Lecture Notes in Computer Science   8428

Commenced Publication in 1973
Founding and Former Series Editors:
Gerhard Goos, Juris Hartmanis, and Jan van Leeuwen

T0235622

Peter Höfner   Peter Jipsen
Wolfram Kahl   Martin Eric Müller (Eds.)

Relational and Algebraic Methods in Computer Science

14th International Conference, RAMiCS 2014
Marienstatt, Germany, April 28 – May 1, 2014
Proceedings

Springer

Volume Editors

Peter Höfner
NICTA and UNSW
Anzac Parade 223, Kensington, NSW 2033, Australia
E-mail: peter.hoefner@nicta.com.au

Peter Jipsen
Chapman University, School of Computational Sciences
One University Drive, Orange, CA 92866, USA
E-mail: jipsen@chapman.edu

Wolfram Kahl
McMaster University, Department of Computing and Software
1280 Main Street West, Hamilton, ON L8S 4K1, Canada
E-mail: kahl@mcmaster.ca

Martin Eric Müller
University of Augsburg, Department of Computer Science
Universitätsstraße 6a, 86159 Augsburg, Germany
E-mail: m.e.mueller@acm.org

ISSN 0302-9743   e-ISSN 1611-3349
ISBN 978-3-319-06250-1   e-ISBN 978-3-319-06251-8
DOI 10.1007/978-3-319-06251-8
Springer Cham Heidelberg New York Dordrecht London

Library of Congress Control Number: Applied for

LNCS Sublibrary: SL 1 – Theoretical Computer Science and General Issues

Typesetting: Camera-ready by author, data conversion by Scientific Publishing Services, Chennai, India

Printed on acid-free paper

Springer is part of Springer Science+Business Media (www.springer.com)

Preface

This volume contains the proceedings of the 14th International Conference on Relational and Algebraic Methods in Computer Science (RAMiCS 2014). The conference took place in Marienstatt, Germany, from April 27 to May 1, 2014, and was the third conference using the RAMiCS title, after Cambridge, UK, 2012, and Rotterdam, The Netherlands, 2011, but the 14th in a series that started out using the name "Relational Methods in Computer Science" with the acronym RelMiCS. From 2003 to 2009, the 7th through 11th RelMiCS conferences were held as joint events with Applications of Kleene Algebras (AKA) conferences, motivated by the substantial common interests and overlap of the two communities. The purpose of the RAMiCS conferences continues to be bringing together researchers from various subdisciplines of computer science, mathematics, and related fields who use the calculus of relations and/or Kleene algebra as methodological and conceptual tools in their work.

The call for papers invited submissions in the general area of relational and algebraic methods in computer science, placing special focus on formal methods for software engineering, logics of programs, and links with neighboring disciplines. This focus was also realized in the choice of the following three invited talks: "Developments in Concurrent Kleene Algebra" by Tony Hoare, "Preparing Relation Algebra for 'Just Good Enough' Hardware" by José Nuno Oliveira, and "Relation Lifting" by Alexander Kurz.

The body of this volume is made up of invited papers accompanying the invited talks by Hoare and Oliveira, and of 25 contributions by researchers from all over the world. The papers have been arranged into five groups, with the invited talks closely related to the first three:

Concurrent Kleene Algebras and Related Formalisms,
  including Kleene algebras with tests and Kleene algebras with converse, both in theoretical investigations and in applications to program correctness.
Reasoning about Computations and Programs,
  with considerations of faults and imperfect hardware, separation logics, infinite computations, process calculi, and program verification.
Heterogeneous and Categorical Approaches,
  including "relation-categorical" studies of topology, concept lattices, semilattice categories, and fuzzy relations.
Applications of Relational and Algebraic Methods,
  including to voting systems, databases and data learning, optimization, and mereotopology.
Developments Related to Modal Logics and Lattices,
  with papers related to domain operators for homogeneous fuzzy relations, accessibility relation semantics and tableau proving for tense operators,

representation theorems for nominal sets, and fixed-point theory of lattice μ-calculus.

The contributed papers were selected by the Program Committee from 37 relevant submissions. Each submission was reviewed by at least three Program Committee members; the Program Committee did not meet in person, but had over one week of intense electronic discussions.

We are very grateful to the members of the Program Committee and the subreviewers for their care and diligence in reviewing the submitted papers. We would like to thank the members of the RAMiCS Steering Committee for their support and advice especially in the early phases of the conference organization. We are grateful to the Bonn Rhein Sieg University of Applied Sciences and especially Nadine Kutz for generously providing administrative support. We gratefully appreciate the excellent facilities offered by the EasyChair conference administration system. Last but not least, we would like to thank the Deutsche Forschungsgemeinschaft for their generous financial support.

February 2014   Peter Höfner
                                                                 Peter Jipsen
                                                               Wolfram Kahl
                                                             Martin E. Müller

Organization

Organizing Committee

Local Organizer

Martin E. Müller   Universität Augsburg, Germany

Programme Co-chairs

Peter Jipsen   Chapman University, USA
Wolfram Kahl   McMaster University, Canada

Publicity

Peter Höfner   NICTA, Australia

Program Committee

Rudolf Berghammer   Christian-Albrechts-Universität zu Kiel,
                                       Germany
Harrie de Swart   Erasmus University Rotterdam,
                                       The Netherlands
Jules Desharnais   Université Laval, Canada
Marc Frappier   University of Sherbrooke, Canada
Hitoshi Furusawa   Kagoshima University, Japan
Timothy G. Griffin   University of Cambridge, UK
Robin Hirsch   University College of London, UK
Peter Höfner   NICTA Ltd., Australia
Ali Jaoua   Qatar University, Qatar
Peter Jipsen   Chapman University, USA
Wolfram Kahl   McMaster University, Canada
Tadeusz Litak   FAU Erlangen-Nürnberg, Germany
Larissa Meinicke   The University of Queensland, Australia
Szabolcs Mikulas   University of London, UK
Bernhard Möller   Universität Augsburg, Germany
Martin E. Müller   Universität Augsburg, Germany
José Nuno Oliveira   Universidade do Minho, Portugal
Ewa Orłowska   National Institute of Telecommunications,
                                       Warsaw, Poland
Matthew Parkinson   Microsoft Research Cambridge, UK
Damien Pous   CNRS, Lab. d'Informatique de Grenoble,
                                       France
Ingrid Rewitzky   Stellenbosch University, South Africa

Holger Schlingloff	Fraunhofer FIRST and Humboldt University, Germany
Gunther Schmidt	Universität der Bundeswehr München, Germany
Renate Schmidt	University of Manchester, UK
Georg Struth	University of Sheffield, UK
George Theodorakopoulos	University of Derby, UK
Michael Winter	Brock University, Canada

Steering Committee

Rudolf Berghammer	Christian-Albrechts-Universität zu Kiel, Germany
Harrie de Swart	Erasmus University Rotterdam, The Netherlands
Jules Desharnais	Université Laval, Canada
Ali Jaoua	Qatar University, Qatar
Bernhard Möller	Universität Augsburg, Germany
Ewa Orłowska	National Institute of Telecommunications, Warsaw, Poland
Gunther Schmidt	Universität der Bundeswehr München, Germany
Renate Schmidt	University of Manchester, UK
Michael Winter	Brock University, Canada

Additional Reviewers

Musa Al-hassy	Koki Nishizawa
Daniela Da Cruz	Rasmus Lerchedahl Petersen
Han-Hing Dang	Patrick Roocks
Nikita Danilenko	Agnieszka Rusinowska
Stéphane Demri	David Rydeheard
Ernst-Erich Doberkat	Thomas Sewell
Zoltan Esik	Paulo F. Silva
Roland Glück	John Stell
Dirk Hofmann	Insa Stucke
Andrzej Indrzejczak	Norihiro Tsumagari
Yannis Kassios	Stephan van Staden
Dexter Kozen	Andreas Zelend
Steffen Lösch	

Sponsor

Deutsche Forschungsgemeinschaft

Table of Contents

Developments in Concurrent Kleene Algebra

Tony Hoare[1], Stephan van Staden[2], Bernhard Möller[3], Georg Struth[4],
Jules Villard[5], Huibiao Zhu[6], and Peter O'Hearn[7]

[1] Microsoft Research, Cambridge, United Kingdom
[2] ETH Zurich, Switzerland
[3] Institut für Informatik, Universität Augsburg, Germany
[4] Department of Computer Science, The University of Sheffield, United Kingdom
[5] Department of Computing, Imperial College London, United Kingdom
[6] Software Engineering Institute, East China Normal University, China
[7] Facebook, United Kingdom

Abstract. This report summarises recent progress in the research of its co-authors towards the construction of links between algebraic presentations of the principles of programming and the exploitation of concurrency in modern programming practice. The research concentrates on the construction of a realistic family of partial order models for Concurrent Kleene Algebra (aka, the Laws of Programming). The main elements of the model are objects and the events in which they engage. Further primitive concepts are traces, errors and failures, and transferrable ownership. In terms of these we can define other concepts which have proved useful in reasoning about concurrent programs, for example causal dependency and independence, sequentiality and concurrency, allocation and disposal, synchrony and asynchrony, sharing and locality, input and output.

1 Introduction

Concurrency has many manifestations in computer system architecture of the present day. It is provided in networks of distributed systems and mobile phones on a world-wide scale; and on a microscopic scale, it is implemented in the multi-core hardware of single computer chips. In addition to these differences of scale, there are many essential differences in detail. As in other areas of basic scientific research, we will initially postpone consideration of these differences, and try to construct a mathematical model which captures the essence of concurrency at every scale and in all its variety.

Concurrency also has many manifestations in modern computer programming languages. It has been embedded into the structure of numerous new and experimental languages, and in languages for specialised applications, including hardware design. It is provided in more widely used languages by a choice of thread packages and concurrency libraries. Further variation is introduced by a useful range of published concurrency design patterns, from which a software architect can select one that reconciles the needs of a particular application with the particular hardware available.

P. Höfner et al. (Eds.): RAMiCS 2014, LNCS 8428, pp. 1–18, 2014.

Concurrency is also a pervasive phenomenon of the real world in which we live. A general mathematical model of concurrency shares with the real world the concept of an object engaging together with other objects in events that occur at various points in space and at various instants in time. It also shares the principle of causality, which states that no event can occur before an event on which it causally depends, as well as a principle of separation, which states that separate objects occupy separate regions of space. It is these principles that guide definitions of sequential and concurrent composition of programs in a model of CKA. They provide evidence for a claim that CKA is an algebraic presentation of common-sense (non-metric) spatio-temporal reasoning, similar to that formalised in various ancient and modern modal logics.

1.1 Domain of Discourse

The construction of a model of a set of algebraic laws consists of three tasks. The first task is the definition of the domain of discourse (carrier set of the algebra). Any element of the domain is a value that may be attributed to any of the variables occurring in any of the laws. It is a mathematically defined structure, describing at some level of abstraction the relevant aspects of the real or conceptual world to which the algebra is applied. The second task is the interpretation of each operator of the algebra as a mathematical function, with its arguments and its result in the domain of discourse. Finally, there is the proof that the laws of the algebra are true for any attribution of values in the domain of discourse to all the variables in each of the equations.

It is instructive and convenient to introduce a series of three domains, each one including all the elements of the next. The most comprehensive domain is that of specifications, which describe properties and behaviour of a computer system while executing a program. They may be desirable properties of programs that are not yet written, or undesirable properties of a program that is still under test. Formally, a specification is just a set containing traces of all executions that satisfy the property. It may be expressed in any meaningful mathematical notation, including arbitrary set unions, and arbitrary intersections, and even complementation. A most important quality of a specification is that it should be comprehensible; and therefore it should be accompanied by an informal explanation that makes it so. That is a precept that we hope to follow in this presentation.

The second domain consists of programs. A program can be regarded as a precise description of the exact range of all its own possible behaviours when executed by computer. It is expressed in a highly restricted notation, namely its programming language. The language excludes negation and other operators that are incomputable in the sense of Turing and Church. As a result, a program text can be input directly by computer, and (after various mechanised transformations) it can be directly executed to cause the computer to produce just one of the behaviours specified. Inefficiency of implementation is another good reason for omission from a programming language of the more general operators useful in a specification. For example, intersection is usually excluded, even though it is the most useful operator for assembling large sets of design requirements into a specification.

At the third level, a single trace, produced for example by a single program test, describes just one particular execution of a particular program at a particular time on a particular computer system or network. The execution itself is also modelled as a lower-level set, consisting of events that occurred during that execution, including events in the real world environment immediately surrounding the computer system. A trace is effectively a singleton set in the domain of programs, so it cannot be composed by union; but it can still be composed either sequentially or concurrently with another trace, and (in our partial order model) the result is still a single trace. The composition operators are easily proved to satisfy the laws of CKA.

There is close analogy between this classification of domains for concurrent programming and the classification of the standard number systems of arithmetic - reals, rationals, and integers. For example, the operators of a programming language share the same kind of elementary algebraic properties as arithmetic operators - distribution, association and commutation, with units and zeroes.

The analogy can be pursued further: our modelling methods are also similar to those used in the foundations of arithmetic. For example, the reals are defined as downward-closed sets of rationals (Dedekind cuts). The operators at each level are then defined in terms of the operators at the lower level. For example, an operator like addition on reals is defined as the set obtained by adding each rational from the first real operand to each rational from the second operand. This construction is known as 'lifting to sets', and we will use it to lift individual traces to the domain of programs and specifications that describe them.

The constructions at the foundations of arithmetic show that the operators of all the number systems obey the same algebraic laws, or nearly so. Our models are designed to do the same for the laws of programming, as expressed in a Concurrent Kleene Algebra.

1.2 Contracts and Counterexamples

Models play an essential role in the development of theories and the practical use of an algebra in mathematics. They provide evidence (a counterexample) for the invalidity of an inaccurately formulated conjecture, explaining why it can never be proved from a given set of algebraic laws. In pure logical and algebraic research, such evidence proves the independence of each axiom of the algebra from all the others. For purposes of counterexample generation, appropriate selection from a family of simple models can be more useful and more efficient than repeated use of a single realistic model that is more complicated. An experienced mathematician is familiar with a wide range of models, and selects the one most likely to serve current purposes. However, the research reported here seeks realism rather than simplicity of its model.

Discovery of counterexamples is also a primary role of models of programming. A counterexample consists in a trace of program execution which contains an error; it thereby demonstrates falsity of the conjecture that a program is correct.

An automatic test case generator should obviously concentrate on finding such counterexamples. It should also indicate where the errors have been detected, and where they cannot have occurred. The information should be provided in a form that guides human judgement in diagnosing the error, tracing where it has occurred, and deciding whether it should be corrected or worked around.

The definition of what counts as an error, and of where it is to be attributed, can be formalised as a contract at the interface between one part of the program and another. For each of its participants, a contract has two sides. One side is a description of the obligations, which any of the other participants may expect to be fulfilled. An example is the post-condition of a method body, which every call of the method will rely on to be true afterwards. The other side is a description of the requirements which each participant may expect of the behaviour of all the other participants taken together. An example is the precondition of the method body. Every calling program is required to make this true before the call, and the method body may rely on this as an assumption.

In addition to violation of contracts, there are various kinds of generic error, which are universally erroneous, independent of the purposes of the program. Examples familiar to sequential programmers are undefined operations, overflows, and resource leakages. Concurrency has introduced into programming several new classes of generic error, for example, deadlock, livelock, and races (interference). To deal with these errors, we need new kinds of contract, formulated in terms of new concepts such as dependency, resource sharing, ownership, and ownership transfer. We also need to specify dynamic interactions (by synchronisation, input, or output) between a component of a concurrent program and its surrounding environment.

A full formal definition (semantics) of a programming language will specify the range of generic errors which programs in the language are liable to commit. The semantics itself may be regarded as a kind of contract between the implementer and the user of the language, and they will often allocate responsibility for errors that occur in a running program. For example, syntax errors and violation of type security are often avoided by compile-time checks, and the implementer undertakes to ensure that a program which contains any such errors will not be released for execution, even in a test.

Conversely, for certain intractable errors, the programmer must accept the responsibility to avoid them. In the case of a violation occurring at run time, the language definition may state explicitly that the implementer is freed of all responsibility for what does or does not happen afterwards. For example, in the case of deadlock, nothing more will happen. Or worse, the error may even make the program susceptible to malware attack, with totally unpredictable and usually unpleasant consequences.

The inclusion of contractual obligations in a model lends it an aspect of deontic logic, which has no place in the normal pursuit of pure scientific knowledge. However, it plays a vital role in engineering applications of the discoveries of science.

1.3   Semantics

There are four well-known styles for formalising the definition of the meaning of a programming language. They are denotational, algebraic, operational and deductive (originally called axiomatic). They are all useful in defining a common understanding for the design and exploitation of various software engineering tools in an Integrated Development Environment (IDE), and for defining sound contracts between them.

A model of the laws of programming plays the role of a *denotational* semantics (due to Scott and Strachey) of a language which obeys the laws. The denotation of each program component is a mathematical structure, which describes program behaviour at a suitable level of abstraction. The first examples of such a model were mathematical functions, mapping an input value to an output value. Later examples included concurrent behaviour, modelled as sets of traces of events. We follow the later examples, and extend them to support the discovery and attribution of errors in a program. The denotations therefore provide a conceptual basis for the design and implementation of testing tools, including test case generators, test trace explorers, and error analysers.

The laws themselves present an *algebraic* semantics (advocated, for example by Bergstra and his colleagues) of the same abstract programming language. Algebra is useful in all forms of reasoning about programs, and the proofs are often relatively simple, both for man and for machine. The most obvious example is the use of algebra to validate the transformation of a program into one with the same meaning, but with more efficient executions. An algebraic semantics is therefore a good theoretical foundation for program translators, synthesisers and optimisers.

The rules of an *operational* semantics (due to Plotkin and Milner) show how to derive, from the text of a program plus its input data, the individual steps of just a single execution of the program. This is exactly what any implementation of the language has to do. The rules thereby provide a specification of the correctness of more efficient methods of implementation, for example, by means of interpreters written in the same or a different language, or compilers together with their low-level run-time support.

The *deductive* semantics (attributed to Hoare) gives proof rules for constructing a proof that a program is correct. Correctness means that no possible execution of the program contains an error. Some of the errors, like an overflow, a race condition or a deadlock, are generic errors. Others are violations of some part of a contract, for example an assertion, written in the program itself. A deductive semantics is most suitable as a theoretical basis for program verifiers, analysers and model checkers, whose function is to prove correctness of programs.

When the full range of tools, based on the four different formalisations of semantics, are assembled into an IDE, it is obviously important that they should be mutually consistent, and provably so. It is common for the tools to communicate with each other by passing annotated programs between them. The programs are often expressed in a common verification-oriented intermediate language like Boogie designed and implemented by Leino. The semantics of this common language must obviously be rigorously formalised and understood by the designers of all the tools. As described above, the semantics needs to be formalised in different ways, to suit the purposes of different classes of tool. The mutual consistency of all the forms of semantics establishes confidence in the successful integration of the tools, by averting errors at the interface between them. Ideally, this proof can be presented and checked, even before the individual tools are written.

An easy way to prove consistency of two different formalisations is to prove one of them on an assumption of the validity of the other. For example, Hoare and Wehrman describe how the laws of the algebraic semantics can be derived rather simply from a graphical denotational model. Similarly, Hoare and van Staden show that the rules of the operational semantics, as well as the rules of the deductive semantics, can be derived from the same algebraic laws of programming. In combination, these proofs ensure that all the models of the laws satisfy the rules of all three of the other kinds of semantics.

In fact, most of the laws can themselves be derived in the other direction, either from the rules of the operational or from the rules of the deductive semantics, or even from both. For example, the principal law of concurrency (the exchange law) is derivable either from the deductive separation logic rules for concurrency formalised by O'Hearn, or from the Milner transition rules which define concurrency in CCS. This direction of derivation gives convincing evidence that our laws for concurrency are consistent with well-established understanding of the principles of concurrent programming. Similar mutual derivations are familiar from the study of propositional logic, where the rules of natural deduction are derived from Boolean algebra, and vice versa.

2   The Laws of Programming

The laws of programming are an amalgam of laws obtained from many sources: relational algebra (Tarski), regular languages (Kleene), process algebras (Brookes and Roscoe, Milner, Bergstra), action algebra (Pratt) and Concurrent Kleene Algebra (Hoare et al.). The pomset models of Gischer and others have also provided inspiration.

An earlier introduction to the laws for sequential programming is (Hoare et al., Laws of Programming). This was written for general computer scientists and professional software developers. It contains simple proofs that the laws are satisfied by a relational model of program execution. Unfortunately, the relational model does not extend easily to concurrency.

The purpose of this section is to list a comprehensive (but not complete) selection of the laws applicable to concurrent programming. The laws are motivated informally by describing their consequences and utility. The informal description of the operators gives the most general meaning of each of them, when applied to programs and specifications. Several of them do not apply to traces.

The model described in section 3 offers a choice of definitions for many of the operators. Any combination of the choices will satisfy the laws. The choice is usually made by a programming language definition; but in principle, the choice could be left as a parameter of an individual test run of a program.

2.1   The Basic Operators

Basic Commands

1 (skip) does nothing, because there is nothing it has been asked to do.

$\top$ is a program whose behaviour is totally undetermined. For example, it might be under control of an undetected virus. Other names for this behaviour are abort (Dijkstra), CHAOS (in CSP) and havoc in Boogie.

$\bot$ is a program with no executions. For example, it might contain a type error, which the compiler is required to detect. As a result, the program is prevented from running.

Binary Operators

Sequential composition $p \,; q$ executes both p and q, where p can finish before q starts. It is associative with unit 1, and has $\bot$ as zero.

Concurrent composition $p \,|\, q$ executes both p and q, where p and q can start together, proceed together with mutual interactions, and finally they can finish together. The operator is associative and commutative with unit 1 , and has $\bot$ as zero.

Choice $(p \cup q)$ executes just one of p or q. The choice may be determined or influenced by the environment, or it may be left wholly indeterminate. The operator is associative, commutative and idempotent, with $\bot$ as unit.

Refinement

The refinement relation $p \Rightarrow q$ is reflexive and transitive, i.e., a pre-order. It means that p is comparable to q in some relevant respect. For example, p may have less traces, so its behaviour is more deterministic than q, and therefore easier to predict and control. The three operators listed above are covariant (also called monotone or isotone) in both arguments with respect to this ordering. The ordering has $\bot$ as bottom, $\top$ as top and $\cup$ as lub. For further explanation of refinement, see section 2.2.

Distribution

All three binary operators distribute through choice.

Sequential and concurrent composition distribute through each other, as described by the following analogue of the exchange (or interchange) law of Category Theory:

$$(p \,|\, q) \,; (p' \,|\, q') \Rightarrow (p \,; p') \,|\, (q \,; q')$$

For further explanation of the exchange law, see section 2.3.

Iterations

The sequential iteration $p*$ performs a finite sequential repetition of p, zero or more times.

The concurrent iteration $p!$ performs a finite concurrent repetition of p, zero or more times.

Residuals

The weakest precondition q -; r (Dijkstra) is the most general specification of a program p which can be executed before q in order to satisfy specification r. The weakest precondition consequently cancels sequential composition (and vice-versa), but the cancellation is only approximate in the refinement ordering:

$$(q\text{ -; }r)\,;\,q \Rightarrow r \quad\text{and}\quad p \Rightarrow (q\text{ -;}(p\,;\,q))$$

The specification statement p ;- r (due to Back and Morgan) is the most general specification of a program q that can be executed after p in order to satisfy specification r.

p -| r the magic wand (due to O'Hearn and Pym) is similar to the above for concurrent composition.

Notes:

1. the result of the residuals is a specification rather than a program. Residuals are in general incomputable. That is why the residual operations are excluded from programming languages.
2. The constants $\top$ and $\bot$, and the operators of iteration and choice, are not available in the algebra of traces.

2.2   Refinement

The refinement relation $p \Rightarrow q$ expresses an engineering judgement, comparing the quality of two products p and q. By convention, the better product is on the left, and the worse one on the right. For example, the better operand p on the left may be a program with less possible executions than q. Consequently, if q is also a program, it is more non-deterministic than p, and so more difficult to predict and control. If p is a program and q is a specification, the refinement relation means that p meets the specification q, in the sense that everything that p can do is described by the specification q. And if they are both specifications, it means that p logically implies q. Consequently p places stronger constraints on an implementation, which can be more difficult to meet.

Refinement between programs may also account for failures and errors. For example, if p is a program that has the same observable behaviour as q, but q contains a generic programming error that is not present in p, this may be the grounds for a judgment that p refines q. In other words, a program can be improved by removing its programming errors, but otherwise leaving its behaviour unchanged. Dually, if p is a specification, it is made weaker (easier to meet) by strengthening the obligation which it places on its environment. Thus the meaning of refinement is relative to the contracts between the components of a program, and between the whole program and its environment.

A more precise interpretation for refinement is usually made in a programming language definition. But a testing tool might allow the definition to be changed, to reflect exactly the purposes of each test.

2.3 The Exchange Law: $(p \,|\, q) \,;\, (p' \,|\, q') \Rightarrow (p \,;\, p') \,|\, (q \,;\, q')$

The purpose of this law is made clear by describing its consequences, which are to relate a concurrent composition to one of its possible implementations by interleaving. The law also permits events to occur concurrently, and requires dependent events like input output to occur in the right order. Inspection of the form of the law shows that the left hand side of the law describes a subset of the possible interleavings of the atomic actions from the two component threads $(p \,;\, p')$ and $(q \,;\, q')$ on the right hand side. This subset results from a scheduling decision that the two semicolons shown on the right hand side will be 'synchronised' as the single semicolon on the left.

The algorithm for finding an interleaving uses the recursive principle of 'divide-and conquer'. The interleavings $(p \,|\, q)$ before the semicolon on the left are formed from the two first operands p and q of the two semicolons on the right. The interleavings $(p' \,|\, q')$ after the semicolon on the left are formed from the two second operands p' and q' of the two semicolons on the right. Every execution of the left hand side is the sequential composition of a pair of executions, one from each of $(p \,|\, q)$ and $(p' \,|\, q')$. Each such execution achieves synchronisation of the two semicolons on the right, but it places no other constraint on the interleavings. (The constraints are specified in the definition of sequential composition).

By introduction and elimination of the unit 1, the exchange law can be adapted to cases where the term to be transformed has only two or three operands. The following three theorems are called frame laws:

$$p \,;\, q \Rightarrow p \,|\, q \qquad \text{(frame law 0)}$$

$$p \,;\, (q \,|\, r) \Rightarrow (p \,;\, q) \,|\, r \qquad \text{(frame law 1)}$$

$$(p \,|\, q) \,;\, r \Rightarrow p \,|\, (q \,;\, r) \qquad \text{(frame law 2)}$$

By commuting the operands of concurrent composition, the first frame law gives a weak principle of sequential consistency:

$$q \,;\, p \Rightarrow p \,|\, q, \qquad \text{from which, by covariance and idempotence,}$$

$$q \,;\, p \,\cup\, p \,;\, q \Rightarrow p \,|\, q$$

If p and q are atomic commands, then the left hand side of the above conclusion shows the only two possible interleavings of their concurrent combination on the right hand side. A strong principle of sequential consistency would allow the conclusion to be strengthened to an equation (but only in the case of atomic commands). However, we will continue to exploit the weaker formulation of the principle.

When there are larger numbers of atomic commands in a formula, the exchange law can be used, in conjunction with commutation, association and distribution, to reduce the formula to a normal form in which the outer operator is union and the inner operator is sequential composition. The technique is to use the exchange law to drive the occurrences of concurrent composition to the atoms, and then apply the weak principle of sequential consistency given above.

For example, from the frame laws we get:

$$p\,;\,(q\,;\,r\ \cup\ r\,;\,q) \Rightarrow (p\,;\,q)\,|\,r \quad \text{and} \quad (p\,;\,r\ \cup\ r\,;\,p)\,;\,q \Rightarrow r\,|\,(p\,;\,q)$$

By commutation, distribution, covariance and idempotence, we can combine these to an analogue of Milner's expansion theorem for CCS:

$$p\,;\,q\,;\,r\ \cup\ p\,;\,r\,;\,q\ \cup\ r\,;\,p\,;\,q \Rightarrow (p\,;\,q)\,|\,r$$

This theorem remains valid when there are synchronised interactions between the concurrent commands. When an interleaving $\cdots p\,;\,q\,;\,\cdots$ violates a synchronisation constraint that p must follow q, the definition of sequential composition will ensure that this interleaving takes the value $\bot$, which is the unit of choice. This particular interleaving is thereby excluded from the left hand side of the theorem.

3   A Diagrammatic Model

The natural sciences often model a real-world system as a diagram (graph) drawn in two dimensions: a space dimension extends up and down the vertical axis, and a time dimension extends along the horizontal axis. We model what happens inside a computer in the same way. The fundamental components of the model are objects, which are represented by lines (trajectories) drawn from left to right on the diagram. An object has a unique identifier (name or address) associated with it at its allocation. This may be used just like a numeric value in assignments and communications.

Examples of objects are variables (local or shared), semaphores (for exclusion or synchronisation), communication channels (buffered or synchronised), and threads. These classes of object are often built into a programming language; but in an object-oriented language they can be supplemented by programmed class declarations.

The lines representing two or more objects may intersect at a point, which represents an atomic event or action in which the given objects participate simultaneously. Examples of events are allocations and disposals of an object, assignments or fetches of a variable, input or output of a message, seizures or releases of a semaphore, and forking or joining of threads. An example of multiple participation is an atomic assignment, which involves fetches from many variables and assignments to one or more target variables, together with the thread that contains the assignment.

The line for each object passes through the time-ordered sequence of events in which the object engages. In the case of a thread object, the ordering of events within a thread is often called program order. It seems reasonable to require that no event can occur without participation of exactly one thread.

In the diagram for a Petri net, an event is drawn as a transition, in the form of a box or a bold vertical line. Extending the same convention, we will represent participation of an object in the transition as a line which passes straight through the transition. This contrasts with an allocation of a new object whose line begins at the transition, or with a disposal in the case of a line which ends at it. The other Petri net component (a place) represents choice; it is therefore not needed in the diagram for a single trace, for which all choices have already been made.

An arrow is defined as a pair of consecutive points on the same line. It is drawn with its source on the left and its target on the right. An arrow is labelled by a primitive constant predicate of an assertion language. For example, in standard separation logic, the primitive is a pair written (say) $101 \mapsto 27$, where the constant 101 is the unique identifier of the object, and 27 is the value that is held by the object between the event at the source of the arrow and the event at its target.

Arrows are classified as either local or global. The source and target of a local arrow must be events that involve the same thread, called its current owner: violation of this rule of locality is an error. In a language like occam, the compiler is responsible for detecting this error, and making sure that the program is not executed. However in a language like C this check would be too difficult, and it is not required. Instead, violation of locality is attributed as an error of the program.

An object is defined to be local if all its arrows are local, and volatile (shared) if all its arrows are global. An object with both kinds of arrow is one whose ownership may change between the event at the tail and the event at the head of any one of its global arrows. The distinction between local and global arrows is familiar from Message Sequence Charts. The local arrows representing concurrent tasks are drawn downwards, and global arrows are drawn between points on the vertical lines. The points represent calls, call-backs, returns, and other communications between the tasks.

In a diagram of program execution, an instant of time (real or virtual) can be drawn as a vertical coordinate which crosses just one arrow in the line for each object that is allocated but not yet disposed at that instant. The collection of labels on the arrows which cross a vertical coordinate describes the state of the entire system at the given instant. A global arrow denotes a message which has been buffered between the tail event of the arrow and its head. An arrow of a volatile object is effectively a special case of a message. A local arrow crossing the coordinate represents the value held in the computer memory allocated to the owning object. The state of the entire local memory at the relevant instant is the relation whose pairs are (loc, val), where the label on the crossing arrow is $loc \mapsto val$. This is necessarily a function, because no line can cross a coordinate twice: that would involve a backward crossing somewhere in between.

In a diagram of program execution, a point in space can be drawn as a horizontal coordinate separating the threads above it from the threads below it. The set of global arrows which cross the coordinate in either direction give a complete account of the dynamic interactions between the threads that reside on either side of the coordinate. They must all be global arrows. There is no significance attached to the vertical ordering of the horizontal coordinates. That is why concurrent composition commutes.

The important concept of causal dependency (happens before) is defined in terms of arrows. A causal chain is a sequence of arrows (taken usually from different object lines), in which the head of each arrow is the same point as the tail of the next arrow (if any). Occurrence of an event on a causal chain is a (necessary) cause of all subsequent events on the same chain; and it is dependent on all earlier events on the chain. Obviously, no event can occur before an event which it depends on, or after an event that depends on it. This is represented by the left-to-right direction of drawing the arrows.

In summary, the primitive concepts of our geometry are lines and points at which the lines meet. Arrows are defined and classified as local or global, and they are given labels. In terms of these primitives we define vertical and horizontal coordinates, system state, ownership and transfer of ownership, and causal dependency. The concept of synchrony can be defined as mutual dependency, and the concept of 'true' concurrency can be defined in the usual way as causal independence.

3.1   Decomposition of Diagrams

A diagram in plane geometry can be decomposed into segments in two ways, either horizontally or vertically. A horizontal segment (slice) contains the entire lines of a group of related objects, which interact by participating jointly with each other in their events. This segmentation is useful in analysing the behaviour of individual objects from the same class, or groups of interacting objects from the same package of classes.

A vertical segment is similarly separated from its left and right neighbours by two vertical coordinates, representing the initial and the final instants of time. The segment contains all events in the diagram which occurred between the two instants. This form of segmentation is useful in analysing everything that happens during a particular phase in the execution, for example, a method call.

A third form of decomposition mirrors the syntactic structure of the program whose execution is recorded in a trace. Each segment (called a tracelet) contains all the events that occurred during execution of one of the branches of the abstract syntax tree of the program; the tracelets for two syntactically disjoint commands of the program will have disjoint sets of events, as they do in reality.

Inside a diagram, the tracelet is surrounded by a perimeter, with vertical and horizontal sides. The west and east sides of the perimeter are segments from two vertical coordinates, and the north and south sides are segments of two horizontal coordinates. Consequently, the tracelet for a sequential composition $p\,;q$ is split vertically into two smaller tracelets, one for p and one for q, with no dependency of any event in p on any event in q; similarly, the tracelet for a concurrent composition is split horizontally, with no local arrows crossing the split. The whole plane is tiled by these splits, like a crazy paving. The tiles are often drawn as rectangles, but this is not necessary.

An arrow with its source outside the box and its target inside is defined as an input arrow; and an output arrow is defined similarly. The local input arrows represent the portion of local state (called the initial statelet) which is passed to the tracelet on entry. By convention, these arrows enter the box on the west side. Similarly, the local output arrows represent the final statelet, and leave the box on the right side. The global arrows may cross the north or the south sides of the box, as convenient. They represent dynamic interactions that take place with the environment of the tracelet between the start and the finish of its execution.

A fourth form of segmentation splits a tracelet into three segments, sharing just a single event. One segment contains all events that the shared event causally depends on. A second segment contains all events that depend on the shared event; and the remaining segment contains all remaining events, which are irrelevant to its occurrence.

The diagrammatic representation of the trace described in this section is intended to be helpful to the user of a visual debugging tool, by conveying an understanding of what has gone wrong in a failed test, and what can be done about it. For example, the segmentation into tracelets will give the closest possible indication of where in the source program an error has been detected. In a visual tool, a hover of the mouse on the perimeter of the tracelet should highlight the command in the original source program whose execution is recorded in the tracelet.

Similarly, the causal segmentation gives clear access to the events which may have caused the error: to prevent the error, at least one of these will have to be changed. When the culprit has been detected and corrected, the segment that is dependent on it contains all the events that may have been affected by the change. The remaining events in the third segment that are causally independent of the error could be greyed out on a display of the trace.

3.2   Refinement

We represent an error that is detected inside a tracelet by colouring its perimeter. We attribute the errors as described in section 2.2. If the error is attributed to the program being executed, the perimeter is drawn in red, or if it is attributed to the environment of the tracelet, it is marked blue. Where necessary, a single point can be coloured. For example, evaluation of an assertion to false is marked red, whereas a false assumption is marked blue. If no error is detected, a normal black perimeter is drawn. A black perimeter with no points inside represents the execution of the SKIP command, which literally does nothing.

The refinement relation $p \Rightarrow q$ between tracelets p and q is defined by looking only at their events, and also at the colour of their two perimeters. The definition deliberately ignores the internal structure of tracelets within p or q. Validity of the refinement means that the diagram of one operand are just an isomorphic copy of the diagram in the other, and that the perimeter of p has a lower colour than that of q in the natural ordering, with blue below black and red above it. The definition of the isomorphism can be weakened by ignoring much of the internal content of the tracelet. However, the labels on the arrows that cross the perimeter must be preserved, and so must the causal dependencies between these arrows.

The laws of programming require observance of the following principles in colouring of perimeters. The first three principles state the obvious fact that a tracelet inherits all the errors that are contained in any of its subtracelets; but if it contains both a red and a blue error, a somewhat arbitrary decision states that the blue dominates. This is required by the zero laws for $\perp$: it is certainly justified when the bottom denotes a program with no executions.

1. If a tracelet contains a tracelet with a blue perimeter, it also has a blue perimeter.
2. Otherwise, if it contains a tracelet with a red perimeter, it also has a red perimeter.
3. Otherwise, both operands are black, and the whole tracelet has a black perimeter too.

Further rules are introduced in the definition of the two composition operators.

4. Any failure to observe the rules of sequential composition colours the perimeter blue.
5. Any failure to observe the rules of concurrent composition colours the perimeter red, except in the case that principle 1 requires it to be blue.

There is a choice of reasonable meanings for sequential composition. In the strongest variant, every event of the second operand must be dependent on every event of the first operand. This is an appropriate definition for sequential compositions which occur within a single thread. In most programming languages, this is the only kind of sequential composition that can be written in the program. But if strong composition is applied to a multi-threaded trace, it requires that all the threads pass the semicolon together, as in PRAM model of lock-step program execution.

The weakest variant of sequential composition involves the minimum of synchronisation. The principle is simply that no event of the first operand can depend on any event of the second operand. Violation of this principle would make it impossible to complete execution of the first operand before the second operand starts: this was quoted informally in section 2 as the general defining condition for sequential composition.

This definition is weak enough to allow the reordering optimisations that are commonly made in modern compilers for widely used languages. When applied to multi-threaded programs, it allows each thread to pass the semicolon at a different time.

Turning to concurrent composition, its weakest definition imposes only the condition that no local arrow can cross its north or south sides. A more realistic definition has to make the occurrence of deadlock into a programming error. More formally, the condition states that there is no dependency cycle that crosses from an event of one operand to an event of the other. An exception may be made to allow synchronised communication between an outputting and an inputting thread, as in CSP.

The principle that a local arrow cannot cross between threads ensures that in any correct trace there is no interference by one thread with the values of a local variable of another thread. Thus separation logic is a valid method of reasoning about concurrent programs, even in the presence of extra features like synchronisation, atomicity, and communication. This claim still needs to be checked in detail.

That concludes our informal description of a diagrammatic model for the algebra of traces. Our description has been analytic (decompositional). It is presented as a set of principles that are applied to test whether a given fully decomposed and annotated trace has been correctly decomposed, and whether its errors have been correctly attributed, according to the five principles above. This is in contrast to the usual approach of denotational semantics, which is compositional (synthetic): the denotation of the result of each operation is fully defined in terms of the denotations of its operands. The contrast between the decompositional and compositional interpretations is similar to the analytic and synthetic readings of a set of recursive syntactic equations of a context-free language.

The problem with such denotational definitions is that they are too prescriptive of all the details of the model. This is because every needed property of every aspect of the operator has to be deducible from its definition. In a decompositional presentation, each aspect of an operator can be described separately, and as weakly as desired. Indeed, the weakness is often desirable, because interesting variations of the operator can be identified, classified, and left for later choice.

The problem with the analytic approach is to decide when enough principles have been given. We suggest that the relevant criterion is simply that all the laws of programming are provably satisfied by the given collection of principles. Section 1.3 has presented evidence that the laws are sufficient as a foundation for reasoning about programs, and for the design of programming tools, which analyse, implement and verify them.

4   Sketch of a Formal CKA Model

4.1   Graphs and Tracelets

Definition 4.1. Given a set EV of *events* and a set AR of *arrows*, a *graph* is a structure $H = (E, A, s, t)$ where $E \subseteq \text{EV}$, $A \subseteq \text{AR}$ and $s, t : A \to E$ are total functions yielding *source* and *target* of an arrow. A *tracelet* is a pair $tr = (H, F)$ where $H = (E, A, s, t)$ is a graph, called the *overall trace*, and $F \subseteq E$ is a distinguished set of events, called the *focus* of tr and denoted by $foc(tr)$.

The pairwise disjoint sets of *input, output* and *internal* arrows of the tracelet are given by

$$a \in in(tr) \Leftrightarrow_{df} t(a) \in F \wedge s(a) \notin F ,$$
$$a \in out(tr) \Leftrightarrow_{df} s(a) \in F \wedge t(a) \notin F ,$$
$$a \in int(tr) \Leftrightarrow_{df} s(a) \in F \wedge t(a) \in F .$$

As mentioned in Section 3, arrows are classified as local and global, but in this section we ignore the distinction.

We want to combine tracelets by connecting outputs of one tracelet to inputs of another. For separation we require the events of tr_1 and tr_2 to be disjoint. Moreover, the combination is meaningful only if both tracelets have the same overall trace. More precisely, consider tracelets tr_1, tr_2 with disjoint focuses but same overall trace H, and an arrow a in $out(tr_1) \cap in(tr_2)$. Then a is automatically an internal arrow of the tracelet $(H, foc(tr_1) \cup foc(tr_1))$. If a carries values of some kind, we view the combination as transferring these values from the source event of a in tr_1 to the target event of a in tr_2.

Definition 4.2. Two tracelets tr_1, tr_2 are *combinable* if $H_1 = H_2$ and $foc(tr_1) \cap foc(tr_2) = \emptyset$. Then their *join* is $tr_1 \sqcup tr_2 =_{df} (H_1, F_1 \cup F_2)$. Clearly, $tr_1 \sqcup tr_2$ is a tracelet again.

Since disjoint union is associative, also $\sqcup$ is associative. In the set of all tracelets with a common overall trace H the *empty tracelet* $\square_H =_{df} (H, \emptyset)$ with empty focus is the unit of $\sqcup$.

4.2 Tracelets and Colours

In Sect. 3.2 we have presented the idea of accounting for errors by colours. Formally we use *tiles*, i.e. pairs (tp, c) with a tracelet tp and a colour c. For abbreviation we represent the colours red, black and blue by the values $\bot, 1, \top$ ordered by $\bot \le 1 \le \top$.

For combining scores, we use the *summary* operator $\circ$ defined by the table at the right. Obviously, this operator is commutative and idempotent and has 1 as its unit which is indivisible, i.e., $c \circ c' = 1$ implies $c = 1 = c'$. The operator is also covariant w.r.t. $\le$. Finally, it is also associative since it coincides with the supremum operator

$\circ$	$\bot$	1	$\top$
$\bot$	$\bot$	$\bot$	$\bot$
1	$\bot$	1	$\top$
$\top$	$\bot$	$\top$	$\top$

on the lattice induced by the second ordering $1 \preceq \top \preceq \bot$ (we are not going to use that ordering any further, though).

The *refinement relation*, a partial order between tiles, is given by

$$(p, c) \Rightarrow (p', c') \Leftrightarrow_{df} p = p' \wedge c \le c' .$$

Tiles are composed by joining their tracelet parts and summarising their colours together with additional error information the joined trace may provide. For sequential and parallel composition ; and | this information is computed by two operators $\downarrow$ and $\uparrow$ mapping pairs of traces to colours.

We can realise $\downarrow$ and $\uparrow$ using binary relations R, R' that must hold between the events of joined traces. For combinable traces p, q we set

$$p \downarrow q =_{df} \begin{cases} 1 \text{ if } foc(p) \times foc(q) \subseteq R \text{ ,} \\ \bot \text{ otherwise ,} \end{cases} \qquad p \uparrow q =_{df} \begin{cases} 1 \text{ if } foc(p) \times foc(q) \subseteq R' \text{ ,} \\ \top \text{ otherwise .} \end{cases}$$

Here R and R' are any of the relations listed for sequential and parallel composition, respectively, at the end of Section 3.

By definition, these operators satisfy $p \downarrow q \leq 1$ and $1 \leq p \uparrow q$. Moreover, $\uparrow$ is commutative. Using the indivisibility of 1 one sees that $\downarrow$ and $\uparrow$ distribute through trace join, i.e., $(p \sqcup q) \downarrow r = p \downarrow r \circ q \downarrow r$ etc.

Definition 4.3. Sequential and parallel composition of tiles with combinable traces are defined as follows:

$$(p, s) ; (p', s') =_{df} (p \sqcup p', s \circ s' \circ p \downarrow p') \text{ ,}$$
$$(p, s) \mid (p', s') =_{df} (p \sqcup p', s \circ s' \circ p \uparrow p') \text{ .}$$

Theorem 4.4. *The operators $;$ and $\mid$ are associative and $\mid$ is commutative. Moreover they satisfy the frame and exchange laws. Under the additional assumptions $\square_H \downarrow p = 1 = p \downarrow \square_H = p \uparrow \square_H$, the tile $(\square_H, 1)$ is a shared unit of $;$ and $\mid$ in the set of all tracelets with common overall trace H.*

The latter assumptions mean that the empty tracelet is error-free, which is reasonable. By this Theorem we have provided a recipe for constructing specific models to meet specific purposes, with a prior guarantee that the model will satisfy the laws.

4.3   Programs and Lifting

A *program* is a set of tiles that is downward closed w.r.t. refinement $\Rightarrow$.

We already have presented the idea that operators on programs should arise by pointwise lifting of the corresponding operators on tiles. Of course, this makes sense only if also the laws for tiles lift to programs.

A sufficient condition for this is bilinearity, viz. that every variable occurs exactly once on both sides of the law. Examples are associativity, commutativity and neutrality in the case of equations and the frame and exchange laws in the case of refinement laws.

While it is clear what equality means for programs, i.e., downward closed sets of tiles, there are several ways to extend refinement to sets. We choose the following definition:

$$p \Rightarrow p' \quad \text{iff} \quad \forall t \in p : \exists t' \in p' : t \Rightarrow t' \text{ .}$$

By this, a program p refines a specification p' if each of its tiles refines a tile admitted by the specification. Downward closure implies that $\Rightarrow$ in fact coincides with inclusion $\subseteq$ between programs. Hence the set of programs forms a complete lattice w.r.t. the inclusion ordering; it has been called the *Hoare power domain* in the theory of denotational semantics.

The operators at the tile level can be lifted to downward closed sets by forming all possible combinations of the tiles in the operands and closing the result set downward. For instance, $p\,;p'$ is defined as the downward closure $dc(\{t\,;t'\mid t \in p, t' \in p'\}) =_{df} \{r\mid r \Rightarrow t\,;t' \text{ for some } t \in p, t' \in p'\}$, and analogously for the other operators. The lifted versions of covariant tile operators are covariant again, but even distribute through arbitrary unions of programs. Therefore, by the Tarski-Knaster fixed point theorem, recursion equations have least and greatest solutions.

Moreover, it can be shown that with this construction bilinear refinement laws lift to programs. We illustrate this for the case of the frame law $p\,;p' \Rightarrow p\mid p'$.

Assume $r \in p\,;p'$. By the above definition there are $t \in p, t' \in p'$ such that $r \Rightarrow t\,;t'$. Since the frame law holds at the tile level, we have $t\,;t' \Rightarrow t\mid t'$. Moreover, $t\mid t'$ is in $dc(\{t\mid t'\mid t \in p, t' \in p'\}) = p\mid p'$ and we are done.

By this and the second part of Theorem 4.4 the program

$$1 =_{df} dc(\{(\Box_H, 1)\mid H \text{ a graph}\})$$

is a shared unit of the liftings of ; and | to programs.

4.4 Residuals

By the distributivity of lifted covariant operators and completeness of the lattice of downward closed programs the residuals mentioned in Section 2.1 are guaranteed to exist. They can be defined by the Galois connections

$$\begin{aligned} p \Rightarrow q\,\text{-}\,;\,r & \quad \text{iff} \quad p\,;q \Rightarrow r\,, \\ q \Rightarrow p\,;\text{-}\,r & \quad \text{iff} \quad p\,;q \Rightarrow r\,. \end{aligned}$$

This independent characterisation is necessary, since these operators cannot reasonably be defined as the liftings of corresponding ones at the tile level. An analogous definition can be given for the magic wand -|. The semi-cancellation laws of Section 2.1 are immediate consequences of these definitions. Residuals enjoy many more useful properties, but we forego the details.

Endowing Concurrent Kleene Algebra with Communication Actions

Jason Jaskolka, Ridha Khedri, and Qinglei Zhang

Department of Computing and Software, Faculty of Engineering
McMaster University, Hamilton, Ontario, Canada
{jaskolj,khedri,zhangq33}@mcmaster.ca

Abstract. Communication is integral to the understanding of agent
interactions in concurrent systems. In this paper, we propose a mathematical framework for communication and concurrency called Communicating
Concurrent Kleene Algebra (C^2KA). C^2KA extends concurrent Kleene algebra with the notion of communication actions. This extension captures
both the influence of external stimuli on agent behaviour as well as the communication and concurrency of communicating agents.

Keywords: concurrency, communication, concurrent Kleene algebra, semimodules, specification, algebraic approaches to concurrency.

1 Introduction

Systems interact with other systems resulting in the development of patterns of
stimuli-response relationships. Therefore, models for concurrency are commonly
constructed upon the assumption of uninterruptible system execution or atomic
events. Models for concurrency differ in terms of how they capture this notion. A coarse-grained classification categorises models for concurrency as either
state-based models or *event-based* models [4]. State-based models describe the
behaviour of a system in terms of the properties of its states. Typical state-based
approaches consist of representing system properties as formulae of temporal logics, for example, such as LTL [26], CTL [2], or CTL* [5], and model-checking the
state space of the system against them. Conversely, event-based models represent
systems via structures consisting of atomic events. There is an extensive variety
of examples of event-based models for concurrency including labelled transition
systems [17], Petri nets [25], process calculi (e.g., CCS [22], CSP [7], ACP [1],
and π-calculus [24]), Hoare traces [8], Mazurkiewicz traces [21], synchronisation
trees [22], pomsets [27], and event structures [31].

Recently, Hoare et al. [9–12] proposed a formalism for modelling concurrency
called *Concurrent Kleene Algebra* (CKA). CKA extends the algebraic framework
provided by Kleene algebra by offering, aside from choice and finite iteration,
operators for sequential and concurrent composition.

In this paper, we propose a mathematical framework for communication and
concurrency called *Communicating Concurrent Kleene Algebra* (C^2KA). It
extends the algebraic model of concurrent Kleene algebra and allows for the

P. Höfner et al. (Eds.): RAMiCS 2014, LNCS 8428, pp. 19–36, 2014.
© Springer International Publishing Switzerland 2014

separation of communicating and concurrent behaviour in a system and its environment. With C²KA, we are able to express the influence of external stimuli on the behaviours of a system of agents resulting from the occurrence of external events either from communication among agents or from the environment of a particular agent. In this way, we can think about concurrent and communicating systems from two different perspectives: a behavioural perspective and an external event (stimulus) perspective. We can obtain a behavioural perspective by focussing on the behaviour of a particular agent in a communicating system and considering the influence of stimuli, from the rest of the world in which the agent resides, as transformations of the agent's behaviour. Similarly, we can obtain an external event perspective by considering the influence of agent behaviours as transformations of external stimuli. It provides a framework which presents a different view of communication and concurrency than what is traditionally given by existing process calculi.

The remainder of the paper is organised as follows. Section 2 discusses the notions of external stimuli and induced behaviours and introduces a hybrid view of agent communication. Section 3 provides the mathematical preliminaries needed for the remainder of this paper. Section 4 presents the proposed mathematical framework for communication and concurrency and the related results. Section 5 discusses the proposed framework and related work. Finally, Section 6 draws conclusions and points to the highlights of our current and future work.

2   Stimuli and Induced Behaviours

An essential aspect of concurrent systems is the notion of communication. As presented in [9–12], communication in CKA is not directly captured. Variables and communication channels are modelled as sets of traces. Communication can be perceived only when programs are given in terms of the dependencies of shared events [13]. One needs to instantiate the low-level model of programs and traces for CKA in order to define any sort of communication. We would like to have a way to specify communication in CKA without the need to articulate the state-based system of each action (i.e., at a convenient abstract level).

Furthermore, CKA does not directly deal with describing how the behaviours of agents in a system are influenced by external stimuli. From the perspective of behaviourism [30], a stimulus constitutes the basis for behaviour. In this way, agent behaviour can be explained without the need to consider the internal states of an agent. When examining the effects of external stimuli on agent behaviours, it is important to note that every external stimulus *invokes a response* from an agent. When the behaviour of an agent changes as a result of the response, we say that the external stimulus *influences* the behaviour of the agent. Moreover, it is important to have an understanding of how agent behaviours may evolve due to the influence of external stimuli. In particular, it is often useful to have an idea of the possible influence that any given external stimulus may have on a particular agent. We call these possible influences, the *induced behaviours* via external stimuli.

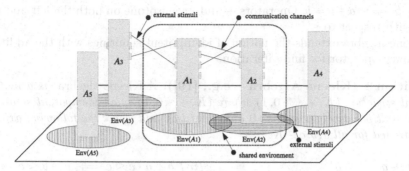

Fig. 1. A hybrid view of agent communication

Agents can communicate via their shared environment and through their local communication channels, but they may also be influenced by external stimuli. For example, if we consider agents A_1 and A_2 (dotted box) depicted in Figure 1, they have a shared environment through which they can communicate. Additionally, they have some communication channels at their disposal for sending and receiving messages. However, the behaviour of A_1 and A_2 can be influenced by the external stimuli coming from A_3, for example. The system formed by A_5 alone is a closed system and does not communicate with the rest of the world neither by external stimuli nor a shared environment. Consider the case where A_1 is subjected to an external stimulus from A_3. Then, A_1 may respond to the stimulus by changing its behaviour which can affect the communication between it and A_2. Currently, this notion cannot be directly handled with CKA. We would like to have a mathematical framework for systems of communicating agents which can capture both the influence of external stimuli on agent behaviour, as well as the communication and concurrency of agents at the abstract algebraic level.

3   Mathematical Background

In this section, we provide the mathematical preliminaries of monoids, semirings, Kleene algebras, and semimodules, and we introduce concurrent Kleene algebra.

3.1   Monoids, Semirings, Kleene Algebras, and Semimodules

A *monoid* is a mathematical structure $(S, \cdot, 1)$ consisting of a nonempty set S, together with an associative binary operation $\cdot$ and a distinguished constant 1 which is the identity with respect to $\cdot$. A monoid is called *commutative* if $\cdot$ is commutative and a monoid is called *idempotent* if $\cdot$ is idempotent.

A *semiring* is a mathematical structure $(S, +, \cdot, 0, 1)$ where $(S, +, 0)$ is a commutative monoid and $(S, \cdot, 1)$ is a monoid such that operator $\cdot$ distributes over operator $+$. We say that element 0 is *multiplicatively absorbing* if it annihilates S with respect to $\cdot$. We say that a semiring is *idempotent* if operator $+$ is idempotent. Every idempotent semiring has a natural partial order $\leq$ on S defined by $a \leq b \iff a + b = b$. Operators $+$ and $\cdot$ are isotone on both the left and the right with respect to $\leq$.

Kleene algebra extends the notion of idempotent semirings with the addition of a unary operator for finite iteration.

Definition 1 (Kleene Algebra – e.g., [19]). *A Kleene algebra is a mathematical structure* $(K, +, \cdot, ^*, 0, 1)$ *where* $(K, +, \cdot, 0, 1)$ *is an idempotent semiring with a multiplicatively absorbing* 0 *and identity* 1 *and where the following axioms are satisfied for all* $a, b, c \in K$:

(i) $1 + a \cdot a^* = a^*$ (iii) $b + a \cdot c \leq c \implies a^* \cdot b \leq c$

(ii) $1 + a^* \cdot a = a^*$ (iv) $b + c \cdot a \leq c \implies b \cdot a^* \leq c$

An important notion required for the proposed framework for communication and concurrency is that of semimodules.

Definition 2 (Left S-semimodule – e.g., [6]). *Let* $S = (S, +, \cdot, 0_S, 1)$ *be a semiring and* $K = (K, \oplus, 0_K)$ *be a commutative monoid. We call* $({}_S K, \oplus)$ *a left S-semimodule if there exists a mapping* $S \times K \to K$ *denoted by juxtaposition such that for all* $s, t \in S$ *and* $a, b \in K$

(i) $s(a \oplus b) = sa \oplus sb$

(ii) $(s + t)a = sa \oplus sb$

(iii) $(s \cdot t)a = s(ta)$

(iv) $({}_S K, \oplus)$ *is called* unitary *if it also satisfies* $1a = a$

(v) $({}_S K, \oplus)$ *is called* zero-preserving *if it also satisfies* $0_S a = 0_K$

A right S-semimodule can be defined analogously. From Definition 2, it is easy to see that each unitary left S-semimodule $({}_S K, \oplus)$ has an embedded left S-act ${}_S K$ with respect to the monoid $(S, \cdot, 1)$. We say that ${}_S K$ is a left S-act if there exists a mapping satisfying Axioms (iii) and (iv) of Definition 2 [18].

3.2   Concurrent Kleene Algebra

Concurrent Kleene algebra is an algebraic framework extended from Kleene algebra offering operators for sequential and concurrent composition, along with those for choice and finite iteration. The operators for sequential and concurrent composition are related by an inequational form of the exchange axiom.

Definition 3 (Concurrent Kleene Algebra – e.g., [9]). *A concurrent Kleene algebra (CKA) is a structure* $(K, +, *, ;, ^\circledast, ^\odot, 0, 1)$ *such that* $(K, +, *, ^\circledast, 0, 1)$ *and* $(K, +, ;, ^\odot, 0, 1)$ *are Kleene algebras linked by the* exchange axiom *given by* $(a * b) ; (c * d) \leq (b ; c) * (a ; d)$.

A selection of laws for CKA which are needed for the remainder of this paper are found in [9] and are given in Proposition 1. An additional useful law is given in Proposition 2.

Proposition 1 (e.g., [9]). *For all* $a, b, c, d \in K$,

(i) $a * b = b * a$
(ii) $(a * b) \,;\, (c * d) \leq (a \,;\, c) * (b \,;\, d)$
(iii) $a \,;\, b \leq a * b$

(iv) $(a * b) \,;\, c \leq a * (b \,;\, c)$
(v) $a \,;\, (b * c) \leq (a \,;\, b) * c$

Proposition 2. *For all* $a \in K$, $a^{\circledcirc} \leq a^{\circledast}$.

Proof. The proof involves the application of Definition 1(iii), Definition 1(i), and Proposition 1(iii). The detailed proof can be found in Appendix A of [16].

4   The Proposed Framework

In the following subsections, we first articulate the algebraic structures which capture agent behaviours and external stimuli. After that, we use the aforementioned algebraic structures for agent behaviours and external stimuli to develop the proposed framework for communication and concurrency. Throughout this section, all omitted proofs can be found in Appendix A of [16].

4.1   A Simple Example of a System of Communicating Agents

We adapt a simple illustrative example from [23] to illustrate the basic notions of specifying a system of communicating agents using the proposed framework. Consider the behaviour of a one-place buffer. Suppose that the buffer uses two flags to indicate its current status. Let $flag_1$ denote the empty/full status of the buffer and let $flag_2$ denote the error status of the buffer. In this simple system of communicating agents, assume that there are two basic system agents, **P** and **Q**, which are responsible for controlling the buffer state flags $flag_1$ and $flag_2$, respectively. Throughout the following subsections, we illustrate how we can utilise the proposed framework to specify the communicating and concurrent behaviours of the agents **P** and **Q**, as well as the overall system behaviour of the one-place buffer.

4.2   Structure of Agent Behaviours

In [9–12], Hoare et al. presented the framework of concurrent Kleene algebra which captures the concurrent behaviour of agents. In this paper, we adopt the framework of CKA in order to describe agent behaviours in systems of communicating agents. In what follows, let $\mathcal{K} \overset{\text{def}}{=} (K, +, *, \,;\, , ^{\circledast}, ^{\circledcirc}, 0, 1)$ be called a CKA.

It is important to note that throughout this paper, the term *agent* is used in the sense used by Milner in [23] to mean any system whose behaviour consists of discrete actions. In this way, an agent can be defined by simply describing its behaviour. Because of this, we may use the terms agents and behaviours interchangeably. With this understanding of agents, the support set K of the CKA $\mathcal{K}$ represents a set of possible behaviours. The operator $+$ is interpreted as a choice between two behaviours, the operator ; is interpreted as a sequential composition of two behaviours, and the operator $*$ is interpreted as a parallel composition of two behaviours. The element 0 represents the behaviour of the *inactive agent* and the element 1 represents the behaviour of the *idle agent* just as in many process calculi. Moreover, associated with a CKA is a natural ordering relation $\leq_{\mathcal{K}}$ representing the sub-behaviour relation. For behaviours $a, b \in K$, $a \leq_{\mathcal{K}} b$ indicates that a is a sub-behaviour of b if and only if $a + b = b$.

For the one-place buffer example of Section 4.1, we consider the following set of events which are simple assignments to the buffer status flags:

$$P_1 \overset{\text{def}}{=} (\mathit{flag}_1 := \mathit{off}) \qquad Q_1 \overset{\text{def}}{=} (\mathit{flag}_2 := \mathit{off})$$
$$P_2 \overset{\text{def}}{=} (\mathit{flag}_1 := \mathit{on}) \qquad Q_2 \overset{\text{def}}{=} (\mathit{flag}_2 := \mathit{on})$$

In this way, K is generated by the set of basic behaviours $\{P_1, P_2, Q_1, Q_2, 0, 1\}$ where 0 is interpreted as abort and 1 is interpreted as skip.

4.3   Structure of External Stimuli

As mentioned in Section 2, a stimulus constitutes the basis for behaviour. Because of this, each discrete, observable event introduced to a system, such as that which occurs through the communication among agents or from the system environment, is considered to be an external stimulus which invokes a response from each system agent.

Definition 4 (Stimulus Structure). *Let* $\mathcal{S} \overset{\text{def}}{=} (S, \oplus, \odot, \eth, \mathfrak{n})$ *be an idempotent semiring with a multiplicatively absorbing* $\eth$ *and identity* $\mathfrak{n}$. *We call* $\mathcal{S}$ *a stimulus structure.*

Within the context of external stimuli, S is the set of external stimuli which may be introduced to a system. The operator $\oplus$ is interpreted as a choice between two external stimuli and the operator $\odot$ is interpreted as a sequential composition of two external stimuli. The element $\eth$ represents the *deactivation stimulus* which influences all agents to become inactive and the element $\mathfrak{n}$ represents the *neutral stimulus* which has no influence on the behaviour of all agents. Furthermore, each stimulus structure has a natural ordering relation $\leq_{\mathcal{S}}$ representing the sub-stimulus relation. For external stimuli $s, t \in S$, we write $s \leq_{\mathcal{S}} t$ and say that s is sub-stimulus of t if and only if $s \oplus t = t$.

Continuing with the one-place buffer example of Section 4.1, suppose that the behaviour of each agent in the one-place buffer system is influenced by a number of external stimuli which either place an item in the buffer, remove an item from the buffer, or generate an error. We denote these stimuli by *in*, *out*, and *error* respectively. These external stimuli form a stimulus structure $\mathcal{S}$ where S is generated by the set of basic external stimuli $\{in, out, error, \eth, \mathfrak{n}\}$ where we interpret $\eth$ as a kill signal and $\mathfrak{n}$ as any stimulus with no influence that belongs to the complement of the set of external stimuli which may be introduced to a system.

4.4 Communicating Concurrent Kleene Algebra (C²KA)

C²KA extends the algebraic foundation of CKA with the notions of semimodules and stimulus structures to capture the influence of external stimuli on the behaviour of system agents.

Definition 5 (Communicating Concurrent Kleene Algebra). *A Communicating Concurrent Kleene Algebra (C²KA) is a system $(S, \mathcal{K})$, where $S = (S, \oplus, \odot, \mathfrak{d}, \mathfrak{n})$ is a stimulus structure and $\mathcal{K} = (K, +, *, ;, \circledast, \odot, 0, 1)$ is a CKA such that $({}_S K, +)$ is a unitary and zero-preserving left S-semimodule   with mapping $\circ : S \times K \to K$ and $(S_\mathcal{K}, \oplus)$ is a unitary and zero-preserving right $\mathcal{K}$-semimodule with mapping[1] $\lambda : S \times K \to S$, and where the following axioms are satisfied for all $a, b, c \in K$ and $s, t \in S$:*

(i) $s \circ (a;b) = (s \circ a) ; (\lambda(s,a) \circ b)$ *(iii)* $\lambda(s \odot t, a) = \lambda(s, (t \circ a)) \odot \lambda(t, a)$
(ii) $c \leq_\mathcal{K} a \ \lor \ (s \circ a) ; (\lambda(s,c) \circ b) = 0$

In essence, a C²KA consists of two semimodules which describe how the stimulus structure S and the CKA $\mathcal{K}$ mutually act upon one another in order to characterise the response invoked by an external stimulus on the behaviour of an agent as a next behaviour and a next stimulus.

First, the left S-semimodule $({}_S K, +)$ describes how the stimulus structure S acts upon the CKA $\mathcal{K}$ via the mapping $\circ$. We call $\circ$ the *next behaviour mapping* since it describes how an external stimulus invokes a behavioural response from a given agent. From $({}_S K, +)$, we have that the next behaviour mapping $\circ$ distributes over $+$ and $\oplus$. Additionally, since $({}_S K, +)$ is unitary, we have that the neutral stimulus has no influence on the behaviour of all agents and since it is zero-preserving, the deactivation stimulus influences all agents to become inactive. Second, the right $\mathcal{K}$-semimodule $(S_\mathcal{K}, \oplus)$ describes how the CKA $\mathcal{K}$ acts upon the stimulus structure S via the mapping λ. We call λ the *next stimulus mapping* since it describes how a new stimulus is generated as a result of the response invoked by a given external stimulus on an agent behaviour. From $(S_\mathcal{K}, \oplus)$, we have that the next stimulus mapping λ distributes over $\oplus$ and $+$. Also, since $(S_\mathcal{K}, \oplus)$ is unitary, we have that the idle agent forwards any external stimulus that acts on it and since $(S_\mathcal{K}, \oplus)$ is zero-preserving, the inactive agent always generates the deactivation stimulus.

In Definition 5, Axiom (i) describes the interaction of the next behaviour mapping $\circ$ with the sequential composition operator $;$ for agent behaviours. This axiom corresponds to the definition of the transition function for the cascading product (or synchronous serial composition) of Mealy automata [14]. Axiom (ii), which we call the *cascading output law*, states that when an external

---

[1] We use an infix notation for the next behaviour mapping $\circ$ and a prefix notation for the next stimulus mapping λ. We adopt these notations in an effort to reach out to those in the communities of monoid acts and Mealy automata since they adopt a similar non-uniform notation.

stimulus is introduced to the sequential composition $(a\,;b)$, then the cascaded stimulus must be generated by a sub-behaviour of a. In this way, the cascading output law ensures consistency between the next behaviour and next stimulus mappings with respect to the sequential composition of agent behaviours. It allows distributivity of $\circ$ over $;$ to be applied indiscriminately. Finally, Axiom (iii) describes the interaction of the next stimulus mapping λ with the sequential composition operator $\odot$ for external stimuli. This can be viewed as the analog of Axiom (i) with respect to the next stimulus mapping λ when considering the action of $(S_\mathcal{K}, \oplus)$.

In a given system of communicating agents, agent behaviour can be initiated in two ways. The first way to initiate agent behaviour in a system of communicating agents is by reactivation. We say that a $\mathrm{C}^2\mathrm{KA}$ is *with reactivation* if $s \circ 1 \neq 1$ for some $s \in S\backslash\{\mathfrak{d}\}$. Consider the case where the idle agent 1 is not fixed with respect to some given external stimulus. Then, the passive idle agent could be influenced to behave as any active agent. In this case, we say that the agent has been *reactivated* as it then begins to actively participate in the system operation. If a $\mathrm{C}^2\mathrm{KA}$ is without reactivation, then the idle agent 1 reflects an idle behaviour that is not influenced by any external stimulus other than the deactivation stimulus. In this case, the idle agent does not actively participate in the operation of a system and it cannot initiate agent behaviours. The second way in which agent behaviour can be initiated in a system of communicating agents is by external stimuli. In a $\mathrm{C}^2\mathrm{KA}$, we say that an agent $a \in K\backslash\{0, 1\}$ is a *stimulus initiator* if and only if $\lambda(\mathfrak{n}, a) \neq \mathfrak{n}$. When an agent is a stimulus initiator then that agent may generate a new stimulus without outside influence. Because $(S_\mathcal{K}, \oplus)$ is unitary and zero-preserving, the inactive agent 0 and the idle agent 1 cannot be stimulus initiators. Intuitively, the inactive agent is not a stimulus initiator since it can only generate the deactivation stimulus to influence all other agents to cease their behaviours and become inactive. Likewise, the idle agent is not a stimulus initiator since it can be seen as having no state-changing observed behaviour and therefore it cannot generate any stimuli.

A Comment on a Model for $\mathrm{C}^2\mathrm{KA}$. In [9–12], we find the following model for CKA. Let EV be a set of event occurrences. A trace is a set of events and a program is a set of traces. The set of all traces over EV is denoted by $TR(EV) \overset{\text{def}}{=} \mathcal{P}(EV)$ and the set of all programs is denoted by $PR(EV) \overset{\text{def}}{=} \mathcal{P}(TR(EV))$. Obviously, $(PR(EV), \cup, *, \,;, ^\circledast, ^\odot, \emptyset, \{\emptyset\})$ is a CKA [9–12]. Moreover, the structure of external stimuli is modelled by sets of strings. In this way, it is easy to see that $(\mathcal{P}(\Lambda), \cup, \bullet, \emptyset, \{\epsilon\})$ is a stimulus structure where Λ is a set of alphabet symbols, $\bullet$ denotes set concatenation, and ϵ is the empty string.

In a $\mathrm{C}^2\mathrm{KA}$, the semimodules $(_\mathcal{S}K, +)$ and $(S_\mathcal{K}, \oplus)$ contain a left $\mathcal{S}$-act $_\mathcal{S}K$ and a right $\mathcal{K}$-act $K_\mathcal{S}$, respectively. It is well known that monoid acts can be considered as semiautomata [18, pg. 45]. By combining these two semiautomata, we obtain a Mealy automaton. A Mealy automaton is given by a five-tuple $(Q, \Sigma, \Theta, F, G)$ [14]. The set of states Q is a subset of $PR(EV)$ (i.e., the set K). In this way, each state of the Mealy automaton represents a possible program that can be executed by the system as a reaction to the stimulus (input) leading to the state. The input alphabet Σ and output alphabet Θ are given by the stimulus structure such that $\Sigma = \Theta = S$. Finally, the transition function $F : \Sigma \times Q \to Q$ and the output function $G : \Sigma \times Q \to \Theta$ correspond to the next behaviour mapping $\circ : S \times K \to K$ and next stimulus mapping $\lambda : S \times K \to S$, respectively. These mappings respectively correspond to the transition functions from the semiautomata representations of $_SK$ and K_S.

The proposed model is also equipped with two operations for Mealy automata. The operation ; is associative and the operation $+$ is associative, idempotent, and commutative. The ; operation corresponds to the *cascading product* of Mealy automata and the operation $+$ corresponds to the *full direct product* of Mealy automata [14].

Proposition 3. *Let* (S, K) *be a* C^2KA. *For all* $a, b \in K$ *and* $s, t \in S$:

(i) $a \leq_K b \wedge s \leq_S t \implies s \circ a \leq_K t \circ b$
(ii) $a \leq_K b \wedge s \leq_S t \implies \lambda(s, a) \leq_S \lambda(t, b)$

The isotonicity laws of Corollary 1 follow immediately from Proposition 3. In [9], an idempotent semiring is called a *quantale* if the natural order induces a complete lattice and multiplication distributes over arbitrary suprema.

Corollary 1. *In a* C^2KA *where the underlying* CKA *and stimulus structure are built up from quantales, the following laws hold:*

(i) $a \leq_K b \implies s \circ a \leq_K s \circ b$ *(vi)* $s \leq_S t \implies \lambda(s, a) \leq_S \lambda(t, a)$
(ii) $s \leq_S t \implies s \circ a \leq_K t \circ a$ *(vii)* $a \leq_K b \implies \lambda(s, a) \leq_S \lambda(s, b)$
(iii) $s \circ (a ; b + b ; a) \leq_K s \circ (a * b)$ *(viii)* $\lambda(s, (a ; b + b ; a)) \leq_S \lambda(s, (a * b))$
(iv) $s \circ a^\odot \leq_K s \circ a^\circledast$ *(ix)* $\lambda(s, a^\odot) \leq_S \lambda(s, a^\circledast)$
(v) $s \circ a^\odot = +(n \mid n \geq 0 : s \circ a^n)$ *(x)* $\lambda(s, a^\odot) = \oplus(n \mid n \geq 0 : \lambda(s, a^n))$

4.5 Specifying Systems of Communicating Agents with C^2KA

In order to specify a system of communicating agents using C^2KA, we have three levels of specification. Using the illustrative example of the one-place buffer from Section 4.1, we show how to specify the system agents using the proposed framework.

The *stimulus-response specification of agents* level gives the specification of the next behaviour mapping $\circ$ and the next stimulus mapping λ for each agent in the system. Assuming that we have a C^2KA without reactivation, the agent behaviours of **P** and **Q** are compactly specified as shown in Table 1. By composing the behaviours of **P** and **Q**, we are able to obtain the complete behaviour of the one-place buffer. The full stimulus-response specification of the buffer agent can be found in Table 3 in [16].

The *abstract behaviour specification* level restricts the specification to the desired behaviour of an agent in the communicating system by computing the responses to the external stimuli that can be introduced into the system in the given context. In the one-place buffer example, consider a context in which we

Table 1. Stimulus-Response Specification for Agents **P** and **Q**

$\mathbf{P} \overset{\text{def}}{=} P_1 + P_2$ $\qquad\qquad\qquad\qquad\qquad$ $\mathbf{Q} \overset{\text{def}}{=} Q_1 + Q_2$

$\circ_{\mathbf{P}}$	$\mathfrak{n}$	in	out	$error$
P_1	P_1	P_2	P_1	P_1
P_2	P_2	P_2	P_1	P_2

$\circ_{\mathbf{Q}}$	$\mathfrak{n}$	in	out	$error$
Q_1	Q_1	Q_1	Q_1	Q_2
Q_2	Q_2	Q_2	Q_2	Q_2

$\lambda_{\mathbf{P}}$	$\mathfrak{n}$	in	out	$error$
P_1	$\mathfrak{n}$	$\mathfrak{n}$	$error$	$\mathfrak{n}$
P_2	$\mathfrak{n}$	$error$	$\mathfrak{n}$	$\mathfrak{n}$

$\lambda_{\mathbf{Q}}$	$\mathfrak{n}$	in	out	$error$
Q_1	$\mathfrak{n}$	$\mathfrak{n}$	$\mathfrak{n}$	$\mathfrak{n}$
Q_2	$\mathfrak{n}$	$\mathfrak{n}$	$\mathfrak{n}$	$\mathfrak{n}$

$$\forall(P_i, Q_i \mid 1 \leq i \leq 2 : \mathfrak{d} \circ P_i = 0 \land \mathfrak{d} \circ Q_i = 0 \land \lambda(\mathfrak{d}, P_i) = \mathfrak{d} \land \lambda(\mathfrak{d}, Q_i) = \mathfrak{d})$$

only consider the buffer as behaving either as an empty buffer or as a full buffer. Furthermore, assume that the behaviour of the buffer may only be influenced by the introduction of in and out stimuli since these are the only stimuli that another external agent may have control over. This is to say that an external agent cannot issue an $error$ since this is an uncontrollable stimulus which cannot be issued at will. In this way, after simple computation, we find that the abstract behaviour of the one-place buffer is given by $P_1 ; Q_1 + P_1 ; Q_2 + P_2 ; Q_1 + P_2 ; Q_2$. At the abstract behaviour specification level, C²KA can be viewed as an event-based model of communication. In C²KA, the left S-semimodule $\left({}_{\mathcal{S}}K, +\right)$ and the right $\mathcal{K}$-semimodule $\left(S_{\mathcal{K}}, \oplus\right)$ allow us to specify how the external stimuli influence the behaviour of each agent in a given system. For this reason, this level of specification is best suited for describing message passing communication where agents transfer information explicitly through the exchange of data structures, either synchronously or asynchronously.

Finally, the *concrete behaviour specification* level provides the state-level specification of each agent behaviour (i.e., each program). At this level, we define the concrete programs for each of the CKA terms which specify each agent behaviour. The concrete behaviour specification provides the following state-level programs for each behaviour of the one-place buffer.

$$\text{EMPTY} \overset{\text{def}}{=} P_1 ; Q_1 = \quad (flag_1 := off \; ; \; flag_2 := off)$$
$$\text{FULL} \overset{\text{def}}{=} P_2 ; Q_1 = \quad (flag_1 := on \; ; \; flag_2 := off)$$
$$\text{UNDERFLOW} \overset{\text{def}}{=} P_1 ; Q_2 = \quad (flag_1 := off \; ; \; flag_2 := on)$$
$$\text{OVERFLOW} \overset{\text{def}}{=} P_2 ; Q_2 = \quad (flag_1 := on \; ; \; flag_2 := on)$$

Fig. 2. Concrete behaviour specification of the one-place buffer

Since C²KA extends concurrent Kleene algebra, it inherits this model of communication from CKA. Just as in CKA, the instantiation of a low-level model of programs and traces for C²KA affords the ability to specify communication through shared events and the dependencies between them. Because of this, this level of specification is best suited for shared-variable communication where agents transfer information through a shared medium such as variables, memory locations, etc.

Depending on which level of specification we are working at, the model can be viewed as either event-based or state-based. This gives flexibility in allowing us to choose which level is most suitable for the given problem. The context of the given problem will help to dictate at which level we need to work. For a full treatment of the illustrative example of the one-place buffer, the reader is referred to [16].

4.6 C²KA and Orbits, Stabilisers, and Fixed Points

Orbits, stabilisers, and fixed points are notions that allow us to perceive a kind of topology of a system with respect to the stimulus-response relationships among the system agents. Because of this, we are able to gain some insight into the communication channels that can be established among system agents. For example, with C²KA, we are able to compute the strong orbits (presented below) of the agent behaviours in a given system. The strong orbits represent the strongly connected agent behaviours in the system and therefore can provide some insight into the abilities of the agents in the same strong orbit to influence one another's behaviour through communication. Furthermore, having an idea of the topology of the system allows for the abstraction of components of the overall system behaviour. This kind of abstraction can aid in separating the communicating and concurrent behaviour in a system and its environment. Moreover, computing the orbits and stabilisers of agent behaviours can aid in the analysis and verification of systems of communicating agents, since it allows us to model the possible reaction of a system to a stimulus. Also, they allow us, in some cases, to reduce the analysis to only some relevant orbits of a system. Similarly, stabilisers allow us to reduce the analysis to studying only the stimuli that influence the behaviour of an agent. We conjecture that such reduction could, for example, alleviate the state explosion problem in model checking.

Since a C²KA consists of two semimodules $\left(_S K, +\right)$ and $\left(S_{\mathcal{K}}, \oplus\right)$ for which we have a left S-act $_S K$ and a right $\mathcal{K}$-act $S_{\mathcal{K}}$, we have two complementary notions of orbits, stabilisers, and fixed points within the context of agent behaviours and external stimuli, respectively. In this way, one can use these notions to think about concurrent and communicating systems from two different perspectives, namely the behavioural perspective provided by the action of external stimuli on agent behaviours described by $\left(_S K, +\right)$ and the external event (stimulus) perspective provided by the action of agent behaviours on external stimuli described by $\left(S_{\mathcal{K}}, \oplus\right)$. In this section, we focus only on the treatment of these notions with respect to the left S-semimodule $\left(_S K, +\right)$ and agent behaviours. In a very similar way, we can present the same notions for the right $\mathcal{K}$-semimodule $\left(S_{\mathcal{K}}, \oplus\right)$ and external stimuli.

Definition 6 recalls the notions of orbits, stabilisers, and fixed points from the mathematical theory of monoids acting on sets [18].

Definition 6. Let $\left(_S K, +\right)$ be the unitary and zero-preserving left S-semimodule of a C²KA and let $a \in K$.

(i) The orbit of a in S is the set given by $\mathrm{Orb}(a) = \{s \circ a \mid s \in S\}$.

(ii) The strong orbit of a in S is the set given by $\mathrm{Orb}_S(a) = \{b \in K \mid \mathrm{Orb}(b) = \mathrm{Orb}(a)\}$.

(iii) *The stabiliser of a in S is the set given by* $\text{Stab}(a) = \{s \in S \mid s \circ a = a\}$.
(iv) *An element* $a \in K$ *is called a* fixed point *if* $\forall (s \mid s \in S \backslash \{\eth\} : s \circ a = a)$.

We can define a preorder on K as $a \preceq_{\mathcal{K}} b \iff \text{Orb}(a) \subseteq \text{Orb}(b)$. Given this preorder, we can obtain an equivalence relation $\sim_{\mathcal{K}}$ from the intersection of $\preceq_{\mathcal{K}}$ and $\succeq_{\mathcal{K}}$. The equivalence classes of $\sim_{\mathcal{K}}$ give the strong orbits [20]. The strong orbits can also be viewed as the strongly connected components of a directed graph [29]. Additionally, when $a \in K$ is a fixed point, $\text{Orb}(a) = \{0, a\}$ and $\text{Stab}(a) = S \backslash \{\eth\}$. It is important to note that since $({}_{s}K, +)$ is zero-preserving, every agent behaviour becomes inactive when subjected to the deactivation stimulus $\eth$. Because of this, we exclude this special case when discussing fixed agent behaviours.

Before we discuss the interplay between C^2KA and the notions of orbits, stabilisers, and fixed points, we first extend the partial order of sub-behaviours $\leq_{\mathcal{K}}$ to sets in order to express sets of agent behaviours encompassing one another.

Definition 7 (Encompassing Relation). *Let* $A, B \subseteq K$ *be two subsets of agent behaviours. We write* $A \lessdot_{\mathcal{K}} B$ *and say that* A *is encompassed by* B *(or* B *encompasses* A *) if and only if* $\forall (a \mid a \in A : \exists (b \mid b \in B : a \leq_{\mathcal{K}} b))$.

The encompassing relation $\lessdot_{\mathcal{S}}$ for external stimuli can be defined similarly.

Orbits. The orbit of an agent $a \in K$ represents the set of all possible behavioural responses from an agent behaving as a to any external stimulus from $\mathcal{S}$. In this way, the orbit of a given agent can be perceived as the set of all possible future behaviours for that agent. With regard to the specification of the one-place buffer, we can compute the orbits of each of the buffer behaviours. For instance, $\text{Orb}(\text{EMPTY}) = \{\text{EMPTY}, \text{FULL}, \text{UNDERFLOW}, \text{OVERFLOW}\}$.

Proposition 4 provides an isotonicity law with respect to the orbits and the encompassing relation for agent behaviours.

Proposition 4. *Let* $(\mathcal{S}, \mathcal{K})$ *be a* C^2KA. *Then,* $a \leq_{\mathcal{K}} b \implies \text{Orb}(a) \lessdot_{\mathcal{K}} \text{Orb}(b)$ *for all* $a, b \in K$.

A selection of additional properties follow immediately from Proposition 4 and are given in Corollary 2.

Corollary 2. *In a* C^2KA *the following laws hold for all* $a, b, c \in K$:

  (i) $\text{Orb}(a) \lessdot_{\mathcal{K}} \text{Orb}(a + b)$
 (ii) $\text{Orb}((a * b) \,;(c * d)) \lessdot_{\mathcal{K}} \text{Orb}((a \,;c) * (b \,;d))$
(iii) $\text{Orb}(a \,;b) \lessdot_{\mathcal{K}} \text{Orb}(a * b)$
 (iv) $\text{Orb}(a \,;b + b \,;a) \lessdot_{\mathcal{K}} \text{Orb}(a * b)$
  (v) $\text{Orb}((a * b) \,;c) \lessdot_{\mathcal{K}} \text{Orb}(a * (b \,;c))$
 (vi) $\text{Orb}(a \,;(b * c)) \lessdot_{\mathcal{K}} \text{Orb}((a \,;b) * c)$
(vii) $\text{Orb}(a^{\oplus}) \lessdot_{\mathcal{K}} \text{Orb}(a^{\circledast})$
(viii) $\text{Orb}(a) \lessdot_{\mathcal{K}} \text{Orb}(c) \wedge \text{Orb}(b) \lessdot_{\mathcal{K}} \text{Orb}(c) \iff \text{Orb}(a) \cup \text{Orb}(b) \lessdot_{\mathcal{K}} \text{Orb}(c)$

As stated before, without discussing the properties derived from the right $\mathcal{K}$-semimodule $(S_{\mathcal{K}}, \oplus)$, due to the cascading output law (see Definition 5 (ii)), we also have that $\text{Orb}((s \circ a) \,;(\lambda(s, c) \circ b)) = \{0\}$ for any $(a \,;b) \in K$ and $\neg(c \leq_{\mathcal{K}} a)$.

Another Interpretation of Orbits. As mentioned in Section 2, we call the influence of external stimuli on agent behaviours the *induced behaviours* via external stimuli. The notion of induced behaviours allows us to make some predictions about the evolution of agent behaviours in a given system by providing some insight into the topology of the system and how different agents can respond to any external stimuli. Here, we provide a formal treatment of the notion of induced behaviours. While studying induced behaviours, we focus particularly on the next behaviour mapping ∘ and the effects of external stimuli on agent behaviours since we are interested in examining the evolution of agent behaviours via the influence of external stimuli in a given system of communicating agents.

Definition 8 (Induced Behaviour). *Let $a, b \in K$ be agent behaviours such that $a \neq b$. We say that b is* induced by a *via external stimuli (denoted by $a \lhd b$) if and only if $\exists(s \mid s \in S : s \circ a = b)$.*

Equivalently, we can express $a \lhd b \iff b \in \mathrm{Orb}(a)$ for $a \neq b$. In this way, it can be seen that the orbit of a behaviour a represents the set of all behaviours which are induced by a via external stimuli. Considering the one-place buffer example, it is plain to see, for instance, that EMPTY $\lhd$ UNDERFLOW via the external stimulus *out* and EMPTY $\lhd$ OVERFLOW via the external stimulus $in \odot in$.

Strong Orbits. Two agents are in the same strong orbit, denoted $a \sim_{\mathcal{K}} b$ for $a, b \in K$, if and only if their orbits are identical. This is to say when $a \sim_{\mathcal{K}} b$, if an agent behaving as a is influenced by an external stimulus to behave as b, then there exists an external stimulus which influences the agent, now behaving as b, to revert back to its original behaviour a. Furthermore, if $a \sim_{\mathcal{K}} b$, then $\exists(s, t \mid s, t \in S : s \circ a = b \wedge t \circ b = a)$. In this case, the external stimuli s and t can be perceived as *inverses* of one another and allow us to revert an agent back to its original behaviour since $t \circ s \circ a = a$ and $s \circ t \circ b = b$ (i.e., $s \odot t \in \mathrm{Stab}(a)$ and $t \odot s \in \mathrm{Stab}(b)$). In the specification of the one-place buffer, we have two strong orbits, namely, those given by {EMPTY, FULL} and {UNDERFLOW, OVERFLOW} which represent the behaviours from agents **P** and **Q**, respectively. This is to say that we have (EMPTY $\sim_{\mathcal{K}}$ FULL) and (UNDERFLOW $\sim_{\mathcal{K}}$ OVERFLOW).

Stabilisers. For any agent $a \in K$, the stabiliser of a represents the set of external stimuli which have no observable influence (or act as neutral stimuli) on an agent behaving as a. In the illustrative example of the one-place buffer, we can compute the stabilisers of each of the buffer behaviours from the specification of the buffer agent. For example, $\mathrm{Stab}(\text{EMPTY})$ is generated by {*error*, $in \odot out$}.

By straightforward calculation and the definition of the encompassing relation $\leqslant_{\mathcal{S}}$ for external stimuli, we have that $\mathrm{Stab}(a) \cap \mathrm{Stab}(b) \leqslant_{\mathcal{S}} \mathrm{Stab}(a + b)$ for $a, b \in K$. However, consider a case where $\exists(s \mid s \in S : s \circ a = b \wedge s \circ b = a)$. Then, $s \notin \mathrm{Stab}(a)$ and $s \notin \mathrm{Stab}(b)$ but $s \in \mathrm{Stab}(a + b)$. Therefore, it is easy to see that in general $\neg(\mathrm{Stab}(a + b) \leqslant_{\mathcal{S}} (\mathrm{Stab}(a) \cap \mathrm{Stab}(b)))$ and $\neg(\mathrm{Stab}(a + b) \leqslant_{\mathcal{S}} (\mathrm{Stab}(a) \cup \mathrm{Stab}(b)))$.

Fixed Points. Depending on the given specification of a system of communicating agents, there may be any number of fixed points with respect to the next behaviour mapping ∘. When an agent behaviour is a fixed point, it is not influenced by any external stimulus other than the deactivation stimulus $\eth$. For example, with regard to the specification of agents for the one-place buffer example, it is easy to see that the behaviour Q_2 is a fixed point. The existence of fixed point behaviours is important when considering how agents can communicate via external stimuli. For instance, an agent that has a fixed point behaviour, does not have any observable response to any external stimuli (except for the deactivation stimulus) and therefore it can be seen that such an agent cannot be a receiver in any sort of communication via external stimuli.

Proposition 5 gives a selection of properties regarding fixed agent behaviours.

Proposition 5. *Let* $(\mathcal{S}, \mathcal{K})$ *be a* C^2KA *and let* $a, b \in K$ *such that a and b are fixed points. We have:*

(i) 0 *is a fixed point*
(ii) $a + b$ *is a fixed point*
(iii) $a \,;b$ *is a fixed point*
(iv) $a^{\odot}$ *is a fixed point if additionally* $(\mathcal{S}, \mathcal{K})$ *is without reactivation*

In Proposition 5, Identity (i) states that the inactive agent 0 is a fixed point with respect to the next behaviour mapping ∘. In this way, the inactive agent is not influenced by any external stimulus. Similarly, we can see that the deactivation stimulus $\eth$ is a fixed point with respect to the next stimulus mapping λ if we consider the notion of a fixed point in terms of external stimuli. Identity (ii) (resp. (iii) and (iv)) state that the choice (resp. sequential composition and sequential iteration) of fixed point behaviours results in a fixed point behaviour. In general, even if $a, b \in K$ are both fixed points, we are unable to say anything about $(a * b)$ as a fixed point.

Proposition 6 provides further insight into how the topology of a system of communicating agents can be perceived using C^2KA and the notion of induced behaviours.

Proposition 6. *Let* $a, b, c \in K$ *be agent behaviours.*

(i) a *is a fixed point* $\implies$ $\forall (b \mid b \in K \,\wedge\, b \neq 0 \,\wedge\, b \neq a : \neg(a \lhd b))$
(ii) $a \sim_{\mathcal{K}} b \implies a \lhd b \,\wedge\, b \lhd a$
(iii) $a \sim_{\mathcal{K}} b \implies (a \lhd c \iff b \lhd c)$

Proposition 6(i) states that if an agent has a fixed point behaviour, then it does not induce any agent behaviours via external stimuli besides the inactive behaviour 0. This is a direct consequence of the fact that an agent with a fixed point behaviour is not influenced by any external stimuli (except for the deactivation stimulus $\eth$) and therefore remains behaving as it is. Proposition 6(ii) states that all agent behaviours which belong to the same strong orbit are mutually induced via some (possibly different) external stimuli. This is to say that if two agent behaviours are in the same strong orbit, then there exists inverse stimuli for each agent behaviour in a strong orbit allowing an agent to revert back to its original behaviour. Finally, Proposition 6(iii) states that if two agent behaviours are in the same strong orbit, then a third behaviour can be induced via external stimuli by either of the behaviours within the strong orbit. This is to say that each behaviour in a strong orbit can induce the same set of behaviours (perhaps via different external stimuli). Therefore, the strong orbit to which these behaviours belong can be abstracted and perceived as an equivalent agent behaviour with respect to the behaviours which it can induce via external stimuli.

5   Related Work and Discussion

Existing state-based and event-based formalisms for communication and concurrency such as temporal logics, labelled transition systems, Petri nets, and process calculi are primarily interested in modelling the behaviour of a system either in terms of the properties of its states or in terms of the observability of events. However, they do not directly, if at all, provide a hybrid model of communication and concurrency which encompass the characteristics of both state-based and event-based models. Concurrent Kleene algebra is perhaps the closest formalism to providing such a hybrid model. While CKA can be perceived as a hybrid model for concurrency, the same cannot be said for communication since communication in CKA is not directly evident.

C^2KA offers an algebraic setting which can capture both the influence of external stimuli on agent behaviour as well the communication and concurrency of agents at the abstract algebraic level. It uses notions from classical algebra to extend the algebraic foundation provided by CKA. If we consider a C^2KA with a trivial stimulus structure (i.e., $S = \{\mathfrak{n}\}$), then the next behaviour and next stimulus mappings are trivial and the C^2KA reduces to a CKA.

In the past, communication has been studied in process algebras such as CCS and CSP. As discussed in [9, 11, 12], some analogies can be made between relating CKA with process algebras. Therefore, if we consider the case where we have a trivial stimulus structure, then we can make the same kind of analogies relating C^2KA with existing process algebras.

In [9–12], Hoare et al. have taken steps towards investigating some aspects of communication through the derivation of rules for a simplified rely/guarantee calculus using CKA. However, this kind of communication is only captured via shared events. Since the proposed framework provides an extension of CKA, it is also capable of achieving these results. Furthermore, C^2KA supports the ability to work in either a state-based model (as illustrated by Figure 2) or an event-based model (as illustrated by Table 1) for the specification of concurrent and communicating systems. It gives us the ability to separate the communicating and concurrent behaviour in a system and its environment. This separation of concerns allows us to consider the influence of stimuli from the world in which the agent resides as transformations of agent behaviour and yields the three levels of specification offered by C²KA. With these levels of specification, C²KA is able to capture the notions of message passing communication and shared-variable communication consistent with the hybrid view of agent communication depicted in Figure 1. Specifically, at the abstract behaviour specification level, we are interested only in the behaviour of an agent as dictated by the stimulus-response relationships that exist in the given system. In this way, the behaviour of an agent is dictated by its responses to external stimuli without the need to articulate the internal state-based system of each behaviour. On the other hand, by instantiating a concrete model of agent behaviour, such as that of programs and traces similar to what is done with CKA [9–12] at the concrete behaviour specification level, we have the ability to define the state-based model of agent behaviour. In this way, if the given problem requires insight into how external stimuli are processed by an agent, the concrete behaviour specification level affords the ability to specify such internal states of agent behaviours in terms of programs on concrete state variables. Because of this, C²KA is flexible in allowing the context of the given problem to dictate which level of abstraction is most suitable. For example, if the given problem need not worry about the internal states of agent behaviours, then we can specify the system at the abstract behaviour specification level without any modifications to the proposed framework. Moreover, C²KA inherits the algebraic foundation of CKA with all of its models and theory.

6   Conclusion and Future Work

In this paper, we proposed a mathematical framework for communication and concurrency called Communicating Concurrent Kleene Algebra (C²KA). C²KA extends the algebraic setting of concurrent Kleene algebra with semimodules in order to capture the influence of external stimuli on the behaviour of system agents in addition to the communication among agents through shared variables and communication channels. C²KA supports the ability to work in either a state-based or event-based model for both the specification of communicating and concurrent behaviour by providing three levels of specification which reflect different levels of abstraction for the behaviour of agents in a given system. To the best of our knowledge, such a formalism does not currently exist in the literature and is required for dealing with problems such as studying the necessary conditions for covert channel existence [15]. A hybrid view of communication among agents and the influence of external stimuli on agent behaviour needs to be considered when examining the potential for communication condition for covert channels. Because of the separation of communicating and concurrent behaviour, we expect that C²KA can aid in designing and analysing systems which are robust against covert communication channels. Since it provides a means for specifying systems of communicating agents, C²KA can be an integral part of verifying the necessary conditions for covert channels [15]. We are using it to formalise and verify the potential for communication condition for covert channel existence. Also, we are developing a prototype tool using the Maude term rewriting system [3] to support the automated computation and specification of systems of communicating agents using C^2KA. In future work, we aim to examine the ability to adapt C^2KA for use in solving interface equations (e.g., [28]) which can allow for implicit agent behaviour specifications in a variety of application domains. Furthermore, we intend to further investigate the theory and use of C^2KA to capture and explain the influence of external stimuli on agent behaviour in social networking environments.

Acknowledgements. This research is supported by the Natural Sciences and Engineering Research Council of Canada (NSERC) through the grant RGPIN227806-09 and the NSERC PGS D program. We would also like to thank the anonymous reviewers for their valuable comments which helped us considerably improve the quality of the paper.

References

1. Bergstra, J., Klop, J.: Process algebra for synchronous communication. Information and Control 60(1-3), 109–137 (1984)
2. Clarke, E.M., Emerson, E.A.: Design and synthesis of synchronization skeletons using branching time temporal logic. In: Kozen, D. (ed.) Logic of Programs 1981. LNCS, vol. 131, pp. 52–71. Springer, Heidelberg (1982)
3. Clavel, M., Durán, F., Eker, S., Lincoln, P., Martí-Oliet, N., Meseguer, J., Talcott, C.: The Maude 2.0 System. In: Nieuwenhuis, R. (ed.) RTA 2003. LNCS, vol. 2706, pp. 76–87. Springer, Heidelberg (2003)
4. Cleaveland, R., Smolka, S.: Strategic directions in concurrency research. ACM Computing Surveys 28(4), 607–625 (1996)
5. Emerson, E., Halpern, J.: "sometimes" and "not never" revisited: On branching versus linear time temporal logic. Journal of the ACM 33(1), 151–178 (1986)
6. Hebisch, U., Weinert, H.: Semirings: Algebraic Theory and Applications in Computer Science. Series in Algebra, vol. 5. World Scientific (1993)
7. Hoare, C.: Communicating sequential processes. Communications of the ACM 21(8), 666–677 (1978)
8. Hoare, C.: Communicating Sequential Processes. Prentice-Hall (1985)
9. Hoare, C., Möller, B., Struth, G., Wehrman, I.: Concurrent Kleene algebra. In: Bravetti, M., Zavattaro, G. (eds.) CONCUR 2009. LNCS, vol. 5710, pp. 399–414. Springer, Heidelberg (2009)
10. Hoare, C., Möller, B., Struth, G., Wehrman, I.: Foundations of concurrent Kleene algebra. In: Berghammer, R., Jaoua, A.M., Möller, B. (eds.) RelMiCS/AKA 2009. LNCS, vol. 5827, pp. 166–186. Springer, Heidelberg (2009)
11. Hoare, C., Möller, B., Struth, G., Wehrman, I.: Concurrent Kleene algebra and its foundations. Tech. Rep. CS-10-04, University of Sheffield, Department of Computer Science, Sheffield, UK (August 2010), http://www.dcs.shef.ac.uk/~georg/ka/
12. Hoare, C., Möller, B., Struth, G., Wehrman, I.: Concurrent Kleene algebra and its foundations. Journal of Logic and Algebraic Programming 80(6), 266–296 (2011)
13. Hoare, C., Wickerson, J.: Unifying models of data flow. In: Broy, M., Leuxner, C., Hoare, C. (eds.) Proceedings of the 2010 Marktoberdorf Summer School on Software and Systems Safety, pp. 211–230. IOS Press (August 2011)

14. Holcombe, W.: Algebraic Automata Theory. Cambridge Studies in Advanced Mathematics. Cambridge University Press (2004)
15. Jaskolka, J., Khedri, R., Zhang, Q.: On the necessary conditions for covert channel existence: A state-of-the-art survey. Procedia Computer Science 10, 458–465 (2012); Proceedings of the 3rd International Conference on Ambient Systems, Networks and Technologies, ANT 2012 (2012)
16. Jaskolka, J., Khedri, R., Zhang, Q.: Foundations of communicating concurrent Kleene algebra. Tech. Rep. CAS-13-07-RK, McMaster University, Hamilton, Ontario, Canada (November 2013),
    http://www.cas.mcmaster.ca/cas/0template1.php?601
17. Keller, R.: Formal verification of parallel programs. Communications of the ACM 19(7), 371–384 (1976)
18. Kilp, M., Knauer, U., Mikhalev, A.: Monoids, Acts And Categories With Applications to Wreath Products and Graphs: A Handbook for Students and Researchers. De Gruyter Expositions in Mathematics Series, vol. 29. Walter de Gruyter (2000)
19. Kozen, D.: Automata and Computability. Undergraduate Texts in Computer Science. Springer (1997)
20. Linton, S., Pfeiffer, G., Robertson, E., Ruškuc, N.: Computing transformation semigroups. Journal of Symbolic Computation 33(2), 145–162 (2002)
21. Mazurkiewicz, A.: Trace theory. In: Brauer, W., Reisig, W., Rozenberg, G. (eds.) APN 1986. LNCS, vol. 255, pp. 278–324. Springer, Heidelberg (1987)
22. Milner, R.: A Calculus of Communication Systems. LNCS, vol. 92. Springer, Heidelberg (1980)
23. Milner, R.: Communication and Concurrency. Prentice-Hall International Series in Computer Science. Prentice Hall (1989)
24. Milner, R., Parrow, J., Walker, D.: A calculus of mobile processes part I. Information and Computation 100(1), 1–40 (1992)
25. Petri, C.: Kommunikation mit Automaten. Ph.D. thesis, Institut für instrumentelle Mathematik, Bonn, Germany (1962), English translation available as: Communication with Automata, Technical Report RADC-TR-65-377, vol. 1, supplement 1, Applied Data Research, Princeton, NJ (1966)
26. Pnueli, A.: The temporal logic of programs. In: Proceedings of the 18th Annual Symposium on Foundations of Computer Science, pp. 46–57 (1977)
27. Pratt, V.: Modeling concurrency with partial orders. International Journal of Parallel Programming 15(1), 33–71 (1986)
28. Shields, M.: Implicit system specification and the interface equation. The Computer Journal 32(5), 399–412 (1989)
29. Steinberg, B.: A theory of transformation monoids: Combinatorics and representation theory. The Electronic Journal of Combinatorics 17(1) (2010)
30. Watson, J.: Behaviorism. University of Chicago Press (1930)
31. Winskel, G.: Event structures. In: Brauer, W., Reisig, W., Rozenberg, G. (eds.) APN 1986. LNCS, vol. 255, pp. 325–392. Springer, Heidelberg (1987)

Concurrent Kleene Algebra with Tests

Peter Jipsen

Chapman University, Orange, California 92866, USA
jipsen@chapman.edu

Abstract. Concurrent Kleene algebras were introduced by Hoare, Möller, Struth and Wehrman in [HMSW09, HMSW09a, HMSW11] as idempotent bisemirings that satisfy a concurrency inequation and have a Kleene-star for both sequential and concurrent composition. Kleene algebra with tests (KAT) were defined earlier by Kozen and Smith [KS97]. *Concurrent Kleene algebras with tests* (CKAT) combine these concepts and give a relatively simple algebraic model for reasoning about operational semantics of concurrent programs. We generalize guarded strings to *guarded series-parallel strings*, or gsp-strings, to provide a concrete language model for CKAT. Combining nondeterministic guarded automata [Koz03] with branching automata of Lodaya and Weil [LW00] one obtains a model for processing gsp-strings in parallel, and hence an operational interpretation for CKAT. For gsp-strings that are simply guarded strings, the model works like an ordinary nondeterministic guarded automaton. If the test algebra is assumed to be $\{0, 1\}$ the language model reduces to the regular sets of bounded-width sp-strings of Lodaya and Weil.

Since the concurrent composition operator distributes over join, it can also be added to relation algebras with transitive closure to obtain the variety CRAT. We provide semantics for these algebras in the form of coalgebraic arrow frames expanded with concurrency.

Keywords: Concurrent Kleene algebras, Kleene algebras with tests, parallel programming models, series-parallel strings, relation algebras with transitive closure.

1 Introduction

Relation algebras and Kleene algebras with tests have been used to model specifications and programs, while automata and coalgebras have been used to model state based systems and object-oriented programs. To compensate for plateauing processor speed, multi-core architectures and cluster-computing are becoming widely available. However there is little agreement on how to efficiently develop software for these technologies or how to model them with suitably abstract and simple principles. The recent development of concurrent Kleene algebra [HMSW09, HMSW09a, HMSW11] builds on a computational model that is well understood and has numerous applications. Hence it is useful to explore which aspects of Kleene algebras can be lifted fairly easily to the concurrent setting, and whether the simplicity of regular languages and guarded strings can be preserved along the way. For the nonguarded case many interesting results have

P. Höfner et al. (Eds.): RAMiCS 2014, LNCS 8428, pp. 37–48, 2014.

been obtained by Lodaya and Weil [LW00] using labeled posets (or pomsets) of Pratt [Pra86] and Gisher [Gis88], but restricted to the class of series-parallel pomsets called sp-posets. This is a special case of the set-based traces and dependency relation used in [HMSW09, HMSW09a, HMSW11] to motivate the laws of CKA. Here we investigate how to extend guarded strings to handle concurrent composition with the same approach as for sp-posets in [LW00].

Recall from [KS97] that a *Kleene algebra with tests* (KAT) is an idempotent semiring with a Boolean subalgebra of tests and a unary Kleene-star operation that plays the role of reflexive-transitive closure. More precisely, it is a two-sorted algebra of the form $\mathbf{A} = (A, A', +, 0, \cdot, 1, ^-, ^*)$ where A' is a subset of A, $(A, +, 0, \cdot, 1, ^*)$ is a Kleene algebra and $(A', +, 0, \cdot, 1, ^-)$ is a Boolean algebra (the complementation operation is only defined on A').

Let Σ be a set of *basic program symbols* $p, q, r, p_1, p_2, \ldots$ and T a set of *basic test symbols* $t, t_1, t_2, \ldots$, where we assume that $\Sigma \cap T = \emptyset$. Elements of T are Boolean generators, and we write 2^T for the set of *atomic tests*, given by characteristic functions on T and denoted by $\alpha, \beta, \gamma, \alpha_1, \alpha_2, \ldots$

The collection of guarded strings over $\Sigma \cup T$ is $GS_{\Sigma,T} = 2^T \times \bigcup_{n<\omega} (\Sigma \times 2^T)^n$, and a typical guarded string is denoted by $\alpha_0 p_1 \alpha_1 p_2 \alpha_2 \ldots p_n \alpha_n$, or by $\alpha_0 w \alpha_n$ for short, where $\alpha_i \in 2^T$ and $p_i \in \Sigma$. Note that for finite T the members of $2^T \subseteq GS_{\Sigma,T}$ can be identified with the atoms of the free Boolean algebra generated by T.

Concatenation of guarded strings is via the coalesced product: $w\alpha \diamond \beta w' = w\alpha w'$ if $\alpha = \beta$ and undefined otherwise. For subsets L, M of $GS_{\Sigma,T}$ define

- $L + M = L \cup M$,
- $LM = \{v \diamond w : v \in L, w \in M \text{ and } v \diamond w \text{ is defined}\}$,
- $0 = \emptyset$, $1 = 2^T$, $\bar{L} = GS_{\Sigma,T} \setminus L$ and
- $L^* = \bigcup_{n<\omega} L^n$ where $L^0 = L$ and $L^n = LL^{n-1}$ for $n > 0$.

Then $\mathcal{P}(GS_{\Sigma,T})$ is a KAT under these operations, and one defines a map G from KAT terms over $\Sigma \cup T$ to this concrete model by

- $G(t) = \{\alpha \in 2^T : \alpha(t) = 1\}$ for $t \in T$,
- $G(p) = \{\alpha p \beta : \alpha, \beta \in 2^T\}$ for $p \in \Sigma$,
- $G(p + q) = G(p) + G(q)$, $G(pq) = G(p)G(q)$, $G(p^*) = G(p)^*$, for any terms p, q and
- $G(0) = 0$, $G(1) = 1$, $G(\bar{b}) = \overline{G(b)}$ for any Boolean term b.

The *language theoretic model* $\mathbf{G}_{\Sigma,T}$ is the subalgebra of $\mathcal{P}(GS_{\Sigma,T})$ generated by $\{G(t) : t \in T\} \cup \{G(p) : p \in \Sigma\}$. In fact $\mathbf{G}_{\Sigma,T}$ is the free KAT and its members are the *regular guarded languages*. Subsets of 2^T are called Boolean tests, and other members of $\mathbf{G}_{\Sigma,T}$ are called programs.

A *nondeterministic guarded automaton* is a coalgebra

$$\mathcal{A} : X \to \mathcal{P}(X)^{\Sigma \cup \mathcal{P}(2^T)} \times 2$$

where X is a set of states, $\mathcal{A}_0(x)(y)$ is the *set of successor states* of $x \in X$ for symbol $y \in \Sigma \cup \mathcal{P}(2^T)$, and $F = \{x : \mathcal{A}_1(x) = 1\}$ is the set of *final states*.

Alternatively one can describe these automata in the more traditional way as a tuple $\mathcal{A}' = (X, \delta, F)$ where $\delta \subseteq X \times (\Sigma \cup \mathcal{P}(2^T)) \times X$ is the transition relation and $F \subseteq X$ is the set of final states. Acceptance of a guarded string w by $\mathcal{A}$ starting from initial state x_0 and ending in state x_f is defined recursively by:

- If $w = \alpha \in 2^T$ then w is accepted iff for some $n \geq 1$ there is a path $x_0 t_1 x_1 t_2 \ldots x_{n-1} t_n x_f$ in $\mathcal{A}$ of n test transitions $t_i \in \mathcal{P}(2^T)$ such that $\alpha \in t_i$ for $i = 1, \ldots, n$.
- If $w = \alpha p v$ then w is accepted iff there exist states x_1, x_2 such that α is accepted ending in state x_1, there is a transition labeled p from x_1 to x_2 (i.e., $x_2 \in \mathcal{A}_0(x_1)(p)$) and v is accepted by $\mathcal{A}$ starting from initial state x_2.

Finally, w is *accepted by $\mathcal{A}$ starting from* x_0 if the ending state x_f is indeed a final state, i.e., satisfies $x_f \in F$.

Kozen [Koz03] proved that the equational theory of KAT is decidable in PSPACE. Moreover KAT is much more versatile that Kleene algebra since it can faithfully express "if b then p else q" by the term $bp + \bar{b}q$ and "while b do p" using $(bp)^*\bar{b}$, as well as several other standard programming constructs. It also interprets Hoare logic and properly distinguishes between simple Boolean tests and complex assertions.

2 Adding Concurrency

After this rather brief discussion of the language semantics and operational semantics of KAT, we now describe how these definitions generalize to handle concurrency. Intuitively, elements P, Q of a concurrent Kleene algebra with tests can be thought of as programs or program fragments, and they are represented by sets of "computation paths". The operation that needs to be added to KAT is the concurrent composition $P \| Q$. Whereas in the sequential model the computation paths are guarded strings, we now need to be able to place two such sequential strings "next to each other", and then we also need to be able to sequentially compose such "concurrent strings" etc. A convenient way to visualize the semantic objects that we would like to construct is to view sequential composition as vertical concatenation (top to bottom) and concurrent composition as horizontal concatenation.

So for example, given two guarded strings $\alpha_0 v \alpha_m$ and $\beta_0 w \beta_n$ we would like to construct

$$
\begin{array}{c|c}
\alpha_0 & \beta_0 \\
v & w \\
\alpha_m & \beta_n
\end{array}
$$

As with sequential composition, this operation is not always defined. In order for these type of objects to be sequentially (vertically) composable, we impose the condition that $\alpha_0 = \beta_0$ and $\alpha_m = \beta_n$. So in fact we have $\alpha_0 v \alpha_m \| \alpha_0 w \alpha_m$ and the resulting object is denoted by $\alpha_0 \{\!| v, w |\!\} \alpha_m$ or vertically by

$$\begin{array}{c} \alpha_0 \\ v|w \\ \alpha_m \end{array}$$

In particular, if α, β are distinct atomic tests then $\alpha||\beta$ is undefined and $\alpha||\alpha = \alpha$. Similarly, $\alpha||\beta w\gamma$ is undefined for all atomic tests α, β, γ. Also, we define concurrent composition to be commutative, which is already reflected in our choice of notation: $\{v, w\} = \{w, v\}$ is a multiset. Moreover it is associative, which means that in these "strings", multisets are not members of multisets, i.e., $\{\{u, v\}, w\}$ is normalized to $\{u, v, w\}$. This ensures that $(\alpha p\beta||\alpha q\beta)||\alpha r\beta = \alpha\{p, q, r\}\beta = \alpha p\beta||(\alpha q\beta||\alpha r\beta)$. Via successive concurrent and sequential compositions we obtain *guarded series-parallel strings*, or *gsp-strings* for short. Formally the set of gsp-strings generated by Σ, T is the smallest set $GSP_{\Sigma,T}$ that has 2^T and $2^T \times \Sigma \times 2^T$ as subsets and is closed under the coalesced product $\diamond$ as well as the concurrent product $||$. For example, if $\Sigma = \{p, q\}$ and $T = \{t\}$ then, abbreviating 2^T by $\{\alpha, \beta\}$, the following expressions are gsp-strings: α, $\alpha p\alpha$, $\alpha p\beta$, $\alpha\{p, q\}\alpha$, $\alpha\{p, q\}\alpha q\beta$, $\alpha\{p, \{p, q\}\alpha q\}\beta$, ...

The language model over gsp-strings is defined as in the case of guarded strings, except that we now have an additional operation. For $L, M \in \mathcal{P}(GSP_{\Sigma,T})$ let

- $L||M = \{v||w : v \in L, w \in M \text{ and } v||w \text{ is defined}\}$.

This makes $\mathcal{P}(GSP_{\Sigma,T})$ into a complete bisemiring with a Kleene-star for sequential composition. The map G from the previous section is extended to all terms of KAT with $||$, by defining $G(p||q) = G(p)||G(q)$. The bi-Kleene algebra of *series-rational gsp-languages*[1], denoted by $\mathbf{C}_{\Sigma,T}$, is the subalgebra generated by $\{G(t) : t \in T\} \cup \{G(p) : p \in \Sigma\}$.

Note that for $b \in \mathcal{P}(2^T)$ and for any subset p of $GSP_{\Sigma,T}$ the concurrent composition $b||p$ is equal to $b \cap p$. In particular, concurrent and sequential composition coincide on tests. However, in general $||$ is not idempotent for sets of gsp-strings and the identity 1 of sequential composition is not an identity of concurrent composition.

With this language model as guide, we now define a *concurrent Kleene algebra with tests* (CKAT) as an algebra $\mathbf{A} = (A, A', +, 0, ||, \cdot, 1, ^*, \bar{\ })$ where

- $(A, A', +, 0, \cdot, 1, ^*, \bar{\ })$ is a Kleene algebra with tests,
- $(A, +, 0, ||)$ is a commutative semiring with 0 (but possibly no unit), and
- $b||c = bc$ for all $b, c \in A'$.

We do not include iterated parallel composition (i.e., parallel star) in the definition of a CKAT since this operation prevents the generalization of Kleene's theorem to gsp-languages ([LW00], see Section 3 for further discussion).

The language model also shows that the concurrency inequation $(x||y)(z||w) \leq (xz)||(yw)$ of CKA is not satisfied under the present definition of CKAT. Take for

---

[1] Lodaya and Weil used the name series-rational sp-language for the members of their language model

example $x = \{\alpha p\beta\}$, $y = \{\alpha q\beta\}$, $z = \{\beta p\gamma\}$, and $w = \{\beta q\gamma\}$, then $(x\|y)(z\|w) = \{\alpha\{\!|p,q|\!\}\beta\{\!|p,q|\!\}\gamma\}$ whereas $(xz)\|(yw) = \{\alpha\{\!|p\beta p, q\beta q|\!\}\gamma\}$. So each expression produces a singleton set, but the two elements are distinct, hence the two expressions are not comparable. However one can impose the concurrency inequation on the generators of the regular gsp-languages to obtain a homomorphic image that satisfies this condition. Not all forms of concurrency satisfy this inequation (in some cases the reverse inequality is applicable), so having a more general axiomatization could be advantageous.

3   Automata over Guarded Series-Parallel Strings

The notion of nondeterministic automaton for gsp-strings is based on the one for guarded strings, but it is expanded with fork and join transitions taken from the branching automata of Lodaya and Weil [LW00]. Specifically a *guarded branching automaton* is a coalgebra for the functor $F(X) = \mathcal{P}(X)^{\Sigma \cup \mathcal{P}(2^T)} \times \mathcal{P}(\mathcal{M}(X)) \times \mathcal{P}(\mathcal{M}(X)) \times 2$ defined on the category of sets, where $\mathcal{M}(X)$ is the collection of multisets of X with more than one element. This means that an automaton is a map $\alpha : X \to F(X)$ where X is the set of states. As for guarded automata, the transition function is given by the first component of α and the set of final states is given by the last component in the form of a characteristic function. The second and third component are the *fork* and *join* relations respectively. In traditional notation, the automaton can also be specified by the tuple $\alpha' = (X, \delta, \delta_{\text{fork}}, \delta_{\text{join}}, F)$, where

- (X, δ, F) is a guarded automaton,
- $\delta_{\text{fork}} \subseteq X \times \mathcal{M}(X)$ and
- $\delta_{\text{join}} \subseteq \mathcal{M}(X) \times X$.

Fork transitions in δ_{fork} are denoted $(x, \{\!|x_1, x_2, \ldots, x_n|\!\})$, and if the multiset has n elements they are called forks of arity n. The join transitions of arity n are defined similarly, but with the order of the two components reversed.

While coalgebraic automata do not have an explicit initial state, they can be augmented with such a state whenever this is required. The advantages of the coalgebraic point of view is that it turns the class of all automata for this functor into a concrete category, and allows many standard results on bisimulation and coalgebraic modal logic to be applied to this setting. We will not make use of it at this point, but in the later part of this paper we again use the coalgebraic perspective to define frame semantics for concurrent relation algebras with transitive closure.

The acceptance condition for gsp-strings does have to be defined carefully since it substantially extends the one for guarded strings. Intuitively one can think of an automaton as evaluating the acceptance condition for parallel parts of the input string concurrently on separate processors. In many cases, when large scale parallel programs are run on a distributed cluster of computers, (part of) the program code is distributed to all the available processors and executes in separate environments until at an appropriate point results are communicated back to a subset of the processors (perhaps a single one) and combined into a new state. This fork and join paradigm is of course a fairly restricted model of concurrent programming, but it has the merit of being quite simple and algebraic since it avoids syntactic annotations for named channels and other more architecture-dependent features. It also meshes well with our generalization of guarded strings and with the laws of concurrent Kleene algebra.

For the actual definition of acceptance we do not need to have separate copies of automata, instead we simply map the parallel parts of a gsp-string into the same automaton. Looking back at the recursive definition of acceptance for a (non-concurrent) guarded string relative to an initial state x_0, it is apparent that this condition is equivalent to finding a path from x_0 to some final state x_f such that the atomic program symbols in the string match with symbols along the path in the same order, and if $p_{i-1}\alpha_i p_i$ occurs in the guarded string then there is a path $\beta_1 \ldots \beta_{n_i}$ of Boolean tests $\beta_k \geq \alpha_i$ along edges of the automaton that lie between the edges matched by p_{i-1} and p_i. For gsp-strings we define a similar "embedding" into the automaton where parallel branches correspond to a fork transition, followed by parallel (not necessarily disjoint) paths along matching edges until they reach a join transition. The precise recursive definition is as follows: A *weak guarded series parallel string* (or wgsp-string for short) is a gsp-string but possibly without the first and/or last atomic test. Acceptance of a wgsp-string w by $\mathcal{A}$ starting from initial state x_0 and ending at state x_f, is defined recursively by:

- If $w = \alpha \in 2^T$ then w is accepted iff for some $n \geq 1$ there is a sequential path $x_0 t_1 x_1 t_2 \ldots x_{n-1} t_n x_f$ in $\mathcal{A}$ (i.e., (x_{i-1}, t_i, x_i) is an edge in $\mathcal{A}$) of n test transitions $t_i \in \mathcal{P}(2^T)$ such that $\alpha \in t_i$ for $i = 1, \ldots, n$.
- If $w = p \in \Sigma$ then w is accepted iff there exist a transition labelled p from x_0 to x_f.
- If $w = \{u_1, \ldots, u_m\}v$ for $m > 1$ then w is accepted iff there exist a fork $(x_0, \{x_1, \ldots, x_m\})$ and a join $(\{y_1, \ldots, y_m\}, y_0)$ in $\mathcal{A}$ such that u_i is accepted starting from x_i and ending at y_i for all $i = 1, \ldots, m$, and furthermore βv is accepted by $\mathcal{A}$ starting at y_0 and ending at x_f.
- If $w = uv$ then w is accepted iff there exist a state x such that u is accepted ending in state x and v is accepted by $\mathcal{A}$ starting from initial state x and ending at x_f.

Finally, w is *accepted by $\mathcal{A}$ starting from x_0* if the ending state x_f is indeed a final state, i.e., satisfies $\mathcal{A}_2(x_f) = 1$.

In the second recursive clause the fork transition corresponds to the creation of n separate processes that can work concurrently on the acceptance of the wgsp-strings $u_1, \ldots, u_n$. The matching join-operation then corresponds to a communication or merging of states that terminates these processes and continues in a single thread.

The sets of gsp-strings that are accepted by a finite automaton are called *regular gsp-languages*. For sets of (unguarded) strings, the regular languages and the series-rational languages (i.e., those built from Kleene algebra terms) coincide.

However, Loyala and Weil pointed out that this is not the case for sp-posets (defined like gsp-strings except without using atomic tests), since for example the language $\{p, p||p, p||p||p, \ldots\}$ is regular, but not a series-rational language. The *width* of an sp-poset or a gsp-string is the maximal cardinality of an antichain in the underlying poset. A (g)sp-language is said to be of *bounded width* if there exists $n < \omega$ such that every member of the language has width less than n. Intuitively this means that the language can be accepted by a machine that has no more than n processors. The series-rational languages are of bounded width since concurrent iteration was not included as one of the operations of CKAT. For languages of bounded width we regain familiar results such as Kleene's theorem which states that a language is series-rational if and only if it is regular (i.e., accepted by a finite automaton) and has bounded width.

We now use a method from Kozen and Smith [KS97] to relate the bounded-width regular languages of Lodaya and Weil [LW00] to guarded bounded-width regular languages. Let $\overline{T} = \{\bar{t} : t \in T\}$ be the set of negated basic tests. From now on we will assume that $T = \{t_1, \ldots, t_n\}$ is finite, and we consider atomic tests α to be (sequential) strings of the form $b_1 b_2 \ldots b_n$ where each b_i is either the element t_i or $\bar{t}_i$. Every term p can be transformed into a term p' in negation normal form using DeMorgan laws and $\bar{\bar{b}} = b$, so that negation only appears on t_i.

Hence the term p' is also a CKA term over the set $\Sigma \cup T \cup \overline{T}$. Let $R(p')$ be the result of evaluating p' in the set of sp-posets of Lodaya and Weil. In [KS97] it is shown how to transform p' further to a sum $\hat{p}$ of *externally guarded* terms such that $p = p' = \hat{p}$ in KAT and $R(\hat{p}) = G(\hat{p})$. This argument extends to terms of CKAT since $||$ distributes over $+$. Therefore the completeness result of Lodaya and Weil [LW00] can be lifted to the following result.

Theorem 1. *CKAT* $\models p = q \iff G(p) = G(q)$

It follows that $\mathbf{C}_{\Sigma,T}$ is indeed the free algebra of CKAT. With the same approach one can also deduce the next result from [LW00].

Theorem 2. *A set of gsp-strings is series-rational (i.e. an element of $\mathbf{C}_{\Sigma,T}$) if and only if it is accepted by a finite guarded branching automaton and has bounded width.*

The condition of bounded width can be rephrased as a restriction on the automaton. A run of $\mathcal{A}$ is called *fork-acyclic* if a matching fork-join pair never occurs as a matched pair nested within itself. The automaton is fork-acyclic if all the accepted runs of $\mathcal{A}$ are fork-acyclic. Lodaya and Weil prove that if a language is accepted by a fork-acyclic automaton then it has bounded width, and their proof applies equally well to gsp-languages.

At this point it is not clear whether this correspondence can be used as a decision procedure for the equational theory of concurrent Kleene algebras with tests.

4   Trace Semantics for Concurrent Kleene Algebras with Tests

Kozen and Tiuryn [KT03] (see also [Koz03]) show how to provide trace semantics for programs (i.e. terms) of Kleene algebra with tests. This is based on an elegant connection between computation traces in a Kripke structure and guarded strings. Here we point out that this connection extends very simply to the setting of concurrent Kleene algebras with tests, where traces are related to labeled Hasse diagrams of posets and these objects in turn are associated with guarded series-parallel strings.

Exactly as for KAT, a Kripke frame over Σ, T is a structure (K, m_K) where K is a set of *states*, $m_K : \Sigma \to \mathcal{P}(K \times K)$ and $m_K : T \to \mathcal{P}(K)$. An sp-trace τ in K is essentially a gsp-string with the atomic guards replaced by states in K, such that whenever a triple $spt \in K \times \Sigma \times K$ is a subtrace of τ then $(s, t) \in m_K(p)$. As with gsp-strings one can form the coalesced product $\sigma \diamond \tau$ of two sp-traces σ, τ (if σ ends at the same state as where τ starts) as well as the parallel product $\sigma \| \tau$ (if σ and τ start at the same state and end at the same state). These partial operations lift to sets X, Y of sp-traces by

- $XY = \{\sigma \diamond \tau : \sigma \in X, \tau \in Y \text{ and } \sigma \diamond \tau \text{ is defined}\}$
- $X\|Y = \{\sigma\|\tau : \sigma \in X, \tau \in Y \text{ and } \sigma\|\tau \text{ is defined}\}$.

Programs (terms of CKAT) are interpreted in K using the inductive definition of Kozen and Tiuryn [KT03] extended by a clause for $\|$:

- $[\![p]\!]_K = \{spt | (s, t) \in m_K(p)\}$ for $p \in \Sigma$
- $[\![0]\!]_K = \emptyset$ and $[\![b]\!]_K = m_K(b)$ for $b \in T$
- $[\![\bar{b}]\!]_K = K \setminus m_K(b)$ and $[\![p + q]\!]_K = [\![p]\!]_K \cup [\![q]\!]_K$
- $[\![pq]\!]_K = ([\![p]\!]_K)([\![q]\!]_K)$ and $[\![p^*]\!]_K = \bigcup_{n < \omega} [\![p]\!]_K^n$
- $[\![p\|q]\!]_K = [\![p]\!]_K \| [\![q]\!]_K$.

Each sp-trace τ has an associated gsp-string $\mathrm{gsp}(\tau)$ obtained by replacing every state s in τ with the corresponding unique atomic test $\alpha \in 2^T$ that satisfies $s \in [\![\alpha]\!]_K$. It follows that $\mathrm{gsp}(\tau)$ is the unique guarded string over Σ, T such that $\tau \in [\![\mathrm{gsp}(\tau)]\!]_K$. As a result the connection between sp-trace semantics and gsp-strings is the same as in [KT03] (the proof is also by induction on the structure of p).

Theorem 3. *For a Kripke frame K, program p and sp-trace τ, we have $\tau \in [\![p]\!]_K$ if and only if $\mathrm{gsp}(\tau) \in G(p)$, whence $[\![p]\!]_K = \mathrm{gsp}^{-1}(G(p))$. In fact gsp^{-1} is a CKAT homomorphism from the free algebra $\mathbf{C}_{\Sigma,T}$ to the algebra of series-rational sets of sp-traces over K.*

The trace model for guarded strings has many applications since each trace in $[\![p]\!]_K$ can be interpreted as a sequential run of the program p starting from the first state of the trace. The sp-trace model provides a similar interpretation for programs that fork and join threads during their runs. Each sp-trace in $[\![p]\!]_K$ is a representation of the basic programs and tests that were performed during the possibly concurrent execution of the program p. Note that there are no explicit fork and join transitions in an sp-trace since, unlike a gsp-automaton (which has to allow for nondeterministic choice), whenever a state in an sp-trace has several immediate successor states, this is the result of a fork, and similarly states with several immediate predecessors represent a join.

While series-parallel traces are more complex than linear traces, they can, like the gsp-strings in Section 2, still be represented by planar diagrams where parallel composition is denoted by placing traces next to each other (with only one copy of the start state and end state), and sequential composition is given by placing traces vertically above each other (with only one connecting state between them).

The sp-trace semantics are useful for analysing the behavior of threads that communicate only indirectly with other concurrent threads via joint termination in a single state. While this is a restricted model of concurrency, it has a simple algebraic model based on Kleene algebras with tests, and it satisfies most of the laws of concurrent Kleene algebra.

5   Expanding Relation Algebras with Concurrency

Kleene algebra with tests provides a reasonable operational semantics for imperative programs, but for specification purposes it would be useful to also have the full language of binary relations available when reasoning about concurrent software. In this section we show how coalgebraic arrow frames of relation algebras can be augmented with an additional component that corresponds to the $\|$ operation. Recall that a relation algebra is of the form $\mathbf{A} = (A, +, 0, \wedge, \top, ^-, ;, 1, ^\smile)$ where $(A, +, 0, \wedge, \top, ^-)$ is a Boolean algebra, $(A, ;, 1)$ is a monoid and for all $x, y, z \in A$

$$ x; y \leq \bar{z} \quad \Longleftrightarrow \quad x^\smile; z \leq \bar{y} \quad \Longleftrightarrow \quad z; y^\smile \leq \bar{x}. $$

It follows that both ; and $^\smile$ distribute over the Boolean join, and that $^\smile$ is an involution, i.e., $x^{\smile\smile} = x$ and $(x; y)^\smile = y^\smile; x^\smile$. Jónsson and Tarski showed that every relation algebra $\mathbf{A}$ can be embedded in a complete and atomic relation algebra, and one can define a relational structure on the set of atoms from which the algebra can be reconstructed as a complex (powerset) algebra. The structure is known as *atom structure* or *ternary Kripke frame* or *arrow frame*, but it is in fact a coalgebra. Hence we define an *arrow coalgebra* to be of the form $\gamma : X \to \mathcal{P}(X^2) \times X \times 2$ such that for all $x, y, z \in X$,

- $(x \circ y) \circ z = x \circ (y \circ z)$ where $x \circ y = \gamma_0^{-1}\{(x, y)\}$ and $A \circ z = \{a \circ z : a \in A\}$,
- $I \circ x = x = x \circ I$ where $I = \gamma_2^{-1}\{1\}$ and
- $(x, y) \in \gamma_0(z) \quad \Longleftrightarrow \quad (x^\smile, z) \in \gamma_0(y) \quad \Longleftrightarrow \quad (z, y^\smile) \in \gamma_0(x)$ where $x^\smile = \gamma_1(x)$.

For $A, B \subseteq X$, define $A; B = \{a \circ b : a \in A, b \in B\}$ and $A^\smile = \{a^\smile : a \in A\}$ and $1 = I$. Then the *complex algebra* over γ, denoted

$$ \mathcal{C}m(\gamma) = (\mathcal{P}(X), \cup, \emptyset, \cap, X, ^-, ;, ^\smile, 1') $$

is a complete relation algebra and $;,\smile$ distribute over arbitrary unions. Hence we can expand this algebra to a relation algebra with reflexive transitive closure (or RAT for short):

- $x^* = \bigcup_{n<\omega} x^n$, where $x^0 = 1'$ and $x^n = x; x^{n-1}$ for $n > 0$.

The variety generated by these algebras has a finite equational axiomatization, and has been studied by Tarski and Ng [NT77, Ng84]. We now expand arrow coalgebras further by adding another factor $\mathcal{P}(X^2)$ to the type functor. A *concurrent arrow coalgebra* is of the form $\gamma : X \to \mathcal{P}(X^2) \times X \times 2 \times \mathcal{P}(X^2)$ such that the projection onto the first three components is an arrow coalgebra and for all $x, y \in X$,

- $(x||y)||z = x||(y||z)$ and $x||y = y||x$ where $x||y = \gamma_3^{-1}\{x, y\}$
- $x \in \gamma_2^{-1}(1)$ implies $x||x = x$ and if $x \neq y$ then $x||y$ is undefined.

The complex algebra of a concurrent arrow coalgebra is a relation algebra with an additional binary operation $||$ defined on subsets A, B of X by $A||B = \{a||b : a \in A, b \in B\}$. Adding reflexive transitive closure is done as before. Based on this concrete model we have the following definition:

A *concurrent relation algebra with reflexive transitive closure* (or CRAT) is an algebra of the form

$$\mathbf{A} = (A, +, 0, \wedge, \top, \bar{\ }, ||, ;, 1, \smile, {}^*)$$

where $\mathbf{A} = (A, +, 0, \wedge, \top, \bar{\ }, ;, 1, \smile, {}^*)$ is a RAT, $(A, +, 0, ||)$ is a commutative semiring with zero and $(x \wedge 1)||y = x \wedge y \wedge 1$ holds for all $x, y \in A$. The result below follows from the theory of Boolean algebras with operators.

Theorem 4. *The complex algebra of a concurrent arrow coalgebra is a complete and atomic CRAT, and every CRAT can be embedded into such a complex algebra.*

The next result establishes a connection between CRAT and concurrent Kleene algebras with test.

Theorem 5. *Let $\mathbf{A} = (A, +, 0, \wedge, \top, \bar{\ }, ||, ;, 1, \smile, {}^*)$ be a CRAT and define $A' = \{b \in A : b \leq 1\}$. Then $\mathbf{A}'' = (A, A', +, 0, ||, \cdot, 1, \bar{\ }, {}^*)$ is a CKAT.*

The proof is simply a matter of checking that the axioms of CKAT hold for $\mathbf{A}''$. It is currently not known if every CKAT is embeddable into an algebra of the form $\mathbf{A}''$. Some related results about KAT can be found in [Koz06].

The concurrency inequality $(x||y); (z||w) \leq (x;z)||(y;w)$ can be added to CRAT and defines a proper subvariety. In the language of concurrent arrow coalgebras the inequality takes the following form: for all $t, u, v, w, x, y, z \in X$

- $t \in u \circ v$ and $u \in x||y$ and $v \in z||w \implies \exists r, s \in X \ (t \in r||s$ and $r \in x \circ z$ and $s \in y \circ w)$.

Other inequations that could be considered are $x||x = x$ or $x; y \leq x||y$ or $x||y \leq x; y$.

Unlike Kleene algebras with tests, the equational theory of relation algebras is known to be undecidable. This is a consequence of having complementation defined on the whole algebra, together with the associativity of a join-preserving operation (see [KNSS93] for such general results). However Andreka, Mikulas and Nemeti [AMN11] have recently proved that the theory of Kleene lattices is decidable. It is an interesting question whether their result can be extended to Kleene lattices with tests or concurrent Kleene lattices (with tests).

6 Conclusion

Many theoretical models of concurrency have been proposed and studied during the last five decades. Here we have taken an algebraic approach starting from Kleene algebras with tests and adapting them to concurrent Kleene algebras of Hoare et. al. and bounded-width series-parallel language models. This provides semantics for concurrency based on standard notions such as regular languages and automata. The addition of tests allows KAT to express standard imperative programming constructs such as if-then-else and while-do. Adding concurrency into this elegant algebraic model is likely to lead to new applications such as verifying compiler optimizations targeting multicore architectures or modeling computations on large distributed clusters. In the last section we have also shown how to add concurrency to relation algebras with reflexive and transitive closure, thus making concurrent composition part of this well-known and expressive algebraic setting.

References

[AMN11]   Andréka, H., Mikulás, S., Németi, I.: The equational theory of Kleene lattices. Theoret. Comput. Sci. 412(52), 7099–7108 (2011)

[Gis88]   Gisher, L.: The equational theory of pomsets. Theoretical Computer Science 62, 224–299 (1988)

[HMSW11]   Hoare, C.A.R., Möller, B., Struth, G., Wehrman, I.: Concurrent Kleene algebra and its foundations. J. Log. Algebr. Program. 80(6), 266–296 (2011)

[HMSW09]   Hoare, C.A.R., Möller, B., Struth, G., Wehrman, I.: Foundations of concurrent Kleene algebra. In: Berghammer, R., Jaoua, A.M., Möller, B. (eds.) RelMiCS/AKA 2009. LNCS, vol. 5827, pp. 166–186. Springer, Heidelberg (2009)

[HMSW09a]   Hoare, C.A.R., Möller, B., Struth, G., Wehrman, I.: Concurrent Kleene algebra. In: Bravetti, M., Zavattaro, G. (eds.) CONCUR 2009. LNCS, vol. 5710, pp. 399–414. Springer, Heidelberg (2009)

[KNSS93]   Kurucz, Á., Németi, I., Sain, I., Simon, A.: Undecidable varieties of semilattice-ordered semigroups, of Boolean algebras with operators, and logics extending Lambek calculus. Logic Journal of IGPL 1(1), 91–98 (1993)

[Koz03]   Kozen, D.: Automata on guarded strings and applications. In: 8th Work-
            shop on Logic, Language, Informations and Computation WoLLIC 2001
            (Braslia). Mat. Contemp., vol. 24, pp. 117–139 (2003)
[Koz06]   Kozen, D.: On the representation of Kleene algebras with tests. In:
            Královič, R., Urzyczyn, P. (eds.) MFCS 2006. LNCS, vol. 4162, pp. 73–83.
            Springer, Heidelberg (2006)
[KS97]   Kozen, D., Smith, F.: Kleene algebra with tests: Completeness and decid-
            ability. In: van Dalen, D., Bezem, M. (eds.) CSL 1996. LNCS, vol. 1258,
            pp. 244–259. Springer, Heidelberg (1997)
[KT03]   Kozen, D., Tiuryn, J.: Substructural logic and partial correctness. ACM
            Trans. Computational Logic 4(3), 355–378 (2003)
[LW00]   Lodaya, K., Weil, P.: Series-parallel languages and the bounded-width
            property. Theoret. Comput. Sci. 237(1-2), 347–380 (2000)
[Ng84]   Ng, K.C.: Relation Algebras with Transitive Closure. PhD thesis,
            University of California, Berkeley (1984)
[NT77]   Ng, K.C., Tarski, A.: Relation algebras with transitive closure, Abstract
            742-02-09. Notices Amer. Math. Soc. 24, A29–A30 (1977)
[Pra86]   Pratt, V.: Modelling concurrency with partial orders. Internat. J. Parallel
            Prog. 15(1), 33–71 (1986)

Algebras for Program Correctness in Isabelle/HOL

Alasdair Armstrong, Victor B.F. Gomes, and Georg Struth

Department of Computer Science, University of Sheffield
{a.armstrong,v.gomes,g.struth}@shefield.ac.uk

Abstract. We present a reference formalisation of Kleene algebra and demonic refinement algebra with tests in Isabelle/HOL. It provides three different formalisations of tests. Our structured comprehensive libraries for these algebras extend an existing Kleene algebra library. It includes an algebraic account of Hoare logic for partial correctness and several refinement and concurrency control laws in a total correctness setting. Formalisation examples include a complex refinement theorem, a generic proof of a loop transformation theorem for partial and total correctness and a simple prototypical verification tool for while programs, which is itself formally verified.

1 Introduction

This article documents the formalisation of computationally important algebraic concepts and structures within a larger project of making variants of Kleene algebras and relation algebras available in the Isabelle proof assistant. It presents variants of test semirings [17] in Isabelle together with their expansions to Kleene algebras and demonic refinement algebras with tests [15, 22, 23]. The latter two algebras have been applied in the verification and correctness of sequential programs; the first one in partial correctness, the second one in a total correctness setting. Demonic refinement algebras have also been used for concurrency verification with action systems [7].

Implementing these algebras with their most important models—the binary relation model for Kleene algebras with tests and the conjunctive predicate transformer model for demonic refinement algebras—yields a basis for building lightweight tools for program verification and correctness in Isabelle. The general approach is quite simple. The algebraic layer captures part of reasoning about programs, in particular their control flow, abstractly and concisely. Other aspects, such as data flow, however, are performed within more concrete models, for example the relational model of program store. In addition, algebra helps at the meta-level to derive inference, refinement or transformation rules and implement tactics or decision procedures. Isabelle allows one to reason seamlessly across these layers, for instance, by programming algebra-driven tactics which automatically generate verification conditions for concrete models or data types or by inferring abstract properties of assignment commands. All these features are provided by our implementation and are illustrated in this article.

P. Höfner et al. (Eds.): RAMiCS 2014, LNCS 8428, pp. 49–64, 2014.
© Springer International Publishing Switzerland 2014

More concretely, the main contributions of our formalisation are as follows.

First, based on a comprehensive reference library for variants of dioids and Kleene algebras [4], we have implemented demonic refinements algebras [22, 23] as an extension of a variant of Kleene algebra via Isabelle's type class mechanism, the standard tool for such formalisations. While Kleene algebras provide operations for the programming concepts of abort, skip, sequential composition, nondeterministic choice and finite iteration, demonic refinement algebras add an operation of potentially infinite iteration. We have implemented a library for equational reasoning in this algebra which contains more than 50 facts from the literature.

Second, we have formalised three different approaches to tests for variants of dioids (idempotent semirings) and developed comprehensive libraries for these. The first one is one-sorted. It implements functions for tests and antitests (boolean test complements) as in the standard approach to domain semirings [11]. The second, two-sorted one follows the approach of embedding a boolean algebra of tests into the dioid [17]. Both approaches are purely axiomatic; they do not mention an underlying carrier set. While such axiomatic versions often suffice for verification applications, a third variant with explicit carrier sets is provided as a basis for mathematical investigations. From variants of test dioids, Kleene and demonic refinement algebras with tests are obtained as straightforward expansions. The libraries for these structures contain more than 350 facts. In particular the third variant required implementing a range of background theories.

Third, we illustrate our formalisation through two classical examples from the literature, which any formalisation of these structures should feature. We have formalised proofs of three variants of Back's atomicity refinement theorem for action systems [22, 9, 13] in demonic refinement algebras with tests. We also present a new generic proof of Kozen's transformation theorem for while loops in Kleene algebras with tests [15] and demonic refinement algebras [21]. It is based on an axiomatisation of regular algebras by Conway [10] in which the iteration axioms are too weak to distinguish finite from potentially infinite iteration.

Fourth, we demonstrate how simple prototypical tools for the verification, refinement and transformation of sequential programs can be obtained in a generic way from the algebras considered. By developing the tool in Isabelle, it is itself formally verified. We first derive the inference rules of Hoare logic except the assignment rule in Kleene algebras with tests. To capture assignment in concrete models, we then formalise the relational model of Kleene algebra with tests and the predicate transformer model of demonic refinement algebra and specialise the first model further to program stores. We can then derive assignment laws easily within this model. We have also implemented a tactic for automatically generating the usual verification conditions for while programs and show the approach at work on a simple verification example.

In sum, this formalisation spans the gulf between abstract algebras of programs and concrete tools for program correctness and verification in a simple, coherent, principled way. We are using it as a template for developing more sophisticated and applicable tools for sequential and concurrent programs.

2   Algebraic Preliminaries

A *dioid*, or *idempotent semiring* is a structure $(S, +, \cdot, 0, 1)$ such that $(S, \cdot, 1)$ is a monoid, $(S, +, 0)$ is a semilattice with least element 0, and the distributivity laws $x \cdot (y + z) = x \cdot z + y \cdot z$ and $(x + y) \cdot z = x \cdot z + y \cdot z$ as well as the annihilation laws $0 \cdot x = 0$ and $x \cdot 0 = 0$ hold. Addition and multiplication are isotone with respect to the semilattice order defined by $x \leq y \longleftrightarrow x + y = y$, that is, $x \leq y$ implies $z + x = z + y$, $z \cdot x \leq z \cdot y$ and $x \cdot z \leq y \cdot z$.

A *right Kleene algebra* is a dioid expanded by a Kleene star which satisfies the unfold axiom $1 + x^* \cdot x \leq x^*$ and the iteration axiom $z + y \cdot x \leq y \longrightarrow z \cdot x^* \leq y$. The dual unfold law $1 + x \cdot x^* \leq x^*$ is derivable. A right Kleene algebra is a *Kleene algebra* if the left induction axiom $z + x \cdot y \leq y \longrightarrow x^* \cdot z \leq y$ holds too. For an overview of variants of dioids and Kleene algebras, their most useful laws and their most important models see [4].

In this context it is important to know that binary relations form Kleene algebras. This relational model is discussed further in Section 7. Binary relations, in turn, yield a standard semantics for sequential programs. Addition models nondeterministic choice, multiplication models sequential composition, 1 models **skip**, 0 models **abort**, * models finite iteration.

For modelling conditionals and while loops according to the relational partial correctness semantics, a notion of *test* needs to be added. In the relational model a test is simply an element between the empty and the identity relation. Abstractly, a *test dioid* [17] is a structure (S, B) such that S is a dioid and B a boolean algebra which is embedded into the subalgebra of elements between 0 and 1 of the dioid. There are the following correspondences between operations of the dioid and those of the boolean algebra: 0 corresponds to the least element of the boolean algebra, 1 to its greatest element, + corresponds to join and $\cdot$ to meet. Complementation $-$ has no counterpart in the dioid, it exists only for the subalgebra of tests. A *Kleene algebra with tests* is a test dioid which is also a Kleene algebra. We write x, y, z for general Kleene algebra elements and p, q, r for tests. A technical development can be found in Section 4. Using tests, an abstract algebraic semantics for conditionals and while loops is given by

$$\textbf{if } p \textbf{ then } x \textbf{ else } y = p \cdot x + (-p) \cdot y, \qquad \textbf{while } p \textbf{ do } x = (p \cdot x)^* \cdot (-p).$$

Multiplying a program x with a test p from the left means restricting the input of the program to those states where the test holds; multiplying from the right means an output restriction.

Total program semantics require another variant of Kleene algebra [22, 23]. A *demonic refinement algebra* is a Kleene algebra in which the right annihilation axiom $x \cdot 0 = 0$ is absent and which is expanded by an operation for possibly infinite iterations which satisfies the unfold axiom $1 + x \cdot x^\infty = x^\infty$, the coinduction axiom $y \leq x \cdot y + z \longrightarrow y \leq x^\infty \cdot z$ and the isolation axiom $x^\infty = x^* + x^\infty \cdot 0$.

This captures total correctness where an agent has no control over termination; $(p \cdot x)^\infty \cdot (-p)$, for instance, models a while loop which may not terminate. For similar reasons, $x \cdot 0 = 0$ is invalid due to potentially infinite processes.

In the isolation axiom, $x^* \cdot 0$ annihilates if all processes in x are finite whereas $x^\infty \cdot 0$ projects on the strictly infinite processes in x^∞.

Demonic refinement algebra can be expanded by tests in the obvious way. The semantics of choice, in this case, is a predicate transformer algebra, which we discuss in detail in Section 7.

The refinement community's notation unfortunately deviates from the regular algebra notation. Their refinement order $\sqsubseteq$ is the converse of $\leq$; the symbols $\top$, $\sqcap$, ; and $^\omega$ are used instead of 0, $+$, $\cdot$ and ∞. Finally tests are known as *guards*.

3   Demonic Refinement Algebra in Isabelle

We now sketch our formalisation of demonic refinement algebra in the theorem proving environment Isabelle/HOL [18]. We introduce some basic features of Isabelle while discussing our formalisation. For additional information about Isabelle we refer to its excellent documentation[1]. The complete Isabelle code of our implementation can be found online[2]. A reference formalisation is available from the Archive of Formal Proofs [3]. We recommend reading these in parallel.

Isabelle is an interactive proof assistant with embedded automatic theorem provers and counterexample generators. It is based on a small logical core to guarantee correctness. It has been used to formalise a wide range of mathematical theories and applied in numerous computing applications, including program correctness and verification. Isabelle/HOL, in particular, is based on a typed higher-order logic which supports reasoning with sets, polymorphic data types, inductive definitions and recursive functions.

Algebraic hierarchies, like those in the previous section, are usually formalised with Isabelle's type class and locale infrastructure. Type classes typically suffice for simple structures with one single type parameter. More advanced formalisations often require locales. Both mechanisms support theory expansion and the formalisation of subclass relationships. Theorems proved for reducts or superclasses thus become automatically available in expansions or subclasses. Within this infrastructure, algebras can be linked formally with their models by instantiation or interpretation statements.

We have integrated our formalisation of demonic refinement algebra into the existing Kleene algebra hierarchy [4]. More precisely, we have formalised demonic refinement algebra as an expansion of Kleene algebra with a left annihilator, adding simply the unfold, coinduction and isolation axiom for ∞. By this expansion, all facts proved for this variant of Kleene algebra become automatically available in demonic refinement algebra.

class $dra = kleene\text{-}algebra\text{-}zerol + strong\text{-}iteration\text{-}op +$
  assumes $iteration\text{-}unfoldl$: $1 + x \cdot x^\infty = x^\infty$
  and $coinduction$: $y \leq z + x \cdot y \longrightarrow y \leq x^\infty \cdot z$
  and $isolation$: $x^\infty = x^* + x^\infty \cdot 0$

---

[1] http://isabelle.in.tum.de
[2] http://www.dcs.shef.ac.uk/~victor/ramics2014

We have developed a comprehensive library of theorems of demonic refinement algebra from the literature. Isabelle offers various ways of proving such facts. First, there is a range of built-in tactics, provers and simplifiers. These are generally insufficient for automating algebraic reasoning, but quite powerful for higher-order reasoning with models. Second, Isabelle's Sledgehammer tactic calls external automated theorem provers and SMT solvers and reconstructs their output internally to increase trustworthiness. In this way, many equational algebraic theorems can be proved fully automatically, but the approach is limited to first-order reasoning. Finally, Isabelle offers different modes of interactive reasoning, notably the proof scripting language Isar which supports human-readable proofs, as in the following example.

lemma *iteration-sim*: $z \cdot y \leq x \cdot z \longrightarrow z \cdot y^\infty \leq x^\infty \cdot z$
proof
  assume *assms*: $z \cdot y \leq x \cdot z$
  have $z \cdot y^\infty = z + z \cdot y \cdot y^\infty$
    by (*metis distrib-left mult-assoc mult-oner iteration-unfoldl*)
  also have ... $\leq z + x \cdot z \cdot y^\infty$
    by (*metis assms add-commute add-iso mult-isor*)
  finally show $z \cdot y^\infty \leq x^\infty \cdot z$
    by (*metis mult-assoc coinduction*)
qed

In this proof, individual proof steps have been proved automatically by Sledgehammer and internally verified by the metis prover. Isar links these steps into a complete proof. In total we have proved 57 theorems about demonic refinement algebra, 43 of which were fully automatic. The remaining 14 facts required user intervention at the level of the previous example.

Isabelle also offers counterexample generators such as *nitpick* and *quickcheck*, which is very important for exploring mathematical theories. The dual simulation law $y \cdot z \leq z \cdot x \longrightarrow y^\infty \cdot z \leq z \cdot x^\infty$, for instance, has been refuted by nitpick with a three-element counterexample, whereas both simulation laws—with ∞ replaced by *—hold in Kleene algebra.

The most interesting and difficult theorems come from demonic refinement algebra with tests. Before discussing these in Section 5 we present three alternative formalisations of test dioids in the following section.

4   Three Formalisations of Tests

The embedding of a boolean test algebras into a Kleene algebra can be formalised in different ways in Isabelle. Our first implementation is based on an unpublished manuscript by Jipsen and Struth. It is inspired by the axiomatisation of domain semirings [11]. The main idea is to add a function t to a semiring or dioid S and axiomatise it in such a way that the image $t(S)$ forms a boolean subalgebra of tests. The function t is assumed to be a retraction, that is, $t \circ t = t$, since then $p \in t(S)$ if and only if $t(p) = p$. We can use this fixpoint property for typing tests or verifying closure conditions.

For encoding test complementation, however, it is more suitable to axiomatise an *antitest function* n which satisfies $t = n \circ n$:

class *dioid-tests-zerol = dioid-one-zerol + comp-op +*
  assumes *test-one*: $n\,n\,1 = 1$
  and   *test-mult*: $n\,n\,(n\,n\,x \cdot n\,n\,y) = n\,n\,y \cdot n\,n\,x$
  and   *test-mult-comp*: $n\,x \cdot n\,n\,x = 0$
  and   *test-de-morgan*: $n\,x + n\,y = n\,(n\,n\,x \cdot n\,n\,y)$

We then abbreviate $t\,x \equiv n\,(n\,x)$ and define *test* $p \equiv t\,p = p$. In fact, if these axioms are added to an arbitrary semiring, idempotence is enforced. It is straightforward to verify that tests satisfy the boolean algebra axioms, but the fact that $t(S)$ forms a boolean algebra cannot be expressed explicitly in Isabelle by a subclass or sublocale statement, simply because the carrier set S is not explicit in an type class. Thus we cannot formally integrate Isabelle's library for boolean algebra and had to build up our own one with the most important boolean theorems for tests. We provide an alternative implementation where this problem can be circumvented.

The expansion of test dioids to Kleene algebras with tests is straightforward and therefore not shown in this paper. We have also verified that our test axioms are independent, using nitpick for finding counterexamples when trying to prove each individual axiom from the remaining ones. Despite its limitations, this formalisation is simple and yields a high degree of automation. Overall, 122 theorems about Kleene algebras with tests and boolean algebra were proved, all of which fully automatically.

Our second formalisation of test dioids integrates Isabelle's boolean algebra type class. In contrast to the previous one-sorted implementation it is therefore two-sorted. This requires locales instead of type classes.

locale *dioid-tests-zerol =*
  fixes *test* :: $'a{::}boolean\text{-}algebra \Rightarrow 'b{::}dioid\text{-}one\text{-}zerol$
  and *not* :: $'b{::}dioid\text{-}one\text{-}zerol \Rightarrow 'b{::}dioid\text{-}one\text{-}zerol$
  assumes *test-sup*: *test* $(sup\ p\ q) = \,'p + q\,'$
  and *test-inf*: *test* $(inf\ p\ q) = \,'p \cdot q\,'$
  and *test-top*: *test top* $= 1$
  and *test-bot*: *test bot* $= 0$
  and *test-not*: *test* $(-p) = \,'{-}p\,'$
  and *test-iso-eq*: $p \leq q \longleftrightarrow \,'p \leq q\,'$

Now the function *test* embeds the boolean algebra into the dioid as usual. A boolean complementation is also defined on the dioid. The other axioms of this locale link the boolean operations with the dioid ones, as described in Section 2. To obtain the typical Kleene algebra with test notation, where the embedding is implicit, we have implemented a syntax translation which automatically recognises tests in formulas. Hence one can write '$p + q$' for the join of two tests.

With this two-sorted approach, Isabelle's libraries for boolean algebras become automatically available. From an automation point of view, however, we noted little difference between the two approaches. As before, we do not explicitly show the expansion of test dioids to Kleene algebras with tests.

Our third implementation of test dioids provides explicit carrier sets. It follows the general Isabelle recipe for setting up such algebras.

record $'a$ *test-dioid-structure* $=$ $'a$ *dioid* $+$ *test* $::$ $'a$ *ord*

abbreviation *tests* $A \equiv$ *carrier* $($*test* $A)$

locale *dioid-tests-zerol* $=$
  fixes $A ::$ $'a$ *test-dioid-structure* (**structure**)
  assumes *is-dioid*: *dioid-tests-zerol* A
  and *test-subset*: *tests* $A \subseteq$ *carrier* A
  and *test-le*: *le* $($*test* $A) =$ *dioid.nat-order* A
  and *test-ba*: *boolean-algebra* $($*test* $A)$
  and *test-one*: *top* $($*test* $A) = 1$
  and *test-zero*: *bot* $($*test* $A) = 0$
  and *test-join*: $[\![x \in$ *tests* $A; y \in$ *tests* $A]\!] \implies$ *join* $($*test* $A)$ x $y = x + y$
  and *test-meet*: $[\![x \in$ *tests* $A; y \in$ *tests* $A]\!] \implies$ *meet* $($*test* $A)$ x $y = x \cdot y$

This formalisation expands carrier-based formalisations of dioids and boolean algebras. In this setting, algebraic signatures are specified in records. In this case it is said that tests have a pre-defined order type. The axioms yield a dioid without left annihilation where the carrier set of tests is a subset of the main carrier and the operations are embedded as usual.

To support this approach we had to implement several background theories from scratch with more than 250 theorems about lattices, dioids, Kleene algebra and Kleene algebras with tests. Because of the additional constraints, Sledgehammer may struggle to automate simple proofs. Hence there is a trade-off between mathematical precision and automation. This approach has previously been used to implement schematic Kleene algebra with tests and derive flow chart equivalence as well as simple program verification proofs in this setting [5].

In sum, our three formalisations all have their advantages and disadvantages. The one-sorted and two-sorted implementation offer comparable proof automation and might be superior for program verification applications. Which one is preferable in practice remains to be seen. The carrier-based implementation leads to less automatic proofs, but for investigations in universal algebra, for instance, this price needs to be paid.

5   A Program Refinement Example

All axiomatisations from the previous section have been given for dioids without the axiom $x \cdot 0 = 0$. This makes all three formalisations compatible with demonic refinement algebra. The one-sorted formalisation of tests, for instance, is

class *dra-tests* $=$ *dioid-tests-zerol* $+$ *dra*

An expansion to proper test dioids is, of course, given in our Isabelle theory files.

The addition of tests or guards make demonic refinement algebra suitable for program development applications. We have also formalised the dual notion of *assertion*. Assertions are used as context information for weakest precondition reasoning [22, 23] in guarded command languages. We have formalised assertions as $p^o = (-p) \cdot \top + 1$. The constant $\top$ denotes the greatest element of the demonic refinement algebra, which exists in this class and is equal to 1^∞. Intuitively, an assertion p^o aborts when p is false and skips when p is true. We have verified that guards and assertions are adjoints of Galois connections, $p \cdot x \leq y \longleftrightarrow x \leq p^o \cdot y$ and $x \cdot p^o \leq y \longleftrightarrow x \leq y \cdot p$, as well as further properties from the literature.

Demonic refinement algebra is also interesting for modelling concurrency in Back's *action system* framework [7]. As a complex example we have verified three algebraic versions of Back's *atomicity refinement theorem* [6, 22, 23, 9, 13]. For an explanation we refer to these articles. Here we only discuss algebraic aspects and proof automation. Von Wright's variant states that the identity

$$x \cdot (y + z + v + w)^\infty \cdot p \leq x \cdot (yz^\infty p + v + w)^\infty$$

can be derived from the 12 assumptions

$$t\, p = p, \qquad x = x \cdot p, \qquad y = p \cdot y, \qquad p \cdot z = 0, \qquad v \cdot z \leq z \cdot v,$$

$$v \cdot w \leq w \cdot v, \qquad v \cdot p \leq p \cdot v, \qquad y \cdot w \leq w \cdot y, \qquad z \cdot w \leq w \cdot z,$$

$$p \cdot w \leq w \cdot p, \qquad z^\infty = z^*, \qquad v^\infty = v^*.$$

Note that $z^\infty = z^*$ and $w^\infty = w^*$ express that z and w are finite. Von Wright's original proof covers about 3 pages. Our Isabelle proof essentially translates this proof at this level of granularity; a more coarse grained automation seems difficult for metis. The main reason is that the terms appearing in this proof are quite long and many rules can match. This combinatorics is difficult to handle in particular for metis, which is inferior to Sledgehammer's external provers. In fact, a more general proof of this theorem with Prover9 [13] was much more coarse grained but required excessive running times. Theorems like this provide interesting benchmarks for Sledgehammer in particular and automated theorem provers in general. This general version can also be found in our Isabelle files.

Finally, we have verified Cohen's simplified version of the atomicity refinement theorem [9] which derives the equation

$$(x + y + z)^\infty = (p \cdot z)^\infty \cdot (x + (-p) \cdot z + y \cdot (-p))^\infty \cdot (y \cdot p)^\infty$$

from the assumptions $t\, p = p$, $x \cdot 0 = 0$, $y \cdot 0 = 0$, $p \cdot y \cdot (-p) = 0$, $p \cdot z \cdot (-p) = 0$, $y \cdot p \cdot x \leq x \cdot y$, $x \cdot p \cdot z \leq z \cdot x$ and $y \cdot p \cdot z \leq z \cdot y$. Cohen assumes partial correctness, so we must explicitly express that x and y must terminate: $x \cdot 0 = 0$ and $y \cdot 0 = 0$. Our proof requires 10 particular steps with Isar.

The results in this section show that libraries that support program refinement can be developed quite easily at the algebraic level with Isabelle. Demonic refinement algebra is part of more powerful calculi which have been described, for instance, in the book of Back and von Wright [8]. Their approach is based on lattice and fixpoint theory. It can easily be obtained by theory expansion from our formalisation of demonic refinement algebra. This is left for future work.

6   A Program Transformation Example

We now consider a classical program transformation example which has first been considered in the partial correctness setting of Kleene algebra with tests. We formalise Kozen's loop transformation theorem in Kleene algebra with tests: *Every sequential while program, appropriately augmented with subprograms of the form $z \cdot (p \cdot q + (-p) \cdot (-q))$, can be viewed as a while program with at most one loop under certain preservation assumptions* [15]. Hence any while program, suitably augmented with finitely many new *dummy* subprograms, is equivalent to a simple while program of the form x; **while** p **do** y, where x and y do not contain any nested loops.

A key ingredient of Kozen's approach are commutativity conditions of the form $p \cdot x = x \cdot p$. We use preservation conditions instead, which are of the form $p \cdot x = p \cdot x \cdot p$ and $(-p) \cdot x = (-p) \cdot x \cdot (-p)$. In Kleene algebra with tests, these two conditions are equivalent. However we prove the transformation theorem in the weaker setting of *pre-Conway algebras*, where the former imply the latter, but not vice versa (according to nitpick). Pre-Conway algebras are defined as

class *pre-conway = pre-dioid-one-zerol + dagger-op +*
  assumes *dagger-denest*: $(x + y)^\dagger = (x^\dagger \cdot y)^\dagger \cdot x^\dagger$
  and *dagger-prod-unfold*: $(x \cdot y)^\dagger = 1 + x \cdot (y \cdot x)^\dagger \cdot y$
  and *dagger-simr*: $z \cdot x \leq y \cdot z \longrightarrow z \cdot x^\dagger \leq y^\dagger \cdot z$

As the first line shows, they are based on *pre-dioids* with only a left-annihilating zero [4]. In these structures, the left distributivity law $x \cdot (y + z) = x \cdot y + x \cdot z$ is weakened to sub-distributivity $x \cdot y + x \cdot z \leq x \cdot (y + z)$ which is equivalent to isotonicity $x \leq y \longrightarrow z \cdot x \leq z \cdot y$. Furthermore, the right annihilation law $x \cdot 0 = 0$ is absent. To avoid confusion we use the operator † instead of *. The denest and product-unfold axioms are part of Conway's *classical axioms* for regular algebra [10], but several other axioms, including the idempotency axiom $x^{\dagger\dagger} = x^\dagger$, are absent. In particular, Conway's classical axioms are based on a full dioid. In fact, the dioid-based version plus dagger idempotence is equivalent to the axioms of right Kleene algebra; and complete with respect to the equational theory of regular expressions (see [12] for an overview).

In preparation to the proof of the loop transformation theorem we have verified a number of laws about the dagger in pre-Conway algebra, for instance isotonicity of dagger, $x \leq y \longrightarrow x^\dagger \leq y^\dagger$, a slide rule, $x \cdot (y \cdot x)^\dagger = (x \cdot y)^\dagger \cdot x$, unfold laws for the dagger, $x^\dagger = 1 + x \cdot x^\dagger$ and $x^\dagger = 1 + x^\dagger \cdot x$, along with some preservation properties, such as that $p \cdot x \cdot p = p \cdot x$ implies $p \cdot x^\dagger = (p \cdot x)^\dagger \cdot p$ and $p \cdot (p \cdot x + (-p) \cdot y)^\dagger = (p \cdot x)^\dagger \cdot p$.

The proof itself is by structural induction on while programs. This can be formalised in Isabelle by defining a grammar for programs and imposing the quotient of pre-Conway algebra identities, using Isabelle's quotient package. We only discuss the individual cases of this inductive argument. For each program construct, an inner loop is moved to the outside of a program and these program transformations are verified in pre-Conway algebra with tests. Programs can be augmented by dummy subprograms under preservation assumptions. We follow Kozen's case analysis, but proofs for individual cases are different due to our more general assumptions and the weaker axioms of pre-Conway algebras. To save space we write xy instead of $x \cdot y$ and $\overline{x}$ instead of $-x$. Following Kozen, we take the sequential composition operator to be of lower precedence than the other program constructs.

For conditionals, Kozen shows that the following programs are equivalent:

$pq + \overline{pq}$; **if** p **then** $(x_1;$ **while** r_1 **do** $y_1)$ **else** $(x_2;$ **while** r_2 **do** $y_2)$,

$pq + \overline{pq}$; **if** q **then** x_1 **else** x_2; **while** $qr_1 + \overline{q}r_2$ **do** (**if** q **then** y_1 **else** y_2).

Translated into pre-Conway algebra we must prove that

$$(pq + \overline{pq})(px_1(r_1y_1)^\dagger\overline{r_1} + \overline{p}x_2(r_2y_2)^\dagger\overline{r_2}) =$$
$$(pq + \overline{pq})(qx_1 + \overline{q}x_2)((qr_1 + \overline{q}r_2)(qy_1 + \overline{q}y_2))^\dagger\overline{qr_1 + \overline{q}r_2}.$$

This consists of two phases. First, the two terms are simplified by right distributivity, yielding two subterms each. Second, we proved this by verifying the following two equations between these subterms, using preservation:

$$pqx_1(r_1y_1)^\dagger\overline{r_1} = pqx_1(qr_1y_1 + \overline{q}r_2y_2)^\dagger(q\overline{r_1} + \overline{q}\overline{r_2})$$
$$p\overline{q}x_2(r_2y_2)^\dagger\overline{r_2} = p\overline{q}x_2(qr_1y_1 + \overline{q}r_2y_2)^\dagger(q\overline{r_1} + \overline{q}\overline{r_2})$$

For nested loops, Kozen proves the following two programs equivalent:

while p **do** $(x;$ **while** q **do** $y)$

if p **then** $(x;$ **while** $p + q$ **do** (**if** q **then** y **else** $x))$

The corresponding proof in pre-Conway algebra was fully automatic.

$$(px(qy)^\dagger\overline{q})^\dagger\overline{q} = px((p+q)(qy + \overline{q}x))^\dagger(\overline{p+q} + \overline{p})$$

The case of sequential composition has two subcases. The first one—called *postcomputation*—composes a while loop with a loop-free program:

(**while** p **do** x); y

if $\overline{p}$ **then** y **else** (**while** p **do** $(x;$ **if** $\overline{p}$ **then** $y))$

The corresponding identity in Conway algebra is

$$(px)^\dagger\overline{p}y = \overline{p}y + p(px(\overline{p}y + p))^\dagger\overline{p}.$$

Due to the weaker setting, our proof differs from Kozen's.

$$p(px(\overline{p}y + p))^\dagger\overline{p} = p\overline{p} + ppx((\overline{p}y + p)px)^\dagger(\overline{p}y + p)\overline{p}$$
$$= px(\overline{p}ypx + px)^\dagger\overline{p}y\overline{p}$$
$$= px(\overline{p}y0 + px)^\dagger\overline{p}y$$
$$= px(px)^\dagger(\overline{p}y0)^\dagger\overline{p}y$$
$$= px(px)^\dagger\overline{p}y(0\overline{p}y)^\dagger$$
$$= px(px)^\dagger\overline{p}y.$$

The first step uses the product unfold law. The second step uses right distributivity and boolean algebra. The third step uses the preservation assumption $\overline{p}y = \overline{p}y\overline{p}$. The forth step uses denesting and right annihilation. The fifth step uses the sliding rule. The last step uses right annihilation and the rule $0^\dagger = 1$, which can be derived from the left unfold law. Finally, adding the term $\overline{p}y$ to both sides and applying unfold yields the desired identity.

The second subcase is the composition of two while loops, which leads to the equivalence of

$$\textbf{while } p \textbf{ do } x; \textbf{ while } q \textbf{ do } y$$

$$\textbf{if } \overline{p} \textbf{ then } (\textbf{while } q \textbf{ do } y) \textbf{ else } (\textbf{while } p \textbf{ do } (x; \textbf{ if } \overline{p} \textbf{ then } (\textbf{while } q \textbf{ do } y)))$$

and the identity $(px)^\dagger \overline{p}(qy)^\dagger \overline{q} = \overline{p}(qy)^\dagger \overline{q} + p(px(\overline{p}(qy)^\dagger \overline{q} + p))^\dagger \overline{p}$.

Its proof has two steps. We first prove that $(qy)^\dagger \overline{q}$ preserves p, that is, $p(qy)^\dagger \overline{q} = p(qy)^\dagger \overline{q}p$ and $\overline{q}(qy)^\dagger \overline{q} = \overline{p}(qy)^\dagger \overline{q}p$. Then we prove the identity by applying the previous subcase. This finishes the case analysis.

We have formally shown that every Kleene algebra with tests is a pre-Conway algebra where we interpret † as *.

sublocale *kat* $\subseteq$ *pre-conway star* ⟨*proof*⟩

Thus our proof generalises Kozen's result; and Isabelle makes our theorem automatically available in Kleene algebra with tests. We have also shown that every demonic refinement algebra is a pre-Conway algebra when interpreting † as $^\infty$.

sublocale *dra-tests* $\subseteq$ *pre-conway strong-iteration* ⟨*proof*⟩

Hence our result holds in demonic refinement algebra as well; our proof generalises a previous result by Solin [21].

Finally, Rabehaja and Sanders [20] have further generalised the loop refinement theorem to a probabilistic demonic refinement algebra in which the star and the isolation axiom are absent and the left distributivity axiom is weakened to general left sub-distributivity and to a special left distributivity axiom $p \cdot (x + y) = p \cdot x + p \cdot y$ for tests p. We have adapted our proof so that it covers all three cases. We do not display this most generic result here since probabilistic variants are not the subject of this article. Our Isabelle file contains all relevant details. Note that left distributivity does not hold in pre-Conway algebras and that the product unfold axiom and simulation axiom cannot be derived from Rabehaja and Sanders' axioms. The decision whether the Conway-style axiom set is appropriate for probabilistc reasoning depends on probabilistic semantics.

In pre-Conway algebras, the dagger axioms are too weak to distinguish between finite and potentially infinite iteration. Conway's axiom $x^{\dagger\dagger} = x^\dagger$, which we have dropped, holds of *, but not of $^\infty$, since $x^{\infty\infty} = \top$. Conway has analysed the relevance of this axiom for regular algebras and remarked that it is equivalent to $1^\dagger = 1$. In demonic refinement algebra, however, $1^\infty = \top$.

7   Relational and Predicate Transformer Semantics

This section presents the formalisation of the two most important models of
Kleene algebra with tests and demonic refinement algebra: the relational model
for the first and the predicate transformer model for the second. We restrict our
attention to the one-sorted formalisation.

It is well known that, for each set A, the structure $(2^{A \times A}, \cup, ; , \emptyset, Id, ^*)$ forms a
Kleene algebra; the *full relation Kleene algebra* over A. Here, $\cup$ corresponds to $+$,
relational composition $;$ to $\cdot$, $\emptyset$ to 0, the identity relation Id to 1 and the reflexive
transitive closure operation to *. In addition, every subalgebra of a full relation
Kleene algebra forms a *relation Kleene algebra*. In the one-sorted approach to
the relational model, tests are subidentities and, for each relation x, $n\, x$ is the
complement of x intersected with the identity relation: $n\, x = Id \cap (-x)$. In
Isabelle we have formalised the fact that binary relations form Kleene algebras
with tests by an interpretation statement:

interpretation *rel-kat*: *kat*
  "op $\cup$" "op O" "Id" "op $\subseteq$" "op $\subset$" "rtrancl" "$\lambda x.\ Id \cap (-x)$"
  $\langle proof \rangle$

The proof is fully automatic because binary relations have already been shown
to form Kleene algebras [4], hence only the axioms for n need to be checked.
Moreover, Isabelle's libraries for binary relations are very well developed.

The formalisation of Kleene algebra in Isabelle contains additional models,
including formal languages and regular languages, sets of paths in digraphs,
sets of traces and matrices. For languages there are only two tests: the empty
language and the empty word language. Linking these structures with Kleene
algebra with tests is therefore uninteresting. The other models have a richer test
structures. Interpretation statements with respect to Kleene algebra with tests
seem straightforward. This is left for future work.

The intended model of demonic refinement algebras is formed by conjunctive
predicate transformers [23]. Abstractly speaking these are functions $f : B \to B$
over boolean algebras that distribute over arbitrary meets. Boolean algebras
with such functions are also known as *boolean algebras with operators* [14]. We
have formalised the isomorphic case where B is a field of sets and functions are
strict and additive. In this model, multiplication is function composition and 1
is the identity function; the other dioid operations are

definition $f + g \equiv \lambda\sigma.\ f\,\sigma \cup g\,\sigma$

definition $0 \equiv \lambda\sigma.\ \{\}$

definition $f \leq g \equiv \forall\sigma.\ f\,\sigma \subseteq g\,\sigma$

The iterations * and $^\infty$ correspond to least and greatest fixpoints of the function
$\lambda\sigma.\ 1 + \rho \cdot \sigma$. To characterise the boolean subalgebra, we have defined the adjoint
of a function f, following Jónsson and Tarski, as *adjoint* $f \equiv (\lambda\sigma.\ -\,f\,(-\sigma))$.

We could then define the operation n in this model as

definition $n\ f \equiv (adjoint\ f\ \cdot\ 0) + 1$

Finally, we have created an Isabelle type for the set of strict additive functions—or boolean operators—and proved that, along with the operators defined above, these functions form a demonic refinement algebra with tests.

typedef $'a\ bool\text{-}op = \{f::'a\ set \Rightarrow\ 'a\ set.\ (\forall g\ h.\ f\cdot(g + h) = f\cdot g + f\cdot h \wedge 0\cdot f = 0)\}$

instantiation $bool\text{-}op :: (type)\ dioid\text{-}tests\text{-}zerol\ \langle proof\rangle$

instantiation $bool\text{-}op :: (dioid\text{-}tests\text{-}zerol)\ dra\text{-}tests\ \langle proof\rangle$

A dual statement for multiplicative functions or conjunctive predicate transformers could be obtained similarly. The characterisation of more general function spaces can also be achieved along these lines. We have not pursued this any further since Preoteasa [19] has already formalised an isotone predicate transformer model for demonic refinement algebra. Hence our main contribution lies in the formalisation of the function n. An integration of Preoteasa's model into the Kleene algebra hierarchy is certainly desirable for applications.

8 A Prototypical Verification Tool

We have already explained that Kleene algebras with tests provide an algebraic semantics of while programs in a partial correctness setting. It is also well known [16] that validity of Hoare triples $\vdash \{\!|p|\!\}x\{\!|q|\!\}$ can be encoded as $p \cdot x \cdot (-q) = 0$. This formula states that there are no successful executions of program x from states in p into the complement of q. In other words, if x is executed from precondition p, then its output will satisfy postcondition q upon termination. We have formalised validity of Hoare triples in Kleene algebras with tests. We have also derived all inference rules of propositional Hoare logic without the assignment rule. The derivations were fully automatic.

We now demonstrate how the relational model can be used to derive assignment rules in Isabelle and how the algebraic layer can be extended to a simple, formally verified tool prototype for program verification and correctness. This semantic approach is in contrast to a previous axiomatic treatment of assignment [5] with schematic Kleene algebra with tests [1]. Within this tool, Kleene algebra with tests also allows us to automatically generate verification conditions which completely eliminate the control structure of programs. By our formal linkage of the relation model with abstract Kleene algebra with tests, we can of course use all abstract theorems in this particular model. Although this is not needed for verification, it is important for program transformation.

In the standard relational semantics of imperative programs, a command is a relation between states and a state is a function from variables to values. We provide a prototypical implementation of states as functions from strings to natural numbers and have defined an Isabelle type for this:

type-synonym $state = string \Rightarrow nat$

We have also defined assignment commands as functions from variable names, update functions and states. They return a new state in which the value of the variable has been updated. This is defined in Isabelle as follows, where *lift-fn* lifts the assignment function into the relational model.

definition *lift-fn f* $\equiv$ *Abs-relation* $\{(x, f\,x) \mid x.\ True\}$

definition *assign-fn x f* $\sigma \equiv (\lambda y.\ if\ x = y\ then\ f\ \sigma\ else\ \sigma\ y)$

definition $x := e \equiv$ *lift-fn* (*assign-fn x e*)

Subsets of the identity relation represent tests.

definition *assert P* $\equiv$ *Abs-relation* (*Id-on P*)

This set-up allows us to derive assignment axioms, for instance,

$$P[x|e] \subseteq Q \longrightarrow \{\!|assert\ P|\!\}\ x := e\ \{\!|assert\ Q|\!\},$$

where $P[x|e]$ is the set of states in P in which the variable x has value e.

For convenience we have added a notion of loop invariant for while loops,

$$\textbf{while } p \textbf{ inv } i \textbf{ do } x = (p \cdot x)^* \cdot (-p).$$

Invariants are tests or assertions. They are used for generating verification conditions according to the rule

$$(p \leq i) \wedge (i \cdot (-b) \leq q) \wedge \{\!|i \cdot b|\!\}x\{\!|i|\!\} \longrightarrow \{\!|p|\!\} \textbf{ while } b \textbf{ inv } i \textbf{ do } x\ \{\!|q|\!\},$$

which can be derived easily from the original Hoare rule for the while loop.

Finally, we have adapted the simple proof tactic *hoare-auto* from [5] for generating verification conditions in Isabelle. It applies Isabelle's simplifiers together with the rules of Hoare logic. This works in practice since Hoare logic provides precisely one rule per programming construct. Resolving verification conditions then depends on Isabelle's libraries for the underlying data domains; algebra is no longer needed at this level. We verify Euclid's algorithm as an illustration.

lemma *euclids-algorithm*:
  $\{\!|\{\sigma.\ \sigma\ ''x'' = x \wedge \sigma\ ''y'' = y\}|\!\}$ -- *states σ where $''x'' = x$ and $''y'' = y$*
    while $\{\sigma.\ \sigma\ ''y'' \neq 0\}$ -- *while the state has $''y'' \neq 0$*
    inv $\{\sigma.\ gcd\ (\sigma\ ''x'')\ (\sigma\ ''y'') = gcd\ x\ y\}$
    do (
        $''z'' := ''y'';\ ''y'' := ''x''\ mod\ ''y'';\ ''x'' := ''z''$
)
  $\{\!|\{\sigma.\ \sigma\ ''x'' = gcd\ x\ y\}|\!\}$ -- *states σ where $''x'' = gcd\ x\ y$*
by *hoare-auto* (*metis gcd-red-nat*)

In this simple case, *hoare-auto* presents only $gcd\ x\ y = gcd\ y\ (x\ mod\ y)$ as a single verification condition; the other ones have been discharged by simplification.

Invoking Sledgehammer discharges this condition automatically, using the fact *gcd-red-nat* which been drawn from Isabelle's library for natural numbers.

This simple prototype of a verification tool yields a general template for algebra-based program analysis systems. It can readily be adapted and extended for complex applications. Refinement and transformation tools using the predicate transformer semantics can be built along the same lines.

9   Conclusion

We have extended a reference formalisation for variants of Kleene algebras in Isabelle by two algebras that are important for program verification and correctness applications: Kleene algebras with tests and demonic refinement algebras. We provide more than 10 algebraic structures, hundreds of theorems and two important models. We have demonstrated the applicability of the implementation by two main examples, and have shown how trustworthy tools for program construction and verification can be implemented from such algebras. A coherent integration of algebraic methods into program analysis tools has thereby been achieved. The associated Isabelle theories in the Archive of Formal Proofs [3] serve as a reference for extensions and further applications.

Main applications of our formalisation lie in the development of tools for program verification and correctness. Our technique for integrating the control flow into the algebraic layer is generic. We have already extended it to arbitrary data types beyond natural numbers and verified additional algorithms. An integration of data flow into predicate transformer semantics and the extension of our tool to refinement or program transformation are topics for future work. Finally, we have applied our approach in the context of shared variable concurrency verification [2], with similar algebras, but trace-based semantics.

Acknowledgements. The authors are grateful to Jordan Milner for a preparatory implementation and to Peter Jipsen for joint work on the one-sorted test axiomatisation. They also acknowledge funding from CNPq and EPSRC.

References

[1] Angus, A., Kozen, D.: Kleene algebra with tests and program schematology. Technical Report TR2001-1844, Computer Science Department, Cornell University (July 2001)

[2] Armstrong, A., Gomes, V.B.F., Struth, G.: Algebraic principles for rely-guarantee style concurrency verification tools. CoRR, abs/1312.1225 (2013)

[3] Armstrong, A., Gomes, V.B.F., Struth, G.: Kleene algebras with tests and demonic refinement algebras. Archive of Formal Proofs (2014)

[4] Armstrong, A., Struth, G., Weber, T.: Kleene algebra. Archive of Formal Proofs (2013)

[5] Armstrong, A., Struth, G., Weber, T.: Program analysis and verification based on Kleene algebra in Isabelle/HOL. In: Blazy, S., Paulin-Mohring, C., Pichardie, D. (eds.) ITP 2013. LNCS, vol. 7998, pp. 197–212. Springer, Heidelberg (2013)

[6] Back, R.-J.R.: A method for refining atomicity in parallel algorithms. In: Odijk, E., Rem, M., Syre, J.-C. (eds.) PARLE 1989. LNCS, vol. 366, pp. 199–216. Springer, Heidelberg (1989)

[7] Back, R.-J., Kurki-Suonio, R.: Distributed cooperation with action systems. ACM TOPLAS 10(4), 513–554 (1988)

[8] Back, R.-J., von Wright, J.: Refinement Calculus: A Systematic Introduction. Springer (1998)

[9] Cohen, E.: Separation and reduction. In: Backhouse, R., Oliveira, J.N. (eds.) MPC 2000. LNCS, vol. 1837, pp. 45–59. Springer, Heidelberg (2000)

[10] Conway, J.H.: Regular Algebra and Finite Machines. Chapman and Hall (1971)

[11] Desharnais, J., Struth, G.: Internal axioms for domain semirings. Science of Computer Programming 76(3), 181–203 (2011)

[12] Foster, S., Struth, G.: Automated analysis of regular algebra. In: Gramlich, B., Miller, D., Sattler, U. (eds.) IJCAR 2012. LNCS, vol. 7364, pp. 271–285. Springer, Heidelberg (2012)

[13] Höfner, P., Struth, G., Sutcliffe, G.: Automated verification of refinement laws. Ann. Mathematics and Artificial Intelligence 55(1-2), 35–62 (2009)

[14] Jónsson, B., Tarski, A.: Boolean algebras with operators, part 1. American Journal of Mathematics 73(4), 891–939 (1951)

[15] Kozen, D.: Kleene algebra with tests. ACM TOPLAS 19(3), 427–443 (1997)

[16] Kozen, D.: On Hoare logic and Kleene algebra with tests. ACM TOCL 1(1), 60–76 (2000)

[17] Manes, E.G., Benson, D.B.: The inverse semigroup of a sum-ordered semiring. Semigroup Forum 31(1), 129–152 (1985)

[18] Nipkow, T., Paulson, L.C., Wenzel, M.T.: Isabelle/HOL - A Proof Assistant for Higher-Order Logic. LNCS, vol. 2283. Springer, Heidelberg (2002)

[19] Preoteasa, V.: Algebra of monotonic boolean transformers. In: Simao, A., Morgan, C. (eds.) SBMF 2011. LNCS, vol. 7021, pp. 140–155. Springer, Heidelberg (2011)

[20] Rabehaja, T.M., Sanders, J.W.: Refinement algebra with explicit probabilism. In: Chin, W.-N., Qin, S. (eds.) TASE, pp. 63–70. IEEE Comp. Soc. (2009)

[21] Solin, K.: Normal forms in total correctness for while programs and action systems. J. Logic and Algebraic Programming 80(6), 362–375 (2011)

[22] von Wright, J.: From Kleene algebra to refinement algebra. In: Boiten, E.A., Möller, B. (eds.) MPC 2002. LNCS, vol. 2386, pp. 233–262. Springer, Heidelberg (2002)

[23] von Wright, J.: Towards a refinement algebra. Science of Computer Programming 51(1-2), 23–45 (2004)

Completeness Theorems for Bi-Kleene Algebras and Series-Parallel Rational Pomset Languages

Michael R. Laurence and Georg Struth

Department of Computer Science, University of Sheffield, UK
{m.laurence,g.struth}@sheffield.ac.uk

Abstract. The congruence on series-parallel rational pomset expressions induced by series-parallel rational pomset language identity is shown to be axiomatised by the Kleene algebra axioms plus those of commutative Kleene algebra. A decision procedure is extracted from this proof. On the way to this result, series-parallel rational pomset languages are proved to be closed under the operations of co-Heyting algebras and homomorphisms.

1 Introduction

Pomsets, or partially ordered multisets, form a standard and widely studied model of true concurrency [1–6]. They are essentially partial orders which model the causal dependencies between events in a concurrent system. Vertices are labelled by letters from some given alphabet to capture the actions taking place at particular events. Pomsets generalise both words, which are linear pomsets, and commutative words, which are discrete ones. Words of the former kind can be generated from singleton pomsets using a non-commutative sequential composition, whereas commutative words are generated using parallel composition which is commutative. In pomsets, both operations are present.

Pomset languages are sets of pomsets. Several classes have been studied in the literature, since equivalent ways of defining regularity become distinct when generalising from word to pomset languages. A word language X over an alphabet Σ is regular if it satisfies any, hence all, of the following conditions. (i) It is defined as the image of some Σ-term t with regular operations under a regular homomorphism h, that is, $X = h(t)$. (ii) There is a homomorphism from the free monoid M with basis Σ into a finite monoid such that X is the preimage of a subset of the target monoid. (iii) X is accepted by a finite automaton.

Lodaya and Weil [7, 8] have generalised these conditions as follows. A pomset language is *series rational* if it is definable by a term with operations $0, 1, +, \cdot, *$ and $\|$; it is *series-parallel rational* if it is definable by a term with operations in $0, 1, +, \cdot, *, \|$ and $^{(*)}$, where $^{(*)}$ denotes the parallel star. It is *recognisable* if is definable using preimages of homomorphisms from a free algebra with signature $1, \cdot, \|$ into a finite algebra, by analogy with the special case for word languages. Lastly, it is *regular* if it is accepted by a *branching automaton*, a generalisation of finite automata. Lodaya and Weil have shown that the class of series-rational

P. Höfner et al. (Eds.): RAMiCS 2014, LNCS 8428, pp. 65–82, 2014.
© Springer International Publishing Switzerland 2014

languages is a strict subset of both the classes of recognisable and regular languages and that the class of series-parallel-rational languages is a strict subset of the class of regular languages.

Algebraic axioms for pomset languages have been studied by Gischer [9]. He proved that additively idempotent bi-semirings, which possess both a sequential and a commutative parallel multiplication, are sound and complete with respect to series-rational pomset languages—but without both stars.

The main contribution of this paper is a completeness result for series-parallel rational pomset languages, where both stars are defined. The algebras employed are bi-Kleene algebras consisting of a Kleene algebra and a commutative Kleene algebra with shared addition and units. As in the star-free case, no further axioms are needed for the interaction between sequential and concurrent behaviours in the pomset language model.

Completeness is proved relative to Kozen's completeness result for Kleene algebras and regular languages [10], as well as Pilling and Conway's completeness result for commutative Kleene algebra and regular sets of commutative words [11]. These special cases are used for computing a finite atom structure for series-parallel rational pomset expressions recursively with respect to the alternation depth of their serial and parallel operations. Subexpression memoisation is used for propagating relationships between expressions across these layers. Equations between series-parallel rational pomset expressions can then be decided by comparing finite sets of atoms.

Further contributions are a much simpler completeness proof for bi-Kleene algebras with a certain continuity property, and a proof that series-parallel rational pomset languages are closed under the operations of co-Heyting algebras and homomorphisms. At least the proofs for differences of languages and homomorphisms seem to be new.

Bi-Kleene algebras have been proposed as tools for the verification and correctness of concurrent programs [12]. Our completeness and decidability result can make reasoning about such programs simpler and more automatic.

2 Pomsets, Pomset Languages, Pomset Algebras

A Σ-*labelled poset* over a finite alphabet Σ is a structure $(P, \leq, \lambda)$, where $(P, \leq)$ is a poset and $\lambda : P \to \Sigma$ a function labelling vertices. We call them *labelled posets* when the alphabet is clear. The labelled posets $(P \leq_P, \lambda_P)$ and $(Q, \leq_Q, \lambda_Q)$ are isomorphic if there is a bijection preserving the ordering and labelling.

A Σ-*pomset* or *partially ordered multiset* over Σ is an isomorphism class of Σ-labelled posets. We tacitly assume that all labelled posets and pomsets are finite. We write ε for the empty pomset and $\mathsf{Pom}(\Sigma)$ for the set of all (finite) Σ-pomsets. A *pomset language* over Σ is a subset of $\mathsf{Pom}(\Sigma)$.

There are two extremal cases: any pomset over a linear order is isomorphic to a finite word or string; any pomset over the discrete partial order is isomorphic to a finite commutative word or finite multiset. We refer to the literature cited in the introduction for more information and examples.

Pomsets can be composed sequentially (or serially) and concurrently (or parallely). The labelled poset $P \cdot Q$ is obtained by taking the disjoint union of two labelled posets P and Q and making every element of P precede every element in Q. The labelled poset $P||Q$ is obtained by simply taking the disjoint union of P and Q (and leaving the ordering unchanged). We assume that labelled posets are always made disjoint before performing these operations. Formally,

$$P \cdot Q = (P \cup Q, \leq_P \cup \leq_Q \cup P \times Q, \lambda_P \cup \lambda_Q),$$
$$P||Q = (P \cup Q, \leq_P \cup \leq_Q, \lambda_P \cup \lambda_Q).$$

Since the domains of λ_P and λ_Q are disjoint, $\lambda_P \cup \lambda_Q$ is indeed a function. These definitions are lifted to pomsets in the obvious way.

The algebra of these operations is captured as follows. A *bimonoid* is a structure $(M, \cdot, ||, 1)$ such that $(M, \cdot, 1)$ is a monoid and $(M, ||, 1)$ a commutative monoid [13]. The following fact is routine.

Lemma 1. $(\mathsf{Pom}(\Sigma), \cdot, ||, \varepsilon)$ *forms a bimonoid.*

Strictly speaking, this holds for finite and for infinite pomsets; but in particular for finite ones since finiteness is preserved by the pomset operations.

The bimonoidal structure on individual pomsets is lifted to the powerset level of pomset languages as usual by defining complex products, for pomset languages $X, Y \subseteq \mathsf{Pom}(\Sigma)$, as

$$X \cdot Y = \{P \cdot Q : P \in X \wedge Q \in Y\}, \qquad X||Y = \{P||Q : P \in X \wedge Q \in Y\}.$$

The corresponding algebraic structure is defined as follows. A *dioid* or idempotent semiring is a structure $(S, +, \cdot, 0, 1)$ such that $(S, \cdot, 1)$ is a monoid and $(S, +, 0)$ a semilattice with least element 0 that satisfies the distributivity laws $x \cdot (y + z) = x \cdot y + x \cdot z$ and $(x + y) \cdot z = x \cdot z + y \cdot z$ as well as the annihilation laws $x \cdot 0 = 0$ and $0 \cdot x = 0$. A dioid is *commutative* if $x \cdot y = y \cdot x$. A *trioid* is a structure $(S, +, \cdot, ||, 0, 1)$ such that $(S, +, \cdot, 0, 1)$ is a dioid and $(S, +, ||, 0, 1)$ is a commutative dioid. The following fact is again routine.

Lemma 2. $(2^{\mathsf{Pom}(\Sigma)}, \cup, \cdot, ||, \emptyset, \{\varepsilon\})$ *forms a trioid.*

At the level of pomset languages, stars can be defined as for word languages:

$$X^* = \bigcup_{i \geq 0} X^i, \qquad X^{(*)} = \bigcup_{i \geq 0} X^{(i)},$$

where powers of products are defined as for word languages. A *$*$-continuous bi-Kleene algebra* is a trioid expanded by two star operations that satisfy

$$xy^*z = \sum_{i \geq 0} xy^i z, \qquad xy^{(*)}z = \sum_{i \geq 0} xy^{(i)}z,$$

where powers x^i and $x^{(i)}$ are defined like those on pomsets.

Proposition 3. $(2^{\mathsf{Pom}(\Sigma)}, \cup, \cdot, ||, \emptyset, \{\varepsilon\}, ^*, ^{(*)})$ *forms a* *-*continuous bi-Kleene algebra.*

In other words, pomset languages form *-continuous bi-Kleene algebras.

As special cases, a *-*continuous Kleene algebra* [10] is a dioid expanded by a sequential star that satisfies the first star axiom above, whereas a *-*continuous commutative Kleene algebra* is a commutative dioid expanded by a concurrent star that satisfies the second star axiom.

The above star axioms are first-order infinitary. Every dioid is ordered by the semilattice order $x \leq y \Leftrightarrow x + y = y$. The axiom $x \cdot y^* \cdot z = \sum_{i \geq 0} x \cdot y^i \cdot z$ is therefore equivalent to the conjunction of the two infinitary Horn formulas $\forall i \in \mathbb{N}.x \cdot y^i \cdot z \leq x \cdot y^* \cdot z$ and $(\forall i \in \mathbb{N}.x \cdot y^i \cdot z \leq w) \Rightarrow x \cdot y^* \cdot z \leq w$. Hence the class of star-continuous pomset algebras has free algebras and is closed under products and subalgebras.

Finally, a *Kleene algebra* is a dioid where the star is a least pre-fixpoint:

$$1 + x \cdot x^* \leq x^*, \qquad z + x \cdot y \leq y \Rightarrow x^* \cdot z \leq y, \qquad z + y \cdot x \leq y \Rightarrow z \cdot x^* \leq y.$$

The law $1 + x^* \cdot x \leq x^*$ is then derivable. A *commutative Kleene algebra* is a Kleene algebra which is also a commutative dioid. A *bi-Kleene algebra* is a trioid, a Kleene algebra and a commutative Kleene algebra.

We write T for the class of trioids and the trioid axioms, KA for the class of Kleene algebras and the Kleene algebra axioms, KA_* for the *-continuous variants, cKA and cKA_* for the commutative variants and bKA as well as bKA_* for the cases of bi-Kleene algebras.

3   Series-Parallel Rational Expressions and Languages

Series-parallel rational expressions (spr-expressions) [8] over Σ are defined as

$$T_{\mathsf{bKA}}(\Sigma) ::= 0 \mid 1 \mid a \in \Sigma \mid e \cdot e \mid e||e \mid e + e \mid e^* \mid e^{(*)}.$$

Obviously, spr-expressions correspond to the ground terms over the bKA signature with constants from Σ, which explains the notation $T_{\mathsf{bKA}}(\Sigma)$. It is therefore appropriate to study these expressions in the context of bi-Kleene algebras. The operations in $(+, \cdot, ^*, 0, 1)$ are the usual *regular operations*; the expressions over that signature are the *regular expressions* or ground KA-terms. The operations in $(+, ||, ^{(*)}, 0, 1)$ are called *commutative regular operations*; expressions over that signature are called *commutative regular expressions* or ground cKA-terms. We write $T_{\mathsf{T}}(\Sigma)$, $T_{\mathsf{KA}}(\Sigma)$ and $T_{\mathsf{cKA}}(\Sigma)$ for the sets of trioid ground terms, regular expressions and commutative regular expressions over Σ.

The *canonical homomorphism* $h : T_{\mathsf{bKA}}(\Sigma) \to 2^{\mathsf{Pom}(\Sigma)}$ from spr-expressions to pomset languages is given, for all $a \in \Sigma$ and $s, t \in T_{\mathsf{bKA}}(\Sigma)$, by

$$h(0) = \emptyset, \qquad h(1) = \{\varepsilon\}, \qquad h(a) = \{a\}, \qquad h(s + t) = h(s) \cup h(t),$$
$$h(s \cdot t) = h(s) \cdot h(t), \qquad h(s||t) = h(s)||h(t),$$
$$h(s^*) = h(s)^*, \qquad h(s^{(*)}) = h(s)^{(*)}.$$

The *series-parallel rational pomset languages* (*spr-languages*) over Σ [8] are the homomorphic images of $T_{\mathsf{bKA}}(\Sigma)$ under h. Alternatively they form the inductive subset of $\mathsf{Pom}(\Sigma)$ which contains $\emptyset$, $\{\varepsilon\}$ and the singleton sets $\{a\}$ for all $a \in \Sigma$ and which is closed under finite applications of the operations $+, \cdot, ^*, ||, ^{(*)}$. The set of all spr-languages over Σ is denoted $\mathsf{SPR}(\Sigma)$.

The kernel of h induces a congruence on $T_{\mathsf{bKA}}(\Sigma)$. For all $s, t \in T_{\mathsf{bKA}}(\Sigma)$,

$$s \sim t \Leftrightarrow h(s) = h(t).$$

The main contribution of this paper is to show that $\sim$ is axiomatised precisely by the bi-Kleene algebra axioms.

It is well known that the elements of spr-languages over Σ are special pomsets, called *series-parallel pomsets* (*sp-pomsets*) over Σ. We write $\mathsf{SP}(\Sigma)$ for the set of all sp-pomsets. These have been characterised in two different ways.

First, they are built inductively from the empty pomset ε and the singleton pomsets (isomorphism classes of one-element posets) labelled by $a \in \Sigma$ by closing under finite sequential and concurrent compositions. Thus all sp-pomsets are finite and all pomsets occurring in $\mathsf{SPR}(\Sigma)$ are sp-pomsets; $\mathsf{SPR}(\Sigma) \subseteq 2^{\mathsf{SP}(\Sigma)}$.

For the second characterisation consider the poset N with Hasse diagram

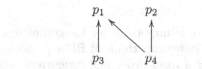

It has been shown that this pomset cannot be constructed as $P \cdot Q$ or $P||Q$ under any labelling. A po(m)set is called *N-free* if it does not contain a subpo(m)set isomorphic to N. More generally, a finite pomset is series-parallel if and only if it is N-free [1, 14]. Hence not all pomsets are series-parallel and the definition of sp-pomset is meaningful.

4   Bimonoids and Series-Parallel Pomsets

Let BM denote the bimonoid axioms and the class of all bimonoids. Let $T_{\mathsf{BM}}(\Sigma)$ denote the ground bimonoid terms, or monomials, with constants from Σ. Grabowski and Gischer [1, 9] have shown the following fact.

Proposition 4. *Every finite N-free pomset has a unique decomposition as an sp-pomset modulo associativity of sequential composition and associativity and commutativity of concurrent composition.*

So, for all monomials $s, t \in T_{\mathsf{BM}}(\Sigma)$, we define the congruence

$$s \sim_{\mathsf{BM}} t \Leftrightarrow \mathsf{BM} \vdash s = t$$

and write $[s]_{\mathsf{BM}}$ for the equivalence class of s with respect to $\sim_{\mathsf{BM}}$. The bimonoid operations are lifted in the standard way as $[s]_{\mathsf{BM}} \cdot [t]_{\mathsf{BM}} = [s \cdot t]_{\mathsf{BM}}$ and likewise. The canonical bimonoid homomorphism $h : T_{\mathsf{BM}}(\Sigma) \to \mathsf{Pom}(\Sigma)$ is lifted as $h([s]_{\mathsf{BM}}) = \{h(t) : t \sim_{\mathsf{BM}} s\}$. Grabowski's result can then be rephrased as follows.

Proposition 5.

(1) $h([s]_{\mathsf{BM}}) = h(s)$ *holds for every* $s \in T_{\mathsf{BM}}(\varSigma)$.
(2) h *is injective:* $h([s]_{\mathsf{BM}}) = h([t]_{\mathsf{BM}}) \Rightarrow [s]_{\mathsf{BM}} = [t]_{\mathsf{BM}}$.

This immediately implies the following soundness and completeness result which links pomsets with bimonoids.

Theorem 6. *If* $s, t \in T_{\mathsf{BM}}(\varSigma)$, *then* $h(s) = h(t) \Leftrightarrow \mathsf{BM} \vdash s = t$.

Proof. Soundness, the right-to-left implication, follows from Lemma 1.

For completeness suppose $h(s) = h(t)$. Then $h([s]_{\mathsf{BM}}) = h([t]_{\mathsf{BM}})$ by Proposition 5(1) and $[s]_{\mathsf{BM}} = [t]_{\mathsf{BM}}$ by Proposition 5(2), whence $s \sim_{\mathsf{BM}} t$ and therefore $\mathsf{BM} \vdash s = t$ by definition of $\sim_{\mathsf{BM}}$. □

Corollary 7. *The bimonoid homomorphism* h *is a bijection between equivalence classes of monomials and sp-pomsets.*

Proof. By Proposition 5(1), each equivalence class $[s]_{\mathsf{BM}}$ is mapped to the singleton language $h(s)$, which, by the isomorphism between these singleton sets and sp-pomsets, yields a map to a unique pomset. By Proposition 5(2), this map is injective, hence invertible. Since every sp-pomset is the image of some monomial, the map is bijective. □

This bijective correspondence is important for the completeness results in Sections 5 and and 7. Due to it we can think of BM-equivalence classes as trees where sequential edges have a fixed order and concurrent edges are commutative. The monomial $a||(b \cdot (c||c) \cdot a)||(b \cdot b)||c$, or rather its equivalence class, and therefore the unique associated pomset, is represented in Figure 4.

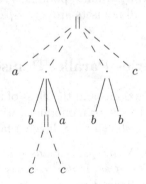

Fig. 1. Tree representation of $a||(b \cdot (c||c) \cdot a)||(b \cdot b)||c$

In Section 6 we define an alternation depth which measures the maximal number of alternations between the sequential and the concurrent layers of an sp-pomset tree. The alternation depth of the above monomial and pomset is three. We say that a monomial is sequential if its top symbol is sequential composition, and similar for the concurrent case. The above monomial is of course concurrent.

As usual, the completeness result for bimonoids can be turned into a result about free algebras [13].

Lemma 8. $(\mathsf{SP}(\Sigma), \cdot, ||, \varepsilon)$ *is freely generated in the variety* BM *by* Σ.

Proof. Let $g : T_{\mathsf{BM}}(\Sigma) \to B$ be a mapping from monomials to some bimonoid B; let h be the canonical homomorphism from monomials to sp-pomsets. We must show that $g = f \cdot h$ and that f is uniquely defined. We first show that $f = g \cdot h^{-1}$ is a map. We have $\ker h \subseteq \ker g$ because $h(s) = h(t)$ implies $\mathsf{BM} \vdash s = t$ and therefore $g(s) = g(t)$. Thus, if h^{-1} relates an sp-pomset to different spr-expressions, then these are in $\ker g$ and identified by g. This yields indeed a function. Now suppose $f' \cdot h = g = f \cdot h$. Then $f' = f$ since h is surjective.   □

The proofs of the following well known facts, which are due to Gischer, Kozen and Pilling and Conway, are similar to this one.

Lemma 9.

(1) $(\mathsf{SPR}_f(\Sigma), \cup, \cdot, ||, \emptyset, \{\varepsilon\})$ *is freely generated in the variety* T *by* Σ.
(2) $(\mathsf{Reg}(\Sigma), \cup, \cdot, \emptyset, \{\varepsilon\}, ^*)$ *is freely generated in the variety* KA *by* Σ.
(3) $(\mathsf{cReg}(\Sigma), \cup, ||, \emptyset, \{\varepsilon\}, ^{(*)})$ *is freely generated in the variety* cKA *by* Σ.

Here, $\mathsf{SPR}_f(\Sigma)$ denotes the finite series-parallel rational languages over Σ; $\mathsf{Reg}(\Sigma)$ and $\mathsf{cReg}(\Sigma)$ denote the regular and commutative regular languages over Σ.

5   Completeness of bKA$_*$

We now prove that bKA$_*$ is complete for the equational theory of spr-expressions. This is a simple consequence of the following proposition, which extends a well known fact about Kleene algebras and regular languages.

Proposition 10. *For all* $r, s, t \in T_{\mathsf{bKA}}(\Sigma)$,

$$\mathsf{bKA}_* \vdash r \cdot s \cdot t = \sum_{P \in h(s)} r \cdot \epsilon(h^{-1}(\{P\})) \cdot t,$$

where ϵ *chooses some representative* p *from every equivalence class* $[p]_{\mathsf{BM}}$.

Proof. By induction on s. Since, by Corollary 7, the correspondence between p and $\{P\} = h(p)$ is bijective, we identify $[p]_{\mathsf{BM}}$, its representative p as well as $\{P\} = h(p)$; we write P instead of $\{P\}$.

- $s = 0$. Then $h(s) = h(0) = \emptyset$ and the right-hand side is $\sum 0 = 0$ as well.
- $s = 1$. Then $h(s) = h(1) = \{\varepsilon\}$, and $\sum r \cdot 1 \cdot t = r \cdot t$.
- $s = a \in \Sigma$. Then $h(a) = \{a\}$ and $\sum r \cdot a \cdot t = r \cdot a \cdot t$.

$- s = s_1 + s_2$. Then

$$r \cdot (s_1 + s_2) \cdot t = r \cdot s_1 \cdot t + r \cdot s_2 \cdot t = r \cdot \left(\sum_{p \in h(s_1)} p \right) \cdot t + r \cdot \left(\sum_{q \in h(s_2)} q \right) \cdot t$$

$$= \sum_{p \in h(s_1)} (r \cdot p \cdot t) + \sum_{q \in h(s_2)} (r \cdot q \cdot t) = \sum_{p \in h(s_1) \cup h(s_2)} (r \cdot p \cdot t)$$

$$= \sum_{p \in h(s_1 + s_2)} (r \cdot p \cdot t).$$

$- s = s_1 \cdot s_2$. Then

$$r \cdot s_1 \cdot s_2 \cdot t = r \cdot \left(\sum_{p \in h(s_1)} p \right) \cdot s_2 \cdot t = \sum_{p \in h(s_1)} (r \cdot p \cdot s_2 \cdot t)$$

$$= \sum_{p \in h(s_1)} \left(r \cdot p \cdot \left(\sum_{q \in h(s_2)} q \right) \cdot t \right) = \sum_{p \in h(s_1)} \sum_{q \in h(s_2)} (r \cdot p \cdot q \cdot t)$$

$$= \sum_{p \in h(s_1), q \in h(s_2)} (r \cdot p \cdot q \cdot t) = \sum_{p \in h(s_1 \cdot s_2)} (r \cdot p \cdot t).$$

$- s = u^*$. Then

$$r \cdot u^* \cdot t = r \cdot \left(\sum_{i \geq 0} u^i \right) \cdot t = \sum_{i \geq 0} (r \cdot u^i \cdot t)$$

$$= \sum_{i \geq 0} \left(r \cdot \left(\sum_{p \in h(u^i)} p \right) \cdot t \right) = \sum_{i \geq 0} \sum_{p \in h(u^i)} (r \cdot p \cdot t)$$

$$= \sum_{p \in \sum_{i \geq 0} h(u^i)} (r \cdot p \cdot t) = \sum_{p \in h(u^*)} (r \cdot p \cdot t).$$

$- s = s_1 \| s_2$ and $s = u^{(*)}$ are similar to the sequential cases. $\qquad \square$

Completeness of bKA_* is now straightforward.

Theorem 11. $h(s) = h(t) \Leftrightarrow \mathsf{bKA}_* \vdash s = t$ for all $s, t \in T_{\mathsf{bKA}}(\Sigma)$.

Proof. Soundness follows from Proposition 3. For completeness assume that $h(s) = h(t)$. Then, by Proposition 10,

$$\mathsf{bKA}_* \vdash s = \sum_{P \in h(s)} \epsilon(h^{-1}(P)) = \sum_{P \in h(t)} \epsilon(h^{-1}(P)) = t.$$

$\qquad \square$

Theorem 11 generalises the following well known results.

Theorem 12.

(1) $h(s) = h(t) \Leftrightarrow \mathsf{T} \vdash s = t$, for all $s, t \in T_{\mathsf{T}}(\Sigma)$.
(2) $h(s) = h(t) \Leftrightarrow \mathsf{KA}_* \vdash s = t \Leftrightarrow \mathsf{KA} \vdash s = t$, for all $s, t \in T_{\mathsf{KA}}(\Sigma)$.
(3) $h(s) = h(t) \Leftrightarrow \mathsf{cKA}_* \vdash s = t \Leftrightarrow \mathsf{cKA} \vdash s = t$, for all $s, t \in T_{\mathsf{cKA}}(\Sigma)$.

For the sake of simplicity, we overload notation and do not notationally distinguish between homomorphisms in the categories of trioids, Kleene algebras, commutative Kleene algebras and bi-Kleene algebras. (1) is due to Gischer, (2) to Kozen; (3) is new for cKA$_*$, but not surprising. For cKA, the result is due to Conway and Pilling. In fact, Conway proves completeness of a finite equational axiom system based on commutative dioids which is different from KA. On the one hand, the axiom $xy^{(*)}z = \sum_{i \geq 0} xy^i z$ entails Conway and Pilling's star axioms. On the other hand, their axioms can be derived from those of cKA.

6 Alternation Depth

We prove our main theorems by induction on the alternation depth of spr-expressions. This is based on a semantic definition of alternation depth for pomsets [15], which we extend to spr-expressions. The alternation depth measures the number of layers of sequential and concurrent operations in sp-pomsets or spr-expressions (cf. Section 4).

Let P be an sp-pomset. The *alternation depth* $\delta : \mathsf{SP}(\Sigma) \to \mathbb{N}$ is defined recursively as follows, using unique decomposability (Proposition 4).

- if P is a singleton pomset or $P = \epsilon$, then $\delta(P) = 0$;
- if $P = P_1 \parallel \cdots \parallel P_n$ then $\delta(P) = \max_{i \leq n} \delta(P_i) + 1$;
- if $P = P_1 \cdots \cdots P_n$ then $\delta(P) = \max_{i \leq n} \delta(P_i) + 1$;

where all decompositions are maximal and $n \geq 2$. The operations of addition and star are incorporated by the following syntactic definition on spr-expressions, which is justified by the next lemma.

Lemma 13. *For all $t \in T_{\mathsf{bKA}}(\Sigma)$ the set $\{\delta(P) : P \in h(t)\}$ has a finite upper bound.*

Proof. Let $\Delta(t) = \{\delta(P) : P \in h(t)\}$

- If $t = 0$, then $h(t) = \emptyset$ and we can chose 0 as an upper bound.
- If $t = 1$, then $h(t) = \{\varepsilon\}$ and 0 is an upper bound.
- If $t = a$ for $a \in \Sigma$, then $h(t) = \{a\}$ and 0 is an upper bound.
- If $t = t_1 + t_2$, then max $\Delta(t_1) \leq m$ and max $\Delta(t_2) \leq n$ for some $m, n \in \mathbb{N}$. Therefore max $\Delta(t_1 + t_2) = \max(\Delta(t_1) \cup \Delta(t_2)) = \max\{m, n\}$.
- If $t = t_1 \cdot t_2$, then max $\Delta(t_1) = m$ and max $\Delta(t_2) = n$ for some $m, n \in \mathbb{N}$. Then

$$\max \Delta(t_1 \cdot t_2) = \max \{\delta(P_1 \cdot P_2) : P_1 \in h(t_1) \wedge P_2 \in h(t_2)\} \leq \max \{m, n\} + 1.$$

- If $t = u^*$, then max $\Delta(u) \leq n$ for some $n \in \mathbb{N}$.

$$\max \Delta(u^*) = \max \{\delta(P) : P \in \bigcup_{i \geq 0} h(u)^i\} = \max \{\delta(P) : P \in h(u)^2\} \leq n + 1.$$

- For $t = t_1 \| t_2$ or $t = u^{(*)}$ the proofs are similar to the sequential cases. □

In particular, $\delta(t^*) = \delta(t \cdot t)$ and $\delta(t^{(*)}) = \delta(t\|t)$. This allows us to define a syntactic notion of alternation depth for spr-expressions for which we overload notation. For $t \in T_{\mathsf{bKA}}(\Sigma)$,

$$\delta(t) = \max\{\delta(P) : P \in h(t)\}$$

is well defined owing to Lemma 13. For convenience, if T is a finite subset of $T_{\mathsf{bKA}}(\Sigma)$, we write $\delta(T) = \max\{\delta(t) : t \in T\}$. Intuitively, $\delta(t)$ can be computed by replacing all subterms s^* of t by $s \cdot s$ and all subterms $s^{(*)}$ by $s\|s$ in t. By definition, the alternation depth is preserved by distributivity laws, e.g., $\delta(a \cdot (b+c)) = \delta(a \cdot b + a \cdot c)$, hence the alternation depth of any spr-expression can be computed as the maximal alternation depth of all monomials in a finite sum, according to these transformations.

The following fact is an immediate consequence of the definition.

Lemma 14. *If $s,t \in T_{\mathsf{bKA}}(\Sigma)$, then $h(s) = h(t) \Rightarrow \delta(s) = \delta(t)$.*

The next lemma is again obvious from the definition of δ and the above observation on computing alternation depths.

Lemma 15. *Let $s,t \in T_{\mathsf{bKA}}(\Sigma)$ such that t is a strict subterm of s and s is not a sum. Then $\delta(t) < \delta(s)$.*

7   Closure Properties

The proof of the closure properties of spr-languages and the completeness proof for bKA are based on subterm memoisation, and we need the following definition. A *substitution morphism* is a function $f : \Sigma \to T_{\mathsf{bKA}}(\Sigma)$ which replaces constants by spr-expressions. We lift f to a function of type $T_{\mathsf{bKA}}(\Sigma) \to T_{\mathsf{bKA}}(\Sigma)$ as usual:

$$f(1) = 1, \qquad f(x \cdot y) = f(x) \cdot f(y), \qquad f(x\|y) = f(x)\|f(y),$$
$$f(x+y) = f(x) + f(y), \qquad f(x^*) = f(x)^*, \qquad f(x^{(*)}) = f(x)^{(*)}.$$

We next show that the set $\mathsf{SPR}(\Sigma)$ of series-parallel rational languages is closed under the union, intersection and difference operations, and also under homomorphisms. Closure under unions is trivial.

Lemma 16. *If $X,Y \in \mathsf{SPR}(\Sigma)$, then $X \cup Y \in \mathsf{SPR}(\Sigma)$.*

Proof. Let $s,t \in T_{\mathsf{bKA}}(\Sigma)$ with $X = h(s)$ such that $Y = h(t)$. Then $X \cup Y = h(s) \cup h(t) = h(s+t)$ is the image of $s+t$ under h and therefore in $\mathsf{SPR}(\Sigma)$.   □

Closure under homomorphisms requires some work. A *pomset homomorphism* is a function $f_l : \mathsf{Pom}(\Sigma) \to \mathsf{Pom}(\Sigma)$ such that

$$f_l(\varepsilon) = \{\varepsilon\}, \qquad f_l(P \cdot Q) = f_l(P) \cdot f_l(Q), \qquad f_l(P\|Q) = f_l(P)\|f_l(Q)$$

for all $P, Q \in \mathsf{Pom}(\Sigma)$. We lift f_l to languages by stipulating

$$f_l(X) = \bigcup \{f_l(P) : P \in X\}.$$

In particular, $f_l(\emptyset) = \emptyset$.

Obviously every substitution homomorphism f defines a homomorphism f_l on spr-languages by $f_l(\{a\}) = h(f(a)) = \{f(a)\}$ for all $a \in \Sigma$. More generally, on spr-expressions, we obtain the following two facts.

Lemma 17. *Let f be a substitution homomorphism and f_l the corresponding pomset language homomorphism. Then, for all $t \in T_{\mathsf{bKA}}(\Sigma)$,*

$$h(f(t)) = \bigcup_{P \in h(t)} f_l(\{P\}).$$

Proof. By induction on t. The proof is similar to that of Proposition 10.

- $t = 0$. Then $f(t) = 0$ and $h(f(0)) = h(0) = \emptyset$, hence both sides reduce to $\emptyset$.
- $t = 1$. Then $f(t) = 1$ and $h(f(1)) = h(1) = \{\varepsilon\} = f_l(\{\varepsilon\})$.
- $t = a \in \Sigma$. Suppose that $f(a) = p$ and that $h(p) = \{P\}$. Then $f_l(\{a\}) = \{P\}$ and both sides are equal.
- $t = t_1 + t_2$. Then

$$h(f(t_1 + t_2)) = h(f(t_1)) \cup h(f(t_2)) = \bigcup_{P \in h(t_1)} f_l(\{P\}) \cup \bigcup_{P \in h(t_2)} f_l(\{P\})$$

$$= \bigcup_{P \in h(t_1) \cup h(t_2)} f_l(\{P\}) = \bigcup_{P \in h(t_1 + t_2)} f_l(P).$$

- $t = t_1 \cdot t_2$. Then

$$h(f(t_1 \cdot t_2)) = h(f(t_1)) \cdot h(f(t_2))$$

$$= \left(\bigcup_{P_1 \in h(t_1)} f_l(\{P_1\}) \right) \cdot h(f_l(t_2))$$

$$= \bigcup_{P_1 \in h(t_1)} \left(f_l(\{P_1\}) \cdot h(f_l(t_2)) \right)$$

$$= \bigcup_{P_1 \in h(t_1)} \left(f_l(\{P_1\}) \cdot \left(\bigcup_{P_2 \in h(t_2)} f_l(\{P_2\}) \right) \right)$$

$$= \bigcup_{P_1 \in h(t_1)} \bigcup_{P_2 \in h(t_2)} \left(f_l(\{P_1\}) \cdot f_l(\{P_2\}) \right)$$

$$= \bigcup_{P_1 \in h(t_1), P_2 \in h(t_2)} \left(f_l(\{P_1\}) \cdot f_l(\{P_2\}) \right)$$

$$= \bigcup_{P \in h(t_1 \cdot t_2)} f_l(\{P\}).$$

$-\ t = u^*$. Then

$$h(f(u^*)) = h(f(u))^* = \bigcup_{i \geq 0} h(f(u))^i = \bigcup_{i \geq 0} \bigcup_{P \in h(u^i)} f_l(\{P\})$$

$$= \bigcup_{P \in \bigcup_{i \geq 0} h(u^i)} f_l(\{P\}) = \bigcup_{P \in h(u^*)} f_l(\{P\}).$$

$-\ t = t_1 \| t_2$ and $t = u^{(*)}$ are similar to the sequential cases.

□

Lemma 18. $f_l(h(t)) = h(f(t))$.

Proof. $f_l(h(t)) = \bigcup_{P \in h(t)} f_l(\{P\}) = h(f(t))$ by definition of f_l and Lemma 17.   □

This shows that the following diagram commutes.

$$
\begin{array}{ccc}
t & \xrightarrow{\ f\ } & t' \\
{\scriptstyle h}\downarrow & & \downarrow{\scriptstyle h} \\
X & \xrightarrow[\ f_l\ ]{} & X'
\end{array}
$$

Closure under homomorphisms then follows.

Proposition 19. $X \in \mathsf{SPR}(\Sigma)$ *implies* $f_l(X) \in \mathsf{SPR}(\Sigma)$ *for all pomset homomorphisms* f_l.

Proof. By Lemma 18, $f_l(h(t)) = h(f(t))$. Thus, if $X = h(t)$ for some spr-expression t, then f maps t to an spr-expression $t' = f(t)$ such that $h(t') = f_l(X)$. Hence $f_l(X)$ is the image of some spr-expression under h and thus in $\mathsf{SPR}(\Sigma)$.

□

Closure under the intersection and difference operators requires a further few lemmas.

Lemma 20. *If f is a monomorphism, then* $f(X \cap Y) = f(X) \cap f(Y)$.

Proof. The map f has an inverse, so

$$x \in f(X \cap Y) \Leftrightarrow f^{-1}(x) \in X \wedge f^{-1}(x) \in Y \Leftrightarrow x \in f(X) \cap f(Y).$$

□

Lemma 21. *If f is a substitution homomorphism, then*

$$h(f(s)) \cap h(f(t)) = \emptyset \Rightarrow h(s) \cap h(t) = \emptyset.$$

Proof. Suppose $h(f(s)) \cap h(f(t)) = \emptyset$. By Lemma 17 the sets $\bigcup_{P \in h(s)} f_l(\{P\})$ and $\bigcup_{Q \in h(t)} f_l(\{Q\})$ are disjoint, thus $f_l(\{P\}) \neq f_l(\{Q\})$ holds for all $P \in h(s)$ and $Q \in h(t)$. Therefore $\{P\} \neq \{Q\}$ holds for all $P \in h(s)$ and all $Q \in h(t)$, so $\{P : P \in h(s)\}$ and $\{Q : Q \in h(t)\}$ are disjoint, whence $h(s) \cap h(t) = \emptyset$.   □

Proposition 22. *If f is an injective substitution homomorphism, then*

$$h(s) \cap h(t) = \emptyset \Leftrightarrow h(f(s)) \cap h(f(t)) = \emptyset.$$

Proof. The right-to-left implication holds by Lemma 21. For its converse, observe that f_l is injective whenever f is. Now suppose $h(s) \cap h(t) = \emptyset$. Then

$$h(f(s)) \cap h(f(t)) = f_l(h(s)) \cap f_l(h(t)) = f_l(h(s) \cap h(t)) = f_l(\emptyset) = \emptyset$$

by Lemma 18 and Lemma 20.   □

We now prepare for closure under intersection and difference.

Lemma 23. *For all (commutative) regular expressions r and s there exists (commutative) regular expressions t and t' such that*

$$h(t) = h(r) - h(s), \qquad h(t') = h(r) \cap h(s).$$

For regular languages the automata-based constructions are well known; a proof for commutative regular languages can be found in Conway's book [11].

Lemma 24. *For every finite set T of (commutative) regular expressions, the set $h(T) = \{h(t) : t \in T\}$ generates a finite boolean algebra of (commutative) regular languages.*

The construction is standard. For $|T| = n$ there are 2^n atoms which are obtained by computing intersections and set differences between the $h(t)$ with $t \in T$. They are given by the meets of all elements of $h(T)$ and their complements. In the regular case the atoms are regular sets. Hence there are terms $s_1, \ldots, s_{2^n}$ which generate them and which can be computed effectively. By definition, $h(s_i) \cap h(s_j) = \emptyset$ for $s_i \neq s_j$. Due to atomicity, every language $X \in h(T)$ satisfies

$$X = \bigcup \{h(s_i) : h(s_i) \subseteq X\};$$

it is the union of the atomic languages below it. Thus $X = h(t)$ implies

$$t = \sum \{s : \mathsf{KA} \vdash s \le t \wedge h(s) \text{ is an atom}\}.$$

This sum is finite since the underlying boolean algebra is.

In the commutative regular case, the constructions are similar, but can no longer be based on automata. Language intersections and complementations are based on normal form constructions explained in Conway's book [11, p.92–97]

Proposition 25. *For every finite set $T \subseteq T_{\mathsf{bKA}}(\Sigma)$, the set*

$$h(T) = \{h(t) : t \in T\}$$

generates a finite boolean algebra, each of whose elements is representable as $h(t)$ for some $t \in T_{\mathsf{bKA}}(\Sigma)$.

Proof. By induction on the alternation depth $\delta(T)$.

- Suppose $\delta(T) = 1$ and let $T = \{t_1, \ldots, t_n\}$. Then every t_i can be decomposed as $t_i = u_i + v_i$, where u_i is a possibly empty regular expression and v_i a possibly empty commutative regular expression, all of which are not sums. By Lemma 24, the sets $h(\{u_i : 1 \leq i \leq n\})$ and $h(\{v_i : 1 \leq i \leq n\})$ form finite boolean algebras with atoms generated by the regular expressions $r_1^u, \ldots, r_{2^n}^u$ and the commutative regular expressions $r_1^v, \ldots, r_{2^n}^v$. We also have that $h(r_i^u) \cap h(r_j^v) = \emptyset$, because every element of $h(r_i^u)$ is a word, whereas every element of $h(r_j^v)$ is a commutative word or finite multiset—as far as constants are concerned, we assume that they are part of the regular sets. Hence the union of the two sets of expressions generates the atoms of a larger finite boolean algebra. Each of its elements is generated by a union of atoms and therefore a regular pomset language according to Lemma 16.

- Suppose the claim holds for $\delta(T) = n$ and consider $\delta(T) = n+1$. Let U be the set of all maximal subterms of terms in T which are commutative at top level if the element of T is sequential and sequential at top level if the element of T is commutative (if a term is both, split into smaller terms). It follows that $\delta(U) < \delta(T)$ by Lemma 15, hence we can apply the induction hypothesis. Accordingly, the elements of U generate a finite boolean algebra and there are terms $u_1, \ldots, u_n$ which generate their atoms. Hence $h(u_i) \cap h(u_j) = \emptyset$ whenever $i \neq j$.

Now introduce constants c_u for every term generating an atom $h(u)$ and define a substitution homomorphism f by $f(c_u) = u$. Then f is injective by atomicity. For each $t \in T$ we obtain a term t' by replacing each subterm from U that occurs in it by the sum of the constants c_u representing the atoms below it. Let T' be the resulting set of terms.

Obviously, all terms in T' can be decomposed into a a regular and a commutative regular term as in the base case. Hence the sets generated by these terms form a finite boolean algebra with atoms $r_1', \ldots, r_m'$, as in the base case. They satisfy $h(r_i') \cap h(r_j') = \emptyset$ for all $i \neq j$. By injectivity of f and Lemma 22, this is the case if and only if $h(f(r_i')) \cap h(f(r_j')) = \emptyset$ for all $i \neq j$. In other words, the set $\{f(r_i') : 1 \leq i \leq m\}$ generates the set of atoms $\{h(f(r_i')) : 1 \leq i \leq m\}$ of the finite boolean algebra $h(T)$. $\square$

Proposition 26. *For all $r, s \in T_{\mathsf{bKA}}(\Sigma)$ there exist $t, u \in T_{\mathsf{bKA}}(\Sigma)$ such that*

$$h(t) = h(r) - h(s), \qquad h(u) = h(r) \cap h(s).$$

Proof. The spr-expressions t and u can be constructed from the atom structure of the finite boolean algebra generated by r and s, which exists by Proposition 25. t is the sum of the atoms below r which are not below s; u is the sum of the atoms below both r and s. $\square$

Theorem 27. $\mathrm{SPR}(\Sigma)$ *is closed under the operations of co-Heyting algebras.*

Proof. This is a consequence of Lemma 16 and Proposition 26. co-Heyting algebras are (distributive) lattices with an additional operation corresponding to set difference. □

All spr-expressions arising from unions, intersections and differences of spr-languages can be computed effectively. For unions this is straightforward. For set differences and intersections this can be achieved by a recursive algorithm which applies the memoisation construction in the proof of Proposition 26 in a bottom-up fashion. The complexity of set intersections and differences depends on that of determining intersections and differences of commutative regular languages using Pilling and Conway's constructions, which is open.

Peter Jipsen has pointed out that $SPR(\Sigma)$ might not be closed under complementation. In fact, by Lemma 13, the alternation depth of any spr-expression is bounded, whereas the alternation depth of sp-pomsets is unbounded. Hence there is no maximal spr-expression generated by Σ and therefore 0 has no spr-expression as a complement.

8   Completeness of bKA

Finally, we prove completeness of bi-Kleene algebras with respect to the equational theory of spr-expressions, as induced by spr-language identity.

Theorem 28. $h(s) = h(t) \Leftrightarrow \mathsf{bKA} \vdash s = t$, for all $s, t \in T_{\mathsf{bKA}}(\Sigma)$.

Proof. Soundness follows from Proposition 3. For completeness assume that $h(s) = h(t)$. We proceed by induction over the alternation depth. The assumption implies $\delta(s) = \delta(t)$ by Lemma 14, which justifies the approach. We need to show that $\mathsf{bKA} \vdash s = t$.

- $\delta(t) \leq 1$. Then $s = s_1 + s_2$ for some KA-term s_1 (possibly empty) and cKA-term s_2 (possibly empty). Similarly $t = t_1 + t_2$ for some KA-term t_1 (possibly empty) and cKA-term t_2 (possibly empty). We also treat the elements of Σ and the constants 0 and 1 ambivalently as sequential or concurrent terms, as needed, to simplify the presentation. Hence

$$
\begin{aligned}
h(s) = h(t) \ &\Leftrightarrow\ h(s_1) \cup h(s_2) = h(t_1) \cup h(t_2) \\
&\Leftrightarrow\ h(s_1) = h(t_1) \ \wedge\ h(s_2) = h(t_2) \\
&\Leftrightarrow\ \mathsf{KA} \vdash s_1 = t_1 \ \wedge\ \mathsf{cKA} \vdash s_2 = t_2 \\
&\Rightarrow\ \mathsf{bKA} \vdash s = t.
\end{aligned}
$$

  by completeness of KA and cKA (Theorem 12).
- Suppose the claim holds for $\delta(t) = n$ and consider $\delta(t) = n+1$. We decompose $s = s_1 + s_2$ and $t = t_1 + t_2$ into KA and cKA algebra parts as before, whenever we can find a sequential and a concurrent subterm with the same alternation depth, including terms with sequential or concurrent stars at top level. Elements of Σ, 0 and 1 are again considered as ambivalent as needed.

Let U be the set of all maximal concurrent subterms of s_1 and t_1 and of all maximal sequential subterms of s_2 and t_2. Then $\delta(u) < \delta(s) = \delta(t)$ for all $u \in U$ by Lemma 15 and the induction hypothesis yields, for all $u, v \in U$,

$$h(u) = h(v) \Leftrightarrow \mathsf{bKA} \vdash u = v.$$

Next, we calculate the finite set of atoms generating the Boolean algebra of $h(U)$ according to Proposition 25 and assume that $V = \{v_1 \ldots v_k\}$ is the resulting set of spr-expressions generating the atoms $h(v_1), \cdots, h(v_k)$. By definition, the atoms generate disjoint languages: $h(v_i) \cap h(v_j) = \emptyset$ whenever $i \neq j$. This implies that h is injective on V: $v_i \neq v_j$ implies $h(v_i) \neq h(v_j)$. We can then replace each subterm from U that occurs in s or t by the (finite) sum of the atoms below it. This preserves language identity. Then, following Proposition 25, we introduce fresh constants c_v for each $v \in V$ and we replace each atom v by its associated constant c_v in s and t. We also define the subsitution homomorphism f by $f(c_v) = v$ as in Proposition 25. Again, f is injective by atomicity and, due to Lemma 18 and injectivity of h, so is the associated pomset homomorphism f_l. It follows that

$$h(f(c_u)) = h(f(c_v)) \Leftrightarrow \mathsf{bKA} \vdash f(c_u) = f(c_v) \Leftrightarrow c_u = c_v.$$

Let now s' and t' be obtained from s and t by replacing subterms with constants as described. Then

$$
\begin{aligned}
h(s) = h(t) &\Leftrightarrow h(f(s')) = h(f(t')) \\
&\Leftrightarrow h(s') = h(t') \\
&\Leftrightarrow h(s'_1) = h(t'_1) \wedge h(t'_1) = h(t'_2) \\
&\Leftrightarrow \mathsf{cKA} \vdash s'_1 = t'_1 \wedge \mathsf{KA} \vdash s'_2 = t'_2 \\
&\Rightarrow \mathsf{bKA} \vdash s' = t' \\
&\Leftrightarrow \mathsf{bKA} \vdash f(s') = f(t').
\end{aligned}
$$

The first step holds by definition of the subsitution homomorphism, s' and t'. The second step holds by Lemma 18 and injectivity of f_l corresponding to that of f. The third step holds by definition of s'_1, s'_2, t'_1, t'_2 and h. The fourth step holds by completeness of KA and cKA. The fifth step is trivial. The last step holds again because f is injective:

$$\mathsf{bKA} \vdash f(s') = f(t') \Rightarrow \mathsf{bKA} \vdash f^{-1}f(s') = f^{-1}f(t') \Leftrightarrow \mathsf{bKA} \vdash s' = t'.$$

Finally, $\mathsf{bKA} \vdash s = f(s')$ and $\mathsf{bKA} \vdash t = f(t')$, because these terms have been obtained from each other by replacing equals by equals at level $\delta(u) < \delta(t)$ applying the induction hypothesis.

This proves that $h(s) = h(t) \Rightarrow \mathsf{bKA} \vdash s = t$. $\square$

Theorem 29. *The equational theory of* bKA *is decidable.*

Proof. Recursively apply the subterm memoisation procedure from the innermost subterms upwards, using the fact that all equations and inequalities can be decided at each level, because the equational theories of KA and cKA are decidable. KA is decidable by automata-theoretic means due to Kozen's completeness result [10]. Decidability of the equational theory of cKA follows from results of Conway and Pilling, as presented in Conway's book [p.92–97].   □

Theorem 29 says in other words that the congruence ~ is decidable. The complexity of the decision procedure depends on deciding identities in commutative Kleene algebra for which the complexity seems open.

9   Conclusion

We have proved, in this paper, that the class of series-parallel rational pomset languages is closed under the intersection and difference operations, and that every identity that is valid for all those languages is a consequence of the set of valid regular and commutative-regular identities. We have also shown that the problem of establishing whether two series-parallel rational pomset expressions define the same language is decidable. The complexity of this is not clear however. It is known that decidability of equivalence of two regular expressions is PSPACE-complete [16, 17], and can be shown that the analogous problem for commutative-regular expressions lies in PSPACE; hence it is possible that generalising to series-parallel rational pomset expressions does not increase the bound beyond PSPACE. Nevertheless this problem may still be EXPTIME-complete or EXPSPACE-complete. This is worth investigating further.

Gischer [9] has studied another class of series-parallel rational pomset languages, which are closed with respect of series-parallel pomset subsumption, but for the case without both stars. He has shown that trioids which satisfy the additional interchange law

$$(w||x) \cdot (y||z) \leq (w \cdot y)||(x \cdot z)$$

are sound and complete for this class. Parallel composition needs to be redefined in this context to preserve subsumption closure. Bi-Kleene algebras which satisfy this interchange law have recently found applications in the verification and correctness of concurrent programs [12]. An extension of our completeness and decidability results to this class seems therefore interesting.

Finally, the algorithmic aspects of our completeness proofs and decision procedures should be elaborated and a technical more precise and detailed presentation might be achieved by using categories. To this end, recent work on coalgebraic and bialgebraic approaches to completeness proofs for Kleene algebras is certainly worth considering [18, 19].

Acknowledgments. We would like to thank the anonymous referees for their suggestions to improve the presentation of the paper. We are especially grateful to Peter Jipsen for pointing out a mistake in a previous version.

References

1. Grabowski, J.: On partial languages. Fundamenta Informaticae 4(2), 427–498 (1981)
2. Pratt, V.R.: On the composition of processes. In: DeMillo, R.A. (ed.) POPL 1982, pp. 213–223. ACM (1982)
3. Pratt, V.R.: Some constructions for order-theoretic models of concurrency. In: Parikh, R. (ed.) Logic of Programs 1985. LNCS, vol. 193, pp. 269–283. Springer, Heidelberg (1985)
4. Brookes, S.D.: Traces, pomsets, fairness and full abstraction for communicating processes. In: Brim, L., Jančar, P., Křetínský, M., Kučera, A. (eds.) CONCUR 2002. LNCS, vol. 2421, pp. 466–482. Springer, Heidelberg (2002)
5. Gastin, P., Mislove, M.: A truly concurrent semantics for a process algebra using resource pomsets. Theoretical Computer Science 281, 369–421 (2002)
6. Zhao, Y., Wang, X., Zhu, H.: Towards a pomset semantics for a shared-variable parallel language. In: Qin, S. (ed.) UTP 2010. LNCS, vol. 6445, pp. 271–285. Springer, Heidelberg (2010)
7. Lodaya, K., Weil, P.: Series-parallel posets: Algebra, automata and languages. In: Morvan, M., Meinel, C., Krob, D. (eds.) STACS 1998. LNCS, vol. 1373, pp. 555–565. Springer, Heidelberg (1998)
8. Lodaya, K., Weil, P.: Series-parallel languages and the bounded-width property. Theoretical Computer Science 237(1-2), 347–380 (2000)
9. Gischer, J.L.: The equational theory of pomsets. Theoretical Computer Science 61(2-3), 199–224 (1988)
10. Kozen, D.: A completeness theorem for Kleene algebras and the algebra of regular events. Information and Computation 110(2), 366–390 (1994)
11. Conway, J.H.: Regular Algebra and Finite Machines. Chapman and Hall (1971)
12. Hoare, T., Möller, B., Struth, G., Wehrman, I.: Concurrent Kleene algebra and its foundations. Journal of Logical Algebraic Programming 80(6), 266–296 (2011)
13. Bloom, S.L., Ésik, Z.: Free shuffle algebras in language varieties. Theoretical Computer Science 163(1&2), 55–98 (1996)
14. Valdes, J., Tarjan, R.E., Lawler, E.L.: The recognition of series parallel digraphs. SIAM Journal of Computing 11(2), 298–313 (1982)
15. Ésik, Z., Németh, Z.L.: Automata on series-parallel biposets. In: Kuich, W., Rozenberg, G., Salomaa, A. (eds.) DLT 2001. LNCS, vol. 2295, pp. 217–227. Springer, Heidelberg (2002)
16. Stockmeyer, L.J., Meyer, A.R.: Word problems requiring exponential time: Preliminary report. In: Aho, A.V., Borodin, A., Constable, R.L., Floyd, R.W., Harrison, M.A., Karp, R.M., Strong, H.R. (eds.) STOC 1973, pp. 1–9. ACM (1973)
17. Stockmeyer, L.J.: The Complexity of Decision Problems in Automata Theory and Logic. PhD thesis. MIT, Cambridge, Massachusetts, USA (1974)
18. Jacobs, B.: A bialgebraic review of deterministic automata, regular expressions and languages. In: Futatsugi, K., Jouannaud, J.-P., Meseguer, J. (eds.) Algebra, Meaning, and Computation. LNCS, vol. 4060, pp. 375–404. Springer, Heidelberg (2006)
19. Bonsangue, M.M., Milius, S., Silva, A.: Sound and complete axiomatizations of coalgebraic language equivalence. ACM TOCL 14(1) (2013)

A Modified Completeness Theorem of KAT and Decidability of Term Reducibility

Takeo Uramoto

Department of Mathematics, Kyoto University
Sakyo-ku, Kyoto, Japan
takeo-u@math.kyoto-u.ac.jp

Abstract. Kleene algebra with tests (KAT) was introduced by Kozen as an extension of Kleene algebra (KA). The decidability of equational formulas $p = q$ and Horn formulas $\wedge_i p_i = q_i \rightarrow p = q$ in KAT has been studied so far by several researchers. Continuing this line of research, this paper studies the decidability of existentially quantified equational formulas $\exists q \in P.(p = q)$ in KAT, where P is a fixed collection of KAT terms. A new completeness theorem of KAT is proved, and via the completeness theorem, the decision problem of $\exists q \in P.(p = q)$ is reduced to a certain membership problem of regular languages, to which a pseudo-identity-based decision method is applicable. Based on this reduction, an instance of the problem is studied and shown to be decidable.

Keywords: Kleene algebra with tests, completeness theorem, decidability.

1 Introduction

A *Kleene algebra with tests* is a pair $(\mathcal{K}, \mathcal{B})$ of a *Kleene algebra* $\mathcal{K}$ [Koz94] and an embedded Boolean subalgebra $\mathcal{B} \subseteq \mathcal{K}$ whose members are called *tests*. Kleene algebra with tests was introduced by Kozen [Koz97], and the paradigm provides an algebraic approach to program logic and formal verification of program equivalences [Koz97, Koz00, Koz08a, KP00].

From a syntactic viewpoint, the equational logic for Kleene algebra with tests (KAT) is an extension of that for Kleene algebra (KA). Regular expressions (i.e. KA terms) are extended so that, by KAT terms, one can naturally encode simple **while**-programs [Koz97, Koz00] in such a manner that the encoding is compatible with relational semantics of **while**-programs [KT90]. Moreover, KAT has a necessary and sufficient set of axioms for reasoning about relational equivalences of programs. That is, an identity $p = q$ between programs (or KAT terms in general) is valid over all relational interpretations if and only if it is formally deducible from the axioms of KAT [KS96], which we denote by $\text{KAT} \vdash p = q$.

So far, the formal deducibility $\text{KAT} \vdash \phi$ of several forms of formulas ϕ under the axioms of KAT (or other KA variants) and their decision problems have been studied by several authors [Coh94, KCS96, KS96]. It was shown in [KS96] that $\text{KAT} \vdash p = q$ (i.e. the equational theory of KAT) is decidable. In the case

P. Höfner et al. (Eds.): RAMiCS 2014, LNCS 8428, pp. 83–100, 2014.
© Springer International Publishing Switzerland 2014

of universal Horn formulas ($\wedge_i p_i = q_i \rightarrow p = q$), the problem is undecidable in general under the axioms of *-continuous Kleene algebras (KAT*) [Coh94, KS96]. However, the case of universal Horn formulas of the form $r = 0 \rightarrow p = q$ is proved to be decidable for KA [Coh94] and also for KAT [KS96], by effectively reducing it to the decision problem of equational formulas.

Continuing this line of investigations of decision problems, the present paper studies the decidability of existentially quantified equational formulas $\exists q \in$ P.$(p = q)$ in KAT with P being a fixed collection of terms. The problem is to decide if there exists $q \in$ P such that KAT $\vdash p = q$ for a given p. When such $q \in$ P exists, we say that p is *reducible* to the class P. Also we refer to the problem as the *term reducibility problem*, writing as KAT $\vdash \exists q \in$ P.$(p = q)$. See §5 for an instance of this problem.

This form of decision problem arises naturally in connection with program optimizations in particular. In program optimization, one is concerned with whether a program p of interest can be refined to another program q that satisfies some fixed criterion (e.g. has PTIME complexity or uses bounded resources), keeping the equivalence of programs. The decision problem of the formula $\exists q \in$ P.$(p = q)$ concerns the existence of such an equivalent program that satisfies an intended criterion (i.e. the membership in the class P). Finding an equivalent program q for a given program p from a restricted class of programs is a crucial step in KAT-based studies of program optimizations such as [Koz97, KP00]. In the present paper, we discuss the decidability of the existence of a desired equivalent program and develop its decision method.

The method of this paper follows the tradition of the algebraic decision methods in combinatorics of regular languages [Pin86]. The key step of this decision method is *pseudo-identity*-based characterizations of combinatorial properties of regular languages (§3). In order to apply the method to term reducibility problems, however, we need to prove a new completeness theorem of KAT (§4). This theorem plays a central role in the reduction of term reducibility problems to combinatorial problems of regular languages, to which the pseudo-identity-based method is applicable. Based on this reduction, we also study an instance of term reducibility problems and show its decidability (§5).

Related Work. As far as completeness of KAT is concerned, there is a related achievement due to Kozen and Smith [KS96]. Despite of this achievement, we shall present our completeness theorem, because there is a certain technical problem on the relation between Kozen and Smith's completeness theorem and the pseudo-identity-based decision method.

In [KS96], Kozen and Smith have shown that KAT is deductively complete with respect to a model of KAT, denoted $\mathcal{G}_{\Sigma,B}$, consisting of *regular sets of guarded strings* (§2). This completeness theorem shows that there is a term interpretation $p \mapsto G(p)$ that assigns effectively to each term p a regular set $G(p) \in \mathcal{G}_{\Sigma,B}$ of guarded strings in such a manner that $G(p) = G(q)$ if and only if KAT $\vdash p = q$. This theorem provided a counterpart to the well-known completeness theorem of KA: For any regular expressions p and q, $R(p) = R(q)$ if and only if KA $\vdash p = q$, where $R(p)$ is the standard interpretation of the regular expression p as a regular language. In model-theoretic words, $\mathcal{G}_{\Sigma,B}$ is a free Kleene algebra with tests generated over Σ and B (where B is the extended alphabet of KAT representing tests) in the same way that the algebra $\mathcal{R}_\Sigma$ of regular languages over an alphabet Σ is a free Kleene algebra generated over Σ.

Technically speaking, $\mathcal{G}_{\Sigma,B}$ is given as a subclass of $\mathcal{R}_{\Sigma \cup B \cup \bar{B}}$ simply because regular sets of guarded strings are defined as a certain type of regular languages over an alphabet of the form $\Sigma \cup B \cup \bar{B}$. Due to the completeness of KAT with respect to $\mathcal{G}_{\Sigma,B}$, the term reducibility problem KAT $\vdash \exists q \in \mathrm{P}.(p = q)$ is equivalent to the decision problem of the membership $G(p) \in G(\mathrm{P})$. Thus, if one could find a decision method for the membership problem in the class $G(\mathrm{P}) \subseteq \mathcal{R}_{\Sigma \cup B \cup \bar{B}}$ of regular languages (i.e. a method of deciding whether or not $L \in G(\mathrm{P})$ for $L \in \mathcal{R}_{\Sigma \cup B \cup \bar{B}}$), then it would follow that KAT $\vdash \exists q \in \mathrm{P}.(p = q)$ is decidable.

Characterizations of membership by pseudo identities provide a systematic decision method for this type of decision problem. Schützenberger's theorem [Sch65] is a pioneering result in this direction, from which it follows that the membership problem in the class $\mathcal{SF}$ of *star-free languages* is decidable: A regular language L is said to be *star-free* if there exists an extended regular expression q that contains no Kleene star and $L = R(q)$. Schützenberger proved that a regular language L is star-free if and only if its *syntactic monoid* $M(L)$ satisfies the identity $x^\omega = x^{\omega+1}$ ($\Leftrightarrow \exists n \in \mathbb{N}. \forall x \in M(L). \; x^n = x^{n+1}$). A syntactic monoid is a monoid $M(L)$, which is attached canonically to each language L and is finite if and only if L is regular. Since the multiplication table of $M(L)$ is calculable from a regular language L and the equational formula above is decidable by searching the table, it is decidable if a regular language L is star-free (i.e. $L \in \mathcal{SF}$). The key of this decidability proof is that the membership in $\mathcal{SF}$ is characterized by the identity $x^\omega = x^{\omega+1}$ of syntactic monoids. When a class $\mathcal{V}$ of regular languages has such characterizing identities, we say that $\mathcal{V}$ is *definable* by the identities. So far, several decidability results of membership problems were established in a similar way, including the decidabilities for *locally testable languages* [BS72] and *piecewise testable languages* [Sim75].

However, in a sharp contrast to this line, $\mathcal{G}_{\Sigma,B}$ is not definable by any set of pseudo identities when it is regarded as a subclass of $\mathcal{R}_{\Sigma \cup B \cup \bar{B}}$, as shown in §3. Even worse, every non-trivial subclass $\mathcal{V} \subseteq \mathcal{G}_{\Sigma,B}$ is not definable by any set of pseudo identities. In particular, for any class P of terms, the subclass $G(\mathrm{P}) \subseteq \mathcal{G}_{\Sigma,B}$ is not definable. This undefinability implies that it is essentially impossible to apply the above pseudo-identity-based argument to the decision problem KAT $\vdash \exists q \in \mathrm{P}.(p = q)$.

The source of this undefinability is that the class $\mathcal{G}_{\Sigma,B}$ is not closed under quotients by finite strings. More specifically, residuals of a guarded string fail to be guarded strings in general.

To remedy this technical issue caused by the undefinability, we introduce the notion of *weakly guarded strings* that relaxes the definition of guarded strings. While in guarded strings test symbols (i.e. letters in $B \cup \bar{B}$) must occur in a definite order and cannot appear twice adjacently, in weakly guarded strings they can occur in an arbitrary order and may be duplicated and also eliminated. Then we define another Kleene algebra with tests, denoted $\mathcal{W}_{\Sigma,B}$, consisting of certain regular sets of weakly guarded strings. We show that KAT is still deductively complete with respect to this refined model $\mathcal{W}_{\Sigma,B}$ as well as $\mathcal{G}_{\Sigma,B}$, but that the class $\mathcal{W}_{\Sigma,B}$ is definable by a set of identities, unlike $\mathcal{G}_{\Sigma,B}$. There is a term interpretation $p \mapsto W(p)$ that assigns effectively to each term p a regular set $W(p) \in \mathcal{W}_{\Sigma,B}$ of weakly guarded strings. This term interpretation W provides an effective reduction of the term reducibility problem $\text{KAT} \vdash \exists q \in \text{P}.(p = q)$ to the membership problem $W(p) \in W(\text{P})$, to which the pseudo-identity-based method is applicable (§5).

Remark. Our completeness theorem of KAT with respect to $\mathcal{W}_{\Sigma,B}$ does not imply that $\mathcal{G}_{\Sigma,B}$ is no longer interesting. In fact, the model $\mathcal{G}_{\Sigma,B}$ has a close connection to the study of deterministic flowcharts [AK01, Koz03, KT08], and is of independent interest. As far as completeness of KAT is concerned, there is no distinction between $\mathcal{G}_{\Sigma,B}$ and $\mathcal{W}_{\Sigma,B}$ because KAT is equally complete with respect to both of them. The major distinction between them is that $\mathcal{G}_{\Sigma,B}$ is more compatible with the study of deterministic flowcharts, while $\mathcal{W}_{\Sigma,B}$ is more compatible with the decision method using pseudo identities and thus more suitable for the reduction of term reducibility problems $\text{KAT} \vdash \exists q \in \text{P}.(p = q)$ to membership problems of regular languages.

Also the completeness of KAT with respect to $\mathcal{G}_{\Sigma,B}$ plays an important role in the *coinductive proof method* for reasoning about KAT term equivalences [Koz08a], i.e. a proof method based on the idea of *Brzozowski derivative* on KAT terms and *bisimulation* between *automata on guarded strings* [Koz03]. See also [Rut98, CP04]. In the last section, we will mention a contribution of our result to this subject.

2   Kleene Algebras with Tests

In this section, we briefly recall the syntax and semantics (models) of KAT, and the structure of $\mathcal{G}_{\Sigma,B}$ for the sake of reader's convenience. For more information, the reader is referred to [Koz94, KS96].

Terminology. Throughout this paper, alphabets are assumed to be finite and nonempty. For an alphabet A, we denote by A^* the free monoid over A and by ε the unit element (i.e. the empty string). Given a string $w = \mathsf{a_1 a_2 \cdots a_n} \in A^*$, the set $\{\mathsf{a_1}, \mathsf{a_2}, \cdots, \mathsf{a_n}\}$ of letters occuring in w is denoted by $C(w)$ and referred as the *content* of w. The length of a string w is denoted by $|w|$. For a string $u \in A^*$ and a language $L \subseteq A^*$, the *quotient* of L by u *from left* is the language $u \backslash L := \{v \in A^* \mid uv \in L\}$. Similarly, the quotient of L by u *from right* is defined as $L/u := \{v \in A^* \mid vu \in L\}$. Also the *both-sided* quotient of L by u and $v \in A^*$ is denoted as $u \backslash L/v := \{w \in A^* \mid uwv \in L\}$.

2.1   Syntax and Models of KAT

Syntax. Let Σ and B be finite sets of symbols. Elements of Σ are called *primitive actions*, and elements of B are called *primitive tests*. *Boolean terms* and (KAT) *terms* over (Σ, B) are formal expressions defined in BNF format as follows:

$$b ::= 0 \mid 1 \mid \mathsf{b} \in B \mid b+b \mid bb \mid \bar{b} \tag{1}$$

$$p ::= \mathsf{p} \in \Sigma \mid b \mid p+p \mid pp \mid p^* \tag{2}$$

Boolean terms are given by b and terms are given by p. Note that every Boolean term is a term. We denote by $\mathrm{T}_{\Sigma,B}$ the set of all terms over (Σ, B) and by BT_B the set of all Boolean terms over B. By definition, a regular expression is a term in KAT.

KAT is an extension of KA: In addition to the axioms of KA [Koz94], KAT also has the following axioms on Boolean terms $b, b' \in \mathrm{BT}_B$, which just state that Boolean terms form a Boolean algebra:

$$bb = b \qquad\qquad bb' = b'b \tag{3}$$
$$b + \bar{b} = 1 \qquad\qquad b\bar{b} = 0. \tag{4}$$

(These axioms are a proper subset of those given in [KK06], but it is readily seen that the two axiomatizations are equivalent.)

Given two terms $p, q \in \mathrm{T}_{\Sigma,B}$, we denote by KAT $\vdash p = q$ if the equality $p = q$ is formally deducible from the axioms of KAT. It is not difficult to see that KAT is a conservative extension of KA in the sense that, for two regular expressions p, q over Σ (i.e. p, q do not contain symbols from B), then KA $\vdash p = q$ if and only if KAT $\vdash p = q$.

Models. A *Kleene algebra with tests* [Koz97] is a two-sorted algebraic structure $(\mathcal{K}, \mathcal{B}, +, \cdot, *, ^-, 0, 1)$, where $(\mathcal{K}, +, \cdot, *, 0, 1)$ is a Kleene algebra [Koz94]; $\mathcal{B} \subseteq \mathcal{K}$; and $(\mathcal{B}, +, \cdot, ^-, 0, 1)$ is a Boolean algebra.

Let $\mathcal{K}$ be a Kleene algebra with tests. We define a (Σ, B)-*interpretation* over $\mathcal{K}$ to be a pair of maps $I_\Sigma : \Sigma \to \mathcal{K}$ and $I_B : B \to \mathcal{B}$. Given an interpretation over $\mathcal{K}$, it extends uniquely to (and is identified with) a map $I : \mathrm{T}_{\Sigma,B} \to \mathcal{K}$ such that I coincides with I_Σ and I_B on Σ and B respectively, commutes with all algebraic operators, and preserves constants. A *model* of KAT is a pair of a Kleene algebra with tests $\mathcal{K}$ and an interpretation I over $\mathcal{K}$. If $I(p) = I(q)$, then we denote this as $(\mathcal{K}, I) \models p = q$.

The *soundness* of KAT follows directly from the definition: For any model $(\mathcal{K}, I)$ of KAT and terms $p, q \in \mathrm{T}_{\Sigma,B}$, KAT $\vdash p = q$ implies that $(\mathcal{K}, I) \models p = q$.

2.2   Completeness of KAT

In what follows, let $B = \{\mathsf{b}_1, \mathsf{b}_2, \cdots, \mathsf{b}_N\}$ with $1 \leq N$. Also let us denote by $\bar{B}$ the set $\{\bar{\mathsf{b}}_1, \bar{\mathsf{b}}_2, \cdots, \bar{\mathsf{b}}_N\}$ and assume that it is disjoint from Σ and B. In general, we call a string γ over $B \cup \bar{B}$ a *test* of B. An *atom* of B is a test α of B of the form $\alpha = c_1 c_2 \cdots c_N$ where $c_i \in \{b_i, \bar{b}_i\}$ for each $1 \leq i \leq N$. (For instance, if $B = \{b_1, b_2, b_3\}$, then $b_1 b_2 \bar{b}_3$ is an atom, but $b_2 \bar{b}_1 b_3$ and $b_1 b_2 b_2 \bar{b}_3$ are not, because the order of b_1 and b_2 is reversed in the first one and b_2 appears twice in the second one.) We denote by $\text{At}(B)$ the set of all atoms of B.

Definition 1. A *guarded string* over (Σ, B) is a string over $\Sigma \cup B \cup \bar{B}$ of the form $\alpha_0 p_1 \alpha_1 \cdots p_n \alpha_n$, where $\alpha_i \in \text{At}(B)$ and $p_i \in \Sigma$.

In other words, a guarded string is a member of the language $(\text{At}(B) \cdot \Sigma)^* \cdot \text{At}(B)$. We denote by GS the set of all guarded strings over (Σ, B). By definition, GS is a regular language over $\Sigma \cup B \cup \bar{B}$. In particular GS $\subseteq (\Sigma \cup B \cup \bar{B})^*$. We say that a language $L \subseteq (\Sigma \cup B \cup \bar{B})^*$ is a *language of guarded strings* over (Σ, B) if $L \subseteq$ GS.

Let $u = x\alpha$ and $v = \beta y$ be two guarded strings over (Σ, B), where α is the last atom in u and β is the initial atom in v. Then the *coalesced product* $\diamond$ is defined as a partial binary operation on GS:

$$x\alpha \diamond \beta y := \begin{cases} x\alpha y & (\alpha = \beta) \\ \text{undefined} & (\alpha \neq \beta) \end{cases} \tag{5}$$

Moreover, given two languages of guarded strings $L, R \subseteq$ GS, define:

$$L \diamond R := \{u \diamond v \in \text{GS} \mid u \in L,\ v \in R\}. \tag{6}$$

Given a term $p \in T_{\Sigma, B}$, one can define a language of guarded strings $G(p) \subseteq$ GS by induction on the structure of p. For the base case $p \in \Sigma$ and $b \in B$,

$$G(p) := \{\alpha p \beta \in \text{GS} \mid \alpha, \beta \in \text{At}(B)\} \tag{7}$$
$$G(b) := \{\alpha \in \text{At}(B) \mid b \in C(\alpha)\}. \tag{8}$$

For the induction step,

$$G(p+q) := G(p) \cup G(q) \qquad\qquad G(1) := \text{At}(B) \tag{9}$$
$$G(pq) := G(p) \diamond G(q) \qquad\qquad G(0) := \emptyset \tag{10}$$
$$G(p^*) := G(p)^* \qquad\qquad G(\bar{b}) := \text{At}(B) - G(b). \tag{11}$$

Here the star $G(p)^*$ above is intended to be $\sum_{n=0}^{\infty} \overbrace{G(p) \diamond \cdots \diamond G(p)}^{n \text{ times}}$. A language of guarded strings L is said to be a *regular set of guarded strings* if there exists a term p such that $L = G(p)$. Now we define $\mathcal{G}_{\Sigma, B}$ as the set of all regular sets of guarded strings over (Σ, B). As pointed out in [Koz03], a regular set of guarded strings is the same as a regular language consisting of guarded strings only: i.e. $\mathcal{G}_{\Sigma, B} := \{G(p) \in \mathcal{R}_{\Sigma \cup B \cup \bar{B}} \mid p \in T_{\Sigma, B}\} = \{L \in \mathcal{R}_{\Sigma \cup B \cup \bar{B}} \mid L \subseteq \text{GS}\}$. This fact will be used later in the construction of $\mathcal{W}_{\Sigma, B}$ (§4).

Naturally, $\mathcal{G}_{\Sigma, B}$ admits a structure of a Kleene algebra with tests: For $L, L' \in \mathcal{G}_{\Sigma, B}$, their summation is given by the union $L \cup L'$ and the multiplication is given by the coalesced product $L \diamond L'$. The Kleene star of L is by $\sum_{n=0}^{\infty} L \diamond L \diamond \cdots \diamond L$.

The Boolean subalgebra of tests in $\mathcal{G}_{\Sigma,B}$ is $\{L \in \mathcal{G}_{\Sigma,B} \mid L \subseteq \mathrm{At}(B)\}$, which coincides with the image of BT_B under G. Also, the above inductive assignment G defines a map $G : \mathrm{T}_{\Sigma,B} \ni p \mapsto G(p) \in \mathcal{G}_{\Sigma,B}$. By definition, G gives a (Σ, B)-interpretation over $\mathcal{G}_{\Sigma,B}$ and is referred as the *standard interpretation* over $\mathcal{G}_{\Sigma,B}$. Moreover, the standard interpretation G is computable. That is, there exists an effective procedure whose input is a term $p \in \mathrm{T}_{\Sigma,B}$ and output is the regular expression denoting the regular language $G(p) \in \mathcal{G}_{\Sigma,B}$ over $\Sigma \cup B \cup \bar{B}$. The procedure can be obtained, e.g., by using the construction given in Lemma 7 of [KS96], where the authors constructed a regular expression $\hat{p}$ for each term p by induction so that $R(\hat{p}) = G(p)$.

Having prepared these, we can now state the completeness theorem of KAT with respect to the model $(\mathcal{G}_{\Sigma,B}, G)$:

Theorem 1 (Kozen-Smith, [KS96]). *Let $p, q \in \mathrm{T}_{\Sigma,B}$ be two terms over (Σ, B). Then $\mathrm{KAT} \vdash p = q$ if and only if $(\mathcal{G}_{\Sigma,B}, G) \models p = q$, namely iff $G(p) = G(q)$.*

3 Pseudo-identity and Undefinability of $\mathcal{G}_{\Sigma,B}$

In this section, we recall necessary notions and known results from the variety theory of regular languages, including *syntactic semirings* and *pseudo-identities* in particular. We also see that $\mathcal{G}_{\Sigma,B} \subseteq \mathcal{R}_{\Sigma \cup B \cup \bar{B}}$ and its non-trivial subclasses $\{\emptyset\} \neq \mathcal{V} \subseteq \mathcal{G}_{\Sigma,B}$ are not closed under quotients and are thus undefinable.

3.1 Syntactic Semirings

A *syntactic semiring* is an idempotent semiring $S(L)$ attached to each language L and subsumes syntactic monoids. Syntactic semirings were introduced by Polák, who developed the fundamental theory of syntactic semirings. See [Pol03, Pol04a, Pol04b] for more information.

Let A be an alphabet and denote by $\mathrm{F}(A^*)$ the idempotent semiring consisting of all *finite languages* over A (i.e. finite subsets of A^*), where the sum is the set-theoretic union and the multiplication is the element-wise concatenation, i.e. $U \cdot V = \{uv \in A^* \mid u \in U, v \in V\}$ for $U, V \in \mathrm{F}(A^*)$. The unit and zero element of $\mathrm{F}(A^*)$ are the singleton language $\{\varepsilon\}$ and the empty language $\emptyset$ respectively. We usually identify a string $w \in A^*$ with the singleton language $\{w\} \in \mathrm{F}(A^*)$. Thus a finite language $U = \{w_1, \cdots, w_n\} \in \mathrm{F}(A^*)$ can be denoted as $U = w_1 + \cdots + w_n$. Also the singleton language $\{\varepsilon\}$ and the empty language $\emptyset$ are denoted by 1 and 0 respectively when they are regarded as members of $\mathrm{F}(A^*)$.

Definition 2. Let $L \subseteq A^*$ be a language. The *syntactic congruence* $\equiv_L$ and *syntactic quasi-ordering* $\leq_L$ on the idempotent semiring $\mathrm{F}(A^*)$ are defined respectively as follows: For two elements $U, V \in \mathrm{F}(A^*)$,

$$U \equiv_L V \Leftrightarrow \forall x, y \in A^* \ (xUy \cap L \neq \emptyset \Leftrightarrow xVy \cap L \neq \emptyset). \tag{12}$$

$$U \leq_L V \Leftrightarrow \forall x, y \in A^* \ (xUy \cap L \neq \emptyset \Rightarrow xVy \cap L \neq \emptyset). \tag{13}$$

Here, for a finite language $U = \{w_1, \cdots, w_n\} \in \mathrm{F}(A^*)$, we denote by xUy the finite language $\{xw_1y, \cdots, xw_ny\}$.

Definition 3. Let $L \subseteq A^*$ be a language. The *syntactic semiring* $S(L)$ of L is defined as the quotient semiring $\mathrm{F}(A^*)/\equiv_L$. The canonical projection is denoted by $\pi_L : \mathrm{F}(A^*) \twoheadrightarrow S(L)$.

Remark 1. The above syntactic congruence $\equiv_L$ and quasi-ordering $\leq_L$ for $L \subseteq A^*$ coincide with Polák's original syntactic congruence and quasi-ordering for the *complement* L^c. Originally, syntactic semirings were defined so as to be applicable to membership problems for classes of regular languages that are closed under *conjunction* $\wedge$ (called *conjunctive varieties*). Taking complement of languages enables us to use syntactic semirings also for classes of regular languages closed under *disjunction* $\vee$ (called *disjunctive varieties*), which we need in this paper. This dualization using complement was discussed also in §6 of [Pol04a].

Remark 2. The syntactic semiring $S(L)$ contains the syntactic monoid $M(L)$: By definition of syntactic monoids (e.g. [Pin86]), $M(L)$ coincides with the image of $A^* \subseteq \mathrm{F}(A^*)$ under π_L, i.e. $\pi_L(A^*) = M(L)$. Thus $M(L)$ is a submonoid of $S(L)$ with respect to multiplication.

Similarly to the fundamental property that a language L is regular if and only if its syntactic monoid $M(L)$ is finite, Polák proved that the regularity of L is also characterized by the finiteness of $S(L)$:

Proposition 1 (Polák, [Pol04a]). *A language L is regular if and only if $S(L)$ is a finite semiring.*

The structure of $S(L)$ is calculable from a given regular language L, e.g. by using a finite automaton accepting L or a regular expression denoting L, similarly to $M(L)$ [Pin86].

3.2 Pseudo-identities and Definability

Let $\mathcal{S}$ be the class of all finite idempotent semirings. An *n-ary implicit operation* is a family $(\rho_S)_{S \in \mathcal{S}}$ of maps $\rho_S : S^n \to S$ on each $S \in \mathcal{S}$ satisfying the following: For any semiring homomorphism $f : S \to S'$ and $s_i \in S$ $(1 \leq i \leq n)$, we have $\rho_{S'}(f(s_1), \cdots, f(s_n)) = f(\rho_S(s_1, \cdots, s_n))$.

Example 1. The semiring operations $+$ and $\cdot$ (and all algebraic terms composed from $+, \cdot$) define 2-ary (multi-ary) implicit operations, denoted $\rho^+ = (\rho_S^+)$ and $\rho^{\cdot} = (\rho_S^{\cdot})$ and given as follows respectively: $\rho_S^+(s_1, s_2) = s_1 + s_2$ and $\rho_S^{\cdot}(s_1, s_2) = s_1 s_2$. An implicit operation that is obtained from a term is called an *explicit operation*. We sometimes denote them as $x_1 + x_2$ and $x_1 x_2$, where x_1 and x_2 represent distinguished variables.

Example 2. Let S be a finite idempotent semiring and $s \in S$ be an arbitrary element. Then the sub-semigroup generated by s contains the unique multiplicative idempotent, denoted as $s^\omega \in S$. Define a map $x_S^\omega : S \ni s \mapsto s^\omega \in S$, which gives a unary implicit operation $x^\omega = (x_S^\omega)$.

Example 3. Let S be a finite idempotent semiring. Then S is a Kleene algebra. Define a family $x^* = (x_S^*)$ by $x_S^* : S \ni s \mapsto s^* \in S$, where s^* is the Kleene star of $s \in S$. It is readily seen that x^* is a unary implicit operation.

Definition 4. Let us denote by I_n the set of all n-ary implicit operations. An n-ary pseudo-identity is an ordered pair (ρ, τ) of n-ary implicit operations $\rho, \tau \in I_n$. We denote it as $\rho = \tau$ or $\rho(x_1, \cdots, x_n) = \tau(x_1, \cdots, x_n)$ to indicate that ρ and τ are n-ary.

Denote by $V_n = \{x_1, \cdots, x_n\}$ a set of distinguished n variables. We call a map $\sigma : V_n \to \mathrm{F}(A^*)$ an n-ary substitution of $\sigma(x_i) \in \mathrm{F}(A^*)$ to the variables x_i. Let S be a finite idempotent semiring having generators A, i.e. S is equipped with a fixed surjective semiring homomorphism $\pi : \mathrm{F}(A^*) \twoheadrightarrow S$. Then we say that S satisfies the n-ary pseudo-identity $\rho = \tau$ with respect to an (n-ary) substitution $\sigma : V_n \to \mathrm{F}(A^*)$ if $\rho_S(s_1, \cdots, s_n) = \tau_S(s_1, \cdots, s_n)$, where $s_i = \pi \circ \sigma(x_i)$ ($1 \le i \le n$). When this is the case, we denote as $S \models \rho(\sigma) = \tau(\sigma)$.

Example 4. Let S be a finite semiring with generators A (i.e. a surjective homomorphism $\pi : \mathrm{F}(A^*) \twoheadrightarrow S$) and let $a_1, a_2 \in A$ be two generators. Then they commute in S if and only if $S \models x_1 x_2(\sigma) = x_2 x_1(\sigma)$ for the 2-ary substitution σ such that $\sigma(x_i) = a_i$. We sometimes denote it simply as $S \models a_1 a_2 = a_2 a_1$.

Let $L \subseteq A^*$ be a regular language over A, and $\pi_L : \mathrm{F}(A^*) \twoheadrightarrow S(L)$ be the canonical projection onto the syntactic semiring $S(L)$. Also let E be a set of pseudo identities $\rho = \tau$ equipped with substitutions σ, denoted $\rho(\sigma) = \tau(\sigma)$. Then one can define a class $\mathcal{W}(E) \subseteq \mathcal{R}_A$ of regular languages over A by $\mathcal{W}(E) := \{L \in \mathcal{R}_A \mid \forall \rho(\sigma) = \tau(\sigma) \in E. \, S(L) \models \rho(\sigma) = \tau(\sigma)\}$. If a class $\mathcal{V}$ is of the form $\mathcal{W}(E)$ for some E, then we say that $\mathcal{V}$ is definable by E. We also say that L satisfies E if $L \in \mathcal{W}(E)$.

Example 5. Let $L \subseteq A^*$ be a language over an alphabet A and $\pi_L : \mathrm{F}(A^*) \twoheadrightarrow S(L)$ be the canonical projection onto the syntactic semiring of L. Then, by definition, $S(L) \models a_1 a_2 = a_2 a_1$ if and only if $a_1 a_2 \equiv_L a_2 a_1$. That is, $u a_1 a_2 v \in L$ implies $u a_2 a_1 v \in L$ and vice versa for every $u, v \in A^*$. Thus if $E = \{a_1 a_2 = a_2 a_1\}$ for example, then $L \in \mathcal{W}(E)$ exactly when $u a_1 a_2 v \in L \Leftrightarrow u a_2 a_1 v \in L$ for every $u, v \in A^*$.

Example 6 (Polák, [Pol04b]). Let $\mathcal{L}_{k,m,l} \subseteq \mathcal{R}_A$ be the class of regular languages over A, whose member is a finite summation of regular languages of the form $u B_1^* \cdots B_m^* v$ with $u, v \in A^*$, $\mid u \mid \le k$, $\mid v \mid \le l$, $B_1, \cdots, B_m \subseteq A$. In [Pol04b], it was shown that, if $m \le 2$, $\mathcal{L}_{k,m,l}$ is definable by a finite set of certain pseudo identities. Due to this definability, it is decidable whether a regular language $L \in \mathcal{R}_A$ is a member of $\mathcal{L}_{k,m,l}$ for $m \le 2$. In fact, $L \in \mathcal{L}_{k,m,l}$ is equivalent to the property that its syntactic semiring $S(L)$ satisfies the finite number of defining identities, which is decidable.

3.3 Undefinability of $\mathcal{G}_{\Sigma,B}$

Let $\mathrm{P} \subseteq \mathrm{T}_{\Sigma,B}$ be a fixed collection of KAT terms. Then the term reducibility problem $\mathrm{KAT} \vdash \exists q \in \mathrm{P}.(p = q)$ can be effectively reduced to the membership problem in the class $G(\mathrm{P}) \subseteq \mathcal{R}_{\Sigma \cup B \cup \bar{B}}$. If the class $G(\mathrm{P})$ has a finite number of defining identities as Example 6, it follows that $\mathrm{KAT} \vdash \exists q \in \mathrm{P}.(p = q)$ is decidable.

This strategy of reduction is, however, not successful because the class $G(\mathrm{P})$ has no defining identities for any non-trivial $\mathrm{P} \subseteq \mathrm{T}_{\Sigma,B}$. This follows from the following fact:

Proposition 2. *If* $\mathcal{V} \subseteq \mathcal{G}_{\Sigma,B}$ *is definable, then* $\mathcal{V} = \{\emptyset\}$.

Proof. Note that a definable class $\mathcal{W}(E)$ must be closed under quotients by strings due to the general fact that, for every language L and a string u, there are canonical projections $S(L) \twoheadrightarrow S(u \backslash L)$ and $S(L) \twoheadrightarrow S(L/u)$. Assume that $\mathcal{V} \subseteq \mathcal{G}_{\Sigma,B}$ is definable. Then it is closed under quotients. Thus, if $\mathcal{V}$ should contain a non-empty language $L \neq \emptyset$, its quotient $\mathsf{b} \backslash L$ by a primitive test $\mathsf{b} \in B$ must be in $\mathcal{V}$. However, $\mathsf{b} \backslash L$ is not even a language of guarded strings. Thus $\mathsf{b} \backslash L \notin \mathcal{G}_{\Sigma,B}$, when $\mathsf{b} \backslash L \notin \mathcal{V}$ in particular. This is a contradiction. Thus $\mathcal{V}$ cannot contain any non-empty language. On the other hand, the empty language $\emptyset$ satisfies an arbitrary identity, thus $\emptyset \in \mathcal{V}$. Consequently $\mathcal{V} = \{\emptyset\}$.

Corollary 1. $\mathcal{G}_{\Sigma,B}$ *is undefinable.*

4   A Modified Completeness Theorem of KAT

The subject of this section is to construct a subclass $\mathcal{W}_{\Sigma,B} \subseteq \mathcal{R}_{\Sigma \cup B \cup \bar{B}}$ that is definable but isomorphic to $\mathcal{G}_{\Sigma,B}$ by mutually computable isomorphisms. In particular, it follows that the membership problem in $\mathcal{W}_{\Sigma,B}$ is decidable. Also, by the isomorphism with $\mathcal{G}_{\Sigma,B}$, the class $\mathcal{W}_{\Sigma,B}$ forms a free Kleene algebra with tests over the generators Σ and B. To construct such a model, we relax the definition of guarded strings and define instead *weakly guarded strings*.

4.1   Weakly Guarded Strings

We say that a test $\alpha' \in (B \cup \bar{B})^*$ is *consistent* if there is no $\mathsf{b}_i \in B$ such that both of b_i and $\bar{\mathsf{b}}_i$ belong to its content $C(\alpha')$. Thus, for instance, $\bar{\mathsf{b}}_2 \mathsf{b}_1 \mathsf{b}_1$ is consistent, but $\mathsf{b}_2 \mathsf{b}_3 \bar{\mathsf{b}}_2$ is not. Note that consistent tests subsume atoms. However, in general consistent tests, the order of $\mathsf{b}_1, \cdots, \mathsf{b}_N$ is arbitrary, and multiple occurrences and absences of them are permitted. The null string $\varepsilon \in (B \cup \bar{B})^*$ is also a consistent test. We denote by $\mathrm{Cst}(B)$ the set of all consistent tests. Then $\mathrm{At}(B) \subseteq \mathrm{Cst}(B)$. If we use metavariables $\alpha, \beta \cdots$, then we intend that they denote atoms, while we use metavariables with prime $\alpha', \beta' \cdots$ to denote consistent tests (i.e. they may not be atoms in general).

Definition 5. A *weakly guarded string* is a string over $\Sigma \cup B \cup \bar{B}$ of the form $\alpha'_0 \mathsf{p}_1 \alpha'_1 \mathsf{p}_2 \cdots \mathsf{p}_n \alpha'_n$, where $\alpha'_i \in \mathrm{Cst}(B)$ and $\mathsf{p}_i \in \Sigma$.

In other words, a weakly guarded string is a member of the language $(\mathrm{Cst}(B) \cdot \Sigma)^* \cdot \mathrm{Cst}(B)$, which we denote by WGS. By definition, a guarded string is always a weakly guarded string. Thus GS $\subseteq$ WGS.

Given a language $L \subseteq$ GS of guarded strings, its *weakening* wgs$(L) \subseteq$ WGS is defined as follows:

$$\text{wgs}(L) := \{\alpha'_0 \text{p}_1 \alpha'_1 \cdots \text{p}_n \alpha'_n \mid \exists \alpha_0 \text{p}_1 \alpha_1 \cdots \text{p}_n \alpha_n \in L.\ \alpha'_i \in C(\alpha_i)^*\}. \quad (14)$$

Note that, if α is an atom, any element α' of $C(\alpha)^*$ is consistent. Thus we have wgs$(L) \subseteq$ WGS for each $L \subseteq$ GS. Conversely, given a language $L' \subseteq$ WGS, we can define a language of guarded strings by $L' \cap$ GS $\subseteq$ GS, which we denote as gs(L').

Lemma 1. *For any $L \subseteq$ GS, we have gs $\cdot$ wgs$(L) = L$.*

Proof. Note that, for any atom $\alpha \in \text{At}(B)$, the language $C(\alpha)^*$ contains the unique atom α. Thus, if $\alpha_0 \text{p}_1 \alpha_1 \cdots \text{p}_n \alpha_n \in$ gs $\cdot$ wgs(L), then by definition, there exists $\beta_0 \text{p}_1 \beta_1 \cdots \text{p}_n \beta_n \in L$ such that $\alpha_i \in C(\beta_i)^*$, i.e. $\alpha_i = \beta_i$. This proves gs $\cdot$ wgs$(L) \subseteq L$. The inverse inclusion is immediate from the definitions of wgs and gs.

Lemma 2. *For any $L \subseteq$ GS, its weakening wgs$(L) \subseteq$ WGS is regular if and only if $L \in \mathcal{G}_{\Sigma,B}$.*

Proof. The only if part follows from Lemma 1 and the fact that $L \in \mathcal{G}_{\Sigma,B}$ iff L is regular and $L \subseteq$ GS. The if part follows from the next proposition.

Proposition 3. *There is an effective procedure to compute the regular expression of wgs(L) from a given regular expression of $L \in \mathcal{G}_{\Sigma,B}$.*

Proof. Since regular expressions and deterministic finite automata (DFAs) are effectively transformable, it is sufficient to give an effective procedure whose input is a DFA accepting the input regular language $L \in \mathcal{G}_{\Sigma,B}$ and output is the regular expression denoting wgs$(L) \subseteq$ WGS. Informally speaking, the procedure replaces all the occurrences of an atom α_i in $\alpha_0 \text{p}_1 \alpha_1 \cdots \text{p}_n \alpha_n \in L$ by the regular expression $C(\alpha_i)^*$. To put it formally, let L be accepted by a DFA with state set Q.

1. For each atom $\alpha = c_1 c_2 \cdots c_N$ and each state $x \in Q$, add to the input DFA a new edge $x \xrightarrow{\alpha} x'$ labeled by a formal symbol $\underline{\alpha}$ if the input DFA has a path $x \xrightarrow{c_1} x_1 \xrightarrow{c_2} x_2 \cdots x_{N-1} \xrightarrow{c_N} x'$;
2. Remove all edges $x \xrightarrow{c} x'$ labeled by primitive tests $c \in B \cup \bar{B}$;
3. Regarding the resulting graph as a DFA over the new alphabet $A' = \Sigma \cup \{\underline{\alpha} \mid \alpha \in \text{At}(B)\}$, apply the algorithm (e.g. [HMU06]) whose input is a DFA over A' and output is the regular expression over A' denoting the regular language accepted by the automaton;
4. Replace the occurrence of $\underline{\alpha}$ in the resulting regular expression over A' by the regular expression $(c_1 + \cdots + c_N)^*$ over $\Sigma \cup B \cup \bar{B}$, where $\alpha = c_1 \cdots c_N$.

Then the resulting regular expression denotes wgs(L).

We define the class $\mathcal{W}_{\Sigma,B} \subseteq \mathcal{R}_{\Sigma \cup B \cup \bar{B}}$ as follows:

$$\mathcal{W}_{\Sigma,B} := \{L' \in \mathcal{R}_{\Sigma \cup B \cup \bar{B}} \mid \exists L \in \mathcal{G}_{\Sigma,B}.\ L' = \mathrm{wgs}(L)\}. \tag{15}$$

On the one hand, the class $\mathcal{G}_{\Sigma,B}$ is characterized as the set of regular languages L over $\Sigma \cup B \cup \bar{B}$ such that $L \subseteq \mathrm{GS}$. On the other hand, the class $\mathcal{W}_{\Sigma,B}$ is not characterized as the set of regular languages L over $\Sigma \cup B \cup \bar{B}$ such that $L \subseteq \mathrm{WGS}$. In fact, there exists a regular language L such that $L \subseteq \mathrm{WGS}$ but $L \notin \mathcal{W}_{\Sigma,B}$. However, $\mathcal{W}_{\Sigma,B}$ is definable unlike $\mathcal{G}_{\Sigma,B}$ and thus the membership in $\mathcal{W}_{\Sigma,B}$ can be characterized by the defining identities.

4.2   Identities Defining $\mathcal{W}_{\Sigma,B}$

We now show that the class $\mathcal{W}_{\Sigma,B}$ coincides with the class $\mathcal{W}(E_{\Sigma,B})$ defined by the following set $E_{\Sigma,B}$ of identities, which is exactly the axioms added to KA. That is:

$$E_{\Sigma,B} := \langle bb = b,\ bb' = b'b,\ b + \bar{b} = 1,\ b\bar{b} = 0 \mid b, b' \in B \cup \bar{B} \rangle.$$

Note that E is a finite set.

Theorem 2. $\mathcal{W}_{\Sigma,B} = \mathcal{W}(E_{\Sigma,B})$

The proof of Theorem 2 is divided into two proofs of the mutual inclusions $\mathcal{W}_{\Sigma,B} \subseteq \mathcal{W}(E_{\Sigma,B})$ and $\mathcal{W}(E_{\Sigma,B}) \subseteq \mathcal{W}_{\Sigma,B}$.

Claim 1. $\mathcal{W}_{\Sigma,B} \subseteq \mathcal{W}(E_{\Sigma,B})$.

Proof. We need to show that, for any $L \in \mathcal{G}_{\Sigma,B}$, the weakening $\mathrm{wgs}(L)$ satisfies the equations $E_{\Sigma,B}$. Here we prove only that $\mathrm{wgs}(L)$ satisfies $b + \bar{b} = 1$, i.e. the equivalence $b + \bar{b} \equiv_{\mathrm{wgs}(L)} 1$ for $b \in B$. The others are similar.

By definition of the equivalence $b + \bar{b} \equiv_{\mathrm{wgs}(L)} 1$, we need to prove that, for any $u, v \in (\Sigma \cup B \cup \bar{B})^*$, $uv \in \mathrm{wgs}(L)$ if and only if $ubv \in \mathrm{wgs}(L)$ or $u\bar{b}v \in \mathrm{wgs}(L)$. However, note that if either of u or v is not weakly guarded, then none of uv, ubv nor $u\bar{b}v$ can belong to $\mathrm{wgs}(L)$. Thus it is sufficient to consider only the case when u, v are weakly guarded strings. Let $u = u'\alpha', v = \beta'v'$ be weakly guarded strings, where α' is the longest suffixal consistent test in u and β' is the longest prefixal consistent test in v. We first show that $uv \in \mathrm{wgs}(L)$ implies $ubv \in \mathrm{wgs}(L)$ or $u\bar{b}v \in \mathrm{wgs}(L)$. Assume that $uv \in \mathrm{wgs}(L)$. Then there exists $\alpha_0 p_1 \alpha_1 \cdots p_n \alpha_n \in L$ such that $\alpha'\beta' \in C(\alpha_i)^*$ for some i. Since α_i is an atom, either of b or $\bar{b}$ is a member of $C(\alpha_i)$. Thus we have $\alpha'b\beta' \in C(\alpha_i)^*$ or $\alpha'\bar{b}\beta' \in C(\alpha_i)^*$. From this, it follows that $ubv \in \mathrm{wgs}(L)$ or $u\bar{b}v \in \mathrm{wgs}(L)$. Conversely, both of $\alpha'b\beta' \in C(\alpha_i)^*$ and $\alpha'\bar{b}\beta' \in C(\alpha_i)^*$ imply $\alpha'\beta' \in C(\alpha_i)^*$. This proves the claim.

Claim 2. $\mathcal{W}(E_{\Sigma,B}) \subseteq \mathcal{W}_{\Sigma,B}$.

Proof. To show this, it is sufficient to prove that $\mathrm{wgs} \cdot \mathrm{gs}(L') = L'$ for any $L' \in \mathcal{W}(E_{\Sigma,B})$. First we prove the inclusion $L' \subseteq \mathrm{wgs} \cdot \mathrm{gs}(L')$. Let $\alpha'_0 p_1 \alpha'_1 \cdots p_n \alpha'_n \in L'$ and let $C(\alpha'_i) = \{c^i_{j_1}, \cdots, c^i_{j_k}\}$ with $c^i_j \in \{b_j, \bar{b}_j\}$ and $1 \le j_1 < \cdots < j_k \le N$.

(Note that, since L' satisfies $b\bar{b} = 0$ and $bb' = b'b$, elements of L' are of such form.) By $bb \equiv_{L'} b$ and $bb' \equiv_{L'} b'b$, we have the following equivalence for each i:

$$\alpha_i' \equiv_{L'} c_{j_1}^i \cdots c_{j_k}^i. \tag{16}$$

By $b_j + \bar{b}_j \equiv_{L'} 1$ for each $b_j \in B$, if $j_l < j < j_{l+1}$ for some j, one obtains the equivalence: $c_{j_l}^i c_{j_{l+1}}^i \equiv_{L'} c_{j_l}^i (b_j + \bar{b}_j) c_{j_{l+1}}^i$. Thus the right hand side of (16) is equivalent to the following:

$$\sum_{C(\alpha_i') \subseteq C(\alpha_i)} \alpha_i. \tag{17}$$

Here α_i ranges over atoms such that $C(\alpha_i') \subseteq C(\alpha_i)$, equivalently $\alpha_i' \in C(\alpha_i)^*$. Since $\equiv_{L'}$ is a semiring congruence on $\mathrm{F}(A^*)$, one obtains the equivalence:

$$\alpha_0' p_1 \alpha_1' \cdots p_n \alpha_n' \equiv_{L'} \sum_{\alpha_i' \in C(\alpha_i)^*} \alpha_0 p_1 \alpha_1 \cdots p_n \alpha_n. \tag{18}$$

Since $\alpha_0' p_1 \alpha_1' \cdots p_n \alpha_n' \in L'$, this equivalence proves that there exists some tuple $(\alpha_0, \cdots, \alpha_n)$ of atoms α_i such that $\alpha_i' \in C(\alpha_i)^*$ for each $1 \leq i \leq n$ and $\alpha_0 p_1 \alpha_1 \cdots p_n \alpha_n \in \mathrm{gs}(L')$. By definition of wgs, it follows that $\alpha_0' p_1 \alpha_1' \cdots p_n \alpha_n' \in \mathrm{wgs} \cdot \mathrm{gs}(L')$. Thus $L' \subseteq \mathrm{wgs} \cdot \mathrm{gs}(L')$.

Conversely, let $\alpha_0' p_1 \alpha_1' \cdots p_n \alpha_n' \in \mathrm{wgs} \cdot \mathrm{gs}(L')$. Then by definition of wgs, there exists $\alpha_0 p_1 \alpha_1 \cdots p_n \alpha_n \in \mathrm{gs}(L')$ such that $\alpha_i' \in C(\alpha_i)^*$. By the equivalence (18) and $\alpha_0 p_1 \alpha_1 \cdots p_n \alpha_n \in \mathrm{gs}(L') \subseteq L'$, it follows that $\alpha_0' p_1 \alpha_1' \cdots p_n \alpha_n' \in L'$. This concludes the proof.

Note that $\mathrm{wgs} \cdot \mathrm{gs}(L') = L'$ for any $L' \in \mathcal{W}(E_{\Sigma,B}) = \mathcal{W}_{\Sigma,B}$. This means that the maps $\mathrm{gs} : \mathcal{W}_{\Sigma,B} \to \mathcal{G}_{\Sigma,B}$ and $\mathrm{wgs} : \mathcal{G}_{\Sigma,B} \to \mathcal{W}_{\Sigma,B}$ are mutual inverse isomorphisms. Also both of gs and wgs preserve unions of languages and are computable. Thus we have proved:

Proposition 4. *The class* $\mathcal{W}_{\Sigma,B}$ *is isomorphic to* $\mathcal{G}_{\Sigma,B}$ *under the computable isomorphisms* wgs *and* gs.

Remark 3. Since wgs and gs preserve unions of languages, $\mathcal{W}_{\Sigma,B}$ and $\mathcal{G}_{\Sigma,B}$ are isomorphic as join semilattices. Note that, however, wgs does not preserve intersections of languages. We also remark that $\mathcal{G}_{\Sigma,B}$ is closed under intersections of languages, while $\mathcal{W}_{\Sigma,B}$ is not. (Of course, $\mathcal{W}_{\Sigma,B}$ has the greatest lower bounds $L_1' \wedge L_2'$ for any $L_1', L_2' \in \mathcal{W}_{\Sigma,B}$ because so does $\mathcal{G}_{\Sigma,B}$, but the greatest lower bounds are not given as set-theoretic intersections $L_1' \cap L_2'$.)

Since the class $\mathcal{W}_{\Sigma,B}$ has the finite number of defining identities $E_{\Sigma,B}$, the membership problem $L \in \mathcal{W}_{\Sigma,B}$ is decidable by judging if the syntactic semiring $S(L)$ satisfies the identities.

Corollary 2. *It is decidable whether a given regular language over* $\Sigma \cup B \cup \bar{B}$ *belongs to* $\mathcal{W}_{\Sigma,B}$.

Since $\mathcal{G}_{\Sigma,B}$ and $\mathcal{W}_{\Sigma,B}$ are isomorphic as join semilattices, the structure of free Kleene algebra on $\mathcal{G}_{\Sigma,B}$ can be transmitted onto $\mathcal{W}_{\Sigma,B}$ via wgs and gs.

Proposition 5. *The class $\mathcal{W}_{\Sigma,B}$ admits the structure of the free Kleene algebra with tests over the generating sets Σ and B.*

In other words, the equational logic KAT is deductively complete with respect to the computable KAT-term interpretation $W := \text{wgs} \circ G : \text{T}_{\Sigma,B} \to \mathcal{W}_{\Sigma,B}$. That is:

Theorem 3. *Let $p, q \in \text{T}_{\Sigma,B}$ be terms over (Σ, B). Then KAT $\vdash p = q$ if and only if $(\mathcal{W}_{\Sigma,B}, W) \models p = q$, namely iff $W(p) = W(q)$.*

Remark 4. In this paper, we do not describe explicitly the multiplication structure of $\mathcal{W}_{\Sigma,B}$ in the style of that of $\mathcal{G}_{\Sigma,B}$ (i.e. coalesced product on languages of guarded strings) nor present an inductive construction of $p \mapsto W(p)$ in the style of G. However, we have seen that $\mathcal{W}_{\Sigma,B}$ can admit the structure of free Kleene algebra with tests; KAT is complete with respect to $\mathcal{W}_{\Sigma,B}$; and also the term interpretation W is effectively calculable (because W is the composition of G and wgs and both of them are effectively calculable). These properties are the only necessary properties for the reduction of KAT $\vdash \exists q \in \text{P}.(p = q)$ to the membership problem in $W(\text{P})$.

5   Decidability of a Term Reducibility

If the class $W(\text{P})$ is definable by a finite set of identities $E = \{\rho_i(\sigma_i) = \tau_i(\sigma_i)\}_{i=1}^{m}$, then the membership problem $L \in W(\text{P})$ can be reduced to the decision problem of the quantifier-free sentence $\bigwedge_{i=1}^{m}\big(S(L) \models \rho_i(\sigma_i) = \tau_i(\sigma_i)\big)$, which is readily decidable. As shown in §3, a similar reduction by G is never successful for any P because $G(\text{P})$ is never definable. In this section, we construct a non-trivial P_0 for which $W(\text{P}_0)$ is definable by finite identities and thus the term reducibility KAT $\vdash \exists q \in \text{P}_0.(p = q)$ is decidable.

For this aim, recall from [KS96] that an *externally guarded term* is a term of the form $\alpha q \beta$, where α, β are atoms and q is a regular expression (or term) over $\Sigma \cup B \cup \bar{B}$. It was shown in [KS96] that any term p can be transformed inductively to a finite sum $\hat{p} = \sum_i \alpha_i q_i \beta_i$ of externally guarded terms $\alpha_i q_i \beta_i$ such that KAT $\vdash p = \hat{p}$ and $G(\hat{p}) = R(\hat{p})$. In general, the term q_i may still contain test symbols $b \in B \cup \bar{B}$, and in some cases, this is unavoidable.

Consider, for example, a program $r = \textbf{while } b \textbf{ do } p$ with $b \in B$ and $p \in \Sigma$, which can be encoded by $(bp)^*\bar{b}$ as a KAT term. (See [Koz97] for more information on KAT-term encoding of **while**-programs.) Then (as easily follows from the proof in Appendix) one can prove that there is no way to refine this **while**-program r to a sum of externally guarded terms $\sum_{i=0}^{n} \alpha_i q_i \beta_i$ so that every q_i does not contain test symbols (i.e. q_i is a regular expression over Σ). In other words, if one can prove,

$$\text{KAT} \vdash \textbf{while } b \textbf{ do } p = \sum_{i=0}^{n} \alpha_i q_i \beta_i, \tag{19}$$

then some of q_i must contain test symbols.

Now let P_0 be the following collection of KAT terms:

$$P_0 = \Big\{ \sum_{i=0}^{n} \alpha_i q_i \beta_i \mid \alpha_i, \beta_i : \text{atom}, \ q_i \in \text{RExp}(\Sigma) \Big\}. \tag{20}$$

Here $\text{RExp}(\Sigma)$ denotes the set of all regular expressions over Σ. If one finds that a program p is reducible to the class P_0, it intuitively means that the process represented by the program p can be simulated by a process that executes tests only in its initial and terminal parts. As seen by the above example, not all programs p are reducible to the class P_0. We can show that, however, it is decidable if a given term (program) p is reducible to P_0. In fact:

Lemma 3. *The class $W(P_0) \subseteq \mathcal{W}_{\Sigma,B}$ is definable by a finite set of identities.*

Proof. The proof and the set of defining identities are given in the Appendix.

Of course, as seen in §3, the class $G(P_0)$ is undefinable unlike $W(P_0)$. From the definability of $W(P_0)$, one can then prove the decidability of KAT $\vdash \exists q \in P_0.(p = q)$.

Theorem 4. *The term reducibility KAT $\vdash \exists q \in P_0.(p = q)$ is decidable.*

Proof. The reducibility KAT $\vdash \exists q \in P_0.(p = q)$ is equivalent to $W(p) \in W(P_0)$ by the completeness of KAT to W. Also, $W(p) \in W(P_0)$ is decidable because $W(P_0)$ is definable by a finite set of identities. Furthermore, since the map $p \mapsto W(p)$ is effectively calculable, one can conclude that KAT $\vdash \exists q \in P_0.(p = q)$ is decidable.

6   Conclusion

This paper presented a modified completeness theorem of KAT with respect to a new model $(\mathcal{W}_{\Sigma,B}, W)$. Although both models $(\mathcal{G}_{\Sigma,B}, G)$ and $(\mathcal{W}_{\Sigma,B}, W)$ share the same equational theory (i.e. the equational theory of KAT) and both of them consist of regular languages over $\Sigma \cup B \cup \bar{B}$, the characteristic difference between them is that one of them (i.e. $\mathcal{W}_{\Sigma,B}$) is definable and the other (i.e. $\mathcal{G}_{\Sigma,B}$) is undefinable as subclasses of $\mathcal{R}_{\Sigma \cup B \cup \bar{B}}$. This definability of $\mathcal{W}_{\Sigma,B}$ (and its subclasses) is crucial when we study the decidability of the validity of existentially quantified equational formulas $\exists q \in P.(p = q)$ in KAT, i.e. the term reducibility. In fact, as shown by the above instance P_0, our modified completeness theorem provides a pseudo-identity-based decision strategy for term reducibility problem KAT $\vdash \exists q \in P.(p = q)$.

In this paper, however, we left unexplored term reducibility problem for classes P of programs other than P_0. This direction of research would deserve further investigation, which is of interest as KAT-based study of program optimization.

Finally, we would like to mention that our completeness theorem contributes to the development of coinductive proof methods studied in [Rut98, CP04, Koz08a]. Using $(\mathcal{W}_{\Sigma,B}, W)$, one would obtain a coinductive proof method based on the standard deterministic automata only, while Chen and Pucella [CP04] and Kozen [Koz08a] needed nonstandard variants of automata. However, it is yet unknown if our method can improve the complexity of those in [CP04, Koz08a].

Acknowledgements. I thank Prof. Susumu Nishimura for his helpful comments on the draft of this paper. Also I am grateful to the anonymous reviewers, whose comments were valuable for improving this paper.

References

[AK01]   Angus, A., Kozen, D.: Kleene algebra with tests and program schematology. Technical Report TR2001-1844, Computer Science Department, Cornell University (2001)

[BS72]   Brzozowski, J.A., Simon, I.: Characterizations of locally testable events. Discrete Mathematics 4, 243–271 (1972)

[CP04]   Chen, H., Pucella, R.: A coalgebraic approach to Kleene algebra with tests. Theoret. Comput. Sci. 327, 23–44 (2004)

[Coh94]   Cohen, E.: Hypotheses in Kleene algebra (1994),
ftp://ftp.bellcore.com/pub/ernie/research/homepage.html

[HMU06]   Hopcroft, J.E., Motwani, R., Ullman, J.D.: Introduction to Automata Theory, Languages, and Computation, 3rd edn. Prentice Hall (2006)

[KK06]   Kamal, A.-H., Kozen, D.: KAT-ML: An interactive theorem prover for Kleene algebra with tests. Journal of Applied Non-Classical Logics 16, 9–33 (2006)

[KT90]   Kozen, D., Tiuryn, J.: Logics of programs. In: Handbook of Theoretical Computer Science. Elsevier (1990)

[Koz94]   Kozen, D.: A completeness theorem for Kleene algebras and the algebra of regular events. Infor. and Comput. 110, 366–390 (1994)

[KCS96]   Kozen, D., Cohen, E., Smith, F.: The complexity of Kleene algebra with tests. Technical Report TR96-1598, Computer Science Department, Cornell University (1996)

[KS96]   Kozen, D., Smith, F.: Kleene algebra with tests: Completeness and decidability. In: van Dalen, D., Bezem, M. (eds.) CSL 1996. LNCS, vol. 1258, pp. 244–259. Springer, Heidelberg (1997)

[Koz97]   Kozen, D.: Kleene algebra with tests. Transactions on Programming Languages and Systems 19, 427–443 (1997)

[Koz00]   Kozen, D.: On Hoare logic and Kleene algebra with tests. Trans. Computational Logic 1, 60–76 (2000)

[KP00]   Kozen, D., Patron, M.-C.: Certification of compiler optimizations using Kleene algebra with tests. In: Lloyd, J., et al. (eds.) CL 2000. LNCS (LNAI), vol. 1861, pp. 568–582. Springer, Heidelberg (2000)

[Koz03]   Kozen, D.: Automata on guarded strings and applications. Matématica Contemporânea 24, 117–139 (2003)

[Koz08a]   Kozen, D.: Nonlocal flow of control and Kleene algebra with tests. In: Proc. 23rd IEEE Symp. Logic in Computer Science (LICS 2008), pp. 105–117 (2008)

[Koz08b]   Kozen, D.: On the coalgebraic theory of Kleene algebra with tests. Technical Report Computer Science Department, Cornell University (2008),
http://hdl.handle.com/1813/10173

[KT08]   Kozen, D., Tseng, W.-L.D.: The Böhm-Jacopini theorem is false, propositionally. In: Audebaud, P., Paulin-Mohring, C. (eds.) MPC 2008. LNCS, vol. 5133, pp. 177–192. Springer, Heidelberg (2008)

[Pin86]   Pin, J.E.: Varieties of Formal Languages. Foundations of Computer Science. Springer (1986)

[Pol03]   Polák, L.: Syntactic semiring and language equations. In: Champarnaud, J.-M., Maurel, D. (eds.) CIAA 2002. LNCS, vol. 2608, pp. 182–193. Springer, Heidelberg (2003)

[Pol04a]   Polák, L.: A classification of rational languages by semilattice-ordered monoids. Archivum Mathematium 40, 395–406 (2004)

[Pol04b]   Polák, L.: On Pseudovarieties of Semiring Homomorphisms. In: Fiala, J., Koubek, V., Kratochvíl, J. (eds.) MFCS 2004. LNCS, vol. 3153, pp. 635–647. Springer, Heidelberg (2004)

[Rut98]   Rutten, J.J.M.M.: Automata and coinduction (an exercise in coalgebra). In: Sangiorgi, D., de Simone, R. (eds.) CONCUR 1998. LNCS, vol. 1466, pp. 194–218. Springer, Heidelberg (1998)

[Sch65]   Schützenberger, M.-P.: On finite monoids having only trivial subgroups. Information and Control 8, 190–194 (1965)

[Sim75]   Simon, I.: Piecewise testable events. In: Brakhage, H. (ed.) GI-Fachtagung 1975. LNCS, vol. 33, pp. 214–222. Springer, Heidelberg (1975)

Appendix: The Proof of Lemma 3

Let E_0 be the following set of identities:

$$E_0 := \langle \mathsf{p}b\mathsf{q} = \mathsf{pq} \mid \mathsf{p}, \mathsf{q} \in \Sigma, \ b = c_{i_1} c_{i_2} \cdots c_{i_l},$$
$$c_i \in \{\mathsf{b}_i, \bar{\mathsf{b}}_i\}, 1 \leq i_1 < i_2 \cdots < i_l \leq N \rangle.$$

Note that this is a finite set. We now show that:

$$W(\mathsf{P}_0) = \mathcal{W}(E_{\Sigma,B} \cup E_0). \tag{21}$$

Claim 3. $W(\mathsf{P}_0) \subseteq \mathcal{W}(E_{\Sigma,B} \cup E_0)$.

Proof. Since $W(\mathsf{P}_0) \subseteq \mathcal{W}_{\Sigma,B} = \mathcal{W}(E_{\Sigma,B})$ and $\mathcal{W}(E_{\Sigma,B} \cup E_0)$ is closed under summations, it is sufficient to prove that $W(\alpha q \beta)$ satisfies E_0, where $q \in$ RExp(Σ) and $\alpha, \beta \in$ At(B). The language $W(\alpha q \beta)$ can be explicitly specified as follows:

$$W(\alpha q \beta) = \{\alpha_0' \mathsf{p}_1 \alpha_1' \cdots \mathsf{p}_n \alpha_n' \mid \alpha_0' \in C(\alpha)^*, \alpha_i' \in \mathrm{Cst}(B) \ (i \neq 0, n),$$
$$\alpha_n' \in C(\beta)^*, \mathsf{p}_1 \cdots \mathsf{p}_n \in R(q).\}$$

Note that internal consistent tests α_i' $(i \neq 0, n)$ are arbitrary. This implies that if $\alpha_0' \mathsf{p}_1 \alpha_1' \mathsf{p}_2 \cdots \mathsf{p}_n \alpha_n' \in W(\alpha q \beta)$, then $\alpha_0' \mathsf{p}_1 \mathsf{p}_2 \cdots \mathsf{p}_n \alpha_n' \in W(\alpha q \beta)$ and vice versa. By using this fact, it follows that $W(\alpha q \beta)$ satisfies E_0 above.

We now show the difficult part of the proof:

Claim 4. $\mathcal{W}(E_{\Sigma,B} \cup E_0) \subseteq W(\mathsf{P}_0)$.

Proof. Let $L \in \mathcal{W}(E_{\Sigma,B} \cup E_0)$. Since L satisfies $E_{\Sigma,B}$ in particular, we have $\mathbf{b}_i + \bar{\mathbf{b}}_i \equiv_L 1$ for any $\mathbf{b}_i \in B$. This means that, for any string $w \in (\Sigma \cup B \cup \bar{B})^*$, we have the following equivalence:

$$w \equiv_L \sum_{\alpha,\beta \in \mathrm{At}(B)} \alpha w \beta. \tag{22}$$

This implies that $w \in L$ if and only if $\alpha w \beta \in L$ for some $\alpha, \beta \in \mathrm{At}(B)$, i.e. $L = \sum_{\alpha,\beta} \alpha \backslash L / \beta$. Since $\mathcal{W}(E_{\Sigma,B} \cup E_0)$ is closed under quotients, each $\alpha \backslash L / \beta$ belongs to $\mathcal{W}(E_{\Sigma,B} \cup E_0)$. Let us denote as $L_{\alpha,\beta} := \alpha \backslash L / \beta$ for short. It is sufficient to show that each $L_{\alpha,\beta}$ belongs to $W(\mathrm{P}_0)$, because $W(\mathrm{P}_0)$ is closed under sums and $L = \sum L_{\alpha,\beta}$.

Since $L_{\alpha,\beta} \cap \Sigma^*$ is a regular language over Σ, there exists a regular expression $q_{\alpha,\beta} \in \mathrm{RExp}(\Sigma)$ such that $R(q_{\alpha,\beta}) = L_{\alpha,\beta} \cap \Sigma^*$. We now show that $L_{\alpha,\beta} = W(\alpha q_{\alpha,\beta} \beta)$. First, let $\alpha_0' \mathbf{p}_1 \alpha_1' \cdots \mathbf{p}_n \alpha_n' \in L_{\alpha,\beta}$. Then $\mathbf{p}_1 \cdots \mathbf{p}_n \in L_{\alpha,\beta}$ because $L_{\alpha,\beta}$ satisfies $E_{\Sigma,B}$, when $\mathbf{p}_1 \cdots \mathbf{p}_n \in R(q_{\alpha,\beta})$. By the definition of $L_{\alpha,\beta}$ and the assumption $L \in \mathcal{W}(E_{\Sigma,B})$, we have $\alpha \backslash L_{\alpha,\beta} / \beta = L_{\alpha,\beta}$. Thus $\alpha \alpha_0' \mathbf{p}_1 \alpha_1' \cdots \mathbf{p}_n \alpha_n' \beta \in L_{\alpha,\beta}$. If either $\alpha_0' \notin C(\alpha)^*$ or $\alpha_n' \notin C(\beta)^*$, then $\alpha \alpha_0' \equiv_{L_{\alpha,\beta}} 0$ or $\alpha_n' \beta \equiv_{L_{\alpha,\beta}} 0$ because $L_{\alpha,\beta}$ satisfies $E_{\Sigma,B}$, i.e. $b\bar{b} \equiv_{L_{\alpha,\beta}} 0$ and $bb' \equiv_{L_{\alpha,\beta}} b'b$ in particular. This contradicts the fact that $\alpha \alpha_0' \mathbf{p}_1 \alpha_1' \cdots \mathbf{p}_n \alpha_n' \beta \in L_{\alpha,\beta}$. Thus $\alpha_0' \in C(\alpha)^*$ and $\alpha_n' \in C(\beta)^*$. This proves that $\alpha_0' \mathbf{p}_1 \alpha_1' \cdots \mathbf{p}_n \alpha_n' \in W(\alpha q_{\alpha,\beta} \beta)$, and thus, $L_{\alpha,\beta} \subseteq W(\alpha q_{\alpha,\beta} \beta)$.

Conversely, assume that $\alpha_0' \mathbf{p}_1 \alpha_1' \cdots \mathbf{p}_n \alpha_n' \in W(\alpha q_{\alpha,\beta} \beta)$. Then $\mathbf{p}_1 \cdots \mathbf{p}_n \in L_{\alpha,\beta}$ by definition of $q_{\alpha,\beta}$. By $\alpha \backslash L_{\alpha,\beta} / \beta = L_{\alpha,\beta}$, we have $\alpha \mathbf{p}_1 \cdots \mathbf{p}_n \beta \in L_{\alpha,\beta}$. Also since $\alpha_0' \in C(\alpha)^*$, $\alpha_n' \in C(\beta)^*$ and $L_{\alpha,\beta}$ satisfies $E_{\Sigma,B}$, it follows that $\alpha_0' \mathbf{p}_1 \cdots \mathbf{p}_n \alpha_n' \in L_{\alpha,\beta}$. Furthermore, since $L_{\alpha,\beta}$ satisfies E_0, any consistent test $c_{i_1} \cdots c_{i_l}$ can be inserted between $\mathbf{p}_j$ and $\mathbf{p}_{j+1}$ up to equivalence: That is,

$$\alpha_0' \mathbf{p}_1 \cdots \mathbf{p}_j \mathbf{p}_{j+1} \cdots \mathbf{p}_n \alpha_n' \equiv_{L_{\alpha,\beta}} \alpha_0' \mathbf{p}_1 \cdots \mathbf{p}_j c_{i_1} \cdots c_{i_l} \mathbf{p}_{j+1} \cdots \mathbf{p}_n \alpha_n' \tag{23}$$

Again, since $L_{\alpha,\beta}$ satisfies $E_{\Sigma,B}$, the order of $c_{i_1}, \cdots, c_{i_l}$ can be replaced arbitrarily and also each occurrence of c_{i_k} can be copied up to equivalence. Thus, applying this fact to $C(\alpha_j') = \{c_{i_1}^j, \cdots, c_{i_l}^j\}$ for each $1 \le j \le n-1$, we obtain:

$$\alpha_0' \mathbf{p}_1 \mathbf{p}_2 \cdots \mathbf{p}_n \alpha_n' \equiv_{L_{\alpha,\beta}} \alpha_0' \mathbf{p}_1 \alpha_1' \mathbf{p}_2 \alpha_2' \cdots \cdots \mathbf{p}_n \alpha_n'. \tag{24}$$

Thus, by $\alpha_0' \mathbf{p}_1 \mathbf{p}_2 \cdots \mathbf{p}_n \alpha_n' \in L_{\alpha,\beta}$, we obtain $\alpha_0' \mathbf{p}_1 \alpha_1' \mathbf{p}_2 \alpha_2' \cdots \mathbf{p}_n \alpha_n' \in L_{\alpha,\beta}$. Consequently, $W(\alpha q_{\alpha,\beta} \beta) \subseteq L_{\alpha,\beta}$. This concludes the proof.

Kleene Algebra with Converse

Paul Brunet and Damien Pous*

LIP, CNRS, ENS Lyon, INRIA, Université de Lyon, UMR 5668

Abstract. The equational theory generated by all algebras of binary
relations with operations of union, composition, converse and reflexive
transitive closure was studied by Bernátsky, Bloom, Ésik, and Stefanescu
in 1995. We reformulate some of their proofs in syntactic and elementary terms, and we provide a new algorithm to decide the corresponding
theory. This algorithm is both simpler and more efficient; it relies on
an alternative automata construction, that allows us to prove that the
considered equational theory lies in the complexity class PSPACE.

Specific regular languages appear at various places in the proofs. Those
proofs were made tractable by considering appropriate automata recognising those languages, and exploiting symmetries in those automata.

Introduction

In many contexts in computer science and mathematics operations of union, sequence or product and iteration appear naturally. *Kleene Algebra*, introduced by
John H. Conway under the name *regular algebra* [Con71], provides an algebraic
framework allowing to express properties of these operators, by studying the
equivalence of expressions built with these connectives. It is well known that the
corresponding equational theory is decidable [Kle51], and that it is complete for
language and relation models.

As expressive as it may be, one may wish to integrate other usual operations in
such a setting. Theories obtained this way, by addition of a finite set of equations
to the axioms of Kleene Algebra, are called *Extensions of Kleene Algebra*. We
shall focus here on one of these extensions, where an operation of *converse* is
added to Kleene Algebra. The converse of a word is its mirror image (the word
obtained by reversing the order of the letters), and the converse $R^\vee$ of a relation
R is its reciprocal ($xR^\vee y \triangleq yRx$). This natural operation can be expressed
simply as a set of equations that we add to Kleene Algebra's axioms.

The question that arises once this theory is built is its decidability: given
two formal expressions built with the connectives product, sum, iteration and
converse, can one decide automatically if they are equivalent, meaning that
their equality can be proven using the axioms of the theory? Bloom, Ésik,

---

* Work partially funded by the french projects PiCoq (ANR-09-BLAN-0169-01) and
PACE (ANR-12IS02001).

P. Höfner et al. (Eds.): RAMiCS 2014, LNCS 8428, pp. 101–118, 2014.

Stefanescu and Bernátsky gave an affirmative answer to that question in two articles, [BÉS95] and [ÉB95], in 1995.

However, although the algorithm they define proves the decidability result, it is too complicated to be used in actual applications. In this paper, beside some simplifications of the proofs given in [BÉS95], we give a new and more efficient algorithm to decide this problem, which we place in the complexity class PSPACE.

The equational theory of Kleene algebra cannot be finitely axiomatised [Red64]. Krob presented the first purely axiomatic (but infinite) presentation [Kro90]. Several finite quasi-equational characterisations have been proposed [Sal66, Bof90, Kro90, Koz91, Bof95]; here we follow the one from Kozen [Koz91].

A Kleene Algebra is an algebraic structure $\langle K, +, \cdot, ^\star, 0, 1 \rangle$ such that $\langle K, +, \cdot, 0, 1 \rangle$ is an idempotent semi-ring, and the operation * satisfies the following properties

$$1 + aa^\star \leqslant a^\star \tag{1a}$$

$$1 + a^\star a \leqslant a^\star \tag{1b}$$

$$b + ax \leqslant x \Rightarrow a^\star b \leqslant x \tag{1c}$$

$$b + xa \leqslant x \Rightarrow ba^\star \leqslant x \tag{1d}$$

(Here $a \leqslant b$ is a shorthand for $a + b = b$.)

The quasi-variety KA consists in the axioms of an idempotent semi-ring together with axioms and inference rules (1a) to (1d). Kleene Algebras are thus *models* of KA. We shall call *regular expressions over X*, written Reg_X, the expressions built from letters of X, the binary connectives $+$ and $\cdot$, the unary connective * and the two constants 0 and 1.

Two families of such algebras are of particular interest: languages (sets of finite words over a finite alphabet, with union as sum and concatenation as product) and relations (binary relations over an arbitrary set with union and composition). KA is complete for both these models [Kro90, Koz91], meaning that for any $e, f \in \mathrm{Reg}_X$, $\mathrm{KA} \vdash e = f$ if and only if e and f coincide under any language (resp. relational) interpretation. This last property will be written $e \equiv_{\mathrm{Lang}} f$ (resp. $e \equiv_{\mathrm{Rel}} f$).

More remarkably, if we denote by $[\![e]\!]$ the language denoted by an expression e, we have that for any $e, f \in \mathrm{Reg}_X$, $\mathrm{KA} \vdash e = f$ if and only if $[\![e]\!] = [\![f]\!]$. By Kleene's theorem (see [Kle51]) the equality of two regular languages can be reduced to the equivalence of two finite automata, which is easy to compute. Hence, the theory KA is decidable.

Now let us add a unary operation of converse to regular expressions. We shall denote by $\mathrm{Reg}_X^\vee$ the set of regular expressions with converse over a finite alphabet X. While doing so, several questions arise:

1. Can the converse on languages and on relations be encoded in the same theory?
2. What axioms do we need to add to KA to model these operations?
3. Are the resulting theories complete for languages and relations?
4. Are these theories decidable?

There is a simple answer to the first question: no. Indeed the equation $a \leqslant a \cdot a^\vee \cdot a$ is valid for any relation a (because if $(x, y) \in a$, then $(x, y) \in a$, $(y, x) \in a^\vee$, and $(x, y) \in a$, so that $(x, y) \in a \circ a^\vee \circ a$). But this equation is not satisfied for all languages a (for instance, with the language $a = \{x\}$, $a \cdot a^\vee \cdot a = \{xxx\}$ and $x \notin \{xxx\}$). This means that there are two distinct theories corresponding to these two families of models. Let us begin by considering the case of languages.

Theorem 1 (Completeness of KAC⁻ [BÉS95]). *A complete axiomatisation of the variety Lang$^\vee$ of languages generated by concatenation, union, star, and converse consists of the axioms of KA together with axioms (2a) to (2d).*

$$(a + b)^\vee = a^\vee + b^\vee \tag{2a}$$

$$(a \cdot b)^\vee = b^\vee \cdot a^\vee \tag{2b}$$

$$(a^\star)^\vee = (a^\vee)^\star \tag{2c}$$

$$a^{\vee\vee} = a. \tag{2d}$$

We call this theory KAC⁻; it is decidable.

As for relations, we write $e \equiv_{\text{Lang}^\vee} f$ if e and f have the same language interpretations (for a formal definition, see the "Notation" subsection below). To prove this result, one first associates to any expression $c \in \text{Reg}_X^\vee$ an expression $\mathbf{e} \in \text{Reg}_{\mathbf{X}}$, where $\mathbf{X}$ is an alphabet obtained by adding to X a disjoint copy of itself. Then, one proves that the following implications hold.

$$e \equiv_{\text{Lang}^\vee} f \quad \Rightarrow \quad [\![\mathbf{e}]\!] = [\![\mathbf{f}]\!] \tag{3}$$

$$[\![\mathbf{e}]\!] = [\![\mathbf{f}]\!] \quad \Rightarrow \quad \text{KAC}^- \vdash e = f \tag{4}$$

(That $\text{KAC}^- \vdash e = f$ entails $e \equiv_{\text{Lang}^\vee} f$ is obvious; decidability comes from that of regular languages equivalence.) We reformulate Bloom et al.'s proofs of these implications in elementary terms in Section 1.1.

As stated before, the equation $a \leqslant a \cdot a^\vee \cdot a$ provides a difference between languages with converse and relations with converse. It turns out that it is the only difference, in the sense that the following theorem holds:

Theorem 2 (Completeness of KAC [BÉS95, ÉB95]). *A complete axiomatisation of the variety Rel$^\vee$ of relations generated by composition, union, star, and converse consists of the axioms of KAC⁻ together with the axiom (5).*

$$a \leqslant a \cdot a^\vee \cdot a. \tag{5}$$

We call this theory KAC; it is decidable.

The proof of this result also relies on a translation into regular languages. Ésik et al. define a notion of *closure*, written $cl()$, for languages over $\mathbf{X}$, and they prove the following implications:

$$e \equiv_{\text{Rel}^\vee} f \quad \Rightarrow \quad cl([\![\mathbf{e}]\!]) = cl([\![\mathbf{f}]\!]) \tag{6}$$

$$cl([\![\mathbf{e}]\!]) = cl([\![\mathbf{f}]\!]) \quad \Rightarrow \quad \text{KAC} \vdash e = f \tag{7}$$

(Again, that $\mathrm{KAC} \vdash e = f$ entails $e \equiv_{\mathrm{Rel}^\vee} f$ is obvious.) The first implication (6) was proven in [BÉS95]; we give a new formulation of this proof in Section 1.2. The second one (7) was proven in [ÉB95].

The last consideration is the decidability of KAC. To this end, Bloom et al. propose a construction to obtain an automaton recognising $\mathit{cl}(L)$, when given an automaton recognising L. Decidability follows: to decide whether $\mathrm{KAC} \vdash e = f$ one can build two automata recognising $\mathit{cl}(\llbracket e \rrbracket)$ and $\mathit{cl}(\llbracket f \rrbracket)$ and check if they are equivalent. Unfortunately, their construction tends to produce huge automata, which makes it useless for practical application. We propose a new and simpler one in Section 2; by analysing this construction, we show in Section 3 how it leads to a proof that the problem of equivalence in KAC is PSPACE.

Notation

For any word w, $|w|$ is the size of w, meaning its number of letters; for any $1 \leqslant i \leqslant |w|$, we'll write $w(i)$ for the ith letter of w and $w|_i \triangleq w(1)w(2)\cdots w(i)$ for its prefix of size i. Also, $\mathrm{suffixes}(w) \triangleq \{v \mid \exists u : uv = w\}$ is the set of all suffixes of w. A deterministic automaton is a tuple $\langle Q, \Sigma, q_0, T, \delta \rangle$; with Q a set of states, Σ an alphabet, $q_0 \in Q$ an initial state, $T \subseteq Q$ a set of final states and $\delta : Q \times \Sigma \to Q$ a transition function. A non-deterministic automaton is a tuple $\langle Q, \Sigma, I, T, \Delta \rangle$; with Q, Σ and T same as before, $I \subseteq Q$ a set of initial states and $\Delta \subseteq Q \times \Sigma \times Q$ a set of transitions. We write $L(\mathscr{A})$ for the *language recognised by the automaton* $\mathscr{A}$. For any $a \in \Sigma$, we write $\Delta(a)$ for $\{(p,q) \mid (p,a,q) \in \Delta\}$. We also use the compact notation $p \xrightarrow{w}_{\mathscr{A}} q$ to denote that there is in the automaton $\mathscr{A}$ a path labelled by w from the state p to the state q. For a set $E \subseteq Q$ and a relation R over Q, we write $E \cdot R$ for the set $\{y \mid \exists x \in E : xRy\}$.

Given a map σ from a set X to the languages on an alphabet Σ (resp. the relations on a set S), there is a unique extension of σ into a homomorphism from Reg_X to Lang_Σ (resp. Rel_S), which we denote by $\widehat{\sigma}$. The same thing can be done with regular expressions with converse, and we will use the same notation for it. We finally denote by $\equiv_V$ the equality in a variety V (Lang, Rel, Lang$^\vee$ or Rel$^\vee$): $e \equiv_V f \triangleq \forall K, \forall \sigma : X \to V_K, \widehat{\sigma}(e) = \widehat{\sigma}(f)$.

1   Preliminary Material

1.1   Languages with Converse: Theory KAC⁻

We consider regular expressions with converse over a finite alphabet X. The alphabet $\mathbf{X}$ is defined as $X \cup X'$, where $X' \triangleq \{x' \mid x \in X\}$ is a disjoint copy of X. As a shorthand, we use $'$ as an internal operation on $\mathbf{X}$ going from X to X' and from X' to X such that if $x \in X$, $x' \triangleq x' \in X'$ and $(x')' \triangleq x \in X$. An important operation in the following is the translation of an expression $e \in \mathrm{Reg}_X^\vee$ to an expression $\mathbf{e} \in \mathrm{Reg}_{\mathbf{X}}$. We proceed to its definition in two steps.

Let $\tau(e)$ denote the normal form of an expression $e \in \mathrm{Reg}_X^\vee$ in the following convergent term rewriting system:

$$(a+b)^\vee \to a^\vee + b^\vee \qquad 0^\vee \to 0 \qquad (a^\star)^\vee \to (a^\vee)^\star$$
$$(a \cdot b)^\vee \to b^\vee \cdot a^\vee \qquad 1^\vee \to 1 \qquad a^{\vee\vee} \to a$$

The corresponding equations being derivable in KAC^-, one easily obtain that

$$\forall e \in \mathrm{Reg}_X^\vee, \ \mathrm{KAC}^- \vdash \tau(e) = e \tag{8}$$

We finally denote by $\mathbf{e}$ the expression obtained by further applying the substitution $\nu \triangleq [x^\vee \mapsto x', (\forall x \in \mathbf{X})]$, i.e., $\mathbf{e} \triangleq \nu(\tau(e))$. (Note that $\mathbf{e} \in \mathrm{Reg}_\mathbf{X}$: it is regular, all occurrences of the converse operation have been eliminated.) As explained in the introduction, Bloom et al.'s proof [BÉS95] amounts to proving the implications (3) and (4). We include a syntactic and elementary presentation of this proof, for the sake of completeness.

Lemma 3. *For all $e, f \in \mathrm{Reg}_X^\vee$, $e \equiv_{\mathrm{Lang}^\vee} f$ entails $[\![\mathbf{e}]\!] = [\![\mathbf{f}]\!]$.*

Proof. For any $e \in \mathrm{Reg}_X^\vee$, we have $\tau(e) \equiv_{\mathrm{Lang}^\vee} e$ (†) as an immediate consequence of (8). Let us write $X_\bullet \triangleq X \uplus \{\bullet\}$ and consider the following interpretations (which appear in [BÉS95, proof of Proposition 4.3]):

$$\mu : X \longrightarrow \mathcal{P}(X_\bullet^\star) \qquad\qquad \eta : \mathbf{X} \longrightarrow \mathcal{P}(X_\bullet^\star)$$
$$x \longmapsto \{x \cdot \bullet\} \qquad\qquad x \in X \longmapsto \{x \cdot \bullet\}$$
$$x' \in X' \longmapsto \{\bullet \cdot x\}$$

One can check that $\widehat{\eta}$ is injective modulo equality of denoted languages, in the sense that for any expression $e \in \mathrm{Reg}_\mathbf{X}$, we have

$$\widehat{\eta}(e) = \widehat{\eta}(f) \text{ implies that } [\![e]\!] = [\![f]\!] \ . \tag{9}$$

By a simple induction on e, we get $\widehat{\mu}(\tau(e)) = \widehat{\eta}(\nu(\tau(e))) = \widehat{\eta}(\mathbf{e})$. Combined with (†), we deduce that $\widehat{\mu}(e) = \widehat{\eta}(\mathbf{e})$. All in all, we obtain: $e \equiv_{\mathrm{Lang}^\vee} f \Rightarrow \widehat{\mu}(e) = \widehat{\mu}(f) \Rightarrow \widehat{\eta}(\mathbf{e}) = \widehat{\eta}(\mathbf{f}) \Rightarrow [\![\mathbf{e}]\!] = [\![\mathbf{f}]\!]$. □

The second implication is even more immediate, using KA completeness.

Lemma 4. *For all $e, f \in \mathrm{Reg}_X^\vee$, if $[\![\mathbf{e}]\!] = [\![\mathbf{f}]\!]$ then $\mathrm{KAC}^- \vdash e = f$.*

Proof. By completeness of KA [Kro90, Koz91], if $[\![\mathbf{e}]\!] = [\![\mathbf{f}]\!]$, then we know that there is a proof $\pi_1 : \mathrm{KA} \vdash \mathbf{e} = \mathbf{f}$. As KA is contained in KAC^-, the same proof can be seen as $\pi_1 : \mathrm{KAC}^- \vdash \mathbf{e} = \mathbf{f}$. By substituting x' by $(x^\vee)$ everywhere in this proof, we get a new proof $\pi_2 : \mathrm{KAC}^- \vdash \tau(e) = \tau(f)$. By (8) and transitivity we thus get $\mathrm{KAC}^- \vdash e = f$. □

We finally deduce that $e \equiv_{\mathrm{Lang}^\vee} f \Leftrightarrow [\![\mathbf{e}]\!] = [\![\mathbf{f}]\!] \Leftrightarrow \mathrm{KAC}^- \vdash e = f$. Since the regular expressions $\mathbf{e}$ and $\mathbf{f}$ can be easily computed from e and f, the problem of equivalence in KAC^- thus reduces to an equality of regular languages, which makes it decidable.

1.2   Relations with Converse: Theory KAC

We now move to the equational theory generated by relational models. It turns out that this theory will be characterised using "closed" languages on the extended alphabet $\mathbf{X}$. To define this closure operation, we first define a mirror operation $\overline{w}$ on words over $\mathbf{X}$, such that $\overline{\epsilon} \triangleq \epsilon$ and for any $x, w \in \mathbf{X} \times \mathbf{X}^*$, $\overline{wx} = x'\overline{w}$. Accordingly with the axiom (5) of KAC we define a reduction relation $\leadsto$ on words over $\mathbf{X}$, using the following word rewriting rule.

$$w\overline{w}w \leadsto w \ .$$

We call $w\overline{w}w$ a *pattern* of *root* w. The last two thirds of the pattern are $\overline{w}w$. Following [BÉS95, ÉB95], we extend this relation into a closure operation on languages.

Definition 5. The *closure* of a language $L \subseteq \mathbf{X}^*$ is the smallest language containing L that is downward-closed with respect to $\leadsto$:

$$\mathit{cl}(L) \triangleq \{v \mid \exists u \in L : u \leadsto^* v\} \ .$$

Example 6. If $X = \{a, b, c, d\}$, then $\mathbf{X} = \{a, b, c, d, a', b', c', d'\}$, and $\overline{ab'} = ba'$. We have the reduction $cab'ba'ab'd' \leadsto cab'd'$, by triggering a pattern of root ab'. For $L = \{aa'a, b, cab'ba'ab'd'\}$, we have $\mathit{cl}(L) = L \cup \{a, cab'd'\}$.

Now we define a family of languages which play a prominent role in the sequel.

Definition 7. For any word $w \in \mathbf{X}^*$, we define a regular language $\Gamma(w)$ by:

$$\Gamma(\epsilon) \triangleq \{\epsilon\}$$
$$\forall x \in \mathbf{X}, \forall w \in \mathbf{X}^*, \quad \Gamma(wx) \triangleq (x'\Gamma(w)x)^* \ .$$

An equivalent operator called G is used in [BÉS95]: we actually have $\Gamma(w) = G(\overline{w})$, and our recursive definition directly corresponds to [BÉS95, Proposition 5.11.(2)]. By using such a simple recursive definition, we avoid the need for the notion of *admissible maps*, which is extensively used in [BÉS95].

Instead, we just have the following property to establish, which illustrates why these languages are of interest: words in $\Gamma(w)$ reduce into the last two thirds of a pattern compatible with w. Therefore, in the context of recognition by an automaton, $\Gamma(w)$ contains all the words that could potentially be skipped after reading w, in a closure automaton.

Proposition 8. *For all words u and v, $u \in \Gamma(v) \Leftrightarrow \exists t \in suffixes(v) : u \leadsto^* \overline{t}t$.*

Proof. The proof of the implication from left to right is routine but a bit lengthy, we include it in [BP14].

For the converse implication, we first define the following language: $\Gamma'(v) \triangleq \{\overline{t}t \mid t \in \mathrm{suffixes}(v)\}$. We thus have to show that the upward closure of $\Gamma'(v)$ is contained in $\Gamma(v)$. We first check that this language satisfies $\Gamma'(\epsilon) = \epsilon$ and

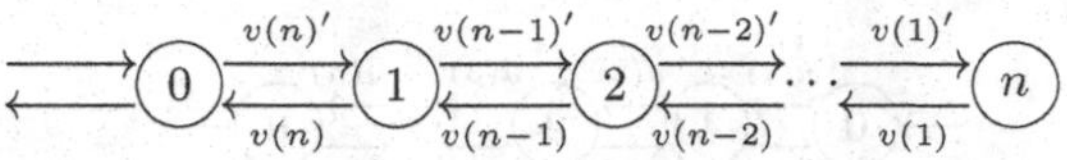

Fig. 1. Automaton $\mathscr{G}(v)$ recognising $\Gamma(v)$, with $|v| = n$

$\Gamma'(vx) = \epsilon + x'\Gamma'(v)x$, which allows us to deduce that $\Gamma'(v) \subseteq \Gamma(v)$ by a straightforward induction.

It thus suffices to show that $\Gamma(v)$ is upward-closed with respect to $\leadsto$. For this, we introduce the family of automata $\mathscr{G}(v)$ depicted in Figure 1. One can check that $\mathscr{G}(v)$ recognises $\Gamma(v)$ by a simple induction on v. One can moreover notice that in this automaton, if $p \xrightarrow{x}_{\mathscr{G}(v)} q$, then $q \xrightarrow{x'}_{\mathscr{G}(v)} p$. More generally, for any word u, if $p \xrightarrow{u}_{\mathscr{G}(v)} q$, then $q \xrightarrow{\overline{u}}_{\mathscr{G}(v)} p$. So if $u_1 w u_2 \in \Gamma(v)$, then by definition of the automaton we have $0 \xrightarrow{u_1}_{\mathscr{G}(v)} q_1 \xrightarrow{w}_{\mathscr{G}(v)} q_2 \xrightarrow{u_2}_{\mathscr{G}(v)} 0$, and thus, by the previous remark:

$$0 \xrightarrow[\mathscr{G}(v)]{u_1} q_1 \xrightarrow[\mathscr{G}(v)]{w} q_2 \xrightarrow[\mathscr{G}(v)]{\overline{w}} q_1 \xrightarrow[\mathscr{G}(v)]{w} q_2 \xrightarrow[\mathscr{G}(v)]{u_2} 0 \quad ,$$

i.e., $u_1 w \overline{w} w u_2 \in \Gamma(v)$. In other words, for any words v and w and any $u \in \Gamma(v)$, if $w \leadsto u$ then w is also in $\Gamma(v)$, meaning exactly that $\Gamma(v)$ is upward-closed with respect to $\leadsto$.

Since $\Gamma'(v) \subseteq \Gamma(v)$, we deduce that $\Gamma(v)$ contains the upward closure of $\Gamma'(v)$, as expected. $\qquad\square$

We now have enough material to embark in the proof of the implication (6) from the introduction, stating that if two expressions $e, f \in \mathrm{Reg}_X^{\vee}$ are equal for all interpretations in all relational models, then $cl(\mathbf{e}) = cl(\mathbf{f})$.

Proof. Bloom et al. [BÉS95] consider specific relational interpretations: for any word $u \in \mathbf{X}^*$ and for any letter $x \in \mathbf{X}$, they define

$$\phi_u(x) \triangleq \{(i-1, i) \mid u(i) = x\} \cup \{(i, i-1) \mid u(i) = x'\} \subseteq \{0, \ldots, n\}^2 \quad ,$$

where $n \triangleq |u|$. The key property of those interpretations is the following:

$$(0, n) \in \widehat{\phi_u}(v) \Leftrightarrow v \leadsto^* u \quad . \tag{10}$$

We give a new proof of this property, by using the automaton $\Phi(u)$ depicted in Figure 2. By definition of $\Phi(u)$ and ϕ_u, we have that

$$(i, j) \in \phi_u(x) \Leftrightarrow i \xrightarrow{x}_{\Phi(u)} j \quad .$$

Therefore, proving (10) amounts to proving

$$v \in L(\Phi(u)) \Leftrightarrow v \leadsto^* u \quad . \tag{11}$$

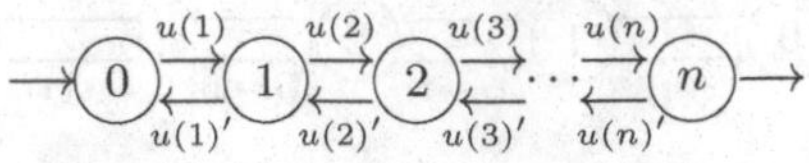

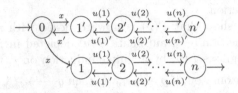

Fig. 2. Automaton $\Phi(u)$, with $|u| = n$

Fig. 3. Automaton $\Phi'(xu)$, with $|u| = n$, language equivalent to $\Phi(xu)$

First notice that $i \xrightarrow{x}_{\Phi(u)} j \Leftrightarrow j \xrightarrow{x'}_{\Phi(u)} i$. We can extend this to paths (as in the proof of Proposition 8) and then prove that if $s \rightsquigarrow t$ and $i \xrightarrow{t}_{\Phi(u)} j$ then $i \xrightarrow{s}_{\Phi(u)} j$. As u is clearly in $L(\Phi(u))$, any v such that $v \rightsquigarrow^* u$ is also in $L(\Phi(u))$.

We proceed by induction on u for the other implication. The case $u = \epsilon$ being trivial, we consider $v \in L(\Phi(xu))$. We introduce a second automaton $\Phi'(xu)$ given in Figure 3, that recognises the same language as $\Phi(xu)$. The upper part of this automaton is actually the automaton $\mathscr{G}(\overline{xu})$ (as given in Figure 1), recognising the language $\Gamma(\overline{xu})$. Moreover, the lower part starting from state 1 is the automaton $\Phi(u)$. This allows us to obtain that $L(\Phi(xu)) = \Gamma(\overline{xu})xL(\Phi(u))$. Hence, for any $v \in L(\Phi(xu))$, there are $v_1 \in \Gamma(\overline{xu})$ and $v_2 \in L(\Phi(u))$ such that $v = v_1xv_2$. By induction, we get $v_2 \rightsquigarrow^* u$, and by Proposition 8 we know that $v_1 \rightsquigarrow^* \overline{w}w$, with $w \in \text{suffixes}(\overline{xu})$. That means that $\overline{xu} = tw$, for some word t, so $xu = \overline{tw} = \overline{w}\,\overline{t}$. If we put everything back together:

$$v = v_1xv_2 \rightsquigarrow^* v_1xu \rightsquigarrow^* \overline{w}wxu = \overline{w}w\overline{w}\,\overline{t} \rightsquigarrow \overline{w}\,\overline{t} = xu \ .$$

This concludes the proof of (11), and thus (10).

We follow Bloom et al.'s proof [BÉS95] to deduce that the implication (6) from the introduction holds: we first prove that for all $e \in \text{Reg}_{\mathbf{X}}$, we have

$$u \in cl(\llbracket e \rrbracket) \Leftrightarrow \exists v \in \llbracket e \rrbracket, v \rightsquigarrow^* u \qquad \text{(by definition)}$$

$$\Leftrightarrow \exists v \in \llbracket e \rrbracket, (0, n) \in \widehat{\phi_u}(v) \qquad \text{(by (10))}$$

$$\Leftrightarrow (0, n) \in \widehat{\phi_u}(e) \ .$$

(For the last line, we use the fact that for any relational interpretation ϕ, we have $\widehat{\phi}(e) = \bigcup_{w \in \llbracket e \rrbracket} \widehat{\phi}(w)$.)

Furthermore, as $\phi_u(x') = \phi_u(x)^\vee$, we can prove that $\widehat{\phi}_u(e) = \widehat{\phi}_u(\mathbf{e})$. There-fore, for all expressions $e, f \in \mathrm{Reg}_X^\vee$ such that $e \equiv_{\mathrm{Rel}^\vee} f$, we have $\widehat{\phi}_u(e) = \widehat{\phi}_u(e) = \widehat{\phi}_u(f) = \widehat{\phi}_u(\mathbf{f})$, and we deduce that $\mathit{cl}(\llbracket e \rrbracket) = \mathit{cl}(\llbracket f \rrbracket)$ thanks to the above characterisation. □

2   Closure of an Automaton

The problem here is the following: given two regular expressions $e, f \in \mathrm{Reg}_X^\vee$, how to decide $\mathit{cl}(\llbracket e \rrbracket) = \mathit{cl}(\llbracket f \rrbracket)$? We follow the approach proposed by Bloom et al.: given an automaton recognising a language L, we show how to construct an automaton recognising $\mathit{cl}(L)$. To solve the initial problem, it then suffices to build two automata recognising $\llbracket e \rrbracket$ and $\llbracket f \rrbracket$, to apply a construction to obtain two automata for $\mathit{cl}(\llbracket e \rrbracket)$ and $\mathit{cl}(\llbracket f \rrbracket)$, and to check those for language equivalence.

As a starting point, we first recall the construction proposed in [BÉS95].

2.1   The Original Construction

This construction uses the transition monoid of the input automaton:

Definition 9 (Transition monoid). Let $\mathscr{A} = \langle Q, \Sigma, q_0, T, \delta \rangle$ be a deterministic automaton. Each word $u \in \Sigma^*$ induces a function $u_{\mathscr{A}} : Q \to Q$ which associates to a state p the state q obtained by following the unique path from p labelled by u. The *transition monoid* of $\mathscr{A}$, written $M_{\mathscr{A}}$, is the set of functions $Q \to Q$ induced by words of Σ^*, equipped with the composition of functions and the identity function.

This monoid is finite, and its subsets form a Kleene Algebra. Bloom et al. then proceed to define the closure automaton in the following way:

Theorem 10 (Closure automaton of [BÉS95]). *Let $L \subseteq \mathbf{X}^*$ be a regular language, recognised by the deterministic automaton $\mathscr{A} = \langle Q, \mathbf{X}, q_0, Q_f, \delta \rangle$. Let $M_{\mathscr{A}}$ be the transition monoid of $\mathscr{A}$. Then the following deterministic automaton recognises $\mathit{cl}(L)$:*

$$\mathscr{B} \triangleq \langle \mathcal{P}(M_{\mathscr{A}}) \times \mathcal{P}(M_{\mathscr{A}}), \mathbf{X}, (\{\epsilon_{\mathscr{A}}\}, \{\epsilon_{\mathscr{A}}\}), T, \delta_1 \rangle$$

$$\text{with } T \triangleq \{(F, G) \mid \exists u_{\mathscr{A}} \in F : u_{\mathscr{A}}(q_0) \in Q_f\} \ ,$$

$$\text{and } \delta_1((F, G), x) \triangleq \left(F \cdot \{x_{\mathscr{A}}\} \cdot ((\{x'_{\mathscr{A}}\} \cdot G \cdot \{x_{\mathscr{A}}\})^*), \ (\{x'_{\mathscr{A}}\} \cdot G \cdot \{x_{\mathscr{A}}\})^* \right) \ .$$

An important idea in this construction, that inspired our own, is the transition rule for the second component above. Let us write $\delta_2(G, x)$ for the expression $(\{x'_{\mathscr{A}}\} \cdot G \cdot \{x_{\mathscr{A}}\})^*$, so that the definition of δ_1 can be reformulated as

$$\delta_1((F, G), x) = (F \cdot \{x_{\mathscr{A}}\} \cdot \delta_2(G, x), \ \delta_2(G, x)).$$

With that in mind, one can see the second component as some kind of *history*, that runs on its own, and is used at each step to enrich the first component. At this point, it might be interesting to notice that the formula for $\delta_2(G, x)$ closely resembles the one for $\Gamma(wx) = (x'\Gamma(w)x)^*$, which we defined in Section 1.2.

2.2 Intuitions

Let us forget the above construction, and try to build a closure automaton. One way would be to simply add transitions to the initial automaton. This idea comes naturally when one realises that if $u \leadsto^* v$, then v is obtained by erasing some subwords from u: at each reduction step $u_1 w \overline{w} w u_2 \leadsto u_1 w u_2$ we just erase $\overline{w} w$. To "erase" such subwords using an automaton, it suffices to allow one to jump along certain paths.

Suppose for instance that we start from the following automaton:

$$\rightarrow \boxed{q_0} \xrightarrow{a} \boxed{q_1} \xrightarrow{b} \boxed{q_2} \xrightarrow{b'} \boxed{q_3} \xrightarrow{a'} \boxed{q_4} \xrightarrow{a} \boxed{q_5} \xrightarrow{b} \boxed{q_6} \rightarrow$$

We can detect the pattern $ab\overline{a}\overline{b}ab$, and allow one to "jump" over it when reading the last letter of the root of the pattern, in this case the b in second position. Our automaton thus becomes:

$$\rightarrow \boxed{q_0} \xrightarrow{a} \boxed{q_1} \xrightarrow{b} \boxed{q_2} \xrightarrow{b'} \boxed{q_3} \xrightarrow{a'} \boxed{q_4} \xrightarrow{a} \boxed{q_5} \xrightarrow{b} \boxed{q_6} \rightarrow$$

However, this approach is too naive, and it quickly leads to errors. If for instance we slightly modify the above example by adding a transition labelled by b' between q_0 and q_1, the same method leads to the following automaton, by detecting the patterns $b'bb'$ between q_0 and q_3 and $abb'a'ab$ between q_0 and q_6.

$$\rightarrow \boxed{q_0} \overset{a}{\underset{b'}{\rightrightarrows}} \boxed{q_1} \xrightarrow{b} \boxed{q_2} \xrightarrow{b'} \boxed{q_3} \xrightarrow{a'} \boxed{q_4} \xrightarrow{a} \boxed{q_5} \xrightarrow{b} \boxed{q_6} \rightarrow$$

The problem is that the word $b'b$ is now wrongly recognised in the produced automaton. What happens here is that we can use the jump from q_1 to q_6, even though we didn't read the prerequisite for doing so, in this case the a constituting the beginning of the root ab of pattern $ab\overline{a}\overline{b}ab$. (Note that the dual idea, consisting in enabling a jump when reading the first letter of the root of the pattern, would lead to similar problems.)

A way to prevent that, which was implicitly introduced in the original construction, consists in using a notion of *history*. The states of the closure automaton will be pairs of a state in the initial automaton and a history. That will allow us to distinguish between the state q_1 after reading a and the state q_1 after reading b', and to specify which jumps are possible considering what has been previously read. In the construction given in [BÉS95], the history is given by an element of $\mathcal{P}(M_{\mathscr{A}})$, in the second component of the states (the "G" part). We will define a history as a set of words allowing for the same jumps, using $\Gamma(w)$.

2.3 Our Construction

We have shown in Section 1.2 that $\forall u \in \Gamma(w), \exists v \in \text{suffixes}(w) : u \rightsquigarrow^* \overline{v}v$, so we do have a characterisation of the words "allowing jumps" after having read some word w. The problem is that we want a finite number of possible histories, and there are infinitely many $\Gamma(w)$ (for instance, all the $\Gamma(a^n)$ are different). To get that, we will project $\Gamma(w)$ on the automaton. Let us consider a non-deterministic automaton $\mathscr{A} = \langle Q, \mathbf{X}, I, T, \Delta \rangle$ recognising a language L.

Definition 11. For any word $w \in \mathbf{X}^*$ we define the relation $\gamma(w)$ between states of $\mathscr{A}$ by $\gamma(\epsilon) \triangleq \text{Id}_Q$ and $\gamma(wx) = (\Delta(x') \circ \gamma(w) \circ \Delta(x))^*$.

One can notice right away the strong relationship between γ and Γ:

Proposition 12. $\forall w, q_1, q_2, \ (q_1, q_2) \in \gamma(w) \ \Leftrightarrow \ \exists u \in \Gamma(w) : q_1 \xrightarrow{u}_{\mathscr{A}} q_2$.

This result is straightforward once one realises that $\gamma(w) = \widehat{\sigma}(\Gamma(w))$ with $\sigma(x) = \Delta(x)$. By composing Propositions 8 and 12 we eventually obtain that $((q_1, q_2) \in \gamma(w))$ iff $\exists u : q_1 \xrightarrow{u}_{\mathscr{A}} q_2$ and $u \rightsquigarrow^* \overline{v}v$, with v a suffix of w.

The set Q being finite, γ has a finite index and one can define a finite set of histories as follows:

Definition 13. Let $\sim_\gamma$ be the kernel of γ: $u \sim_\gamma v$ iff $\gamma(u) = \gamma(v)$. We define the set G as the quotient of $\mathbf{X}^*$ by $\sim_\gamma$. We denote by $[w]$ the elements of G, in such a way that $[u] = [v] \Leftrightarrow u \sim_\gamma v \Leftrightarrow \gamma(u) = \gamma(v)$.

We now have all the tools required for our construction of the closure of $\mathscr{A}$:

Theorem 14 (Closure Automaton). *The closure of the language L is recognised by the automaton* $\mathscr{A}' \triangleq \langle Q \times G, \mathbf{X}, I \times \{[\epsilon]\}, T \times G, \Delta' \rangle$ *with:*

$$\Delta' = \{((q_1, [w]), x, (q_2, [wx])) \mid (q_1, q_2) \in \Delta(x) \circ \gamma(wx)\}.$$

We shall write L' for the language recognised by $\mathscr{A}'$. One can read the set of transitions as "from a state q_1 with an history w, perform a step x in the automaton $\mathscr{A}$, and then a jump compatible with wx, which becomes the new history". One can see, from the definition of Δ' and Proposition 12 that :

$$(q_1, [u]) \xrightarrow{x}_{\mathscr{A}'} (q_2, [ux]) \ \Leftrightarrow \ \exists (q_3, v) \in Q \times \Gamma(ux) : q_1 \xrightarrow{x}_{\mathscr{A}} q_3 \xrightarrow{v}_{\mathscr{A}} q_2. \ (12)$$

Now we prove the correctness of this construction. First recall the notion of *simulation* [Mil89]:

Definition 15 (Simulation). A relation R between the states of two automata $\mathscr{A}$ and $\mathscr{B}$ is a *simulation* if for all $(p, q) \in R$ we have (a) if $p \xrightarrow{x}_{\mathscr{A}} p'$, then there exists q' such that $q \xrightarrow{x}_{\mathscr{B}} q'$ and $(p', q') \in R$, and (b) if $p \in T_{\mathscr{A}}$ then $q \in T_{\mathscr{B}}$.

We say that $\mathscr{A}$ *is simulated by* $\mathscr{B}$ if there is a simulation R such that for any $p_0 \in I_{\mathscr{A}}$, there is $q_0 \in I_{\mathscr{B}}$ such that $p_0 \, R \, q_0$.

The following property of γ is proved by exhibiting such a simulation:

Proposition 16. *For all words $u, v \in \mathbf{X}^\star$ such that $u \rightsquigarrow v$, we have $\gamma(u) \subseteq \gamma(v)$.*

Proof. First, notice that $\Gamma(u) \subseteq \Gamma(v) \Rightarrow \gamma(u) \subseteq \gamma(v)$, using Proposition 12. It thus suffices to prove $u \rightsquigarrow v \Rightarrow \Gamma(u) \subseteq \Gamma(v)$, which can be rewritten as $\Gamma(u_1 w \overline{w} w u_2) \subseteq \Gamma(u_1 w u_2)$. We can drop u_2 (it is clear that $\Gamma(w_1) \subseteq \Gamma(w_2) \Rightarrow \forall x \in \mathbf{X}, \Gamma(w_1 x) \subseteq \Gamma(w_2 x)$, from the definition of Γ): we now have to prove that $\Gamma(u_1 w \overline{w} w) \subseteq \Gamma(u_1 w)$. The proof of this inclusion relies on the fact that the automaton $\mathscr{G}(u_1 w \overline{w} w)$ is simulated by the automaton $\mathscr{G}(u_1 w)$ (see [BP14]). $\square$

We define an order relation $\preccurlyeq$ on the states of the produced automaton ($Q \times G$), by $(p, [u]) \preccurlyeq (q, [v]) \triangleq p = q \wedge \gamma(u) \subseteq \gamma(v)$.

Proposition 17. *The relation $\preccurlyeq$ is a simulation for the automaton $\mathscr{A}'$.*

Proof. Suppose that $(p, [u]) \preccurlyeq (q, [v])$ and $(p, [u]) \xrightarrow{x}_{\mathscr{A}'} (p', [ux])$, i.e., $(p, p') \in \Delta_x \circ \gamma(ux)$. We have $p = q$ and $\gamma(u) \subseteq \gamma(v)$, hence $\gamma(ux) \subseteq \gamma(vx)$, and thus $(p, p') \in \Delta_x \circ \gamma(vx)$ meaning that $(p, [v]) \xrightarrow{x}_{\mathscr{A}'} (p', [vx])$. It remains to check that $(p', [ux]) \preccurlyeq (p', [vx])$, i.e., $\gamma(ux) \subseteq \gamma(vx)$, which we just proved. $\square$

We may now prove that $L' = cl(L)$.

Lemma 18. $L' \subseteq cl(L)$

Proof. We prove by induction on u that for all q_0, q such that $(q_0, [\epsilon]) \xrightarrow{u}_{\mathscr{A}'} (q, [u])$, there exists v such that $v \rightsquigarrow^\star u$ and $q_0 \xrightarrow{v}_{\mathscr{A}} q$. The case $u = \epsilon$ is trivial.

If $(q_0, [\epsilon]) \xrightarrow{u}_{\mathscr{A}'} (q_1, [u]) \xrightarrow{x}_{\mathscr{A}'} (q, [ux])$, by induction one can find v_1 such that $q_0 \xrightarrow{v_1}_{\mathscr{A}} q_1$ and $v_1 \rightsquigarrow^\star u$. We also know (by (12) and Proposition 8) that there are some q_2, v_2 and $v_3 \in \text{suffixes}(ux)$ such that $q_1 \xrightarrow{x}_{\mathscr{A}} q_2$, $v_2 \rightsquigarrow^\star \overline{v}_3 v_3$ and $q_2 \xrightarrow{v_2}_{\mathscr{A}} q$. We thus get

$$q_0 \xrightarrow{v_1}_{\mathscr{A}} q_1 \xrightarrow{x}_{\mathscr{A}} q_2 \xrightarrow{v_2}_{\mathscr{A}} q \text{ and } v_1 x v_2 \rightsquigarrow^\star u x v_2 \rightsquigarrow^\star u x \overline{v}_3 v_3 \rightsquigarrow u x.$$

By choosing $q \in T$, we obtain the desired result. $\square$

Lemma 19. $L \subseteq L'$

Proof. This is actually very simple. First notice that for all u, $\gamma(u)$ is a reflexive relation, hence $q_1 \xrightarrow{x}_{\mathscr{A}} q_2$ entails $\forall u, (q_1, [u]) \xrightarrow{x}_{\mathscr{A}'} (q_2, [ux])$. This means that the relation R defined by $p \ R \ (q, [w]) \Leftrightarrow p = q$ is a simulation between $\mathscr{A}$ and $\mathscr{A}'$, and thus $L = L(\mathscr{A}) \subseteq L(\mathscr{A}') = L'$. $\square$

Lemma 20. L' *is downward-closed for* $\rightsquigarrow$.

A technical lemma is required to establish this closure property:

Lemma 21. *If $(q_1, [uw]) \xrightarrow{x}_{\mathscr{A}'} (q_2, [uwx]) \xrightarrow{\overline{wx} \ wx}_{\mathscr{A}'} (q_3, [uwx \ \overline{wx} \ wx])$, then $(q_1, [uw]) \xrightarrow{x}_{\mathscr{A}'} (q_3, [uwx])$.*

Proof sketch. The proof being quite verbose and dry, we shall only give a sketch of it here, referring to [BP14] for a detailed one. If $|w| = n$ and $|u| = m$, the premise can be equivalently stated:

$$(q_1, [(uw)|_{m+n-1}]) \xrightarrow{\quad w(n) \quad}_{\mathscr{A}'} (q_2, [uw]) \xrightarrow{\quad \overline{w}w \quad}_{\mathscr{A}'} (q_3, [uw\overline{w}w]).$$

(Recall that $u|_i$ denotes the prefix of length i of a word u.) Let us write $\Gamma_i = \Gamma((uw\overline{w}w)|_{m+n+i}) = \Gamma(uw(\overline{w}w)|_i)$ and $x_i = (uw\overline{w}w)(n+m+i)$ for $0 \leqslant i \leqslant 2n$. By Proposition 12 and the definition of $\mathscr{A}'$, we can show that there are $v_i \in \Gamma_i$ such that the execution above can be lifted into an execution in $\mathscr{A}$:

$$q_1 \xrightarrow{\quad x_0 v_0 x_1 v_1 \cdots x_i v_i \cdots x_{2n} v_{2n} \quad}_{\mathscr{A}} q_3.$$

Then one can prove by recurrence on i and using Proposition 8 that:

$$\forall i, \exists t_i \in \Gamma(uw) : (\overline{w}w)|_i v_i \rightsquigarrow^* t_i(\overline{w}w)|_i. \tag{13}$$

We deduce that $v_0 x_1 v_1 \cdots x_i v_i \cdots x_{2n} v_{2n} \rightsquigarrow^* t_0 t_1 \cdots t_{2n}\overline{w}w \in \Gamma(uw)^{2n+2} \subseteq \Gamma(uw)$. By Proposition 8, this means that $v_0 x_1 v_1 \cdots x_i v_i \cdots x_{2n} v_{2n}$ is in $\Gamma(uw)$, so that $(q_1, q_3) \in \Delta(w(n)) \circ \gamma(uw)$, and $(q_1, [uw|_{n-1}]) \xrightarrow{\quad w(n) \quad}_{\mathscr{A}'} (q_2, [uw])$. □

With this intermediate lemma, one can obtain a succinct proof of Lemma 20:

Proof. The statement of the lemma is equivalent to saying that if $u \rightsquigarrow v$ with $u \in L'$ then v is also in L'. Consider $u = u_1 w \cdot \overline{w} \cdot wu_2$ and $v = u_1 wu_2$ with $|w| = n \geqslant 1$ (the case where $w = \epsilon$ doesn't hold any interest since it implies that $u = v$). By combining Lemma 21 and Proposition 17 we can build the following diagram:

$$(q_0, [\epsilon]) \xrightarrow{\quad u_1 w|_{n-1} \quad} (q_1, [u_1 w|_{n-1}]) \xrightarrow{\quad w(n) \quad} (q_2, [u_1 w]) \xrightarrow{\quad \overline{w}w \quad} (q_3, [u_1 w\overline{w}w]) \xrightarrow{\quad u_2 \quad} (q_f, [u])$$

Lem. 21 $\searrow$ $w(n)$ $\stackrel{}{\curlyvee}$ Prop. 16

$$(q_3, [u_1 w]) - - - - - \underset{\text{Prop. 17}}{\overset{u_2}{- - - -}} \to (q_f, [v])$$

□

Lemmas 19 and 20 tell us that L' is closed and contains L, so by definition of the closure of a language, we get $cl(L) \subseteq L'$. Lemma 18 gives us the other inclusion, thus proving Theorem 14.

3   Analysis and Consequences

3.1   Relationship with [BÉS95]'s construction

As suggested by an anonymous referee, one can also formally relate our construction to the one from [BÉS95]: we give below an explicit and rather natural bisimulation relation between the automata produced by both these methods.

This results in an alternative correctness proof of our construction, by reducing it to the correctness of the one from [BÉS95].

We first make the two constructions comparable: the original construction, because it considers the transition monoid, takes as input a deterministic automaton. It returns a deterministic automaton. Instead, our construction does not require determinism in its input, but produces a non-deterministic automaton. We thus have to ask of both methods to accept as their input a *non-deterministic* automaton, and to return a *deterministic* automaton.

For our construction, the straightforward thing to do would be to determinise the automaton afterwards. We can actually do better, by noticing that from a state $(p, [u])$, reading some x, there may be a lot of accessible states, but all of their histories (second components) will be equal to $[ux]$. So in order to get a deterministic automaton, one only has to perform the power-set construction on the first component of the automaton. This way, we get an automaton $\mathscr{A}_1$ with states in $\mathcal{P}(Q) \times G$ and a transition function

$$\delta_1((P, [u]), x) = (P \cdot (\Delta(x) \circ \gamma(ux)), [ux]).$$

The original construction can also be adjusted very easily: first build a deterministic automaton $\mathcal{D}$ with the usual powerset construction, then apply the construction as described in Theorem 10 to get an automaton which we call $\mathscr{A}_2$. An important thing here is to understand the shape of the resulting transition monoid $M_\mathcal{D}$: its elements are functions over sets of states (because of the powerset construction) induced by words; more precisely, they are sup-semilattice homomorphisms, and they are in bijection with binary relations on states.

Define the following KA-homomorphism from $\mathcal{P}(M_\mathcal{D})$ to $\mathcal{P}(Q^2)$:

$$i(F) = \{(p, q) \mid \exists u_\mathcal{D} \in F : q \in u_\mathcal{D}(\{p\})\} .$$

(That i is a KA-homomorphism comes from the fact that the elements of $M_\mathcal{D}$ are themselves sup-semilattice homomorphisms on $\mathcal{P}(Q)$.) We can check that for all $x \in \mathbf{X}$, we have

$$i(\{x_\mathcal{D}\}) = \{(p, q) \mid q \in x_\mathcal{D}(\{p\})\} = \{(p, q) \mid q \in \delta(\{p\}, x)\}$$
$$= \left\{(p, q) \;\middle|\; p \xrightarrow{x}_\mathscr{A} q\right\} = \Delta(x) ,$$

It follows that the following relation is a bisimulation between $\mathscr{A}_1$ and $\mathscr{A}_2$.

$$\{((Q, [u]), (F, G)) \mid Q = I \cdot i(F) \text{ and } \gamma(u) = i(G)\}$$

(See [BP14] for a detailed proof.)

3.2   Complexity

Because we are speaking about algorithms rather than actual programs, it is a bit difficult to give accurate complexity bounds, considering the many possible data structures appearing during the computation. However, one may think that a relevant complexity measure of the final algorithm (for deciding equality in KAC) could be the size of the produced automata. In the following the *size* of an automaton is its number of states. In order to give a fair comparison, we will consider the generic algorithms given in the previous subsection, taking as their input a *non-deterministic* automaton, and returning a *deterministic* automaton.

Let us begin by evaluating the size of the automaton produced by the method in [BÉS95], given a non-deterministic automaton of size n. As explained above, the states of the constructed transition monoid ($M_{\mathcal{D}}$) are in bijection with the binary relations on Q. There are thus at most 2^{n^2} elements in this monoid. We deduce that the final automaton, whose states are pairs of subsets of $M_{\mathcal{D}}$ has at most $2^{2^{n^2}} \times 2^{2^{n^2}} = 2^{2^{n^2}+1}$ states.

Now with the deterministic version of our construction, the states are in the set $\mathcal{P}(Q) \times G$. Since G is the set of equivalence classes of $\sim_\gamma$ and γ has values in the reflexive binary relations over Q, we know that $\sim_\gamma$ has less than $2^{n \times (n-1)}$ elements. Hence we can see that $|\mathcal{P}(Q) \times G| \leqslant 2^n \times 2^{n \times (n-1)} = 2^{n^2}$, which is significantly smaller than the $2^{2^{n^2}+1}$ states we get with the other construction.

3.3   A Polynomial-Space Algorithm

The above upper-bound on the number of states of the automata produced by our construction allows us to show that the problem of equivalence in KAC is in PSPACE (the problem was already known to be PSPACE-hard since KAC is conservative over KA, which is PSPACE-complete [MS73]).

Recall that the equivalence of two deterministic automata $\mathcal{A}$ and $\mathcal{B}$ is in LOGSPACE. The algorithm to show that relies on the fact that $\mathcal{A}$ and $\mathcal{B}$ are different if and only if there is a word w in the difference of $L(\mathcal{A})$ and $L(\mathcal{B})$ such that $|w| \leqslant |\mathcal{A}| \times |\mathcal{B}|$. With that in mind, we can give a non-deterministic algorithm, by simulating a computation in both automata with a letter chosen non-deterministically at each step, with a counter to stop us at size $|\mathcal{A}| \times |\mathcal{B}|$. The resulting algorithm will only have to store the counter of size $\log(|\mathcal{A}| \times |\mathcal{B}|)$ and the two current states.

For our problem, the first step is to compute $\mathbf{e}$ and $\mathbf{f}$ from the regular expressions with converse e and f. It is obvious that such a transformation can be done in linear time and space, by a single sweep of both e and f. Then we have to build automata for $\mathbf{e}$ and $\mathbf{f}$. Once again this is a very light operation: if one considers for instance the position automaton (also called Glushkov's construction [Glu61]), we obtain automata of respective sizes $n = |\mathbf{e}| + 1 = |e| + 1$ and $m = |\mathbf{f}| + 1 = |f| + 1$, where $|\cdot|$ denotes the number of variable leaves of a regular expression (possibly with converse).

Our construction then produces closed automata of size at most 2^{n^2} and 2^{m^2}, so that the non-deterministic algorithm to check their equivalence needs to scan all words of size smaller than by $2^{n^2} \times 2^{m^2} = 2^{n^2+m^2}$. The counter used to bound the recursion depth can thus be stored in polynomial space $(n^2 + m^2)$.

```
    input  : Two regular expressions with converse e, f ∈ Reg_X^∨
    output: A Boolean, saying whether or not KAC ⊢ e = f.
 1  𝒜_1 = ⟨Q_1, X, I_1, T_1, Δ_1⟩ ← Glushkov' automaton recognising [[e]];
 2  𝒜_2 = ⟨Q_2, X, I_2, T_2, Δ_2⟩ ← Glushkov' automaton recognising [[f]];
 3  N ← (2^((|e|+1)^2) × 2^((|f|+1)^2));         /* N gets a value ⩾ |cl(𝒜_1)| · |cl(𝒜_2)| */
 4  ((P_1, R_1), (P_2, R_2)) ← ((I_1, Id_{Q_1}), (I_2, Id_{Q_1}));
 5  while N > 0 do
 6  |   N ← N − 1;                               /* N bounds the recursion depth */
 7  |   f_1 ← is_empty(P_1 ∩ T_1);
 8  |   f_2 ← is_empty(P_2 ∩ T_2);
 9  |   if f_1 = f_2 then
10  |   |   x ←random(X);                        /* Non-deterministic choice */
11  |   |   (R_1, R_2) ← ((Δ_1(x') ∘ R_1 ∘ Δ_1(x))^⋆, (Δ_2(x') ∘ R_2 ∘ Δ_2(x))^⋆);
12  |   |   (P_1, P_2) ← (P_1 · (Δ_1(x) ∘ R_1), P_2 · (Δ_2(x) ∘ R_2));
13  |   else
14  |   |   return false;    /* A difference appeared for some word, e ≠ f */
15  |   end
16  end
17  return true;                                /* There was no difference, KAC ⊢ e = f */
```

Algorithm 1. A PSPACE algorithm for KAC

It is worth mentioning here that with the automata constructed in [BÉS95], the counter would have size $2^{n^2+1} + 2^{m^2+1}$ which is not a polynomial.

Now the last two important things to worry about are the representation of the states of the closure automata, in particular their "history" component, and the way to compute their transition function. Let us focus on the automaton for e and let Q be the set of states of the Glushkov automaton built out of it.

– For the state representation, one needs to represent an equivalence class $[u] \in G$ by its image under γ: while the smallest word $w \in [u]$ may be quite long, $\gamma(u)$ is just a binary relation on Q. We shall thus represent the states in the determinised closure automaton as pairs of a set of states in Q and a binary relation (set of pairs) over Q. Such a pair can be stored in polynomial space (recall that $|Q| = n = |e| + 1$).

– For computing the transition function, the image of a pair $(\{q_1, \cdots, q_k\}, R)$ (with $R \subseteq Q^2$) by a letter $x \in X$ is done in two steps: first the relation becomes $R' = (\Delta(x') \circ R \circ \Delta(x))^\star$, then the set of states becomes $\{q \mid \exists i, 1 \leqslant i \leqslant k : (q_i, q) \in \Delta(x) \circ R'\}$. Those computations take place in PSPACE. (The composition of two relations in Q^2 can be performed in space $\mathcal{O}(|Q|^2)$, and the same holds for the reflexive and transitive closure of a relation R by building the powers $(R + Id_Q)^{2^k}$ and keeping a copy of the previous iteration to stop when the fixed-point is reached.)

Summing up, we obtain Algorithm 1, which is PSPACE.

Conclusion

Starting from the works of Bernátsky, Bloom, Ésik and Stefanescu, we gave a new and more efficient algorithm to decide the theory KAC. This algorithm relies on a new construction for the closure of an automaton, which allowed us to show that the problem was in fact in the complexity class PSPACE.

To prove the correctness of our construction, we used the family of regular languages $\Gamma(w)$ ($G(w^\vee)$ in [BÉS95]), and we establish its main properties using a proper finite automata characterisation. Moreover, this function allowed us to reformulate the proof of the completeness of the reduction from equality in Rel$^\vee$ to equivalence of closed automata (implication (6) from the introduction).

As an exercise, we have implemented and tested the various constructions and algorithms in an OCAML program which is available online[1].

To continue this work, we would like to implement our algorithm in the proof assistant COQ, as a tactic to automatically prove the equalities in KAC—as it has already been done for the theories KA and KAT. The simplifications we propose in this paper give us hope that such a task is feasible. The main difficulty certainly lies in the formalisation of the completeness proof of KAC (implication (7) from the introduction): the proof given in [ÉB95] uses yet another automaton construction for the closure, which is much more complicated than the one used in [BÉS95], and which seems quite difficult to formalise in COQ. We hope to find an alternative completeness proof, by exploiting the simplicity of the presented construction.

Acknowledgements. We are grateful to the anonymous referees who suggested us the alternative proof of correctness which we provide in Section 3.1, and who helped us to improve this paper.

References

[BÉS95] Bloom, S.L., Ésik, Z., Stefanescu, G.: Notes on equational theories of relations. Algebra Universalis 33, 98–126 (1995)

[Bof90] Boffa, M.: Une remarque sur les systèmes complets d'identités rationnelles. Informatique Théorique et Applications 24, 419–428 (1990)

[Bof95] Boffa, M.: Une condition impliquant toutes les identités rationnelles. Informatique Théorique et Applications 29, 515–518 (1995)

[BP14] Brunet, P., Pous, D.: Extended version of this abstract, with omitted proofs. Technical report, LIP - CNRS, ENS Lyon (2014), http://hal.archives-ouvertes.fr/hal-00938235

[Con71] Conway, J.H.: Regular algebra and finite machines. Chapman and Hall Mathematics Series (1971)

[ÉB95] Ésik, Z., Bernátsky, L.: Equational properties of Kleene algebras of relations with conversion. Theoretical Computer Science 137, 237–251 (1995)

[Glu61] Glushkov, V.M.: The abstract theory of automata. Russian Mathematical Surveys 16, 1 (1961)

---

[1] http://perso.ens-lyon.fr/paul.brunet/cka.html

[Kle51]   Kleene, S.C.: Representation of Events in Nerve Nets and Finite Automata. Memorandum. Rand Corporation (1951)
[Koz91]   Kozen, D.: A Completeness Theorem for Kleene Algebras and the Algebra of Regular Events. In: LICS, pp. 214–225. IEEE Computer Society (1991)
[Kro90]   Krob, D.: A Complete System of B-Rational Identities. In: Paterson, M. (ed.) ICALP 1990. LNCS, vol. 443, pp. 60–73. Springer, Heidelberg (1990)
[MS73]   Meyer, A., Stockmeyer, L.J.: Word problems requiring exponential time. In: Proc. ACM Symposium on Theory of Computing, pp. 1–9. ACM (1973)
[Mil89]   Milner, R.: Communication and Concurrency. Prentice Hall (1989)
[Red64]   Redko, V.N.: On defining relations for the algebra of regular events. In: Ukrainskii Matematicheskii Zhurnal, pp. 120–126 (1964)
[Sal66]   Salomaa, A.: Two Complete Axiom Systems for the Algebra of Regular Events. J. ACM 13, 158–169 (1966)

Preparing Relational Algebra
for "Just Good Enough" Hardware

José N. Oliveira

High Assurance Software Laboratory
INESC TEC and University of Minho
Braga, Portugal
jno@di.uminho.pt

Abstract. Device miniaturization is pointing towards tolerating imperfect hardware provided it is "good enough". Software design theories will have to face the impact of such a trend sooner or later.

A school of thought in software design is *relational*: it expresses specifications as relations and derives programs from specifications using relational algebra.

This paper proposes that *linear* algebra be adopted as an evolution of relational algebra able to cope with the quantification of the impact of imperfect hardware on (otherwise) reliable software.

The approach is illustrated by developing a monadic calculus for *component oriented* software construction with a probabilistic dimension quantifying (by linear algebra) the propagation of imperfect behaviour from lower to upper layers of software systems.

1 Introduction

In the trend towards miniaturization of automated systems the size of circuit transistors cannot be reduced endlessly, as these eventually become unreliable. There is, however, the idea that inexact hardware can be tolerated provided it is "good enough" [16].

Good enough has always been the way engineering works as a broad discipline: why invest in a "perfect" device if a less perfect (and less expensive) alternative fits the needs? Imperfect circuits will make a certain number of errors, but these will be tolerated if they nevertheless exhibit *almost* the same performance as perfect circuits. This is the principle behind *inexact circuit design* [16], where accuracy of the circuit is exchanged for cost savings (e.g. energy, delay, silicon) in a controlled way.

If unreliable hardware becomes widely accepted on the basis of fault tolerance guarantees, what will the impact of this be into the software layers which run on top of it in virtually any automated system? Running on less reliable hardware, functionally correct (e.g. proven) code becomes faulty and risky. Are we prepared to handle such risk at the software level in the same way it is tackled by hardware specialists? One needs to know how risk propagates across networks of software components so as to mitigate it.

P. Höfner et al. (Eds.): RAMiCS 2014, LNCS 8428, pp. 119–138, 2014.

The theory of software design by stepwise refinement already copes with some form of "approximation" in the sense that "vague" specifications are eventually realized by precise algorithms by taking design decisions which lead to (deterministic) code. However, there is a fundamental difference: all input-output pairs of a post-condition in a software specification are *equally* acceptable, giving room for the implementer to choose among them. In the case of imperfect design, one is coping with undesirable, possibly catastrophic outputs which one wishes to prove very unlikely.

In the area of safety critical systems, NASA has defined a *probabilistic risk assessment* (PRA) methodology [27] which characterizes risk in terms of three basic questions: *what can go wrong? how likely is it? and what are the consequences?* The PRA process answers these questions by systematically modeling and quantifying those scenarios that can lead to undesired consequences.

Altogether, it seems that (as happened with other sciences in the past) software design needs to become a *quantitative* or *probabilistic* science. Consider concepts such as e.g. *reliability*. From a qualitative perspective, a software system is *reliable* if it can *successfully carry out its own task as specified* [9]. But our italicized text is a inexact quotation of [9], the exact one being: *reliability [is] defined as a probabilistic measure of the system ability to successfully carry out its own task as specified.*

From a functional perspective, this means moving from specifications (input/output relations) and implementations (functions) to something which lives in between, for instance *probabilistic functions* expressing the propensity, or likelihood of multiple, possibly erroneous outputs. Typically, the classic non-deterministic choice between alternative behaviours,

$$bad \cup good \tag{1}$$

has to be replaced by probabilistic choice [19]

$$bad \ _p\diamond \ good \tag{2}$$

and the reasoning should be able to ensure that the probability p of bad behaviour is acceptably small.

Does the above entail abandoning relational reasoning in software design? Interestingly, the same style of reasoning will be preserved provided binary relations are generalized to (typed) matrices, the former being just a special case of the latter. This leads to a kind of *linear algebra of programming* [21]. Technically, in the same way relations can be *transposed* to set valued functions, which rely on the powerset monad to express non-determinism, so do probabilistic matrices, which transpose to distribution-valued functions that rely on the distribution monad to express probabilistic behaviour. It turns out that it is the converse of such transposition which helps, saving explicit set-theoretical constructions in one case and explicit distribution manipulation in the other via pointfree styled, algebraic reasoning.

Contribution. This paper proposes that, similarly to what has happened with the increasing role of *relational algebra* in computer science [7, 5, 25], *linear algebra* be adopted as its natural development where quantitative reasoning is required. Relation algebra and linear algebra share a lot in common once addressed from e.g. a categorial perspective [8, 17]. So it seems that there is room for evolution rather than radical change.

In this setting, this paper contributes (a) with a case study on such an evolution concerning a calculus of software components [2, 3] intended for quantitative analysis of software reliability; (b) with a strategy for reducing the impact of the "probabilistic move" based on re-interpreting software component semantics in linear algebra through a so-called "Kleisli-lifting' which keeps as much of the original semantics definition as possible.

2 Context

Quantitative software reliability analysis is not so easy in practice because, as is well-known, software systems are nowadays built component-wise. Cortellessa and Grassi [9] quantify component-to-component error propagation in terms of a matrix whose entry (i, j) gives the probability of component i transferring control to component j — a kind of probabilistic call-graph. For our purposes, this abstracts too much from the semantics of component-oriented systems, which have been quite successfully formalized under the *components as coalgebras* motto (see e.g. [2]), building on extensive work on automata using coalgebra theory [24, 12].

Coalgebra theory can be regarded as a generic approach to transition systems, described by functions of type

$$f : S \to \mathbb{F}\, S \tag{3}$$

where S is a set of states and $\mathbb{F}\, S$ captures the future behaviour of the system according to evolution "pattern" $\mathbb{F}$ which (technically) is a functor. For $\mathbb{F}\, S = \mathbb{P}\, S$, the powerset functor, f is the *power-transpose* [5] of a binary relation on the state space S. Other instances of $\mathbb{F}$ lead to more sophisticated transition structures, for instance Mealy and Moore machines involving inputs and outputs. Barbosa [2] gives a software component calculus in which components are regarded as such machines, expressed as coalgebras.

In this paper we wish to investigate a (technically) cheap way of promoting the *components as coalgebras* approach from the qualitative, original formulation [2] to a quantitative, probabilistic extension able to cope with the impact of *inexact circuit design* into software. As the survey by Sokolova [26] shows, probabilistic systems have been in the software research agenda for quite some time. From the available literature we focus on a paper [12] which suits our needs: it studies trace semantics of state-based systems with different forms of branching such as e.g. the non-deterministic and the probabilistic, in a categorial setting. This fits with our previous work [22] on probabilistic automata as coalgebras in categories of matrices which shows that the cost of *going quantitative* amounts essentially to changing the underlying category where the reasoning takes place.

3 Motivation

This section presents a brief account of the component algebra [2] which is central to the current paper. For illustrative purposes, we have implemented this algebra of component combinators in Haskell, for the particular situation in which components are regarded as (monadic) Mealy machines. The original algebra has furthermore been extended probabilistically relying on the PFP library written by Erwig and Kollmansberger [10]. On purpose, the examples hide many technical details which are deferred to later sections.

Abstract Mealy machines. An $\mathbb{F}$-branching Mealy machine is a function of type

$$S \times I \to \mathbb{F}\,(S \times O) \tag{4}$$

where S is the machine's internal state space, I is the set of inputs and O the set of outputs. Our main principle is that of regarding a software system as a combination of Mealy machines, from elementary to more complex ones. For this to work, $\mathbb{F}$ in (4) will be regarded as a *monad* capturing effects which are propagated upwards, from component to composite machines.

Functions of type (4), for $\mathbb{F}$ a monad, will be referred to as monadic Mealy machines (MMM) in the sequel. This type (4) can be written in two other equivalent (isomorphic) ways, all useful in component algebra: the coalgebraic $S \to (\mathbb{F}\,(S \times O))^I$ — compare with (3) — and the state-monadic $I \to (\mathbb{F}\,(S \times O))^S$, depending on how currying is applied.

Methods = elementary Mealy machines. Let us see an example which shows how an aggregation of (possibly partial) functions sharing a data type already is a Mealy machine. In the example, a stack is modelled as a (partial) algebra of finite lists written in Haskell syntax as follows

$push\,(s, a) = a : s$		$push :: ([a], a) \to [a]$
$pop = tail$	whose types are	$pop :: [a] \to [a]$
$top = head$		$top :: [a] \to a$
$empty\ s = (0 \equiv length\ s)$		$empty :: [a] \to \mathbb{B}$

Below we show how to write each individual function as an elementary Mealy machine on the shared state space $S = [a]$ before aggregating them all into a single machine (component).

In the case of *push*, $I = a$. (Note that in Haskell syntax type variables are denoted by lower-case letters). What about O? We may regard it as an instance of the singleton type 1, whose unique inhabitant carries the information that the action indeed took place:

$$push' :: ([a], a) \to ([a], 1)$$
$$push' = push \vartriangle \,!$$

This definition relies on the *pairing* operator, $(f \vartriangle g)\ x = (f\ x, g\ x)$ and on the uniquely defined (total and constant) function $! :: b \to 1$, often referred to as the "bang" function.

Note how action ("method") *push'* is pure in the sense that it does not generate any effect. The same happens with

$$empty' :: ([\,a\,], 1) \rightarrow ([\,a\,], \mathbb{B})$$
$$empty' = (id \vartriangle empty) \cdot \pi_1$$

where this time the singleton type is at the input side, meaning a "trigger" for the operation to take place. Functions *id* and π_1 are the identity function and the projection $\pi_1\,(x, y) = x$, respectively, the former ensuring that no state change takes place.

Concerning *pop* and *top* we have a new situation: as these are partial functions, some sort of totalization is required before promoting them to Mealy machines. The cheapest way of totalizing partial functions resorts to the "Maybe" monad $\mathbb{M}$, mapping into an error value $\bot$ the inputs for which the function is undefined and otherwise signaling a successful computation using the monad's unit $\eta :: S \rightarrow \mathbb{M}\,S$:[1]

$$\cdot \Leftarrow \cdot :: (a \rightarrow b) \rightarrow (a \rightarrow \mathbb{B}) \rightarrow a \rightarrow \mathbb{M}\,b$$
$$(f \Leftarrow p)\,a = \textbf{if } p\,a \textbf{ then } (\eta \cdot f)\,a \textbf{ else } \bot$$

Note how $f \Leftarrow p$ "fuses" f with a given pre-condition p, as in the following promotion of *top* to a $\mathbb{M}$-monadic Mealy machine

$$top' :: ([\,a\,], 1) \rightarrow \mathbb{M}\,([\,a\,], a)$$
$$top' = (id \vartriangle top \Leftarrow (\neg \cdot empty)) \cdot \pi_1$$

which, as *empty'*, does not change the state. Opting for the usual semantics of the *pop* method,

$$pop' :: ([\,a\,], 1) \rightarrow \mathbb{M}\,([\,a\,], a)$$
$$pop' = (pop \vartriangle top \Leftarrow (\neg \cdot empty)) \cdot \pi_1$$

we finally go back to pure *push'* and *empty'* making them $\mathbb{M}$-compatible (ie. $\mathbb{M}$-resultric) through the *success* operator:

$$push' :: ([\,a\,], a) \rightarrow \mathbb{M}\,([\,a\,], 1)$$
$$push' = \eta \cdot (push \vartriangle !)$$
$$empty' :: ([\,a\,], 1) \rightarrow \mathbb{M}\,([\,a\,], \mathbb{B})$$
$$empty' = \eta \cdot (id \vartriangle empty) \cdot \pi_1$$

Components $= \sum$ *methods.* Now that we have the *methods* of a stack written as individual Mealy machines over the same monad and shared state space, we *add them up* to obtain the intended *stack* component [2]

$$stack :: ([\,a\,], 1 + 1 + a + 1) \rightarrow \mathbb{M}\,([\,a\,], a + a + 1 + \mathbb{B})$$
$$stack = pop' \oplus top' \oplus push' \oplus empty'$$

---

[1] Symbols $\bot$ and η pretty-print *Nothing* and *Just* of Haskell's concrete syntax, respectively, cf. definition **data** $\mathbb{M}\,a = Nothing \mid Just\,a$.

[2] Notation $x + y$ pretty-prints Haskell's syntax for disjoint union, *Either x y*.

Before giving the details of the binary operator $\oplus$ which binds methods together, note that *stack* is also a (composite) Mealy machine (4), for $I = 1 + 1 + a + 1$ and $O = a + a + 1 + \mathbb{B}$. This I/O interfacing, pictured aside, captures the four alternatives which are available for interacting with a stack. Note how singleton types (1) at the input side mean *"do it!"* and at the output side mean *"done!"*.

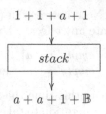

$$1 + 1 + a + 1$$
$$\downarrow$$
$$\boxed{stack}$$
$$\downarrow$$
$$a + a + 1 + \mathbb{B}$$

Components such as *stack* arise as the *sum* of their methods, a MMM binary combinator whose definition in Haskell syntax is given aside. Isomorphism $dr :: (s, i+j) \rightarrow (s, i) + (s, j)$ (resp. its converse dr°) distributes (resp. factorizes) the shared state across the sum of inputs (resp. outputs); $m_1 + m_2$ is the sum of m_1 and

$\cdot \oplus \cdot ::(Functor\ \mathbb{F}) \Rightarrow$
-- input machines
$((s, i) \rightarrow \mathbb{F}\ (s, o)) \rightarrow$
$((s, j) \rightarrow \mathbb{F}\ (s, p)) \rightarrow$
-- output machine
$(s, i + j) \rightarrow \mathbb{F}\ (s, o + p)$
-- definition
$m_1 \oplus m_2 = (\mathbb{F}\ dr^\circ) \cdot \Delta \cdot (m_1 + m_2) \cdot dr$

m_2 and "cozip" operator $\Delta :: \mathbb{F}\ a + \mathbb{F}\ b \rightarrow \mathbb{F}\ (a + b)$ promotes sums through functor $\mathbb{F}$.

Systems = component compositions. Let us now consider the idea of building a system in which two stacks interact with each other, e.g. by popping from one and pushing the outcome onto the other.[3]

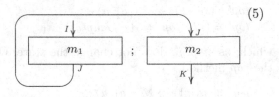

(5)

For this another MMM combinator is needed taking two I/O compatible MMM m_1 and m_2 (with different internal states in general) and building a third one, $m_1 ; m_2$, in which outputs of m_1 are sent to m_2 (5).

The type of this combinator as implemented in Haskell is given aside. It requires $\mathbb{F}$ to be a *strong* monad [15], a topic to be addressed later. Note how the output machine has a composite state pairing the states of the two input machines.

$\cdot ; \cdot ::(Strong\ \mathbb{F}, Monad\ \mathbb{F}) \Rightarrow$
-- input machines
$((s, i) \rightarrow \mathbb{F}\ (s, j)) \rightarrow$
$((r, j) \rightarrow \mathbb{F}\ (r, k)) \rightarrow$
-- output machine
$((s, r), i) \rightarrow \mathbb{F}\ ((s, r), k)$

We defer to a later stage the analysis of the formal definition of this combinator, which is central to the principle of building components out of other components [2]. Instead, we build the composite machine already anticipated above,

---

[3] This interaction will of course fail if the source stack is empty, but this is not our concern — monad $\mathbb{M}$ will take care of such effects.

$$m = pop' \; ; push'$$

which pops from a source stack ($m_1 = pop'$) and pushes onto a target stack ($m_2 = push'$). By running e.g. [4]

```
> m(([1],[2]),())
Just (([],[1,2]),())
```

we obtain the expected output and new state, while

```
> m(([],[2]),())
Nothing
```

fails, because the source stack is empty.

Faulty components. Let us finally consider the possibility of, due to hardware imperfection, pop' behaving in the source stack as expected, with probability p, and unexpectedly like top' with probability $1 - p$,

$$pop'' :: \mathsf{P} \to ([a], 1) \to \mathbb{D}\,(\mathsf{M}\,([a], a))$$
$$pop''\ p = pop'\ {}_p\diamond top'$$

recall (2). P pretty-prints the probability representation data type $\mathtt{ProbRep}$ of the PFP library and $\mathbb{D}$ denotes the (finite) *distribution* monad implemented in the same library. The choice operator $\cdot\ {}_p\diamond\ \cdot$ is the pointfree counterpart of a similar operator in PFP.

Concerning the target stack, the conjectured fault of $push'$ is that it does not push anything with probability $1 - q$:

$$push'' :: \mathsf{P} \to ([a], a) \to \mathbb{D}\,(\mathsf{M}\,([a], 1))$$
$$push''\ q = push'\ {}_q\diamond\ !'$$

where $!' = \eta \cdot (id \times !)$, of generic type $(s, a) \to \mathsf{M}\,(s, 1)$, is the promotion of the *bang* function $!$ to a MMM.

Note how pop'' and $push''$ have become "doubly" monadic in their cascading of the distribution ($\mathbb{D}$) and Maybe (M) monads. To compose them as in $m = pop' \; ; push'$ above we need a more sophisticated version of the semi-

```
· ;_D · ::
-- input probabilistic MMMs
((s, i) → D (M (s, j))) →
((r, j) → D (M (r, k))) →
-- output probabilistic MMM
((s, r), i) → D (M ((s, r), k))
```

colon combinator (aside) whose actual implementation is once again intentionally skipped. Thanks to this new combinator, we can build a faulty version of machine m above [5]

$$m_2 = (pop''\ 0.95) \; ;_D (push''\ 0.8)$$

and test it for the same composite state ($[1], [2]$) as in the first experiment above, obtaining

---

[4] Recall that () is the Haskell notation for the unique inhabitant of type 1.

[5] The probabilities in these examples are chosen with no criterion apart from leading to distributions visible to the naked eye. By all means, 5% would be extremely high risk in realistic PRA [27], where only figures as small as 1.0E-7 become "acceptable".

```
> m2(([1],[2]),())
  Just (([],[1,2]),())    76.0%
    Just (([],[2]),())    19.0%
  Just (([1],[1,2]),())    4.0%
    Just (([1],[2]),())    1.0%
```

The simulation shows that the overall risk of faulty behaviour is 24% $(1 - 0.76)$, structured as 1%: both stacks misbehave; 4%: source stack misbehaves; 19%: target stack misbehaves. As expected, the second experiment

```
> m2(([],[2]),())
Nothing 100.0%
```

is *always catastrophic* (again popping from an empty stack).

Summing up: our animation in Haskell has been able to *simulate* fault propagation between two stack components with different fault patterns arising from conjectured hardware imperfections. In the sequel we will want to *reason* about such fault propagation rather than just simulate it.

4   Related Work and Research Questions

Elsewhere [21, 20] we have shown that fault propagation can be reasoned about for functional programs of a particular kind — they are inductive extensions (termed *folds* or *catamorphisms*) of given *algebras*. In particular, the *linear (matrix) algebra of programming* mentioned in the introduction is used to decide which laws of programming [5] hold probabilistically or to find side-conditions for them to hold.

In the current paper we are faced with the dual situation: our programs are *coalgebras* and we want to *observe* and compare their behaviour expressed by *unfolds* (also known as *anamorphisms*) which tell how likely particular execution traces are. In particular, we want to be able to ascertain which (different) machines exhibit the same (probabilistic) traces for the same starting states (trace equivalence).

Looking at the types of *push''*, m_2 etc. above we realize that our MMMs have become probabilistic, leading to coalgebras of general shape

$$S \to (\mathbb{D}(\mathbb{F}(S \times O)))^I \tag{6}$$

This leads into our main research questions: *How tractable (mathematically) is this doubly-monadic framework? Can $\mathbb{F}$ be any monad?*

Relatives of shape (6) have been studied elsewhere [26], namely *reactive probabilistic automata*, $S \to (\mathbb{M}\,(\mathbb{D}\,S))^I$; *generative probabilistic automata*, $S \to \mathbb{M}\,(\mathbb{D}\,(O \times S))$; *bundle systems*, $S \to \mathbb{D}\,(\mathbb{P}\,(O \times S))$ and so on. In a coalgebraic approach to weighted automata, reference [6] studies coalgebras of functor $S \to \mathbb{K} \times (\mathbb{K}_\omega^S)^I$ for $\mathbb{K}$ a field. Such coalgebras rely on the so-called *field valuation* (exponential) functor $\mathbb{K}_\omega^-$ calling for *vector spaces*.

Inspired by this approach, a similar framework was studied directly in suitable *categories of matrices* [22]. We will follow a similar strategy in the current paper concerning probabilistic MMMs and their combinators.

5 Composition

The essence of the component algebra of [2] is a notion of component *composition* stated in a coalgebraic, categorial setting. Let us briefly review this framework, instantiated for generic $\mathbb{F}$-branching Mealy machines (4).

Let $X \xrightarrow{\eta} \mathbb{F}X \xleftarrow{\mu} \mathbb{F}^2 X$ be a monad and m_1, m_2 be two machines (functions) of types $S \times I \to \mathbb{F}\,(S \times J)$ and $Q \times J \to \mathbb{F}\,(Q \times K)$, respectively. Abstracting from their internal states as in picture (5) above, these machines can be represented by the arrows $I \xrightarrow{m_1} J$ and $J \xrightarrow{m_2} K$, respectively. Their *composition* by $I \xrightarrow{m_1;m_2} K$ is a machine with composite state $S \times Q$ built in the following way: first, m_1 is "wrapped" with the state Q of m_2,

$$\mathbb{F}\,(S \times J) \times Q \xleftarrow{m_1 \times id} (S \times I) \times Q \xleftarrow{xr} (S \times Q) \times I$$

$$\tau_r \Big\downarrow \qquad\qquad \xleftarrow{\quad g \quad}$$

$$\mathbb{F}\,((S \times J) \times Q)$$

where xr is the obvious isomorphism and τ_r is the right *strength* of monad $\mathbb{F}$, $\tau_r : (\mathbb{F}\,A) \times B \to \mathbb{F}\,(A \times B)$, which therefore has to be a *strong* monad. The purpose of xr is to ensure the compound state and input I on the input side. In turn, m_2 is wrapped with the state of m_1,

$$\mathbb{F}\,(S \times (Q \times K)) \xleftarrow{\tau_l} S \times \mathbb{F}\,(Q \times K) \xleftarrow{id \times m_2} S \times (Q \times J) \xleftarrow{xl} (S \times J) \times Q$$

$$\mathbb{F}\,a^\circ \Big\downarrow \qquad\qquad \xleftarrow{\quad f \quad}$$

$$\mathbb{F}\,((S \times Q) \times K)$$

where a° is the converse of isomorphism $a : (A \times B) \times C \to A \times (B \times C)$ and xl is a variant of xr above. Finally, τ_l is the left strength of $\mathbb{F}$, $\tau_l : (B \times \mathbb{F}\,A) \to \mathbb{F}\,(B \times A)$.

Note how $\mathbb{F}\,a^\circ$ ensures the compound state and type K on the output. In spite of the efforts of xl to approximate the input of contribution f to the output of the other (g), they do not match, as the latter is $\mathbb{F}$-*more complex* than the former.

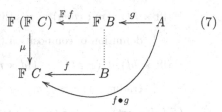

$$(7)$$

This suggests that f and g be composed using the *Kleisli composition* associated with monad $\mathbb{F}$, denoted by $f \bullet g$ and depicted in diagram (7). Thus we obtain the Haskell implementation of machine composition which was left unspecified in section 3:

$$m_1 \,;\, m_2 = ((\mathbb{F}\,a^\circ) \cdot \tau_l \cdot (id \times m_2) \cdot \mathsf{xl}) \bullet (\tau_r \cdot (m_1 \times id) \cdot \mathsf{xr}) \tag{8}$$

An advantage of relying on Kleisli composition (7) is its rich algebra, forming a monoid with η

$$f \bullet (g \bullet h) = (f \bullet g) \bullet h \tag{9}$$

$$f \bullet \eta = f = \eta \bullet f \tag{10}$$

and trading nicely with normal composition, cf. for instance

$$(\mathbb{F}\, f) \cdot (h \bullet k) = (\mathbb{F}\, f \cdot h) \bullet k \tag{11}$$

$$(f \cdot g) \bullet h = f \bullet (\mathbb{F}\, g \cdot h) \tag{12}$$

It turns out that Mealy machines too form a monoid whose binary operator is (8) and whose unit is the machine $J \xrightarrow{copy} J$ which faithfully passes its input to the output, never changing state:

$$copy : 1 \times J \to \mathbb{F}\,(1 \times J)$$
$$copy = \eta$$

However, such an algebraic structure holds up to behavioural equivalence only, denoted by symbol $\simeq$:

$$m \,;\, copy \;\simeq\; m \;\simeq\; copy \,;\, m \tag{13}$$

$$m \,;\, (n \,;\, p) \;\simeq\; (m \,;\, n) \,;\, p \tag{14}$$

Behavioural equivalence can be established by defining *morphisms* between equivalent machines regarded as coalgebras. In general, given two $\mathbb{F}$-Mealy machines m_1 and m_2 (aside), a state transformation $h{:}S \to Q$ is a *morphism* between them if the diagram aside commutes.

$$
\begin{array}{ccc}
\mathbb{F}\,(S \times J) \xleftarrow{\ m_1\ } S \times I & & S \\
{\scriptstyle \mathbb{F}\,(h \times id)}\downarrow \quad & \quad \downarrow{\scriptstyle h \times id} & \downarrow{\scriptstyle h} \\
\mathbb{F}\,(Q \times J) \xleftarrow[m_2]{} Q \times I & & Q
\end{array}
$$

For instance, to establish the first part of (13) it suffices to show that the natural isomorphism $\mathsf{lft}{:}A \times 1 \to A$ is a morphism between $m \,;\, copy$ and m itself: $\mathbb{F}\,(\mathsf{lft} \times id) \cdot (m \,;\, copy) = m \cdot (\mathsf{lft} \times id)$. As example of pointfree calculation typical of $\simeq$ reasoning, the proof of this equality is given next, where $\mathsf{rgt}{:}1 \times A \to A$ is another natural isomorphism:

$$\mathbb{F}\,(\mathsf{lft} \times id) \cdot (m \,;\, copy) = m \cdot (\mathsf{lft} \times id)$$

$\equiv$ { definition of composition (8) }

$$\mathbb{F}\,(\mathsf{lft} \times id) \cdot (((\mathbb{F}\, \mathsf{a}^{\circ}) \cdot \tau_l \cdot (id \times \eta) \cdot \mathsf{xl}) \bullet \tau_r \cdot (m \times id) \cdot \mathsf{xr}) = m \cdot (\mathsf{lft} \times id)$$

$\equiv$ { trading (11); $\tau_l \cdot (id \times \eta) = \eta$; lft commutes with rgt via a° }

$$(\mathbb{F}\,(id \times \mathsf{rgt}) \cdot \eta \cdot \mathsf{xl}) \bullet (\tau_r \cdot (m \times id) \cdot \mathsf{xr}) = m \cdot (\mathsf{lft} \times id)$$

$\equiv$ { naturality of η; $(id \times \mathsf{rgt}) \cdot \mathsf{xl} = \mathsf{lft}$ }

$$(\eta \cdot \mathsf{lft}) \bullet (\tau_r \cdot (m \times id) \cdot \mathsf{xr}) = m \cdot (\mathsf{lft} \times id)$$

$\equiv$ { (12); unit η (10) }

$$(\mathbb{F}\, \mathsf{lft}) \cdot \tau_r \cdot (m \times id) \cdot \mathsf{xr} = m \cdot (\mathsf{lft} \times id)$$

$\equiv$ { $\mathbb{F}\, \mathsf{lft} \cdot \tau_r = \mathsf{lft}$; $\mathsf{lft} \cdot \mathsf{xr} = \mathsf{lft} \times id$ }

$$\mathsf{lft} \cdot (m \times id) \cdot \mathsf{xr} = m \cdot \mathsf{lft} \cdot \mathsf{xr}$$

$$\equiv \qquad \{ \text{ cancel xr; naturality of lft: } \text{lft} \cdot (m \times id) = m \cdot \text{lft } \}$$

$$\textit{true}$$

The component algebra of [2] contains several other combinators which are of interest. For economy of space we will restrict ourselves to composition, as this is enough for our main point in the paper: what is the impact in the component algebra of [2] of having *faulty* Mealy machines as models of components?

6 Composing Non-deterministic Components

Recall from the motivation how we have simulated faulty composition of M-Mealy machines on top of the PFP Haskell library. In general, this means handling machines of type $I \to J$, that is, functions of type $Q \times I \to \mathbb{D}\,(\mathbb{F}\,(Q \times J))$ for some space state Q, where before we had $Q \times I \to \mathbb{F}\,(Q \times J)$. Thus, further to monad $\mathbb{F}$, another monad is around, the *distribution* monad $\mathbb{D}$.

Generalizing even further, we want to consider machines of type

$$Q \times I \to \mathbb{T}\,(\mathbb{F}\,(Q \times J)) \tag{15}$$

where monad $X \xrightarrow{\eta_\mathbb{F}} \mathbb{F}\,X \xleftarrow{\mu_\mathbb{F}} \mathbb{F}^2 X$ caters for *transitional* effects (how the machine evolves) and monad $X \xrightarrow{\eta_\mathbb{T}} \mathbb{T}\,X \xleftarrow{\mu_\mathbb{T}} \mathbb{T}^2 X$ specifies the *branching type* of the system [12]. A typical instance is $\mathbb{T} = \mathbb{P}$ (powerset) and $\mathbb{F} = \mathbb{M} = (1+)$ ('maybe'), that is, we have machines

$$m : Q \times I \to \mathbb{P}\,(1 + Q \times J) \tag{16}$$

which are reactive, non-deterministic finite state automata with explicit termination.

Note, however, that m (16) could alternatively be specified as a *binary relation* R of type $Q \times I \to 1 + Q \times J$ of which m is the *power transpose* [5], following the equivalence

$$R = \lceil m \rceil \quad \equiv \quad \langle \forall b, a :: b\,R\,a \equiv b \in m\,a \rangle \tag{17}$$

which tells that a set-valued function m is uniquely represented by a binary relation $R = \lceil m \rceil$ and vice-versa. Moreover, $\lceil m \bullet n \rceil = \lceil m \rceil \cdot \lceil n \rceil$ holds, where $m \bullet n$ means the Kleisli composition of two *set-valued functions* and $\lceil m \rceil \cdot \lceil n \rceil$ is the *relational* composition of the corresponding *binary relations*:

$$b\,(R \cdot S)\,a \quad \equiv \quad \langle \exists c :: b\,R\,c \wedge c\,S\,a \rangle \tag{18}$$

In categorial-speak, this means that the category of binary relations coincides with the Kleisli category of the powerset monad $X \xrightarrow{sing} \mathbb{P}\,X \xleftarrow{dunion} \mathbb{P}^2 X$ where $sing\,a = \{a\}$ and $dunion\,S = \bigcup_{s \in S} s$.

Back to (16), the advantage of "thinking relationally" is that machine m can be "replaced" by the relation $\lceil m \rceil : Q \times I \to 1 + Q \times J$ from whose (relational) type the powerset has vanished. So, in a sense, it is as if we were back to the situation where only $\mathbb{M} = (1+)$ is present.

How do relational, $\mathbb{M}$-machines compose? Recall from (8) that machine composition relies on Kleisli composition (7) — in this case, of monad $\mathbb{M}\,X = 1 + X$, with structure $X \xrightarrow{\;i_2\;} 1 + X \xleftarrow{\;[i_1, id]\;} 1 + (1 + X)$, where i_1 and i_2 are the injections associated to binary sums. Thus

$$f \bullet g = [i_1, id] \cdot (id + f) \cdot g = [i_1, f] \cdot g \tag{19}$$

where $[f, g]$ is the *junc* combinator satisfying $[f, g] \cdot i_1 = f$ and $[f, g] \cdot i_2 = g$. How about relations? Consider evaluating expression $[i_1, f] \cdot g$ for f replaced by some relation $1 + B \xleftarrow{\;R\;} C$,[6] g replaced by some other relation $1 + C \xleftarrow{\;S\;} A$ and functional composition replaced by relational composition:

$$R \bullet S = [i_1, R] \cdot S \tag{20}$$

Unfolding $[i_1, R]$ to $i_1 \cdot i_1^\circ \cup R \cdot i_2^\circ$ — where R° denotes the converse of R — one obtains, abbreviating by $*$ the application of i_1 to the unique inhabitant of singleton type 1:

$$y\,(R \bullet S)\,a \;\; \equiv \;\; (y = *) \wedge (* \, S \, a) \vee \langle \exists c :: (y \; R \; c) \wedge ((i_2 \; c) \; S \; a) \rangle$$

In words: composition $R \bullet S$ is doomed to fail wherever S fails; otherwise, it will fail where R fails. For the same input, $R \bullet S$ may *both* succeed or fail.

Summing up: we have encoded the Kleisli composition of a monad ($\mathbb{M}$) not in the category of sets and functions but in the Kleisli category of another monad (the powerset) which eventually we found familiar with — we met *relational algebra* there. Back to (8), can think of $\mathbb{M}$-monadic Mealy machines as *binary relations* which compose (as machines) according to definition

$$S\,;\,R \;\; = \;\; [i_1, (id + \mathsf{a}^\circ) \cdot \tau_l \cdot (id \times R) \cdot \mathsf{xl}] \cdot \tau_r \cdot (S \times id) \cdot \mathsf{xr} \tag{21}$$

where all dots mean relational composition (18).[7]

7   Composing Probabilistic Components

For the above constructions to help in reasoning about non-deterministic components we have to check if the properties of monad $\mathbb{M}$ remain intact once encoded relationally, extensive to the properties of the *strength* operators also present in (21).

---

[6] Relations of this type *express the possibility that for some inputs, both termination and nontermination are possible* [that is] *relations from legal states to a "lifted" state set containing all legal states and in addition one "illegal state" standing for nontermination* [13].

[7] As usual, every function symbol f in (21) should be regarded as the homonym relation f such that $b\,f\,a$ holds iff $b = f\,a$.

Prior to this, however, let us not forget that our aim is to *prepare relational algebra for "just good enough" hardware* and its imprecision, calling for a probabilistic treatment of faults. In this direction, we should also check the scenario of the previous section once the *distribution* monad replaces the powerset,

$$m : Q \times I \to \mathbb{D} \left(1 + Q \times J\right) \tag{22}$$

whereupon non-deterministic branching becomes weighted with probabilities indicating the likelihood of state transitions, recall (2). So, m is now a distribution-valued function. We assume below that such distributions have countable support.[8]

It turns out that the strategy to cope with this situation is similar to that of the previous section: distribution-valued functions are adjoint to so-called *column stochastic* (CS) matrices, which represent the inhabitants of the Kleisli category associated with monad $\mathbb{D}$; and, for this monad, Kleisli composition corresponds to matrix *composition*, usually termed matrix *multiplication*:

$$b \left(M \cdot N\right) a \; \equiv \; \left\langle \sum c :: (b \; M \; c) \times (c \; N \; a) \right\rangle \tag{23}$$

In this formula, both M and N are matrices. We prefer to denote the cell in (say) M addressed by row b and column a by the infix notation $b \; M \; a$, rather than the customary $M(b,a)$ or $M_{b,a}$. This stresses on the notational proximity with relations: matrices are just weighted relations.[9]

Summing up: in the same way the "Kleisli lifting" of section 6 makes the *powerset* monad implicit, leading into *relational algebra*, the same lifting now hides the *distribution* monad and leads to the *linear algebra* of CS matrices [21], under the universal correspondence

$$M = \lceil f \rceil \; \equiv \; \langle \forall b, a :: b \; M \; a = (f \; a) \; b \rangle \tag{24}$$

where $f : A \to \mathbb{D} \; B$ is a probabilistic function and $\mathbb{D} \; B$ is the set of all distributions on B with countable support; that is, for every $a \in A, f \; a = \mu$ where $\mu : B \to [0,1]$ is a function such that $\sum_{b \in B} m \; b = 1$.

Correspondence (24) establishes the isomorphism

$$A \to \mathbb{D} \; B \; \cong \; A \to B \tag{25}$$

where on the left hand side we have $\mathbb{D}$-valued functions and on the right hand side $A \to B$ denotes the set of all CS matrices with columns indexed by A, rows indexed by B and cells taking values from the interval $[0,1]$.[10] This *matrices as arrows* approach [17] regards them as *morphisms* of suitable categories (of typed matrices). In the current paper we only consider matrices on the interval $[0,1]$ subject to the *column stochasticity* constraint expressed above.

---

[8] This is reasonable in the sense that they arise from a finite number of applications of the choice (2) operator.

[9] Reference [22] argues in this direction by adapting rules typical of relational algebra to linear algebra.

[10] Each such column represents a distribution and therefore adds up to 1, as written above.

With no further detours let us adapt definition (20) of (relational) $\mathbb{M}$-Kleisli composition to the corresponding definition in linear algebra, where relation R gives place to matrix $1 + B \xleftarrow{\ M\ } C$, relation S to matrix $1 + C \xleftarrow{\ N\ } A$ and the little dot now denotes matrix composition (23):

$$M \bullet N = [i_1 | M] \cdot N \tag{26}$$

The reader may wonder about how does injection i_1 (a function) fit into a linear algebra expression (26). The explanation is the same as for functions in relational expressions: every function $f : A \to B$ is uniquely represented by the homonym matrix f defined by $b \, f \, a = 1$ if $b = f \, a$ and 0 otherwise.[11] Combinator $[M|N]$ occurring in (26) means the juxtaposition of matrices M and N, which therefore have to exhibit the same output type (and thus the same number of rows). Similarly to relations, it decomposes into $[M|N] = M \cdot i_1^\circ + N \cdot i_2^\circ$ where addition of matrices is the obvious cell-wise operation and the converse M° of a matrix M swaps its rows with columns (it is commonly known as the *transpose* of M). Because matrix multiplication is bilinear, we obtain $M \bullet N = i_1 \cdot i_1^\circ \cdot N + M \cdot i_2^\circ \cdot N$ and therefore the following pointwise version of (26)

$$y \, (M \bullet N) \, a \;=\; (y = *) \times (* \, N \, a) + \left\langle \sum c :: (y \, M \, c) \times ((i_2 \, c) \, N \, a) \right\rangle$$

where $*$ is the same abbreviation used before and term $y = *$ evaluates to 1 if the equality holds and to 0 otherwise.[12]

The picture aside shows an example of probabilistic, $\mathbb{M}$-Kleisli composition of two matrices $N : \{a_1, a_2, a_3\} \to 1 + \{c_1, c_2\}$ and $M : \{c_1, c_2\} \to 1 + \{b_1, b_2\}$. Injection $i_1 : 1 \to 1 + \{b_1, b_2\}$ is the leftmost column vector. Note how, for input a_1, there is 60% probability of $M \bullet N$ failing, partly due (50%) to N failing or (50%) to passing output c_1 to M, which for such an input has 20% probability of failing again.

					a1	a2	a3
				*	0.5	0	0
			c1		0.5	1	0.7
		*	c1	c2	0	0	0.3
	*	1	0.2	0	0.6	0.2	0.14
b1	0	0	0	0.6	0	0	0.18
b2	0	0.8	0.4	0.4	0.4	0.8	0.68

As before with relations, we can think of probabilistic $\mathbb{M}$-monadic Mealy machines as column stochastic matrices which compose (matricially) as follows

$$N \; ; \; M \;=\; [i_1 | (id \oplus a^\circ) \cdot \tau_l \cdot (id \otimes M) \cdot \mathsf{xl}] \cdot \tau_r \cdot (N \otimes id) \cdot \mathsf{xr} \tag{27}$$

where relational product becomes matrix Kronecker product

$$(y, x)(M \otimes N)(b, a) = (yMb) \times (xNa) \tag{28}$$

and relational sum gives place to matrix direct sum, $M \oplus N = \begin{bmatrix} M & 0 \\ \hline 0 & N \end{bmatrix}$.

---

[11] See section 8 for a technically more detailed explanation.

[12] See [22, 20] for a number of useful rules interfacing index-free and index-wise matrix notation. Such rules, expressed in the style of the Eindhoven quantifier calculus [1], provide evidence of the safe mix among matrix, predicate and function notation in typed linear algebra.

8 Monads in Relational/Linear Algebra

The evolution from relational to (typed) linear algebra proposed in the previous sections corresponds to moving from non-deterministic choice (1) to probabilistic choice (2). The latter can now be defined matricially, for probabilistic f and g:
$\lceil f_{\,p}\diamond g\rceil = p\otimes\lceil f\rceil + (1-p)\otimes\lceil g\rceil.$

A generic strategy can be identified: having a notion of composition (8) for machines of type $Q\times I \to \mathbb{F}\,(Q\times J)$ (4), where monad $\mathbb{F}$ captures their *transition* pattern, we want to *reuse* such a definition for more sophisticated machines of type $Q \times I \to \mathbb{T}\,(\mathbb{F}\,(Q \times J))$ (15) by porting it "as is" to the Kleisli category of the extra monad $\mathbb{T}$ which captures the *branching* structure.

For this to make sense we must be sure that the *lifting* of monad $\mathbb{F}$ by $\mathbb{T}$ still is a monad in the Kleisli category of $\mathbb{T}$. In general, let $X \xrightarrow{\ \eta_\mathbb{T}\ } \mathbb{T}X \xleftarrow{\ \mu_\mathbb{T}\ } \mathbb{T}^2 X$ and $X \xrightarrow{\ \eta_\mathbb{F}\ } \mathbb{F}X \xleftarrow{\ \mu_\mathbb{F}\ } \mathbb{F}^2X$ be two monads in a category C, and let $\mathsf{C}^\flat$ denote the Kleisli category induced by $\mathbb{T}$. Denote by $B \xleftarrow{\ f^\flat\ } A$ the morphism in $\mathsf{C}^\flat$ corresponding to $\mathbb{T}B \xleftarrow{\ f\ } A$ in C and define:

$$f^\flat \cdot g^\flat = (f \bullet g)^\flat = (\mu_\mathbb{T} \cdot \mathbb{T}f \cdot g)^\flat \qquad\qquad (29)$$

For *any* morphism $B \xleftarrow{\ f\ } A$ in C define its lifting to $\mathsf{C}^\flat$ by $\overline{f} = (\eta_\mathbb{T} \cdot f)^\flat$. As in [12], assume a distributive law $\lambda : \mathbb{F}\mathbb{T} \to \mathbb{T}\mathbb{F}$ and define, for each endofunctor $\mathbb{F}$ in C, its lifting $\overline{\mathbb{F}}$ to $\mathsf{C}^\flat$ by

$$\overline{\mathbb{F}}(f^\flat) \;=\; (\lambda \cdot \mathbb{F}\,f)^\flat \quad \text{cf. diagram} \quad \mathbb{T}\mathbb{F}B \xleftarrow{\ \lambda\ } \mathbb{F}\mathbb{T}B \xleftarrow{\ \mathbb{F}\,f\ } \mathbb{F}\,A \qquad (30)$$

for $\mathbb{T}\,B \xleftarrow{\ f\ } A$. For $\overline{\mathbb{F}}$ to be a functor in $\mathsf{C}^\flat$ two conditions must hold [12]:

$$\lambda \cdot \mathbb{F}\,\eta_\mathbb{T} = \eta_\mathbb{T} \qquad\qquad (31)$$
$$\mu_\mathbb{T} \cdot \mathbb{T}\lambda \cdot \lambda = \lambda \cdot \mathbb{F}\,\mu_\mathbb{T} \qquad\qquad (32)$$

We want to check under what conditions monad $\mathbb{F}$ lifts to a monad in the Kleisli of $\mathbb{T}$, that is, whether

$$X \xrightarrow{\ \overline{\eta_\mathbb{F}}=(\eta_\mathbb{T}\cdot\eta_\mathbb{F})^\flat\ } \overline{\mathbb{F}}X \xleftarrow{\ \overline{\mu_\mathbb{F}}=(\eta_\mathbb{T}\cdot\mu_\mathbb{F})^\flat\ } \overline{\mathbb{F}}^2 X \qquad\qquad (33)$$

is a monad in $\mathsf{C}^\flat$. The standard monadic laws, e.g. $\overline{\mu_\mathbb{F}} \cdot \overline{\eta_\mathbb{F}} = id$, hold by construction.[13] It can be checked that the remaining natural laws, $\overline{\mathbb{F}}\,f^\flat \cdot \overline{\eta_\mathbb{F}} = \overline{\eta_\mathbb{F}} \cdot f^\flat$

---

[13] The general rule is that $\overline{f} = (\eta_\mathbb{T} \cdot f)^\flat$ embeds C in $\mathsf{C}^\flat$. Thus the lifting of e.g. an equality $f \cdot g = h$ in C, that is $\overline{f} \cdot \overline{g} = \overline{h}$ in $\mathsf{C}^\flat$, is $(\eta_\mathbb{T} \cdot f)^\flat \cdot (\eta_\mathbb{T} \cdot f)^\flat = (\eta_\mathbb{T} \cdot h)^\flat$ which, by (29), reduces to the original $f \cdot g = h$. Within the image of the embedding, everything in $\mathsf{C}^\flat$ "works as if" in C. Our previous use of a function symbol f as denotation of the *corresponding* relation or matrix $\overline{f}$ is a very convenient abuse of notation.

and $\overline{\mathbb{F}} \, f^\flat \cdot \overline{\mu_\mathbb{F}} = \overline{\mu_\mathbb{F}} \cdot (\overline{\mathbb{F}}^2 \, f^\flat)$ are ensured by two "monad-monad" compatibility conditions:[14]

$$\lambda \cdot \eta_\mathbb{F} = \mathbb{T}\eta_\mathbb{F} \tag{34}$$

$$\mathbb{T}\mu_\mathbb{F} \cdot \lambda \cdot \mathbb{F}\lambda = \lambda \cdot \mu_\mathbb{F} \tag{35}$$

Recall that, in our component algebra illustration, $\mathbb{F}$ is the *maybe* monad $\mathbb{M}$ and $\mathbb{T}$ is one of either the powerset or distribution monads. From a result in [12] it can immediately be shown that the distributive law $\lambda : 1 + \mathbb{T} \, X \to \mathbb{T} \, (1 + X)$ between $\mathbb{M}$ and any other monad $\mathbb{T}$, $\lambda = [\eta_\mathbb{T} \cdot i_1, \mathbb{T} \, i_2]$, satisfies (31,32) in both cases.[15]

It is also easy to show that $\mathbb{M}$ satisfies (34,35) for any $\mathbb{T}$. [16] So nondeterministic (resp. probabilistic) composition of $\mathbb{M}$-monadic Mealy machines regarded as binary relations (resp. matrices) given by monadic definitions (21) (resp. (27)) is sound, where $\mathbb{M}$ can be *generalized* to any $\mathbb{T}$-liftable monad $\mathbb{F}$ satisfying (34,35).

In retrospect, recall from the motivation that we went as far as simulating probabilistic composition of $\mathbb{M}$-machines in Haskell using the operator $m_1 \, ;_D \, m_2$, the probabilistic evolution of $m_1 ; m_2$ (8). Although we have not seen its actual definition, we can say that fact $\lceil m_1 \, ;_D \, m_2 \rceil = \lceil m_1 \rceil \, ; \, \lceil m_2 \rceil$ holds, where composition $\lceil m_1 \rceil \, ; \, \lceil m_2 \rceil$ is given by (27).[17]

Instead of staying in the *original category* and elaborating the definition to the probabilistic case we have kept the *original definition* by changing category. The advantage is that all *probabilistic accounting* is silently carried out by (monadic) matrix composition and is not our concern.

9 Strong Monads in Relational/Linear Algebra

We are not yet done, however: definition (27) is *strongly* monadic and we need to know in what sense *strength* is preserved through Kleisli-lifting. The question is, which strong monads ($\mathbb{F}$) are still strong once lifted to the Kleisli category of another monad ($\mathbb{T}$)? Recall that the two strengths

$$\tau_l : B \times \mathbb{F} \, A \to \mathbb{F} \, (B \times A)$$
$$\tau_r : \mathbb{F} \, A \times B \to \mathbb{F} \, (A \times B)$$

distribute context (B) across $\mathbb{F}$-data structures. Their basic properties, $\mathbb{F} \, \mathsf{lft} \cdot \tau_r = \mathsf{lft}$ and $\mathbb{F} \, \mathsf{a}^\circ \cdot \tau_r = \tau_r \cdot (\tau_r \times id) \cdot \mathsf{a}^\circ$ (similarly for τ_l) are preserved by their liftings $\overline{\tau_r}$ and $\overline{\tau_l}$, recall footnote 13. So, what may fail is their *naturality*, e.g.

$$\overline{\tau_l} \cdot (N \otimes \overline{\mathbb{F}} \, M) = \overline{\mathbb{F}} \, (N \otimes M) \cdot \overline{\tau_l} \tag{36}$$

---

[14] See e.g. [28], among the literature emerging from [4].

[15] This happens because the powerset and distribution monads are *commutative*.

[16] Concerning (34): $[\eta_\mathbb{T} \cdot i_1, \mathbb{T} \, i_2] \cdot i_2 = \mathbb{T} \, i_2$; concerning (35): $\mathbb{T} \, [i_1, id] \cdot [\eta_\mathbb{T} \cdot i_1, \mathbb{T} \, i_2 \cdot \lambda] = [\eta_\mathbb{T} \cdot [i_1, id] \cdot i_1, \lambda] = [\lambda \cdot i_1, \lambda] = \lambda \cdot [i_1, id]$.

[17] The actual implementation of $\cdot ;_D \cdot$ in the Haskell simulator follows *verbatim* pointfree formula (27) carefully using the encodings of section 8.

where M and N are arbitrary column-stochastic matrices. This is important because strength naturality is essential to many proofs of the component algebra of [2], for instance to that of (14). All proofs go in the style of that given for (13), with matrices in the place of functions.

Let us investigate (36) for $\mathbb{F} = \mathbb{M}$, in which case we have $\overline{\tau_l} = (\overline{!} \oplus id) \cdot \overline{\mathsf{dr}}$, of type $B \times (1 + A) \to 1 + B \times A$, where $\mathsf{dr} : A \times (C + B) \to A \times C + A \times B$ is the obvious isomorphism and $! : A \to 1$ lifts the "bang" function to the row vector of its type wholly filled with 1s.

The naturality of τ_l (hereupon we drop the lifting bars, under the convention we have used before) arises from that of dr and of $! \oplus id$. The naturality of dr is easy to prove from that of its converse using relational/matrix biproducts [17]. Concerning $! \oplus id$:

$$(id \oplus N) \cdot (! \oplus id) = (! \oplus id) \cdot (M \oplus N)$$

$$\equiv \quad \{ \text{ bifunctor } \cdot \oplus \cdot \}$$

$$! \oplus N = (! \cdot M) \oplus N$$

$$\equiv \quad \{ \ ! \cdot M = ! \text{ because } M \text{ is assumed column stochastic } [22] \ \}$$

$$true$$

The calculation for τ_r is similar. Thus $\overline{\mathbb{M}}$ is strong.

Note, however, that not every natural transformation remains natural once "Kleisli lifted". A very simple example is the diagonal function $\Delta : A \to A \times A$, $\Delta\, a = (a, a)$, shown aside for $A = \mathbb{B}$. Its natural property, $(M \otimes M) \cdot \Delta = \Delta \cdot M$ does not hold because $(y, z)\ ((M \otimes M) \cdot \Delta)\ x = (y\ M\ x) \times (z\ M\ x)$ on the

	F	T
(F, F)	1	0
(F, T)	0	0
(T, F)	0	0
(T, T)	0	1

left hand side, and $(y, z)\ (\Delta \cdot M)\ x = (\text{if } y = z \text{ then } y\ M\ x \text{ else } 0)$ on the right hand side. Thus the distributions captured by $(M \otimes M) \cdot \Delta$ have, in general, larger support. The same happens, of course, for relations: $(R \otimes R) \cdot \Delta \supseteq \Delta \cdot R$ holds but the converse inclusion does not.

In the terminology of categorial physics, Δ fails to be a *uniform copying operation* [8]. This has to do with the fact that the *pairing* operator $(f \vartriangle g)\ a = (f\ a, g\ a)$ (note that $\Delta = id \vartriangle id$) does not form a categorial product once Kleisli-lifted [20]. The corresponding matrix operation is the so-called Khatri-Rao matrix product, defined by $(b, c)\ (M \vartriangle N)\ a = (b\ M\ c) \times (c\ N\ a)$. In relational algebra it is known as (strict) *fork* [11, 25]. [18]

10 Conclusions and Future Work

Faced with the need to quantify software (un)reliability in presence of faults arising from (intentionally) inexact hardware, the semantics of software systems has to evolve towards weighted nondeterminism, for instance in a probabilistic way.

[18] Both *Khatri-Rao* and *fork* can be regarded as the lifting of the pairing operator, $f^\flat \,\overline{\vartriangle}\, g^\flat = (dstr \cdot (f \vartriangle g))^\flat$ where $dstr$ denotes the "double strength" of a commutative monad [12], a class of monads which includes both $\mathbb{D}$ and $\mathbb{P}$.

This paper proposes that such *semantics evolution* be obtained without sacrificing the simplicity of the original (qualitative) semantics definition. The idea is to keep quantification *implicit* rather than explicit, the trick being a change of category: instead of the category of sets where traditional (e.g. coalgebraic) semantics is expressed, we change to a suitable category (e.g. of matrices) tuned to the specific quantitative (e.g. probabilistic) effect.

Technically, this *"keep definition, change category"* approach consists of investing in the Kleisli category of the monad chosen to capture the new (e.g. quantitative) effect. The approach is useful because such a *Kleisli lifting* leads to rich algebraic theories: to *relational algebra* and *linear algebra*[19] in particular, both offering a useful *pointfree styled* calculus.

The approach is illustrated in the paper by enriching an existing software component calculus with fault propagation, by lifting it through a discrete distribution monad. As the original semantics are already monadic and coalgebraic, "keeping the definitions" entails monad-monad lifting.

Ideally, the proposed Kleisli-lifting should preserve theories, not only definitions (the theory of component behavioural equivalence of [2], in our case). But things are not so immediate in presence of *tupling* (cf. strong monads) as products become *weak* once lifted. Weak tupling calls for a wider perspective, interestingly bridging relational algebra to *categorial quantum physics* under the umbrella of *monoidal* categories. Thus the remarks by Coecke and Paquette, in their *Categories for the Practising Physicist* [8]:

> *Rel* [the category of relations] possesses more 'quantum features' than the category *Set* of sets and functions [...] The categories *FdHilb* [of finite dimensional Hilbert spaces] and *Rel* moreover admit a categorical matrix calculus.

Future work. This paper is part of a research line aiming at promoting linear algebra as the "natural" evolution of (pointfree) relational algebra towards quantitative reasoning in the software sciences. Work in this direction is still in its infancy [21, 17, 22, 20].

A full-fledged coalgebraic trace semantics for probabilistic, component oriented software systems will call for sub-distributions and, more generally, to measure theory [12, 23, 14]. The main result of [12] — that the (final) behaviour coalgebra in a Kleisli category is given by an initial algebra in sets — is central to the approach.

The connection to categorial quantum physics and monoidal categories [8] should be exploited, in particular concerning partial orders defined for quantum states which could be used to support a notion of refinement.

On the applications side, it would be interesting to address case studies such as that of [18], the verification of a persistent memory manager (in IBM's 4765 secure coprocessor) in face of restarts and hardware failures, using probabilistic component algebra. As the authors of [18] write, the inclusion of hardware failures incurs a significant jump in system complexity.

---

[19] This carries over to more sophisticated algebras and monads, for instance that of *stochastic relations*, the "Kleisli lifting" of the Giry monad [23].

Acknowledgments. This work is funded by ERDF - European Regional Development Fund through the COMPETE Programme (operational programme for competitiveness) and by National Funds through the *FCT - Fundação para a Ciência e a Tecnologia* (Portuguese Foundation for Science and Technology) within project FCOMP-01-0124-FEDER-020537.

Feedback and exchange of ideas with Tarmo Uustalu, Alexandra Silva and Luís Barbosa are gratefully acknowledged.

References

[1] Backhouse, R., Michaelis, D.: Exercises in quantifier manipulation. In: Uustalu, T. (ed.) MPC 2006. LNCS, vol. 4014, pp. 69–81. Springer, Heidelberg (2006)

[2] Barbosa, L.: Towards a Calculus of State-based Software Components. JUCS 9(8), 891–909 (2003)

[3] Barbosa, L., Oliveira, J.: Transposing Partial Components — An Exercise on Coalgebraic Refinement. Theor. Comp. Sci. 365(1), 2–22 (2006), http://dx.doi.org/10.1016/j.tcs.2006.07.030

[4] Beck, J.: Distributive laws. In: Eckmann, B. (ed.) Seminar on Triples and Categorical Homology Theory. Lecture Notes in Mathematics, vol. 80, pp. 119–140. Springer (1969)

[5] Bird, R., de Moor, O.: Algebra of Programming. Series in Computer Science. Prentice-Hall International (1997)

[6] Bonchi, F., Bonsangue, M., Boreale, M., Rutten, J., Silva, A.: A coalgebraic perspective on linear weighted automata. Inf. & Comp. 211, 77–105 (2012)

[7] Brink, C., Kahl, W., Schmidt, G. (eds.): Relational methods in computer science. Springer-Verlag New York, Inc., New York (1997)

[8] Coecke, B. (ed.): New Structures for Physics. Lecture Notes in Physics, vol. 831. Springer (2011)

[9] Cortellessa, V., Grassi, V.: A modeling approach to analyze the impact of error propagation on reliability of component-based systems. In: Schmidt, H.W., Crnković, I., Heineman, G.T., Stafford, J.A. (eds.) CBSE 2007. LNCS, vol. 4608, pp. 140–156. Springer, Heidelberg (2007)

[10] Erwig, M., Kollmansberger, S.: Functional pearls: Probabilistic functional programming in Haskell. J. Funct. Program. 16, 21–34 (2006)

[11] Frias, M., Baum, G., Haeberer, A.: Fork algebras in algebra, logic and computer science. Fundam. Inform., 1–25 (1997)

[12] Hasuo, I., Jacobs, B., Sokolova, A.: Generic trace semantics via coinduction. Logical Methods in Computer Science 3(4), 1–36 (2007)

[13] Kahl, W.: Refinement and development of programs from relational specifications. ENTCS 44(3), 4.1–4.43 (2003)

[14] Kerstan, H., König, B.: Coalgebraic trace semantics for probabilistic transition systems based on measure theory. In: Koutny, M., Ulidowski, I. (eds.) CONCUR 2012. LNCS, vol. 7454, pp. 410–424. Springer, Heidelberg (2012)

[15] Kock, A.: Strong functors and monoidal monads. Archiv der Mathematik 23(1), 113–120 (1972), http://dx.doi.org/10.1007/BF01304852

[16] Lingamneni, A., Enz, C., Palem, K., Piguet, C.: Synthesizing parsimonious inexact circuits through probabilistic design techniques. ACM Trans. Embed. Comput. Syst. 12(2s), 93:1–93:26 (2013)

[17] Macedo, H., Oliveira, J.: Typing linear algebra: A biproduct-oriented approach. Science of Computer Programming 78(11), 2160–2191 (2013)

[18] Marić, O., Sprenger, C.: Verification of a transactional memory manager under hardware failures and restarts (2013), conference paper (submitted)

[19] McIver, A., Morgan, C.: Abstraction, Refinement and Proof for Probabilistic Systems. Monographs in Computer Science. Springer (2005)

[20] Murta, D., Oliveira, J.N.: Calculating risk in functional programming. CoRR abs/1311.3687 (2013)

[21] Oliveira, J.: Towards a linear algebra of programming. Formal Aspects of Computing 24(4-6), 433–458 (2012)

[22] Oliveira, J.: Weighted automata as coalgebras in categories of matrices. Int. Journal of Found. of Comp. Science 24(06), 709–728 (2013)

[23] Panangaden, P.: Labelled Markov Processes. Imperial College Press (2009)

[24] Rutten, J.: Universal coalgebra: A theory of systems. Theor. Comp. Sci. 249(1), 3–80 (2000); Revised version of CWI Techn. Rep. CS-R9652 (1996)

[25] Schmidt, G.: Relational Mathematics. Encyclopedia of Mathematics and its Applications, vol. 132. Cambridge University Press (November 2010)

[26] Sokolova, A.: Probabilistic systems coalgebraically: A survey. Theor. Comput. Sci. 412(38), 5095–5110 (2011)

[27] Stamatelatos, M., Dezfuli, H.: Probabilistic Risk Assessment Procedures Guide for NASA Managers and Practitioners, NASA/SP-2011-3421, 2nd edn (December 2011)

[28] Tanaka, M.: Pseudo-Distributive Laws and a Unified Framework for Variable Binding. Ph.D. thesis, School of Informatics, University of Edinburgh (2005)

Extended Conscriptions Algebraically

Walter Guttmann

Department of Computer Science and Software Engineering,
University of Canterbury, New Zealand
walter.guttmann@canterbury.ac.nz

Abstract. Conscriptions are a model of sequential computations with
assumption/commitment specifications in which assumptions can refer
to final states, not just to initial states. We show that they instantiate
existing algebras for iteration and infinite computations. We use these algebras to derive an approximation order for conscriptions and one for extended conscriptions, which additionally represent aborting executions.
We give a new computation model which generalises extended conscriptions and apply the algebraic techniques for a unified treatment.

1   Introduction

Various relational models have been proposed for sequential computations, for
example, in [15,4,13,9,11]. The most precise of these models can represent finite, infinite and aborting executions independently of each other. Less precise
models ignore certain kinds of executions if others are present, but have simpler
descriptions. All of these models have two properties in common:

- They represent only the initial and final states of computations and disregard
  the intermediate states.
- Whenever they represent infinite or aborting executions, they only record if
  such executions are present or absent from each starting state.

The latter in particular means that there is no notion of a final state when it
comes to infinite or aborting executions. For several reasons it is desirable to
eliminate this restriction:

- Its removal provides a basis for more detailed models that involve time.
- Different interpretations of a model may replace infinite or aborting execu-
  tions with other kinds of executions for which final states are observable.
- A final state may be observable for a blocking computation which waits for
  external input that never arrives.
- A final state may be observable for an infinite execution, for example, if it
  stabilises and continues as an endless loop that does not change the state.
- A final state may be observable for an aborting execution, namely the last
  state before the execution aborts.
- The restriction is a technical constraint on the model: some entries in the
  matrix of relations that represent a computation have to be vectors, which
  are special relations. It is natural to generalise them to arbitrary relations.

P. Höfner et al. (Eds.): RAMiCS 2014, LNCS 8428, pp. 139–156, 2014.

Two models that lift the restriction have been described in [5]. *Conscriptions* represent infinite and finite executions independently, but have no notion of aborting executions. There is no restriction on the final states in the case of infinite executions. *Extended conscriptions* represent aborting, infinite and finite executions independently. There is no restriction on the final states in the case of aborting executions, but the restriction to vectors still applies for infinite executions. For both models, basic algebraic properties of sequential composition and non-deterministic choice have been derived.

The present paper extends this investigation by considering the following questions:

- How is recursion defined for the new computation models?
- Can extended conscriptions be further generalised by lifting the restriction on infinite executions?
- What algebraic structures underlie iteration in these models?
- Can all of these models be captured by a unifying algebraic theory?

We provide the following contributions:

- Several new computation models, the most precise of which represents finite, infinite and aborting executions independently of each other with no restrictions on the final states for any kind of execution.
- Instances of previously introduced algebras for iteration and infinite computations for each of these models. Consequences are a unified description of recursion and applications including separation and refinement theorems and various program transformations for the new models.
- An approximation order for the new models, which is used to define the semantics of recursion. The approximation orders known from previous models do not generalise in a direct way. Instead, the new models turn out to satisfy axioms of algebras previously developed for non-strict computations [12]; we use these algebras to obtain the approximation order.

Section 2 gives the basic algebraic structures referred to in the remainder of this paper. Section 3 derives an approximation order for conscriptions and shows how conscriptions instantiate algebraic structures to describe iterations and infinite executions. Section 4 applies the method of Section 3 to extended conscriptions. Section 5 introduces the most precise computation model, applies the method of Section 3 to it and shows how to specialise it to obtain other models.

2   Algebraic Structures for Sequential Computations

In this section we axiomatise the operations of non-deterministic choice, conjunction and sequential composition, the infinite executions of a computation and various forms of iteration featured by many computation models.

A *lattice-ordered semiring* is an algebraic structure $(S, +, \curlywedge, \cdot, 0, 1, \top)$ such that the following axioms hold:

$$x + (y + z) = (x + y) + z \qquad\qquad x \curlywedge (y \curlywedge z) = (x \curlywedge y) \curlywedge z$$
$$x + y = y + x \qquad\qquad x \curlywedge y = y \curlywedge x$$
$$x + x = x \qquad\qquad x \curlywedge x = x$$
$$0 + x = x \qquad\qquad \top \curlywedge x = x$$
$$x + (y \curlywedge z) = (x + y) \curlywedge (x + z) \qquad x \curlywedge (y + z) = (x \curlywedge y) + (x \curlywedge z)$$
$$x + (x \curlywedge y) = x \qquad\qquad x \curlywedge (x + y) = x$$
$$1 \cdot x = x \qquad\qquad x \cdot (y + z) = (x \cdot y) + (x \cdot z)$$
$$x \cdot 1 = x \qquad\qquad (x + y) \cdot z = (x \cdot z) + (y \cdot z)$$
$$x \cdot (y \cdot z) = (x \cdot y) \cdot z \qquad\qquad 0 \cdot x = 0$$

The axioms not involving $\cdot$ make up a *bounded distributive lattice* $(S, +, \curlywedge, 0, \top)$. The axioms not involving $\curlywedge$ make up an idempotent semiring without right annihilator $(S, +, \cdot, 0, 1)$, simply called *semiring* in the remainder of this paper. In particular, $x \cdot 0 = 0$ is not an axiom. The *lattice order* $x \leq y \Leftrightarrow x + y = y \Leftrightarrow x \curlywedge y = x$ has least element 0, greatest element $\top$, least upper bound $+$ and greatest lower bound $\curlywedge$. The operations $+$, $\curlywedge$ and $\cdot$ are $\leq$-isotone. We abbreviate $x \cdot y$ as xy.

In many computation models the operation $+$ represents non-deterministic choice, the operation $\curlywedge$ conjunction, the operation $\cdot$ sequential composition, 0 the computation with no executions, 1 the computation that does not change the state, $\top$ the computation with all executions, and $\leq$ the refinement relation.

The following algebras capture various fixpoints of the function $\lambda x.yx + z$, which are useful to describe iteration.

A *Kleene algebra* $(S, +, \cdot, {}^*, 0, 1)$ adds to a semiring an operation * with the following unfold and induction axioms [16]:

$$1 + yy^* \leq y^* \qquad\qquad z + yx \leq x \Rightarrow y^*z \leq x$$
$$1 + y^*y \leq y^* \qquad\qquad z + xy \leq x \Rightarrow zy^* \leq x$$

It follows that y^*z is the $\leq$-least fixpoint of $\lambda x.yx + z$ and that zy^* is the $\leq$-least fixpoint of $\lambda x.xy + z$. The operation * is $\leq$-isotone.

An *omega algebra* $(S, +, \cdot, {}^*, {}^\omega, 0, 1)$ adds to a Kleene algebra an operation $^\omega$ with the following unfold and induction axioms [2,17]:

$$yy^\omega = y^\omega \qquad\qquad x \leq yx + z \Rightarrow x \leq y^\omega + y^*z$$

It follows that $y^\omega + y^*z$ is the $\leq$-greatest fixpoint of $\lambda x.yx + z$. In particular, $\top = 1^\omega$ is the $\leq$-greatest element. The operation $^\omega$ is $\leq$-isotone.

For computation models that require different fixpoints of $\lambda x.yx + z$, we use the following generalisations of Kleene algebras.

An *extended binary itering* $(S, +, \cdot, \star, 0, 1)$ adds to a semiring a binary operation $\star$ with the following axioms [10]:

$$(x + y) \star z = (x \star y) \star (x \star z) \qquad x \star (y + z) = (x \star y) + (x \star z)$$
$$(xy) \star z = z + x((yx) \star (yz)) \qquad (x \star y)z \leq x \star (yz)$$

$$zx \leq y(y \star z) + w \Rightarrow z(x \star v) \leq y \star (zv + w(x \star v))$$
$$xz \leq z(y \star 1) + w \Rightarrow x \star (zv) \leq z(y \star v) + (x \star (w(y \star v)))$$
$$w(x \star (yz)) \leq (w(x \star y)) \star (w(x \star y)z)$$

It follows that $y \star z$ is a fixpoint of $\lambda x.yx + z$. The operation $\star$ is $\leq$-isotone. The element $y \star z$ corresponds to iterating y an unspecified number of times, followed by a single occurrence of z. This may involve an infinite number of iterations of y.

In models that satisfy $(x \star y)z = x \star (yz)$, the binary itering operation specialises to a unary operation $^\circ$ with the following simpler axioms. An *itering* $(S, +, \cdot, ^\circ, 0, 1)$ adds to a semiring an operation $^\circ$ with the sumstar and product-star equations of [3] and two simulation axioms [9]:

$$(x + y)^\circ = (x^\circ y)^\circ x^\circ \qquad\qquad zx \leq yy^\circ z + w \Rightarrow zx^\circ \leq y^\circ(z + wx^\circ)$$
$$(xy)^\circ = 1 + x(yx)^\circ y \qquad\qquad xz \leq zy^\circ + w \Rightarrow x^\circ z \leq (z + x^\circ w)y^\circ$$

It follows that $y^\circ z$ is a fixpoint of $\lambda x.yx + z$ and that zy° is a fixpoint of $\lambda x.xy + z$. The operation $^\circ$ is $\leq$-isotone.

Every Kleene algebra is an itering using $x^\circ = x^*$. Every omega algebra is an itering using $x^\circ = x^\omega 0 + x^*$. Every itering is an extended binary itering using $x \star y = x^\circ y$. Further instances and consequences of iterings are given in [9,10].

We finally describe the set of states $n(x)$ from which a computation x has infinite executions. Sets are represented as tests, that is, as elements ≤ 1. The axioms have been developed in [12] for a unified treatment of strict and non-strict computations; the latter can produce defined outputs from undefined inputs. An axiomatisation for strict computations has been given in [9].

An *n-algebra* $(S, +, \curlywedge, \cdot, n, 0, 1, \mathsf{L}, \top)$ adds to a lattice-ordered semiring an operation $n : S \to S$ and a constant L with the following axioms:

$(n1)$	$n(x) + n(y) = n(n(x)\top + y)$	$(n6)$	$n(x) \leq n(\mathsf{L}) \curlywedge 1$
$(n2)$	$n(x)n(y) = n(n(x)y)$	$(n7)$	$n(x)\mathsf{L} \leq x$
$(n3)$	$n(x)n(x + y) = n(x)$	$(n8)$	$n(\mathsf{L})x \leq xn(\mathsf{L})$
$(n4)$	$n(\mathsf{L})x = (x \curlywedge \mathsf{L}) + n(\mathsf{L}0)x$	$(n9)$	$xn(y)\top \leq x0 + n(xy)\top$
$(n5)$	$x\mathsf{L} = x0 + n(x\mathsf{L})\mathsf{L}$	$(n10)$	$x\top y \curlywedge \mathsf{L} \leq x\mathsf{L}y$

An *n-omega algebra* $(S, +, \curlywedge, \cdot, n, *, \omega, 0, 1, \mathsf{L}, \top)$ adds the following axioms to an n-algebra $(S, +, \curlywedge, \cdot, n, 0, 1, \mathsf{L}, \top)$ and an omega algebra $(S, +, \cdot, *, \omega, 0, 1)$:

$$(n11) \quad n(\mathsf{L})x^\omega \leq x^* n(x^\omega)\top \qquad\qquad (n12) \quad x\mathsf{L} \leq x\mathsf{L}x\mathsf{L}$$

The constant L represents the endless loop, that is, the computation with all infinite executions. A constant for the computation with all aborting executions is not provided.

3   Conscriptions

The state space A of a sequential computation is given by the values of the program variables. A computation is thus represented as a relation R on A, that is, as a subset of the Cartesian product $A \times A$. A pair $(x, x') \in R$ signifies that there is a finite execution of the program which starts in state x and ends in state x'; in other words, x' is a possible output for input x. Several outputs for the same input indicate non-determinism.

This simple model is sufficient for partial correctness, but does not provide means to represent infinite executions. For the latter, extended models can be used which represent computations as assumption/commitment pairs. The assumption part specifies the conditions under which termination is guaranteed and the commitment part specifies the effect if the program terminates. The conditions of termination traditionally refer only to the pre-state x of the computation, not to its post-state x'. As discussed in the introduction, conscriptions are introduced in [5] to eliminate this restriction.

A *conscription* is a 2×2 matrix whose entries are relations over A. The matrix has the following form:

$$\begin{pmatrix} 1 & 0 \\ Q & R \end{pmatrix}$$

The entries in the top row are the identity relation 1 and the empty relation 0, respectively, for each conscription. Only the entries in the bottom row can vary: the relation Q represents the complement of the assumption (the infinite executions) and the relation R represents the commitment (the finite executions). No restrictions are placed on Q and R. This is in contrast to other models such as the designs of [15], the prescriptions of [4] and the extended designs of [13]. They require Q to be a vector, that is, in those models Q relates every state x either to all states x' or to no state.

Sequential composition and non-deterministic choice of conscriptions are given by matrix product and componentwise union, respectively. The refinement order on conscriptions is the componentwise set inclusion order. The computation which does not change the state and the endless loop are represented by the conscriptions

$$\mathsf{skip} = \begin{pmatrix} 1 & 0 \\ 0 & 1 \end{pmatrix} \qquad \mathsf{L} = \begin{pmatrix} 1 & 0 \\ \top & 0 \end{pmatrix}$$

where $\top$ is the universal relation. The conscription skip is a neutral element of sequential composition and L is a left annihilator.

We wish to define the semantics of recursion by least fixpoints in a suitable approximation order $\sqsubseteq$. Because the endless loop L has to be the $\sqsubseteq$-least element, the refinement order cannot be used for approximation.

3.1   An Approximation Order for Conscriptions: Two Attempts

In the following we discuss two attempts to define an approximation order for conscriptions and the reasons why they fail. The first attempt is to take the approximation order of prescriptions [7,6]. Prescriptions correspond to a subset of conscriptions in which the assumption component Q is a vector. It is therefore natural to assume that the approximation order for conscriptions specialises to the approximation order for prescriptions when restricted to this subset. The approximation order $\sqsubseteq_1$ on conscriptions would accordingly be defined as

$$\begin{pmatrix} 1 & 0 \\ Q_1 & R_1 \end{pmatrix} \sqsubseteq_1 \begin{pmatrix} 1 & 0 \\ Q_2 & R_2 \end{pmatrix} \Leftrightarrow Q_2 \subseteq Q_1 \wedge R_1 \subseteq R_2 \subseteq R_1 \cup Q_1$$

The intuition underlying $\sqsubseteq_1$ is that in states with infinite executions, finite executions can be added but only such that have the same output. A problem with $\sqsubseteq_1$ is that sequential composition from the right is not $\sqsubseteq_1$-isotone. Namely,

$$\begin{pmatrix} 1 & 0 \\ 1 & 0 \end{pmatrix} \sqsubseteq_1 \begin{pmatrix} 1 & 0 \\ 0 & 1 \end{pmatrix}$$

but

$$\begin{pmatrix} 1 & 0 \\ 1 & 0 \end{pmatrix}\begin{pmatrix} 1 & 0 \\ \top & \top \end{pmatrix} = \begin{pmatrix} 1 & 0 \\ 1 & 0 \end{pmatrix} \not\sqsubseteq_1 \begin{pmatrix} 1 & 0 \\ \top & \top \end{pmatrix} = \begin{pmatrix} 1 & 0 \\ 0 & 1 \end{pmatrix}\begin{pmatrix} 1 & 0 \\ \top & \top \end{pmatrix}$$

The second attempt converts conscriptions to prescriptions and takes their order:

$$\begin{pmatrix} 1 & 0 \\ Q_1 & R_1 \end{pmatrix} \sqsubseteq_2 \begin{pmatrix} 1 & 0 \\ Q_2 & R_2 \end{pmatrix} \Leftrightarrow \begin{pmatrix} 1 & 0 \\ Q_1\top & R_1 \end{pmatrix} \sqsubseteq_1 \begin{pmatrix} 1 & 0 \\ Q_2\top & R_2 \end{pmatrix}$$

The assumptions are converted to vectors by composing them with $\top$. Hence the resulting prescription has an infinite execution from state x if the original conscription has any infinite execution starting in x. The intuition underlying $\sqsubseteq_2$ is that in states with any infinite executions, any finite execution can be added. A problem with $\sqsubseteq_2$ is that it is not antisymmetric. Namely,

$$\begin{pmatrix} 1 & 0 \\ 1 & 0 \end{pmatrix} \sqsubseteq_2 \begin{pmatrix} 1 & 0 \\ \top & 0 \end{pmatrix} \quad \text{and} \quad \begin{pmatrix} 1 & 0 \\ \top & 0 \end{pmatrix} \sqsubseteq_2 \begin{pmatrix} 1 & 0 \\ 1 & 0 \end{pmatrix}$$

More generally, it is inconsistent to assume all of the following four properties of an approximation relation $\sqsubseteq$ for conscriptions:

- $\sqsubseteq$ is a partial order,
- sequential composition from the right is $\sqsubseteq$-isotone,
- $\begin{pmatrix} 1 & 0 \\ \top & 0 \end{pmatrix} \sqsubseteq \begin{pmatrix} 1 & 0 \\ 0 & 1 \end{pmatrix}$,
- $\begin{pmatrix} 1 & 0 \\ 1 & 0 \end{pmatrix} \sqsubseteq \begin{pmatrix} 1 & 0 \\ 0 & 1 \end{pmatrix}$.

This is because they would imply

$$\begin{pmatrix} 1 & 0 \\ 1 & 0 \end{pmatrix} = \begin{pmatrix} 1 & 0 \\ 1 & 0 \end{pmatrix}\begin{pmatrix} 1 & 0 \\ \top & 0 \end{pmatrix} \sqsubseteq \begin{pmatrix} 1 & 0 \\ 0 & 1 \end{pmatrix}\begin{pmatrix} 1 & 0 \\ \top & 0 \end{pmatrix} = \begin{pmatrix} 1 & 0 \\ \top & 0 \end{pmatrix}$$

and

$$\begin{pmatrix} 1 & 0 \\ \top & 0 \end{pmatrix} = \begin{pmatrix} 1 & 0 \\ \top & 0 \end{pmatrix}\begin{pmatrix} 1 & 0 \\ 1 & 0 \end{pmatrix} \sqsubseteq \begin{pmatrix} 1 & 0 \\ 0 & 1 \end{pmatrix}\begin{pmatrix} 1 & 0 \\ 1 & 0 \end{pmatrix} = \begin{pmatrix} 1 & 0 \\ 1 & 0 \end{pmatrix}$$

and therefore

$$\begin{pmatrix} 1 & 0 \\ 1 & 0 \end{pmatrix} = \begin{pmatrix} 1 & 0 \\ \top & 0 \end{pmatrix}$$

The approximation relation we subsequently derive satisfies the first three properties of the above list, but not the last one.

3.2 An Approximation Order for Conscriptions

To obtain a suitable approximation order for conscriptions, we use the algebraic method developed in [9,12]. We first observe the following basic structure.

Theorem 1. *Conscriptions form a lattice-ordered semiring with the operations*

$$\begin{pmatrix} 1 & 0 \\ Q_1 & R_1 \end{pmatrix} + \begin{pmatrix} 1 & 0 \\ Q_2 & R_2 \end{pmatrix} = \begin{pmatrix} 1 & 0 \\ Q_1 \cup Q_2 & R_1 \cup R_2 \end{pmatrix} \qquad 0 = \begin{pmatrix} 1 & 0 \\ 0 & 0 \end{pmatrix}$$

$$\begin{pmatrix} 1 & 0 \\ Q_1 & R_1 \end{pmatrix} \curlywedge \begin{pmatrix} 1 & 0 \\ Q_2 & R_2 \end{pmatrix} = \begin{pmatrix} 1 & 0 \\ Q_1 \cap Q_2 & R_1 \cap R_2 \end{pmatrix} \qquad \top = \begin{pmatrix} 1 & 0 \\ \top & \top \end{pmatrix}$$

$$\begin{pmatrix} 1 & 0 \\ Q_1 & R_1 \end{pmatrix} \cdot \begin{pmatrix} 1 & 0 \\ Q_2 & R_2 \end{pmatrix} = \begin{pmatrix} 1 & 0 \\ Q_1 \cup R_1 Q_2 & R_1 R_2 \end{pmatrix} \qquad 1 = \begin{pmatrix} 1 & 0 \\ 0 & 1 \end{pmatrix}$$

Proof. The claims follow by simple calculations since $+$ and $\curlywedge$ are defined componentwise and $\cdot$ is the matrix product. $\qquad\square$

The lattice order $\leq$ on conscriptions therefore amounts to componentwise inclusion. The algebraic approach to approximation is based on the operation n that represents the infinite executions of a computation as a test, that is, as an element ≤ 1. For conscriptions, tests take the form

$$\begin{pmatrix} 1 & 0 \\ 0 & R \end{pmatrix}$$

where $R \subseteq 1$. The operation n that maps semiring elements to tests is characterised by the Galois connection

$$n(x)\mathsf{L} \leq y \Leftrightarrow n(x) \leq n(y)$$

Hence $n(y)$ is the greatest test whose composition with L is below y. We use this Galois connection to obtain a definition of n for conscriptions. Because the result of n is a test, assume that n is given by the general form

$$n\begin{pmatrix} 1 & 0 \\ Q & R \end{pmatrix} = \begin{pmatrix} 1 & 0 \\ 0 & f(Q,R) \end{pmatrix}$$

using a function f that maps its argument relations Q and R to a relation below the identity 1, that is, $f(Q,R) \subseteq 1$. By the Galois connection,

$$f(Q_1,R_1) \subseteq f(Q_2,R_2) \Leftrightarrow \begin{pmatrix} 1 & 0 \\ 0 & f(Q_1,R_1) \end{pmatrix} \leq \begin{pmatrix} 1 & 0 \\ 0 & f(Q_2,R_2) \end{pmatrix}$$

$$\Leftrightarrow n\begin{pmatrix} 1 & 0 \\ Q_1 & R_1 \end{pmatrix} \leq n\begin{pmatrix} 1 & 0 \\ Q_2 & R_2 \end{pmatrix} \Leftrightarrow n\begin{pmatrix} 1 & 0 \\ Q_1 & R_1 \end{pmatrix}\mathsf{L} \leq \begin{pmatrix} 1 & 0 \\ Q_2 & R_2 \end{pmatrix}$$

$$\Leftrightarrow \begin{pmatrix} 1 & 0 \\ 0 & f(Q_1,R_1) \end{pmatrix}\begin{pmatrix} 1 & 0 \\ \top & 0 \end{pmatrix} \leq \begin{pmatrix} 1 & 0 \\ Q_2 & R_2 \end{pmatrix} \Leftrightarrow \begin{pmatrix} 1 & 0 \\ f(Q_1,R_1)\top & 0 \end{pmatrix} \leq \begin{pmatrix} 1 & 0 \\ Q_2 & R_2 \end{pmatrix}$$

$$\Leftrightarrow f(Q_1,R_1)\top \subseteq Q_2 \Leftrightarrow f(Q_1,R_1) \subseteq \overline{Q_2\top} \Leftrightarrow f(Q_1,R_1) \subseteq \overline{Q_2\top} \cap 1$$

The operation $\bar{}$ is the relational complement. The next-to-last step is a consequence of a Schröder equivalence.

The above calculation suggests the definition $f(Q,R) = \overline{\overline{QT}} \cap 1$. A simple rearrangement of the calculation shows that this satisfies the Galois connection. We therefore define

$$n\begin{pmatrix} 1 & 0 \\ Q & R \end{pmatrix} = \begin{pmatrix} 1 & 0 \\ 0 & \overline{\overline{QT}} \cap 1 \end{pmatrix}$$

In [9] we give an approximation order in terms of n that covers a range of models of strict computations including prescriptions and extended designs. The present definition of n for conscriptions does not satisfy the axioms $n(x+y) = n(x)+n(y)$ and $xn(y)L = x0 + n(xy)L$ used there. However, n satisfies the weaker axioms given in [12], which have been developed to uniformly describe strict and non-strict computations.

Theorem 2. *Conscriptions form an n-algebra with the operations*

$$n\begin{pmatrix} 1 & 0 \\ Q & R \end{pmatrix} = \begin{pmatrix} 1 & 0 \\ 0 & \overline{\overline{QT}} \cap 1 \end{pmatrix} \qquad L = \begin{pmatrix} 1 & 0 \\ T & 0 \end{pmatrix}$$

Proof. Observe that $n(L) = 1$ and $Lx = L$; this implies axioms $(n4)$, $(n6)$ and $(n8)$. The remaining axioms are shown as follows. See [20] for properties of relations used in the calculations.

$(n1)$

$$n\left(n\begin{pmatrix} 1 & 0 \\ Q_1 & R_1 \end{pmatrix}\begin{pmatrix} 1 & 0 \\ T & T \end{pmatrix} + \begin{pmatrix} 1 & 0 \\ Q_2 & R_2 \end{pmatrix} \right)$$

$$= n\left(\begin{pmatrix} 1 & 0 \\ 0 & \overline{\overline{Q_1 T}} \cap 1 \end{pmatrix}\begin{pmatrix} 1 & 0 \\ T & T \end{pmatrix} + \begin{pmatrix} 1 & 0 \\ Q_2 & R_2 \end{pmatrix} \right) = n\begin{pmatrix} 1 & 0 \\ \overline{\overline{Q_1 T}} \cup Q_2 & \overline{\overline{Q_1 T}} \cup R_2 \end{pmatrix}$$

$$= \begin{pmatrix} 1 & 0 \\ 0 & \overline{\overline{\overline{\overline{Q_1 T}} \cup Q_2 T}} \cap 1 \end{pmatrix} = \begin{pmatrix} 1 & 0 \\ 0 & \overline{\overline{Q_1 T}} \cap \overline{\overline{Q_2 T}} \cap 1 \end{pmatrix} = \begin{pmatrix} 1 & 0 \\ 0 & (\overline{\overline{Q_1 T}} \cup \overline{\overline{Q_2 T}}) \cap 1 \end{pmatrix}$$

$$= \begin{pmatrix} 1 & 0 \\ 0 & \overline{\overline{Q_1 T}} \cap 1 \end{pmatrix} + \begin{pmatrix} 1 & 0 \\ 0 & \overline{\overline{Q_2 T}} \cap 1 \end{pmatrix} = n\begin{pmatrix} 1 & 0 \\ Q_1 & R_1 \end{pmatrix} + n\begin{pmatrix} 1 & 0 \\ Q_2 & R_2 \end{pmatrix}$$

$(n2)$

$$n\left(n\begin{pmatrix} 1 & 0 \\ Q_1 & R_1 \end{pmatrix}\begin{pmatrix} 1 & 0 \\ Q_2 & R_2 \end{pmatrix} \right) = n\left(\begin{pmatrix} 1 & 0 \\ 0 & \overline{\overline{Q_1 T}} \cap 1 \end{pmatrix}\begin{pmatrix} 1 & 0 \\ Q_2 & R_2 \end{pmatrix} \right)$$

$$= n\begin{pmatrix} 1 & 0 \\ (\overline{\overline{Q_1 T}} \cap 1)Q_2 & (\overline{\overline{Q_1 T}} \cap 1)R_2 \end{pmatrix} = n\begin{pmatrix} 1 & 0 \\ \overline{\overline{Q_1 T}} \cap Q_2 & \overline{\overline{Q_1 T}} \cap R_2 \end{pmatrix}$$

$$= \begin{pmatrix} 1 & 0 \\ 0 & \overline{\overline{\overline{Q_1 T}} \cap Q_2 T} \cap 1 \end{pmatrix} = \begin{pmatrix} 1 & 0 \\ 0 & \overline{\overline{Q_1 T}} \cap \overline{\overline{Q_2 T}} \cap 1 \end{pmatrix} = \begin{pmatrix} 1 & 0 \\ 0 & (\overline{\overline{Q_1 T}} \cap 1)(\overline{\overline{Q_2 T}} \cap 1) \end{pmatrix}$$

$$= \begin{pmatrix} 1 & 0 \\ 0 & \overline{\overline{Q_1 T}} \cap 1 \end{pmatrix}\begin{pmatrix} 1 & 0 \\ 0 & \overline{\overline{Q_2 T}} \cap 1 \end{pmatrix} = n\begin{pmatrix} 1 & 0 \\ Q_1 & R_1 \end{pmatrix} n\begin{pmatrix} 1 & 0 \\ Q_2 & R_2 \end{pmatrix}$$

(n3) The calculation uses that $\overline{\overline{Q_1\mathsf{T}}} \subseteq \overline{\overline{Q_1 \cup Q_2 \mathsf{T}}}$:

$$n\begin{pmatrix} 1 & 0 \\ Q_1 & R_1 \end{pmatrix} n\left(\begin{pmatrix} 1 & 0 \\ Q_1 & R_1 \end{pmatrix} + \begin{pmatrix} 1 & 0 \\ Q_2 & R_2 \end{pmatrix}\right) = n\begin{pmatrix} 1 & 0 \\ Q_1 & R_1 \end{pmatrix} n\begin{pmatrix} 1 & 0 \\ Q_1 \cup Q_2 & R_1 \cup R_2 \end{pmatrix}$$

$$= \begin{pmatrix} 1 & 0 \\ 0 & \overline{Q_1\mathsf{T}} \cap 1 \end{pmatrix}\begin{pmatrix} 1 & 0 \\ 0 & \overline{Q_1 \cup Q_2 \mathsf{T}} \cap 1 \end{pmatrix} = \begin{pmatrix} 1 & 0 \\ 0 & (\overline{Q_1\mathsf{T}} \cap 1)(\overline{Q_1 \cup Q_2 \mathsf{T}} \cap 1) \end{pmatrix}$$

$$= \begin{pmatrix} 1 & 0 \\ 0 & \overline{Q_1\mathsf{T}} \cap \overline{Q_1 \cup Q_2 \mathsf{T}} \cap 1 \end{pmatrix} = \begin{pmatrix} 1 & 0 \\ 0 & \overline{Q_1\mathsf{T}} \cap 1 \end{pmatrix} = n\begin{pmatrix} 1 & 0 \\ Q_1 & R_1 \end{pmatrix}$$

(n5) The calculation uses that $\overline{\overline{Q\mathsf{T}}} \subseteq Q$:

$$\begin{pmatrix} 1 & 0 \\ Q & R \end{pmatrix}\begin{pmatrix} 1 & 0 \\ 0 & 0 \end{pmatrix} + n\left(\begin{pmatrix} 1 & 0 \\ Q & R \end{pmatrix}\begin{pmatrix} 1 & 0 \\ \mathsf{T} & 0 \end{pmatrix}\right)\begin{pmatrix} 1 & 0 \\ \mathsf{T} & 0 \end{pmatrix}$$

$$= \begin{pmatrix} 1 & 0 \\ Q & 0 \end{pmatrix} + n\begin{pmatrix} 1 & 0 \\ Q \cup R\mathsf{T} & 0 \end{pmatrix}\begin{pmatrix} 1 & 0 \\ \mathsf{T} & 0 \end{pmatrix} = \begin{pmatrix} 1 & 0 \\ Q & 0 \end{pmatrix} + \begin{pmatrix} 1 & 0 \\ 0 & \overline{Q \cup R\mathsf{T}\mathsf{T}} \cap 1 \end{pmatrix}\begin{pmatrix} 1 & 0 \\ \mathsf{T} & 0 \end{pmatrix}$$

$$= \begin{pmatrix} 1 & 0 \\ Q & 0 \end{pmatrix} + \begin{pmatrix} 1 & 0 \\ \overline{\overline{Q \cup R\mathsf{T}\mathsf{T}}} & 0 \end{pmatrix} = \begin{pmatrix} 1 & 0 \\ Q \cup \overline{\overline{Q\mathsf{T}}} \cap \overline{\overline{R\mathsf{T}}} & 0 \end{pmatrix} = \begin{pmatrix} 1 & 0 \\ Q \cup \overline{\overline{Q\mathsf{T}}} \cup R\mathsf{T} & 0 \end{pmatrix}$$

$$= \begin{pmatrix} 1 & 0 \\ Q \cup R\mathsf{T} & 0 \end{pmatrix} = \begin{pmatrix} 1 & 0 \\ Q & R \end{pmatrix}\begin{pmatrix} 1 & 0 \\ \mathsf{T} & 0 \end{pmatrix}$$

(n7)

$$n\begin{pmatrix} 1 & 0 \\ Q & R \end{pmatrix}\begin{pmatrix} 1 & 0 \\ \mathsf{T} & 0 \end{pmatrix} = \begin{pmatrix} 1 & 0 \\ 0 & \overline{Q\mathsf{T}} \cap 1 \end{pmatrix}\begin{pmatrix} 1 & 0 \\ \mathsf{T} & 0 \end{pmatrix} = \begin{pmatrix} 1 & 0 \\ \overline{\overline{Q\mathsf{T}}} & 0 \end{pmatrix} \leq \begin{pmatrix} 1 & 0 \\ Q & R \end{pmatrix}$$

(n9) The calculation uses that $\overline{\overline{R_1 \overline{Q_2\mathsf{T}}}} \subseteq \overline{\overline{R_1 Q_2 \mathsf{T}}} \subseteq \overline{\overline{Q_1 \cup R_1 Q_2 \mathsf{T}}}$:

$$\begin{pmatrix} 1 & 0 \\ Q_1 & R_1 \end{pmatrix} n\begin{pmatrix} 1 & 0 \\ Q_2 & R_2 \end{pmatrix}\begin{pmatrix} 1 & 0 \\ \mathsf{T} & \mathsf{T} \end{pmatrix} = \begin{pmatrix} 1 & 0 \\ Q_1 & R_1 \end{pmatrix}\begin{pmatrix} 1 & 0 \\ 0 & \overline{Q_2\mathsf{T}} \cap 1 \end{pmatrix}\begin{pmatrix} 1 & 0 \\ \mathsf{T} & \mathsf{T} \end{pmatrix}$$

$$= \begin{pmatrix} 1 & 0 \\ Q_1 & R_1 \end{pmatrix}\begin{pmatrix} 1 & 0 \\ \overline{Q_2\mathsf{T}} & \overline{Q_2\mathsf{T}} \end{pmatrix} = \begin{pmatrix} 1 & 0 \\ Q_1 \cup R_1\overline{Q_2\mathsf{T}} & R_1\overline{Q_2\mathsf{T}} \end{pmatrix}$$

$$= \begin{pmatrix} 1 & 0 \\ Q_1 & 0 \end{pmatrix} + \begin{pmatrix} 1 & 0 \\ R_1\overline{Q_2\mathsf{T}} & R_1\overline{Q_2\mathsf{T}} \end{pmatrix} \leq \begin{pmatrix} 1 & 0 \\ Q_1 & 0 \end{pmatrix} + \begin{pmatrix} 1 & 0 \\ \overline{\overline{R_1 Q_2\mathsf{T}}} & \overline{\overline{R_1 Q_2\mathsf{T}}} \end{pmatrix}$$

$$\leq \begin{pmatrix} 1 & 0 \\ Q_1 & 0 \end{pmatrix} + \begin{pmatrix} 1 & 0 \\ \overline{\overline{Q_1 \cup R_1 Q_2\mathsf{T}}} & \overline{\overline{Q_1 \cup R_1 Q_2\mathsf{T}}} \end{pmatrix}$$

$$= \begin{pmatrix} 1 & 0 \\ Q_1 & 0 \end{pmatrix} + \begin{pmatrix} 1 & 0 \\ 0 & \overline{Q_1 \cup R_1 Q_2\mathsf{T}} \cap 1 \end{pmatrix}\begin{pmatrix} 1 & 0 \\ \mathsf{T} & \mathsf{T} \end{pmatrix}$$

$$= \begin{pmatrix} 1 & 0 \\ Q_1 & 0 \end{pmatrix} + n\begin{pmatrix} 1 & 0 \\ Q_1 \cup R_1 Q_2 & R_1 R_2 \end{pmatrix}\begin{pmatrix} 1 & 0 \\ \mathsf{T} & \mathsf{T} \end{pmatrix}$$

$$= \begin{pmatrix} 1 & 0 \\ Q_1 & R_1 \end{pmatrix}\begin{pmatrix} 1 & 0 \\ 0 & 0 \end{pmatrix} + n\left(\begin{pmatrix} 1 & 0 \\ Q_1 & R_1 \end{pmatrix}\begin{pmatrix} 1 & 0 \\ Q_2 & R_2 \end{pmatrix}\right)\begin{pmatrix} 1 & 0 \\ \mathsf{T} & \mathsf{T} \end{pmatrix}$$

(n10)

$$
\begin{pmatrix} 1 & 0 \\ Q_1 & R_1 \end{pmatrix} \begin{pmatrix} 1 & 0 \\ \top & \top \end{pmatrix} \begin{pmatrix} 1 & 0 \\ Q_2 & R_2 \end{pmatrix} \curlywedge \begin{pmatrix} 1 & 0 \\ \top & 0 \end{pmatrix}
$$

$$
= \begin{pmatrix} 1 & 0 \\ Q_1 \cup R_1 \top & R_1 \top \end{pmatrix} \begin{pmatrix} 1 & 0 \\ Q_2 & R_2 \end{pmatrix} \curlywedge \begin{pmatrix} 1 & 0 \\ \top & 0 \end{pmatrix}
$$

$$
= \begin{pmatrix} 1 & 0 \\ Q_1 \cup R_1 \top \cup R_1 \top Q_2 & R_1 \top R_2 \end{pmatrix} \curlywedge \begin{pmatrix} 1 & 0 \\ \top & 0 \end{pmatrix}
$$

$$
= \begin{pmatrix} 1 & 0 \\ Q_1 \cup R_1 \top & 0 \end{pmatrix} = \begin{pmatrix} 1 & 0 \\ Q_1 & R_1 \end{pmatrix} \begin{pmatrix} 1 & 0 \\ \top & 0 \end{pmatrix} = \begin{pmatrix} 1 & 0 \\ Q_1 & R_1 \end{pmatrix} \begin{pmatrix} 1 & 0 \\ \top & 0 \end{pmatrix} \begin{pmatrix} 1 & 0 \\ Q_2 & R_2 \end{pmatrix}
$$

$\square$

As a consequence, we can use the approximation order given in [12]:

$$
x \sqsubseteq y \iff x \leq y + \mathsf{L} \wedge n(\mathsf{L})y \leq x + n(x)\top
$$

For conscriptions, $n(\mathsf{L}) = 1$ holds, whence the order elaborates to

$$
\begin{pmatrix} 1 & 0 \\ Q_1 & R_1 \end{pmatrix} \sqsubseteq \begin{pmatrix} 1 & 0 \\ Q_2 & R_2 \end{pmatrix}
$$

$$
\iff \begin{pmatrix} 1 & 0 \\ Q_1 & R_1 \end{pmatrix} \leq \begin{pmatrix} 1 & 0 \\ Q_2 & R_2 \end{pmatrix} + \begin{pmatrix} 1 & 0 \\ \top & 0 \end{pmatrix} = \begin{pmatrix} 1 & 0 \\ \top & R_2 \end{pmatrix} \wedge
$$

$$
\begin{pmatrix} 1 & 0 \\ Q_2 & R_2 \end{pmatrix} \leq \begin{pmatrix} 1 & 0 \\ Q_1 & R_1 \end{pmatrix} + n \begin{pmatrix} 1 & 0 \\ Q_1 & R_1 \end{pmatrix} \begin{pmatrix} 1 & 0 \\ \top & \top \end{pmatrix}
$$

$$
= \begin{pmatrix} 1 & 0 \\ Q_1 & R_1 \end{pmatrix} + \begin{pmatrix} 1 & 0 \\ 0 & \overline{Q_1 \top} \cap 1 \end{pmatrix} \begin{pmatrix} 1 & 0 \\ \top & \top \end{pmatrix}
$$

$$
= \begin{pmatrix} 1 & 0 \\ Q_1 & R_1 \end{pmatrix} + \begin{pmatrix} 1 & 0 \\ \overline{Q_1 \top} & \overline{Q_1 \top} \end{pmatrix} = \begin{pmatrix} 1 & 0 \\ Q_1 & R_1 \cup \overline{Q_1 \top} \end{pmatrix}
$$

$$
\iff Q_2 \subseteq Q_1 \wedge R_1 \subseteq R_2 \subseteq R_1 \cup \overline{Q_1 \top}
$$

The relation $\overline{Q_1 \top}$ is a vector that represents the states where all infinite executions are present. The intuition underlying the approximation order is that in states with all infinite executions, any finite execution can be added. In states where at least one infinite execution is missing, no finite execution can be added. Infinite executions can only be removed.

It follows that all properties shown in [12, Theorems 1–4] hold for conscriptions. This includes various properties of the operation n and of $\sqsubseteq$ and representations of $\sqsubseteq$-least fixpoints in terms of $\leq$-least and $\leq$-greatest fixpoints. For reference they are reproduced in the appendix of this paper.

3.3 Iteration

For instantiating further results concerning iteration, we show that conscriptions form a Kleene algebra and an omega algebra. The Kleene star is derived by the standard automata-based matrix construction [3], according to which

$$\begin{pmatrix} a & b \\ c & d \end{pmatrix}^* = \begin{pmatrix} e^* & a^*bf^* \\ d^*ce^* & f^* \end{pmatrix} \quad \text{where} \quad \begin{pmatrix} e \\ f \end{pmatrix} = \begin{pmatrix} a \cup bd^*c \\ d \cup ca^*b \end{pmatrix}$$

The Kleene star of a relation is its reflexive-transitive closure. For conscriptions, $e = 1 \cup 0R^*Q = 1$ and $f = R \cup Q1^*0 = R$, so the Kleene star elaborates to

$$\begin{pmatrix} 1 & 0 \\ Q & R \end{pmatrix}^* = \begin{pmatrix} 1^* & 1^*0R^* \\ R^*Q1^* & R^* \end{pmatrix} = \begin{pmatrix} 1 & 0 \\ R^*Q & R^* \end{pmatrix}$$

The result is a conscription, whence the operation satisfies the Kleene algebra axioms.

The standard automata-based construction does not work for the omega operation as the resulting matrix is not a conscription. This problem, which arises also for prescriptions and extended designs, is solved by typed omega algebras as detailed in [8]. The resulting operation is

$$\begin{pmatrix} 1 & 0 \\ Q & R \end{pmatrix}^\omega = \begin{pmatrix} 1 & 0 \\ R^\omega \cup R^*Q & R^\omega \end{pmatrix}$$

It satisfies the omega algebra axioms. The operation $^\omega$ on relations describes the states from which infinite transition paths exist. We obtain the following result.

Theorem 3. *Conscriptions form an n-omega algebra with the operations*

$$\begin{pmatrix} 1 & 0 \\ Q & R \end{pmatrix}^* = \begin{pmatrix} 1 & 0 \\ R^*Q & R^* \end{pmatrix} \qquad \begin{pmatrix} 1 & 0 \\ Q & R \end{pmatrix}^\omega = \begin{pmatrix} 1 & 0 \\ R^\omega \cup R^*Q & R^\omega \end{pmatrix}$$

Proof. Axiom (n12) follows since $\mathsf{L}x = \mathsf{L}$. As $n(\mathsf{L}) = 1$, axiom (n11) follows by

$$\begin{pmatrix} 1 & 0 \\ Q & R \end{pmatrix}^\omega = \begin{pmatrix} 1 & 0 \\ R^\omega \cup R^*Q & R^\omega \end{pmatrix} \le \begin{pmatrix} 1 & 0 \\ R^\omega \cup R^*Q & R^\omega \cup \overline{R^*Q\mathsf{T}} \end{pmatrix}$$

$$= \begin{pmatrix} 1 & 0 \\ R^*R^\omega \cup R^*Q \cup R^*\overline{\overline{R^*Q}\mathsf{T}} & R^*R^\omega \cup R^*\overline{\overline{R^*Q}\mathsf{T}} \end{pmatrix}$$

$$= \begin{pmatrix} 1 & 0 \\ R^*Q \cup R^*(R^\omega \cup \overline{R^*Q\mathsf{T}}) & R^*(R^\omega \cup \overline{R^*Q\mathsf{T}}) \end{pmatrix}$$

$$= \begin{pmatrix} 1 & 0 \\ R^*Q & R^* \end{pmatrix}\begin{pmatrix} 1 & 0 \\ R^\omega \cup \overline{R^*Q\mathsf{T}} & R^\omega \cup \overline{R^*Q\mathsf{T}} \end{pmatrix}$$

$$= \begin{pmatrix} 1 & 0 \\ Q & R \end{pmatrix}^*\begin{pmatrix} 1 & 0 \\ \overline{R^\omega \cup R^*Q\mathsf{T}} & \overline{R^\omega \cup R^*Q\mathsf{T}} \end{pmatrix}$$

$$= \begin{pmatrix} 1 & 0 \\ Q & R \end{pmatrix}^*\begin{pmatrix} 1 & 0 \\ 0 & \overline{R^\omega \cup R^*Q\mathsf{T}} \cap 1 \end{pmatrix}\begin{pmatrix} 1 & 0 \\ \mathsf{T} & \mathsf{T} \end{pmatrix}$$

$$= \begin{pmatrix} 1 & 0 \\ Q & R \end{pmatrix}^*_n\begin{pmatrix} 1 & 0 \\ R^\omega \cup R^*Q & R^\omega \end{pmatrix}\begin{pmatrix} 1 & 0 \\ \mathsf{T} & \mathsf{T} \end{pmatrix} = \begin{pmatrix} 1 & 0 \\ Q & R \end{pmatrix}^*_n\left(\begin{pmatrix} 1 & 0 \\ Q & R \end{pmatrix}^\omega\right)\begin{pmatrix} 1 & 0 \\ \mathsf{T} & \mathsf{T} \end{pmatrix}$$

The calculation uses that R^ω is a vector and that $R^*\overline{R^*Q\mathsf{T}} = \overline{R^*Q\mathsf{T}} \subseteq R^*Q$. $\quad\square$

Therefore all properties shown in [12, Theorems 5–6] hold for conscriptions. This includes further properties of the operation n and representations of iteration in terms of the Kleene star and omega operations; see again the appendix. In particular, conscriptions form an extended binary itering and an itering as follows.

Corollary 1. *Conscriptions form an extended binary itering and an itering with*

$$\begin{pmatrix} 1 & 0 \\ Q_1 & R_1 \end{pmatrix} \star \begin{pmatrix} 1 & 0 \\ Q_2 & R_2 \end{pmatrix} = \begin{pmatrix} 1 & 0 \\ R_1^\omega \cup R_1^*(Q_1 \cup Q_2) & R_1^* R_2 \end{pmatrix}$$

$$\begin{pmatrix} 1 & 0 \\ Q & R \end{pmatrix}^\circ = \begin{pmatrix} 1 & 0 \\ R^\omega \cup R^* Q & R^* \end{pmatrix}$$

Proof. The extended binary itering instance follows by [12, Theorem 6]. Moreover, conscriptions satisfy the property $(x \star y)z = x \star (yz)$:

$$\left(\begin{pmatrix} 1 & 0 \\ Q_1 & R_1 \end{pmatrix} \star \begin{pmatrix} 1 & 0 \\ Q_2 & R_2 \end{pmatrix} \right) \begin{pmatrix} 1 & 0 \\ Q_3 & R_3 \end{pmatrix}$$

$$= \begin{pmatrix} 1 & 0 \\ R_1^\omega \cup R_1^*(Q_1 \cup Q_2) & R_1^* R_2 \end{pmatrix} \begin{pmatrix} 1 & 0 \\ Q_3 & R_3 \end{pmatrix}$$

$$= \begin{pmatrix} 1 & 0 \\ R_1^\omega \cup R_1^*(Q_1 \cup Q_2) \cup R_1^* R_2 Q_3 & R_1^* R_2 R_3 \end{pmatrix}$$

$$= \begin{pmatrix} 1 & 0 \\ R_1^\omega \cup R_1^*(Q_1 \cup Q_2 \cup R_2 Q_3) & R_1^* R_2 R_3 \end{pmatrix}$$

$$= \begin{pmatrix} 1 & 0 \\ Q_1 & R_1 \end{pmatrix} \star \begin{pmatrix} 1 & 0 \\ Q_2 \cup R_2 Q_3 & R_2 R_3 \end{pmatrix}$$

$$= \begin{pmatrix} 1 & 0 \\ Q_1 & R_1 \end{pmatrix} \star \left(\begin{pmatrix} 1 & 0 \\ Q_2 & R_2 \end{pmatrix} \begin{pmatrix} 1 & 0 \\ Q_3 & R_3 \end{pmatrix} \right)$$

Hence conscriptions form an itering with $x^\circ = x \star 1$.   □

It follows that all consequences of iterings and binary iterings shown in [9,10] hold for conscriptions. They include separation theorems generalised from omega algebras and Back's atomicity refinement theorem.

4   Extended Conscriptions

Extended conscriptions combine aspects of three computation models:

- They represent aborting executions in addition to finite and infinite executions; so do extended designs [13].
- They represent aborting, infinite and finite executions independently; so does the model introduced in [9].
- Aborting executions can refer to final states; so do infinite executions in conscriptions.

Infinite executions of extended conscriptions are restricted to refer to initial states only.

An *extended conscription* is a 3×3 matrix whose entries are relations over the state space A. The matrix has the following form

$$\begin{pmatrix} 1 & 0 & 0 \\ 0 & \top & 0 \\ P & Q & R \end{pmatrix}$$

where Q is a vector, that is, $Q\top = Q$. The relation P represents the aborting executions, Q represents the states from which infinite executions exist and R represents the finite executions. Hence the endless loop is represented by the extended conscription

$$L = \begin{pmatrix} 1 & 0 & 0 \\ 0 & \top & 0 \\ 0 & \top & 0 \end{pmatrix}$$

Sequential composition and non-deterministic choice of extended conscriptions are given by matrix product and componentwise union, respectively. The operation n for extended conscriptions is derived by the method applied to conscriptions in Section 3.2. The result is

$$n \begin{pmatrix} 1 & 0 & 0 \\ 0 & \top & 0 \\ P & Q & R \end{pmatrix} = \begin{pmatrix} 1 & 0 & 0 \\ 0 & \top & 0 \\ 0 & 0 & Q \cap 1 \end{pmatrix}$$

The simpler form $Q \cap 1$ is due to the fact that Q is a vector. This operation satisfies the axioms given in [9] for models of strict computations and the axioms given in [12]. The approximation order instantiates to

$$\begin{pmatrix} 1 & 0 & 0 \\ 0 & \top & 0 \\ P_1 & Q_1 & R_1 \end{pmatrix} \sqsubseteq \begin{pmatrix} 1 & 0 & 0 \\ 0 & \top & 0 \\ P_2 & Q_2 & R_2 \end{pmatrix} \Leftrightarrow \begin{array}{c} P_1 \subseteq P_2 \subseteq P_1 \cup Q_1 \wedge \\ Q_2 \subseteq Q_1 \wedge \\ R_1 \subseteq R_2 \subseteq R_1 \cup Q_1 \end{array}$$

The intuition is that in states with an infinite execution, any aborting and finite executions can be added. In states with no infinite execution, no executions can be added. Infinite executions can only be removed.

The standard matrix construction for the Kleene star and the typed matrix construction for the omega operation yield the following operations. Moreover extended conscriptions form an itering that does not satisfy $x° = x^\omega 0 + x^*$ in general:

$$\begin{pmatrix} 1 & 0 & 0 \\ 0 & \top & 0 \\ P & Q & R \end{pmatrix}^* = \begin{pmatrix} 1 & 0 & 0 \\ 0 & \top & 0 \\ R^*P & R^*Q & R^* \end{pmatrix}$$

$$\begin{pmatrix} 1 & 0 & 0 \\ 0 & \top & 0 \\ P & Q & R \end{pmatrix}^\omega = \begin{pmatrix} 1 & 0 & 0 \\ 0 & \top & 0 \\ R^\omega \cup R^*P & R^\omega \cup R^*Q & R^\omega \end{pmatrix}$$

$$\begin{pmatrix} 1 & 0 & 0 \\ 0 & \top & 0 \\ P & Q & R \end{pmatrix}° = \begin{pmatrix} 1 & 0 & 0 \\ 0 & \top & 0 \\ R^*P & R^\omega \cup R^*Q & R^* \end{pmatrix}$$

The following result summarises the algebraic properties of extended conscriptions. Hence all properties shown in [9,12] hold for extended conscriptions.

Theorem 4. *Extended conscriptions form a lattice-ordered semiring, an n-algebra, an n-omega algebra, an itering and an extended binary itering.*

Proof. The lattice-ordered semiring, n-algebra and n-omega algebra instances follow by calculations as in the proof of Theorems 1–3. The itering and extended binary itering instances follow as in the proof of Corollary 1.   □

5   Further Computation Models

Comparing the various computation models – designs, prescriptions, extended designs, conscriptions, extended conscriptions – it is natural to further generalise extended conscriptions by eliminating the restriction placed on the infinite executions. This is done in a similar way as for conscriptions and for the aborting executions of extended conscriptions.

A computation in the resulting model is a 3×3 matrix of the following form:

$$(P|Q|R) = \begin{pmatrix} 1 & 0 & 0 \\ 0 & 1 & 0 \\ P & Q & R \end{pmatrix}$$

There are no restrictions on P, Q or R. The calculations to obtain the operation n, the approximation order, the Kleene star, the omega operation, the itering operation and the binary itering operation follow the method of Section 3. The following result summarises the algebraic structure.

Theorem 5. *Let A be a set and let $S = \{(P|Q|R) \mid P, Q, R \subseteq A \times A\}$. Then S is a lattice-ordered semiring, an n-algebra, an n-omega algebra, an itering and an extended binary itering using the following operations:*

$$(P_1|Q_1|R_1) + (P_2|Q_2|R_2) = (P_1 \cup P_2|Q_1 \cup Q_2|R_1 \cup R_2)$$
$$(P_1|Q_1|R_1) \curlywedge (P_2|Q_2|R_2) = (P_1 \cap P_2|Q_1 \cap Q_2|R_1 \cap R_2)$$
$$(P_1|Q_1|R_1) \cdot (P_2|Q_2|R_2) = (P_1 \cup R_1 P_2|Q_1 \cup R_1 Q_2|R_1 R_2)$$
$$(P_1|Q_1|R_1) \star (P_2|Q_2|R_2) = (R_1^*(P_1 \cup P_2)|R_1^\omega \cup R_1^*(Q_1 \cup Q_2)|R_1^* R_2)$$
$$n(P|Q|R) = (0|0|\overline{\overline{Q}\top} \cap 1)$$
$$(P|Q|R)^* = (R^* P|R^* Q|R^*)$$
$$(P|Q|R)^\omega = (R^\omega \cup R^* P|R^\omega \cup R^* Q|R^\omega)$$
$$(P|Q|R)^\circ = (R^* P|R^\omega \cup R^* Q|R^*)$$
$$0 = (0|0|0)$$
$$\top = (\top|\top|\top)$$
$$1 = (0|0|1)$$
$$L = (0|\top|0)$$

The approximation order on S is

$$(P_1|Q_1|R_1) \sqsubseteq (P_2|Q_2|R_2) \iff P_1 \subseteq P_2 \subseteq P_1 \cup \overline{\overline{Q_1\top}} \wedge Q_2 \subseteq Q_1 \wedge$$
$$R_1 \subseteq R_2 \subseteq R_1 \cup \overline{\overline{Q_1\top}}$$

Proof. The lattice-ordered semiring, n-algebra and n-omega algebra instances follow by calculations as in the proof of Theorems 1–3. The itering and extended binary itering instances follow as in the proof of Corollary 1.   □

This computation model is the most precise among those considered in this paper: it can represent finite, infinite and aborting executions independently and without any restrictions. Previously investigated computation models are isomorphic to substructures of this model:

- extended conscriptions: $\{(P|Q|R) \mid Q = Q\top\}$,
- the model of [9]: $\{(P|Q|R) \mid P = P\top \wedge Q = Q\top\}$,
- extended designs: $\{(P|Q|R) \mid P = P\top \wedge Q = Q\top \wedge P \subseteq Q \wedge P \subseteq R\}$,
- conscriptions: $\{(P|Q|R) \mid P = 0\}$,
- prescriptions: $\{(P|Q|R) \mid P = 0 \wedge Q = Q\top\}$,
- designs: $\{(P|Q|R) \mid P = 0 \wedge Q = Q\top \wedge Q \subseteq R\}$.

Other restrictions lead to further computation models which can be represented by matrices, for example,

- $\{(P|Q|R) \mid P = P\top\}$ requires that aborting executions do not refer to final states;
- $\{(P|Q|R) \mid P = P\top \wedge Q = Q\top \wedge Q \subseteq P \wedge Q \subseteq R\}$ requires that aborting and infinite executions do not refer to final states and that in the presence of infinite executions, aborting or finite executions cannot be distinguished.

Further combinations are possible, but in each case it has to be verified that the subset is closed under operations such as sequential composition.

6   Conclusion

In this paper we have derived approximation orders for new computation models based on a Galois connection for infinite executions and on algebras previously introduced for other models. Once more this shows that the algebraic approach can essentially contribute to the development of computation models. Additionally we inherit a multitude of results that have been proved for the previously introduced algebras. Future work will be concerned with computation models involving time, such as those studied in [13,14,5].

Acknowledgement. I thank the anonymous referees for helpful comments.

References

1. Blanchette, J.C., Böhme, S., Paulson, L.C.: Extending Sledgehammer with SMT solvers. In: Bjørner, N., Sofronie-Stokkermans, V. (eds.) CADE 2011. LNCS, vol. 6803, pp. 116–130. Springer, Heidelberg (2011)
2. Cohen, E.: Separation and reduction. In: Backhouse, R., Oliveira, J.N. (eds.) MPC 2000. LNCS, vol. 1837, pp. 45–59. Springer, Heidelberg (2000)

3. Conway, J.H.: Regular Algebra and Finite Machines. Chapman and Hall (1971)
4. Dunne, S.: Recasting Hoare and He's Unifying Theory of Programs in the context of general correctness. In: Butterfield, A., Strong, G., Pahl, C. (eds.) 5th Irish Workshop on Formal Methods. Electronic Workshops in Computing. The British Computer Society (2001)
5. Dunne, S.: Conscriptions: A new relational model for sequential computations. In: Wolff, B., Gaudel, M.-C., Feliachi, A. (eds.) UTP 2012. LNCS, vol. 7681, pp. 144–163. Springer, Heidelberg (2013)
6. Dunne, S.E., Hayes, I.J., Galloway, A.J.: Reasoning about loops in total and general correctness. In: Butterfield, A. (ed.) UTP 2008. LNCS, vol. 5713, pp. 62–81. Springer, Heidelberg (2010)
7. Guttmann, W.: General correctness algebra. In: Berghammer, R., Jaoua, A.M., Möller, B. (eds.) RelMiCS/AKA 2009. LNCS, vol. 5827, pp. 150–165. Springer, Heidelberg (2009)
8. Guttmann, W.: Towards a typed omega algebra. In: de Swart, H. (ed.) RAMiCS 2011. LNCS, vol. 6663, pp. 196–211. Springer, Heidelberg (2011)
9. Guttmann, W.: Algebras for iteration and infinite computations. Acta Inf. 49(5), 343–359 (2012)
10. Guttmann, W.: Unifying lazy and strict computations. In: Kahl, W., Griffin, T.G. (eds.) RAMiCS 2012. LNCS, vol. 7560, pp. 17–32. Springer, Heidelberg (2012)
11. Guttmann, W.: Extended designs algebraically. Sci. Comput. Program. 78(11), 2064–2085 (2013)
12. Guttmann, W.: Infinite executions of lazy and strict computations (2013) (submitted)
13. Hayes, I.J., Dunne, S.E., Meinicke, L.: Unifying theories of programming that distinguish nontermination and abort. In: Bolduc, C., Desharnais, J., Ktari, B. (eds.) MPC 2010. LNCS, vol. 6120, pp. 178–194. Springer, Heidelberg (2010)
14. Hayes, I.J., Dunne, S.E., Meinicke, L.A.: Linking Unifying Theories of Program refinement. Sci. Comput. Program. 78(11), 2086–2107 (2013)
15. Hoare, C.A.R., He, J.: Unifying theories of programming. Prentice Hall Europe (1998)
16. Kozen, D.: A completeness theorem for Kleene algebras and the algebra of regular events. Information and Computation 110(2), 366–390 (1994)
17. Möller, B.: Kleene getting lazy. Sci. Comput. Program. 65(2), 195–214 (2007)
18. Nipkow, T., Paulson, L.C., Wenzel, M.: Isabelle/HOL: A Proof Assistant for Higher-Order Logic. LNCS, vol. 2283. Springer, Heidelberg (2002)
19. Paulson, L.C., Blanchette, J.C.: Three years of experience with Sledgehammer, a practical link between automatic and interactive theorem provers. In: Sutcliffe, G., Ternovska, E., Schulz, S. (eds.) Proceedings of the 8th International Workshop on the Implementation of Logics, pp. 3–13 (2010)
20. Schmidt, G., Ströhlein, T.: Relationen und Graphen. Springer (1989)

Appendix: Consequences of n-Algebras

Because the models discussed in this paper form n-omega algebras, they satisfy all of the following results, which appear as Theorems 1–6 in [12]. The results have been verified in Isabelle/HOL [18], making heavy use of its integrated automated theorem provers and SMT solvers [19,1]. The proofs can be found in the theory files at http://www.csse.canterbury.ac.nz/walter.guttmann/algebra/.

Proposition 1. *Let S be an n-algebra. Then $(n(S), +, \cdot, n(0), n(\top))$ is a semi-ring with right annihilator $n(0)$ and a bounded distributive lattice with meet $\cdot$. Moreover, n is $\leq$-isotone and the following properties hold for $x, y \in S$:*

1. $n(x)n(y) = n(y)n(x)$
2. $n(x)n(x) = n(x)$
3. $n(x)n(y) \leq n(x)$
4. $n(x)n(y) \leq n(y)$
5. $n(x) \leq n(x + y)$
6. $n(x) \leq 1$
7. $n(x)0 = 0$
8. $n(x)n(0) = n(0)$
9. $n(x) \leq x + n(x0)$
10. $n(x + n(x)\top) = n(x)$
11. $n(n(x)\mathsf{L}) = n(x)$
12. $n(x)n(\mathsf{L}) = n(x)$
13. $n(x) \leq n(\mathsf{L})$
14. $n(x) \leq n(x\mathsf{L})$
15. $n(x)\mathsf{L} \leq x\mathsf{L}$
16. $n(0)\mathsf{L} = 0$
17. $n(\mathsf{L}) = n(\top)$
18. $n(x\top) = n(x\mathsf{L})$
19. $n(x)\top = n(x)\mathsf{L} + n(x0)\top$
20. $n(xn(y)\mathsf{L}) \leq n(xy)$

21. $xn(y)\top \leq xy + n(xy)\top$
22. $n(x)\top y \leq xy + n(xy)\top$
23. $xn(y)\mathsf{L} = x0 + n(xn(y)\mathsf{L})\mathsf{L}$
24. $xn(y)\mathsf{L} \leq x0 + n(xy)\mathsf{L}$
25. $n(\mathsf{L})x \leq x0 + n(x\mathsf{L})\top$
26. $n(\mathsf{L})\mathsf{L} = \mathsf{L}n(\mathsf{L}) = \mathsf{L}$
27. $\mathsf{LL} = \mathsf{L}\top = \mathsf{L}\top\mathsf{L} = \mathsf{L}$
28. $\mathsf{L}x \leq \mathsf{L}$
29. $x\mathsf{L} \leq x0 + \mathsf{L}$
30. $x\top \curlywedge \mathsf{L} \leq x\mathsf{L}$
31. $x\top y \curlywedge \mathsf{L} = x\mathsf{L}y \curlywedge \mathsf{L}$
32. $x\top y \curlywedge \mathsf{L} \leq x0 + \mathsf{L}y$
33. $(x \curlywedge \mathsf{L})0 \leq x0 \curlywedge \mathsf{L}$
34. $n(x) = n(x \curlywedge \mathsf{L}) = (n(x) \curlywedge \mathsf{L}) + n(x0)$
35. $n(x)\mathsf{L} \leq x \curlywedge \mathsf{L} \leq n(\mathsf{L})x$
36. $n(x) \curlywedge \mathsf{L} \leq (n(x) \curlywedge \mathsf{L})\top \leq n(x)\mathsf{L} \leq x$
37. $x \leq y \Leftrightarrow x \leq y + \mathsf{L} \wedge n(\mathsf{L})x \leq y + n(y)\top$
38. $x \leq y \Leftrightarrow x \leq y + \mathsf{L} \wedge x \leq y + n(y)\top$
39. $n(y)x \leq xn(y) \Leftrightarrow n(y)x = n(y)xn(y)$
40. $n(x) \leq n(y) \Leftrightarrow n(x)\mathsf{L} \leq y$

In an n-algebra S the approximation relation $\sqsubseteq$ is defined by

$$x \sqsubseteq y \Leftrightarrow x \leq y + \mathsf{L} \wedge n(\mathsf{L})y \leq x + n(x)\top$$

where $x, y \in S$. The following result gives properties of $\sqsubseteq$.

Proposition 2. *Let S be an n-algebra.*

1. *The relation $\sqsubseteq$ is a partial order with least element L.*
2. *The operations $+$ and $\cdot$ and $\lambda x.x \curlywedge \mathsf{L}$ and $\lambda x.n(x)\mathsf{L}$ are $\sqsubseteq$-isotone.*
3. *If S is an itering, the operation $^\circ$ is $\sqsubseteq$-isotone.*
4. *If S is a Kleene algebra, the operation * is $\sqsubseteq$-isotone.*

Further results concern fixpoints of a function $f : S \to S$. Provided they exist, the $\leq$-least, $\leq$-greatest and $\sqsubseteq$-least fixpoints of f are denoted by μf, νf and κf, respectively:

$$
\begin{aligned}
f(\mu f) &= \mu f & f(x) = x &\Rightarrow \mu f \leq x \\
f(\nu f) &= \nu f & f(x) = x &\Rightarrow \nu f \geq x \\
f(\kappa f) &= \kappa f & f(x) = x &\Rightarrow \kappa f \sqsubseteq x
\end{aligned}
$$

We abbreviate $\kappa(\lambda x.f(x))$ by $\kappa x.f(x)$. Provided it exists, the $\sqsubseteq$-greatest lower bound of $x, y \in S$ is denoted by $x \sqcap y$:

$$x \sqcap y \sqsubseteq x \qquad x \sqcap y \sqsubseteq y \qquad z \sqsubseteq x \wedge z \sqsubseteq y \Rightarrow z \sqsubseteq x \sqcap y$$

Proposition 3. *Let S be an n-algebra, let $f : S \to S$ be $\leq$- and $\sqsubseteq$-isotone, and assume that μf and νf exist. Then the following are equivalent:*

1. *κf exists.*
2. *κf and $\mu f \sqcap \nu f$ exist and $\kappa f = \mu f \sqcap \nu f$.*
3. *κf exists and $\kappa f = (\nu f \curlywedge L) + \mu f$.*
4. *$n(L)\nu f \leq (\nu f \curlywedge L) + \mu f + n(\nu f)\top$.*
5. *$n(L)\nu f \leq (\nu f \curlywedge L) + \mu f + n((\nu f \curlywedge L) + \mu f)\top$.*
6. *$(\nu f \curlywedge L) + \mu f \sqsubseteq \nu f$.*
7. *$\mu f \sqcap \nu f$ exists and $\mu f \sqcap \nu f = (\nu f \curlywedge L) + \mu f$.*
8. *$\mu f \sqcap \nu f$ exists and $\mu f \sqcap \nu f \leq \nu f$.*

Condition 4 of this proposition characterises the existence of κf in terms of μf and νf. Condition 3 shows how to obtain κf from μf and νf. This simplifies calculations as $\leq$ is less complex than $\sqsubseteq$. Further characterisations generalise to n-algebras as shown in the following result.

Proposition 4. *Let S be an n-algebra, let $f : S \to S$ be $\leq$- and $\sqsubseteq$-isotone, and assume that μf and νf exist. Then the following are equivalent and imply the statements of Proposition 3:*

1. *κf exists and $\kappa f = n(\nu f)L + \mu f$.*
2. *$n(L)\nu f \leq \mu f + n(\nu f)\top$.*
3. *$n(\nu f)L + \mu f \sqsubseteq \nu f$.*
4. *$\mu f \sqcap \nu f$ exists and $\mu f \sqcap \nu f = n(\nu f)L + \mu f$.*

Proposition 5. *Let S be an n-omega algebra and $x, y, z \in S$. Then the following properties hold:*

1. $Lx^* = L$
2. $(xL)^* = 1 + xL$
3. $(xL)^\omega = xL = xLxL$
4. $(xL)^*y \leq y + xL$
5. $(xL + y)^* = y^* + y^*xL$
6. $(xL + y)^\omega = y^\omega + y^*xL$
7. $n(x) \leq n(x^\omega)$
8. $n(y^\omega + y^*z) = n(y^\omega) + n(y^*z)$
9. $x^* + n(x^\omega)L = x^* + x^*n(x^\omega)L$
10. $x^* + n(x^\omega)L = x^* + xn(x^\omega)L$
11. $yx^* + n(yx^\omega)L = yx^* + yn(x^\omega)L$
12. $x^*0 + n(x^\omega)L = x^*0 + x^*n(x^\omega)L$
13. $xx^*0 + n(x^\omega)L = xx^*0 + xn(x^\omega)L$
14. $yx^*0 + n(yx^\omega)L = yx^*0 + yn(x^\omega)L$
15. $n(L)x^\omega \leq x^*0 + n(x^\omega)\top$
16. $n(L)(y^\omega + y^*z) \leq y^*z + n(y^\omega + y^*z)\top$

Proposition 6. *Let S be an n-omega algebra, let $x, y, z \in S$, and let $f : S \to S$ be given by $f(x) = yx + z$.*

1. *The $\sqsubseteq$-least fixpoint of f is $\kappa f = (y^\omega \curlywedge L) + y^*z = n(y^\omega)L + y^*z$.*
2. *The operations $^\omega$ and $\lambda y.(\kappa x.yx + z)$ and $\lambda z.(\kappa x.yx + z)$ are $\sqsubseteq$-isotone.*
3. *S is an extended binary itering using $x \star y = n(x^\omega)L + x^*y$.*

Abstract Dynamic Frames

Han-Hing Dang

Institut für Informatik, Universität Augsburg, 86159 Augsburg, Germany
h.dang@informatik.uni-augsburg.de

Abstract. Based on a former relation-algebraic approach to separation logic we present an abstraction of the theory of dynamic frames and algebraically describe concepts, properties and behaviour of that theory in a pointfree fashion. Moreover, relationships to abstract concepts of separation logic are given to pave the way for a unified treatment of both approaches. In particular, we also sketch the main ideas within the framework of local actions.

Keywords: Frame problem, local actions, relational semantics, separation algebra.

1 Introduction

For obtaining a methodology that guarantees modularity and hence scalability in specification and correctness proofs of computer programs, an adequate solution to the frame problem [MH69] is required. The frame problem asks for a methodology that allows specifying which resources of a program can be changed and which ones are left unchanged without naming them explicitly. A popular approach to this problem is the *theory of dynamic frames* [Kas11] that provides the mentioned modularity while still being expressive enough to handle a variety of useful programs. Further variations of the theory address the automation of program verification (e.g., [SJP09, Lei10, GGN11]).

Another approach to the frame problem is given by *separation logic* [Rey02] which allows, due to its popular frame rule, modular reasoning about parts of a program without the need to construct a program proof in a larger context anew. For this logic there exist a few abstract and algebraic approaches that are used to extract and formalise general behaviour [COY07, DHM11, HHM⁺11, DM12], also in the case of concurrency. Unfortunately, for the theory of dynamic frames such approaches and considerations barely exist.

In the present paper we revisit a former relation-based algebraic calculus [DHM11, DM12] that was used as a formal base for pointfree proofs of inference rules of separation logic and combine it with the theory of dynamic frames.

The *contributions* of this work comprise an abstract treatment of the resources and locations dealt with in that theory, based on separation algebras. Moreover we give point-free characterisations and proofs of crucial concepts within the extended relational approach of [DM12] and explain their concrete meaning. By this, the relational calculus extends towards a unifying approach for dynamic frames and separation logic.

P. Höfner et al. (Eds.): RAMiCS 2014, LNCS 8428, pp. 157–172, 2014.
© Springer International Publishing Switzerland 2014

The structure of this paper is as follows. First, we present all required basic definitions of separation algebras and the extended relational structure. In Section 3 we give pointfree variants of framing requirements and consequences of this. By this, Section 4 abstractly clarifies the relationship between locality principles and their application in accumulating frames. We conclude this work with a discussion on the relationship to so-called local actions.

2   Basics of the Algebraic Structure

This section provides the formal background to abstractly characterise dynamic frames. Framing requirements are defined in [Kas11] using a relational style. This motivates the idea to use the relation-algebraic structures of [DHM11, DM12] as an abstract base.

2.1   Separation Algebras

Before giving basic definitions and direct consequences of the algebra we start with the concept of *separation algebras* that provides a general way to characterise the structure and properties of resources [COY07].

Definition 2.1. A *separation algebra* is a cancellative and partial commutative monoid that we denote by $(\Sigma, \bullet, u)$. Elements of the algebra are called *states* and denoted by $\sigma, \tau, \ldots \in \Sigma$. Due to partiality two terms are defined to be equal iff both are defined and equal or both terms are undefined. This induces a *combinability* relation $\#$ defined by

$$\sigma_0 \# \sigma_1 \Leftrightarrow_{df} \sigma_0 \bullet \sigma_1 \text{ is defined}$$

and a *substate* relation given for $\sigma_0, \sigma_1 \in \Sigma$ by

$$\sigma_0 \preceq \sigma_1 \Leftrightarrow_{df} \exists \sigma_2. \, \sigma_0 \bullet \sigma_2 = \sigma_1.$$

When writing $\sigma \bullet \tau$ for states σ, τ we will implicitly assume $\sigma \# \tau$ in the following.

The *empty state* u is the unit of the partial binary operator $\bullet$ which, additionally satisfies cancellativity, i.e., $\sigma_1 \bullet \tau = \sigma_2 \bullet \tau \Rightarrow \sigma_1 = \sigma_2$ for arbitrary states σ_1, σ_2, τ.

A concrete instance of a separation algebra can be found in the dynamic frames setting. Resources or states in that approach are finite mappings from an infinite set of locations Loc to an infinite set of values Val that comprises at least integers and Booleans. Formally. we use the concrete dynamic frames separation algebra DFSA $=_{df}$ (Loc $\rightsquigarrow$ Val, $\dot{\cup}$, $\emptyset$) where $\dot{\cup}$ denotes union of location-disjoint functions, $\emptyset$ the completely undefined function and $\sigma \# \tau \Leftrightarrow dom(\sigma) \cap dom(\tau) = \emptyset$. We write $dom(\sigma)$ for a mapping or state σ to denote its domain or more concretely all of its allocated locations, i.e., a subset of Loc. Moreover, we define the substate $\sigma|_X$ that restricts the domain of the state σ to a set of locations X.

Lemma 2.2. *For a state τ assume $dom(\tau) = X$. Then for arbitrary σ we have $(\sigma \bullet \tau)|_X = \tau$.*

We continue to characterise and manage several central properties of the dynamic frames approach within the abstraction to separation algebras. For this we require additional assumptions given in [DHA09] and basically follow the approach of that work. A separation algebra $(\Sigma, \bullet, u)$ satisfies *disjointness* iff for all σ, τ

$$\sigma \bullet \sigma = \tau \Rightarrow \sigma = \tau \tag{1}$$

and it satisfies *cross-split* iff for arbitrary σ_i with $i \in \{1, 2, 3, 4\}$

$$\sigma_1 \bullet \sigma_2 = \sigma_3 \bullet \sigma_4 \Rightarrow \exists \sigma_{13}, \sigma_{14}, \sigma_{23}, \sigma_{24}.\ \sigma_1 = \sigma_{13} \bullet \sigma_{14} \ \wedge\ \sigma_2 = \sigma_{23} \bullet \sigma_{24}$$
$$\wedge\ \sigma_3 = \sigma_{13} \bullet \sigma_{23} \ \wedge\ \sigma_4 = \sigma_{14} \bullet \sigma_{24}. \tag{2}$$

Disjointness in the presence of cancellativity implies that the only element that can be combined with itself is the neutral element u, i.e.,

$$\sigma \# \sigma \Rightarrow \sigma = u. \tag{3}$$

Equivalently, non-unit elements cannot be combined with themselves since any allocated resources will overlap in such products. Therefore, the condition of (1) is called disjointness.

For a proof of (3) assume a state σ that satisfies $\sigma \# \sigma$. By definition of $\#$, Equation (1), and a logic step:

$$\sigma \# \sigma \Leftrightarrow (\exists \tau.\ \sigma \bullet \sigma = \tau) \Rightarrow (\tau = \sigma) \Rightarrow (\sigma \bullet \sigma = \sigma).$$

Now, by cancellativity we can infer $u \bullet \sigma = \sigma \bullet \sigma \Rightarrow u = \sigma$.

To explain the idea of the cross-split assumption, assume that a state can be combined in two ways or that there exist two possible splits of a state. Then there need to exist four substates that represent a partition of the original state w.r.t. the mentioned splits. The partitions of the state can be depicted as follows:

For the remaining sections we assume separation algebras that satisfy disjointness and cross-split. A concrete example of such a separation algebra can be found in [HV13]. The assumptions are required there to establish basic properties of operators for reasoning about sharing within data structures. Note that the separation algebra DFSA also satisfies disjointness and cross-split.

2.2   The Relational Structure

In what follows we define a relational structure enriched by an operator $*$ that is also called *separating conjunction*. It ensures disjointness of program states or executions on disjoint states (cf. [DM12]).

Definition 2.3. Assume a separation algebra $(\Sigma, \bullet, u)$. A *command* is a relation $P \subseteq \Sigma \times \Sigma$. Relational composition of commands is denoted by ; . Its unit skip $=_{df} \{(\sigma, \sigma) : \sigma \in \Sigma\}$ is the identity relation while the universal relation is denoted by $\top$. *Tests* are special commands p, q, r that satisfy $p \subseteq$ skip. As particular tests we define emp $=_{df} \{(u, u)\}$ that characterises the empty state u and $\ulcorner P$ that represents the domain of a command P. It is characterised by the universal property

$$\ulcorner P \subseteq q \;\Leftrightarrow\; P \subseteq q \,; P \tag{4}$$

where q is an arbitrary test. In particular, $P \subseteq \ulcorner P \,; P$ and hence $P = \ulcorner P \,; P$. Moreover, we have $\ulcorner P = (P \,; \top) \cap$ skip.

Note that tests form a Boolean algebra with skip as its greatest and $\emptyset$ as its least element w.r.t. $\subseteq$. Moreover, on tests $\cup$ coincides with join and ; with meet. In particular, tests are idempotent and commute under composition, i.e., $p \,; p = p$ and $p \,; q = q \,; p$.

We now come to the definitions to introduce state separation relationally. Separation of commands can be interpreted either as their parallel execution on disjoint portions of states or, in the special case of tests as assertions characterising disjoint resources. For both cases we need a concept to be able to reason independently on disjoint portions of resources relationally.

First we introduce the *Cartesian product* $P \times Q$ of commands P, Q by

$$(\sigma_1, \sigma_2)\,(P \times Q)\,(\tau_1, \tau_2) \;\Leftrightarrow_{df}\; \sigma_1 \, P \, \tau_1 \wedge \sigma_2 \, Q \, \tau_2 \,.$$

Union, inclusion and intersection of such relations on pairs are straightforward while composition is defined componentwise.

We assume that ; binds tighter than $\times$ and $\cap$. It is clear that skip $\times$ skip is the identity of ; on products. Note that $\times$ and ; satisfy an *equational* exchange law:

$$P \,; Q \times R \,; S \;=\; (P \times R) \,; (Q \times S) \,. \tag{5}$$

Pairs of tests are subidentities w.r.t. skip $\times$ skip and thus are idempotent and commute under ; . A special test is given by the *combinability check* # [DM12], on pairs of states:

$$(\sigma_1, \sigma_2)\,\#\,(\tau_1, \tau_2) \;\Leftrightarrow_{df}\; \sigma_1 \,\#\, \sigma_2 \wedge \sigma_1 = \tau_1 \wedge \sigma_2 = \tau_2 \,.$$

It main usage is to rule out pairs of incompatible states that can occur within products $P \times Q$ for arbitrary commands P, Q.

As in [DHM11], we connect the pairs of states with single states using the so-called *split* relation $\lhd$ and its converse *join* $\rhd$ defined by

$$\sigma \lhd (\sigma_1, \sigma_2) \;\Leftrightarrow_{df}\; (\sigma_1, \sigma_2) \rhd \sigma \;\Leftrightarrow_{df}\; \sigma_1 \,\#\, \sigma_2 \wedge \sigma = \sigma_1 \bullet \sigma_2 \,.$$

Corollary 2.4. # ; $\rhd = \rhd$ *and symmetrically* $\lhd$; # = $\lhd$. *Moreover,* # $\subseteq$ $\rhd$; $\lhd$.

Corollary 2.5 (Forward/Backward Compatiblity). *For tests p, q we have* $\# \, ; (p \times q) = (p \times q) \, ; \# \, .$

The intuition for this inequation is that tests do not change states as subidentities and hence starting from compatible states they return the same compatible ones. Finally, the $*$-*composition* of commands P, Q is defined by

$$P * Q =_{df} \lhd \, ; (P \times Q) \, ; \rhd \, . \tag{6}$$

For states σ, τ we have $\sigma \, (P * Q) \tau$ iff σ can be split into states σ_P, σ_Q on which P and Q can act and produce results τ_P, τ_Q that are again combinable to $\tau = \tau_P \bullet \tau_Q$. Hence $P * Q$ also provides a possibility to characterise the structure of commands and hence their behaviour on parts of a state.

Moreover, $P * Q$ can also be interpreted as the concurrent execution of programs P, Q running on combinable or disjoint sets of resources [DM12, DM13].

Note that for tests p, q the command $p * q$ is also a test and in particular, skip $*$ skip $=$ skip. Moreover, $*$ is associative and commutative and emp is its unit.

For readers familiar with *fork algebras* (e.g., [FBH97]) we remark that using the pairing operation $\star(\sigma, \tau) = (\sigma, \tau)$ one has the relationship

$$\lhd = ((\succeq) \, \underline{\nabla} \, (\succeq)) \, ; \# \cap (\bullet)$$

where $\underline{\nabla}$ denotes the fork operator, $\succeq$ the converse of $\preceq$ and $\sigma \, (\bullet) \, (\sigma_1, \sigma_2) \Leftrightarrow \sigma = \sigma_1 \bullet \sigma_2$. Moreover, the Cartesian products coincides with the direct product, i.e., $P \times Q = P \otimes Q$. For the sake of simplicity we stay with the above given definitions, since the additional constructs provided by fork algebras are not required for our purposes.

3   Abstracting Dynamic Frames

Dynamic frames are represented in concrete program specifications as specification variables, i.e., variables that serve only for verification purposes and hence are not physically visible in the program itself. Their usage is to cover a set of locations of a state σ ranging over variables or allocated objects. By this mechanism one obtains the expressiveness to specify what a program or a method is allowed to modify and what remains untouched during its execution.

For an abstraction of the theory of dynamic frames we start by considering the concrete separation algebra DFSA. Frequently used examples in the theory of dynamic frames are the auxiliary specification variables

$$\text{used} = \text{used}_\sigma =_{df} dom(\sigma) \quad \text{and} \quad \text{unused} =_{df} \text{Loc} - \text{used} \, .$$

The former denotes the set of locations to which the state σ assigns values while the latter corresponds to all unallocated ones in that state. A *dynamic frame* f at a state σ is defined as a subset of Loc satisfying $f \subseteq$ used. Hence, dynamic frames are state dependent and may vary with state transitions, i.e., considering $\sigma \, P \, \sigma'$ for a command P and a dynamic frame f in σ then generally f in σ' will capture a different set of locations. Following the notation in [Kas11] a dynamic frame f in a final state σ' is denoted by f', i.e., it would correspond to $f_{\sigma'}$.

Our central goal is to derive an abstract and pointfree relational treatment of dynamic frames. Therefore, we are mainly interested in extracting behavioural patterns and aspects or effects of these. For a relational treatment we use a constant set of locations representing an initial dynamic frame f. The dynamic behaviour within state transitions $\sigma P \sigma'$ will be represented by relational and pointfree formalisations rather than using functions or expressions that depend on the states σ or σ'. This will allow more concise structural characterisations and pointfree proofs of basic properties involving dynamic frames.

Concretely, assuming an initial dynamic frame f to be a fixed set of locations we define

$$[\![f]\!] =_{df} \{(\sigma, \sigma) : f = dom(\sigma)\},$$

i.e., embedding f as a relation yields a subidentity which characterises all states where the allocated set of locations equals f. Note that $[\![f]\!] \neq \emptyset$, even if $f = \emptyset$, because then $[\![f]\!] = \{(u, u)\}$. For better readability we will omit the $[\![\_]\!]$ brackets in the following. The context will disambiguate the usage.

This embedding of f implies that the corresponding test satisfies a special behaviour which coincides with a pointfree characterisation of so-called *precise* tests [DM13]:

$$(f \times \mathsf{skip}) \,;\, \triangleright \,;\, \triangleleft \,;\, (f \times \mathsf{skip}) \subseteq f \times \mathsf{skip}. \tag{7}$$

In a pointwise form it reads for arbitrary states $\sigma, \sigma_1, \sigma_2$

$$(\sigma_1 \in f \wedge \sigma_2 \in f \wedge \sigma_1 \preceq \sigma \wedge \sigma_2 \preceq \sigma) \;\Rightarrow\; \sigma_1 = \sigma_2,$$

where $\tau \in f \Leftrightarrow_{df} \tau f \tau$ for arbitrary states τ and test f. This means that in any state τ a unique substate w.r.t. $\preceq$ that contains exactly the locations of f can always be pointed out.

As the next step we introduce pointfree relational variants of framing requirements that are crucial for the theory of dynamic frames [Kas11].

Definition 3.1 (Framing Requirements). Assume a dynamic frame f. Then the *modification* command $\Delta(\_)$ and *preservation* command $\Xi(\_)$ are defined by

$$\Delta f =_{df} \{(\sigma, \sigma') : \sigma|_{used-f} = \sigma'|_{used-f}\},$$
$$\Xi f =_{df} \{(\sigma, \sigma') : \sigma|_{f} = \sigma'|_{f}\}.$$

The modification requirement Δf intuitively asserts that at most resources captured by the frame f can be changed while any other resources remain untouched and hence are not modified. In particular, Δf allows the allocation of fresh storage. Conversely, Ξf asserts that at least the state parts characterised by f are not changed while anything else can be changed arbitrarily.

Theorem 3.2. *Assume a dynamic frame f. Then*

$$\Delta f = (f \,;\, \top) * \mathsf{skip} \qquad and \qquad \Xi f = f * \top.$$

Proof. By definition of Δ_-, definition of skip, by set theory and definition of $\top$, using f is a test, definition of $;$, and definition of $*$:

$$\sigma \, (\Delta f) \, \sigma'$$
$$\Leftrightarrow \sigma|_{used-f} = \sigma'|_{used-f}$$
$$\Leftrightarrow \sigma|_{used-f} \text{ skip } \sigma'|_{used-f}$$
$$\Leftrightarrow \sigma|_{used-f} \text{ skip } \sigma'|_{used-f} \wedge \sigma|_f \top \sigma'|_{used'-(used-f)}$$
$$\Leftrightarrow \sigma|_{used-f} \text{ skip } \sigma'|_{used-f} \wedge \sigma|_f \top \sigma'|_{used'-(used-f)} \wedge \sigma|_f \, f \, \sigma|_f$$
$$\Leftrightarrow \sigma|_{used-f} \text{ skip } \sigma'|_{used-f} \wedge \sigma|_f \, (f\,;\top) \, \sigma'|_{used'-(used-f)}$$
$$\Rightarrow \sigma \, ((f\,;\top)*\text{skip}) \, \sigma'\,.$$

For the reverse implication assume states $\sigma_f, \sigma_{skip}, \sigma_\top$ with $\sigma_f \in f \wedge \sigma = \sigma_f \bullet \sigma_{skip} \wedge \sigma' = \sigma_\top \bullet \sigma_{skip}$. Using Lemma 2.2 we get $\sigma|_f = (\sigma_f \bullet \sigma_{skip})|_f = \sigma_f$. Hence, $\sigma = \sigma|_f \bullet \sigma|_{used-f}$ and cancellativity implies $\sigma_{skip} = \sigma|_{used-f}$. Moreover, we can infer $\sigma'|_{used-f} = (\sigma_\top \bullet \sigma_{skip})|_{used-f} = (\sigma_\top \bullet \sigma|_{used-f})\big|_{used-f} = \sigma|_{used-f}$. Now Lemma 2.2 implies $\sigma_\top = \sigma'|_{used'-(used-f)}$.

By definition of Ξ_-, f is a test, set theory and definition of $\top$, and definition of $*$:

$$\sigma \, (\Xi f) \, \sigma'$$
$$\Leftrightarrow \sigma|_f = \sigma'|_f$$
$$\Leftrightarrow \sigma|_f \, f \, \sigma'|_f$$
$$\Leftrightarrow \sigma|_f \, f \, \sigma'|_f \wedge \sigma|_{used-f} \top \sigma'|_{used'-f}$$
$$\Rightarrow \sigma \, (f*\top) \, \sigma'\,.$$

The reverse implication can be proved analogously to the above case. $\qquad\square$

The algebraic embedding of dynamic frames as precise tests and their use in pointfree characterisations of the framing requirements yields the abstraction from the concrete DFSA separation algebra to arbitrary ones mentioned in Section 2.1. Moreover this allows calculational proofs of fundamental properties that establish the theory as a solution to tackle the frame problem (cf. Section 1). We begin with the following result: Assume two initial disjoint sets of locations f, g where only locations of f can be modified, then all locations of g will remain unchanged. The general idea of this is that expressions depending on locations of f will not affect expressions that depend only on locations in g.

Lemma 3.3. *Assume dynamic frames f, g. Then*

$$(f * g * \text{skip})\,;\Delta f \subseteq g * \Delta f\,.$$

Proof. By Theorem 3.2, definition of $*$, neutrality of skip and Equation (5), f is precise (Equation (7)), skip is neutral and Equation (5), definition of $*$, commutativity of $*$ and Theorem 3.2,

$$(f * g * \text{skip})\,;\Delta f$$
$$= (f * g * \text{skip})\,;((f\,;\top)*\text{skip})$$
$$= \vartriangleleft\,;(f \times (g * \text{skip}))\,;\vartriangleright\,;\vartriangleleft\,;(f\,;\top \times \text{skip})\,;\vartriangleright$$

$$= \quad \vartriangleleft ; (\mathsf{skip} \times (g * \mathsf{skip})) ; (f \times \mathsf{skip}) ; \vartriangleright ; \vartriangleleft ; (f \times \mathsf{skip}) ; (\top \times \mathsf{skip}) ; \vartriangleright$$
$$\subseteq \quad \vartriangleleft ; (\mathsf{skip} \times (g * \mathsf{skip})) ; (f \times \mathsf{skip}) ; (\top \times \mathsf{skip}) ; \vartriangleright$$
$$= \quad \vartriangleleft ; (f ; \top \times (g * \mathsf{skip})) ; \vartriangleright$$
$$= \quad (f ; \top) * g * \mathsf{skip}$$
$$= \quad g * \Delta f .$$

□

Since a dynamic frame f covers a set of locations on a state, it can be concluded that as long as f is not changed then all variables and expressions that depend on its locations will also remain unchanged. Expressions E can be abstracted relationally to tests that only include the states that assign values to at least all free variables occurring in E. Abstractly we define that a dynamic frame f *frames a test E* iff

$$(E * \mathsf{skip}) ; \Xi f \subseteq E * \top . \tag{8}$$

Ξf states that dynamic frame f is preserved while the test $E * \mathsf{skip}$ assumes a starting state σ that contains at least the required locations of E. Now by the relation $E * \top$ we can conclude that these locations will not be modified in a final state σ' since E is a test.

Altogether we can now prove a central theorem of the dynamic frames theory, stating that a dynamic frame will preserve its values while modifications on a disjoint frame are performed.

Lemma 3.4 (Value preservation). *Assume dynamic frames f, g. If g frames a test E then*

$$(E * \mathsf{skip}) ; (f * g * \mathsf{skip}) ; \Delta f \subseteq E * \top .$$

Proof. By Lemma 3.3, isotony, Theorem 3.2 and g frames E (Equation (8)),

$$(E * \mathsf{skip}) ; (f * g * \mathsf{skip}) ; \Delta f \subseteq (E * \mathsf{skip}) ; (g * \Delta f) \subseteq (E * \mathsf{skip}) ; (g * \top) \subseteq E * \top .$$

□

The abstraction of dynamic frames to sets of locations and representing them relationally as precise tests implies that they already come with the so-called *self-framing* property. It is used in the program specifications of [Kas11] to maintain that initial disjointness of dynamic frames is preserved in final states. Concretely it characterises a dynamic frame to be preserved whenever the environment does not change its value.

Lemma 3.5. *Dynamic frames are self-framing.*

Proof. Follows directly from $f * \mathsf{skip} \subseteq \mathsf{skip}$, isotony of $;$ and Theorem 3.2.   □

Basically, dynamic frames in concrete verification applications are always defined to be self-framing. Hence, this does not impose a restriction on the theory.

We continue with an auxiliary result that is required for later calculations.

Lemma 3.6. *For a dynamic frame f we have $\ulcorner(\Delta f) = f * \mathsf{skip} = \ulcorner(\Xi f)$.*

A proof can be found in the appendix.

4   Locality and Frame Accumulation

The relational structure of modification commands (cf. Theorem 3.2) reveals that they are related to so-called *local* commands [DM12, HHM^{+}11]. These commands have the following special behaviour: at most resources in the footprint[1] of such a command are modified while all other resources are left unchanged. Relationally, local commands P are simply characterised by the equation $P * \mathsf{skip} = P$ [DM12]. For modifications we can immediately conclude

Lemma 4.1. *Modifications Δf are local commands.*

Proof. By Theorem 3.2, associativity of $*$, $\mathsf{skip} * \mathsf{skip} = \mathsf{skip}$,

$$\Delta f * \mathsf{skip} = ((f\,;\top) * \mathsf{skip}) * \mathsf{skip} = (f\,;\top) * (\mathsf{skip} * \mathsf{skip}) = (f\,;\top) * \mathsf{skip} = \Delta f.$$

$\square$

Basically, pairs of commands within $*$ - compositions operate separately on disjoint portions of states. In [DM12] it turned out that due to the angelic behaviour of relations, an additional assumption is required for pointfree calculations on the footprint and the resources that remain untouched in $*$-products. The assumption can be encoded relationally by the *frame property* [DHM11], i.e.,

$$(\ulcorner P \times \mathsf{skip}) \,;\, \rhd \,;\, P \subseteq (P \times \mathsf{skip}) \,;\, \rhd. \tag{9}$$

In pointwise form it reads as follows, considering arbitrary $\sigma_P, \sigma_{\mathsf{skip}}, \sigma'$ in a pair $((\sigma_P, \sigma_{\mathsf{skip}}), \sigma')$ of the left-hand side:

$$\sigma_P \in \ulcorner P \wedge (\sigma_P \bullet \sigma_{\mathsf{skip}}) \, P \, \sigma' \;\Rightarrow\; \exists \sigma'_P. \;\; \sigma_P \, P \, \sigma'_P \wedge \sigma' = \sigma'_P \bullet \sigma_{\mathsf{skip}}.$$

This implies that the state portion σ_{skip} above does not contain any resources that P would need for a successful execution and hence is not affected by the execution of P. Equation (9) is named after the frame property of separation logic, since it basically reflects similar behaviour [DHM11]. It can be shown that local commands with a precise footprint satisfy this inequation as, e.g., in the case of modifications Δf.

Lemma 4.2. *Modifications Δf have the frame property.*

A proof can be found in the appendix.

In the present work, Equation (9) will be applied to prove a relational version of the *frame accumulation* law of [Kas11]. For a better intuition we start by providing the logical version of that law and describe its semantics. It is originally given as an imperative specification, i.e., a Boolean expression that is relationally evaluated on arbitrary pairs (σ, σ') where σ denotes the initial and σ' the final state of an arbitrary execution. The accumulation law reads as follows

$$(\Delta f \wedge g' \subseteq f \cup \mathsf{unused})\,;\, \Delta g \;\Rightarrow\; \Delta f. \tag{10}$$

---

[1] The minimal set of resources required for non-aborting executions.

The relational version of the accumulation law is to be understood pointwise on arbitrary pairs (σ, σ') by

$$(\exists \sigma''.\ \sigma\, \Delta f\, \sigma'' \wedge g(\sigma'') \subseteq f(\sigma) \cup \mathsf{unused}(\sigma) \wedge \sigma''\, \Delta g\, \sigma') \ \Rightarrow\ \sigma\, \Delta f\, \sigma'$$

where σ denotes an initial state and σ' a final state. Note that the dynamic frame g' of Equation (10) denotes the final value of g on the intermediate state σ'' instead of σ'.

The law means that whenever g in the intermediate state is bounded by f and can only increase by initially unallocated resources then the overall effect is that at most locations in f are changed in the composition $\Delta f\,;\Delta g$. Or equivalently, all allocated resources initially from f disjoint are preserved. For an algebraic proof we need a pointfree variant to characterise bounds for dynamic frames within modifications, which is of course not trivial to achieve since dynamic frames are state-dependent.

Definition 4.3. For dynamic frames f, g we say that g is bounded by f iff

$$\#\,;(f\,;\top \times \mathsf{skip})\,;\triangleright\,;(g * \mathsf{skip}) \subseteq (f\,;\top\,;(g * \mathsf{skip}) \times \mathsf{skip})\,;\triangleright.$$

To understand the intuition of this formula within the dynamic frames theory we describe its meaning in the concrete separation algebra DFSA. Of course it can be interpreted in other adequate separation algebras, too. Assume an arbitrary pair $((\sigma_f, \sigma_{\mathsf{skip}}), \sigma')$ from the left-hand side of the above inequation. In a pointwise form the premise then reads

$$\exists \sigma_\top, \sigma_g, \tau_{\mathsf{skip}}.\ \sigma_f \in f \wedge \sigma_f \# \sigma_{\mathsf{skip}} \wedge \sigma_\top \bullet \sigma_{\mathsf{skip}} = \sigma_g \bullet \tau_{\mathsf{skip}} = \sigma' \wedge \sigma_g \in g.$$

Intuitively the substate σ_f represents that part of the complete state $\sigma_f \bullet \sigma_{\mathsf{skip}}$ that can be changed while σ_{skip} corresponds to the untouched part in which any changes to resources are not permitted. By assuming $\exists \sigma'.\ \sigma' = \sigma_\top \bullet \sigma_{\mathsf{skip}}$ we also know $\sigma_\top \# \sigma_{\mathsf{skip}}$ and hence σ_{skip} is also disjoint from any additionally allocated resources, i.e., $dom(\sigma_{\mathsf{skip}})$ is disjoint from any locations of $\mathsf{unused}(\sigma_f \bullet \sigma_{\mathsf{skip}})$.

Now, the right-hand side states that

$$\exists \sigma_{\mathsf{rem}}.\ \sigma_f \in f \wedge \sigma' = (\sigma_g \bullet \sigma_{\mathsf{rem}}) \bullet \sigma_{\mathsf{skip}} \wedge \sigma_g \in g.$$

This means by cancellativity of the underlying separation algebra that $\sigma_\top = \sigma_g \bullet \sigma_{\mathsf{rem}}$ and $\tau_{\mathsf{skip}} = \sigma_{\mathsf{rem}} \bullet \sigma_{\mathsf{skip}}$. Hence, $\sigma_g \preceq \sigma_\top$ and $\sigma_{\mathsf{skip}} \preceq \tau_{\mathsf{skip}}$. In particular, we get $\sigma_g \# \sigma_{\mathsf{skip}}$, i.e., σ_g is disjoint from σ_{skip} which in turn implies that its allocated locations can only cover locations of f and initially unallocated ones in $\mathsf{unused}(\sigma_f \bullet \sigma_{\mathsf{skip}})$. The above state partitions can be depicted as follows:

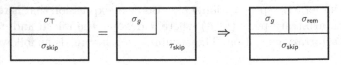

Conversely, we can show using cross-split and disjointness that the underlying separation algebra satisfies the inequation of Definition 4.3, assuming $\sigma_g \# \sigma_{\mathsf{skip}}$.

To see this, note that the premise asserts $\sigma_\top \bullet \sigma_{\mathsf{skip}} = \sigma_g \bullet \tau_{\mathsf{skip}}$ and hence $\sigma_\top \# \sigma_{\mathsf{skip}}$. By cross-split, i.e., Equation (2) we infer

$$\exists \sigma_1, \sigma_2, \sigma_3, \sigma_4 \,.\; \sigma_\top = \sigma_1 \bullet \sigma_2 \wedge \sigma_{\mathsf{skip}} = \sigma_3 \bullet \sigma_4 \;\wedge$$
$$\sigma_g = \sigma_1 \bullet \sigma_3 \;\wedge\; \tau_{\mathsf{skip}} = \sigma_2 \bullet \sigma_4 \,.$$

Thus, $\sigma_g \# \sigma_{\mathsf{skip}} \Leftrightarrow \sigma_1 \bullet \sigma_3 \# \sigma_3 \bullet \sigma_4 \Rightarrow \sigma_3 \# \sigma_3$ and Equation (3) implies that $\sigma_3 = u$. By this we immediately have $\sigma_g = \sigma_1 \wedge \sigma_{\mathsf{skip}} = \sigma_4$ and therefore $\sigma_\top = \sigma_g \bullet \sigma_2 \wedge \tau_{\mathsf{skip}} = \sigma_2 \bullet \sigma_{\mathsf{skip}}$. Since $\sigma_\top \# \sigma_{\mathsf{skip}}$ we can instantiate σ_{rem} as σ_2.

Unfortunately, Definition 4.3 is more complex than its logical variant which is due to implicitly expressing the particular restriction of g to unallocated resources w.r.t. f. However, with Definition 4.3 we now have the possibility to abstractly relate dynamic frames among each other and can continue by reasoning in an (in)equational style. By this we can summarise a central result of dynamic frames within modifications.

Theorem 4.4. *Assume dynamic frames where g is bounded by f then*

$$\Delta f \,;\, \Delta g \subseteq (f \,;\, \top \,;\, \Delta g) * \mathsf{skip} \,.$$

Proof. By Theorem 3.2, Corollary 2.4 and Lemma 3.6, g is bounded by f, $\mathsf{skip} = \mathsf{skip} \,;\, \mathsf{skip}$ and Equation (5), Lemma 3.6, Δg has the frame property and Equation (5) again, and definition of $*$:

$$
\begin{aligned}
&\quad \Delta f \,;\, \Delta g \\
&= \; \triangleleft \,;\, (f \,;\, \top \times \mathsf{skip}) \,;\, \triangleright \,;\, \Delta g \\
&= \; \triangleleft \,;\, \# \,;\, (f \,;\, \top \times \mathsf{skip}) \,;\, \triangleright \,;\, (g * \mathsf{skip}) \,;\, \Delta g \\
&\subseteq \; \triangleleft \,;\, (f \,;\, \top \,;\, (g * \mathsf{skip}) \times \mathsf{skip}) \,;\, \triangleright \,;\, \Delta g \\
&= \; \triangleleft \,;\, (f \,;\, \top \times \mathsf{skip}) \,;\, ((g * \mathsf{skip}) \times \mathsf{skip}) \,;\, \triangleright \,;\, \Delta g \\
&\subseteq \; \triangleleft \,;\, (f \,;\, \top \times \mathsf{skip}) \,;\, (\ulcorner \Delta g \times \mathsf{skip}) \,;\, \triangleright \,;\, \Delta g \\
&\subseteq \; \triangleleft \,;\, (f \,;\, \top \,;\, \Delta g \times \mathsf{skip}) \,;\, \triangleright \\
&= \; (f \,;\, \top \,;\, \Delta g) * \mathsf{skip} \,.
\end{aligned}
$$

$\square$

This characterises the behaviour that only the changes on the execution within f need to be considered for Δg if g is bounded by f, while all other allocated locations w.r.t a starting state will remain unchanged.

Corollary 4.5 (Frame Accumulation). *Assume dynamic frames f, g where g is bounded by f. Then*

$$\Delta f \,;\, \Delta g \subseteq \Delta f \,.$$

Proof. By Theorem 4.4, isotony and definition of $\top$, and Theorem 4.4:

$$\Delta f \,;\, \Delta g \subseteq (f \,;\, \top \,;\, \Delta g) * \mathsf{skip} \subseteq (f \,;\, \top) * \mathsf{skip} = \Delta f \,.$$

$\square$

This result can be interpreted as a pointfree variant of the frame accumulation theorem of [Kas11] (cf. Equation (10)). Its general application is to simplify correctness proofs of specifications by eliminating occurrences of sequential composition in combination with framing requirements.

In [Kas11] the concept of *strong dynamic frames* is also defined. Such frames f come with the additional restriction on a final state σ' that $f(\sigma')$ can only contain locations of $f(\sigma)$ for a starting state σ or unallocated ones w.r.t. σ. Since the given abstractions of dynamic frames in this work imply that they are always self-framing, the modifications Δf are only able to extend f in σ' by previously unallocated locations as in [Kas11]. Hence, simple modifications Δf already coincide with the stronger variant within our abstraction.

As a final result we present another treatment of the abstracted theory in the context of related work.

5   A Related Approach: Local Actions

In [COY07] an abstract approach to separation logic was presented that is built on separation algebras and provides a model of programs in terms of so-called *local actions*. By contrast with the relational approach of Section 2.2 this concept works pointwise. We show in the following by the use of previous ideas about abstracting dynamic frames that formalisations about modifications in that approach satisfy a similar locality condition and allow a calculational proof of the frame accumulation law within the separation algebra DFSA.

Basically, local actions are special state transformers, i.e., special functions that map from states to sets of states or to a distinguished element $\top$ [2]. The element $\top$ is used to denote program abortion, e.g., due to dereferencing of non-allocated resources.

There is also an order $\sqsubseteq$ defined on sets of states and $\top$. For arbitrary sets of states $p, q \in \mathcal{P}(\Sigma)$ it is defined by $p \sqsubseteq q =_{df} p \subseteq q$. Moreover, $\top$ is the greatest element w.r.t. the order $\sqsubseteq$, i.e., for arbitrary $p \in \mathcal{P}(\Sigma) \cup \{\top\}$ we have $p \sqsubseteq \top$. One can extend $\sqsubseteq$ pointwise to state transformers f, g by $f \sqsubseteq g \Leftrightarrow_{df} \forall \sigma.\ f(\sigma) \sqsubseteq g(\sigma)$.

Separating conjunction $*$ on sets of states is given by

$$p * q =_{df} \begin{cases} \{\sigma_1 \bullet \sigma_2 :\ \sigma_1 \# \sigma_2,\ \sigma_1 \in p,\ \sigma_2 \in q\} & \text{if } p, q \in \mathcal{P}(\Sigma) \\ \top & \text{otherwise}. \end{cases}$$

A proper definition of $*$ on strongest postcondition state transformers might lead to problems with associativity. Hence we stay with the definitions of the original approach. A state transformer definition for modifications can be obtained for a fixed set of locations f with the same ideas as in Section 3 by

$$(\Delta f)(\sigma) =_{df} \begin{cases} \Sigma * \{\sigma|_{\mathsf{used}-f}\} & \text{if } f \subseteq \mathsf{used}(\sigma) \\ \top & \text{otherwise}. \end{cases}$$

Intuitively, whenever all locations of f are allocated then all other used locations in σ are preserved. Otherwise, an erroneous execution is signalled by the output $\top$. Analogously, in the case of Ξf we can define

---

[2] $\top$ does not denote the universal relation in this context.

$$(\Xi f)(\sigma) \ =_{df} \ \begin{cases} \Sigma * \{\sigma|_f\} & \text{if } f \subseteq \mathsf{used}(\sigma) \\ \top & \text{otherwise}. \end{cases}$$

According to Lemma 4.2, the relational version of Δf satisfies the frame property, i.e., Equation (9). Similar behaviour is obtained for the state transformer definition of Δf by the *locality* property of [COY07], i.e.,

$$\sigma_1 \# \sigma_2 \ \Rightarrow \ (\Delta f)(\sigma_1 \bullet \sigma_2) \sqsubseteq (\Delta f)(\sigma_1) * \{\sigma_2\}. \tag{11}$$

State transformers that satisfy Equation (11) are called *local actions*. The locality property has similar behaviour as the relational version of the frame property. The state σ_2 represents that part of the state $\sigma_1 \bullet \sigma_2$ that will remain unchanged while σ_1 contains the footprint of Δf.

For a proof of Equation (11) a case distinction is needed. First assume $\sigma_1 \# \sigma_2$. If $f \not\subseteq \mathsf{used}(\sigma_1)$ then $(\Delta f)(\sigma_1) * \{\sigma_2\} = \top * \{\sigma_2\} = \top$ and the inequation holds. Now assume $f \subseteq \mathsf{used}(\sigma_1)$, then

$$\begin{aligned} (\Delta f)(\sigma_1 \bullet \sigma_2) &= \Sigma * \{\sigma_1 \bullet \sigma_2|_{\mathsf{used}(\sigma_1 \bullet \sigma_2)-f}\} \\ &\sqsubseteq \Sigma * \{\sigma_1|_{\mathsf{used}(\sigma_1)-f} \bullet \sigma_2\} \\ &= \Sigma * \{\sigma_1|_{\mathsf{used}(\sigma_1)-f}\} * \{\sigma_2\} \\ &= \Delta f * \{\sigma_2\}. \end{aligned}$$

Next we show that a treatment of the frame accumulation law is also possible using local actions. For a translation of the frame accumulation law into that setting we need to define a local action that models the restricted modification given in its logical variant (cf. Equation (10))

$$\Delta(f,g) \ =_{df} \ \Delta f \wedge g' \subseteq f \cup \mathsf{unused}(\sigma).$$

Note that $g' = g(\sigma')$ generally implies the existence of a set of locations g in each state σ' in the result set $(\Delta f)(\sigma)$, interpreting modifications as a local action. By this we need to restrict the local action definition of modification Δf as follows to get a local action for $\Delta(f,g)$

$$(\Delta(f,g))(\sigma) \ =_{df} \ \begin{cases} \{\sigma' : \mathsf{used}(\sigma') = g\} * \Sigma * \{\sigma|_{\mathsf{used}-f}\} & \text{if } f \subseteq \mathsf{used}(\sigma) \\ \top & \text{otherwise}. \end{cases}$$

The general idea with this is to restrict the output of Δf to involve a fixed set of locations g. Another possibility would be to define another local action that sequentially composed with Δf restricts its output adequately. The above local action for $\Delta(f,g)$ includes the behaviour described in Definition 4.3 in which a bounding between dynamic frames g and f is characterised. Analogously to Δf, the state transformer is also a local action. Now, the frame accumulation law in that setting can be stated as follows

$$\forall \sigma. \ (\Delta(f,g) \,;\, \Delta g)(\sigma) \sqsubseteq \Delta f(\sigma),$$

where for arbitrary local actions f, g one pointwise lifts $(f\,;g)(\sigma) =_{df} \bigsqcup \{ g(\sigma') : \sigma' \in f(\sigma)\}$ if $f(\sigma) \neq \top$ and otherwise $f\,;g$ also equals $\top$. For a proof of the above inequation we assume $f \subseteq \mathsf{used}(\sigma)$ and $g \subseteq f \cup \mathsf{unused}(\sigma)$ and calculate

$$
\begin{aligned}
(\Delta(f,g)\,;\Delta g)\,(\sigma) &= \bigsqcup\{\,\Delta g\,(\sigma'') : \sigma'' \in \{\sigma' : \ \mathsf{used}(\sigma') = g\ \} * \Sigma * \{\sigma|_{\mathsf{used}-f}\}\} \\
&= \bigsqcup\{\,\Delta g\,(\sigma' \bullet \tau \bullet \sigma|_{\mathsf{used}-f}) : \mathsf{used}(\sigma') = g,\, \tau \in \Sigma\} \\
&\sqsubseteq \bigsqcup\{\,\Delta g\,(\sigma') * \{\tau \bullet \sigma|_{\mathsf{used}-f}\} : \mathsf{used}(\sigma') = g,\, \tau \in \Sigma\} \\
&= \bigsqcup\{\,\Sigma * \{\tau \bullet \sigma|_{\mathsf{used}-f}\} : \tau \in \Sigma\} \\
&\sqsubseteq \bigsqcup\{\,\Sigma * \{\sigma|_{\mathsf{used}-f}\}\} \\
&= \Sigma * \{\sigma|_{\mathsf{used}-f}\} \\
&= \Delta f\,(\sigma)\,.
\end{aligned}
$$

6 Conclusion and Outlook

We explored algebraic and abstract calculi for the theory of dynamic frames. It turned out that an extended relational approach, originally used as an algebraic base for separation logic, can also be used to generally formalise effects of the dynamic frames theory. Since definitions in that theory were given in [Kas11] in a relational style, a direct translation to relational pointfree variants was possible by a few abstractions. This yields a step towards a unifying calculus for abstractly capturing crucial behaviours of dynamic frames and separation logic.

As further work it would be interesting to include the *overlapping conjunction* of [HV13] into this setting. Applied to assertions it allows an unspecified portion of resources to be shared among two predicates. For the presented calculus, it would enable an abstract treatment of dynamic frames that share certain parts of their locations as e.g., in the situation when two iterators are attached to the same list as described in [Kas11]. Another possibility for this can be considering separation algebras that involve permissions [BCOP05].

Moreover, the relationships to concrete approaches [DYDG+10, PS11, JB12] and their integration into this framework has to be investigated.

Acknowledgements. I am grateful to Eric C. R. Hehner and Bernhard Möller for drawing my attention to this particular topic. Moreover, I thank Bernhard Möller for valuable remarks and all anonymous reviewers for their helpful feedback and comments that significantly helped to improve the paper. This research was partially funded by the DFG project *MO 690/9-1* ALGSEP — *Algebraic Calculi for Separation Logic*.

References

[BCOP05]   Bornat, R., Calcagno, C., O'Hearn, P.W., Parkinson, M.J.: Permission Accounting in Separation Logic. In: Palsberg, J., Abadi, M. (eds.) Proc. of the 32nd ACM SIGPLAN-SIGACT Symposium on Principles of Programming Languages, pp. 259–270. ACM Press (2005)

[COY07]   Calcagno, C., O'Hearn, P.W., Yang, H.: Local Action and Abstract Separation Logic. In: Proc. of the 22nd Symposium on Logic in Computer Science, pp. 366–378. IEEE Press (2007)

[DHM11]   Dang, H.-H., Höfner, P., Möller, B.: Algebraic Separation Logic. Journal of Logic and Algebraic Programming 80, 221–247 (2011)

[DM12]   Dang, H.-H., Möller, B.: Reverse Exchange for Concurrency and Lo-
            cal Reasoning. In: Gibbons, J., Nogueira, P. (eds.) MPC 2012. LNCS,
            vol. 7342, pp. 177–197. Springer, Heidelberg (2012)

[DM13]   Dang, H.-H., Möller, B.: Concurrency and Local Reasoning under Re-
            verse Exchange. Science of Computer Programming (2013)

[DYDG$^+$10] Dinsdale-Young, T., Dodds, M., Gardner, P., Parkinson, M.J., Vafeiadis,
            V.: Concurrent Abstract Predicates. In: D'Hondt, T. (ed.) ECOOP 2010.
            LNCS, vol. 6183, pp. 504–528. Springer, Heidelberg (2010)

[DHA09]   Dockins, R., Hobor, A., Appel, A.W.: A Fresh Look at Separation Al-
            gebras and Share Accounting. In: Hu, Z. (ed.) APLAS 2009. LNCS,
            vol. 5904, pp. 161–177. Springer, Heidelberg (2009)

[FBH97]   Frias, M.F., Baum, G., Haeberer, A.M.: Fork Algebras in Algebra, Logic
            and Computer Science. Fundam. Inform. 32, 1–25 (1997)

[GGN11]   Garbervetsky, D., Gorín, D., Neisen, A.: Enforcing Structural Invariants
            using Dynamic Frames. In: Abdulla, P.A., Leino, K.R.M. (eds.) TACAS
            2011. LNCS, vol. 6605, pp. 65–80. Springer, Heidelberg (2011)

[HHM$^+$11] Hoare, C.A.R., Hussain, A., Möller, B., O'Hearn, P.W., Petersen, R.L.,
            Struth, G.: On Locality and the Exchange Law for Concurrent Processes.
            In: Katoen, J.-P., König, B. (eds.) CONCUR 2011. LNCS, vol. 6901,
            pp. 250–264. Springer, Heidelberg (2011)

[HV13]   Hobor, A., Villard, J.: The Ramifications of Sharing in Data Struc-
            tures. In: Giacobazzi, R., Cousot, R. (eds.) Proc. of the 40th annual
            ACM SIGPLAN-SIGACT Symposium on Principles of Programming
            Languages, POPL, pp. 523–536. ACM Press (2013)

[JB12]   Jensen, J.B., Birkedal, L.: Fictional Separation Logic. In: Seidl, H. (ed.)
            ESOP 2012. LNCS, vol. 7211, pp. 377–396. Springer, Heidelberg (2012)

[Kas11]   Kassios, I.T.: The Dynamic Frames Theory. Formal Aspects of Comput-
            ing 23, 267–289 (2011)

[Lei10]   Leino, K.R.M.: Dafny: An Automatic Program Verifier for Functional
            Correctness. In: Clarke, E.M., Voronkov, A. (eds.) LPAR-16 2010. LNCS,
            vol. 6355, pp. 348–370. Springer, Heidelberg (2010)

[MH69]   McCarthy, J., Hayes, P.J.: Some Philosophical Problems from the Stand-
            point of Artificial Intelligence. In: Meltzer, B., Michie, D. (eds.) Machine
            Intelligence 4, pp. 463–502. Edinburgh University Press (1969)

[PS11]   Parkinson, M.J., Summers, A.J.: The Relationship between Separation
            Logic and Implicit Dynamic Frames. In: Barthe, G. (ed.) ESOP 2011.
            LNCS, vol. 6602, pp. 439–458. Springer, Heidelberg (2011)

[Rey02]   Reynolds, J.C.: Separation Logic: A Logic for Shared Mutable Data
            Structures. In: Proc. of the 17th Annual IEEE Symposium on Logic
            in Computer Science, pp. 55–74. IEEE Computer Society (2002)

[SJP09]   Smans, J., Jacobs, B., Piessens, F.: Implicit Dynamic Frames: Combin-
            ing Dynamic Frames and Separation Logic. In: Drossopoulou, S. (ed.)
            ECOOP 2009. LNCS, vol. 5653, pp. 148–172. Springer, Heidelberg (2009)

Appendix: Deferred Proofs

Proof of Lemma 3.6.

By Theorem 3.2 and def. of $*$, f is a test, $\mathsf{skip} = \mathsf{skip}\,;\mathsf{skip}$ and Equation (5), Corollary 2.4 and Corollary 2.5, again Corollary 2.4, Theorem 3.2, def. of $*$,

$$
\begin{aligned}
\Delta f &= \lhd\,;(f\,;\top\times\mathsf{skip})\,;\rhd \\
&= \lhd\,;(f\,;f\,;\top\times\mathsf{skip})\,;\rhd \\
&= \lhd\,;(f\times\mathsf{skip})\,;(f\,;\top\times\mathsf{skip})\,;\rhd \\
&= \lhd\,;(f\times\mathsf{skip})\,;\#\,;(f\,;\top\times\mathsf{skip})\,;\rhd \\
&\subseteq \lhd\,;(f\times\mathsf{skip})\,;\rhd\,;\lhd\,;(f\,;\top\times\mathsf{skip})\,;\rhd \\
&= (f*\mathsf{skip})\,;\Delta f\,.
\end{aligned}
$$

Hence, by Equation (4) we have $\ulcorner(\Delta f) \subseteq f*\mathsf{skip}$. For the converse we calculate $f*\mathsf{skip} = \ulcorner(f*\mathsf{skip}) \subseteq \ulcorner((f\,;\top)*\mathsf{skip}) = \ulcorner(\Delta f)$. Analogous calculations show the result for Ξf. $\qquad\square$

Proof of Lemma 4.2.

$$
\sigma_1\,\ulcorner(\Delta f)\,\sigma_1 \wedge \sigma_1\bullet\sigma_2\,\Delta f\,\sigma'
$$

$\Leftrightarrow$ { Lemma 4.1 and Lemma 3.6 }

$$
\sigma_1\,f*\mathsf{skip}\,\sigma_1 \wedge \sigma_1\bullet\sigma_2\,\Delta f*\mathsf{skip}\,\sigma'
$$

$\Leftrightarrow$ { definition of $*$ }

$$
\exists\,\sigma_f,\sigma_\mathsf{skip},\tau_1,\tau_2,\tau_1'.\ \ \sigma_1 = \sigma_f\bullet\sigma_\mathsf{skip} \wedge \sigma_f \in f \wedge \sigma_1\bullet\sigma_2 = \tau_1\bullet\tau_2 \\
\wedge\ \tau_1\,\Delta f\,\tau_1' \wedge \sigma' = \tau_1'\bullet\tau_2
$$

$\Leftrightarrow$ { Theorem 3.2 }

$$
\exists\,\sigma_f,\sigma_\mathsf{skip},\tau_1,\tau_2,\tau_1',\tau_f,\tau_f',\tau_\mathsf{skip}.\ \ \sigma_1 = \sigma_f\bullet\sigma_\mathsf{skip} \wedge \sigma_f \in f \wedge \\
\sigma_1\bullet\sigma_2 = \tau_1\bullet\tau_2 \wedge\ \tau_1 = \tau_f\bullet\tau_\mathsf{skip} \wedge \tau_1' = \tau_f'\bullet\tau_\mathsf{skip} \wedge \tau_f \in f \wedge \\
\sigma' = \tau_1'\bullet\tau_2
$$

$\Rightarrow$ { Equation (7) implies $\sigma_f = \tau_f$, logic }

$$
\exists\,\sigma_f,\sigma_\mathsf{skip},\tau_1,\tau_2,\tau_f',\tau_\mathsf{skip}.\ \ \sigma_1 = \sigma_f\bullet\sigma_\mathsf{skip} \wedge \sigma_f \in f \wedge \\
\sigma_1\bullet\sigma_2 = \tau_1\bullet\tau_2 \wedge\ \tau_1 = \sigma_f\bullet\tau_\mathsf{skip} \wedge \sigma' = \tau_f'\bullet\tau_\mathsf{skip}\bullet\tau_2
$$

$\Rightarrow$ { cancellativity implies $\sigma_\mathsf{skip}\bullet\sigma_2 = \tau_\mathsf{skip}\bullet\tau_2$, logic }

$$
\exists\,\sigma_f,\sigma_\mathsf{skip},\tau_f'.\ \ \sigma_1 = \sigma_f\bullet\sigma_\mathsf{skip} \wedge \sigma_f \in f\ \wedge \sigma' = (\tau_f'\bullet\sigma_\mathsf{skip})\bullet\sigma_2
$$

$\Leftrightarrow$ { definition of $;$, $*$ and Theorem 3.2 }

$$
\exists\,\sigma_1'.\ \ \sigma_1\,\Delta f\,\sigma_1' \wedge \sigma' = \sigma_1'\bullet\sigma_2\,.
$$

$\qquad\square$

Automated Verification
of Relational While-Programs

Rudolf Berghammer[1], Peter Höfner[2,3], and Insa Stucke[1]

[1] Institut für Informatik, Christian-Albrechts-Universität zu Kiel, Germany
[2] NICTA, Australia
[3] Computer Science and Engineering, University of New South Wales, Australia

Abstract. Software verification is essential for safety-critical systems. In this paper, we illustrate that some verification tasks can be done fully automatically. We show how to automatically verify imperative programs for relation-based discrete structures by combining relation algebra and the well-known assertion-based verification method with automated theorem proving. We present two examples in detail: a relational program for determining the reflexive-transitive closure and a topological sorting algorithm. We also treat the automatic verification of the equivalence of common-logical and relation-algebraic specifications.

1   Introduction

Many discrete structures of mathematics and computer science, such as orders, lattices, certain classes of graphs, Petri nets, and games, are relations or can easily be modelled by means of relations. In such cases computational tasks frequently reduce computations on relations and the correctness proofs of the corresponding algorithms to proofs of statements over relations.

In the past, various techniques for programming with relations have been proposed. In this paper, we follow an approach that considers relations only as data structures and manipulates them with a simple, imperative programming language. It is straightforward to translate the relational programs into more efficient programming languages such as Java or C. The approach also bears methodical advantages: if problem specifications are expressed via relation-algebraic formulae, then the correctness proofs allow to intertwine approved program verification steps with formal and precise relation-algebraic calculations. This mathematical rigour drastically reduces errors in the programs. Moreover, this approach is supported by tools for (a) prototyping and testing, (b) interactive theorem proving, and (c) automatic theorem proving. An example for prototyping and testing relation-algebraic specifications is RELVIEW (cf. [30]), which allows the evaluation of relation-algebraic expressions and the formulation of relational programs. With regard to interactive theorem proving either special purpose systems, such as RALF (see [14]), can be used, or relation-algebraic techniques can be integrated into existing provers (for example into Isabelle/HOL as described in [28,12]). Full automatisation of proofs can frequently be achieved

P. Höfner et al. (Eds.): RAMiCS 2014, LNCS 8428, pp. 173–190, 2014.

by off-the-shelf automated theorem provers, such as Prover9 (see [31]). We refer to [16] for such an application. In the present paper, we will follow the latter approach and use Prover9 for automated program verification of while-programs.

Formal verification of imperative programs is often done by use of pre- and post-conditions as problem specifications, and loop-invariants; see e.g., [10,11,13]. This so-called assertion-based technique is particularly useful for while-loops, where it is sufficient to show that the loop-invariant is established and maintained (under the assumption that the pre-condition holds) and that the post-condition is valid as soon as the while-loop terminates. The combination of program verification and relation algebra we are going to use is not new; it was applied in several case studies, for instance in [1,2,3,4].

Encouraged by the practicability and elegance of the latter results and the positive experiences of [16], the combination of assertion-based program verification and relation algebra was combined with automated theorem proving using Prover9; see [5]. This paper is a continuation as well as a step further of this idea. We consider two new and more sophisticated examples, viz. the computation of reflexive-transitive closures by means of decomposition and the computation of topological sortings in case of cycle-free relations. We further demonstrate how the equivalences of the logical specifications and their relation-algebraic counterparts can automatically be verified using Prover9. The paper closes with a short discussion on the lessons we have learned from the two case studies.

2   Preliminaries

In this paper, we formalise data structures and assertions of imperative programs by homogeneous relation algebra, as axiomatised by Tarski in [26]. The pre-conditions, post-conditions, loop-invariants and proof obligations will be formalised via expressions and formulae in relation algebra and implemented in Prover9. In this section, we recapitulate the basic concepts of the relational calculus and its automation via automated theorem proving, which are needed later on. For more details we refer to [22,23] concerning relation algebra, to [31] concerning Prover9, and to [7,24] concerning the use of automated theorem proving in general software engineering.

2.1   (Homogeneous) Relation Algebra

Homogeneous relation algebra was first axiomatised in [26] and further developed in [8,27]. A relation R over a set $\mathcal{X}$, the *universe*, is a subset of the direct product $\mathcal{X} \times \mathcal{X}$. Relation algebra offers five operations on relations, viz. $R \cup S$ (union), $R \cap S$ (intersection), $\overline{R}$ (complement), $R;S$ (composition) and R^{T} (transposition), two predicates to compare relations, viz. $R \subseteq S$ (inclusion) and $R = S$ (equality), and three special relations: O (empty relation), L (universal relation), and I (identity relation). Except composition, transposition and the identity relation all concepts are defined by standard set theory. The composition $R;S$ of two relations R and S is the set of all pairs $(x,y) \in \mathcal{X} \times \mathcal{X}$ such that

$(x, z) \in R$ and $(z, y) \in S$ for some $z \in \mathcal{X}$, the transposition R^T is the set of all pairs $(x, y) \in \mathcal{X} \times \mathcal{X}$ with $(y, x) \in R$, and the identity relation I is the set of all pairs $(x, y) \in \mathcal{X} \times \mathcal{X}$ with $x = y$.

These definitions form the base of *concrete relation algebras*. An *(abstract) relation algebra* abstracts from set theory and is axiomatised as follows, where we follow the axiomatisation of [22] instead of [8,26,27].

1. With regard to $^-$, $\cup, \cap$, the order $\subseteq$, and the constants O and L the relations form a Boolean algebra.
2. With regard to composition and the identity relation I the relations form a monoid.
3. The *Dedekind rule* holds, i.e., for all relations Q, R and S we have

$$Q;R \cap S \subseteq (Q \cap S;R^\mathsf{T});(R \cap Q^\mathsf{T};S) . \tag{1}$$

Since all axioms are first-order, it is easy to encode them in any off-the-shelf automated theorem prover.

From the Dedekind rule we obtain the so-called *Schröder equivalences* (also known as "Theorem K" of de Morgan). They state that

$$Q;R \subseteq S \iff Q^\mathsf{T};\overline{S} \subseteq \overline{R} \qquad\qquad Q;R \subseteq S \iff \overline{S};R^\mathsf{T} \subseteq \overline{Q} \tag{2}$$

for all relations Q, R, and S. The Schröder equivalences are equivalent to the Dedekind rule (see e.g., [22]).

Using relation algebra, we now recapitulate some fundamental classes of relations. These will be used in the remainder of the paper.

A relation R is called *reflexive* if $\mathsf{I} \subseteq R$ and *transitive* if $R;R \subseteq R$. The least reflexive and transitive relation containing R is its *reflexive-transitive closure* R^*, specified by the laws $\mathsf{I} \cup R;R^* = R^*$ and $R;Q \cup S \subseteq Q \Rightarrow R^*;S \subseteq Q$ or, equivalently, by the laws $\mathsf{I} \cup R^*;R = R^*$ and $Q;R \cup S \subseteq Q \Rightarrow S;R^* \subseteq Q$ to hold for all relations Q, R, and S. A relation R is *antisymmetric* if $R \cap R^\mathsf{T} \subseteq \mathsf{I}$ and in combination with the above formulae this allows to characterise *partial order relations* R by $\mathsf{I} \subseteq R$, $R;R \subseteq R$, and $R \cap R^\mathsf{T} \subseteq \mathsf{I}$. A partial order relation R is called a *linear order relation* if additionally $R \cup R^\mathsf{T} = \mathsf{L}$ holds. A relation v satisfying $v = v;\mathsf{L}$ is called a *vector*. In case of a set-theoretic (i.e., concrete) relation $v \subseteq \mathcal{X} \times \mathcal{X}$ this equation means that an element $x \in \mathcal{X}$ is either in relationship to none of the elements of $\mathcal{X}$ or to all elements of $\mathcal{X}$. Due to this property, vectors can be used to model subsets of the universe $\mathcal{X}$. We say that $v \subseteq \mathcal{X} \times \mathcal{X}$ models the subset Y of $\mathcal{X}$ if for all $x, y \in \mathcal{X}$ we have that $x \in Y$ iff $(x, y) \in v$. By definition, a *point* is an injective and surjective vector, i.e., a vector p such that the two properties $\mathsf{L};p = \mathsf{L}$ and $p;p^\mathsf{T} \subseteq \mathsf{I}$ hold. In case of a set-theoretic point $p \subseteq \mathcal{X} \times \mathcal{X}$ these properties mean that it models a singleton subset $\{x\}$ of $\mathcal{X}$, i.e., the element x of the universe if we identify the singleton set $\{x\}$ with the only element x it contains.

2.2 Automating Relation Algebra

Automated/mechanised reasoning is not a new challenge, but has been performed since more than 20 years. Interactive theorem provers for relation algebras have been implemented (see e.g., [28,17]) and relational techniques have been integrated into various proof checkers for B or Z. Special purpose first-order proof systems for relation algebras, including tableaux and Rasiowa-Sikorski calculi, have been proposed as well (e.g., in [20]). Translations of relation-algebraic formulae into (undecidable) fragments of predicate logics have been implemented (see [25]) and integrated into the theorem prover SPASS (see [29]). However, it has been shown that automated reasoning with relation algebra does not need special-purpose tools nor interaction. As demonstrated in [16,5], an off-the-shelf automated theorem prover, such as Prover9, is often sufficient.

In this paper, we follow the latter approach and encode relation algebra in Prover9, which is a saturation-based automated theorem prover for first-order logic with equality. An evaluation of various automated theorem provers has shown that in our context Prover9 is currently best suited for verifying properties in relation algebra; see [9]. We also have experimented with the interactive theorem prover Isabelle/HOL. However, for our specific purpose the proof-effort of interactive theorem provers presently seems to be too high. Moreover, we believe that they often require a rather deep understanding of the used tool and hence experienced user, whereas our approach also can be used by people mainly interested in relation algebra and not in theorem proving.

Prover9 implements a first-order resolution and paramodulation calculus. Equalities are handled via rewriting rules and Knuth-Bendix completion. The tool suite also offers the counterexample generator Mace4, which is very useful in practice. The encoding of relation-algebraic formulae in Prover9 is straight-forward. For example the Dedekind rule (1) can be written as follows:

```
all Q all R all S (Q * R /\ S <= (Q /\ R * S^) * (R /\ Q^ * S)).
```

Since Prover9 allows only ASCII symbols as input, we use the symbols \/, /\, *, _^, _', <= and rtc(_) for union, intersection, composition, transposition, complement, inclusion, and the operator for reflexive-transitiv closure, respectively. An entire input template can be found in the appendix.

Prover9 does not support types. Hence we define the following two predicates to characterise relations as vectors and points; they are nothing else than the translations of the definitions of Section 2.1 into the language of Prover9:

```
vector(R) <-> R = R*L.
point(R)  <-> (R = R*L & L*R = L & R*R^ <= I).
```

Prover9, as any other automated theorem proving system, heavily depends on the axioms given as input. In case one only uses the few axioms of relation algebra given in Section 2.1, an automated theorem prover has to derive each and every relation-algebraic fact used in a proof. For example, if a distributivity law is needed for a proof, Prover9 has to derive it first. This fact does not only increase the running times of the theorem prover, but sometimes even yields failure in the proof search. Due to this, suitable and well-known facts, such as the following distributivity laws, should be added as axioms.

```
all R all S all T ((R \/ S)*T = R*T \/ S*T).
all R all S all T (T*(R \/ S) = T*R \/ T*S).
```

Other examples for useful relation-algebraic facts concern transposition, such as the following formulae:

```
all R (R^^ = R).
all R all S ((R * S)^ = S^ * R^).
```

For the proof automatisation we use a suitable (fixed) set of axioms. In case we need some special fact as additional input, we will state it. All input files can be found at the webpage `http://hoefner-online.de/ramics14/`. Running times presented in this paper are w.r.t. a standard desktop PC equipped with a 3.1 GHz Intel Pentium 5 CPU, 16 GB main memory, running a Mac OS operating system.

3 Automation of Proof Obligations

We start with a description of our general approach to the automation of the assertion-based verification of relational programs. Then we consider two examples. All formulae appearing in the verifications are relation-algebraic ones, but usually the notions in question are specified by predicate-logical means. To connect these two kinds of specifications, we finally show how to automatically verify the equivalence of the relation-algebraic and the common-logical specifications.

3.1 Verification of Relational While-Programs

In the present paper, we treat imperative programs with relations as data type. Concretely this means that the constants, operations and predicates of relation algebra, as introduced in Section 2.1, are available. Furthermore, we consider while-programs of the following specific form only:

$$x := I(\alpha);$$
$$\textbf{while } B(\alpha, x) \textbf{ do} \qquad\qquad\qquad (W)$$
$$x := E(\alpha, x) \textbf{ od}$$

This specific form is only chosen for simplifying program verification. There are no problems on the conceptional side to handle more complicated programs, like those of [3]. Whether the presented approach scales to larger programs, i.e., whether automated theorem provers are able to automatically verify larger relational programs, is part of future work. However, at this place it should be mentioned that relational programs are often small. This is due to the fact that relation-algebraic expressions frequently allow concise descriptions of computations which in conventional programming languages usually are expressed by, for example, (nested) loops.

In the while-program (W) x denotes a non-empty list $x_1, \ldots, x_n$ of variables for relations. Furthermore, α denotes a list of input relations and by $I(\alpha)$ and $I_1(\alpha), \ldots, I_n(\alpha)$ a list of relation-algebraic expressions over the input relations. So, the collateral assignment $x := I(\alpha)$ describes the initialisation of the variables. $E(\alpha, x)$ denotes a list $E_1(\alpha, x), \ldots, E_n(\alpha, x)$ of relation-algebraic expressions, but now over the input relations and the variables. Finally, $B(\alpha, x)$ denotes a quantifier-free formula built over the vocabulary of relation algebra, the input relations, and the variables, usually an inclusion or an equation. It is called the loop-condition. As long as it evaluates to true, the loop-body $x := E(\alpha, x)$, again a collateral assignment, is executed.

A problem specification consists of a pre-condition $Pre(\alpha)$ and a post-condition $Post(\alpha, x)$. The pre-condition describes the input restrictions and the post-condition describes the result(s) which should be computed. In our case both conditions are formulated within the language of relation algebra, frequently as conjunctions of relation-algebraic inclusions and equations. A given algorithm (a while-program) is *partially correct* if it satisfies the post-condition after termination, in case that the pre-condition holds. It is *totally correct* if it is partially correct and also guarantees termination, provided the pre-condition holds.

To prove that a program of the presented form (W) is totally correct w.r.t. a given problem specification, we use the inductive assertion method (see e.g., [10,11,13]). This method consists of three major steps: (a) the identification of a *loop-invariant* $Inv(\alpha, x)$, (b) the verification of three *proof obligations*, viz. that the loop-invariant is established by the initialisation, maintained by the loop-body, and that the loop-invariant together with the negated loop-condition implies the post-condition, and (c) the *termination* of the program.

Since we are looking at relational while-programs, the loop-invariant $Inv(\alpha, x)$ is also formulated within the language of relation algebra. The three proof obligations of (b) then may be formalised by three implications over $Pre(\alpha)$, $Post(\alpha, x)$, $Inv(\alpha)$ and $B(\alpha, x)$. The first one,

$$Pre(\alpha) \Rightarrow Inv(\alpha, I(\alpha)) \qquad \text{(PO1)}$$

says that, if the pre-condition holds, then the loop-invariant has to be established by the initialisation of the variables. After it has been shown that the loop-invariant is established, it needs to be maintained during all runs through the loop. This is formally expressed by the implication

$$Inv(\alpha, x) \wedge B(\alpha, x) \Rightarrow Inv(\alpha, E(\alpha, x)) . \qquad \text{(PO2)}$$

The implication that formalises the third proof obligation is

$$Inv(\alpha, x) \wedge \neg B(\alpha, x) \Rightarrow Post(\alpha, x) . \qquad \text{(PO3)}$$

It expresses that if the while-loop terminates, i.e., $B(\alpha, x)$ does not hold any longer, then the loop-invariant has to imply the post-condition. Since we are interested in total correctness, we also want to prove (correct) termination, i.e.

$$Pre(\alpha) \Rightarrow \text{the program yields a defined value.} \qquad \text{(T)}$$

Usually, (correct) termination of (W) means that its while-loop terminates after a finite number of iterations. However, our instantiations of (W) use a specific partial operation *point* on relations, as we will see in later sections. Therefore, a proof of (T) requires, besides the termination of the while-loop, the verification that each application of *point* yields a defined value.

Unfortunately, it is well known that termination is undecidable. However, in specific cases the termination of while-loops can be proven by measure functions. A measure function δ maps program states into a Noetherian pre-order such that, with the above notations, $Pre(\alpha)$ and $B(\alpha, x)$ imply $\delta(E(\alpha, x)) < \delta(x)$. By this, every execution of the loop-body strictly decreases the measure and, hence, termination of the while-loop is guaranteed. In this paper, we will not only show how proofs of partial correctness can be automatised, we will also show that, under some circumstances, total correctness proofs can be supported.

3.2 Reflexive-Transitive Closure

The first algorithm we verify with the help of Prover9 is an algorithm for computing the reflexive-transitive closure R^* for a given relation R. It is obtained by transforming the functional program of [6] into the following while-program (P1). In the program (P1) a (partial) operation *point* is assumed to be at hand that selects a point from a non-empty vector. The operation *point* is deterministic. In RELVIEW the deterministic selection of a point via the pre-defined operation *point* is done using the internal enumeration of the universe $\mathcal{X}$.

$$
\begin{aligned}
&C, v := \mathsf{I}, \mathsf{O}; \\
&\textbf{while } v \neq R;\mathsf{L} \textbf{ do} \\
&\quad \textbf{let } p = point(R;\mathsf{L} \cap \overline{v}); \\
&\quad C, v := C \cup C;p;p^{\mathsf{T}};R;C, v \cup p \textbf{ od}
\end{aligned}
\tag{P1}
$$

The program (P1) uses two variables: C for computing the result and v, a vector, for looping through all points of the range $R;\mathsf{L}$ of R. To enhance readability, it uses a let-clause.[1]. The selected point $point(R;\mathsf{L} \cap \overline{v})$ is denominated with the letter p for its threefold use in the subsequent assignment. If, in case of set-theoretic relations, v models the subset V of the universe $\mathcal{X}$, then the chosen point p models an element x of the set $\mathcal{X} \setminus V$ that possesses at least one successor w.r.t. R, and the subrelation $p;p^{\mathsf{T}};R$ of R consists precisely of those pairs (y, z) of R, for which $y = x$ holds. Although the program (P1) is deterministic, in principle it does not matter which element x is chosen, as long as it was not handled before and has at least one successor. For the verification we only need the following properties (3) specifying p as a point contained in the vector $R;\mathsf{L} \cap \overline{v}$.

$$
p;\mathsf{L} = p \qquad \mathsf{L};p = \mathsf{L} \qquad p;p^{\mathsf{T}} \subseteq \mathsf{I} \qquad p \subseteq R;\mathsf{L} \cap \overline{v} .
\tag{3}
$$

There is no requirement on the input relation R. So, the pre-condition $Pre(R)$ equals **true**. The post-condition $Post(R, C)$ depends on the input R and the (output) variable C and is $C = R^*$, since we want to compute the reflexive-transitive closure of R. Transferring an idea of [6] to the imperative paradigm, we obtain the conjunction of the two equations of (4) as loop-invariant $Inv(R, C, v)$.

$$
C = (R \cap v)^* \qquad\qquad v = v;\mathsf{L}
\tag{4}
$$

---

[1] We consider the let-clause as syntactical suger only, since the replacement of each occurrence of p in the body of the while-loop of (P1) by $point(R;\mathsf{L} \cap \overline{v})$ and the removal of the let-clause transforms (P1) into the schematic form (W).

Table 1. Auxiliary Facts for Verification

Formula	Running Time
$p;\mathsf{L} = p \wedge \mathsf{L}; p = \mathsf{L} \wedge p; p^\mathsf{T} \subseteq \mathsf{I} \Rightarrow R \cap p = p; p^\mathsf{T}; R$	73 s
$p;\mathsf{L} = p \wedge \mathsf{L}; p = \mathsf{L} \wedge p; p^\mathsf{T} \subseteq \mathsf{I} \Rightarrow (R \cap p); \mathsf{L}; (R \cap p) \subseteq R \cap p; \mathsf{L}$	248 s
$S;\mathsf{L}; S \subseteq S \Rightarrow (R \cup S)^* = R^* \cup R^*; S; R^*$	184 s

The first equation is best described in the Boolean matrix model of relations. It says that C equals the reflexive-transitive closure of the relation (matrix), that is obtained from R by replacing those rows by zero-rows (all entries are zero) where v consists of zeros only.

As discussed in the previous section, it suffices to verify the proof obligations (PO1) to (PO3) to show the partial correctness of the program (P1). Prover9 shows the corresponding instantiation of (PO1) in no time (0 s). Proving the corresponding instantiation of (PO3) is as simple and does not cost time either. In contrast to these cases proving the corresponding instantiation of (PO2), i.e., the maintenance of the loop-invariant, is more complicated. Here, the main goal is to show that under the assumptions of (3) and $v \neq R;\mathsf{L}$ it holds

$$C = (R \cap v)^* \Rightarrow C \cup C; p; p^\mathsf{T}; R; C = (R \cap (v \cup p))^* . \tag{5}$$

Unfortunately, Prover9 is not able to prove the implication (5) from scratch within 1000 s. It does not have sufficient knowledge about the Kleene star. The theorem prover needs additional properties of this operation as input. Adding auxiliary laws, such as star-monotonicity does not help. One needs further specific knowledge about the Kleene star in relation algebra. In [6] the laws listed in Table 1 are used to prove the correctness of the functional program. If these three laws are added, then Prover9 proves (5) and the entire instantiation of (PO2) within 1 s. Luckily, the additional laws can all be proven fully automatically. The running times are presented in Table 1.

The proof of (5) is by far not trivial (even with the additional properties), but definitely shows some limitations of our approach. It cannot be expected that all proofs can be automated. In fact, it is well known that theorem proving in the area of relation algebra is undecidable. However, Prover9 (or any other automated theorem prover) can assist to get rid of proofs of low or medium complexity. The user can then concentrate on the more complicated proofs, such as the maintenance of the loop-invariant of the algorithm under consideration. As we will show in the next section, sometimes even all proofs can be automated. A longer discussion about lessons learned is given in Section 4.

So far we have established partial correctness only. However, we can even show total correctness with the help of Prover9. To prove total correctness, we have to show that the while-loop of the program (P1) terminates and each of its *point*-calls is defined. We use the values of the variable v as measure function and use Prover9 to verify the inclusion

$$v \subseteq R;\mathsf{L} \tag{6}$$

as well as the universally quantified implication

$$\forall p: \ p;\mathsf{L} = p \wedge \mathsf{L}; p = \mathsf{L} \wedge p; p^\mathsf{T} \subseteq \mathsf{I} \wedge p \subseteq R;\mathsf{L} \cap \overline{v} \Rightarrow v \subset v \cup p . \tag{7}$$

With the help of the inclusion (6) and contraposition it is easy to prove that $R;L \cap \overline{v} \neq O$ if $v \neq R;L$: From $R;L \cap \overline{v} = O$ we get $R;L \subseteq v \subseteq R;L$ and this yields $v = R;L$. As a consequence, each call $point(R;L \cap \overline{v})$ in the program (P1) is defined. The formula (7) states that under the assumptions of (3) the vector v grows strictly. If the universe $\mathcal{X}$ is finite, then (6) and (7) together imply that the while-loop terminates, since v is strictly enlarged by every execution of its body but it cannot exceed $R;L$. The two properties (6) and (7) constitute again a loop-invariant and, using Prover9, it can successfully be treated in the same fashion as the previous loop-invariant (4). We summarise the results in Table 2.

3.3   Topological Sorting of Cycle-Free Relations

A *topological sorting* of a cycle-free relation R is a linear order relation that contains R. The relational program we consider in this section stems from [4] and is the relational version of Kahn's well-known algorithm for computing topological sortings (see [18]). It uses two variables, S for computing the result and v as auxiliary vector variable for the while-loop, and looks as follows:

$$S, v := I, O;$$
$$\textbf{while } v \neq L \textbf{ do}$$
$$\quad \textbf{let } p = point(\overline{v} \cap \overline{(R^\mathsf{T} \cap \overline{I});\overline{v}}); \qquad \text{(P2)}$$
$$\quad S, v := S \cup v;p^\mathsf{T}, v \cup p \textbf{ od}$$

Similar to program (P1), program (P2) also uses a let-clause to improve readability. It introduces p as a name for the point chosen from the vector $\overline{v} \cap \overline{(R^\mathsf{T} \cap \overline{I});\overline{v}}$ via the operation *point*. If R and v are set-theoretic relations and the vector v models the subset V of the universe $\mathcal{X}$, then the vector $\overline{v} \cap \overline{(R^\mathsf{T} \cap \overline{I});\overline{v}}$ models the set of minimal elements of the set $\mathcal{X} \setminus V$. So, via the variable v the program (P2) constructs a chain

$$\emptyset \subset \{x_1\} \subset \{x_1, x_2\} \subset \{x_1, x_2, x_3\} \subset \ldots \subset \{x_1, x_2, \ldots, x_n\} = \mathcal{X} \qquad (8)$$

of subsets of $\mathcal{X}$, where for all $i \in \{0, \ldots, n-1\}$ the set $\{x_1, \ldots, x_{i+1}\}$ is obtained from the set $\{x_1, \ldots, x_i\}$ by adding a minimal element of $\mathcal{X} \setminus \{x_1, \ldots, x_i\}$. As

Table 2. Running Times for Termination Proofs

Formula	Running Time
$O \subseteq R;L$	0 s
$(3) \wedge v \subseteq R;L \Rightarrow v \cup p \subseteq R;L$	1 s
$(3) \wedge v \neq R;L \Rightarrow v \subseteq v \cup p$ [2]	0 s
$(3) \wedge v \neq R;L \Rightarrow v \neq v \cup p$	0 s

---

[2] In this example Prover9 finds a proof, but outputs "SEARCH FAILED" followed by "Exiting with 1 proof". A close inspection of the proof logs shows that such situations occur if negative clauses are included in the goals. Then the output is misleading, since in such cases Prover9 *did* find a proof, but thought it had to keep searching.

Table 3. Invariants for Topological Sorting

Name	Invariant	Name	Invariant
$Inv_0(v)$	$v;L \subseteq v$	$Inv_4(S)$	$I \subseteq S$
$Inv_1(S,v)$	$S;v \subseteq v$	$Inv_5(S)$	$S \cap S^T \subseteq S$
$Inv_2(S,v)$	$S \cup S^T = v;v^T \cup I$	$Inv_6(S)$	$S;S \subseteq S$
$Inv_3(R,S,v)$	$R \cap v;v^T \subseteq S$	$Inv_7(R,v)$	$R;v \subseteq v$

before this chain can be used to prove termination later on, if $\mathcal{X}$ is finite. Simultaneously to the chain (8) the program (P2) creates another chain

$$I = S_0 \subset S_1 \subset \ldots \subset S_n \tag{9}$$

of relations, using the variable S. For all $i \in \{0, \ldots, n\}$, the relation S_i is a topological sorting of the input R if both are restricted to $\{x_1, \ldots, x_i\}$. Because of the initialisation of S, outside of this set S_i consists of loops (x, x) only.

Before we treat the automated verification of the above program, we have to be more precise about the choice of p. If p is a point satisfying $p \subseteq \overline{(R^T \cap \overline{I});\overline{v}}$, then $R;p \subseteq v \cup p$ follows by the use of the Schröder equivalences (2) (see [4]). As a consequence, we assume the following properties for p:

$$p;L = p \qquad L;p = L \qquad p;p^T \subseteq I \qquad p \subseteq \overline{v} \qquad R;p \subseteq v \cup p \tag{10}$$

A topological sorting requires a cycle-free relation as input. Since cycle-freeness of R relation-algebraically can be specified as $R;R^* \subseteq \overline{I}$, we take this formula as pre-condition $Pre(R)$. In case of a finite universe $\mathcal{X}$ we then get that the relation $R^T \cap \overline{I}$ is *progressively finite* in the sense of [22]. Hence, $\overline{v} \subseteq (R^T \cap \overline{I});\overline{v}$ implies $\overline{v} = O$. By contraposition we obtain that $\overline{v} \neq O$ implies $\overline{v} \nsubseteq (R^T \cap \overline{I});\overline{v}$ and this is equivalent to the fact that $v \neq L$ implies $\overline{v} \cap \overline{(R^T \cap \overline{I});\overline{v}} \neq O$. So, in the finite case or, more generally, the Noetherian case (since Noetherian relations are precisely those the transposes of which are progressively finite), there exists a p that satisfies the properties of (10), i.e., all calls of *point* in the program (P2) are defined.

The conjunction of the formulae $R \subseteq S$, $I \subseteq S$, $S;S \subseteq S$, $S \cap S^T \subseteq I$, and $S \cup S^T = L$ forms the post-condition $Post(R,S)$ and the loop-invariant $Inv(R, S, v)$ consists of a conjunction of eight formulae, which are shown in Table 3.

The formula $Inv_0(v)$ specifies v as a vector and $Inv_2(S,v)$ to $Inv_6(S)$ constitute the relation-algebraic formalisation of the above described relationship between the sets of the chain (8) and the relations of the chain (9). The remaining two formulae $Inv_1(S,v)$ and $Inv_7(R,v)$ specify that the set, modelled by v, is predecessor-closed w.r.t. S and R, respectively.

As before, we use Prover9 to verify all proof obligations. This time there are no problems at all, and all verification tasks could be fully automated without interactions. The establishment of the loop-invariant as well as the verification of the post condition (proof obligations (PO1) and (PO3)) takes no time; the running times of the maintenance of the invariance are shown in Table 4. This finishes the proof of partial correctness.

We can again use Prover9 to verify total correctness. In fact the proofs are nearly identical to the ones for the program (P1).

3.4 Equivalence of Logical and Relation-Algebraic Specifications

In the previous two sections we have automatically proven the total correctness of two relational while-programs. One reason why we could use automated theorem provers is that we are able to write program specifications and loop-invariants as relation-algebraic formulae. However, often specifications and program properties are *not* given in a relation-algebraic manner, but in predicate logic. For example, the post-condition of the program (P2), which characterises a topological sorting S of R, in first-order logic is the conjunction of the following formulae:

$$\forall x, y : (x, y) \in R \Rightarrow (x, y) \in S$$
$$\forall x : (x, x) \in S$$
$$\forall x, y, z : (x, y) \in S \wedge (y, z) \in S \Rightarrow (x, z) \in S \qquad (11)$$
$$\forall x, y : (x, y) \in S \wedge (y, x) \in S \Rightarrow x = y$$
$$\forall x, y : (x, y) \in S \vee (y, x) \in S .$$

These formulae are standard predicate logic in combination with set theory. For example, the latter three characterise S as transitive, antisymmetric, and total (sometimes also called complete); hence as a linear order. The formulae of (11) are, in the same order, equivalent to $R \subseteq S$, $\mathsf{I} \subseteq S$, $S;S \subseteq S$, $S \cap S^{\mathsf{T}} \subseteq \mathsf{I}$, and $S \cup S^{\mathsf{T}} = \mathsf{L}$, respectively.

In this section we show that Prover9 can also be used to verify such equivalences. By this we close the gap between specifications written in predicate logic and specifications written in relation algebra, as we used them earlier. To do so, we have to define some fragments of set theory in Prover9. We define a new predicate $\mathtt{in}(x, y, R)$, where R is a relation and x, y range over the universe of R. Semantically, we want to have that in(x,y,R) iff $(x, y) \in R$. Hence $\mathtt{in}(x, y, R)$ models the membership property. The predicate in needs additional axioms for the relation-algebraic operations $\cup, \cap, \bar{}, ;, ^{\mathsf{T}}$ and the constants L, O, and I; all being straight forward. For example, union and transposition are defined as

```
all R all S (in(x,y,(R \/ S)) <-> in(x,y,R) | in(x,y,S)).
all R         (in(x,y,R^)      <-> in(y,x,R)).
```

Table 4. Running Times for Proof Obligation (PO2)

Formula	Running Time
$v \neq \mathsf{L} \wedge (10) \wedge Inv(R, S, v) \Rightarrow Inv_0(v \cup p)$	0 s
$v \neq \mathsf{L} \wedge (10) \wedge Inv(R, S, v) \Rightarrow Inv_1(S \cup v; p^{\mathsf{T}}, v \cup p)$	22 s
$v \neq \mathsf{L} \wedge (10) \wedge Inv(R, S, v) \Rightarrow Inv_2(S \cup v; p^{\mathsf{T}}, v \cup p)$	3 s
$v \neq \mathsf{L} \wedge (10) \wedge Inv(R, S, v) \Rightarrow Inv_3(R, S \cup v; p^{\mathsf{T}}, v \cup p)$	1 s
$v \neq \mathsf{L} \wedge (10) \wedge Inv(R, S, v) \Rightarrow Inv_4(S \cup v; p^{\mathsf{T}})$	0 s
$v \neq \mathsf{L} \wedge (10) \wedge Inv(R, S, v) \Rightarrow Inv_5(S \cup v; p^{\mathsf{T}})$	43 s
$v \neq \mathsf{L} \wedge (10) \wedge Inv(R, S, v) \Rightarrow Inv_6(S \cup v; p^{\mathsf{T}})$	24 s
$v \neq \mathsf{L} \wedge (10) \wedge Inv(R, S, v) \Rightarrow Inv_7(R, v \cup p)$	0 s

Table 5. Running Times for the Verification of the Formulae of (11)

Formula	Running Time
$(\forall x, y : (x,y) \in R \Rightarrow (x,y) \in S) \Leftrightarrow R \subseteq S$	0 s
$(\forall x : (x,x) \in S) \Leftrightarrow I \subseteq S$	0 s
$(\forall x, y, z : (x,y) \in S \wedge (y,z) \in S \Rightarrow (x,z) \in S) \Leftrightarrow S;S \subseteq S$	0 s
$(\forall x, y : (x,y) \in S \wedge (y,x) \in S \Rightarrow x = y) \Rightarrow S \cap S^\mathsf{T} \subseteq I$	1 s
$(\forall x, y : (x,y) \in S \wedge (y,x) \in S \Rightarrow x = y) \Leftarrow S \cap S^\mathsf{T} \subseteq I$	0 s
$(\forall x, y : (x,y) \in S \vee (y,x) \in S) \Leftrightarrow S \cup S^\mathsf{T} = \mathsf{L}$	2.5 s

The universal relation L can be defined as `in(x,y,L)` and similar the other two constants O and I can be defined. This combines the algebraic and the logical point of view on relations. In the same manner we can specify inclusion and equality on relations:

```
all R all S (R <= S <-> (all x all y (in(x,y,R)  -> in(x,y,S)))).
all R all S (R == S <-> (all x all y (in(x,y,R) <-> in(x,y,S)))).
```

With an input file containing all facts about the predicate `in` – the full input file can be found again in the appendix – we have verified that the five logical formulae in (11) in fact are equivalent to their relation-algebraic counterparts. Unfortunately, our experiments show that Prover9 does not always find a proof or needs long running times. This is due to two reasons: (a) proving equivalences is often hard, not only for theorem provers, but also for human beings, and (b) Prover9 does not have further knowledge about the operators (as above). Hence, Prover9 needs to derive all facts needed, but it might also derive useless facts, such as "towers" of transpositions. By the latter we mean that Prover9 searches the search space and derives formulae such as

$$(x,y) \in R \Leftrightarrow (x,y) \in R^{\mathsf{T}^\mathsf{T}} \Leftrightarrow (x,y) \in R^{\mathsf{T}^{\mathsf{T}^{\mathsf{T}^\mathsf{T}}}} \Leftrightarrow \dots$$

Splitting equivalences into two implications is an easy solution for problem (a). Moreover, this strategy can easily be automated by a preparation step while generating the input file for Prover9. Often this improves the running times of Prover9 drastically. For problem (b) there are two different approaches. The first one requires the addition of auxiliary lemmas, as we did in Section 3.2. However, for the proofs presented in this section there is a more generic way. When aiming at the proof for the formula

$$(\forall x, y : (x,y) \in R \Rightarrow (x,y) \in S) \Leftrightarrow R \subseteq S,$$

it is unlikely that Prover9 requires facts about the operations $\cup$, $\cap$, $\overline{}$, or $^\mathsf{T}$. Hence the corresponding axioms can be dropped. A quick check with Mace4 can show whether one of the skipped equivalences is needed – this is not the case for our experiments. Using the latter strategy, all but one of the above mentioned five equivalences can be proven in nearly no time. Only one equivalence needs to be split into implications. The results are summarised in Table 5.

4 Lessons Learned

In the present paper, we aimed at proof automation and proof assistance for the assertion-based verification of simple relational while-programs. Overall the experiments performed have been successful and our experience was often positive. However, there are some lessons to be learned when following this approach. The most important ones are discussed in this section.

All automated theorem proving systems depend on the axioms, given as input. If there are too few, many auxiliary facts need to be derived on the fly; if there are too many, the search space explodes and the system probably will not terminate. When starting our experiments, we used a minimal set of axioms only. We noticed that this set was far too small. So, we added a couple of further well-known laws, such as monotonicity and (sub-)distributivity laws – all these facts can be proven automatically by Prover9; see [16]. It turned out that we found a good set of axioms. With this extended set, our second example could be verified fully automatically and the first one only failed for one goal, which could be proven after we added the three laws of Table 1.

Although Prover9 helps a lot, it cannot be expected that a proof for every (true) fact can be found. It is well known that full automatisation is undecidable for relation algebra. Moreover, many researcher often spent years to find single proofs of difficult theorems – how could an automatic tool like Prover9 do it within a couple of minutes? However, theorem provers can help in verifying proofs of low or medium complexity. Those proofs often occur if induction on the structure of certain objects is used. As a consequence, a researcher can leave the easy theorems to the tool and can concentrate on "hard" tasks and the basic strategies for their proofs.

When experimenting, often hypotheses appear which are supposed to be true, but in fact are false. If, as in our case, counterexample generators (here Mace4) work on the same input files, they can be used to falsify hypotheses. We sometimes believed that a loop-invariant or another property is true, but in fact a certain formula was missing – this saves time of the researcher.

If a property is defined by a list of formulae, such as in our examples to be a point, then the definition of a corresponding predicate makes things much more readable. The same holds for the definition of auxiliary operations via certain properties. During our investigations we noticed that Prover9 unfolds such definitions rather late to keep the sets of formulae it has to treat small. Unfortunately, this strategy may lead to very large running times and a proof even may fail since certain rules cannot be applied. In such situations an unfolding of the definitions by the user (or by a preprocessing tool) led to success.

Since Prover9 does not support types, during all our experiments we have been responsible for the correct typing. We usually work within heterogeneous relation algebra in the sense of [22,23]. Therefore, typing was no problem and the rare typing errors immediately have been discovered and corrected with the help of Mace4 or RELVIEW experiments.

5   Conclusion and Outlook

In this paper, we have shown that program verification can sometimes be achieved by the use of automated theorem provers. In particular, we have followed an approach that automatically verifies imperative programs for relation-based discrete structures by combining relation algebra, the assertion-based technique and the automated theorem prover Prover9. By this, we have been able to prove the correctness of a relational program for determining the reflexive-transitive closure and of a relational topological sorting program. We have also treated the automatic verification of the equivalence of the common logical and the relation-algebraic specifications of the properties used in our example (and elsewhere in a similar context).

So far we have only considered relations as data structures. These data structures can be used for algorithms working on many discrete structures, for instance those mentioned in the introduction. However, relation algebra is limited and cannot, for example, reason about words, regular expressions, paths in graphs, and weighted graphs. Reasoning on these structures is often done by calculations on (variants of) Kleene algebra. Since Kleene algebra is also suited for automated theorem provers (see e.g., [15]), we plan to extend our class of algorithms to Kleene-algebraic data structures.

Presently, we manually generate the loop-invariants. In doing so, the main formulae (e.g. $Inv_2(S, v)$ to $Inv_6(S)$ in case of topological sorting) constitute formalisations of the ideas behind the algorithms and are frequently obtained via suitable generalisations of the post-conditions. Based on them, the auxiliary formulae (the remaining ones in case of topological sorting) are usually discovered when trying to verify that the main formulae are maintained by the loop-bodies. For the latter, the tools Mace4 (for generating counterexamples) and RELVIEW (for program evaluation, animation, and visualisation of relations) proved to be very useful. Under this point of view, our approach consists in the computer-supported application of the fundamental principle that "a program and its correctness proof should be developed hand-in-hand with the proof usually leading the way" (cf. [13], p. 164).

If program verification is done using the level of informality common to usual human-produced mathematical proofs, then the facts specified by the above mentioned auxiliary formulae may be overlooked and this may lead to subtle errors. We believe that our approach, as all computer-aided formal methods of programming, leads to results with a much greater mathematical certainty. Hence it increases the confidence.

Although the generation of loop-invariants is in general hard (or even infeasible), techniques for automatically testing and generating loop-invariants and intermediate assertions have been developed since the middle of the 1970s. They are tailored to specific applications and assume a specific structure of the programs and the used assertions. The applied techniques frequently stem from program analysis and computer algebra; see e.g., [21,19]. Apart from automatically testing loop-invariants via Mace4 and RELVIEW, presently we do not use such ideas. However, the automated testing and generation of loop-invariants and intermediate assertions in case of relational programs is part of future work. We hope that algebraic expressions support such tasks, in particular in cases where algebra leads to nice properties and clear structures.

As we have mentioned in Section 4, we usually work within heterogeneous relation algebra where each relation has a distinct type. For reasons of efficiency, in such a setting a vector usually has a type $X \leftrightarrow 1$, with 1 as a specific singleton set, i.e., corresponds to a Boolean column vector. To get along with such situations and to benefit from the advantages of types, w.r.t. the preventation and detection of errors, the extension of our approach to heterogeneous relation algebra is planned for the future, too.

Acknowledgement. We thank the anonymous referees for their valuable comments that helped to improve the paper. NICTA is funded by the Australian Government through the Department of Communications and the Australian Research Council through the ICT Centre of Excellence Program.

References

1. Berghammer, R.: Combining relational calculus and the Dijkstra-Gries method for deriving relational programs. Information Sciences 119, 155–171 (1999)
2. Berghammer, R., Hoffmann, T.: Deriving relational programs for computing kernels by reconstructing a proof of Richardson's theorem. Science of Computer Programming 38, 1–25 (2000)
3. Berghammer, R., Hoffmann, T.: Relational depth-first-search with applications. Information Sciences 139, 167–186 (2001)
4. Berghammer, R.: Applying relation algebra and REL VIEW to solve problems on orders and lattices. Acta Informatica 45, 211–236 (2008)
5. Berghammer, R., Struth, G.: On automated program construction and verification. In: Bolduc, C., Desharnais, J., Ktari, B. (eds.) MPC 2010. LNCS, vol. 6120, pp. 22–41. Springer, Heidelberg (2010)
6. Berghammer, R., Fischer, S.: Simple rectangle-based functional programs for computing reflexive-transitive closures. In: Kahl, W., Griffin, T.G. (eds.) RAMiCS 2012. LNCS, vol. 7560, pp. 114–129. Springer, Heidelberg (2012)
7. Bibel, W., Schmitt, P.: Automated deduction: A basis for applications. Applied Logic Series. Kluwer (1998)
8. Chin, L.H., Tarski, A.: Distributive and modular laws in the arithmetic of relation algebras. Univ. of California Publ. Math. (new series) 1, 341–384 (1951)
9. Dang, H.H., Höfner, P.: First-order theorem prover evaluation w.r.t. relation- and Kleene algebra. In: Berghammer, R., Möller, B., Struth, G. (eds.) Relations and Kleene Algebra in Computer Science – Ph.D. Programme at RelMiCS 10/AKA 05. Technical Report 2008-04, Institut für Informatik, Universität Augsburg, 48-52 (2008)
10. Dijkstra, E.W.: Guarded commands, nondeterminacy and formal derivation of programs. Communications of the ACM 18, 453–457 (1975)
11. Dijkstra, E.W.: A discipline of programming. Prentice-Hall (1976)
12. Foster, S., Struth, G., Weber, T.: Automated engineering of relational and algebraic methods in Isabelle/HOL (invited Tutorial). In: de Swart, H. (ed.) RAMiCS 2011. LNCS, vol. 6663, pp. 52–67. Springer, Heidelberg (2011)

13. Gries, D.: The science of computer programming. Springer (1981)
14. Hattensperger, C., Berghammer, R., Schmidt, G.: RALF – A relation-algebraic formula manipulation system and proof checker. In: Nivat, M., Rattray, C., Rus, T., Scollo, G. (eds.) Algebraic Methodology and Software Technology. Workshops in Computing, pp. 407–408. Springer (1993)
15. Höfner, P., Struth, G.: Automated reasoning in Kleene Algebra. In: Pfenning, F. (ed.) CADE 2007. LNCS (LNAI), vol. 4603, pp. 279–294. Springer, Heidelberg (2007)
16. Höfner, P., Struth, G.: On automating the calculus of relations. In: Armando, A., Baumgartner, P., Dowek, G. (eds.) IJCAR 2008. LNCS (LNAI), vol. 5195, pp. 50–66. Springer, Heidelberg (2008)
17. Kahl, W.: Calculational relation-algebraic proofs in Isabelle/Isar. In: Berghammer, R., Möller, B., Struth, G. (eds.) RelMiCS/Kleene-Algebra Ws 2003. LNCS, vol. 3051, pp. 178–190. Springer, Heidelberg (2004)
18. Kahn, A.B.: Topological sorting of large networks. Communications of the ACM 5, 558–562 (1962)
19. Kovács, L.: Invariant generation for P-solvable loops with assignments. In: Hirsch, E.A., Razborov, A.A., Semenov, A., Slissenko, A. (eds.) CSR 2008. LNCS, vol. 5010, pp. 349–359. Springer, Heidelberg (2008)
20. MacCaull, W., Orłowska, E.: Correspondence results for relational proof systems with application to the Lambek calculus. Studia Logica 71(3), 389–414 (2002)
21. Müller-Olm, M., Seidl, H.: Computing polynomial program invariants. Information Processing Letters 91(5), 233–244 (2004)
22. Schmidt, G., Ströhlein, T.: Relations and graphs, Discrete mathematics for computer scientists. EATCS Monographs on Theoretical Computer Science. Springer (1993)
23. Schmidt, G.: Relational mathematics. Encyclopedia of Mathematics and its Applications, vol. 132. Cambridge University Press (2010)
24. Schumann, J.: Automated theorem proving in software engineering. Springer (2001)
25. Sinz, C.: System description: ARA – An automated theorem prover for relation algebras. In: McAllester, D. (ed.) CADE-17. LNCS (LNAI), vol. 1831, pp. 177–182. Springer, Heidelberg (2000)
26. Tarski, A.: On the calculus of relations. Journal of Symbolic Logic 6(3), 73–89 (1941)
27. Tarski, A., Givant, S.: A formalization of set theory without variables, vol. 41. AMS Colloquium Publications (1987)
28. von Oheimb, D., Gritzner, T.F.: RALL: Machine-supported proofs for relation algebra. In: McCune, W. (ed.) CADE 1997. LNCS (LNAI), vol. 1249, pp. 380–394. Springer, Heidelberg (1997)
29. Weidenbach, C., Schmidt, R.A., Hillenbrand, T., Rusev, R., Topic, D.: System description: SPASS version 3.0. In: Pfenning, F. (ed.) CADE 2007. LNCS (LNAI), vol. 4603, pp. 514–520. Springer, Heidelberg (2007)
30. REL VIEW homepage: http://www.informatik.uni-kiel.de/~progsys/relview/ (accessed April 30, 2013)
31. McCune, W.W.: Prover9 and Mace4., http://www.cs.unm.edu/~mccune/prover9 (accessed April 30, 2013)

A   Prover9 Templates

This appendix contains two templates to be used with Prover9. The first one specifies relation algebra.

```
% LANGUAGE SPECIFICATION
  op(500, infix,    "\/" ).        % union
  op(490, infix,    "/\" ).        % intersection
  op(700, infix,    "<=").         % inclusion
  op(480, postfix, "*" ).          % composition (not Kleene star)
  op(300, postfix, "'").           % complementation
  op(300, postfix, "^").           % transposition
% AXIOMS
  formulas(sos).
    % axioms of Boolean algebra %
      %commutativity
        x \/ y = y \/ x.
        x /\ y = y /\ x.
      %associativity
        x \/ (y \/ z) = (x \/ y) \/ z.
        x /\ (y /\ z) = (x /\ y) /\ z.
      %absorpotion
        x \/  (y /\ x) = x.
        x /\  (y \/ x) = x.
      % ordering
        x <= y  <->  x \/ y = y.
        x <= y  <->  x /\ y = x.
      %distributivity
        x /\ (y \/ z) = (x /\ y) \/ (x /\ z).
        x \/ (y /\ z) = (x \/ y) /\ (x \/ z).
      %constants
        L = x \/ x'.
        O = x /\ x'.
    % composition %
      x * (y * z) = (x * y) * z.
      x * I = x.
      I * x = x.
    % Schroeder/Dedekind %
      x* y /\ z <= (x /\ z* y^) * (y /\ x^* z).
      x* y <= z <-> x^ * z'<= y'.
      x* y <= z <-> z' * y^ <= x'.
    % standard axioms for finite iteration (Kleene star) %
      %unfold laws
        I \/ x * rtc(x) = rtc(x).
        I \/ rtc(x) * x = rtc(x).
      %induction
        x * y \/ z <= y  ->  rtc(x) * z <= y.
        y * x \/ z <= y  ->  z * rtc(x) <= y.
  end_of_list.


% CONJECTURE
  formulas(goals).
    %lemma to be proved
  end_of_list.
```

190   R. Berghammer, P. Höfner, and I. Stucke

Although Prover9 accepts capital letters as variable symbols, such as Q, R, and S, this template uses the small letters x, y, and z for variables. The reason for this renaming is that the latter variable names are automatically qualified by Prover9, i.e., they can be used without using the keyword all.

The second template establishes the relation between local and relation-algebraic specifications (see Section 3.4).

```
% LANGUAGE SPECIFICATION      %--as above--%
% AXIOMS
  formulas(sos).
    %  in()-predicate
      %operations
        all R all S (in(x,y,R \/ S) <-> (in(x,y,R) | in(x,y,S))).
        all R all S (in(x,y,R /\ S) <-> (in(x,y,R) & in(x,y,S))).
        all R all S (in(x,y,R * S)  <-> exists z (in(x,z,R) & in(z,y,S))).
        all R         (in(x,y,R')    <-> -(in(x,y,R))).
        all R         (in(x,y,R^)    <-> in(y,x,R)).
      %constants
        in(x,y,I) <-> x=y.
        -(in(x,y,O)).
        in(x,y,L).
      %inclusion and equality
        all R all S (R == S <-> (R <= S & S <= R)).
        all R all S (R <= S <-> (all x all y (in(x,y,R) -> in(x,y,S)))).
  end_of_list.
% CONJECTURE      %--as above--%
```

On Faults and Faulty Programs*

Ali Mili[1], Marcelo F. Frias[2], and Ali Jaoua[3]

[1] New Jersey Institute of Technology, USA
[2] Instituto Tecnológico de Buenos Aires (ITBA), and CONICET, Argentina
[3] Qatar University, Qatar
ali.mili@njit.edu, mfrias@itba.edu.ar, jaoua@qu.edu.qa

Abstract. A fault is an attribute of a program that precludes it from satisfying its specification; while this definition may sound clear-cut, it leaves many details unspecified. An incorrect program may be corrected in many different ways, involving different numbers of modifications. Hence neither the location nor the number of of faults may be defined in a unique manner; this, in turn, sheds a cloud of uncertainty on such concepts as fault density, and fault forecasting. In this paper, we present a more precise definition of a program fault, that has the following properties: it recognizes that the same incorrect behavior may be remedied in more than one way; it recognizes that removing a fault does not necessarily make the program correct, but may make it less incorrect (in a sense to be defined); it characterizes fault removals that make the program less incorrect, as opposed to fault removals that may remedy one aspect of program behavior at the expense of others; it recognizes that isolating a fault in a program is based on implicit assumptions about the remaining program parts; it identifies instances when a fault may be localized in a program with absolute certainty.

Keywords: faults, faulty programs, correctness, relative correctness, refinement, contingent fault, definite fault, fault removal, monotonic fault removal.

1 Introduction: Faults, an Evasive Concept

In [ALRL04, Lap04], Avizienis et. al. present a comprehensive survey of important concepts in dependable and secure computing, including definitions of these concepts, an investigation of their relevant attributes, and a discussion of the means that can be deployed to affect these attributes; this refines earlier work produced by Laprie [Lap91, Lap95]. At the center of these studies lies a hierarchy of concepts that includes: *failure*, *error*, and *fault*. A *failure* of a system is the event when the system fails to deliver the services that it is intended to provide; an *error* of the system is a deviation of the system's state from its intended value (a precursor to a possible failure); a *fault* of the system is the

* Acknowledgement: This publication was made possible by a grant from the Qatar National Research Fund, NPRP04-1109-1-174. Its contents are solely the responsibility of the authors and do not necessarily represent the official views of the QNRF.

P. Höfner et al. (Eds.): RAMiCS 2014, LNCS 8428, pp. 191–207, 2014.
© Springer International Publishing Switzerland 2014

adjudged or hypothesized cause of an error. Interpreting these definitions in the specific context of software, we find that we can easily define a software failure, as the event when a program fails to behave according to its specification. From this definition, we can define software errors and software faults as follows:

- A software error is the event when the state of the program deviates from its expected value at a given stage in the execution of the program.
- A software fault is a feature of a program that causes it to generate errors in some circumstances.

The definition of an error assumes that we have a precise characterization of correct states at each stage of the computation; and the definition of a fault assumes that we have a precise specification of each program part. Neither of these assumptions is legitimate, as they both refer to a detailed hypothetical design of the program, that has no official existence (except possibly in the mind of the designer), hence has not been vetted and documented. In practice all we have in general is the overall specification of the software product.

Also, we find that in practice, the same program malfunction can be blamed on a number of possible faulty statements, and can be corrected in a number of different ways, involving different numbers of modifications. This makes it difficult to define what a fault is, and difficult to assign precise meanings to such common concepts as fault density, fault forecasting, fault removal, etc. In this paper, we propose an approach to define faults, on the basis of the following premises:

- The definition of a fault in a program depends primarily on two parameters:
  - The specification of the program, which determines the standard that we use to judge correct behavior;
  - The structure of the program, which determines, depending on its level of granularity, the precision with which we wish to localize / isolate faults.
- To the extent that what precludes a program from being correct may be a combination of statements, a single statement, or even no statement at all (e.g. a missing statement), we ought to define a fault in such a way that it does not refer to a single statement, but may refer in general to any program part or a combination of program parts.
- To the extent that a faulty program behavior may be remedied in more than one way, designating a program part as faulty is often a discretionary decision, contingent upon assumptions about other parts of the program.
- There are cases when a program part can be deemed to be faulty regardless of other parts; we refer to these situations as definite faults, and we characterize them mathematically.
- Implicit in the concept of a fault is the idea that the program would be better off without it; to give meaning to the property of being *better off*, we introduce the concept of relative correctness, which we define formally; also, we highlight interesting relationships between relative correctness and traditional refinement.
- Removing a fault does not necessarily make a program correct, since there may be other faults elsewhere; but it ought to make it *less incorrect/ or more*

correct, as we alluded above. Using this property, we introduce the concept of monotonic fault removal.

In section 2 we briefly present elements of relational mathematics, then a simple framework for program analysis. In sections 3 and 4 we define, respectively, contingent faults, i.e. faults that are contingent upon hypotheses on other parts of the program, and definite faults, i.e. faults that preclude the correctness of the program regardless of what assumptions one makes about other program components. We conclude in section 5 with some insights and prospects.

2   A Framework for Program Analysis

2.1   Relational Notations

In this section, we introduce some elements of relational mathematics that we use in the remainder of the paper to carry out our discussions. Dealing with programs, we represent sets using a programming-like notation, by introducing variable names and associated data type (sets of values). For example, if we represent set S by the variable declarations

$x : X; y : Y; z : Z,$

then S is the Cartesian product $X \times Y \times Z$. Elements of S are denoted in lower case s, and are triplets of elements of X, Y, and Z. Given an element s of S, we represent its X-component by $x(s)$, its Y-component by $y(s)$, and its Z-component by $z(s)$. A relation on S is a subset of the Cartesian product $S \times S$; given a pair (s, s') in R, we say that s' is an *image* of s by R. Special relations on S include the *universal* relation $L = S \times S$, the *identity* relation $I = \{(s, s')|s' = s\}$, and the *empty* relation $\phi = \{\}$. Operations on relations (say, R and R') include the set theoretic operations of *union* $(R \cup R')$, *intersection* $(R \cap R')$, *difference* $(R \setminus R')$ and *complement* $(\overline{R})$. They also include the *relational product*, denoted by $(R \circ R')$, or $(RR'$, for short) and defined by:

$$RR' = \{(s, s')|\exists s'' : (s, s'') \in R \wedge (s'', s') \in R'\}.$$

The *power* of relation R is denoted by R^n, for a natural number n, and defined by $R^0 = I$, and for $n > 0$, $R^n = R \circ R^{n-1}$. The *reflexive transitive closure* of relation R is denoted by R^* and defined by $R^* = \{(s, s')|\exists n \geq 0 : (s, s') \in R^n\}$. The *converse* of relation R is the relation denoted by $\widehat{R}$ and defined by

$$\widehat{R} = \{(s, s')|(s', s) \in R\}.$$

Finally, the *domain* of a relation R is defined as the set $dom(R) = \{s|\exists s' : (s, s') \in R\}$, and the *range* of relation R is defined as the domain of $\widehat{R}$. A relation R is said to be *reflexive* if and only if $I \subseteq R$, *antisymmetric* if and only if $(R \cap \widehat{R}) \subseteq I$, and *transitive* if and only if $RR \subseteq R$. A relation is said to be a *partial ordering* if and only if it is reflexive, antisymmetric, and transitive. Also, a relation R is said to be *total* if and only if $I \subseteq R\widehat{R}$, and a relation R is said to be *deterministic* (or: a *function*) if and only if $\widehat{R}R \subseteq I$. A relation R is said to be a *vector* if and only if $RL = R$; a vector on space S is a relation of the form $R = A \times S$, for some subset A of S; we use vectors to represent subsets of S, and we may by abuse of notation write $s \in R$ to mean $s \in A$.

2.2   Relational Semantics

For the purposes of our discusions, we consider a simple programming notation that includes variable declarations, in the syntax discussed above, as well as a number of C-like executable statements. The semantics of variables declarations are simply to define the state space of the program, in the way we discussed above; as for the semantics of executable statements, we define them by means of a relation that captures the effect of the execution on the state of the program. Given a program or program part g, we let its semantics be represented by $[g]$ and defined by: $[g] = \{(s, s') |$ if execution of g starts in state s then it terminates normally in state $s'\}$.

We present the following executable statements, along with their semantic definition.

- *Assignment Statements* have the form s=E(s), where s is a shorthand for the program variables, and $E(s)$ is an expression that involves the program variables. We define the semantics of assignment statements as follows:
  $[\mathtt{s} = \mathtt{E(s)}] = \{(s, s') | s \in def(E) \wedge s' = E(s)\}$,
  where $def(E)$ is the set of states where expression $E(s)$ can be evaluated.
- *Sequence* has the form {g1; g2}; its semantics is defined by:
  $[\mathtt{g1; g2}] = [g1] \circ [g2]$.
- *Alternation* has the form {if (t) {g1} else {g2}} and the following semantic definition:
  $[\{\mathtt{if(t)\{g1\}else\{g2\}}\}] = T \cap [g1] \cup \overline{T} \cap [g2]$,
  where $T = \{(s, s') | t(s)\}$.
- *Conditional* has the form {if (t) {g1}} and the following semantic definition: $[\{\mathtt{if\ (t)\ \{g1\}}\}] = T \cap [g1] \cup \overline{T} \cap I$.
- *Iteration* has the form {while (t) {b}} and the following semantic definition:
  $[\{\mathtt{while(t)\{b\}}\}] = (T \cap [b])^* \cap \widehat{\overline{T}}$.
- The *skip* statement is written as skip and its semantics is defined as the identity relation I on S. This statement is useful for our purposes as a placeholder for faults that result from missing statements.

As a notational convention, we use lower case letters (possibly indexed) to represent programs or program parts, and we use the same letters in upper case to represent the relational semantic denotation of these programs or progrma parts. Using these semantic rules, we can represent a C-like program by a relational expression, at an arbitrary level of abstraction. For illustration, we consider the following program, taken from [GSAGvG11] (with some modifications):

```
#include <iostream>   ...   ...   ...                        line  1
void count (char q[]) {int let, dig, other, i, l; char c;          2
    i=0; let=0; dig=0; other=0; l=strlen(q);  // body init         3
    while (i<l) {                    // cond t                     4
      c = q[i];                      // body b0                    5
      if ('A'<=c && 'Z'>c) let+=2;   // cond c1, body b1           6
      else                                                         7
      if ('a'<=c && 'z'>=c) let+=1;  // cond c2, body b2           8
      else                                                         9
      if ('0'<=c && '9'>=c) dig+=1;  // cond c3, body b3          10
      else                                                        11
          other+=1;                  // body b4                   12
      i++;}                          // body inc                  13
    printf ("%d %d %d\n", let, dig, other);}  // body p           14
```

At a top level, we can view this as an initialized loop, and represent it by the following relational expression:

$$COUNT = INIT \circ ((T \cap B)^* \cap \widehat{T}) \circ P.$$

At this level of abstraction, the whole loop body is viewed as a monolith; if we want finer grained fault localization, we may want to look into the structure of B, and we find:

$$B = B0 \circ NEST \circ INC,$$

where $NEST$ represents the semantics of the nested if-then-else statement in the middle of the loop body. If we want to refine the fault localization to a finer grain scale, we further decompose relation $NEST$, as follows:

$$NEST = (C1 \cap B1) \cup \overline{C1} \cap ((C2 \cap B2) \cup \overline{C2} \cap ((C3 \cap B3) \cup \overline{C3} \cap B4)).$$

We can refine this decomposition further if we wish to distinguish between the conjuncts of the "if" conditions; for example, if we let $C11$ and $C12$ be the vectors defined by the conditions ('A'<=c) and ('Z'>c), then we can write:

$$C1 = C11 \cap C12.$$

More generally, we consider that we can capture the semantics of any program g by means of a relational expression of the form

$$G = \theta(G_1, G_2, G_3, ...G_n),$$

where each term Gi of the expression represents a program part (statement, condition, compound statement, etc), whose scale depends on the precision we wish to achieve in localizing faults. Before we define the concept of fault, we need to discuss correctness and relative correctness; these are the subject of the next section.

2.3 Correctness and Relative Correctness

We define a partial ordering between relations under the name *refinement* as follows.

Definition 2.1. Refinement, due to [BEM92]. *Let R and R' be two relations on set S. We say that R refines relation R' (and we write: $R \sqsupseteq R'$) if and only if: $RL \cap R'L \cap (R \cup R') = R'$.*

Intuitively, R refines R' if and only if it has a larger domain than R', and for all elements in the domain of R', the set of images by R is a subset of the set of images by R'. We admit without proof that this relation is a partial ordering between relations, and we refer to it as the *refinement ordering*. The following Proposition stems readily from the definition, hence is presented without proof.

Proposition 2.2. *If R and R' have the same domain, then R refines R' if and only if $R \subseteq R'$; on the other hand, if R and R' are deterministic, then R refines R' if and only if $R' \subseteq R$.*

In the following definition, we use the refinement ordering to define the property of correctness, from which we infer (a first approximation) of the concept of fault. For the sake of simplicity, we limit our discussions in this paper to deterministic programs, whose relational semantics are captured by a function.

Definition 2.3. Correctness. *Given a program* g *on space S, whose function is denoted by G, and given a relation R on S, we say that* g *is* correct *with respect to R if and only if G refines R. Program* g *is said to be* faulty *with respect to R if and only if it is not correct with respect to R.*

Of course, defining faulty programs as non-correct programs is incomplete: first, we are interested in ways to localize faults on as small a scale as possible; second, we want to characterize the stepwise progression of a faulty program from a faulty state to a fault-free state. The following definition gives us the means to these effects.

Definition 2.4. Relative Correctness. *Given a relation R on space S and two programs* g *and* g' *on space S, we say that* g *is* more-correct *than* g' *with respect to R if and only if*

$$(G \cap R)L \supseteq (G' \cap R)L.$$

Also, we say that g *is* strictly-more-correct *than* g' *with respect to R if and only if*

$$(G \cap R)L \supset (G' \cap R)L.$$

Intuitive interpretation of this definition: Given that G represents the function of a deterministic program, $(G \cap R)L$ represents (by a vector) the set of (initial) states for which G delivers a correct (with respect to R) final state. A program is all the more correct with respect to a specification R that this set is larger;

a simple illustration is given in Figure 1. Note that being more-correct does not (necessarily) mean being a superset, nor (necessarily) having a larger domain, as shown by this Figure. Note also that the relation *more-correct* if reflexive and transitive but is not antisymmetric: two programs g and g' can be in relation with each other and still be distinct; in particular, all correct programs with respect to a specification R are in *more-correct* relation with each other.

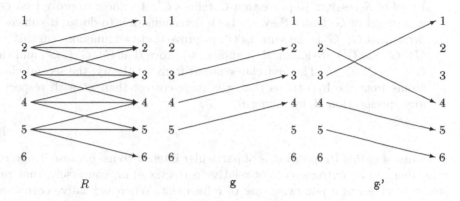

Fig. 1. Program g is more-correct with respect to R than program g'

The following Proposition presents some results about relative correctness, and its relation to correctness and refinement.

Proposition 2.5. *Let R be a relation on set S and let g and g' be programs on space S. We have the following propositions:*

1. *Program g is correct with respect to R if and only if $(R \cap G)L = RL$.*
2. *If program g is correct with respect to R, then it is more-correct than any program g' with respect to R.*
3. *If program g is correct with respect to R, and program g' is not, then g is strictly-more-correct than g' with respect to R.*
4. *If and only if G refines G', program g is more-correct than program g' with respect to any specification R.*

Proof. We consider the clauses of this Proposition in turn.

1. This result is due to [MDM94]. Its interest, for our purposes, is to show that correctness is an extreme case of relative correctness: Among all candidate programs for correctness with respect to R, those that maximize the expression $(R \cap G)L$ are correct; because $(R \cap G)L$ is by construction a subset of RL, the maximum value it may take is RL, and it does so for correct programs.
2. This results immediately from the first item: From $(R \cap G)L = RL$ (by hypothesis) and from $(R \cap G')L \subseteq RL$ (by construction) we infer $(R \cap G)L \subseteq (R \cap G')L$, whence g is more-correct than g'.

3. If **g'** is not correct, then (by item 1), $(R \cap G')L \neq RL$, hence (by construction) $(R \cap G')L \subset RL$; substituting RL by $(R \cap G)L$, we find that **g** is strictly-more-correct than **g'**.

4. Proof of Sufficiency: If G refines G' then, by Proposition 2.2, G is a super-set of G' (since we assume that G and G' are deterministic). We infer, by construction, that $(R \cap G)L \supseteq (R \cap G')L$, for any relation R; whence **g** is more-correct than **g'** with respect to R.

   Proof of Necessity: To prove that G refines G', it suffices to prove that G is a superset of G' (since they are both deterministic); to do so, it suffices to prove that $G \cap G'$ is the same as G'; to prove that two functions (in this case $G \cap G'$ and G') are identical, it suffices to prove that $G \cap G' \subseteq G'$, and that $G'L \subseteq (G \cap G')L$. The first clause stems from set theory; the second clause stems from the hypothesis that **g** is more-correct than **g'** with respect to any specification R, by taking $R = G'$.

qed

Clause 4 of this Proposition is of particular interest to us, because it portrays refinement as an extreme case of relative correctness; or, conversely, that relative correctness is a pointwise case of refinement. Whereas relative correctness is a tripartite relationship (involving G, G' and R), refinement is a bipartite relationship (involving only G and G', and quantifying universally over R).

Whereas so far we have analyzed programs as monoliths, in the next section we consider their finer grained structure, so as to analyze their faults at an arbitrarily small scale.

3   Faults and Fault Removals

3.1   Faults, An Evasive Concept

We use the sample program introduced in section 2.2 to illustrate the difficulty of characterizing and enumerating faults.

```
#include <iostream>  ... ... ...                        line 1
void count (char q[]) {int let, dig, other, i, l; char c;    2
   i=0; let=0; dig=0; other=0; l=strlen(q);  // body init     3
   while (i<l) {                   // cond t                   4
      c = q[i];                    // body b0                  5
      if ('A'<=c && 'Z'>c) let+=2; // cond c1, body b1         6
      else                                                     7
      if ('a'<=c && 'z'>=c) let+=1; // cond c2, body b2        8
      else                                                     9
      if ('0'<=c && '9'>=c) dig+=1; // cond c3, body b3       10
      else                                                    11
         other+=1;                 // body b4                 12
      i++;}                        // body inc                13
   printf ("%d %d %d\n", let, dig, other);}  //  body p       14
```

Upon looking at this program, one may be tempted to consider that condition c1 and assignment b1 are faulty, but as we discussed in section 1, faults are defined with respect to specifications. Hence, we begin by writing down some sample specifications, then discuss faults accordingly. We define the following sets:

- $\alpha_A =' A' ...' Z'$.
- $\alpha_a =' a' ...' z'$.
- $\nu =' 0' ...' 9'$.
- $\sigma = \{'+', '-', '=', ...'/'\}$, the set of all the ascii symbols.

We let $list\langle T\rangle$ denote the set of lists of elements of type T, and we let $\#_A$, $\#_a$, $\#_\nu$ and $\#_\sigma$ be the functions that to each list l assign (respectively) the number of upper case alphabetic characters, lower case alphabetic characters, numeric digits and symbols; also, we let $\#_\alpha$ be defined as $\#_\alpha(l) = \#_a(l) + \#_A(l)$, for an arbitrary list l. Before we write possible specifications against which we may analyze this program, we must first define its space. We let the space of this program be defined by all the variables declared in line 2. Also, by virtue of the include statement of line 1, we add a variable of type stream, that serves as the stream variable of the output file. We let this variable be named os (for output stream), we assume (for the purposes of our example) that the stream is a sequence of natural numbers, and we define the following operations:

- *Tail*: if os is not empty, then $tail(os)$ is the most recently added number written onto os.
- *Rest*: if os is not empty, then $rest(os)$ is the stream obtained by removing the tail of os.
- *n-th Tail*: if os has at least n elements, then $tail_n(os)$ is defined as $tail(rest^{n-1}(os))$.
- *Append*: given a stream os and a natural number n, we let $os \oplus n$ be the stream obtained by appending n to os.

Using these notations, we can write the following specifications on S:

- $R_0 = \{(s, s') | q \in list\langle\alpha_A \cup \alpha_a \cup \nu \cup \sigma\rangle \wedge os' = os \oplus \#_\alpha(q) \oplus \#_\nu(q) \oplus \#_\sigma(q)\}$.
- $R_1 = \{(s, s') | q \in list\langle\alpha_a \cup \nu \cup \sigma\rangle \wedge os' = os \oplus \#_\alpha(q) \oplus \#_\nu(q) \oplus \#_\sigma(q)\}$.
- $R_2 = \{(s, s') | q \in list\langle\alpha_A \cup \nu \cup \sigma\rangle \wedge os' = os \oplus \#_\alpha(q) \oplus \#_\nu(q) \oplus \#_\sigma(q)\}$.
- $R_3 = \{(s, s') | q \in list\langle\alpha_A \cup \nu \cup \sigma \setminus \{'Z'\}\rangle \wedge os' = os \oplus \#_\alpha(q) \oplus \#_\nu(q) \oplus \#_\sigma(q)\}$.
- $R_4 = \{(s, s') | q \in list\langle\alpha_A \cup \nu \cup \sigma\rangle \wedge tail^2(os') = \#_\nu(q) \wedge tail(os') = \#_\sigma(q)\}$.
- $R_5 = \{(s, s') | q \in list\langle\alpha_A \cup \alpha_a \cup \nu \cup \sigma\rangle \wedge tail^2(os') = \#_\nu(q)\}$.

We briefly analyze this program with respect to the specifications put forth above:

- R_0: Condition c1 ought to be changed to ('A'<=c && 'Z'>=c) and statement b1 ought to be changed to let+=1. Alternatively, we can change condition c1 to ('A'<=c && 'Z'>=c), change statement b2 to let+=2 and change statement p (line 14) to printf("%d %d %d n", let/2, dig, other);. Alternatively, we can change condition c1 to ('A'<=c && 'Z'>=c), change statement b2 to let+=2 and add the missing statement let=let/2; between lines 13 and 14.

- R_1: The program is correct with respect to specification R_1, hence it has no faults.
- R_2: Condition $c1$ ought to be changed to ('A'<=c && 'Z'>=c) and statement $b1$ ought to be changed to let+=1. Alternatively, we can change statement p (line 14) to printf("%d %d %d n", let/2, dig, other);. Alternatively, we can change condition $c1$ to ('A'<=c && 'Z'>=c), and add the missing statement let=let/2; between lines 13 and 14.
- R_3: Statement $b1$ ought to be changed to let+=1. Alternatively, we can change statement p (line 14) to printf("%d%d %d n",let/2,dig,other);. Alternatively, we can add the missing statement let=let/2; between lines 13 and 14.
- R_4: Condition $c1$ ought to be changed to ('A'<=c && 'Z'>=c). Alternatively, we can change statement $b4$ (line 12) to if (symbol(c)) other+=1;; this prevents occurrences of 'Z' from being counted as symbols.
- R_5: The program is correct with respect to specification R_5, hence it has no faults.

The foregoing discussion highlights the extent to which fault diagnosis is dependent on the specification; most importantly, it highlights the fact that, for a given specification, neither the number nor the location of the possible corrections is determined. A definition of faults must acknowledge this non-determinacy and make provisions for it.

3.2   Contingent Faults

Definition 3.1. Contingent Faults. *Let g be a program on space S, and let $\theta(G_1, G_2, G_3, ...G_n)$ be a relational representation of program g at a given level of granularity. We say that G_i is a fault of program g with respect to specification R if and only if there exists a relation G_i' on S such that $\theta(G_1, G_2, G_3, ..., G_i', ...G_n)$ is strictly-more-correct with respect to R than $\theta(G_1, G_2, G_3, ..., G_i, ...G_n)$.*

We refer to G_i as a *contingent* fault because, as we have discussed in the previous section, the same faulty program behavior may be remedied in more ways than one; hence the choice of replacing G_i by G_i' is to some extent discretionary. We choose to consider G_i faulty if we assume for the time being that the other terms of the expression are not under suspicion (a somewhat arbitrary decision). Another reason why faults, as defined herein, are deemed contingent: the structure of the program, as represented by the expression $\theta(G_1, G_2, G_3, ..., G_i, ...G_n)$, is not under suspicion; rather individual terms of the expression are. If we wanted to question the structure of the program rather than its components, then we need to represent the program at a coarser level of granularity, in such a way as to encompass the program structures that we are questioning. Note that by replacing G_i by G_i' we are not necessarily making the program correct, since the other components too may be faulty; we are merely making it strictly-more-correct. As to the question of why we choose the strict ordering (strictly-more-correct) rather than the loose ordering (more-correct): choosing the latter would make

every component of the relational expression a potential fault (taking $G'_i = G_i$);
even a correct program would be full of faults.

As an illustration of this definition, we consider the sample program g introduced in section 2.2 and specification R_0 introduced in section 3.1. Given
that $(R_0 \cap G)L$ represents the set of initial states on which program g satisfies
specification R_0, we find:

$$(R_0 \cap G)L = \{(s, s')|q \in list\langle \alpha_a \cup \nu \cup \sigma\rangle\}.$$

Indeed, program g satisfies specification R_0 only so long as there are no upper case alphabetic characters in q: indeed, all occurences of upper case letters
between 'A' and 'Y' increase let by the wrong value (2 instead of 1) and all
occurences of 'Z' erroneously increase variable $other$ rather than variable let. If
we let g' be the program obtained from g by replacing statement $b1$ (let+=2)
by statement $b1'$ (let+=1), then we find:

$$(R_0 \cap G')L = \{(s, s')|q \in list\langle (\alpha_A \setminus \{'Z'\}) \cup \alpha_a \cup \nu \cup \sigma\rangle\}.$$

Clearly, $(R_0 \cap G')L \supset (R_0 \cap G)L$. Hence statement $b1$ (let+=1) is a fault in g
with respect to specification R_0.

3.3   Monotonic Fault Removal

Just because there exists a relation G'_i that makes $\theta(G_1, G_2, G_3, ..., G'_i, ...G_n)$
strictly-more-correct with respect to R than $\theta(G_1, G_2, G_3, ..., G_i, ...G_n)$ does not
mean that we can find it easily. It is all too common for an analyst to identify
a fault, replace it by another statement, only to find that the program is not
better off; it may have improved the program's behavior on some inputs, but
made it worse on others. Whence the following definition.

Definition 3.2. Monotonic Fault Removal. *Let g be a program on space S,
whose expression is $\theta(G_1, G_2, G_3, ..., G_i, ...G_n)$ and let G_i be a contingent fault in
g. We say that the substitution of G_i by G'_i is a monotonic fault removal if and
only if program g' defined by $\theta(G_1, G_2, G_3, ..., G'_i, ...G_n)$ is strictly-more-correct
than g.*

We consider again program g introduced in section 2.2, and specification R_0
introduced in section 3.1. We consider the following versions of program g:

g_{01} The program obtained from g when we replace (let+=2) by (let+=1).
g_{10} The program obtained from g when we replace ('Z'>c) by ('Z'>=c).
g_{11} The program obtained from g when we replace (let+=2) by (let+=1) and
('Z'>c) by ('Z'>=c).

We compute the expression $(R_0 \cap G)L$ for each candidate program, and find the
following:

- $(R_0 \cap G)L = \{(s, s')|q \in list\langle \alpha_a \cup \nu \cup \sigma\rangle\}.$
- $(R_0 \cap G_{01})L = \{(s, s')|q \in list\langle (\alpha_A \setminus \{'Z'\}) \cup \alpha_a \cup \nu \cup \sigma\rangle\}.$

– $(R_0 \cap G_{10})L = \{(s, s') | q \in list \langle \alpha_a \cup \nu \cup \sigma \rangle\}$.
– $(R_0 \cap G_{11})L = \{(s, s') | q \in list \langle \alpha_A \cup \alpha_a \cup \nu \cup \sigma \rangle\}$.

Figure 2 illustrates the more-correct relationships between progrms g, g_{01}, g_{10}, and g_{11}, with respect to specification R_0; each ordering relationship is labeled with the corresponding substitution. Note that the transition from g to g_{11} via g_{01} is uniformly monotonic whereas the transition from g to g_{11} via g_{10} is not uniformly monotonic, even though the final result (g_{11}) is the same. Do all substitutions have to be monotonic? We argue that while it would be ideal to ensure that each substitution is monotonic, it is sufficient to ensure that any substitution that is not monotonic is part of a sequence of substitutions that, together, produce a monotonic transition from a faulty program to a strictly-more-correct program.

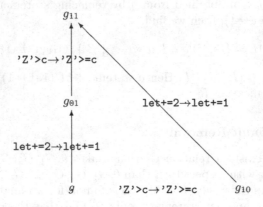

Fig. 2. Monotonic and Non Monotonic Fault Removals

4   Definite Faults

So far, we have defined contingent faults, which are considered faulty under two assumptions: first, that the structure of the program is not in question, but the components of the structure are; second, that the selected (as faulty) component is in question but the other components are not. Yet there are cases when we do not get to choose which components to question and which components to absolve. Whence the following definition.

Definition 4.1. Definite Faults. *We let g be a program on space S, and we let $\theta(G_1, G_2, G_3, ..., G_i, ...G_n)$ be its relational expression. We say that G_i is a definite fault in g with respect to specification R if and only if for all G_1, G_2, G_3, ... G_{i-1}, G_{i+1}, ... G_n, $\theta(G_1, G_2, G_3, ..., G_i, ...G_n)$ does not refine R.*

In other words, the only way to satisfy specification R is to alter G_i (hence G_i is definitely faulty). We know of two cases that are easy to characterize: a pre-processor component that loses injectivity by destroying critical state information whose preservation is mandated by the specification; and a post-processor component whose range of images is smaller than what the specification mandates.

4.1   Definite Faults: Loss of Injectivity

We consider space S defined as the set of naturals, and we let R be the following relation on S:

$$R = \{(s, s')|s' = 5 + s \ mod \ 6\} \ .$$

We let g be a program of the form $g = \{g_1; \ g_2; \}$, and we consider a number of possibilities for g_1:

- $g_1 = \{s = s \ mod \ 6;\}$. Then, g_2 may be $\{s = s+5;\}$.
- $g_1 = \{s = s+6;\}$. Then, g_2 may be $\{s = 5 + s \ mod \ 6;\}$.
- $g_1 = \{s = s+5;\}$. Then, g_2 may be $\{s = 5 + (s-5) \ mod \ 6;\}$.
- $g_1 = \{s = s \ mod \ 12;\}$. Then, g_2 may be $\{s = 5 + s \ mod \ 6;\}$.

But if we choose $g_1 = \{s = s \ mod \ 3;\}$, then no program g_2 can salvage the state of the program and produce a correct outcome. Unlike the first four examples, the fifth example of g_1 has caused a loss of injectivity beyond what R can tolerate. In this case, we find that g_1 is definitely faulty. To establish this result, we introduce a lemma, that is due to [DJM$^+$93, FDM96].

Lemma 4.2. Right Divisibility. *The relational equation in X: $QX \sqsupseteq R$, admits a solution in X if and only if R and Q satisfy the following condition:*

$$RL \subseteq QL \wedge \widehat{Q}(\overline{R} \cap RL)L = L \ .$$

We refer to this condition as the condition of *right divisibility* of R by Q; i.e., it is the condition under which the relational equation above (which seeks to divide R by Q on the right) admits a solution in X. The Proposition below stems readily from this Lemma.

Proposition 4.3. Definite Fault, for loss of injectivity. *We consider a relation R on space S and a program g on S of the form $g = \{g_1; \ g_2\}$. If R and G_1 do not satisfy the right divisibility condition (with G_1 as Q), then g_1 is definitely faulty with respect to R.*

We check briefly that $G_1 = \{(s, s')|s' = s \ mod \ 3\}$ does not, indeed, satisfy the right divisibility condition with specification $R = \{(s, s')|s' = 5 + s \ mod \ 6\}$. Because G_1 is total, the first conjunct is vacuously satisfied, hence we focus on the second conjunct. Because R is total, the second conjunct can be written as: $\widehat{G_1 \overline{R}}L = L$. We find:

$$\widehat{G_1} = \{(s, s')|s = s' \ mod \ 3\}. \ \overline{R} = \{(s, s')|s' \neq 5 + s \ mod \ 6\}.$$

$$\widehat{G_1}\overline{R} = \{(s, s')|\exists t : s = t \ mod \ 3 \wedge s' \neq 5 + t \ mod \ 6\}.$$

$$\overline{\widehat{G_1}\overline{R}} = \{(s, s')|\forall t : s \neq t \ mod \ 3 \vee s' = 5 + t \ mod \ 6\}$$
$$= \{(s, s')|\forall t : s = t \ mod \ 3 \Rightarrow s' = 5 + t \ mod \ 6\}.$$

We find that for $s = 0$, no s' can be found that satisfies the condition of this relation; hence $s = 0$ is not in the domain of this relation. Hence $\widehat{G_1 \overline{R} L} \neq L$. Relation R is not right divisible by G_1; therefore g_1 is definitely faulty. Function G_1 causes a loss of injectivity because whereas R divides the input domain into six equivalence classes (according to the value of $s\ mod\ 6$), G_1 divides it into three classes, thereby causing an irretrievable loss of state information: knowing the mod of s by 3 does not enable us to compute the mod of s by 6.

4.2   Definite Faults: Loss of Surjectivity

We consider space S defined as the set of naturals, and we let R be the following relation on S:

$$R = \{(s, s') | s' = s^2\ mod\ 6\}\ .$$

We let g be a program of the form $g = \{g_1; g_2;\}$, and we consider a number of possibilities for g_2:

- $g_2 = \{$s = s mod 6;$\}$. Then g_1 may be $\{$s = s*s;$\}$.
- $g_2 = \{$s = (s+5) mod 6;$\}$. Then g_1 may be $\{$s = s*s - 5;$\}$.

But if we choose $g_2 = \{$s = s mod 3$\}$, then there is nothing that function g_1 can do to make up for the loss of surjectivity inflicted by g_2. Unlike the first two examples, the third example of g_2 has caused a loss of surjectivity beyond what R can tolerate (the range of R is the interval $[0..5]$ whereas the range of G_2 is $[0..2]$). In this case, we find that g_2 is definitely faulty. To establish this result, we use a Lemma, due to [DJM+93, FDM96].

Lemma 4.4. *Left Divisibility.* *The relational equation in X: $XQ \sqsupseteq R$, $\widehat{X}L \subseteq QL$, admits a solution in X if and only if R and Q satisfy the following condition:*

$$RL \subseteq (\overline{\widehat{RQ}} \cap L\widehat{Q})L\ .$$

We refer to this condition as the condition of *left divisibility* of R by Q; i.e., it is the condition under which the relational equation above (which seeks to divide R by Q on the left) admits a solution in X. The Proposition below stems readily from this Lemma.

Proposition 4.5. *Definite Fault, for loss of surjectivity.* *We consider a relation R on space S and a program g on S of the form $g = \{g_1; g_2;\}$. If R and G_2 do not satisfy the right divisibility condition (with G_2 as Q), then g_2 is definitely faulty with respect to R.*

We check briefly that in program $g = \{g_1; g_2;\}$, component g_2 defined by $g_2 = \{$s = s mod 3;$\}$ is definitely faulty with respect to specification $R = \{(s, s') | s' = s^2\ mod\ 6\}$. To this effect, we check the condition of Lemma 4.4, with $Q = G_2 = \{(s, s')|s' = s \bmod 3\}$. Because R and G_2 are both total, the condition of left divisibility can be simplified into:

$$\overline{\widehat{RG_2}}L = L \ .$$

We find:

$$\overline{R} = \{(s, s')|s' \neq s^2 \bmod 6\}. \ \widehat{G_2} = \{(s, s')|s = s' \bmod 3\}.$$

$$\widehat{\overline{R}G_2} = \{(s, s')|\exists t : t \neq s^2 \bmod 6 \wedge t = s' \bmod 3\}.$$

$$\overline{\widehat{\overline{R}G_2}} = \{(s, s')|\forall t : t = s^2 \bmod 6 \vee t \neq s' \bmod 3\}.$$

Let s be 2; then the predicate defining this relation can be written as:

$$\forall t : t = 4 \vee t \neq s' \bmod 3 \ .$$

The only way to make this formula valid is to find s' such that $s' \bmod 3 = 4$. But the function $\bmod$ 3 takes only values 0, 1 and 2. Hence no s' satisfies this condition. We have found an element s that has no image by relation $\overline{\widehat{\overline{R}G_2}}$. Hence $\overline{\widehat{RG_2}}L \neq L$. Hence g_2 is definitely faulty, due to a loss of surjectivity (the range of G_2 is so small thet G_1G_2 cannot possibly refine R).

5   Conclusion

5.1   Summary

In this paper, we have commented on the evasive nature of faults, and have proposed a definition of faults that reflects most of the attributes that we associate with the concept. To this effect, we have introduced a number of related notions, that are relevant to the analysis of faults in programs, namely:

- The concept of relative correctness which characterizes a program that is more correct than an original faulty program, while not necessarily being correct.
- The concept of a contingent fault, that designates a component of a program as being faulty while absolving other parts of the program.
- The condition of a monotonic fault removal, that ensures that a substitution of a faulty component by a modified component brings the program closer to being correct.
- The concept of a definite fault, that characterizes a program component that single-handedly precludes the program from being correct, regardless of other program components.
- Two formally characterizable definite faults, one that results from loss of injectivity, and one that results from loss of surjectivity, with respect to the specification.

5.2   Theoretical Extensions

We envision two theoretical extensions to this work:

- Extending the results of this paper to non-deterministic programs. A proposition given in [DJM$^+$93] provides that if relation G refines R (i.e. if a non-dterministic program g is correct with respect to specification R) then $RL = \kappa(R, G)L$, where $\kappa(R, G) = \overline{\overline{RG}} \cap L\widehat{G}$. It turns out that for deterministic relation G, $\kappa(R, G)L = (R \cap G)L$. This may be a starting point for revisiting our results, using $\kappa(R, G)L$ instead of $(R \cap G)L$.
- Generalizing the definition of a fault to multiple components rather than a single component. There are cases when a program is incorrect (hence has faults), yet no single relation in its expression may be modified to obtain a more-correct program.

5.3   Applications

Within the field of automated fault removal a number of techniques have been identified. These techniques include some sort of mutation procedure to generate fix candidates, which are subsequently analyzed in order to determine whether they constitute an actual fix. An exponent of this class is the article [DW10], where mutators like those offered by the tool muJava [MOK05] are used in order to generate mutants of the source faulty code. A test suite is used in order to remove those mutants that fail tests from the suite and accept mutants in which all tests pass. In [GMK11] Gopinath et. al. propose a technique based on constraint satisfaction using SAT (instead of testing as in [DW10]). Mutations are employed (actually a few program mutators are supported), and mutants are accepted whenever no faults can be detected resorting to bounded verification using the analysis tool JForge [DYJ08]. In [JM11] Jose et. al. reduce fault localization to a MAX-SAT (maximal satisfiable subset) problem, and mutations of the located faults are used in order to find fix candidates. A candidate is accepted if the execution of the fault localization algorithm on the candidate does not locate any faults. All these articles have something in common: they do not elaborate on the notion of fault, but rather directly reduce fault removal to generating *any* correct mutation. In particular, potential fixes are those that first introduce new faults, and subsequently fix both the old and the newly added faults. We are exploring ways to streamline the selection of the fix candidates by means of the criterion of relative correctness; the main challenge here is to find ways to determine whether a mutant is more-correct than the original, using local information; this is currently under investigation.

Acknowledgement. The authors are very grateful to the anonymous reviewers for their valuable insights and feedback, which have contributed to the content and presentation of this paper.

References

[ALRL04]   Avizienis, A., Laprie, J.C., Randell, B., Landwehr, C.E.: Basic Concepts
             and Taxonomy of Dependable and Secure Computing. IEEE Transac-
             tions on Dependable and Secure Computing 1, 11–33 (2004)
[BEM92]   Boudriga, N., Elloumi, F., Mili, A.: The Lattice of Specifications: Ap-
             plications to a Specification Methodology. Formal Aspects of Comput-
             ing 4, 544–571 (1992)
[DW10]   Debroy, V., Wong, W.E.: Using Mutations to Automatically Suggest
             Fixes for Faulty Programs. In: Proceedings, ICST, pp. 65–74 (2010)
[DYJ08]   Dennis, G., Yessenov, K., Jackson, D.: Bounded Verification of Voting
             Software. In: Shankar, N., Woodcock, J. (eds.) VSTTE 2008. LNCS,
             vol. 5295, pp. 130–145. Springer, Heidelberg (2008)
[DJM+93]   Desharnais, J., Jaoua, A., Mili, F., Boudriga, N., Mili, A.: A Relational
             Division Operator: The Conjugate Kernel. Theoretical Computer Sci-
             ence 114, 247–272 (1993)
[FDM96]   Frappier, M., Desharnais, J., Mili, A.: A Calculus of Program Construc-
             tion by Parts. Science of Computer Programming 6, 237–254 (1996)
[GSAGvG11]   Gonzalez-Sanchez, A., Abreu, R., Gross, H.-G., van Gemund, A.: Pri-
             oritizing Tests for Fault Localization through Ambiguity Group Reduc-
             tion. In: Proceedings, Automated Software Engineering, Lawrence, KS
             (2011)
[GMK11]   Gopinath, D., Malik, M.Z., Khurshid, S.: Specification Based Program
             Repair Using SAT. In: Abdulla, P.A., Leino, K.R.M. (eds.) TACAS
             2011. LNCS, vol. 6605, pp. 173–188. Springer, Heidelberg (2011)
[JM11]   Jose, M., Majumdar, R.: Cause Clue Clauses: Error Localization Using
             Maximum Satisfiability. In: Procedings, PLDI (2011)
[Lap95]   Laprie, J.C.: Dependability —Its Attributes, Impairments and Means.
             In: Predictably Dependable Computing Systems, pp. 1–19. Springer
             (1995)
[Lap91]   Laprie, J.C.: Dependability: Basic Concepts and Terminology: in English,
             French, German, Italian and Japanese. Springer, Heidelberg (1991)
[Lap04]   Laprie, J.C.: Dependable Computing: Concepts, Challenges, Directions.
             In: Proceedings, COMPSAC (2004)
[MOK05]   Ma, Y.S., Offutt, J., Kwon, Y.R.: Mu Java: An Automated Class Mu-
             tation System. Software Testing, Verification and Reliability 15, 97–133
             (2005)
[MDM94]   Mili, A., Desharnais, J., Mili, F.: Computer Program Construction.
             Oxford University Press, New York (1994)

Parameterised Bisimulations: Some Applications

S. Arun-Kumar and Divyanshu Bagga

Department of Computer Science and Engineering
Indian Institute of Technology Delhi
Hauz Khas, New Delhi 110 016, India
{sak,divyanshu}@cse.iitd.ac.in

Abstract. In [AK06] the first author had generalised the notion of bisimulation on labelled transition systems to that of a parameterised relation whose parameters were a pair of relations on the observables of a system. In this paper we present new results which show that notions of parameterised bisimilarity may be defined to capture congruences in process algebras. In particular, we show that observational congruence may be obtained as a parameterised bisimulation, thereby providing a co-inductive characterisation for the same. In another application, parameterisation is employed to prove that amortised bisimilarity is preserved under recursion in CCS by resorting to a generalisation of the so-called "upto"-technique. We then extend the framework to a name passing calculus and show that one can capture (hyper-)bisimulations in the fusion calculus [Vic98] as a parameterised (hyper-)bisimulation. However this involves giving a behaviourally equivalent alternative semantics for the fusion calculus, which is necessary for defining parameterised bisimulations in the fusion calculus and also allows for more natural definitions of bisimulations.

1 Introduction

In [AK06] the notion of bisimilarity was generalised to a bisimilarity relation induced by a pair of relations on the underlying set of observables. The notion was referred to as parameterised bisimilarity. Many of the well-known bisimilarity and pre-bisimilarity relations in the literature are special cases of this generalised notion. Further it was also shown that many of the nice properties that these bisimilarity relations exhibited were essentially inherited from the corresponding properties in the inducing relations. In particular, it was shown that a parameterised bisimilarity relation is a preorder (resp. equivalence) if and only if the inducing relations are themselves preorders (resp. equivalences). A generalised version of Park's induction principle also holds. Finally an efficient on-the-fly algorithm was described for computing parameterised bisimilarity for finite-state labelled transition systems.

In this paper we explore compatible parameterised bisimilarity relations (e.g. congruences and precongruences) in the context of process algebras. We present three very different applications using formulations of parameterised bisimilarity.

Inspired by the axiomatization of observational congruence by Bergstra and Klop [BK85] it was shown in [SAK09] that all parameterised bisimilarities which are preorders (resp. equivalences) are also precongruences (resp. congruences) on process graphs provided there are no "empty observables". Loosely speaking, in the context

P. Höfner et al. (Eds.): RAMiCS 2014, LNCS 8428, pp. 208–225, 2014.

of weak-bisimilarity, the silent action τ of CCS is an empty observable whereas in the context of strong bisimilarity it is not. The proofs in [SAK09] which axiomatized (pre-)congruences that did not involve empty observables, were as lengthy as those in [BK85]. But what eluded a solution therein was a coinductive characterization of observational congruence. The first application we present in this paper is the characterization of observational congruence as a parameterised bisimilarity. The characterization requires a careful analysis and definition of a certain kind of weak transition to capture observational congruence in the presence of empty observables.

In [KAK05] a cost-based notion called amortised bisimilarity was defined on a CCS-like language. The set of actions was augmented with a set of visible actions to which costs were associated. While it was possible to show fairly easily that amortised bisimilarity was preserved by most of the operators of CCS, the issue of whether it is preserved under recursion was left open. The second application we present is a proof that recursion does preserve amortised bisimilarity. However this proof requires casting amortised bisimilarity in the form of an equivalent parameterised bisimilarity and using a generalisation of the "upto"-technique used by Milner and Sangiorgi [SM92] to prove that the recursion operator preserves the equivalent parameterised bisimilarity.

We devote the last section of this paper to extending the theory of parameterised bisimulations to a name-passing calculus. We argue that the meaning of the actions and thus the transitions in name-passing calculi change according to the names being passed. This dynamic update in the meaning of actions needs to be incorporated in the definition of parameterised bisimulations for name-passing calculi. We will use the fusion calculus [Vic98] to develop a general theory of parameterised bisimulations. The notion of "fusion" as an equivalence relation on names comes in quite handy while defining parameterised versions of bisimilarity in a name-passing calculus. We will however, need to give an alternative but equivalent operational semantics for the fusion calculus which allows for a more natural definition of bisimulations and which, we argue is necessary in order to define parameterised bisimulations.

2 Parameterised Bisimulations

A *labelled transition system (LTS)* $\mathcal{L}$ is a triple $\langle \mathbf{P}, \mathcal{O}, \longrightarrow \rangle$, where $\mathbf{P}$ is a set of *process states* or *processes*, $\mathcal{O}$ is a set of *observables* and $\longrightarrow \subseteq \mathbf{P} \times \mathcal{O} \times \mathbf{P}$ is the *transition relation*. We use the notation $p \xrightarrow{a} q$ to denote $(p, a, q) \in \longrightarrow$ and refer to q as a *(strong) a-successor* of p. The set of a-successors of p is denoted Succ_a^p. q is a *successor* of p if it is an a-successor for some observable a. A state q is *reachable* from p if either $p = q$ or q is reachable from some successor of p. An LTS of the form $\langle \mathbf{P}, \mathcal{O}, \longrightarrow \rangle$ may also be thought of as one of the form $\langle \mathbf{P}, \mathcal{O}^+, \longrightarrow \rangle$ such that for any $as \in \mathcal{O}^+$, $p \xrightarrow{as} q$ iff for some p', $p \xrightarrow{a} p'$ and $p' \xrightarrow{s} q$. The notion of successor may be appropriately defined. Further by introducing the transition $p \xrightarrow{\epsilon} p$ we may think of $\mathcal{L}$ also as an LTS of the form $\langle \mathbf{P}, \mathcal{O}^*, \longrightarrow \rangle$. A *rooted* labelled transition system is a 4-tuple $\langle \mathbf{P}, \mathcal{O}, \longrightarrow, p_0 \rangle$ where $\langle \mathbf{P}, \mathcal{O}, \longrightarrow \rangle$ is an LTS and $p_0 \in \mathbf{P}$ a distinguished *initial state*. In general we will consider the set of states of such an LTS as consisting only of those states that are reachable from the initial state. The term "process" will be used to refer to a process state in an LTS, as also to the sub-LTS rooted at that state and containing all the states and transitions reachable from that given state. Since an arbitrary disjoint union of LTSs is also an LTS, we shall often refer to $\mathbf{P}$ as the set of all processes. For each $p \in \mathbf{P}$, Reach(p) denotes the set of all reachable states of p.

Other notational conventions we use are the following.

- $\equiv$ denotes the identity relation on a set. It may be used in the context of observables, processes and also sets of processes.
- $\circ$ denotes relational composition i.e. for $R \subseteq A \times B$ and $S \subseteq B \times C$, $R \circ S = \{(a, c) \mid \exists b : aRbSc\}$.
- R^{-1} denotes the converse of the relation R.
- 2^U denotes the powerset of a set U.
- $|s|$ denotes the length of a sequence s.
- $\mathbf{0} \in \mathbf{P}$ is a process that is incapable of performing any observable action.
- Substitutions are applied in prefix form; two substitutions are composed using the relational composition operator $\circ$ so that $\{x/y\} \circ \{y/z\} = \{x/z\}$.

2.1 (ρ, σ)-Bisimulations

Definition 1. *Let $\mathbf{P}$ be the set of processes and let ρ and σ be binary relations on $\mathcal{O}$. A binary relation $R \subseteq \mathbf{P} \times \mathbf{P}$ is a (ρ, σ)-**induced bisimulation** or simply a (ρ, σ)-**bisimulation** if pRq implies the following conditions for all $a, b \in \mathcal{O}$.*

$$p \xrightarrow{a} p' \Rightarrow \exists b, q'[a\rho b \wedge q \xrightarrow{b} q' \wedge p'Rq']$$

$$q \xrightarrow{b} q' \Rightarrow \exists a, p'[a\sigma b \wedge p \xrightarrow{a} p' \wedge p'Rq']$$

*The largest (ρ, σ)-bisimulation (under set containment) is called (ρ, σ)-**bisimilarity** and denoted $\underline{\sqcup}_{(\rho,\sigma)}$. A $(\equiv, \equiv)$-induced bisimulation will sometimes be called a **natural bisimulation**[1]. $\mathsf{B}_{(\rho,\sigma)}$ denotes the set of all (ρ, σ)-bisimulations.*

Proposition 1. (from [AK06]). *Let ρ and σ be binary relations on $\mathcal{O}$ and let R and S be binary relations on the set $\mathbf{P}$ of processes.*

1. *If R is a (ρ, σ)-bisimulation and pRq then so is $S = R \cap (\text{Reach}(p) \times \text{Reach}(q))$.*
2. *$p \underline{\sqcup}_{(\rho,\sigma)} q$ iff pRq for some $R \in \mathsf{B}_{(\rho,\sigma)}$.* □

In Proposition 2 and Theorem 1 we quote important properties of (ρ, σ)-bisimulations. The reader is referred to [AK06] for some of the proofs.

Proposition 2. (Properties). *Let $R: p \underline{\sqcup}_{(\rho,\sigma)} q$ denote that R is a (ρ, σ)-bisimulation containing the pair (p, q).*

1. **Point-wise Extension** *Let ρ^* and σ^* on $\mathcal{O}^*$ be respectively the point-wise extensions of the relations ρ and σ on $\mathcal{O}$. Then R is a (ρ, σ)-bisimulation iff it is a (ρ^*, σ^*)-bisimulation. Further,*

---

[1] A strong bisimulation on CCS processes with $\mathcal{O} = Act$ is an example of a *natural bisimulation*.

- $R : p \sqsubseteq_{(\rho,\sigma)} q$ iff $R : p \sqsubseteq_{(\rho^*,\sigma^*)} q$ and
- $\sqsubseteq_{(\rho,\sigma)} = \sqsubseteq_{(\rho^*,\sigma^*)}$.

2. **Monotonicity.** *If $\rho \subseteq \rho'$ and $\sigma \subseteq \sigma'$ then every (ρ,σ)-bisimulation is also a (ρ',σ')-bisimulation and hence $\sqsubseteq_{(\rho,\sigma)} \subseteq \sqsubseteq_{(\rho',\sigma')}$.*

3. **Inversion.**
  - $R : p \sqsubseteq_{(\rho,\sigma)} q$ *implies* $R^{-1} : q \sqsubseteq_{(\sigma^{-1},\rho^{-1})} p$
  - $\sqsubseteq_{(\rho,\sigma)}^{-1} = \sqsubseteq_{(\sigma^{-1},\rho^{-1})}$.

4. **Composition.** $\sqsubseteq_{(\rho_1,\sigma_1)} \circ \sqsubseteq_{(\rho_2,\sigma_2)} \subseteq \sqsubseteq_{(\rho_1 \circ \rho_2, \sigma_1 \circ \sigma_2)}$ *since $R_1 : p \sqsubseteq_{(\rho_1,\sigma_1)} q$ and $R_2 : q \sqsubseteq_{(\rho_2,\sigma_2)} r$ implies $R_1 \circ R_2 : p \sqsubseteq_{(\rho_1 \circ \rho_2, \sigma_1 \circ \sigma_2)} r$.*

5. **Reflexivity.** *If ρ and σ are both reflexive then the identity relation $\equiv$ on $\mathbf{P}$ is a (ρ,σ)-bisimulation and consequently $\sqsubseteq_{(\rho,\sigma)}$ is reflexive.*

6. **Symmetry.** *If ρ and σ are both symmetric, the converse of each (ρ,σ)-bisimulation is a (σ,ρ)-bisimulation. In addition, if $\rho = \sigma$ then $\sqsubseteq_{(\rho,\sigma)}$ is a symmetric relation.*

7. **Transitivity.** *If ρ and σ are both transitive then the relational composition of (ρ,σ)-bisimulations is a (ρ,σ)-bisimulation, and $\sqsubseteq_{(\rho,\sigma)}$ is also transitive.*

8. **Preorder characterisation.** $\sqsubseteq_{(\rho,\sigma)}$ *is a preorder iff ρ and σ are both preorders.*

9. **Equivalence characterisation.** $\sqsubseteq_{(\rho,\sigma)}$ *is an equivalence iff ρ and σ are both preorders and $\sigma = \rho^{-1}$.*

10. *If ρ is a preorder then $\sqsubseteq_{(\rho,\rho^{-1})}$ is an equivalence.* $\qquad\square$

Theorem 1. *Let $2^{\mathbf{P} \times \mathbf{P}}$ be the set of all binary relations on processes. Then*

1. $\langle \mathsf{B}_{(\rho,\sigma)}, \cup, \emptyset \rangle$ *is a commutative submonoid of $\langle 2^{\mathbf{P} \times \mathbf{P}}, \cup, \emptyset \rangle$.*
2. $\sqsubseteq_{(\rho,\sigma)}$ *is a preorder if $\langle \mathsf{B}_{(\rho,\sigma)}, \circ, \equiv \rangle$ is a submonoid of $\langle 2^{\mathbf{P} \times \mathbf{P}}, \circ, \equiv \rangle$.* $\qquad\square$

3   On Observational Congruence

In this section we present a characterization of Milner's observational congruence relation $\approx^+$ for divergence-free finite-state CCS agents[2] as a parameterised bisimilarity. We achieve this by deriving an LTS which is observationally equivalent to the original LTS and show that observational congruence on CCS processes is a parameterised bisimilarity on the new LTS. In doing so we obtain a coinductive characterization of observational congruence.

Let $\mathcal{L} = \langle \mathbf{P}, Act, \rightarrow \rangle$ be the usual LTS defined by divergence-free finite-state CCS agents. A process q is a μ-*derivative* (or a *weak μ-successor*) of p if $p \stackrel{\mu}{\Longrightarrow} q$ i.e. for some $m, n \geq 0$, $p \stackrel{\tau^m \mu \tau^n}{\longrightarrow} q$ for any $\mu \in Act$. Similarly, q is a *derivative* of p if it is a μ-derivative for some $\mu \in Act$. For any $s \in Act^*$, let $\hat{s} \in A$ denote the sequence of visible actions obtained by removing all occurrences of τ in s. Given $s, t \in Act^*$, we define $s \cong t$ if $\hat{s} = \hat{t}$. Recall from [Mil89] that

- Observational Equivalence (denoted $\approx$) is the largest symmetric relation on $\mathbf{P}$ such that for all $\mu \in Act$, if $p \stackrel{\mu}{\longrightarrow} p'$ then there exists q' such that $q \stackrel{\hat{\mu}}{\Longrightarrow} q'$ and $p' \approx q'$. Moroever $\approx = \sqsubseteq_{(\cong,\cong)}$ (from [AK06]).

---

[2] That is pure CCS agents without replication.

– Observational Congruence (denoted as $\approx^+$) is the largest symmetric relation on **P** such that for all $\mu \in Act$, if $p \xrightarrow{\mu} p'$ then there exists q' such that $q \xRightarrow{\mu} q'$ and $p' \approx q'$.

It may be seen from the definition that $\approx^+$ is contained in $\approx$. While $\approx$ is easily shown to be the parameterised bisimilarity $\sqsubseteq_{(\triangleq, \triangleq)}$ on $\mathcal{L}$, a formulation of $\approx^+$ as parameterised bisimilarity is trickier. This difficulty comes from the fact that any parameterised bisimulation is defined coinductively while observational congruence is defined in literature everywhere in terms of observational equivalence (Note that in the above definition, observational congruence($\approx^+$) is defined in terms of observational equivalence($\approx$)). It is well known that $\tau.\tau.0 \approx^+ \tau.0$ since they both have observationally equivalent τ-successors. However $\tau.0 \not\approx^+ 0$ because the preemptive power of the τ action may be used to distinguish them in choice contexts. We present a coinductive characterization of observational congruence as our first application.

Keeping the above requirement in mind, we define an LTS $\mathcal{L}_\dagger = \langle \mathbf{P}, Act.\tau^*, \longrightarrow_\dagger \rangle$ derived from $\mathcal{L}$ such that $Act.\tau^* = \{\mu\tau^n \mid \mu \in Act, n \geq 0\}$ and $p \xrightarrow{\mu\tau^n}_\dagger p'$ iff there exists $n \geq 0$ and processes $p_0, \cdots, p_n$ such that $p \xrightarrow{\mu} p_0 \approx\xrightarrow{\tau} p_1 \approx\xrightarrow{\tau} \cdots \approx\xrightarrow{\tau} p_n \equiv p'$ and there does not exists any p_{n+1} such that $p' \approx\xrightarrow{\tau} p_{n+1}$. (Note: We use $p \approx\xrightarrow{\tau} q$ to denote $p \xrightarrow{\tau} q$ and $p \approx q$ here). Since $Act^+ = (Act.\tau^*)^+$ we may identify $\mathcal{L}_\dagger$ with $\mathcal{L}_\dagger^+ = \langle \mathbf{P}, Act^+, \longrightarrow_\dagger^+ \rangle$ where $\longrightarrow_\dagger^+$ denotes one or more transitions via $\longrightarrow_\dagger$. We can prove that the LTS $\mathcal{L}_\dagger$ is observationally equivalent to the original LTS $\mathcal{L}$ for divergence-free[3] finite state processes through the following lemma. The proof of the lemma uses induction on the number of derivatives modulo observational equivalence that a process can have, where the finite state assumption helps. We refer the reader to Appendix A for the detailed proof.

Lemma 1. *If p is a finite state agent then*

1. *For all $\alpha \in Act^+$ if $p \xrightarrow{\alpha}_\dagger^+ p'$ then $p \xrightarrow{\alpha} p'$.*
2. *For all $\alpha \in Act^+$ if $p \xrightarrow{\alpha} p'$ then there exists α' with $\alpha' \triangleq \alpha$ such that $p \xrightarrow{\alpha'}_\dagger^+ p''$ and $p' \approx p''$.* □

The main idea behind defining $\mathcal{L}_\dagger$ was to always ensure that no μ-derivative of any process p in $\mathcal{L}_\dagger$ has any τ-derivative which could be weakly bisimilar to itself. This helps us in ensuring that for any two observationally congruent processes, their derivatives in $\mathcal{L}_\dagger$ are observationally congruent to each other as well. For example, both $\tau.\tau.0$ and $\tau.0$ have 0 as their only derivative in $\mathcal{L}_\dagger$. Thus, we can now define a parameterised bisimulation on $\mathcal{L}_\dagger$ and show that it defines observational congruence. We refer the reader to Appendix A for the detailed proof of the following theorem.

Theorem 2. *For all divergence-free finite-state CCS agents p, q we have $p \sqsubseteq_{(\triangleq_+, \triangleq_+)} q$ in $\mathcal{L}_\dagger^+$ iff $p \approx^+ q$, where $\triangleq_+$ is the restriction of $\triangleq$ to Act^+.* □

---

[3] Divergence-freeness gurantees that there are no infinite τ chains so that we do not lose any behaviour when defining $L_\dagger$.

4 Amortised Bisimulations [KAK05]

In [KAK05] the notion of amortised bisimulations was introduced. The bisimilarity so defined uses priced actions to compare "functionally related" processes in terms of the costs incurred in the long run. The notion generalises and extends the "faster than" preorder defined by Vogler and Lüttgen in [LV06].

Amortised bisimulation is defined on the language of CCS, where in addition to the normal set of actions Act there is a set of priced actions CA as well. Priced actions cannot be restricted or relabelled and (since they do not have complements) cannot take part in synchronisation. They are assigned a cost by a function $c : CA \to \mathbb{N}$. This cost function is extended to $\mathcal{A} = CA \cup Act$ by assigning a zero cost to all actions in Act. For any $a \in \mathcal{A}$, c_a denotes the cost of a. The usual interleaving semantics of CCS is assumed.

Definition 2. *Let $\rho \subseteq \mathcal{A} \times \mathcal{A}$ such that ρ is the identity relation when restricted to Act. A family of relations $\mathscr{R} = \{R_i | \ i \in \mathbb{N}\}$ is called a* **strong amortised ρ-bisimulation**, *if whenever $(p, q) \in R_i$ for some $i \in \mathbb{N}$,*

$$p \xrightarrow{a} p' \Rightarrow \exists b, q'[a\rho b \wedge q \xrightarrow{b} q' \wedge p' R_j q']$$

$$q \xrightarrow{b} q' \Rightarrow \exists a, p'[a\rho b \wedge p \xrightarrow{a} p' \wedge p' R_j q']$$

where $j = i + c_b - c_a$. Process p is said to be **amortised cheaper (more cost efficient)** *than q (denoted $p \preccurlyeq_0^\rho q$ or simply $p \preccurlyeq^\rho q$) if pR_0q for some strong amortised ρ-bisimulation $\mathscr{R}$. Further, p is said to be* **amortised cheaper** *than q* **up to credit** *i (denoted $p \preccurlyeq_i^\rho q$) if $(p, q) \in R_i$. The index i gives the maximum credit which p requires to bisimulate q.*

One problem that has vexed the authors of [KAK05] is that of proving that amortised ρ-bisimilarity is preserved under recursion when ρ is known to be a preorder (it may not work otherwise), since standard techniques are not easily available for bisimulations defined as families of relations on processes.

We therefore characterise amortised bisimilarity as a parameterised one. We define C to be the set of states where each state is of the form $m : p$ where $m \in \mathbb{N}$. The intuition is that the state remembers the total cost incurred so far in reaching the current state. The following rule defines state transitions ($\longrightarrow_C$) in terms of the transitions of a process.

$$\boxed{p \xrightarrow{a} p' \Rightarrow m : p \xrightarrow{(a,n)}_C n : p', \text{ where } n = m + c_a}$$

The set of observables is $\mathcal{O} = \mathcal{A} \times \mathbb{N}$ and the LTS of interest is $\langle C, \mathcal{O}, \longrightarrow_C \rangle$. The following theorem provides the required characterisation of amortised ρ-bisimilarity.

Theorem 3. *Let $\gamma_\rho = \{((a, m), (b, n)) \mid a\rho b, m \leq n\}$. Then $m : p \sqsubseteq_{(\gamma_\rho, \gamma_\rho)} m : q$ iff $p \preccurlyeq^\rho q$, for all $m \in \mathbb{N}$.* □

The use of Theorem 3 in conjunction with the inheritance properties (Proposition 2) simplifies various proofs of properties of $\prec^\rho$ by rendering them in a more convenient form in terms of $\square_{(\gamma_\rho,\gamma_\rho)}$ on $\mathbf{C}$. Some notable examples are parts of Proposition 3, Lemma 4 and Proposition 5 in [KAK05]. For instance when ρ is a preorder, so is γ_ρ and hence $\square_{(\gamma_\rho,\gamma_\rho)}$ is a preorder too.

By Proposition 1.2, to show that a pair of processes is bisimilar it is necessary and sufficient to find a bisimulation containing the pair. However, it actually suffices to find a small relation (containing the pair) which by itself is not a bisimulation, but could be completed by relational composition with bisimilarity to yield a bisimulation. Such relations have been (awkwardly) called "upto"-relations [SM92]. But such a completion may not exist unless the underlying relations are preorders on the observables (see fact 4 and Theorem 5 below).

Definition 3. *Let ρ, σ be relations on observables. A relation $S \subseteq \mathbf{P} \times \mathbf{P}$ is said to be a* **potential** *(ρ, σ)-bisimulation if $\square_{(\rho,\sigma)} \circ S \circ \square_{(\rho,\sigma)}$ is a (ρ, σ)-bisimulation.*

Fact 4.

1. *If ρ and σ are both transitive then every (ρ, σ)-bisimulation is also a potential (ρ, σ)-bisimulation.*
2. *If ρ and σ are both preorders and R is a potential (ρ, σ)-bisimulation, then so are $R \circ \square_{(\rho,\sigma)}$ and $\square_{(\rho,\sigma)} \circ R$.* $\qquad\square$

A generalisation of a sufficiency condition which may be used in proving that recursion preserves bisimilarity is the following (see [Mil89] and [AKH92]).

Theorem 5. *Let ρ and σ both be preorders and R a relation such that $(p, q) \in R$ implies the following conditions for all $a, b \in \mathcal{O}$,*

$$- \; p \xrightarrow{a} p' \Rightarrow \exists b \in \mathcal{O}, q'[a\rho b \wedge q \xrightarrow{b} q' \wedge (p',q') \in \square_{(\rho,\sigma)} \circ R \circ \square_{(\rho,\sigma)}],$$
$$- \; q \xrightarrow{b} q' \Rightarrow \exists a \in \mathcal{O}, p'[a\sigma b \wedge p \xrightarrow{a} p' \wedge (p',q') \in \square_{(\rho,\sigma)} \circ R \circ \square_{(\rho,\sigma)}].$$

Then R is a potential (ρ, σ)-bisimulation. $\qquad\square$

For CCS expressions e and f, $e \prec^\rho f$ and $m : e \; \square_{(\gamma_\rho,\gamma_\rho)} \; n : f$ if under any uniform substitution of processes for the free process variables the resulting processes (respectively states) are related likewise (see [Mil89] for a technically more accurate definition).

Theorem 3 may now be used in conjunction with Theorem 5 to render the problem as one of preserving the $\square_{(\gamma_\rho,\gamma_\rho)}$ relation under recursion on $\mathbf{C}$.

Theorem 6. *Let e and f be guarded CCS expressions and x a free process variable. If ρ is a preorder and $e \prec^\rho f$ then $\underline{\text{rec}}\, x[e] \prec^\rho \underline{\text{rec}}\, x[f]$.*

Proof outline: For simplicity we assume x may be the only process variable free in e and f. Let $E_m = m : e$, $F_n = n : f$, $p = \underline{\text{rec}}\, x[e]$, $q = \underline{\text{rec}}\, x[f]$, $P_0 = 0 : p$ and $Q_0 = 0 : q$. We have (by Theorem 3) $\underline{\text{rec}}\, x[e] \prec^\rho \underline{\text{rec}}\, x[f]$ iff $P_0 \; \square_{(\gamma_\rho,\gamma_\rho)} \; Q_0$. Further since ρ is a preorder, both γ_ρ and $\square_{(\gamma_\rho,\gamma_\rho)}$ are also preorders on $\mathcal{A}$ and $\mathbf{C}$ respectively.

Consider the following relation on **C**.

$$S = \{(m : \{p/x\}g, n : \{q/x\}g) \mid FV(g) \subseteq \{x\}, m \leq n\}$$

$P_0 \equiv 0 : \{p/x\}x$ and $Q_0 \equiv 0 : \{q/x\}x$. So $(P_0, Q_0) \in S$. It then suffices to show that S is a potential $(\gamma_\rho, \gamma_\rho)$-bisimulation. It may be shown by transition induction ([Mil89], [AKH92]) that for all $(P, Q) \in S$ if $P \xrightarrow{\alpha}_C P'$, $\alpha \in \mathcal{O}$, then there exist Q' and β such that $\alpha\gamma_\rho\beta$, $Q \xrightarrow{\beta}_C Q'$ and $P' \: S \circ \sqsubseteq_{(\gamma_\rho, \gamma_\rho)} Q'$. Similarly if $Q \xrightarrow{\beta}_C Q'$, $\beta \in \mathcal{O}$, then we have for some P' and α with $\alpha\gamma_\rho\beta$, that $P \xrightarrow{\alpha}_C P'$ and $P' \sqsubseteq_{(\gamma_\rho, \gamma_\rho)} \circ S \: Q'$. By theorem 5, S is a potential $(\gamma_\rho, \gamma_\rho)$-bisimulation.   □

5   Parameterised Bisimulations in Name-Passing Calculi

Extending parameterised bisimulations to a value passing calculus requires more work in the theory of parameterised bisimulations. One cannot simply define a static relation on the labels of the transitions in a value passing calculus to define parameterised bisimulations over it. This is because the meanings of the labels/actions of the processes in a value passing calculus such as the π-calculus change dynamically based on the values passed over the input and output ports. For example, consider an agent $p \equiv \overline{u}(x)|(u(y).y.0)$ in the π-calculus[MPW92a, MPW92b], which is one of the most well known name-passing calculi. Suppose we wish to define a parameterised bisimulation using a relation ρ which relates the action y with z, i.e. $y \: \rho \: z$. Then by the semantics of the π-calculus, the value x is passed for y over the port u, which is given in the form of the transition $\overline{u}(x)|(u(y).y.0) \xrightarrow{\tau} \{x/y\}y.0 \equiv x.0$. Since the name y has now been identified with x according to the semantics of the π-calculus, the parameterised bisimulation must take this into account and should relate action x with z, since y was related to z. Formulating a theory of parameterised bisimulations which allows the dynamic update of the parameter relations on actions in accordance with the semantics of the value passing calculus is the challenging part which we will address in this section.

In this paper we will use the Fusion Calculus[Vic98] as our name-passing calculus of choice to describe the theory of parameterised bisimulations for value passing calculi. The most important reason in doing so is the explicit identification of names in the fusion calculus using equivalence relations called "fusions" instead of using substitutions to reflect the impact of communications. This explicit equivalence makes it possible for us to give a generalized theory of parameterised bisimulations which is not possible using substitution effects. To see this, consider a simple example of an action $t(u, a)$. Suppose we have a relation on names ρ such that $u \: \rho \: v$, $a \: \rho \: b$ and $t \: \rho \: t$. Then one can extend the relation ρ to relate the action $t(u, a)$ with $t(v, b)$. However, consider the case where some name w has been identified with both u and a, via some interactions between communication actions. Then the action $t(w, w)$ should be ρ-related to $t(v, b)$, however it is not possible to obtain this relationship by applying a substitutive effect on $t(w, w)$, which will replace w with a unique name, which may be either u or a. We need to consider the equivalence relation on names which identifies w with both u and a, in order to be able to relate $t(w, w)$ with $t(u, a)$ under ρ. Thus representing the identification of names as a result of communication by an equivalence relation on names is necessary to develop a general theory of parameterised bisimulations for name-passing calculi. The fusion calculus helps us in this regard as it represents the effects of communications by "fusion" actions which define equivalence relations on names.

Although the semantics of the fusion calculus does define equivalence relations on names via "fusion" actions, bisimulations in the fusion calculus are still defined in terms of the substitutive effects of fusions. As mentioned above, we would like the definitions of parameterised bisimulations to be independent of the substitutive effects. We therefore provide a modified operational semantics of the fusion calculus, which makes the effect of fusion actions explicit and allows for the definitions of bisimulations in a more natural manner without the use of any substitutive effects. We devote the next subsection to providing the details of the modified semantics for the fusion calculus with a brief comparision with the original semantics [Vic98].

5.1 An Alternative Operational Semantics for the Fusion Calculus

Assume an infinite set of names $\mathcal{N}$ ranged over by $u, v, \ldots, z$. Let $\tilde{x}$ denote a (possibly empty) finite sequence of names $x_1, \ldots, x_n$. Then the syntax of *fusions* (ranged over by φ), *free actions* (ranged over by α) and that of *agents* (*processes*) in the fusion calculus is given by the following BNF.

$$\varphi \quad ::= \{\tilde{x} = \tilde{y}\}$$
$$\alpha \quad ::= u\tilde{x} \mid \bar{u}\tilde{x} \mid \varphi$$
$$p, q, r ::= \mathbf{0} \mid \alpha.q \mid q + r \mid q|r \mid (x)q$$

where $u\tilde{x}$ and $\bar{u}\tilde{x}$ are polyadic versions of input and output respectively (collectively called *communication actions* with u and $\bar{u}$ called *subject*s and $\tilde{x}$ the *objects* of the communication) and φ is a *fusion action*. A fusion represented by $\{\tilde{x} = \tilde{y}\}$ (for sequences of names of equal length) is the smallest equivalence relation such that $x_i = y_i$ for $1 \leq i \leq n$. Thus φ ranges over total equivalence relations over $\mathcal{N}$ with only finitely many non-singular classes. For any name z and equivalence φ, $\varphi \backslash z$ is the equivalence $(\varphi \cap (\mathcal{N} - \{z\})^2) \cup \{(z, z)\}$ obtained from φ by deleting all occurrences of the name z. We denote the empty fusion (the identity relation) by $\mathbf{1}$. The set of names occurring in an action α is denoted as $n(\alpha)$. The name x is said to be *bound* in $(x)q$. We write $(\tilde{x})q$ for $(x_1) \cdots (x_n)q$. An *action* is either a fusion action or a communication action of the form $(\tilde{z})u\tilde{x}$ where $\tilde{z} = z_1, \ldots, z_n, n \geq 0$ and each $z_i \in \tilde{x}$. A communication action is *bound* if $n > 0$. There are no bound fusion actions. We will denote the set of all actions by *Act* (ranged over by α, β, γ) and the set of all agents $\mathbf{P}$. We denote sequences of actions by bold Greek letters $\boldsymbol{\alpha}, \boldsymbol{\beta}, \boldsymbol{\gamma}$.

So far, the above description given for actions and processes in the fusion calculus is identical to that given in Victor's thesis[Vic98]. In a departure from Victor's treatment, we associate an *Environment* or "shared state" in the operational semantics of a process. This environment may be seen as the accumulation of the "side effects" of the fusion actions during transitions. Every action and hence a sequence of actions has a side effect, which is the creation of an equivalence on names. We define a function on action sequences which captures this side effect.

Definition 4.

$$E(\alpha) = \begin{cases} \varphi \oplus E(\alpha') & \text{if } \alpha = \varphi.\alpha' \\ E(\alpha') & \text{if } \alpha = \alpha.\alpha' \\ 1 & \text{if } \alpha = \epsilon \end{cases}$$

where α is a communication action and $\varphi \oplus \varphi' = (\varphi \cup \varphi')^*$ denotes the smallest equivalence relation containing both φ and φ'.

We formally define the environment or the shared state which is created during process execution in the following definition. We also define their substitutive effects which allow us to reduce an environment and process pair to a single process. Substitutive effects are important when we need to compare our bisimulations with the one's defined by Victor's semantics.

Definition 5. *Let Env, referred to as the set of* environments, *be the set of all equivalence relations on $\mathcal{N}$ defined by a finite set of pairs of non-identical names. Thus an environment $\psi \in Env$ is an equivalence relation on names and it is extended to actions as follows.*

- $(\tilde{y_1}).u.\tilde{x_1} \ \psi \ (\tilde{y_2}).v.\tilde{x_2}$ *iff* $|\tilde{y_1}| = |\tilde{y_2}|, |\tilde{x_1}| = |\tilde{x_2}|, \tilde{y_1} \ \psi \ \tilde{y_2}, \tilde{x_1} \ \psi \ \tilde{x_2}$ *and* $u \ \psi \ v$.
- $\{\tilde{y_1} = \tilde{x_1}\} \ \psi \ \{\tilde{y_2} = \tilde{x_2}\}$ *iff* $|\tilde{y_1}| = |\tilde{y_2}|, |\tilde{x_1}| = |\tilde{x_2}|, \tilde{y_1} \ \psi \ \tilde{y_2}$ *and* $\tilde{x_1} \ \psi \ \tilde{x_2}$.

Definition 6. *A **substitutive effect** of an environment ψ is a substitution θ such that $\forall x, y$ we have $x\psi y$ if and only if $\theta(x) = \theta(y)$ and $\forall x, y$ if $\theta(x) = y$ then $x\psi y$.*

$$\text{PREFL} \frac{\overline{}}{\alpha.p \xrightarrow{\alpha} p}$$

$$\text{SUML} \frac{p \xrightarrow{\alpha} p'}{p + q \xrightarrow{\alpha} p'}$$

$$\text{PARL} \frac{p \xrightarrow{\alpha} p'}{p|q \xrightarrow{\alpha} p'|q}$$

$$\text{PASSL} \frac{p \xrightarrow{\alpha} p'}{(z)p \xrightarrow{\alpha} (z)p'}, z \notin n(\alpha)$$

$$\text{SCOPEL} \frac{p \xrightarrow{\varphi} p', z\varphi x, x \neq z}{(z)p \xrightarrow{\varphi \backslash z} p'\{x/z\}}$$

$$\text{OPENL} \frac{p \xrightarrow{(\tilde{y})a\tilde{x}} p', z \in \tilde{x} - \tilde{y}, a \notin \{z, \bar{z}\}}{(z)p \xrightarrow{(z\tilde{y})a\tilde{x}} p'}$$

$$\text{COML} \frac{p \xrightarrow{u\tilde{x}} p', q \xrightarrow{\bar{u}\tilde{y}} q', |\tilde{x}| = |\tilde{y}|}{p|q \xrightarrow{\{\tilde{x}=\tilde{y}\}} p'|q'}$$

$$\text{PREFR} \frac{}{(\psi, \alpha.p) \xrightarrow{\alpha} (\psi \oplus \psi(\alpha), p)}$$

$$\text{SUMR} \frac{(\psi, p) \xrightarrow{\alpha} (\psi', p')}{(\psi, p + q) \xrightarrow{\alpha} (\psi', p')}$$

$$\text{PARR} \frac{(\psi, p) \xrightarrow{\alpha} (\psi', p')}{(\psi, p|q) \xrightarrow{\alpha} (\psi', p'|q)}$$

$$\text{PASSR} \frac{(\psi, p) \xrightarrow{\alpha} (\psi', p')}{(\psi, (z)p) \xrightarrow{\alpha} (\psi', (z)p')}, z \notin n(\alpha)$$

$$\text{SCOPER} \frac{(\psi, p) \xrightarrow{\varphi} (\psi \oplus \varphi, p'), z\varphi x, x \neq z}{(\psi, (z)p) \xrightarrow{\varphi \backslash z} (\psi \oplus (\varphi \backslash z), p'\{x/z\})}$$

$$\text{OPENR} \frac{(\psi, p) \xrightarrow{(\tilde{y})a\tilde{x}} (\psi, p'), z \in \tilde{x} - \tilde{y}, a \notin \{z, \bar{z}\}}{(\psi, (z)p) \xrightarrow{(z\tilde{y})a\tilde{x}} (\psi, p')}$$

$$\text{COMR} \frac{(\psi, p) \xrightarrow{u\tilde{x}} (\psi, p'), (\psi, q) \xrightarrow{\bar{v}\tilde{y}} (\psi, q'), \ |\tilde{x}| = |\tilde{y}|, u \ \psi \ v}{(\psi, p|q) \xrightarrow{\{\tilde{x}=\tilde{y}\}} (\psi \oplus \{\tilde{x} = \tilde{y}\}, p'|q')}$$

Fig. 1. Original (L) and alternative (R) SOS rules (modulo structural congruence)

The left half of figure 1 shows the original operational semantics of the fusion calculus (modulo structural congruence)[Vic98]. The right half is the alternative operational semantics based on our notion that the state of process execution should be represented by an environment-process pair. An α denotes a free action in the rules. Every agent according to this semantics executes in an environment (possibly the identity denoted by 1) defined and regulated by scope. It should be noted that all the transitions allowed by the rules in the original semantics hold in the alternative semantics as well, except that the new semantics defines the transition on an environment-process pair. Thus, any transition that an agent p can perform under the rules of the original semantics also holds for (ψ, p) for any environment ψ. In fact, the COM rule in figure 1 allows more transitions in our semantics by specifying possible synchronizations between previously fused names. We illustrate this difference with the following example.

Example 1. Consider a process $r \equiv ua.a.0 | \bar{u}b.\bar{b}.0$. Then by the semantics of figure 1 the following sequence of transitions can be derived starting with the identity environment 1.

$$(1, ua.a.0 | \bar{u}b.\bar{b}.0) \xrightarrow{\{a=b\}} (\{a = b\}, a.0 | \bar{b}.0) \xrightarrow{1} (\{a = b\}, 0)$$

The second transition obtained by the interaction of a with $\bar{b}$ was made possible by the COMR rule as a and b have been fused, but it would not have been possible in the original semantics. However while defining behavioral relations the second transition is indeed taken into account by virtue of a substitutive effect of the first transition (which will either substitute a for b or vice-versa)[Vic98]. Thus even though the new semantics yields more transitions for the processes when compared with the original semantics, it still models the same intended behavior with the added advantage that we do not have to rely on any substitutions while defining bisimulations.

The following lemmas help in establishing the equivalence of behavior as given by original semantics using substitutive effect and the modified semantics using environments.

Lemma 2. ([Bag11]) *Let θ_1 be a substitutive effect of ψ_1 and s be a substitutive effect of $\theta_1(\psi_2)$. Then $s \circ \theta_1$ is a substitutive effect of $\psi_1 \oplus \psi_2 = \psi$.* □

Lemma 3. ([Bag11]) *For any $\psi \in Env$ and $x, z \in \mathcal{N}$ such that $x\psi z$, we have*
$(\psi, p) \xrightarrow{\alpha} (\psi \oplus E(\alpha), p')$ *iff* $(\psi \backslash x, \{z/x\}p) \xrightarrow{\{z/x\}\alpha} (\psi \oplus E(\alpha) \backslash x, \{z/x\}p')$ □

Since the main focus of this paper is on the results concerning parameterised bisimulations, we will limit our discussion to behaviours as described by the modified semantics. We refer the reader to [Bag11] for a formal proof of equivalence upto bisimilarity of the two semantics.

5.2   Parameterised Bisimulations in the Fusion Calculus

We first give the definition of bisimulations for the fusion calculus according to the original semantics[Vic98] before defining parameterised bisimulations for the modified semantics. In order to do so, we must first extend the transition relation for processes defined by Victor's semantics to Act^* by incorporating substitutive effects of the transitions.

Definition 7. *For all $p \in \mathbf{P}$, we have $p \xrightarrow{\epsilon} p$ and for all $\alpha = \alpha.\alpha'$ where $\alpha, \alpha' \in Act^*$, we have $p \xrightarrow{\alpha} p'$ iff there exists some $p'' \in \mathbf{P}$ such that $p \xrightarrow{\alpha} p''$ and $(\theta p'') \xrightarrow{\theta \alpha'} p'$ where θ is some substitutive effect of $E(\alpha)$.*

Definition 8. *A relation $\mathcal{R}$ is a strong bisimulation if $p\mathcal{R}q$ implies $\forall \alpha, \beta \in Act^*$,*

$$p \xrightarrow{\alpha} p' \Rightarrow \exists q' : q \xrightarrow{\alpha} q' \wedge (\theta_\alpha p')\mathcal{R}(\theta_\alpha q')$$
$$q \xrightarrow{\beta} q' \Rightarrow \exists p' : p \xrightarrow{\beta} p' \wedge (\theta_\beta p')\mathcal{R}(\theta_\beta q')$$

*where θ_π is a substitutive effect of $E(\pi)$ with $\pi = \alpha, \beta$. $\dot\sim$ denotes **strong bisimilarity**.*

As we argued before, the meanings of the actions in value passing calculi change in accordance with the names identified by transitions. In the fusion calculus, this identification of names is represented by the environment. This equivalence on names is taken into account using substitutive effects when defining bisimulations for the fusion calculus. With the modified semantics we can incorporate the effect of the environment more easily by arguing that given any state (ψ, p), any computation α performed by process p should be considered exactly identical to a computation α' if $\alpha\psi\alpha'$ holds, where ψ is extended point-wise to actions. Let ρ be a relation on actions which determines which computations should be considered ρ-related for the purpose of defining parameterised bisimulations. Then an action sequence β should be considered ρ-related to another action sequence α if there exist action sequences α' and β' such that $\alpha\psi\alpha'$, $\beta\psi\beta'$ and $\alpha'\rho\beta'$. Equivalently, given an environment ψ, an action sequence β should be considered ρ-related to another action sequence α iff $\alpha \; \psi \circ \rho \circ \psi \; bm\beta$ holds. With this formalisation of "ρ-relatedness" we modify our standard definition of (ρ, σ)-bisimulations to define a generalised (ρ, σ)-bisimulation for the fusion calculus.

Definition 9. *Let $\rho, \sigma \subseteq Act^2$. A relation $\mathcal{G} \subseteq (Env \times \mathbf{P})^2$ is a **generalised (ρ, σ)-bisimulation** if $(\psi, p) \; \mathcal{G} \; (\omega, q)$ implies for all $\alpha_1, \beta_2 \in Act^*$,*

$$(\psi, p) \xrightarrow{\alpha_1} (\psi', p') \Rightarrow \exists \alpha_2 : \alpha_1 \rho' \alpha_2, (\omega', q') : (\omega, q) \xrightarrow{\alpha_2} (\omega', q') \wedge (\psi', p') \; \mathcal{G} \; (\omega', q')$$
$$(\omega, q) \xrightarrow{\beta_2} (\omega', q') \Rightarrow \exists \beta_1 : \beta_1 \sigma' \beta_2, (\psi', p') : (\psi, p) \xrightarrow{\beta_1} (\psi', p') \wedge (\psi', p') \; \mathcal{G} \; (\omega', q')$$

where $\rho' = \psi \circ \rho \circ \omega$ and $\sigma' = \psi \circ \sigma \circ \omega$.

While comparing processes, we may claim that they are related only if they display related behaviours under the same environment. This leads to the following definition of bisimulation.

Definition 10. *A generalised (ρ, σ)-bisimulation $\mathcal{G}$ is a (ρ, σ)-**bisimulation** if for all $\psi, \omega \in Env$ and $p, q \in \mathbf{P}$, $(\psi, p) \; \mathcal{G} \; (\omega, q)$ implies $\psi = \omega$. We refer the largest (ρ, σ)-bisimulation as (ρ, σ)-**bisimilarity** (denoted $\square_{(\rho, \sigma)}$).*

To be able to relate the parameterised bisimulations (Definition 10) with the bisimulations already defined for fusion calculus (Definition 8), we need to define a mapping from states to processes since our bisimulations are defined for states. This motivates the definition of the following translation relation.

Definition 11. *Let the relation* $\mathbf{T} \subseteq States \times \mathbf{P}$ *called the* **translation relation** *be defined by* $(\psi, p)\mathbf{T}p'$ *if and only if there exists* θ *a substitutive effect of* ψ, *such that* $\theta p = p'$.

Proposition 1. *Let* $\mathbf{T}$ *be the translation relation given in Definition 11. Then* $\mathbf{T} \circ \dot{\sim} \circ \mathbf{T}^{-1}$ *is a generalised* $(\equiv, \equiv)$-*bisimulation (Definition 10), where* $\dot{\sim}$ *is strong bisimilarity (Definition 8).*

Proof. Let $(\psi, p)\mathbf{T} \circ \dot{\sim} \circ \mathbf{T}^{-1}(\omega, q)$. Then by definition of the translation relation, for some substitutive effect θ_ψ of ψ and θ_ω of ω we must have $\theta_\psi p \dot{\sim} \theta_\omega q$. Let $(\psi, p) \xrightarrow{\alpha} (\psi', p')$ where $\psi' = \psi \oplus E(\alpha)$. Then by Lemma 3, $\theta_\psi p \xrightarrow{\theta_\psi(\alpha)} \theta_\psi p'$ which implies that there exists $q'' \in \mathbf{P}$ such that $\theta_\omega q \xrightarrow{\theta_\psi(\alpha)} q''$ and $\theta(\theta_\psi(p'))\dot{\sim}\theta(q'')$, where θ is a substitutive effect of $\theta_\psi(\alpha)$. Now by converse of Lemma 3 there must exist $q' \in \mathbf{P}$ and $\beta \in Act^*$ such that $(\omega, q) \xrightarrow{\beta} (\omega', q')$ where $\omega' = \omega \oplus E(\beta)$, $\theta_\omega(\beta) = \theta_\psi(\alpha)$ and $\theta_\omega(q') = q''$. Now by Lemma 2, $\theta \circ \theta_\psi$ is a substitutive effect of ψ' and $\theta \circ \theta_\omega$ is a substitutive effect of ω'. Hence we have $(\omega, q) \xrightarrow{\beta} (\omega', q')$ where $\alpha\psi\circ \equiv \circ\omega\beta$ and $(\psi', p')\mathbf{T} \circ \dot{\sim} \circ \mathbf{T}^{-1}(\omega', q')$. A similar proof may be given for a transition of q. $\square$

Proposition 2. *Let* δ *be a one to one function mapping environments to substitutions such that* $\delta(\psi)$ *is a substitutive effect of* ψ, *for any* $\psi \in Env$. *Given* δ, *we define a sub-relation* $\mathbf{S}$ *of the translation relation* $\mathbf{T}$ *such that* $(\psi, p) \mathbf{S} q$ *iff* $(\delta(\psi))p = q$. *Then* $\mathbf{S}^{-1} \circ \square_{(\equiv, \equiv)} \circ \mathbf{S} \subseteq \dot{\sim}$.

Proof. Let $p\mathbf{S}^{-1} \circ \square_{(\equiv, \equiv)} \circ \mathbf{S}q$. Then for some ψ, p', q' we have $(\psi, p')\square_{(\equiv, \equiv)}(\psi, q')$ where $(\delta(\psi))p' = p$ and $(\delta(\psi))q' = q$ (by definition of $\mathbf{S}$). Suppose $p \xrightarrow{\gamma} p_d$. Then by Lemma 3, we have $(\psi, p') \xrightarrow{\alpha} (\psi', p'')$ where $(\delta(\psi))p'' = p_d$, $\delta(\psi)(\alpha) = \gamma$ and $\psi' = \psi \oplus E(\alpha)$. Then by definition of $\square_{(\equiv, \equiv)}$, there must exist β such that $(\psi, q') \xrightarrow{\beta} (\psi', q'')$ where $\alpha\psi\circ \equiv \circ\psi\beta$ and $(\psi', p'')\square_{(\equiv, \equiv)}(\psi', q'')$. Since $\alpha\psi\circ \equiv \circ\psi\beta$, we have $(\delta(\psi))(\alpha) = \gamma = (\delta(\psi))(\beta)$. Thus we have $q \xrightarrow{\gamma} q_d$ where $q_d = (\delta(\psi))q''$. Now let s be any substitutive effect of $\gamma = \delta(\psi)(\alpha)$, then by Lemma 2, $s \circ \delta(\psi)$ is a substitutive effect of $\psi' = \psi \oplus E(\alpha)$. Furthermore we can choose s and $\delta(\psi')$ to be such that $s \circ \delta(\psi) = \delta(\psi')$. We thus have $s(p_d) \mathbf{S}^{-1} \circ \square_{(\equiv, \equiv)} \circ \mathbf{S} s(q_d)$ (as $(\delta(\psi))p'' = p_d$, $(\psi', p'')\square_{(\equiv, \equiv)}(\psi', q'')$ and $(\delta(\psi))q'' = q_d$). A similar proof may be given for a transition of q. $\square$

Corollary 1. $\square_{(\equiv, \equiv)} = \mathbf{T} \circ \dot{\sim} \circ \mathbf{T}^{-1}$, *where* $\dot{\sim}$ *is strong bisimilarity (Definition 8).*

Proof. It follows from Proposition 2 that $\mathbf{S}^{-1} \circ \square_{(\equiv, \equiv)} \circ \mathbf{S} \subseteq \dot{\sim}$ for some $\mathbf{S} \subseteq \mathbf{T}$. Since $\circ$ is monotonic in each argument with respect to the $\subseteq$ ordering, we get $\square_{(\equiv, \equiv)} \subseteq \mathbf{S} \circ \dot{\sim} \circ \mathbf{S}^{-1} \subseteq \mathbf{T} \circ \dot{\sim} \circ \mathbf{T}^{-1}$. The reverse containment follows from Proposition 1. $\square$

5.3   Parameterised Hyperbisimulations

A (ρ, σ)-bisimulation as defined above only compares two states under identical environments. However we are actually interested in comparing processes and not the states in which they operate. Intuitively speaking, two processes may be considered equivalent only if they are equivalent under all environments. Hence we need to extend (ρ, σ)-bisimulations to a bisimulation based ordering defined over processes.

Definition 12. *A relation $\mathcal{H} \subseteq \mathbf{P}^2$ is a (ρ, σ)-**hyperbisimulation** iff for all $p, q \in \mathbf{P}$, $p\mathcal{H}q$ implies for all $\psi \in Env$, there is a (ρ, σ)-bisimulation $\mathcal{G}$ such that $(\psi, p)\mathcal{G}(\psi, q)$. The largest (ρ, σ)-hyperbisimulation called (ρ, σ)-**hyperbisimilarity** is denoted $\sqsubseteq_{(\rho,\sigma)}$.*

The concept of hyperbisimulations is unique to fusion calculus and it was originally defined by Victor in his work as the largest congruence contained within bisimulation. A very important property of interest for hyperbisimulations in the fusion calculus is the property of substitution closure which is necessary if one wishes to prove that hyperbisimilarity is a congruence.

Definition 13. *A relation ρ is **conservative** iff $\forall \alpha, \beta$ if $\alpha\rho\beta$ then $E(\alpha) = E(\beta)$. It is **substitution-closed** iff $\forall x, y, \alpha, \beta$ if $\alpha\rho\beta$ then $(\{x/y\}\alpha)\rho(\{x/y\}\beta)$.*

The following result (see [Bag11] for a proof) shows that substitution-closure on processes can also be derived from certain properties of the relations on actions for parameterised hyperbisimulations.

Corollary 2. *([Bag11]) If ρ and σ are both conservative and substitution closed relations on actions then for all $\psi \in Env$ and θ such that θ is a substitutive effect of ψ we have $(\psi, p)\sqsubseteq_{(\rho,\sigma)}(\psi, q)$ if and only if $(1, \theta p)\sqsubseteq_{(\rho,\sigma)}(1, \theta q)$. Furthermore if $p\sqsubseteq_{(\rho,\sigma)}q$ then for all substitutions θ we have $(\theta p)\sqsubseteq_{(\rho,\sigma)}(\theta q)$.* $\square$

Our motivation in defining hyperbisimulations is the same as Victor's, i.e. hyperbisimulations should relate processes which have the same behaviour in all contexts. However we have defined hyperbisimulations as the natural lifting of bisimulations, which are relations defined over states, to relations defined over processes, whereas in [Vic98] hyperbisimulation is used to define the largest congruence contained in bisimulations defined on original semantics and is obtained by closing the relation under all substitutions.

Definition 14. *A strong bisimulation (Definition 8) $\mathcal{R}$ is a **strong hyperbisimulation** iff it is substitution-closed i.e. for all substitutions θ, if $p\mathcal{R}q$ then $(\theta p)\mathcal{R}(\theta q)$. We denote the largest strong hyperbisimulation, called **strong hyperbisimilarity**, by $\sim$.*

The reason we choose to call the relation defined in Definition 12 as parameterised hyperbisimulation is because the relation defined by us turns out to be identical to the hyperbisimulations defined by Victor using the original semantics, as shown by the following result.

Corollary 3. *$\sqsubseteq_{(\equiv,\equiv)} =\sim$ i.e. strong hyperbisimilarity (Definition 14), equals parameterised $(\equiv, \equiv)$-hyperbisimilarity (Definition 12).*

Proof. By Definition 14, $p \sim q$ iff for all substitutions θ we have $\theta p \sim \theta q$. By Corollary 1 and noting that $(1, p)\mathbf{T}p$ holds for all processes p, we have $\theta p \sim \theta q$ if and only if $(1, \theta p)\sqsubseteq_{(\equiv,\equiv)}(1, \theta q)$. Let ψ be any environment such that θ is a substitutive effect of ψ, then by Corollary 2, we have $(\psi, p)\sqsubseteq_{(\equiv,\equiv)}(\psi, q)$. Since the result holds for all substitutions θ and hence for all environments ψ which have θ as its substitutive effect, by Definition 12, we have $p\sqsubseteq_{(\equiv,\equiv)}q$. Each step of the proof is reversible, hence the converse also holds. $\square$

6 Concluding Remarks

Some applications of parameterisation to the algebraic theory of bisimulations on process algebras were presented in this paper. While parameterisation has led to a more general notion of bisimulation, we have gone further in this paper by generalising this notion for name-passing calculi. In a manner similar to our earlier results, one can show that the properties of parameterised bisimulations for the value-passing calculus may be derived from the properties of the relations defined on actions, thereby providing a generalized framework for the study of bisimulations in value-passing calculi. In particular, one can show that the monotonicity, inversion, symmetry and reflexivity properties as shown in Proposition 2 also hold for these bisimulations, by simply noting that if ρ is symmetric or reflexive then so is $\psi \circ \rho \circ \psi$ and $(\psi \circ \rho \circ \psi)^{-1} = \psi \circ \rho^{-1} \circ \psi$. Also if one were to limit oneself to processes which can be represented in a non-value passing calculus like CCS, then the Definition 10 reduces to Definition 1. Therefore it strengthens our confidence that Definition 10 is the correct generalization of parameterised bisimulations to value passing calculus.

A more general proof can be given along the lines of corollary 3 to show that analogous notions such as weak (hyper-)bisimulations, efficiency preorder [AKH92] and elaborations [AKN95] over fusion calculus agents are special cases of $(\rho, \sigma)-$(hyper-)bisimulations by choosing ρ and σ appropriately. Further work may be done on investigating and extending the earlier results given for parameterised bisimulations, for example the axiomatization of parameterised bisimulations given in [SAK09], to the more generalised framework given for name passing calculus presented in this paper.

Acknowledgement. The research presented in this paper was partly sponsored by EADS Corp. We are thankful to the suggestions of anonymous reviewers who helped improve this paper. We are also thankful to Shibashis Guha for his careful review and suggestions.

References

[AKH92]   Arun-Kumar, S., Hennessy, M.: An efficiency preorder for processes. Acta Informatica 29, 737–760 (1992)

[AKN95]   Arun-Kumar, S., Natarajan, V.: Conformance: A precongruence close to bisimilarity. In: Structures in Concurrency Theory, pp. 55–68. Springer (1995)

[AK06]   Arun-Kumar, S.: On bisimilarities induced by relations on actions. In: Fourth IEEE International Conference on Software Engineering and Formal Methods, SEFM 2006, pp. 41–49. IEEE (2006)

[Bag11]   Bagga, D.: Parametrised bisimulations for the fusion calculus. Master's thesis, Department of Computer Science and Engineering, IIT Delhi (2011), http://www.cse.iitd.ac.in/~bagga/bag11.html

[BK85]   Bergstra, J.A., Klop, J.W.: Algebra of communicating processes with abstraction. Theoretical Computer Science 37, 77–121 (1985)

[KAK05]   Kiehn, A., Arun-Kumar, S.: Amortised bisimulations. In: Wang, F. (ed.) FORTE 2005. LNCS, vol. 3731, pp. 320–334. Springer, Heidelberg (2005)

[LV06]   Lüttgen, G., Vogler, W.: Bisimulation on speed: A unified approach. Theoretical Computer Science 360, 209–227 (2006)

[Mil89]   Milner, R.: Communication and concurrency. Prentice-Hall, Inc. (1989)
[MPW92a]   Milner, R., Parrow, J., Walker, D.: A calculus of mobile processes, I. Information
             and Computation 100, 1–40 (1992)
[MPW92b]   Milner, R., Parrow, J., Walker, D.: A calculus of mobile processes, II. Information
             and Computation 100, 41–77 (1992)
[PV98]   Parrow, J., Victor, B.: The fusion calculus: Expressiveness and symmetry in mobile
             processes. In: Proceedings of the Thirteenth Annual IEEE Symposium on Logic in
             Computer Science, pp. 176–185. IEEE (1998)
[SM92]   Sangiorgi, D., Milner, R.: The problem of "Weak Bisimulation up to". In: Cleave-
             land, W.R. (ed.) CONCUR 1992. LNCS, vol. 630, pp. 32–46. Springer, Heidelberg
             (1992)
[SAK09]   Singh, P., Arun-Kumar, S.: Axiomatization of a Class of Parametrised Bisimilarities.
             Perspectives in Concurrency Theory. Universities Press, India (2009)
[Vic98]   Victor, B.: The fusion calculus: Expressiveness and symmetry in mobile processes.
             PhD thesis, Uppsala University (1998)

A Appendix: Observational Congruence Proof

We provide the proofs for the observational equivalence of the derived LTS $\mathcal{L}_\dagger$ with the original LTS $\mathcal{L}$ (see Lemma 1 in paper) and the observational congruence as parameterised bisimulation in $\mathcal{L}_\dagger$ (see Theorem 2 in paper) here. The formal proof of the results mentioned in the paper requires the proof of various other lemmas, which makes the proof quite lengthy but it ensures that we cover all the details, thus making them necessary.

Definition A1. *Let $\mathcal{L} = \langle P, Act, \rightarrow \rangle$ be the usual LTS defined by divergence-free finite-state CCS agents. We define an LTS $\mathcal{L}_\dagger = \langle P, Act.\tau^*, \longrightarrow_\dagger \rangle$ derived from $\mathcal{L}$ such that $Act.\tau^* = \{\mu\tau^n \mid \mu \in Act, n \geq 0\}$ and $p \xrightarrow{\mu\tau^n}_\dagger p'$ iff there exists $n \geq 0$ and processes $p_0, \cdots, p_n$ such that $p \xrightarrow{\mu} p_0 \approx\xrightarrow{\tau} p_1 \approx\xrightarrow{\tau} \cdots \approx\xrightarrow{\tau} p_n \equiv p'$ and there does not exists any p_{n+1} such that $p' \approx\xrightarrow{\tau} p_{n+1}$. (Note: We use $p \approx\xrightarrow{\tau} q$ to denote $p \xrightarrow{\tau} q$ and $p \approx q$ here).*

Since $Act^+ = (Act.\tau^*)^+$ we may identify $\mathcal{L}_\dagger$ with $\mathcal{L}_\dagger^+ = \langle P, Act^+, \longrightarrow_\dagger^+ \rangle$ where $\longrightarrow_\dagger^+$ denotes one or more transitions via $\longrightarrow_\dagger$. We formally define the set of all weak μ-successors of p as the set Der_μ^p, i.e. $\mathrm{Der}_\mu^p = \{p' \mid p \xRightarrow{\mu} p', \mu \in Act\}$. We define the following preorder on these sets.

Definition A2. $\mathrm{Der}_\mu^p \lesssim \mathrm{Der}_\mu^q$ *if for every $p' \in \mathrm{Der}_\mu^p$ there exists $q' \in \mathrm{Der}_\mu^q$ such that $p' \approx q'$ and $\mathrm{Der}_\mu^p \not\gtrsim \mathrm{Der}_\mu^q$ if $\mathrm{Der}_\mu^p \lesssim \mathrm{Der}_\mu^q$ and there exists $q' \in \mathrm{Der}_\mu^q$ such that $p' \not\approx q'$ for every $p' \in \mathrm{Der}_\mu^p$. $\mathrm{Der}_\mu^p \gtrsim\!\!\!\lesssim \mathrm{Der}_\mu^q$ if $\mathrm{Der}_\mu^p \lesssim \mathrm{Der}_\mu^q$ and $\mathrm{Der}_\mu^q \lesssim \mathrm{Der}_\mu^p$.*

From Definition A2 it follows that $p \approx q$ iff $\mathrm{Der}_a^p \gtrsim\!\!\!\lesssim \mathrm{Der}_a^q$ for every $a \in Act \setminus \{\tau\}$ and $p \approx^+ q$ iff $\mathrm{Der}_\mu^p \gtrsim\!\!\!\lesssim \mathrm{Der}_\mu^q$ for each $\mu \in Act$. The following lemma follows from the preemptive power of the τ action.

Lemma A1. *For any processes $p, p' \in P$, if $p \xRightarrow{\tau} p'$ then $\mathrm{Der}_\mu^{p'} \lesssim \mathrm{Der}_\mu^p$ for every $\mu \in Act$. Further, if $p \not\approx p'$ implies $\mathrm{Der}_\nu^{p'} \not\gtrsim \mathrm{Der}_\nu^p$ for some $\nu \in Act$.*

Proof. Let $p'' \in \mathrm{Der}_\mu^{p'}$ for some $\mu \in Act$, then $p \xRightarrow{\tau} p' \xRightarrow{\mu} p''$ (by definition of $\mathrm{Der}_\mu^{p'}$), hence $p \xRightarrow{\mu} p''$. Thus we also have $p'' \in \mathrm{Der}_\mu^{p}$. Since p'' was arbitrary, we have $\mathrm{Der}_\mu^{p'} \subseteq_{\approx} \mathrm{Der}_\mu^{p}$ for all $\mu \in Act$. now if $p \not\approx p'$ implies $\exists \nu \in Act$ such that $p \xRightarrow{\nu} p''$ but $\not\exists p'''$ such that $p' \xRightarrow{\nu} p'''$ and $p'' \approx p'''$ (by definition of $\approx$). Clearly $p'' \in \mathrm{Der}_\nu^{p}$ hence $\mathrm{Der}_\nu^{p'} \subsetneq_{\not\approx} \mathrm{Der}_\nu^{p}$ for some $\nu \in Act$. □

Lemma A2. *If p is a finite state agent then*

1. *$p \xRightarrow{\tau} p'$ implies $p \xrightarrow{\tau^n}_\dagger p'' \approx p'$ for some $n > 0$, and*
2. *for any $a \in Act \setminus \{\tau\}$, $p \xRightarrow{a} p'$ implies $p \xrightarrow{\tau^m a \tau^n}_\dagger^+ p'' \approx p'$ for some $m, n \geq 0$.*

Proof. (part 1)

Let $p \xRightarrow{\tau} p'$ then we must have $p \xrightarrow{\tau^n}_\dagger q$ for some $n > 0$ such that either $q \approx p'$ or $q \xrightarrow{\tau^+}_\dagger q' \xRightarrow{\tau} q''$ such that $q'' \approx p'$ and $q' \not\approx p$. In case 1 the result holds. In case 2 we clearly have by Lemma A2, for all $\mu \in Act$ we have $\mathrm{Der}_\mu^{q'} \subseteq_{\approx} \mathrm{Der}_\mu^{p}$ and there exists $\nu \in Act$ such that $\mathrm{Der}_\nu^{q'} \subsetneq_{\not\approx} \mathrm{Der}_\nu^{p}$. By using finite state agent assumption i.e. $\sum_{\mu \in Act} |\mathrm{Der}_\mu^{p}|$ is finite, applying above logic inductively with $\sum_{\mu \in Act} |\mathrm{Der}_\mu^{p}|$ as induction variable we have our result.

(part 2)

Let $p \xRightarrow{a} p'$ for some $a \in Act \setminus \{\tau\}$. Then we have the following two cases:

- case 1: $p \xrightarrow{\tau} p_1 \xRightarrow{a} p'$

  by part 1, in this case we have $p \xrightarrow{\tau^+}_\dagger q$ such that $q \approx p_1$. Hence $q \xRightarrow{a} q'$ such that $q' \approx p'$ (by definition of $q \approx p_1$). now if $q \xrightarrow{\tau} q'' \xRightarrow{a} q'$ then $q \xrightarrow{\tau^+}_\dagger r$ such that $r \approx q''$ (by part 1) but $r \not\approx p$ (by definition of $\mathcal{L}_\dagger$ and since $p \xrightarrow{\tau^+}{}^2_\dagger r$). Then by claim A2 we have $\sum_{\mu \in Act} |\mathrm{Der}_\mu^{r}| < \sum_{\mu \in Act} |\mathrm{Der}_\mu^{p}|$ and $r \xRightarrow{a} r'$ such that $r' \approx p'$.

  Thus under finite state assumption we can only have this case finitely many times and eventually we will get case 2.
- Case 2: $p \xrightarrow{a} p_1 \Rightarrow p'$

  Then by definition of $\mathcal{L}_\dagger$ we must have a q such that $p \xrightarrow{a.\tau^*}_\dagger q$ such that $q \approx p_1$. Since $q \approx p_1$, we have $q \Rightarrow q' \approx p'$. Now either $q \approx p'$ in which case we are done or else we have $q \xRightarrow{\tau} q' \approx p'$ and $q \not\approx p'$. Then by part 1 we will have $q \xrightarrow{\tau^+}_\dagger r \approx q' \approx p'$. □

The following lemma shows that the LTS $\mathcal{L}_\dagger$ is observationally equivalent to the original LTS $\mathcal{L}$.

Lemma A3. *If p is a finite state agent then*

1. *For all $\alpha \in Act^+$ if $p \xrightarrow{\alpha}_\dagger p'$ then $p \xrightarrow{\alpha} p'$.*
2. *For all $\alpha \in Act^+$ if $p \xrightarrow{\alpha} p'$ then there exists α' with $\alpha' \doteq \alpha$ such that $p \xrightarrow{\alpha'}_\dagger p''$ and $p' \approx p''$.*

Proof. The (1) result follows straightforward from the definition of $\mathcal{L}_\dagger$, since every transition in $\mathcal{L}_\dagger$ is defined only if the corresponding transition exists in $\mathcal{L}$. We prove (2) by induction on the length of α. Base case of induction, i.e. $\alpha = a \in Act$, follows trivially from the definition of $\mathcal{L}_\dagger$. We assume by induction hypothesis that for all α such that $|\alpha| = k$ and $k \geq 1$, if $p \overset{\alpha}{\rightarrow} p'$ then there exists $\alpha' \in Act^+$ such that $\hat{\alpha}' = \hat{\alpha}$ and $p \overset{\alpha'}{\rightarrow_\dagger^+} p''$ and $p' \approx p''$. Now Let $|\alpha| = k + 1$ then $\alpha = \gamma.a$ where $|\gamma| = k, k \geq 1$. Now if $p \overset{\alpha}{\rightarrow} p'$ then there exists a q such that $p \overset{\gamma}{\rightarrow} q$ and $q \overset{a}{\longrightarrow} p'$. By IH now there must exist $\gamma' \in Act^+$ such that $\hat{\gamma}' = \hat{\gamma}$ and $p \overset{\gamma'}{\rightarrow_\dagger^+} q'$ and $q \approx q'$. Since $q \approx q'$ and $q \overset{a}{\longrightarrow} p'$ implies $q' \overset{a}{\Longrightarrow} p'''$ such that $p' \approx p'''$. Then by Lemma A2, there must exist p'' such that $q' \overset{\tau^n.a.\tau^m}{\rightarrow_\dagger^+} p''$ for some $n, m \geq 0$ and $p'' \approx p''' \approx p'$. Thus we have $p \overset{\alpha'}{\rightarrow_\dagger^+} p''$ and $p' \approx p''$ where $\alpha' = \gamma'.\tau^n.a.\tau^m$. $\qquad\square$

Theorem A1. *For all divergence-free finite-state CCS agents p, q we have $p \sqsubseteq_{(\hat{=}_+, \hat{=}_+)} q$ in $\mathcal{L}_\dagger^+$ iff $p \approx^+ q$, where $\hat{=}_+$ is the restriction of $\hat{=}$ to Act^+*

Claim A1. *Observational Congruence, $\approx^+$ is a $(\hat{=}_+, \hat{=}_+)-$bisimulation in $\mathcal{L}_\dagger^+$*

Proof. Let $p \approx^+ q$. Then for some $\alpha \in Act^+$ such that $p \overset{\alpha}{\rightarrow_\dagger^+} p'$ we must have $p \overset{\alpha}{\rightarrow} p'$ (by Lemma A3). Since $p \approx^+ q$, there must exist a q' and $\beta \in Act^+$ such that $\hat{\alpha} = \hat{\beta}$ and $q \overset{\beta}{\rightarrow} q'$ and $p' \approx q'$. Now by Lemma A3 there must exist $\beta' \in Act^+$ such that $\hat{\alpha} = \hat{\beta} = \hat{\beta}'$ and $q \overset{\beta'}{\rightarrow_\dagger^+} q''$ and $q' \approx q''$. i.e. $q \overset{\beta'}{\rightarrow} q''$ and $p' \approx q''$. Now by our definition for $\mathcal{L}_\dagger^+$ there does not exists any τ child of p' and q'' which are bisimilar to them, hence p' or q'' cant do a ϵ transition to match a τ transition for the other process and still reach bisimilar states. Hence $p' \approx^+ q''$. Hence we have $q \overset{\beta'}{\rightarrow_\dagger^+} q''$ and $p' \approx^+ q''$ and $\hat{\alpha} = \hat{\beta}'$. Since α was arbitrary, this holds for all α. We can show the result for all transitions of q in the similar way. Hence we have $\approx^+$ as a $(\hat{=}_+, \hat{=}_+)-$bisimulation by definition.

Claim A2. $(\hat{=}_+, \hat{=}_+)-$*bisimulation in $\mathcal{L}_\dagger^+$ is a observational congruence upto weak bisimulation.*

Proof. Let $p \sqsubseteq_{(\hat{=}_+, \hat{=}_+)} q$ in $\mathcal{L}_\dagger^+$. Then for some $\alpha \in Act^+$ such that $p \overset{\alpha}{\rightarrow} p'$ there must exist $\alpha' \in Act^+$ such that $\hat{\alpha}' = \hat{\alpha}$ and $p \overset{\alpha'}{\rightarrow_\dagger^+} p''$ and $p' \approx p''$ (by Lemma A3)

Hence there exists $\beta \in Act^+$ such that $\hat{\alpha}' = \hat{\beta}$ and $q \overset{\beta}{\rightarrow_\dagger^+} q'$ and $p'' \sqsubseteq_{(\hat{=}_+, \hat{=}_+)} q'$.

Therefore, $q \overset{\beta}{\rightarrow} q'$ and $\hat{\alpha} = \hat{\beta}$ and $p' \approx p'' \sqsubseteq_{(\hat{=}_+, \hat{=}_+)} q'$ where $\beta \in Act^+$. Since α was arbitrary, this holds for all α. We can show the result for all transitions of q in the similar way. Hence proved. $\qquad\square$

A Point-Free Relation-Algebraic Approach to General Topology

Gunther Schmidt

Fakultät für Informatik, Universität der Bundeswehr München
85577 Neubiberg, Germany
gunther.schmidt@unibw.de

Abstract. In advanced functional programming, researchers have investigated the existential image, the power transpose, and the power relator, e.g. It will be shown how the existential image is of use when studying continuous mappings between different topologies relationally. Normally, structures are compared using homomorphisms and sometimes isomorphisms. This applies to group homomorphisms, to graph homomorphisms and many more. The technique of comparison for topological structures will be shown to be quite different. Having in mind the cryptomorphic versions of neighborhood topology, open kernel topology, open sets topology, etc., this seems important.

Lifting concepts to a relational and, thus, algebraically manipulable and shorthand form, shows that existential and inverse images must here be used for structure comparison. Applying the relational language TITU-REL to such topological concepts allows to study and also visualize them.

Keywords: relational mathematics, homomorphism, topology, existential image, continuity.

1 Prerequisites

We will work with heterogeneous relations and provide a general reference to [Sch11a], but also to the earlier [SS89, SS93]. Our operations are, thus, binary union "$\cup$", intersection "$\cap$", composition "$;$", unary negation "$\overline{}$", transposition or conversion "^{T}", together with zero-ary null relations "$\mathbb{\bot\!\!\!\bot}$", universal relations "$\mathbb{T\!\!\!T}$", and identities "$\mathbb{I}$". A *heterogeneous relation algebra*

- is a category wrt. composition "$;$" and identities $\mathbb{I}$,
- has as morphism sets complete atomic boolean lattices with $\cup, \cap, \overline{}, \mathbb{\bot\!\!\!\bot}, \mathbb{T\!\!\!T}, \subseteq$,
- obeys rules for transposition ^{T} in connection with the latter two concepts that may be stated in either one of the following two ways:
  Dedekind rule:
  $$R;S \cap Q \subseteq (R \cap Q;S^{\mathsf{T}});(S \cap R^{\mathsf{T}};Q)$$
  Schröder equivalences:
  $$A;B \subseteq C \iff A^{\mathsf{T}};\overline{C} \subseteq \overline{B} \iff \overline{C};B^{\mathsf{T}} \subseteq \overline{A}$$

The two rules are equivalent in the context mentioned. Many rules follow out of this setting; not least for the concepts of a function, mapping or ordering, e.g. that mappings f may be *shunted*, i.e. that $A;f \subseteq B \iff A \subseteq B;f^{\mathsf{T}}$.

P. Höfner et al. (Eds.): RAMiCS 2014, LNCS 8428, pp. 226–241, 2014.

1.1 Quotient Forming

Whoever has a multiplication operation is inclined to ask for division. Division of relations with common source is indeed possible to the following extent:

$R;X = S$ has a solution X precisely when $S \subseteq R;\overline{R^{\mathsf{T}};\overline{S}}$,

or else when $S = R;\overline{R^{\mathsf{T}};\overline{S}}$. Among all solutions of $R;X = S$ the greatest is $\overline{R^{\mathsf{T}};\overline{S}}$. Often this is turned into the operation $R\backslash S := \overline{R^{\mathsf{T}};\overline{S}}$ of forming the *left residuum* — as in division allegories. An illustration of the left residuum is as follows:

$S = $
	Jan	Feb	Mar	Apr	May	Jun	Jul	Aug	Sep	Oct	Nov	Dec
US	0	0	0	1	0	1	1	1	0	1	0	0
French	1	0	0	1	0	0	1	0	0	1	0	0
German	1	1	0	0	1	1	0	1	0	0	0	1
British	1	1	0	0	0	0	1	0	1	0	1	1
Spanish	0	0	0	1	0	1	1	1	0	0	0	0

$R = $
	A	K	Q	J	10	9	8	7	6	5	4	3	2
US	0	0	0	0	0	0	0	0	0	0	0	0	1
French	0	1	0	0	0	0	0	1	0	0	0	0	0
German	0	0	1	0	0	0	1	1	0	1	0	1	0
British	0	1	1	0	0	0	0	1	0	0	0	0	1
Spanish	0	0	0	1	0	0	1	0	0	1	1	0	1

	Jan	Feb	Mar	Apr	May	Jun	Jul	Aug	Sep	Oct	Nov	Dec
A	1	1	1	1	1	1	1	1	1	1	1	1
K	1	0	0	0	0	0	1	0	0	0	0	0
Q	1	1	0	0	0	0	0	0	0	0	0	1
J	0	0	0	1	0	1	1	1	0	0	0	0
10	1	1	1	1	1	1	1	1	1	1	1	1
9	1	1	1	1	1	1	1	1	1	1	1	1
8	0	0	0	0	0	1	0	1	0	0	0	0
7	1	0	0	0	0	0	0	0	0	0	0	0
6	1	1	1	1	1	1	1	1	1	1	1	1
5	0	0	0	0	0	1	0	1	0	0	0	0
4	0	0	0	1	0	1	1	1	0	0	0	0
3	1	1	0	0	1	1	0	1	0	0	0	1
2	0	0	0	0	0	0	1	0	0	0	0	0

$R\backslash S$

Left residua show how columns of the relation R below the fraction backslash are contained in columns of the relation S above, i.e., some sort of subjunction.

As an often used term built upon residua, the *symmetric quotient* of two relations with common source has been introduced as $\mathsf{syq}(R,S) := \overline{R^{\mathsf{T}};\overline{S}} \cap \overline{\overline{R}^{\mathsf{T}};S}$. The illustration of the symmetric quotient is as follows:

$R = $
	A	K	Q	J	10	9	8	7	6	5	4	3	2
US	0	0	0	0	0	0	0	0	0	0	0	0	1
French	0	1	0	0	0	0	0	1	0	0	0	0	0
German	0	0	1	0	0	0	1	1	0	1	0	1	0
British	0	1	1	0	0	0	0	1	0	0	0	0	1
Spanish	0	0	0	1	0	0	1	0	0	1	1	0	1

$S = $
	Jan	Feb	Mar	Apr	May	Jun	Jul	Aug	Sep	Oct	Nov	Dec
US	0	0	0	1	0	1	1	1	0	1	0	0
French	1	0	0	1	0	0	1	0	0	1	0	0
German	1	1	0	0	1	1	0	1	0	0	0	1
British	1	1	0	0	0	0	1	0	1	0	1	1
Spanish	0	0	0	1	0	1	1	1	0	0	0	0

	Jan	Feb	Mar	Apr	May	Jun	Jul	Aug	Sep	Oct	Nov	Dec
A	0	0	1	0	0	0	0	0	0	0	0	0
K	0	0	0	0	0	0	0	0	0	0	0	0
Q	0	1	0	0	0	0	0	0	0	0	0	1
J	0	0	0	0	0	0	0	0	0	0	0	0
10	0	0	1	0	0	0	0	0	0	0	0	0
9	0	0	1	0	0	0	0	0	0	0	0	0
8	0	0	0	0	0	0	0	0	0	0	0	0
7	1	0	0	0	0	0	0	0	0	0	0	0
6	0	0	1	0	0	0	0	0	0	0	0	0
5	0	0	0	0	0	0	0	0	0	0	0	0
4	0	0	0	0	0	0	0	0	0	0	0	0
3	0	0	0	0	1	0	0	0	0	0	0	0
2	0	0	0	0	0	0	0	0	0	0	0	0

$\mathsf{syq}(R,S)$

The symmetric quotient shows which columns of the left are equal to columns of the right relation in $\mathsf{syq}(R,S)$, with S conceived as the denominator.

It is extremely helpful that the symmetric quotient enjoys certain cancellation properties. These are far from being broadly known. Just minor side conditions have to be observed. In any of the following propositions correct typing is assumed. What is more important is that one may calculate with the symmetric quotient in a fairly traditional algebraic way. Proofs may be found in [Sch11a].

1.1 Proposition. Arbitrary relations A, B satisfy in analogy to $a \cdot \frac{b}{a} = b$

i) $A_;\mathsf{syq}(A,B) = B \cap \mathbb{T}_;\mathsf{syq}(A,B)$,
ii) $\mathsf{syq}(A,B)$ surjective $\implies A_;\mathsf{syq}(A,B) = B$. $\square$

The analogy holds except for the fact that certain columns are "cut out" or are annihilated when the symmetric quotient fails to be surjective — meaning that certain columns of the first relation fail to have counterparts in the second.

1.2 Proposition. Arbitrary relations A, B, C satisfy in analogy to $\frac{b}{a} \cdot \frac{c}{b} = \frac{c}{a}$

i) $\mathsf{syq}(A,B)_;\mathsf{syq}(B,C) = \mathsf{syq}(A,C) \cap \mathsf{syq}(A,B)_;\mathbb{T}$
      $= \mathsf{syq}(A,C) \cap \mathbb{T}_;\mathsf{syq}(B,C)$
ii) If $\mathsf{syq}(A,B)$ is total, **or** if $\mathsf{syq}(B,C)$ is surjective, then
      $\mathsf{syq}(A,B)_;\mathsf{syq}(B,C) = \mathsf{syq}(A,C)$. $\square$

1.2 Domain Construction

The relational language TITUREL (see [Sch03, Sch11b]) makes use of characterizations *up to isomorphism* and bases domain constructions on these. This applies to the obvious cases of direct products (tuple forming) with projections named π, ρ and direct sums (variant handling). If any two heterogeneous relations π, ρ with common source are given, they are said to form a **direct product** if

$$\pi^\mathsf{T}_;\pi = \mathbb{I}, \quad \rho^\mathsf{T}_;\rho = \mathbb{I}, \quad \pi_;\pi^\mathsf{T} \cap \rho_;\rho^\mathsf{T} = \mathbb{I}, \quad \pi^\mathsf{T}_;\rho = \mathbb{T}.$$

Thus, the relations π, ρ are mappings, usually called **projections**. Along with the direct product, we automatically have the Kronecker product of any two relations and (when sources coincide) the strict fork operator for relations,

$$(R \otimes S) := \pi_;R_;\pi'^\mathsf{T} \cap \rho_;S_;\rho'^\mathsf{T} \quad \text{and} \quad (P \otimes Q) := P_;\pi'^\mathsf{T} \cap Q_;\rho'^\mathsf{T}.$$

In a similar way, any two heterogeneous relations ι, κ with common target are said to form the left, respectively right, **injection** of a **direct sum** if

$$\iota_;\iota^\mathsf{T} = \mathbb{I}, \quad \kappa_;\kappa^\mathsf{T} = \mathbb{I}, \quad \iota^\mathsf{T}_;\iota \cup \kappa^\mathsf{T}_;\kappa = \mathbb{I}, \quad \iota_;\kappa^\mathsf{T} = \mathbb{L}.$$

TITUREL then enables the construction of natural projections to a quotient modulo an equivalence and the extrusion of a subset out of its domain, so as to have both of them as "first-class citizens" among the domains considered — not just as "dependent types".

Here, we include the **direct power**. *Any* relation ε satisfying

$$\mathsf{syq}(\varepsilon,\varepsilon) \subseteq \mathbb{I}, \quad \mathsf{syq}(\varepsilon, R) \text{ surjective for every relation } R \text{ starting in } X$$

is called a **membership relation** and its codomain the *direct power* of X.

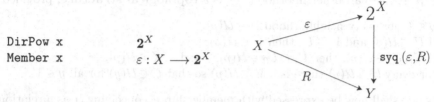

DirPow x 2^X

Member x $\varepsilon : X \longrightarrow 2^X$

Above it is indicated how this is represented in the language. One will observe the 1st-order style of the definition — however quantifying over relations. Classically, the characterisation of the powerset requires 2nd-order.

One will also observe the fractal structure in the following example of a membership relation, together with an interesting interplay between subsets U, their ordering $\Omega := \overline{\varepsilon^{\mathsf{T}}.\overline{\varepsilon}}$, and elements like e in the powerset:

$$U = \varepsilon_{;}e \qquad e = \mathsf{syq}(\varepsilon, U)$$

	{}	{a}	{b}	{a,b}	{c}	{a,c}	{b,c}	{a,b,c}	{d}	{a,d}	{b,d}	{a,b,d}	{c,d}	{a,c,d}	{b,c,d}	{a,b,c,d}	
{}	1	1	1	1	1	1	1	1	1	1	1	1	1	1	1	1	0
{a}	0	1	0	1	0	1	0	1	0	1	0	1	0	1	0	1	0
{b}	0	0	1	1	0	0	1	1	0	0	1	1	0	0	1	1	0
{a,b}	0	0	0	1	0	0	0	1	0	0	0	1	0	0	0	1	0
{c}	0	0	0	0	1	1	1	1	0	0	0	0	1	1	1	1	0
{a,c}	0	0	0	0	0	1	0	1	0	0	0	0	0	1	0	1	0
{b,c}	0	0	0	0	0	0	1	1	0	0	0	0	0	0	1	1	0
{a,b,c}	0	0	0	0	0	0	0	1	0	0	0	0	0	0	0	1	0
{d}	0	0	0	0	0	0	0	0	1	1	1	1	1	1	1	1	0
{a,d}	0	0	0	0	0	0	0	0	0	1	0	1	0	1	0	1	0
{b,d}	0	0	0	0	0	0	0	0	0	0	1	1	0	0	1	1	1
{a,b,d}	0	0	0	0	0	0	0	0	0	0	0	1	0	0	0	1	0
{c,d}	0	0	0	0	0	0	0	0	0	0	0	0	1	1	1	1	0
{a,c,d}	0	0	0	0	0	0	0	0	0	0	0	0	0	1	0	1	0
{b,c,d}	0	0	0	0	0	0	0	0	0	0	0	0	0	0	1	1	0
{a,b,c,d}	0	0	0	0	0	0	0	0	0	0	0	0	0	0	0	1	0

	{}	{a}	{b}	{a,b}	{c}	{a,c}	{b,c}	{a,b,c}	{a,d}	{a,d}	{b,d}	{a,b,d}	{c,d}	{a,c,d}	{b,c,d}	{a,b,c,d}	
a	0	1	0	1	0	1	0	1	0	1	0	1	0	1	0	1	0
b	0	0	1	1	0	0	1	1	0	0	1	1	0	0	1	1	1
c	0	0	0	0	1	1	1	1	0	0	0	0	1	1	1	1	0
d	0	0	0	0	0	0	0	0	1	1	1	1	1	1	1	1	1

$$(0\ 0\ 0\ 0\ 0\ 0\ 0\ 0\ 0\ 0\ 0\ 1\ 0\ 0\ 0\ 0) = e^{\mathsf{T}}$$

Then the direct product together with the direct power allow in particular to define join and meet in the powerset concisely as

$$\mathcal{M} := \mathsf{syq}((\varepsilon \otimes \varepsilon), \varepsilon) \qquad \mathcal{J} := \mathsf{syq}((\overline{\varepsilon} \otimes \overline{\varepsilon}), \overline{\varepsilon}),$$

which then satisfies such nice formulae as

$$(\varepsilon \otimes \varepsilon)\mathcal{M} = \varepsilon$$
$$\mathsf{syq}(X, (\varepsilon \otimes \varepsilon))_{;}\mathcal{M} = \mathsf{syq}(X, \varepsilon).$$

2 Recalling Concepts of Topology

Now we apply the techniques mentioned for topological structures. Topology may be defined via open or closed sets, neighborhoods, transition to open kernels, etc. We show that at least the neighborhood version — in the form given by Felix Hausdorff — shows an inherently "linear" configuration, which is apt to being formulated using relations.

We recall (see [Fra60]) that a set X endowed with a system $\mathcal{U}(p)$ of subsets for every $p \in X$ — called neighborhoods — is a **topological structure**, provided

i) $p \in U$ for every neighborhood $U \in \mathcal{U}(p)$,
ii) if $U \in \mathcal{U}(p)$ and $V \supseteq U$, then $V \in \mathcal{U}(p)$,
iii) if $U_1, U_2 \in \mathcal{U}(p)$, then $U_1 \cap U_2 \in \mathcal{U}(p)$ and $X \in \mathcal{U}(p)$,
iv) for every $U \in \mathcal{U}(p)$ there is a $V \in \mathcal{U}(p)$ so that $U \in \mathcal{U}(y)$ for all $y \in V$.

The same shall now be expressed with membership ε, conceiving $\mathcal{U}$ as a relation

$$\varepsilon : X \longrightarrow \mathbf{2}^X \quad \text{and} \quad \mathcal{U} : X \longrightarrow \mathbf{2}^X.$$

At other occasions, it has been shown that condition (iv), e.g., can semi-formally be lifted step by step to a relational form:

For every $U \in \mathcal{U}(p)$ there is a $V \in \mathcal{U}(p)$ such that $U \in \mathcal{U}(y)$ for all $y \in V$.

$$\forall p, U : U \in \mathcal{U}(p) \rightarrow (\exists V : V \in \mathcal{U}(p) \wedge (\forall y : y \in V \rightarrow U \in \mathcal{U}(y)))$$

$$\forall p, U : \mathcal{U}_{pU} \rightarrow (\exists V : \mathcal{U}_{pV} \wedge (\forall y : \varepsilon_{yV} \rightarrow \mathcal{U}_{yU}))$$

$$\forall p, U : \mathcal{U}_{pU} \rightarrow (\exists V : \mathcal{U}_{pV} \wedge \overline{\exists y : \varepsilon_{yV} \wedge \overline{\mathcal{U}_{yU}}})$$

$$\forall p, U : \mathcal{U}_{pU} \rightarrow (\exists V : \mathcal{U}_{pV} \wedge \overline{\varepsilon^{\mathsf{T}}; \overline{\mathcal{U}}}_{VU})$$

$$\forall p, U : \mathcal{U}_{pU} \rightarrow (\mathcal{U}; \overline{\varepsilon^{\mathsf{T}}; \overline{\mathcal{U}}})_{pU}$$

$$\mathcal{U} \subseteq \mathcal{U}; \overline{\varepsilon^{\mathsf{T}}; \overline{\mathcal{U}}}$$

One could see how the lengthy verbose or the predicate logic formula is traced back to a "lifted" relational version free of quantifiers, that employs a residuum. Such algebraic versions should be preferred in many respects. They support proof assisting systems and may be written down in the language TituRel so as to evaluate terms built with them and, e.g., visualize concepts of this paper. An example of a neighborhood topology $\mathcal{U}$ and the basis of its open sets:

$\mathcal{U} =$

	{}	{a}	{b}	{a,b}	{c}	{a,c}	{b,c}	{a,b,c}	{d}	{a,d}	{b,d}	{a,b,d}	{c,d}	{a,c,d}	{b,c,d}	{a,b,c,d}
a	0	1	0	1	0	1	0	1	0	1	0	1	0	1	0	1
b	0	0	0	0	0	0	1	1	0	0	0	0	0	0	1	1
c	0	0	0	0	1	1	1	1	0	0	0	0	1	1	1	1
d	0	0	0	0	0	0	0	0	0	0	0	0	1	1	1	1

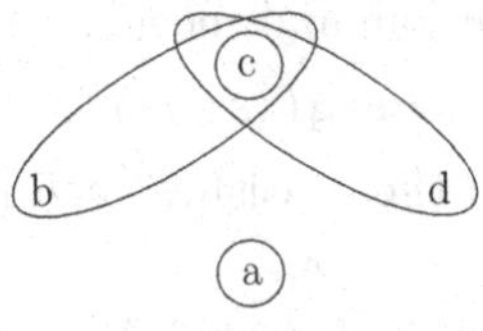

This, together with a transfer of the other properties to the relational level, and using ε derived from the source of $\mathcal{U}$ gives rise to the lifting of the initial Hausdorff definition, thus making it point-free as in:

2.1 Definition. A relation $\mathcal{U} : X \longrightarrow 2^X$ will be called a **neighborhood topology** if the following properties are satisfied:

  i) $\mathcal{U}_{;}\mathbb{T} = \mathbb{T}$ and $\mathcal{U} \subseteq \varepsilon$,
  ii) $\mathcal{U}_{;}\Omega \subseteq \mathcal{U}$,
  iii) $(\mathcal{U} \otimes \mathcal{U})_{;}\mathcal{M} \subseteq \mathcal{U}$,
  iv) $\mathcal{U} \subseteq \mathcal{U}_{;}\varepsilon^{\mathsf{T}}{}_{;}\overline{\overline{\mathcal{U}}}$.   □

Correspondingly, lifting may be executed for various other topology concepts. We start with the mapping to open kernels, assuming $\Omega := \overline{\varepsilon^{\mathsf{T}}{}_{;}\overline{\varepsilon}}$ to represent the powerset ordering.

2.2 Definition. We call a relation $\mathcal{K} : 2^X \longrightarrow 2^X$ a **mapping-to-open-kernel topology**, if

  i) $\mathcal{K}$ is a kernel forming, i.e., $\mathcal{K} \subseteq \Omega^{\mathsf{T}}$, $\Omega_{;}\mathcal{K} \subseteq \mathcal{K}_{;}\Omega$, $\mathcal{K}_{;}\mathcal{K} \subseteq \mathcal{K}$,
  ii) $\varepsilon_{;}\mathcal{K}^{\mathsf{T}}$ is total,
  iii) $(\mathcal{K} \otimes \mathcal{K})_{;}\mathcal{M} = \mathcal{M}_{;}\mathcal{K}$.   □

Conditions (i) obviously request that $\mathcal{K}$ maps to subsets of the original one, is isotonic, and is idempotent. Condition (iii) requires $\mathcal{K}$ and $\mathcal{M}$ to commute: One may obtain kernels of an arbitrary pair of subsets first and then form their intersection, or, equivalently, start intersecting these subsets and then getting the kernel.

2.3 Proposition. The following operations are inverses of one another:

  i) Given any neighborhood topology $\mathcal{U}$, the construct $\mathcal{K} := \mathsf{syq}(\mathcal{U}, \varepsilon)$ is a kernel-mapping topology.
  ii) Given any kernel-mapping topology $\mathcal{K}$, the construct $\mathcal{U} := \varepsilon_{;}\mathcal{K}^{\mathsf{T}}$ results in a neighborhood topology.

We cannot give the full proof for reasons of space, but indicate a part of it: The $\mathcal{K}$ defined in (i) is certainly a mapping, due to cancellation $\mathcal{K}^{\mathsf{T}}{}_{;}\mathcal{K} \subseteq \mathsf{syq}(\varepsilon, \varepsilon) = \mathbb{I}$, and, since forming the symmetric quotient with ε on the right side of syq gives a total relation by definition of a membership relation.

$$\mathcal{U}(\mathcal{K}(\mathcal{U})) = \varepsilon_i [\mathsf{syq}(\mathcal{U},\varepsilon)]^{\mathsf{T}} = \varepsilon_i \mathsf{syq}(\varepsilon,\mathcal{U}) = \mathcal{U} \quad \text{since } \mathsf{syq}(\varepsilon,X) \text{ is surjective}$$
$$\mathcal{K}(\mathcal{U}(\mathcal{K})) = \mathsf{syq}(\varepsilon_i\mathcal{K}^{\mathsf{T}},\varepsilon) = \mathcal{K}_i \mathsf{syq}(\varepsilon,\varepsilon) = \mathcal{K}_i \mathbb{I} = \mathcal{K} \quad \text{since } \mathcal{K} \text{ is a mapping}$$

It remains the obligation to prove

$$\begin{array}{llll}
\mathcal{U}_i\mathbb{T} = \mathbb{T}, & & \mathcal{K} \subseteq \Omega^{\mathsf{T}}, \\
\mathcal{U} \subseteq \varepsilon, & & \Omega_i\mathcal{K} \subseteq \mathcal{K}_i\Omega, \\
\mathcal{U}_i\Omega \subseteq \mathcal{U}, & \Longleftrightarrow & \mathcal{K}_i\mathcal{K} \subseteq \mathcal{K}, \\
(\mathcal{U} \otimes \mathcal{U})_i\mathcal{M} \subseteq \mathcal{U}, & & \varepsilon_i\mathcal{K}^{\mathsf{T}}_i\mathbb{T} = \mathbb{T}, \\
\mathcal{U} \subseteq \mathcal{U}_i\overline{\varepsilon^{\mathsf{T}}_i\overline{\mathcal{U}}}. & & (\mathcal{K} \otimes \mathcal{K})_i\mathcal{M} = \mathcal{M}_i\mathcal{K}.
\end{array}$$

A third form of a topology definition runs as follows:

2.4 Definition. A binary vector $\mathcal{O}_V$ along $\mathbf{2}^X$ will be called an **open set topology** provided

i) $\mathsf{syq}(\varepsilon,\perp\!\!\!\perp) \subseteq \mathcal{O}_V \quad \mathsf{syq}(\varepsilon,\mathbb{T}) \subseteq \mathcal{O}_V$,

ii) $v \subseteq \mathcal{O}_V \implies \mathsf{syq}(\varepsilon,\varepsilon_iv) \subseteq \mathcal{O}_V \quad$ for all vectors $v \subseteq \mathbf{2}^X$,

iii) $\mathcal{M}^{\mathsf{T}}_i(\mathcal{O}_V \otimes \mathcal{O}_V) \subseteq \mathcal{O}_V$. $\qquad\qquad\qquad\qquad\qquad\qquad\qquad\square$

With (i), $\mathbb{T}$ and $\perp\!\!\!\perp$ are declared to be open. The vector v in (ii) determines a set of open sets conceived as points in the powerset. It is demanded that their union be open again. According to (iii), intersection (meet $\mathcal{M}$) applied to two (i.e., finitely many) open sets must be open.

One may also study the membership restricted to open sets $\varepsilon_{\mathcal{O}} := \varepsilon \cap \mathbb{T}_i\mathcal{O}_V^{\mathsf{T}}$.

All these topology concepts are "cryptomorphic". This term has sometimes been used when the "same" concept is defined and axiomatized in quite different ways as here via $\mathcal{U}, \mathcal{K}, \mathcal{O}_V$, e.g. Nevertheless, the transitions below allow to prove equivalence as it is schematically indicated above for $\mathcal{U}, \mathcal{K}$. The transitions below may be written down in TITUREL so as to achieve the version intended. In particular, $\mathcal{O}_V$ and $\mathcal{O}_D$ are distinguished, although they are very similar, namely "diagonal matrix" vs. "column vector" to characterize a subset.

$$\begin{array}{lll}
\mathcal{U} & \mapsto & \mathcal{K} := \mathsf{syq}(\mathcal{U},\varepsilon) : \mathbf{2}^X \longrightarrow \mathbf{2}^X \\
\mathcal{K} & \mapsto & \mathcal{U} := \varepsilon_i\mathcal{K}^{\mathsf{T}} : X \longrightarrow \mathbf{2}^X. \\
\mathcal{O}_D & \mapsto & \mathcal{U} := \varepsilon_i\mathcal{O}_{D_i}\Omega \\
\mathcal{O}_D & \mapsto & \mathcal{O}_V := \mathcal{O}_{D_i}\mathbb{T} \\
\mathcal{K},\mathcal{U},\mathcal{O}_V & \mapsto & \mathcal{O}_D := \mathbb{I} \cap \overline{\varepsilon^{\mathsf{T}}_i\overline{\mathcal{U}}} = \mathbb{I} \cap \mathcal{O}_{V_i}\mathbb{T} = \mathcal{K}^{\mathsf{T}}_i\mathcal{K}
\end{array}$$

One may, thus, obtain the same topology in different forms as it is shown below for the example given before Definition 2.1:

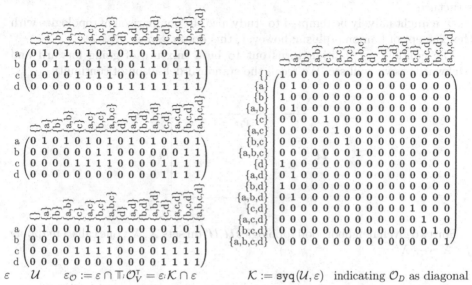

ε $\mathcal{U}$ $\varepsilon_\mathcal{O} := \varepsilon \cap \mathbb{T};\mathcal{O}_V^\mathsf{T} = \varepsilon;\mathcal{K} \cap \varepsilon$ $\mathcal{K} := \mathsf{syq}(\mathcal{U}, \varepsilon)$ indicating $\mathcal{O}_D$ as diagonal

By the way, there exists also a kernel-forming that doesn't lead to a topology; it is not intersection-closed as can be seen from the subsets $\{a, b\}$ and $\{b, d\}$ with intersection $\{d\}$:

$$
\mathcal{K} =
\begin{array}{r}
\{\} \\
\{a\} \\
\{b\} \\
\{a,b\} \\
\{c\} \\
\{a,c\} \\
\{b,c\} \\
\{a,b,c\} \\
\{d\} \\
\{a,d\} \\
\{b,d\} \\
\{a,b,d\} \\
\{c,d\} \\
\{a,c,d\} \\
\{b,c,d\} \\
\{a,b,c,d\}
\end{array}
\left(
\begin{array}{cccccccccccccccc}
1 & 0 & 0 & 0 & 0 & 0 & 0 & 0 & 0 & 0 & 0 & 0 & 0 & 0 & 0 & 0 \\
1 & 0 & 0 & 0 & 0 & 0 & 0 & 0 & 0 & 0 & 0 & 0 & 0 & 0 & 0 & 0 \\
1 & 0 & 0 & 0 & 0 & 0 & 0 & 0 & 0 & 0 & 0 & 0 & 0 & 0 & 0 & 0 \\
1 & 0 & 0 & 0 & 0 & 0 & 0 & 0 & 0 & 0 & 0 & 0 & 0 & 0 & 0 & 0 \\
1 & 0 & 0 & 0 & 0 & 0 & 0 & 0 & 0 & 0 & 0 & 0 & 0 & 0 & 0 & 0 \\
1 & 0 & 0 & 0 & 0 & 0 & 0 & 0 & 0 & 0 & 0 & 0 & 0 & 0 & 0 & 0 \\
1 & 0 & 0 & 0 & 0 & 0 & 0 & 0 & 0 & 0 & 0 & 0 & 0 & 0 & 0 & 0 \\
1 & 0 & 0 & 0 & 0 & 0 & 0 & 0 & 0 & 0 & 0 & 0 & 0 & 0 & 0 & 0 \\
1 & 0 & 0 & 0 & 0 & 0 & 0 & 0 & 0 & 0 & 0 & 0 & 0 & 0 & 0 & 0 \\
0 & 0 & 0 & 0 & 0 & 0 & 0 & 0 & 0 & 1 & 0 & 0 & 0 & 0 & 0 & 0 \\
0 & 0 & 0 & 0 & 0 & 0 & 0 & 0 & 0 & 0 & 1 & 0 & 0 & 0 & 0 & 0 \\
0 & 0 & 0 & 0 & 0 & 0 & 0 & 0 & 0 & 0 & 0 & 1 & 0 & 0 & 0 & 0 \\
1 & 0 & 0 & 0 & 0 & 0 & 0 & 0 & 0 & 0 & 0 & 0 & 0 & 0 & 0 & 0 \\
0 & 0 & 0 & 0 & 0 & 0 & 0 & 0 & 0 & 1 & 0 & 0 & 0 & 0 & 0 & 0 \\
0 & 0 & 0 & 0 & 0 & 0 & 0 & 0 & 0 & 0 & 1 & 0 & 0 & 0 & 0 & 0 \\
0 & 0 & 0 & 0 & 0 & 0 & 0 & 0 & 0 & 0 & 0 & 0 & 0 & 0 & 0 & 1
\end{array}
\right)
$$

3   Continuity

For a mathematical structure, one routinely defines its structure-preserving map-pings. Traditionally, this is handled under the name of a homomorphism; it may be defined for relational structures as well as for algebraic ones (i.e., those where structure is described by *mappings* as for groups, e.g.) in more or less the same standard way; it is available for a homogeneous as well as a heterogeneous structure.

One might naively be tempted to study also the comparison of topologies with the concept of homomorphism; however, this doesn't work.

The continuity condition turns out to be a mixture of going forward and backwards as we will see. We recall the standard definition of continuity.

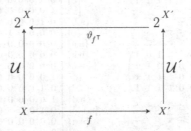

For two given neighborhood topologies $\mathcal{U}, \mathcal{U}'$ on sets X, X', one calls a mapping $f : X \longrightarrow X'$

$$f \text{ continuous} \quad :\Longleftrightarrow \quad \begin{array}{l} \text{For every point } p \in X \text{ and every } U' \in \mathcal{U}'(f(p)), \\ \text{there exists a } U \in \mathcal{U}(p) \text{ such that } f(U) \subseteq U'. \end{array}$$

A first example of a continuous mapping shows two open set bases, arranged as columns of matrices R_1, R_2, and the mapping f:

$$R_1 = \begin{array}{c} 1 \\ 2 \\ 3 \\ 4 \\ 5 \end{array} \begin{pmatrix} 1 & 0 & 0 & 0 \\ 0 & 1 & 0 & 0 \\ 0 & 0 & 0 & 1 \\ 0 & 0 & 0 & 1 \\ 0 & 0 & 1 & 0 \end{pmatrix} \qquad f = \begin{array}{c} 1 \\ 2 \\ 3 \\ 4 \\ 5 \end{array} \begin{pmatrix} 0 & 1 & 0 & 0 & 0 \\ 0 & 0 & 0 & 1 & 0 \\ 1 & 0 & 0 & 0 & 0 \\ 1 & 0 & 0 & 0 & 0 \\ 0 & 0 & 1 & 0 & 0 \end{pmatrix} \qquad R_2 = \begin{array}{c} a \\ b \\ c \\ d \\ e \end{array} \begin{pmatrix} 1 & 0 & 0 \\ 0 & 1 & 0 \\ 0 & 1 & 0 \\ 0 & 0 & 1 \\ 0 & 0 & 1 \end{pmatrix}$$

(columns of R_1: alpha, beta, gamma, delta; columns of f: a, b, c, d, e; columns of R_2: alpha2, beta2, gamma2)

The following is another example of a continuous mapping. Again two open set bases are arranged as columns of matrices R_1, R_2 and shown together with the mapping f:

$$
R_1 = \begin{array}{c} 1 \\ 2 \\ 3 \\ 4 \\ 5 \end{array}
\begin{pmatrix}
1 & 0 & 0 & 0 & 0 \\
0 & 1 & 1 & 0 & 0 \\
0 & 0 & 1 & 0 & 0 \\
0 & 0 & 0 & 1 & 0 \\
0 & 0 & 0 & 1 & 1
\end{pmatrix}
\quad
f = \begin{array}{c} 1 \\ 2 \\ 3 \\ 4 \\ 5 \end{array}
\begin{pmatrix}
0 & 0 & 1 & 0 & 0 \\
0 & 1 & 0 & 0 & 0 \\
1 & 0 & 0 & 0 & 0 \\
0 & 0 & 0 & 1 & 0 \\
0 & 0 & 0 & 0 & 1
\end{pmatrix}
\quad
R_2 = \begin{array}{c} a \\ b \\ c \\ d \\ e \end{array}
\begin{pmatrix}
0 & 0 & 0 & 1 & 0 & 0 \\
1 & 0 & 0 & 1 & 1 & 0 \\
0 & 1 & 0 & 0 & 1 & 0 \\
0 & 0 & 0 & 0 & 0 & 1 \\
0 & 0 & 1 & 0 & 0 & 1
\end{pmatrix}
$$

(columns of R_1: alpha, beta, gamma, delta, epsilon; columns of f: a, b, c, d, e; columns of R_2: alpha2, beta2, gamma2, delta2, epsilon2, eta2)

Now follows a third example of a continuous mapping, using the same style.

$$
R_1 = \begin{array}{c} 1 \\ 2 \\ 3 \\ 4 \\ 5 \end{array}
\begin{pmatrix}
0 & 0 & 0 & 1 & 0 \\
0 & 0 & 0 & 0 & 1 \\
0 & 0 & 1 & 0 & 0 \\
1 & 1 & 1 & 0 & 0 \\
0 & 1 & 0 & 0 & 0
\end{pmatrix}
\quad
f = \begin{array}{c} 1 \\ 2 \\ 3 \\ 4 \\ 5 \end{array}
\begin{pmatrix}
0 & 1 & 0 & 0 \\
1 & 0 & 0 & 0 \\
0 & 0 & 1 & 0 \\
0 & 0 & 0 & 1 \\
0 & 0 & 1 & 0
\end{pmatrix}
\quad
R_2 = \begin{array}{c} a \\ b \\ c \\ d \end{array}
\begin{pmatrix}
1 & 0 & 0 & 0 \\
0 & 0 & 0 & 1 \\
0 & 0 & 1 & 0 \\
0 & 1 & 1 & 0
\end{pmatrix}
$$

(columns of R_1: alpha, beta, gamma, delta, epsilon; columns of f: a, b, c, d; columns of R_2: alpha2, beta2, gamma2, delta2)

According to our general policy, we should try to lift the continuity definition to a point-free relational level. However, one soon sees that this requires that we employ the concept of an existential and of an inverse image.

3.1   Existential and Inverse Image

The lifting of a relation R to a corresponding relation ϑ_R on the powerset level has been called its existential image; cf. [Bd96]. (There exist also the power transpose Λ_R and the power relator ζ_R.)

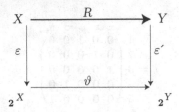

Assuming an arbitrary relation $R : X \longrightarrow Y$ together with membership relations $\varepsilon : X \longrightarrow \mathbf{2}^X$ and $\varepsilon' : Y \longrightarrow \mathbf{2}^Y$ on either side one calls

$$\vartheta := \vartheta_R := \mathsf{syq}\,(R^{\mathsf{T}}{}_{;}\varepsilon, \varepsilon') = \overline{\varepsilon^{\mathsf{T}}{}_{;}R{}_{;}\overline{\varepsilon'}} \cap \overline{\overline{\varepsilon^{\mathsf{T}}{}_{;}R}{}_{;}\varepsilon'},$$

its **existential image**. The **inverse image** is obtained when taking the existential image of the transposed relation.

It turns out, according to [Bd96, Sch11a], that ϑ is

- (lattice-)continuous wrt. the powerset orders $\Omega = \overline{\varepsilon^{\mathsf{T}}{}_{;}\overline{\varepsilon}}$,
- multiplicative: $\vartheta_{Q{;}R} = \vartheta_{Q}{}_{;}\vartheta_{R}$,
- preserves identities: $\vartheta_{\mathbb{I}_X} = \mathbb{I}_{\mathbf{2}^X}$,
- R may be re-obtained from ϑ_R as $R = \overline{\varepsilon{;}\vartheta_R{;}\overline{\varepsilon'}^{\mathsf{T}}}$.

It also satisfies, according to [dRE98, Sch11a], the following simulation property. R and its existential image as well as its inverse image **simulate** each other via $\varepsilon, \varepsilon'$:

$$\varepsilon^{\mathsf{T}}{;}R = \vartheta_R{;}\varepsilon'^{\mathsf{T}} \qquad \varepsilon'^{\mathsf{T}}{;}R^{\mathsf{T}} = \vartheta_{R^{\mathsf{T}}}{;}\varepsilon^{\mathsf{T}}.$$

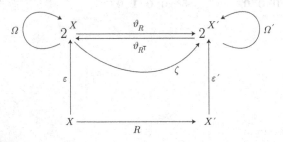

The existential image and the inverse image also satisfy formulae with respect to the powerset orderings:

i) $\Omega'{;}\vartheta_{f^{\mathsf{T}}} \subseteq \vartheta_{f^{\mathsf{T}}}{;}\Omega$ if f is a mapping,

ii) $\Omega{;}\vartheta_{f^{\mathsf{T}}}^{\mathsf{T}} = \vartheta_f{;}\Omega'$ if f is a mapping.

The proof cannot be given in the present limited environment.

$$R = \begin{pmatrix} & a & b & c & d \\ 1 & 0 & 1 & 0 & 1 \\ 2 & 1 & 0 & 0 & 0 \\ 3 & 0 & 0 & 1 & 0 \\ 4 & 0 & 0 & 0 & 1 \\ 5 & 1 & 0 & 1 & 0 \end{pmatrix} \qquad \vartheta_R =$$

	{}	{a}	{b}	{a,b}	{c}	{a,c}	{b,c}	{a,b,c}	{d}	{a,d}	{b,d}	{a,b,d}	{c,d}	{a,c,d}	{b,c,d}	{a,b,c,d}
{}	1	0	0	0	0	0	0	0	0	0	0	0	0	0	0	0
{1}	0	0	0	0	0	0	0	0	0	0	1	0	0	0	0	0
{2}	0	1	0	0	0	0	0	0	0	0	0	0	0	0	0	0
{1,2}	0	0	0	0	0	0	0	0	0	0	0	1	0	0	0	0
{3}	0	0	0	0	1	0	0	0	0	0	0	0	0	0	0	0
{1,3}	0	0	0	0	0	0	0	0	0	0	0	0	0	0	1	0
{2,3}	0	0	0	0	0	1	0	0	0	0	0	0	0	0	0	0
{1,2,3}	0	0	0	0	0	0	0	0	0	0	0	0	0	0	0	1
{4}	0	0	0	0	0	0	0	0	1	0	0	0	0	0	0	0
{1,4}	0	0	0	0	0	0	0	0	0	0	1	0	0	0	0	0
{2,4}	0	0	0	0	0	0	0	0	0	1	0	0	0	0	0	0
{1,2,4}	0	0	0	0	0	0	0	0	0	0	0	1	0	0	0	0
{3,4}	0	0	0	0	0	0	0	0	0	0	0	0	1	0	0	0
{1,3,4}	0	0	0	0	0	0	0	0	0	0	0	0	0	0	1	0
{2,3,4}	0	0	0	0	0	0	0	0	0	0	0	0	0	1	0	0
{1,2,3,4}	0	0	0	0	0	0	0	0	0	0	0	0	0	0	0	1
{5}	0	0	0	0	0	1	0	0	0	0	0	0	0	0	0	0
{1,5}	0	0	0	0	0	0	0	0	0	0	0	0	0	0	0	1
{2,5}	0	0	0	0	0	1	0	0	0	0	0	0	0	0	0	0
{1,2,5}	0	0	0	0	0	0	0	0	0	0	0	0	0	0	0	1
{3,5}	0	0	0	0	0	1	0	0	0	0	0	0	0	0	0	0
{1,3,5}	0	0	0	0	0	0	0	0	0	0	0	0	0	0	0	1
{2,3,5}	0	0	0	0	0	1	0	0	0	0	0	0	0	0	0	0
{1,2,3,5}	0	0	0	0	0	0	0	0	0	0	0	0	0	0	0	1
{4,5}	0	0	0	0	0	0	0	0	0	0	0	0	0	1	0	0
{1,4,5}	0	0	0	0	0	0	0	0	0	0	0	0	0	0	0	1
{2,4,5}	0	0	0	0	0	0	0	0	0	0	0	0	0	1	0	0
{1,2,4,5}	0	0	0	0	0	0	0	0	0	0	0	0	0	0	0	1
{3,4,5}	0	0	0	0	0	0	0	0	0	0	0	0	0	1	0	0
{1,3,4,5}	0	0	0	0	0	0	0	0	0	0	0	0	0	0	0	1
{2,3,4,5}	0	0	0	0	0	0	0	0	0	0	0	0	0	1	0	0
{1,2,3,4,5}	0	0	0	0	0	0	0	0	0	0	0	0	0	0	0	1

The inverse image is obviously not the transpose of the existential image.

$\vartheta_{R^\mathsf{T}} =$

	{}	{1}	{2}	{1,2}	{3}	{1,3}	{2,3}	{1,2,3}	{4}	{1,4}	{2,4}	{1,2,4}	{3,4}	{1,3,4}	{2,3,4}	{1,2,3,4}	{5}	{1,5}	{2,5}	{1,2,5}	{3,5}	{1,3,5}	{2,3,5}	{1,2,3,5}	{4,5}	{1,4,5}	{2,4,5}	{1,2,4,5}	{3,4,5}	{1,3,4,5}	{2,3,4,5}	{1,2,3,4,5}
{}	1	0	0	0	0	0	0	0	0	0	0	0	0	0	0	0	0	0	0	0	0	0	0	0	0	0	0	0	0	0	0	0
{a}	0	0	0	0	0	0	0	0	0	0	0	0	0	0	0	0	0	0	1	0	0	0	0	0	0	0	0	0	0	0	0	0
{b}	0	1	0	0	0	0	0	0	0	0	0	0	0	0	0	0	0	0	0	0	0	0	0	0	0	0	0	0	0	0	0	0
{a,b}	0	0	0	0	0	0	0	0	0	0	0	0	0	0	0	0	0	0	0	1	0	0	0	0	0	0	0	0	0	0	0	0
{c}	0	0	0	0	0	0	0	0	0	0	0	0	0	0	0	0	0	0	0	0	1	0	0	0	0	0	0	0	0	0	0	0
{a,c}	0	0	0	0	0	0	0	0	0	0	0	0	0	0	0	0	0	0	0	0	0	0	1	0	0	0	0	0	0	0	0	0
{b,c}	0	0	0	0	0	0	0	0	0	0	0	0	0	0	0	0	0	0	0	0	0	1	0	0	0	0	0	0	0	0	0	0
{a,b,c}	0	0	0	0	0	0	0	0	0	0	0	0	0	0	0	0	0	0	0	0	0	0	0	1	0	0	0	0	0	0	0	0
{d}	0	0	0	0	0	0	0	0	0	1	0	0	0	0	0	0	0	0	0	0	0	0	0	0	0	0	0	0	0	0	0	0
{a,d}	0	0	0	0	0	0	0	0	0	0	0	0	0	0	0	0	0	0	0	0	0	0	0	0	0	0	0	1	0	0	0	0
{b,d}	0	0	0	0	0	0	0	0	0	1	0	0	0	0	0	0	0	0	0	0	0	0	0	0	0	0	0	0	0	0	0	0
{a,b,d}	0	0	0	0	0	0	0	0	0	0	0	0	0	0	0	0	0	0	0	0	0	0	0	0	0	0	0	1	0	0	0	0
{c,d}	0	0	0	0	0	0	0	0	0	0	0	0	0	0	0	0	0	0	0	0	0	0	0	0	0	0	0	0	0	1	0	0
{a,c,d}	0	0	0	0	0	0	0	0	0	0	0	0	0	0	0	0	0	0	0	0	0	0	0	0	0	0	0	0	0	0	0	1
{b,c,d}	0	0	0	0	0	0	0	0	0	0	0	0	0	0	0	0	0	0	0	0	0	0	0	0	0	0	0	0	0	1	0	0
{a,b,c,d}	0	0	0	0	0	0	0	0	0	0	0	0	0	0	0	0	0	0	0	0	0	0	0	0	0	0	0	0	0	0	0	1

3.2 Lifting the Continuity Condition

With the inverse image, we will manage to lift the continuity definition to a point-free relational level.

3.1 Definition. Consider two neighborhood topologies $\mathcal{U} : X \longrightarrow 2^X$ and $\mathcal{U}' : X' \longrightarrow 2^{X'}$ as well as a mapping $f : X \longrightarrow X'$. We call

$$f \ \ \mathcal{U}\text{-continuous} \quad :\Longleftrightarrow \quad f{;}\mathcal{U}' \subseteq \mathcal{U}{;}\vartheta_{f^\mathsf{T}}^\mathsf{T}. \qquad\qquad \Box$$

An equivalent version $f{;}\mathcal{U}'{;}\vartheta_{f^\mathsf{T}} \subseteq \mathcal{U}$ is obtained by shunting the mapping ϑ_{f^T}.

The semi-formal development of the point-free version out of the predicate-logic form is rather tricky. It is interesting to observe that one must not quantify over subsets U, V; one should always move to quantifying over *elements u, v in the powerset*:

For every $p \in X$, every $V \in \mathcal{U}'(f(p))$, there is a $U \in \mathcal{U}(p)$ so that $f(U) \subseteq V$.

$\forall p \in X : \forall V \in \mathcal{U}'(f(p)) : \exists U \in \mathcal{U}(p) : f(U) \subseteq V$

$\forall p \in X : \forall v \in 2^{X'} : \mathcal{U}'_{f(p),v} \rightarrow \left(\exists u : \mathcal{U}_{pu} \wedge \left[\forall y : \varepsilon_{yu} \rightarrow \varepsilon'_{f(y),v} \right] \right)$

$\forall p : \forall v : (f{;}\mathcal{U}')_{pv} \rightarrow \left(\exists u : \mathcal{U}_{pu} \wedge \left[\forall y : \varepsilon_{yu} \rightarrow (f{;}\varepsilon')_{yv} \right] \right)$

$\forall p : \forall v : (f{;}\mathcal{U}')_{pv} \rightarrow \left(\exists u : \mathcal{U}_{pu} \wedge \overline{\exists y : \varepsilon_{yu} \wedge \overline{(f{;}\varepsilon')_{yv}}} \right)$

$\forall p : \forall v : (f{;}\mathcal{U}')_{pv} \rightarrow \left(\exists u : \mathcal{U}_{pu} \wedge \overline{\varepsilon^\mathsf{T}{;}\overline{f{;}\varepsilon'}}_{uv} \right)$

$\forall p : \forall v : (f{;}\mathcal{U}')_{pv} \rightarrow \left(\mathcal{U}{;}\overline{\varepsilon^\mathsf{T}{;}\overline{f{;}\varepsilon'}} \right)_{pv}$

$f{;}\mathcal{U}' \subseteq \mathcal{U}{;}\overline{\varepsilon^\mathsf{T}{;}\overline{f{;}\varepsilon'}}$

This was the main step. It remains to show relation-algebraically that this sharpens even to $f{;}\mathcal{U}' \subseteq \mathcal{U}{;}\mathsf{syq}(\varepsilon, f{;}\varepsilon') = \mathcal{U}{;}\vartheta_{f^\mathsf{T}}^\mathsf{T}$:

$\mathcal{U}{;}\overline{\varepsilon^\mathsf{T}{;}\overline{f{;}\varepsilon'}} \subseteq \mathcal{U}{;}\overline{\varepsilon^\mathsf{T}{;}\overline{f{;}\varepsilon'}}{;}\vartheta_{f^\mathsf{T}}{;}\vartheta_{f^\mathsf{T}}^\mathsf{T}$ because ϑ_{f^T} is total

$= \mathcal{U}{;}\overline{\varepsilon^\mathsf{T}{;}\overline{f{;}\varepsilon'}}{;}\mathsf{syq}(f{;}\varepsilon', \varepsilon){;}\vartheta_{f^\mathsf{T}}^\mathsf{T}$ by definition of ϑ_{f^T}

$\subseteq \mathcal{U}{;}\overline{\varepsilon^\mathsf{T}{;}\overline{\varepsilon}}{;}\vartheta_{f^\mathsf{T}}^\mathsf{T}$ cancellation

$= \mathcal{U}{;}\overline{\varepsilon^\mathsf{T}{;}\overline{\varepsilon}}{;}\vartheta_{f^\mathsf{T}}^\mathsf{T}$ since ϑ_{f^T} is a mapping

$= \mathcal{U}{;}\Omega{;}\vartheta_{f^\mathsf{T}}^\mathsf{T} = \mathcal{U}{;}\vartheta_{f^\mathsf{T}}^\mathsf{T}$ Def. 2.1.ii

3.3   Remark on Comparison of Structures in General

Comparison of structures via homomorphisms or structure-preserving mappings is omnipresent in mathematics and theoretical computer science, be it for groups, lattices, modules, graphs, or others. Most of these follow a general schema.

Two "structures" of whatever kind shall be given by a relation $R_1 : X_1 \longrightarrow Y_1$ and a relation $R_2 : X_2 \longrightarrow Y_2$. With mappings $\Phi : X_1 \longrightarrow X_2$ and $\Psi : Y_1 \longrightarrow Y_2$

they shall be compared, and we may ask whether these mappings transfer the first structure "sufficiently nice" into the second one.

The standard mechanism is to call the pair Φ, Ψ a **homomorphism** from R_1 to R_2, if $R_1 {}_{;} \Psi \subseteq \Phi {}_{;} R_2$. The two Φ, Ψ constitute an **isomorphism**, if Φ, Ψ as well as $\Phi^{\mathsf{T}}, \Psi^{\mathsf{T}}$ are homomorphisms.

If any two elements x, y are related by R_1, so are their images $\Phi(x), \Psi(y)$ by R_2:

$$\forall x \in X_1 : \forall y \in Y_1 : (x, y) \in R_1 \rightarrow (\Phi(x), \Psi(y)) \in R_2.$$

This concept is also suitable for relational structures; it works in particular for a graph homomorphism Φ, Φ — meaning $X_1 = X_2$, e.g. — as in the following example of a graph homomorphism, i.e., a homomorphism of a non-algebraic structure.

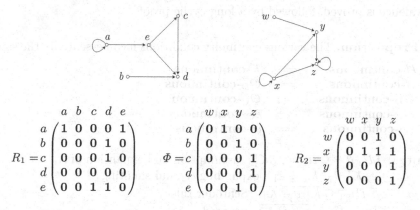

$$R_1 = \begin{array}{c} \\ a \\ b \\ c \\ d \\ e \end{array} \begin{array}{c} a\ b\ c\ d\ e \\ \begin{pmatrix} 1 & 0 & 0 & 0 & 1 \\ 0 & 0 & 0 & 1 & 0 \\ 0 & 0 & 0 & 1 & 0 \\ 0 & 0 & 0 & 0 & 0 \\ 0 & 0 & 1 & 1 & 0 \end{pmatrix} \end{array} \qquad \Phi = \begin{array}{c} \\ a \\ b \\ c \\ d \\ e \end{array} \begin{array}{c} w\ x\ y\ z \\ \begin{pmatrix} 0 & 1 & 0 & 0 \\ 0 & 0 & 1 & 0 \\ 0 & 0 & 0 & 1 \\ 0 & 0 & 0 & 1 \\ 0 & 0 & 1 & 0 \end{pmatrix} \end{array} \qquad R_2 = \begin{array}{c} \\ w \\ x \\ y \\ z \end{array} \begin{array}{c} w\ x\ y\ z \\ \begin{pmatrix} 0 & 0 & 1 & 0 \\ 0 & 1 & 1 & 1 \\ 0 & 0 & 0 & 1 \\ 0 & 0 & 0 & 1 \end{pmatrix} \end{array}$$

We recall the rolling of homomorphisms when Φ, Ψ are mappings as in

$$R_1 {}_{;} \Psi \subseteq \Phi {}_{;} R_2 \iff R_1 \subseteq \Phi {}_{;} R_2 {}_{;} \Psi^{\mathsf{T}} \iff \Phi^{\mathsf{T}} {}_{;} R_1 \subseteq R_2 {}_{;} \Psi^{\mathsf{T}} \iff \Phi^{\mathsf{T}} {}_{;} R_1 {}_{;} \Psi \subseteq R_2$$

If relations Φ, Ψ are not mappings, one cannot fully execute this rolling; there remain different forms of (bi-)simulations as explicated in [dRE98].

This is where the continuity condition fails. One cannot "roll" in this way and has just the two forms given above.

3.4 Cryptomorphy of the Continuity Conditions

Once we have the lifted relation-algebraic form for a neighborhood topology that uses the inverse image, we will immediately extend it to the other topology versions.

3.2 Definition. Given two sets X, X' with topologies, we consider a mapping $f : X \longrightarrow X'$ together with its inverse image $\vartheta_{f^{\mathsf{T}}} : \mathbf{2}^{X'} \longrightarrow \mathbf{2}^X$. Then we say that f is

i) $\mathcal{K}$-**continuous** $:\Longleftrightarrow$ $\mathcal{K}_2^\mathsf{T}\mathbin{;}\vartheta_{f^\mathsf{T}} \subseteq \overline{\varepsilon_2^\mathsf{T}\mathbin{;}f^\mathsf{T}\mathbin{;}\overline{\varepsilon_1}}\mathbin{;}\mathcal{K}_1^\mathsf{T}$,

ii) $\mathcal{O}_D$**continuous** $:\Longleftrightarrow$ $\mathcal{O}_{D2}\mathbin{;}\vartheta_{f^\mathsf{T}} \subseteq \vartheta_{f^\mathsf{T}}\mathbin{;}\mathcal{O}_{D1}$,

iii) $\mathcal{O}_V$-**continuous** $:\Longleftrightarrow$ $\vartheta_{f^\mathsf{T}}^\mathsf{T}\mathbin{;}\mathcal{O}_V' \subseteq \mathcal{O}_V$,

iv) $\varepsilon_\mathcal{O}$-**continuous** $:\Longleftrightarrow$ $f\mathbin{;}\varepsilon_{\mathcal{O}_2}\mathbin{;}\vartheta_{f^\mathsf{T}} \subseteq \varepsilon_{\mathcal{O}_1}$. $\square$

The easiest access is to the open sets version with $\mathcal{O}_V$: Inverse images of open sets have to be open again. Continuity with regard to kernel mapping is an ugly condition — that may be new.

All these versions of continuity can be shown to be equivalent, so that there is an obligation to prove f is $\mathcal{U}$-**continuous** $\Longleftrightarrow$ f is $\mathcal{K}$-**continuous** $\Longleftrightarrow$ f is $\mathcal{O}_D$-**continuous** $\Longleftrightarrow$ f is $\mathcal{O}_V$-**continuous** $\Longleftrightarrow$ f is $\varepsilon_\mathcal{O}$-**continuous**

For economy of proof, we formulate this slightly differently: An immediate equivalence is proved, followed by a long cyclic proof.

3.3 Proposition. The various continuity conditions mean essentially the same:

i) $\mathcal{U}$-**continuous** $\Longleftrightarrow$ $\mathcal{K}$-**continuous**

ii) $\mathcal{U}$-**continuous** $\Longrightarrow$ $\mathcal{O}_D$-**continuous**

iii) $\mathcal{O}_D$-**continuous** $\Longrightarrow$ $\mathcal{O}_V$-**continuous**

iv) $\mathcal{O}_V$-**continuous** $\Longrightarrow$ $\varepsilon_\mathcal{O}$-**continuous**

v) $\varepsilon_\mathcal{O}$-**continuous** $\Longrightarrow$ $\mathcal{U}$-**continuous**

Proof: i) $f\mathbin{;}\mathcal{U}_2\mathbin{;}\vartheta_{f^\mathsf{T}} \subseteq \mathcal{U}_1 = \varepsilon_1\mathbin{;}\mathcal{K}_1^\mathsf{T}$ assumption and expansion of $\mathcal{U}_1$

$\Longleftrightarrow$ $f\mathbin{;}\varepsilon_2\mathbin{;}\mathcal{K}_2^\mathsf{T}\mathbin{;}\vartheta_{f^\mathsf{T}}\mathbin{;}\mathcal{K}_1 \subseteq \varepsilon_1$ expanding $\mathcal{U}_2$ and shunting

$\Longleftrightarrow$ $\varepsilon_2^\mathsf{T}\mathbin{;}f^\mathsf{T}\mathbin{;}\overline{\varepsilon_1} \subseteq \overline{\mathcal{K}_2^\mathsf{T}\mathbin{;}\vartheta_{f^\mathsf{T}}\mathbin{;}\mathcal{K}_1}$ Schröder rule

$\Longleftrightarrow$ $\mathcal{K}_2^\mathsf{T}\mathbin{;}\vartheta_{f^\mathsf{T}}\mathbin{;}\mathcal{K}_1 \subseteq \overline{\varepsilon_2^\mathsf{T}\mathbin{;}f^\mathsf{T}\mathbin{;}\overline{\varepsilon_1}}$ negated

$\Longleftrightarrow$ $\mathcal{K}_2^\mathsf{T}\mathbin{;}\vartheta_{f^\mathsf{T}} \subseteq \overline{\varepsilon_2^\mathsf{T}\mathbin{;}f^\mathsf{T}\mathbin{;}\overline{\varepsilon_1}}\mathbin{;}\mathcal{K}_1^\mathsf{T}$ shunting again

ii) $\overline{\varepsilon_2^\mathsf{T}\mathbin{;}\overline{\mathcal{U}_2}} \subseteq \overline{\varepsilon_2^\mathsf{T}\mathbin{;}f^\mathsf{T}\mathbin{;}f\mathbin{;}\overline{\mathcal{U}_2}} = \overline{\varepsilon_2^\mathsf{T}\mathbin{;}f^\mathsf{T}}\mathbin{;}\overline{f\mathbin{;}\overline{\mathcal{U}_2}} = \vartheta_{f^\mathsf{T}}\mathbin{;}\overline{\varepsilon_1^\mathsf{T}\mathbin{;}\overline{f\mathbin{;}\overline{\mathcal{U}_2}}}$

$\subseteq \vartheta_{f^\mathsf{T}}\mathbin{;}\overline{\varepsilon_1^\mathsf{T}\mathbin{;}\overline{\mathcal{U}_1}\mathbin{;}\vartheta_{f^\mathsf{T}}^\mathsf{T}} = \vartheta_{f^\mathsf{T}}\mathbin{;}\overline{\varepsilon_1^\mathsf{T}\mathbin{;}\overline{\mathcal{U}_1}}\mathbin{;}\vartheta_{f^\mathsf{T}}$

$\Longrightarrow$ $\mathcal{O}_{D2} = \mathbb{I} \cap \overline{\varepsilon_2^\mathsf{T}\mathbin{;}\overline{\mathcal{U}_2}} \subseteq \vartheta_{f^\mathsf{T}}\mathbin{;}\vartheta_{f^\mathsf{T}}^\mathsf{T} \cap \vartheta_{f^\mathsf{T}}\mathbin{;}\overline{\varepsilon_1^\mathsf{T}\mathbin{;}\overline{\mathcal{U}_1}}\mathbin{;}\vartheta_{f^\mathsf{T}}^\mathsf{T}$

$= \vartheta_{f^\mathsf{T}}(\mathbb{I} \cap \overline{\varepsilon_1^\mathsf{T}\mathbin{;}\overline{\mathcal{U}_1}})\mathbin{;}\vartheta_{f^\mathsf{T}}^\mathsf{T} = \vartheta_{f^\mathsf{T}}\mathcal{O}_{D1}\vartheta_{f^\mathsf{T}}^\mathsf{T}$

iii) $\vartheta_{f^\mathsf{T}}^\mathsf{T}\mathbin{;}\mathcal{O}_{V_2} = \vartheta_{f^\mathsf{T}}^\mathsf{T}\mathbin{;}\mathcal{O}_{D2}\mathbb{T} = \vartheta_{f^\mathsf{T}}^\mathsf{T}\mathbin{;}\mathcal{O}_{D2}^\mathsf{T}\mathbb{T} \subseteq \mathcal{O}_{D1}^\mathsf{T}\mathbin{;}\vartheta_{f^\mathsf{T}}^\mathsf{T}\mathbin{;}\mathbb{T} = \mathcal{O}_{D1}\mathbin{;}\vartheta_{f^\mathsf{T}}^\mathsf{T}\mathbin{;}\mathbb{T} \subseteq \mathcal{O}_{D1}\mathbb{T} = \mathcal{O}_{V_1}$

iv) $f\mathbin{;}\varepsilon_{\mathcal{O}_2}\mathbin{;}\vartheta_{f^\mathsf{T}} = f\mathbin{;}(\varepsilon_2 \cap \mathbb{T}\mathbin{;}\mathcal{O}_{V_2}^\mathsf{T})\mathbin{;}\vartheta_{f^\mathsf{T}} = (f\mathbin{;}\varepsilon_2 \cap f\mathbin{;}\mathbb{T}\mathbin{;}\mathcal{O}_{V_2}^\mathsf{T})\mathbin{;}\vartheta_{f^\mathsf{T}}$

$= (\varepsilon_1\mathbin{;}\vartheta_{f^\mathsf{T}}^\mathsf{T} \cap \mathbb{T}\mathbin{;}\mathcal{O}_{V_2}^\mathsf{T})\mathbin{;}\vartheta_{f^\mathsf{T}}$

$= \varepsilon_1 \cap \mathbb{T}\mathbin{;}\mathcal{O}_{V_2}^\mathsf{T}\mathbin{;}\vartheta_{f^\mathsf{T}}$ [Sch11a, Prop. 5.4]

$\subseteq \varepsilon_1 \cap \mathbb{T}\mathbin{;}\mathcal{O}_{V_1}^\mathsf{T} = \varepsilon_{\mathcal{O}_1}$

v) $f\mathbin{;}\mathcal{U}_2\mathbin{;}\vartheta_{f^\mathsf{T}} = f\mathbin{;}\varepsilon_{\mathcal{O}_2}\mathbin{;}\Omega_2\mathbin{;}\vartheta_{f^\mathsf{T}}$

$\subseteq f\mathbin{;}\varepsilon_{\mathcal{O}_2}\mathbin{;}\vartheta_{f^\mathsf{T}}\mathbin{;}\Omega_1$

$\subseteq \varepsilon_{\mathcal{O}_1}\mathbin{;}\Omega_1$ assumption

$= \mathcal{U}_1$ $\square$

4   Conclusion

This article is part of a more extended ongoing research concerning relational methods in topology and in programming. Other attempts are directed towards simplicial complexes, e.g., for pretzels with several holes, the projective plane, or knot decompositions. An important question is whether it is possible to decide orientability, e.g., of a manifold *without* working on it globally. Compare this with the classic philosophers problem. Modelling the actions of the dining philosophers is readily available. One will be able to work on the state space based on 10 philosophers or 15. However, this doesn't scale up, so that local work is necessary. This work is intended to enhance such studies concerning communication and protocols.

Acknowledgement. The author gratefully acknowledges fruitful email discussions with Michael Winter as well as the encouraging remarks of the unknown referees.

Literatur

[Bd96]   Bird, R.S., de Moor, O.: Algebra of Programming. Prentice-Hall International (1996)
[dRE98]   de Roever, W.-P., Engelhardt, K.: Data Refinement: Model-Oriented Proof Methods and their Comparison. Cambridge Tracts in Theoretical Computer Science, vol. 47. Cambridge University Press (1998)
[Fra60]   Franz, W.: Topologie I. Sammlung Göschen 1181. Walter de Gruyter (1960)
[SS89]   Schmidt, G., Ströhlein, T.: Relationen und Graphen. Mathematik für Informatiker. Springer (1989) ISBN 3-540-50304-8, ISBN 0-387-50304-8
[SS93]   Schmidt, G., Ströhlein, T.: Relations and Graphs — Discrete Mathematics for Computer Scientists. EATCS Monographs on Theoretical Computer Science. Springer (1993) ISBN 3-540-56254-0, ISBN 0-387-56254-0
[Sch03]   Schmidt, G.: Relational Language. Technical Report 2003-05, Fakultät für Informatik, Universität der Bundeswehr München, 101 pages (2003), http://mucob.dyndns.org:30531/~gs/Papers/LanguageProposal.html
[Sch11a] Schmidt, G.: Relational Mathematics, Encyclopedia of Mathematics and its Applications, vol. 132, 584 p. Cambridge University Press (2011) ISBN 978-0-521-76268-7
[Sch11b] Schmidt, G.: TITUREL: Sprache für die Relationale Mathematik. Technical Report 132, Arbeitsberichte des Instituts für Wirtschaftsinformatik, Universität Münster, 11 pages (2011), http://mucob.dyndns.org:30531/~gs/Papers/Raesfeld2011ExtendedAbstract.pdf

A Mechanised Abstract Formalisation of Concept Lattices

Wolfram Kahl*

McMaster University, Hamilton, Ontario, Canada
`kahl@cas.mcmaster.ca`

Abstract. Using the dependently-typed programming language Agda, we formalise a category of algebraic contexts with relational homomorphisms presented by [Jip12, Mos13]. We do this in the abstract setting of locally ordered categories with converse (OCCs) with residuals and direct powers, without requiring meets (as in allegories) or joins (as in Kleene categories). The abstract formalisation has the advantage that it can be used both for theoretical reasoning, and for executable implementations, by instantiating it with appropriate choices of concrete OCCs.

1 Introduction

Formal concept analysis (FCA) [Wil05] typically starts from a *context* (E, A, R) consisting of a set E of *entities* (or "objects"), a set A of *attributes*, and an *incidence* relation R from entities to attributes. In such a context, "concepts" arise as "Galois-closed" subsets of E respectively A, and form complete "concept lattices".

In a recent development, M. A. Moshier [Mos13] defined a novel *relational* context homomorphism concept that gives rise to a category of contexts that is dual to the category of complete meet semilattices. This is in contrast with the FCA literature, which typically derives the context homomorphism concept from that used for the concept lattices, as for example in [HKZ06], with the notable exception of Erné, who studied context homomorphisms consisting of pairs of mappings [Ern05].

Jipsen [Jip12] published the central definitions of Moshier's [Mos13] approach, and developed it further to obtain categories of context representations of not only complete lattices, but also different kinds of semirings.

We now set out to mechanise the basis of these developments, and for the sake of reusability we abstract the sets and relations that constitute contexts to objects and morphisms of suitable categories and semigroupoids. Besides the mechanised formalisation itself, our main contribution is the insight that Moshier's relational context category can be formalised in categories of "abstract relations" where neither meet (intersection) nor join (union) are available, and that a large part of this development does not even require the presence of identity relations.

---

* This research is supported by the National Science and Engineering Research Council of Canada, NSERC.

P. Höfner et al. (Eds.): RAMiCS 2014, LNCS 8428, pp. 242–260, 2014.
© Springer International Publishing Switzerland 2014

Overview

We start with an introduction to essential features of the dependently-typed programming language and proof checker Agda2 (in the following just referred to as Agda) and an overview of our RATH-Agda formalisation of categoric abstractions of functions and relations in Sect. 3.

Since formal concept analysis concentrates on subsets of the constituent sets of the contexts we are interested in, we formalise an abstract version of element relations corresponding to the direct powers of [BSZ86, BSZ89] or the power allegories of [FS90], directly in the setting of locally ordered semigroupoids with converse (OSGCs) in Sect. 4. Adding also residuals to that setting (Sect. 5) proves sufficient for the formalisation of the "compatibility conditions" of Moshier's relational context homomorphisms, in Sect. 6. Defining composition of these homomorphisms requires making identity relations available, that is, moving from semigroupoids to categories; locally ordered categories with converse (OCCs) and residuals and a power operator are sufficient to formalise the context category, in Sect. 7. We conclude with additional discussion of the merits of our abstract formalisation.

The Agda source code from this project, including the modules discussed in this paper, are available on-line at http://relmics.mcmaster.ca/RATH-Agda/.

2 Agda Notation

This paper reports on a development in the dependently typed functional programming language and proof assistant Agda [Nor07] based on Martin-Löf type theory. Since Agda has been designed with a clear focus on both readability and writability, we present all mathematical content (except some informal analogies) in Agda notation, as an excerpt of the actual mechanically checked development.

Many Agda features will be explained when they are first used; here we only summarise a few *essential* aspects to make our use of Agda as the mathematical notation in the remainder of this paper more widely accessible.

Syntactically and "culturally", Agda frequently seems quite close to Haskell. However, the syntax of Agda is much more flexible: Almost any sequence of non-space characters is a legal lexeme, permitting the habit of choosing variable names for properties that abbreviate for example their left-hand sides, as for example $\Lambda_{\S\in}^{\smile}$: ... → Λ_0 R $\S$ $\in$ $^{\smile}$ $\approx$ R in Sect. 4 below, or names for proof values that reflect their type, e.g., x≈y : x ≈ y. Infix operators, and indeed mixfix operators of arbitrary arity, have names that contain underscore characters "_" in the positions of the first explicit arguments; below we use a binary infix operator _≈_ for morphism equality in semigroupoids and categories, and the "circumfix" operator [_] in Sect. 7.

Braces "{...}" in a function type "{name : type$_1$} → type$_2$ → ..." declare the first argument to be *implicit*; a function (say, "f") with this type can have this implicit argument supplied in three ways:

- "f v_2" supplies the explicit second argument v_2 explicitly, and thereby supplies the name argument implicitly, which requires that the type checker can determine its value uniquely;
- "f $\{v_1\}$ v_2" supplies the name argument explicitly by position, since it is the first implicit argument;
- "f $\{$name $= v_1\}$ v_2" supplies the name argument explicitly by name, which is useful when earlier implicit arguments can be inferred by the type checker.

Such implicit argument positions are usually declared for arguments that supply the types to other arguments and can therefore be inferred from the latter; the use of implicit arguments in Agda largely corresponds to general mathematical practice.

Since Agda is strongly normalising and has no undefined values, the underlying semantics is quite different from that of Haskell. In particular, since Agda is dependently typed, it does not have Haskell's distinction between terms, types, and kinds (the "types of the types"). The Agda constant Set_0 corresponds to the Haskell kind $*$; it is the type of all "normal" datatypes and is at the bottom of the hierarchy of type-theoretic universes in Agda. Universes Set ℓ are distinguished by universe indices for which we use names like ℓ ℓ_1 ℓa ℓi : Level, where Level is an opaque special-purpose variant of the natural numbers; we write its maximum operator as $\_\cup\_$. This *universe polymorphism* is essential for being able to talk about both "small" and "large" categories or relation algebras, and we always choose our Level parameters so as to enable maximal reusability of our definitions. (We include full Level information in all our types, although it does not essentially contribute to our development.)

Since types in Agda may be uninhabited, predicates use Set ℓ as result type. For example, we define the predicate isIdentity : $\{A : Obj\} \to$ Mor A A $\to$ Set ℓk below so that the application "isIdentity F" denotes the type of proofs that F : Mor A A is an identity morphism, which means that isIdentity F is an inhabited type if and only if F is an identity morphism.

3   Semigroupoids, Categories, OSGCs, OCCs

In [Kah11b], we presented a relatively fine-grained modularisation of sub-theories of division allegories, following our work on using semigroupoids to provide the theory of finite relations between infinite types, as they frequently occur as data structures in programming [Kah08], and on collagories [Kah11a] ("distributive allegories without zero morphisms"). In this section, we present two monolithic definitions that provide appropriate foundations for most of the discussion in this paper. Each of these two definitions bundles a large number of theories of the RATH-Agda libraries summarised in [Kah11b].

We show here a monolithic definition of semigroupoids (i.e., "categories without identity morphisms"), which can be used as an alternative to the one within the fine-grained theory hierarchy of [Kah11b]. We make no provisions for user-defined equality on objects, so Obj is a Set. Morphisms have equality $\_\approx\_$; for

any two objects A and B we have Hom A B : Setoid j k, with the standard
library providing an implementation of the standard type-theoretic concept of
setoid as a carrier set together with an equivalence relation that is considered as
equality on the carrier, much like the equality test == provided by the class Eq
in Haskell.[1] In [Kah11b] and in the Agda development [Kah14] underlying the
Agda theories for the later sections of this paper, only the Levels and Obj are
parameters of the Semigroupoid type; for making the presentation in this section
easier to follow, we choose to have also Hom as a parameter of the Semigroupoid'
type.

record Semigroupoid' $\{\ell i\ \ell j\ \ell k\ :\ \text{Level}\}\ \{\text{Obj}\ :\ \text{Set}\ \ell i\}$
$$(\text{Hom}\ :\ \text{Obj} \to \text{Obj} \to \text{Setoid}\ \ell j\ \ell k)$$
$$:\ \text{Set}\ (\ell i \cup \ell j \cup \ell k)\ \textbf{where}$$
Mor : Obj → Obj → Set ℓj
Mor $= \lambda$ A B → Setoid.Carrier (Hom A B)
infix 4 $\_\approx\_$; **infixr** 9 $\_\,\S\,\_$
$\_\approx\_ = \lambda\ \{A\}\ \{B\} \to \text{Setoid.}\_\approx\_ \text{(Hom A B)}$
field $\_\,\S\,\_$: $\{A\ B\ C\ :\ \text{Obj}\} \to \text{Mor A B} \to \text{Mor B C} \to \text{Mor A C}$
$\S$-cong : $\{A\ B\ C\ :\ \text{Obj}\}\ \{f_1\ f_2\ :\ \text{Mor A B}\}\ \{g_1\ g_2\ :\ \text{Mor B C}\}$
$$\to f_1 \approx f_2 \to g_1 \approx g_2 \to f_1\,\S\,g_1 \approx f_2\,\S\,g_2$$
$\S$-assoc : $\{A\ B\ C\ D\ :\ \text{Obj}\}\ \{f\ :\ \text{Mor A B}\}\ \{g\ :\ \text{Mor B C}\}\ \{h\ :\ \text{Mor C D}\}$
$$\to (f\,\S\,g)\,\S\,h \approx f\,\S\,(g\,\S\,h)$$

Using the infix declarations above, we in particular make morphism composition
$\_\,\S\,\_$ have higher precedence than morphism equality, which allows us to mostly
follow mathematical parenthesisation conventions. Function application, written
as juxtaposition (here only occurring in Mor A B etc.) has higher precedence than
any infix/mixfix operator, and associates to the left: "Mor A B" is "(Mor A) B",
while the **infixr** declaration above specifies that morphism composition associates
to the right: "f $\S$ g $\S$ h" is "f $\S$ (g $\S$ h)".

In semigroupoids, we define the identity property as conjunction (implemented
as pair type $\_\times\_$) of the two one-sided identity properties; the pair components
will later be extracted using the projections proj$_1$ and proj$_2$:

isLeftIdentity isRightIdentity isIdentity : $\{A\ :\ \text{Obj}\}$
$$\to \text{Mor A A} \to \text{Set}\ (\ell i \cup \ell j \cup \ell k)$$
isLeftIdentity $\{A\}$ I = $\{B\ :\ \text{Obj}\}\ \{R\ :\ \text{Mor A B}\} \to I\,\S\,R \approx R$
isRightIdentity $\{A\}$ I = $\{B\ :\ \text{Obj}\}\ \{R\ :\ \text{Mor B A}\} \to R\,\S\,I \approx R$
isIdentity   I = isLeftIdentity I $\times$ isRightIdentity I

---

[1] We can write f $\approx$ g for two morphisms from carrier set Mor A B of the hom-setoid
Hom A B since the object arguments A and B of $\_\approx\_$ are declared implicit and can
be derived from the type of f and g.

The type of congruence of composition, $\S$-cong, is declared using a *telescope* introducing the seven named arguments A, B, C, f_1, f_2, g_1, and g_2 (here all implicit),
which can be referred to in later parts of the type. The resulting *dependent func-*
tion type corresponds to "dependent products" frequently written using Π in other
presentations of type theory.

A Category has the following additional **field**s for identity morphisms and identity properties:

field Id   : {A : Obj} → Mor A A
      leftId : {A : Obj} → isLeftIdentity (Id {A})
      rightId : {A : Obj} → isRightIdentity (Id {A})

As context for the remaining sections, we now show a monolithic definition of ordered semigroupoids with converse (OSGC). As argued in [Kah04], we approach allegories and Kleene categories via common primitives providing a local ordering on homsets. Restricting ourselves to this common core turns out to be sufficient for the current paper, where we will *not* need to add meets or joins in the local homsets orderings.

In locally ordered categories, "homsets" are partial orders, so an OSGC first of all contains a semigroupoid that uses for its "homsets" the underlying setoids. The local poset ordering relations are again collected into a global parameterised relation, ⊑. We also add the involutory converse operator _˘ as a postfix operator, and give it higher precedence than all binary operators.

record OSGC′ {ℓi ℓj ℓk_1 ℓk_2 : Level} {Obj : Set ℓi}
              (Hom : Obj → Obj → Poset ℓj ℓk_1 ℓk_2)
              : Set ($\ell i \cup \ell$suc ($\ell j \cup \ell k_1 \cup \ell k_2$)) **where**
      field semigroupoid : Semigroupoid′ (λ A B → posetSetoid (Hom A B))
      open Semigroupoid′ semigroupoid **hiding** (semigroupoid)
      infix 4 _⊑_ ; **infix** 10 _˘
      _⊑_ = λ {A} {B} → Poset._≤_ (Hom A B)
      field
      ⨾-monotone : {A B C : Obj} {f f′ : Mor A B} {g g′ : Mor B C}
                    → f ⊑ f′ → g ⊑ g′ → f ⨾ g ⊑ f′ ⨾ g′
      _˘   : {A B : Obj}   → Mor A B → Mor B A
      ˘˘   : {A B : Obj} {R : Mor A B} → (R ˘) ˘ ≈ R
      ˘-involution : {A B C : Obj} {R : Mor A B} {S : Mor B C}
                                    → (R ⨾ S) ˘ ≈ S ˘ ⨾ R ˘
      ˘-monotone : {A B : Obj} {R S : Mor A B} → R ⊑ S → R ˘ ⊑ S ˘

Without identities, we frequently need the (one- and two-sided) sub- and super-identity properties, all of type {A : Obj} → (p : Mor A A) → Set ($\ell i \cup \ell j \cup \ell k_2$):

isLeftSubidentity   {A} p = {B : Obj} {R : Mor A B} → p ⨾ R ⊑ R
isRightSubidentity {A} p = {B : Obj} {S : Mor B A} → S ⨾ p ⊑ S
isSubidentity   p = isLeftSubidentity p × isRightSubidentity p
isLeftSuperidentity {A} p = {B : Obj} {R : Mor A B} → R ⊑ p ⨾ R
isRightSuperidentity{A} p = {B : Obj} {S : Mor B A} → S ⊑ S ⨾ p
isSuperidentity   p = isLeftSuperidentity p × isRightSuperidentity p

With these, we can define, already in OSGCs, the following standard relation-algebraic properties, all of type {A B : Obj} → Mor A B → Set ($\ell i \cup \ell j \cup \ell k_2$):

isUnivalent R = isSubidentity $(R \check{\ } \, ; R)$
isTotal   R = isSuperidentity $(R \, ; R \check{\ })$
isMapping R = isUnivalent R × isTotal R
isInjective   R = isSubidentity $(R \, ; R \check{\ })$
isSurjective R = isSuperidentity $(R \check{\ } \, ; R)$
isBijective R = isInjective R × isSurjective R

Total and univalent morphisms (in *Rel*, these are the total functions) are called *mappings*; for morphisms that are known to be mappings we define the dependent sum type Mapping containing the morphism and a proof of its mapping properties:

record Mapping (A B : Obj) : Set $(\ell i \cup \ell j \cup \ell k_2)$ **where**
   field mor : Mor A B
        prf : isMapping mor

The mappings of an OSGC S form a semigroupoid MapSG S where the morphisms from A to B are the Mappings of S, that is, Mor (MapSG S) A B = Mapping S A B.

Adding the identities of Category to an OSGC results in an *ordered category with converse* (OCC); mappings of an OCC C form the category MapCat C, and the OSGC versions of univalence, totality, ... are equivalent to the more habitual OCC versions $R \check{\ } \, ; R \sqsubseteq$ Id etc.

In the remainder of this paper, and in the context of a given OSGS S or OCC C, we append subscript "₁" to material taken from MapSG S, respectively MapCat C, in particular for equality $\_\approx_1\_$ and composition $\_;_1\_$ of mappings.

4 Power Operators in Ordered Semigroupoids with Converse

In the following, the minimal setting is a ordered semigroupoid with converse (OSGC), with equality $\_\approx\_$ and inclusion $\_\sqsubseteq\_$ and composition $\_;\_$ of morphisms. In an OSGC, morphisms are naturally considered as a generalisation of *relations*, not just functions.

Total functions, called *mappings*, are a derived concept in OSGCs (see the end of the previous section); the induced semigroupoid of mappings (base morphisms together with univalence and totality proofs) has equality $\_\approx_1\_$ and composition $\_;_1\_$.

A *power operator* consists of the following items:

$\mathbb{P}$: Obj → Obj   -- power object operator
$\in$: {A : Obj} → Mor A ($\mathbb{P}$ A)   -- membership "relation"
Λ : {A B : Obj} → Mor A B → Mapping A ($\mathbb{P}$ B) -- "power transpose"

"Power transpose" maps a "relation" R : Mor A B to a "set-valued function" Mapping A ($\mathbb{P}$ B).

The following axioms need to be satisfied; these are the two sides of one logical equivalence used by Bird and de Moor [BdM97, Sect. 4.6] to axiomatize the power allegories of Freyd and Scedrov [FS90, 2.4]:

$\Lambda{\Rightarrow}\epsilon$: $\{A\ B\ :\ Obj\}$ $\{R\ :\ Mor\ A\ B\}$ $\{f\ :\ Mapping\ A\ \mathbb{P}B\}$
    $\to f \approx_1 \Lambda\ R$
    $\to$ Mapping.mor $f\ ;\ \epsilon\ \check{}\ \approx R$
$\epsilon{\Rightarrow}\Lambda$: $\{A\ B\ :\ Obj\}$ $\{R\ :\ Mor\ A\ B\}$ $\{f\ :\ Mapping\ A\ \mathbb{P}B\}$
    $\to$ Mapping.mor $f\ ;\ \epsilon\ \check{}\ \approx R$
    $\to f \approx_1 \Lambda\ R$

Throughout this paper, we will use the convention that subscript "$_0$" abbreviates an application of Mapping.mor, but we explicitly show only this first definition following this pattern:

Λ_0 : Mor A B $\to$ Mor A $\mathbb{P}$B
Λ_0 R = Mapping.mor $(\Lambda\ R)$

From the power axioms, we derive the following laws (given objects A and B); the first is used as axiom by Freyd and Scedrov [FS90, 2.4]:

$\Lambda;\epsilon\check{}$: $\{R\ :\ Mor\ A\ B\}$ $\to \Lambda_0\ R\ ;\ \epsilon\ \check{}\ \approx R$
$\Lambda\text{-};\epsilon\check{}$: $\{f\ :\ Mapping\ A\ (\mathbb{P}\ B)\}$ $\to \Lambda\ (Mapping.mor\ f\ ;\ \epsilon\ \check{})\ \approx_1 f$
$\Lambda\text{-cong}$: $\{R_1\ R_2\ :\ Mor\ A\ B\}$ $\to R_1 \approx R_2 \to \Lambda\ R_1 \approx_1 \Lambda\ R_2$
$\check{};\Lambda$: $\{R\ :\ Mor\ A\ B\}$ $\to R\ \check{}\ ;\ \Lambda_0\ R \sqsubseteq \epsilon$

We can define the function that returns, for each "set", the set containing all its "elements":

$\mathsf{Id}\mathbb{P}$: $\{A\ :\ Obj\}$ $\to$ Mapping $(\mathbb{P}\ A)$ $(\mathbb{P}\ A)$
$\mathsf{Id}\mathbb{P} = \Lambda\ (\epsilon\ \check{})$

If there is an identity "relation" on $\mathbb{P}$ A, then $\mathsf{Id}\mathbb{P}\ \{A\}$ is equal to that identity, as one would expect. However, without assuming identities, we only succeeded to show that $\mathsf{Id}\mathbb{P}\ \{A\}$ is a right-identity for mappings.

For any two power operators, we obtain mappings between $\mathbb{P}_1$ A and $\mathbb{P}_2$ A that compose to $\mathsf{Id}\mathbb{P}_1$, respectively $\mathsf{Id}\mathbb{P}_2$. Therefore, if the base OSGC has identities, then that makes the two power operators isomorphic, and more generally, we have that power operators in OCCs are unique up to natural isomorphisms.

In the context of a power operator, a "power order" is an indexed relation on power objects satisfying conditions appropriate for a "subset relation":

record IsPowerOrder $(\Omega\ :\ \{A\ :\ Obj\} \to Mor\ (\mathbb{P}\ A)\ (\mathbb{P}\ A))$
                : Set $(\ell i \cup \ell j \cup \ell k_1 \cup \ell k_2)$ **where**
    field $\epsilon;\Omega$: $\{A\ :\ Obj\}$ $\to \epsilon\ ;\ \Omega\ \{A\} \sqsubseteq \epsilon$
    $\Omega\text{-universal}$: $\{A\ :\ Obj\}$ $\{R\ :\ Mor\ (\mathbb{P}\ A)\ (\mathbb{P}\ A)\} \to \epsilon\ ;\ R \sqsubseteq \epsilon \to R \sqsubseteq \Omega$

(The first condition $\epsilon\ ;\ \Omega \sqsubseteq \epsilon$ could be replaced with the converse implication of the second, namely $R \sqsubseteq \Omega \to \epsilon\ ;\ R \sqsubseteq \epsilon$.) A power operator together with a power order gives rise to existence of all right residuals (which, with converse, in turn implies existence of all left residuals), see Appendix A for the proof.

5 Power Orders via Residuals

The *right-residual* Q \ S, "Q under S", of two morphisms Q : Mor A B and S : Mor A C is the largest solution in X of the inclusion Q ⨾ X ⊑ S; formally, it is defined by:

$$\_\backslash\_ : \{A\ B\ C\ :\ Obj\} \to Mor\ A\ B \to Mor\ A\ C \to Mor\ B\ C$$

\-cancel-outer : {A B C : Obj} {S : Mor A C} {Q : Mor A B} → Q ⨾ (Q \ S) ⊑ S
\-universal :   {A B C : Obj} {S : Mor A C} {Q : Mor A B} {R : Mor B C}
          → Q ⨾ R ⊑ S → R ⊑ Q \ S

The last of these can be understood as an implication axiom \-universal, stating: "If Q ⨾ R ⊑ S, then R ⊑ Q \ S". Technically, it is a function taking, after six implicit arguments, one explicit argument, say "p", of type Q ⨾ R ⊑ S, and then the application \-universal p is of type (i.e., a proof for) R ⊑ Q \ S. For concrete relations, Q \ S relates b with c if and only if for all a with aQb we have aSc.

In the presence of converse, right residuals produce left residuals (S / R, "S over R") and vice versa, so it does not matter whether we assume one or both.

Adding residuals to the base OSGC enables the standard definition of the "set" inclusion relation [BSZ89, FS90] (for which we also easily show IsPowerOrder Ω):

$$\Omega : \{A : Obj\} \to Mor\ (\mathbb{P}\ A)\ (\mathbb{P}\ A)$$
$$\Omega = \epsilon \backslash \epsilon$$

This is transitive, and "as reflexive as we can state" without identities:

Ω-trans : {A : Obj} → Ω ⨾ Ω ⊑ Ω {A}
Ω-trans = \-cancel-middle
Idℙ⊑Ω : {A : Obj} → Mapping.mor Idℙ ⊑ Ω {A}
Idℙ⊑Ω′ = \-universal (⊑-begin
     ε ⨾ Idℙ₀
   ≈˘⟨ ⨾-cong₁ ˘˘ ⟩
     (ε ˘) ˘ ⨾ Λ₀ (ε ˘)
   ⊑⟨ ˘⨾Λ ⟩
     ε
   □)

(Here the argument proof to \-universal is presented in *calculational style*, which technically uses mixfix operators ⊑-begin_, _≈˘⟨_⟩_, _⊑⟨-⟩_, _□ with carefully arranged precedences to produce a fully formal "proof term with type annotations". This technique goes back to Augustsson and Norell [Aug99, Nor07], and in its present form essentially comes from Danielsson's Agda standard library [D⁺13]. The first step above contains an application of ˘˘ : (R ˘) ˘ ≈ R, applied via ⨾-cong₁ at the first argument of the composition, but in *backwards* direction, which is expressed by the ˘ in _≈˘⟨_⟩_. The expansion of the definition of Idℙ₀, that also happens in the first step, does not need to be mentioned, since for Agda, both expressions are the same via normalisation.)

The following property, shown using residual and power properties, will be useful below:

$$\Lambda_0 \mathbin{\raise1pt\hbox{$\scriptstyle\circ$}} \Omega^{\smile} : \{A\ B : \mathsf{Obj}\}\ \{R : \mathsf{Mor}\ A\ B\} \to \Lambda_0\ R \mathbin{\raise1pt\hbox{$\scriptstyle\circ$}} \Omega^{\ \smile} \approx R\ /\ \epsilon^{\ \smile}$$

$$\Lambda_0 \mathbin{\raise1pt\hbox{$\scriptstyle\circ$}} \Omega^{\smile}\ \{R = R\}\ =\ \approx\text{-begin}$$

$$\qquad \Lambda_0\ R \mathbin{\raise1pt\hbox{$\scriptstyle\circ$}} (\epsilon \setminus \epsilon)^{\ \smile}$$

$$\approx (\ \mathbin{\raise1pt\hbox{$\scriptstyle\circ$}}\text{-cong}_2\ \setminus\text{-}^{\smile}\ )$$

$$\qquad \Lambda_0\ R \mathbin{\raise1pt\hbox{$\scriptstyle\circ$}} (\epsilon^{\ \smile} /\ \epsilon^{\ \smile})$$

$$\approx (\ /\text{-outer-}\mathbin{\raise1pt\hbox{$\scriptstyle\circ$}}\text{-}\approx \Lambda\text{-mapping}\ )$$

$$\qquad (\Lambda_0\ R \mathbin{\raise1pt\hbox{$\scriptstyle\circ$}} \epsilon^{\ \smile})\ /\ \epsilon^{\ \smile}$$

$$\approx (\ /\text{-cong}_1\ \Lambda\mathbin{\raise1pt\hbox{$\scriptstyle\circ$}}\epsilon^{\smile}\ )$$

$$\qquad R\ /\ \epsilon^{\ \smile} \hspace{6em} \square$$

6   Contexts in OSGCs with Powers and Residuals

A context in our abstract setting consists of two objects together with a morphism of the base "relation"-OSGC:

```
record AContext : Set (ℓi ⊔ ℓj) where
   field ent : Obj          -- "entities"
         att : Obj          -- "attributes"
         inc : Mor ent att  -- "incidence"
```

In such a context, the incidence "relation" inc induces "concepts" as sets $\mathsf{inc} \uparrow p$ of attributes shared by a set $p : \mathbb{P}$ ent of entities, and sets $\mathsf{inc} \downarrow q$ of entities sharing all attributes in $q : \mathbb{P}$ att, set-theoretically defined in the following way:

$$\mathsf{inc} \uparrow p = \{a : \mathsf{att} \mid \forall e \in p\ .\ e\,\mathsf{inc}\,a\} \qquad \text{and} \qquad \mathsf{inc} \downarrow q = \{e : \mathsf{ent} \mid \forall a \in q\ .\ e\,\mathsf{inc}\,a\}\ .$$

We define the general operators $\_\uparrow$ and $\_\downarrow$ as postfix operators, so they need to be separated from their argument by a space:

$$\_\uparrow : \{A\ B : \mathsf{Obj}\} \to \mathsf{Mor}\ A\ B \to \mathsf{Mapping}\ (\mathbb{P}\,A)\ (\mathbb{P}\,B)$$
$$R\uparrow = \Lambda\,(\epsilon \setminus R)$$

$$\_\downarrow : \{A\ B : \mathsf{Obj}\} \to \mathsf{Mor}\ A\ B \to \mathsf{Mapping}\ (\mathbb{P}\,B)\ (\mathbb{P}\,A)$$
$$R\downarrow = \Lambda\,(\epsilon \setminus (R^{\,\smile}))$$

The fact that these form a Galois connection, set-theoretically

$$p \subseteq R\downarrow q \qquad \Leftrightarrow \qquad q \subseteq R\uparrow p \qquad\qquad \text{for all } p : \mathbb{P}\,A \text{ and } q : \mathbb{P}\,B,$$

can now be stated as a simple morphism equality and shown by algebraic calculation using residual and power properties:

$$\mathsf{Galois}\text{-}\downarrow\text{-}\uparrow : \{A\ B : \mathsf{Obj}\}\ \{R : \mathsf{Mor}\ A\ B\} \to \Omega \mathbin{\raise1pt\hbox{$\scriptstyle\circ$}} (R\downarrow_0)^{\ \smile} \approx R\uparrow_0 \mathbin{\raise1pt\hbox{$\scriptstyle\circ$}} \Omega^{\ \smile}$$

$$\mathsf{Galois}\text{-}\downarrow\text{-}\uparrow\ \{A\}\ \{B\}\ \{R\}\ =\ \approx\text{-begin}$$

$$\qquad \Omega \mathbin{\raise1pt\hbox{$\scriptstyle\circ$}} \Lambda_0\,(\epsilon \setminus R^{\,\smile})^{\ \smile}$$

$\approx\tilde{}\langle\ \tilde{}\text{-involutionRightConv}\ \rangle$
$\quad(\Lambda_0\ (\in\setminus R\ \tilde{})\ \mathring{,}\ \Omega\ \tilde{})\ \tilde{}$
$\approx\langle\ \tilde{}\text{-cong}\ \Lambda_0\mathring{,}\Omega\tilde{}\ \rangle$
$\quad((\in\setminus R\ \tilde{})\ /\ \in\ \tilde{})\ \tilde{}$
$\approx\langle\ /\tilde{}\text{-}\tilde{}\ \rangle$
$\quad\in\setminus(\in\setminus R\ \tilde{})\ \tilde{}$
$\approx\langle\ \setminus\text{-cong}_2\ \setminus\tilde{}\text{-}\tilde{}\ \rangle$
$\quad\in\setminus(R\ /\ \in\ \tilde{})$
$\approx\langle\ \setminus/\text{-}\approx\ \rangle$
$\quad(\in\setminus R)\ /\ \in\ \tilde{}$
$\approx\tilde{}\langle\ \Lambda_0\mathring{,}\Omega\tilde{}\ \rangle$
$\quad\Lambda_0\ (\in\setminus R)\ \mathring{,}\ \Omega\ \tilde{}$ $\square$

For the composed operators $\_\uparrow\downarrow$ and $\_\downarrow\uparrow$, with $R\uparrow\downarrow = R\uparrow \mathring{,}_1 R\downarrow$ and $R\downarrow\uparrow = R\downarrow \mathring{,}_1 R\uparrow$, the closure properties follow via further, partially lengthy calculations.

For a "set-valued *relation*" $R\ :\ \text{Mor}\ X\ (\mathbb{P}\ A)$, the "set-valued function" Lub R can be understood as mapping each "element" x of X to the union of all sets that R relates with x; this union contains an "element" a of A if and only if $R\ \mathring{,}\ \in\ \tilde{}$ relates x with a. From this "relation", Lub R is obtained as its power transpose, and similarly Glb for the intersection instead of union:

Lub Glb $: \{X\ A\ :\ \text{Obj}\}\ (R\ :\ \text{Mor}\ X\ (\mathbb{P}\ A)) \to \text{Mapping}\ X\ (\mathbb{P}\ A)$
Lub $R = \Lambda\ (R\ \mathring{,}\ \in\ \tilde{})$
Glb $R = \Lambda\ (R\ \tilde{}\setminus\in\ \tilde{})$

The properties of "mapping unions to intersections" and vice versa can now be defined as follows, for objects A and B, and $f\ :\ \text{Mapping}\ (\mathbb{P}\ B)\ (\mathbb{P}\ A)$:

Lub-cocontinuous $f = \forall\ \{X\ :\ \text{Obj}\}\ (Q\ :\ \text{Mor}\ X\ (\mathbb{P}\ B))$
$\qquad\qquad\qquad\qquad \to \text{Lub}\ Q\ \mathring{,}_1\ f \approx_1 \text{Glb}\ (Q\ \mathring{,}\ \text{Mapping.mor}\ f)$
Glb-cocontinuous $f = \forall\ \{X\ :\ \text{Obj}\}\ (Q\ :\ \text{Mor}\ X\ (\mathbb{P}\ B))$
$\qquad\qquad\qquad\qquad \to \text{Glb}\ Q\ \mathring{,}_1\ f \approx_1 \text{Lub}\ (Q\ \mathring{,}\ \text{Mapping.mor}\ f)$

Both of the operators $\_\uparrow$ and $\_\downarrow$ are Lub-cocontinuous, as can be shown in somewhat lengthy calculations, of which we show only the first — this contains three levels of nested calculations, indented precisely according to level, and uses in the reasons several different infix transitivity combinators following a common naming pattern, including $\_\langle\approx\tilde{}\sqsubseteq\rangle\_\ :\ \{...\} \to y \approx x \to y \sqsubseteq z \to x \sqsubseteq z$:

$\downarrow$-Lub-cocontinuous $: \{A\ B\ :\ \text{Obj}\}\ (R\ :\ \text{Mor}\ A\ B) \to \text{Lub-cocontinuous}\ (R\downarrow)$
$\downarrow$-Lub-cocontinuous $R\ \{X\}\ Q = \approx\text{-begin}$
$\quad\Lambda_0\ (Q\ \mathring{,}\ \in\ \tilde{})\ \mathring{,}\ \Lambda_0\ (\in\setminus(R\ \tilde{}))$
$\quad\approx\langle\ \in\Rightarrow\Lambda\ \{f = \Lambda\ (Q\ \mathring{,}\ \in\ \tilde{})\ \mathring{,}_1\ \Lambda\ (\in\setminus(R\ \tilde{}))\}\ (\approx\text{-begin}$
$\qquad(\Lambda_0\ (Q\ \mathring{,}\ \in\ \tilde{})\ \mathring{,}\ \Lambda_0\ (\in\setminus(R\ \tilde{})))\ \mathring{,}\ \in\ \tilde{}$
$\qquad\approx\langle\ \mathring{,}\text{-assoc}\ \langle\approx\approx\rangle\ \mathring{,}\text{-cong}_2\ \Lambda_\mathring{,}\in\tilde{}\ \rangle$
$\qquad\Lambda_0\ (Q\ \mathring{,}\ \in\ \tilde{})\ \mathring{,}\ (\in\setminus(R\ \tilde{}))$
$\quad\approx\langle\ \setminus\text{-inner-}\mathring{,}\ \Lambda\text{-mapping}\ \rangle$

$$(\epsilon \mathbin{\fatsemi} \Lambda_0 (Q \mathbin{\fatsemi} \epsilon^{\smallsmile})^{\smallsmile})^{\smallsmile}) \setminus (R^{\smallsmile})$$
$\approx\langle$ \-cong$_1$ ($^{\smallsmile}$-involutionRightConv $\langle\approx^{\smallsmile}\approx\rangle$ $^{\smallsmile}$-cong $\Lambda\mathbin{\fatsemi}\epsilon^{\smallsmile}$) $\langle\approx\approx^{\smallsmile}\rangle$ /-$^{\smallsmile}$ $\rangle$
$$(R / (Q \mathbin{\fatsemi} \epsilon^{\smallsmile}))^{\smallsmile}$$
$\approx\langle$ $^{\smallsmile}$-cong ($\sqsubseteq$-antisym
$\quad$ (/-universal ($\sqsubseteq$-begin
$$(R / (Q \mathbin{\fatsemi} \epsilon^{\smallsmile})) \mathbin{\fatsemi} Q \mathbin{\fatsemi} \Lambda_0 (\epsilon \setminus R^{\smallsmile})$$
$\quad\quad \sqsubseteq\langle$ $\mathbin{\fatsemi}$-assocL $\langle\approx\sqsubseteq\rangle$ $\mathbin{\fatsemi}$-monotone$_1$ /-cancel-$\mathbin{\fatsemi}$-inner $\rangle$
$$(R / \epsilon^{\smallsmile}) \mathbin{\fatsemi} \Lambda_0 (\epsilon \setminus R^{\smallsmile})$$
$\quad\quad \sqsubseteq\langle$ $\mathbin{\fatsemi}$-cong$_1$ \$^{\smallsmile}$-$^{\smallsmile}$ $\langle\approx^{\smallsmile}\sqsubseteq\rangle$ $^{\smallsmile}\mathbin{\fatsemi}\Lambda$ $\rangle$
$$\epsilon$$
$\quad\quad \square$))
$\quad$ (/-universal ($\sqsubseteq$-begin
$$(\epsilon / (Q \mathbin{\fatsemi} \Lambda_0 (\epsilon \setminus R^{\smallsmile}))) \mathbin{\fatsemi} (Q \mathbin{\fatsemi} \epsilon^{\smallsmile})$$
$\quad\quad \sqsubseteq\langle$ $\mathbin{\fatsemi}$-assocL $\langle\approx\sqsubseteq\rangle$ $\mathbin{\fatsemi}$-monotone$_1$ /-cancel-$\mathbin{\fatsemi}$-inner $\rangle$
$$(\epsilon / \Lambda_0 (\epsilon \setminus R^{\smallsmile})) \mathbin{\fatsemi} \epsilon^{\smallsmile}$$
$\quad\quad \sqsubseteq\langle$ $\mathbin{\fatsemi}$-monotone$_2$ (proj$_1$ Λ-total $\langle\sqsubseteq\approx\rangle$ $\mathbin{\fatsemi}$-assoc) $\rangle$
$$(\epsilon / \Lambda_0 (\epsilon \setminus R^{\smallsmile})) \mathbin{\fatsemi} \Lambda_0 (\epsilon \setminus R^{\smallsmile}) \mathbin{\fatsemi} \Lambda_0 (\epsilon \setminus R^{\smallsmile})^{\smallsmile} \mathbin{\fatsemi} \epsilon^{\smallsmile}$$
$\quad\quad \sqsubseteq\langle$ $\mathbin{\fatsemi}$-assocL $\langle\approx\sqsubseteq\rangle$ $\mathbin{\fatsemi}$-monotone$_1$ /-cancel-outer $\rangle$
$$\epsilon \mathbin{\fatsemi} \Lambda_0 (\epsilon \setminus R^{\smallsmile})^{\smallsmile} \mathbin{\fatsemi} \epsilon^{\smallsmile}$$
$\quad\quad \approx\langle$ $\mathbin{\fatsemi}$-assocL $\langle\approx\approx^{\smallsmile}\rangle$ $\mathbin{\fatsemi}$-cong$_1$ $^{\smallsmile}$-involutionRightConv $\rangle$
$$(\Lambda_0 (\epsilon \setminus R^{\smallsmile}) \mathbin{\fatsemi} \epsilon^{\smallsmile})^{\smallsmile} \mathbin{\fatsemi} \epsilon^{\smallsmile}$$
$\quad\quad \approx\langle$ $\mathbin{\fatsemi}$-cong$_1$ ($^{\smallsmile}$-cong $\Lambda\mathbin{\fatsemi}\epsilon^{\smallsmile}$) $\rangle$
$$(\epsilon \setminus R^{\smallsmile})^{\smallsmile} \mathbin{\fatsemi} \epsilon^{\smallsmile}$$
$\quad\quad \sqsubseteq\langle$ $\mathbin{\fatsemi}$-cong$_1$ \$^{\smallsmile}$-$^{\smallsmile}$ $\langle\approx\sqsubseteq\rangle$ /-cancel-outer $\rangle$
$$R$$
$\quad\quad \square$))))
$$(\epsilon / (Q \mathbin{\fatsemi} \Lambda_0 (\epsilon \setminus R^{\smallsmile})))^{\smallsmile}$$
$\approx\langle$ /-$^{\smallsmile}$ $\rangle$
$$(Q \mathbin{\fatsemi} \Lambda_0 (\epsilon \setminus (R^{\smallsmile})))^{\smallsmile} \setminus \epsilon^{\smallsmile}$$
$\square$))
$$\Lambda_0 ((Q \mathbin{\fatsemi} \Lambda_0 (\epsilon \setminus (R^{\smallsmile}))))^{\smallsmile} \setminus \epsilon^{\smallsmile})$$
$\square$

However, the closest we can have to Glb-cocontinuous ($R \uparrow$) is the following (where the final $\rightarrow$ has also been proven in the opposite direction):

$\uparrow$-Glb-cocontinuous : $\{A\ B\ X : Obj\}$ $(R : Mor\ A\ B)$ $(Q : Mor\ X\ (\mathbb{P}\ A))$
$$\rightarrow (\epsilon / Q) \setminus R \approx (Q \mathbin{\fatsemi} (\epsilon \setminus R))$$
$$\rightarrow Glb\ Q \mathbin{\fatsemi_1} (R \uparrow) \approx_1 Lub\ (Q \mathbin{\fatsemi} Mapping.mor\ (R \uparrow))$$

The reason for the general failure of Glb-cocontinuity is that if Q is not total, the resulting empty intersections on the left-hand side may be mapped by $R \uparrow$ to arbitrary sets, but on the right-hand side, the resulting empty unions are always the empty set. In particular in the power-allegory induced by any topos, if we set $Q = \bot$ (the "empty relation") and $R = \top$ (the "universal relation"), then we have:
$$(\epsilon / Q) \setminus R \approx (\epsilon / \bot) \setminus \top \approx \top \not\approx \bot \approx \bot \mathbin{\fatsemi} (\epsilon \setminus \top) \approx Q \mathbin{\fatsemi} (\epsilon \setminus R)$$

For composition of context homomorphisms, we will require Lub-cocontinuity of $G \downarrow_{;1} Y \uparrow_{;1} F \downarrow$ in the following situation:

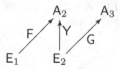

Due to the fact that $Y \uparrow$ is not necessarily Glb-cocontinuous, context homomorphisms require additional "compatibility" conditions to enable the following calculation, which closely follows [Mos13], to go through:

$\downarrow\uparrow\downarrow\text{-Lub-cocontinuous} : \{E_1\ E_2\ A_2\ A_3 : \mathsf{Obj}\}$
$\qquad\qquad \to (F : \mathsf{Mor}\ E_1\ A_2)\ (Y : \mathsf{Mor}\ E_2\ A_2)\ (G : \mathsf{Mor}\ E_2\ A_3)$
$\qquad\qquad \to (\text{F-trgCompat} : Y \downarrow\uparrow_{;1} F \downarrow \approx_1 F \downarrow)$
$\qquad\qquad \to (\text{G-srcCompat} : G \downarrow_{;1} Y \uparrow\downarrow \approx_1 G \downarrow)$
$\qquad\qquad \to \mathsf{Lub\text{-}cocontinuous}\ (G \downarrow_{;1} Y \uparrow_{;1} F \downarrow)$
$\downarrow\uparrow\downarrow\text{-Lub-cocontinuous}\ F\ Y\ G\ \text{F-trgCompat}\ \text{G-srcCompat}\ Q = \approx_1\text{-begin}$
$\quad \mathsf{Lub}\ Q_{;1} G \downarrow_{;1} Y \uparrow_{;1} F \downarrow$
$\quad \approx_1\langle\ _;\text{-assocL}\ \langle\approx\approx\rangle\ _;\text{-cong}_1\ (\downarrow\text{-Lub-cocontinuous}\ G\ Q)\ \rangle$
$\quad \mathsf{Glb}\ (Q_; G \downarrow_0)_{;1} Y \uparrow_{;1} F \downarrow$
$\quad \approx_1\langle\ _;\text{-cong}_1\ (\mathsf{Glb\text{-}cong}\ (_;\text{-cong}_2\ \text{G-srcCompat}\ \langle\approx^\smile\approx\rangle\ _;\text{-assocL}_{3+1}))\ \rangle$
$\quad \mathsf{Glb}\ ((Q_; G \downarrow_0 _; Y \uparrow_0)_; Y \downarrow_0)_{;1} Y \uparrow_{;1} F \downarrow$
$\quad \approx_1\langle\ _;\text{-cong}_1\ (\downarrow\text{-Lub-cocontinuous}\ Y\ (Q_; G \downarrow_0 _; Y \uparrow_0))\ \langle\approx^\smile\approx\rangle\ _;\text{-assoc}\ \rangle$
$\quad \mathsf{Lub}\ (Q_; G \downarrow_0 _; Y \uparrow_0)_{;1} Y \downarrow_{;1} Y \uparrow_{;1} F \downarrow$
$\quad \approx_1\langle\ _;\text{-cong}_2\ (_;\text{-assocL}\ \langle\approx\approx\rangle\ \text{F-trgCompat})\ \rangle$
$\quad \mathsf{Lub}\ (Q_; G \downarrow_0 _; Y \uparrow_0)_{;1} F \downarrow$
$\quad \approx_1\langle\ \downarrow\text{-Lub-cocontinuous}\ F\ (Q_; G \downarrow_0 _; Y \uparrow_0)\ \langle\approx\approx\rangle\ \mathsf{Glb\text{-}cong}\ _;\text{-assoc}_{3+1}\ \rangle$
$\quad \mathsf{Glb}\ (Q_; G \downarrow_0 _; Y \uparrow_0 _; F \downarrow_0)$
$\quad \square_1$

A context homomorphism, following Moshier [Mos13] and Jipsen [Jip12], includes the compatibility properties used above. In order to be able to refer to the **field**s of the source and target contexts X and Y with qualified names like X.ent instead of AContext.ent X, we need to use the "module nature" of **record**s in Agda and define local module names for X and Y. (In the following, we will omit these local module definitions for the sake of brevity, and since there will be no danger of confusion.)

```
record AContextHom (X Y : AContext) : Set (ℓi ⊔ ℓj ⊔ ℓk₁ ⊔ ℓk₂) where
  private module X = AContext X
          module Y = AContext Y
  field   mor : Mor X.ent Y.att
          srcCompat : mor ↓_{;1} X.inc ↑↓ ≈₁ mor ↓
          trgCompat : Y.inc ↓↑_{;1} mor ↓ ≈₁ mor ↓
```

If we now have three contexts X, Y, and Z connected by two context homomorphisms F : AContextHom X Y and G : AContextHom Y Z, then we define:

$G{\downarrow}_§Y{\uparrow}_§F{\downarrow}$: Mapping $(\mathbb{P} Z.att)$ $(\mathbb{P} X.ent)$
$G{\downarrow}_§Y{\uparrow}_§F{\downarrow}$ = G.mor $\downarrow$ $_§{_1}$ Y.inc $\uparrow$ $_§{_1}$ F.mor $\downarrow$

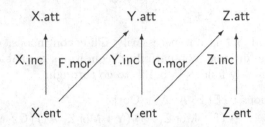

Applying $\downarrow{\uparrow}\downarrow$-Lub-cocontinuous, we obtain that the mapping $G{\downarrow}_§Y{\uparrow}_§F{\downarrow}$ is Lub-cocontinuous just like F.mor $\downarrow$: Mapping $(\mathbb{P} Y.att)$ $(\mathbb{P} X.ent)$ and G.mor $\downarrow$: Mapping $(\mathbb{P} Z.att)$ $(\mathbb{P} Y.ent)$, but we are still missing a way to extract a $\_{\downarrow}$-pre-image of type Mor X.ent Z.att.

7   Abstract Context Categories in OCCs with Powers and Residuals

It turns out that adding identities is sufficient for obtaining a partial inverse to the operator $\_{\downarrow}$. The key is that Λ Id : Mapping A $(\mathbb{P} A)$ can be understood as mapping each "element" a: A to the singleton "set" $\{a\}$: $\mathbb{P} A$.

The "relation" singletons A relates a "subset of A" with all singletons contained in it:

singletons : $\{A : Obj\} \to$ Mor $(\mathbb{P} A)$ $(\mathbb{P} A)$
singletons = $\epsilon^{\smile}$ $_§$ Λ_0 Id

Applying Lub to this produces the identity mapping on $\mathbb{P} A$:

Lub-singletons : $\{A : Obj\} \to$ Lub (singletons $\{A\}$) $\approx_1$ Id$_1$ $\{\mathbb{P} A\}$
Lub-singletons $\{A\}$ = $\approx_1$-begin
    $\Lambda((\epsilon^{\smile} _§ \Lambda_0$ Id$)$ $_§$ $\epsilon^{\smile})$
  $\approx_1\langle$ Λ-cong $(_§$-assoc $\langle\approx\approx\rangle$ $_§$-cong$_2$ $\Lambda_§\epsilon^{\smile})$ $\rangle$
    $\Lambda(\epsilon^{\smile} _§$ Id $\{A\})$
  $\approx_1\langle$ Λ-cong $($rightId $\langle\approx\approx^{\smile}\rangle$ leftId$)$ $\langle\approx\approx\rangle$ Λ-$_§\epsilon^{\smile}$ $\{f = $ Id$_1$ $\{\mathbb{P} A\}\}$ $\rangle$
    Id$_1$ $\{\mathbb{P} A\}$ $\square_1$

The operator $[\_]$ has the opposite type of $\_{\downarrow}$, and $[f]$ relates a with b if and only if $a \in f\{b\}$:

$[\_]$: $\{A\ B : Obj\} \to$ Mapping $(\mathbb{P} B)$ $(\mathbb{P} A) \to$ Mor A B
$[\ f\ ]$ = $(\Lambda_0$ Id $_§$ Mapping.mor f $_§$ $\epsilon^{\smile})^{\smile}$

We always have $[\ R\downarrow\ ]\approx R$:

$[\downarrow]:\{A\ B:Obj\}\ (R:Mor\ A\ B)\to[\ R\downarrow\ ]\approx R$

$[\downarrow]\ R\ =\ \approx\text{-begin}$

$\quad(\Lambda_0\ Id\ \mathring{,}\ \Lambda_0\ (\in\setminus(R\ \check{\ }))\ \mathring{,}\ \in\ \check{\ })\ \check{\ }$

$\approx\langle\ \check{\ }\text{-cong}\ (\mathring{,}\text{-cong}_2\ \Lambda\mathring{,}\in\check{\ })\ \rangle$

$\quad(\Lambda_0\ Id\ \mathring{,}\ (\in\setminus(R\ \check{\ })))\ \check{\ }$

$\approx\langle\ \check{\ }\text{-cong}\ (\setminus\text{-inner-}\mathring{,}\ \Lambda\text{-mapping})\ \rangle$

$\quad((\in\mathring{,}\ (\Lambda_0\ Id)\ \check{\ })\setminus(R\ \check{\ }))\ \check{\ }$

$\approx\langle\ \setminus\check{\ }\text{-}\check{\ }\ \rangle$

$\quad R\ /\ ((\in\mathring{,}\ (\Lambda_0\ Id)\ \check{\ })\ \check{\ })$

$\approx\langle\ /\text{-cong}_2\ \check{\ }\text{-involutionRightConv}\ \rangle$

$\quad R\ /\ (\Lambda_0\ Id\ \mathring{,}\ \in\ \check{\ })$

$\approx\langle\ /\text{-cong}_2\ \Lambda\mathring{,}\in\check{\ }\ \langle\approx\approx\rangle\ /\text{-Id}\ \rangle$

$\quad R$

$\square$

For the opposite composition, $[\ f\ ]\downarrow\approx_1 f$, we need Lub-cocontinuity of f:

$[\ ]\downarrow:\{A\ B:Obj\}\ (f:Mapping\ (\mathbb{P}\ B)\ (\mathbb{P}\ A))$

$\qquad\to Lub\text{-cocontinuous}\ f\to[\ f\ ]\downarrow\approx_1 f$

$[\ ]\downarrow\ f\ f\text{-cocontinuous}\ =\ \approx_1\text{-begin}$

$\quad[\ f\ ]\downarrow$

$\approx_1\langle\ \approx\text{-refl}\ \rangle$

$\quad\Lambda\ (\in\setminus([\ f\ ]\ \check{\ }))$

$\approx_1\langle\ \Lambda\text{-cong}\ (\setminus\text{-cong}_2\ (\check{\ }\check{\ }\ \langle\approx\approx\rangle\ \mathring{,}\text{-assocL}))\ \rangle$

$\quad\Lambda\ (\in\setminus(Mapping.mor\ (\Lambda\ Id\ \mathring{,}_1\ f)\ \mathring{,}\ \in\ \check{\ }))$

$\approx_1\check{\ }\langle\ \Lambda\text{-cong}\ (\setminus\text{-cong}_2\ (\mathring{,}\text{-cong}_1\ \check{\ }\check{\ }))\ \rangle$

$\quad\Lambda\ (\in\setminus((\Lambda_0\ Id\ \mathring{,}\ Mapping.mor\ f)\ \check{\ }\check{\ }\ \mathring{,}\ \in\ \check{\ }))$

$\approx_1\check{\ }\langle\ \Lambda\text{-cong}\ (\setminus\text{-cong}_1\ \check{\ }\text{-involutionLeftConv}$

$\qquad\qquad\langle\approx\approx\rangle\ \setminus\text{-flip}\ (\check{\ }\text{-isBijective}\ (Mapping.prf\ (\Lambda\ Id\ \mathring{,}_1\ f))))\ \rangle$

$\quad\Lambda\ ((\in\ \check{\ }\ \mathring{,}\ \Lambda_0\ Id\ \mathring{,}\ Mapping.mor\ f)\ \check{\ }\setminus\in\ \check{\ })$

$\approx_1\check{\ }\langle\ \Lambda\text{-cong}\ (\setminus\text{-cong}_1\ (\check{\ }\text{-cong}\ \mathring{,}\text{-assoc}))\ \rangle$

$\quad Glb\ (singletons\ \mathring{,}\ Mapping.mor\ f)$

$\approx_1\check{\ }\langle\ f\text{-cocontinuous}\ singletons\ \rangle$

$\quad Lub\ singletons\ \mathring{,}_1\ f$

$\approx_1\langle\ \mathring{,}\text{-cong}_1\ Lub\text{-singletons}\ \langle\approx\approx\rangle\ leftId\ \rangle$

$\quad f$

$\square_1$

The last two steps represent the argument of [Mos13] that "If f sends unions to intersections, its behavior is determined by its behavior on singletons."

Using the instance

$[_{⨾⨾}]↓$: $[$ $G↓_⨾Y↑_⨾F↓$ $]$ $↓$ $≈_1$ $G↓_⨾Y↑_⨾F↓$
$[_{⨾⨾}]↓$ = $[]↓$ $G↓_⨾Y↑_⨾F↓$
($↓↑↓$-Lub-cocontinuous F.mor Y.inc G.mor F.trgCompat G.srcCompat)

for the morphism composition at the end of Sect. 6, we obtain well-definedness:

$\_{⨾⨾}\_$: (F : AContextHom X Y) (G : AContextHom Y Z) → AContextHom X Z
F $_{⨾⨾}$ G = **record**
  {mor = $[$ $G↓_⨾Y↑_⨾F↓$ $]$
  ; srcCompat = $≈_1$-begin
      $[$ $G↓_⨾Y↑_⨾F↓$ $]$ $↓$ $_{⨾1}$ X.inc $↑$ $_{⨾1}$ X.inc $↓$
    $≈_1⟨$ ⨾-cong$_1$ $([_{⨾⨾}]↓$ F $G)$ $⟨≈≈⟩$ ⨾-assoc$_{3+1}$ $⟩$
      G.mor $↓$ $_{⨾1}$ Y.inc $↑$ $_{⨾1}$ F.mor $↓$ $_{⨾1}$ X.inc $↑$ $_{⨾1}$ X.inc $↓$
    $≈_1⟨$ ⨾-cong$_{22}$ F.srcCompat $⟩$
      G.mor $↓$ $_{⨾1}$ Y.inc $↑$ $_{⨾1}$ F.mor $↓$
    $≈_1˘⟨$ $([_{⨾⨾}]↓$ F $G)$ $⟩$
      $[$ $G↓_⨾Y↑_⨾F↓$ $]$ $↓$
    $□_1$
  ; trgCompat = ...   -- analogously
  }

Context homomorphism equality F $≈$ G is defined as the underlying morphism equality F.mor $≈$ G.mor. The left- and right-identity properties of the composition $\_{⨾⨾}\_$ reduce, via $[↓]$, to srcCompat respectively trgCompat due to the fact that the identity context homomorphism on X has X.inc as mor, and the associativity proof also turns into a surprisingly short calculation:

$$X_1 \xrightarrow{\ F\ } X_2 \xrightarrow{\ G\ } X_3 \xrightarrow{\ H\ } X_4$$

ACH-assoc : ... → (F $_{⨾⨾}$ G) $_{⨾⨾}$ H $≈$ F $_{⨾⨾}$ (G $_{⨾⨾}$ H)
ACH-assoc $\{X_1\}$ $\{X_2\}$ $\{X_3\}$ $\{X_4\}$ $\{F\}$ $\{G\}$ $\{H\}$ = $[\,]$-cong
  $\{f_1$ = H.mor $↓$ $_{⨾1}$ X_3.inc $↑$ $_{⨾1}$ FG.mor $↓\}$
  $\{f_2$ = GH.mor $↓$ $_{⨾1}$ X_2.inc $↑$ $_{⨾1}$ F.mor $↓\}$
  ($≈_1$-begin
     H.mor $↓$ $_{⨾1}$ X_3.inc $↑$ $_{⨾1}$ FG.mor $↓$
   $≈_1⟨$ ⨾-cong$_{22}$ $([_{⨾⨾}]↓$ F $G)$ $⟩$
     H.mor $↓$ $_{⨾1}$ X_3.inc $↑$ $_{⨾1}$ G.mor $↓$ $_{⨾1}$ X_2.inc $↑$ $_{⨾1}$ F.mor $↓$
   $≈_1⟨$ ⨾-assocL$_{3+1}$ $⟨≈≈˘⟩$ ⨾-cong$_1$ $([_{⨾⨾}]↓$ G $H)$ $⟩$
     GH.mor $↓$ $_{⨾1}$ X_2.inc $↑$ $_{⨾1}$ F.mor $↓$
   $□_1)$
  where   FG = F $_{⨾⨾}$ G ;   GH = G $_{⨾⨾}$ H

With this, a Category of AContexts with AContextHoms as morphisms is easily defined and checked by Agda.

8   Conclusion

A "natural", more direct formalisation of contexts would allow arbitrary Sets (or possibly Setoids) of entities and attributes, exactly as in the mathematical definition:

> **record** Context (ℓe ℓa ℓr : Level) : Set (ℓsuc ($\ell e \sqcup \ell a \sqcup \ell r$)) **where**
>   **field ent** : Set ℓe -- "entities"
>   att : Set ℓa -- "attributes"
>   inc : $\mathcal{R}el$ ℓr ent att   -- "incidence"

Although this has the advantage of additional universe polymorphism, it appears that the compatibility conditions force all the sets and relations to the same levels:

> **record** Hom {ℓS ℓr : Level} (A B : Context ℓS ℓS ℓr) : Set ($\ell S \sqcup \ell$suc ℓr) **where**
>   **private module** A = Context A
>   **module** B = Context B
>   **field**   mor : $\mathcal{R}el$ ℓr A.ent B.att
>   srcCompat : A.inc $\uparrow\downarrow$ oo mor $\downarrow$ $\approx\lfloor$ $\mathbb{P}$ B.att $\ell r \looparrowright \mathbb{P}$ A.ent _ $\rfloor$ mor $\downarrow$
>   trgCompat : mor $\downarrow$ oo B.inc $\downarrow\uparrow$ $\approx\lfloor$ $\mathbb{P}$ B.att $\ell r \looparrowright \mathbb{P}$ A.ent _ $\rfloor$ mor $\downarrow$

An important disadvantage of this approach is that, for example, quotient contexts will have entity and attribute sets that lack many of the interfaces that would be useful for programming, for example serialisation.

In contrast, using our abstract approach makes it possible to instantiate the base OCC differently for different purposes:

- For theoretical investigations, using OCCs of setoids (or even of Sets) as defined in Relation.Binary.Heterogeneous.Categoric.OCC of the RATH-Agda libraries [Kah11b] provides all the flexibility of the general mathematical setting, but without useful execution mechanisms.
- For data processing applications, that is, "for programming", using implementation-oriented OCCs as for example that of SULists mentioned in [Kah12] provides additional interfaces and correct-by-construction executable implementations.

Beyond the theoretically interesting fact that context categories can be formalised in OCCs with residuals and powers, this paper also demonstrated that such an essentially theoretical development can be fully mechanised and still be presented in readable calculational style, where writing is not significantly more effort than a conventional calculational presentation in LaTeX.

In comparison with similar developments in Isabelle/HOL [Kah03], the use of Agda enables a completely natural mathematical treatment of categories, nested calculational proofs, and direct use of theories as modules of executable programs.

258   W. Kahl

Acknowledgements. I am grateful to the anonymous referees for their constructive comments, and to Musa Al-hassy for numerous useful suggestions for improving readability.

References

[Aug99]   Augustsson, L.: Equality proofs in Cayenne (1999),
          http://tinyurl.com/Aug99eqproof (accessed January 3, 2014)
[BSZ86]   Berghammer, R., Schmidt, G., Zierer, H.: Symmetric Quotients. Technical
          Report TUM-INFO 8620, Technische Universität München, Fakultät für
          Informatik, 18 p. (1986)
[BSZ89]   Berghammer, R., Schmidt, G., Zierer, H.: Symmetric Quotients and Domain
          Constructions. Inform. Process. Lett. 33, 163–168 (1989)
[BdM97]   Bird, R.S., de Moor, O.: Algebra of Programming. International Series in
          Computer Science, vol. 100. Prentice Hall (1997)
[D+13]   Danielsson, N.A. et al.: Agda Standard Library, Version 0.7 (2013),
          http://tinyurl.com/AgdaStdlib
[Ern05]   Erné, M.: Categories of Contexts (2005) (preprint),
          http://www.iazd.uni-hannover.de/~erne/preprints/CatConts.pdf
[FS90]   Freyd, P.J., Scedrov, A.: Categories, Allegories. North-Holland Mathematical Library, vol. 39. North-Holland, Amsterdam (1990)
[HKZ06]   Hitzler, P., Krötzsch, M., Zhang, G.-Q.: A Categorical View on Algebraic
          Lattices in Formal Concept Analysis. Fund. Inform. 74, 301–328 (2006)
[Jip12]   Jipsen, P.: Categories of Algebraic Contexts Equivalent to Idempotent
          Semirings and Domain Semirings. In: [KG12], pp. 195–206
[Kah03]   Kahl, W.: Calculational Relation-Algebraic Proofs in Isabelle/Isar. In:
          Berghammer, R., Möller, B., Struth, G. (eds.) RelMiCS/AKA 2003. LNCS,
          vol. 3051, pp. 178–190. Springer, Heidelberg (2004)
[Kah04]   Kahl, W.: Refactoring Heterogeneous Relation Algebras around Ordered
          Categories and Converse. J. Relational Methods in Comp. Sci. 1, 277–313
          (2004)
[Kah08]   Kahl, W.: Relational Semigroupoids: Abstract Relation-Algebraic Interfaces
          for Finite Relations between Infinite Types. J. Logic and Algebraic Programming 76, 60–89 (2008)
[Kah11a] Kahl, W.: Collagories: Relation-Algebraic Reasoning for Gluing Constructions. J. Logic and Algebraic Programming 80, 297–338 (2011)
[Kah11b] Kahl, W.: Dependently-Typed Formalisation of Relation-Algebraic Abstractions. In: de Swart, H. (ed.) RAMiCS 2011. LNCS, vol. 6663, pp. 230–247.
          Springer, Heidelberg (2011)
[Kah12]   Kahl, W.: Towards Certifiable Implementation of Graph Transformation via
          Relation Categories. In: [KG12], pp. 82–97
[KG12]   Kahl, W., Griffin, T.G. (eds.): RAMiCS 2012. LNCS, vol. 7560. Springer,
          Heidelberg (2012)
[Kah14]   Kahl, W.: Relation-Algebraic Theories in Agda — RATH-Agda-2.0.0. Mechanically checked Agda theories available for download, with 456 pages literate document output (2014), http://RelMiCS.McMaster.ca/RATH-Agda/

[Mos13] Moshier, M.A.: A Relational Category of Polarities (2013) (unpublished draft)

[Nor07] Norell, U.: Towards a Practical Programming Language Based on Dependent Type Theory. PhD thesis, Department of Computer Science and Engineering, Chalmers University of Technology (2007)

[Wil05] Wille, R.: Formal Concept Analysis as Mathematical Theory of Concepts and Concept Hierarchies. In: Ganter, B., Stumme, G., Wille, R. (eds.) Formal Concept Analysis. LNCS (LNAI), vol. 3626, pp. 1–33. Springer, Heidelberg (2005)

A Power Orders Give Rise to Right Residuals

In the context of an OSGC with a power operator, presence of a power implies that all right residuals exist, with $Q \setminus S = \Lambda_0 (Q^{\smile}) \, \mathbin{\S} \, \Omega \, \mathbin{\S} \, (\Lambda_0 (S^{\smile}))^{\smile}$. We use the following additional power operator lemmas:

$\epsilon_{\S}\Lambda^{\smile} : \{R : \mathsf{Mor}\ A\ B\} \to \epsilon \, \mathbin{\S} \, (\Lambda_0\ R)^{\smile} \approx R^{\smile}$

$_{\S}\Lambda\text{-}^{\smile} : \{Q : \mathsf{Mor}\ B\ A\} \to Q \, \mathbin{\S} \, \Lambda_0 (Q^{\smile}) \sqsubseteq \epsilon$

```
module PowerRightRes (Ω : {A : Obj} → Mor (ℙ A) (ℙ A))
(isPowerOrder : IsPowerOrder Ω) where
open IsPowerOrder isPowerOrder
rightResOp : RightResOp orderedSemigroupoid
rightResOp = record
  {_\_ = λ {A} {B} {C} Q S → Λ₀ (Q ˘) ⨾ Ω ⨾ (Λ₀ (S ˘)) ˘
  ;\-cancel-outer = λ {A} {B} {C} {S} {Q} → ⊑-begin
      Q ⨾ Λ₀ (Q ˘) ⨾ Ω ⨾ (Λ₀ (S ˘)) ˘
    ⊑⟨ ⨾-assocL ⟨≈⊑⟩ ⨾-monotone₁ ⨾Λ-˘ ⟩
      ∈ ⨾ Ω ⨾ (Λ₀ (S ˘)) ˘
    ⊑⟨ ⨾-assocL ⟨≈⊑⟩ ⨾-monotone₁ ∈⨾Ω ⟩
      ∈ ⨾ (Λ₀ (S ˘)) ˘
    ≈⟨ ∈⨾Λ˘ ⟨≈≈⟩ ˘˘ ⟩
      S
    □
  ;\-universal = λ {A} {B} {C} {S} {Q} {R} Q⨾R⊑S → ⊑-begin
      R
    ⊑⟨ proj₁ Λ-total ⟨⊑≈⟩ ⨾-assoc ⟩
      Λ₀ (Q ˘) ⨾ (Λ₀ (Q ˘)) ˘ ⨾ R
    ⊑⟨ ⨾-monotone₂₂ (proj₂ Λ-total) ⟩
      Λ₀ (Q ˘) ⨾ (Λ₀ (Q ˘)) ˘ ⨾ R ⨾ Λ₀ (S ˘) ⨾ (Λ₀ (S ˘)) ˘
    ⊑⟨ ⨾-monotone₂ (⨾-assocL₃₊₁ ⟨≈⊑⟩ ⨾-monotone₁ (Ω-universal (⊑-begin
        ∈ ⨾ (Λ₀ (Q ˘)) ˘ ⨾ R ⨾ Λ₀ (S ˘)
      ≈⟨ ⨾-assocL ⟨≈≈⟩ ⨾-cong₁ (∈⨾Λ˘ ⟨≈≈⟩ ˘˘) ⟩
```

$$Q \fatsemi R \fatsemi \Lambda_0 (S \check{\ })$$
$$\sqsubseteq \langle \fatsemi\text{-assocL } \langle \approx\sqsubseteq \rangle \fatsemi\text{-monotone}_1 \; Q\fatsemi R\sqsubseteq S \; \langle \sqsubseteq\sqsubseteq \rangle \fatsemi \Lambda\text{-}\check{\ } \rangle$$
$$\in$$
$$\Box)))\,)$$
$$\Lambda_0 (Q \check{\ }) \fatsemi \Omega \fatsemi (\Lambda_0 (S \check{\ })) \check{\ }$$
$$\Box$$
$$\}$$

The standard definition of the power order via this right residual also returns the given power order: $\epsilon \setminus \epsilon \approx \Omega$.

A Sufficient Condition for Liftable Adjunctions between Eilenberg-Moore Categories

Koki Nishizawa[1] and Hitoshi Furusawa[2]

[1] Department of Information Systems Creation, Faculty of Engineering,
Kanagawa University
nishizawa@kanagawa-u.ac.jp
[2] Department of Mathematics and Computer Science, Kagoshima University
furusawa@sci.kagoshima-u.ac.jp

Abstract. This paper gives a sufficient condition for monads P, P' and T to have an adjunction between the category of P-algebras over T-algebras and the category of P'-algebras over T-algebras. The leading example is an adjunction between the category of idempotent semirings and the category of quantales, where P is the finite powerset monad, P' is the powerset monad, and T is the free monoid monad. The left adjoint of this leading example is given by ideal completion. Applying our result, we show that ideal completion also gives an adjunction between the category of join semilattices over T-algebras and the category of complete join semilattices over T-algebras for a general monad T satisfying certain distributive law.

Keywords: Distributive law, absolute coequalizer, ideal, quantale, idempotent semiring.

1 Introduction

This paper gives a sufficient condition for monads P, P' and T to have an adjunction between the category of P-algebras over T-algebras and the category of P'-algebras over T-algebras.

We have three leading examples. The first one is the adjunction whose right adjoint is the forgetful functor from the category of complete join semilattices to the category of join semilattices. This example can be regarded as the case where P is the finite powerset monad, P' is the powerset monad, and T is the identity monad on Set.

The second one is the adjunction whose right adjoint is the forgetful functor from the category of quantales to the category of idempotent semirings. A quantale is a complete join semilattice together with a monoid structure whose associative multiplication distributes over arbitrary joins. An idempotent semiring is a join semilattice together with a monoid structure whose associative multiplication distributes over finite joins. In this example, P and P' are respectively the same as P and P' of the first example, but T is the free monoid monad.

P. Höfner et al. (Eds.): RAMiCS 2014, LNCS 8428, pp. 261–276, 2014.

In both of the two examples, left adjoints are defined by ideal completion. Our main theorem does not only answer the reason why both of these left adjoints are defined by the same construction, but also shows that ideal completion also gives an adjunction between the category of join semilattices over T-algebras and the category of complete join semilattices over T-algebras for a general monad T satisfying a certain distributive law [NF12].

The third leading example is an adjunction between the category of complete join semilattices and the category of pointed sets, whose right adjoint maps a complete join semilattice to its underlying set with the least element. Its left adjoint maps a set with a point to the set of all subsets containing the point. This example can be regarded as the case where P' is the powerset monad and P is the submonad which returns the set of all singleton subsets and the emptyset.

This paper is organized as follows: Section 2 defines monads for join semilattices. Section 3 shows that a left adjoint between categories of algebras is defined by absolute coequalizer construction. Section 4 shows that the ideal completion gives the absolute coequalizer for join semilattices. Section 5 generalize results in Section 3 for monads combined by distributive laws. Section 6 shows the example for complete join semilattices and pointed sets. Section 7 summarizes this work and discusses future work.

2   Monads for Join Semilattices

Definition 2.1. A *join semilattice* is a tuple $(S, \leq, \bigvee)$ with a partially ordered set $(S, \leq)$ and the join or the least upper bound $\bigvee A$ for a finite subset A of S.

A join semilattice S must have the least element 0 since the empty set $\emptyset$ is a finite subset of S and $\bigvee \emptyset = 0$.

SLat denotes the category whose objects are join semilattices and whose arrows are homomorphisms between them. SLat is equivalent to the Eilenberg-Moore category $\wp_f$-Alg of the finite powerset monad $\wp_f$, whose endofunctor sends a set X to the set of finite subsets $\wp_f(X) = \{A \subseteq X \mid |A| < \omega\}$, whose unit sends an element x in X to $\{x\}$ in $\wp_f(X)$, and whose multiplication sends a subset family α in $\wp_f(\wp_f(X))$ to its union $\bigcup \alpha = \{x \mid \exists X \in \alpha, x \in X\}$ in $\wp_f(X)$. The forgetful functor from $\wp_f$-Alg to Set has the left adjoint which sends a set X to $(\wp_f(X), \bigcup : \wp_f(\wp_f(X)) \to \wp_f(X))$. The counit for $\wp_f$-algebra $(S, \bigvee)$ is $\bigvee : (\wp_f(S), \bigcup) \to (S, \bigvee)$.

Definition 2.2. A *complete join semilattice* is a tuple $(S, \leq, \bigvee)$ with a partially ordered set $(S, \leq)$ and the join or the least upper bound $\bigvee A$ for a subset A of S.

We write CSLat for the category whose objects are complete join semilattices and whose arrows are homomorphisms between them. CSLat is equivalent to the Eilenberg-Moore category $\wp$-Alg of the powerset monad $\wp$, whose endofunctor sends a set X to the set of all subsets $\wp(X) = \{A \mid A \subseteq X\}$, whose unit, multiplication and counit are given in the same way as the finite powerset monad $\wp_f$.

Definition 2.3. Let $P = (P, \mu^P, \eta^P)$ and $P' = (P', \mu^{P'}, \eta^{P'})$ be monads on C. A *monad map* ι from P to P' is a natural transformation from P to P' satisfying the following diagrams.

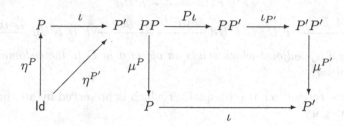

Lemma 2.4. *Let* $P = (P, \mu^P, \eta^P)$ *and* $P' = (P', \mu^{P'}, \eta^{P'})$ *be monads on* C. *If* ι *is a monad map from* P *to* P', *then, the following* G *is a functor from* P'*-Alg to* P*-Alg.*

- *For a* P'*-algebra* (c, p'), $G(c, p') = (c, p' \circ \iota_c)$.
- *For a* P'*-algebra homomorphism* $f\colon (c_1, p'_1) \to (c_2, p'_2)$, $Gf = f$.

We write $- \circ \iota$ for the above functor G.

Example 2.5. Let P be the finite powerset monad $\wp_f$. Let P' be the powerset monad $\wp$. Let ι_X be the inclusion from $\wp_f(X)$ to $\wp(X)$. Then, ι is a monad map from $\wp_f$ to $\wp$. The functor $- \circ \iota$ is the forgetful functor from CSLat to SLat.

3   Left Adjoint by Absolute Coequalizers

The next theorem is a corollary of the Theorem 2(b) of Section 3.7 of the book [BW85]. This theorem mentions the relationship between a monadic functor and coequalizers. A functor is called *monadic* if it is equivalent to the forgetful functor from the Eilenberg-Moore category of a monad.

Theorem 3.1. *Let* C, D, D' *be categories and* G, U, U' *be functors satisfying the following conditions.*

- $G\colon D' \to D$ *is a functor.*
- *A functor* $U\colon D \to C$ *has a left adjoint* F *(call its unit* η *and its counit* ϵ*).*
- *A functor* $U'\colon D' \to C$ *has a left adjoint* F' *(call its unit* η' *and its counit* ϵ'*).*
- $U \circ G$ *is natural isomorphic to* U'.
- U *is monadic.*

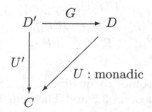

– *For an object d in D, the following parallel pair has a coequalizer in D'.*

$$F'UFUd \xrightarrow{F'U\epsilon_d} F'Ud$$

$$F'UFUd \xrightarrow{F'UF\eta'_{Ud}} F'UFU'F'Ud \xrightarrow{\cong\, \circ F'U\epsilon_{GF'Ud}\circ\, \cong} F'U'F'Ud \xrightarrow{\epsilon'_{F'Ud}} F'Ud$$

Then, G has a left adjoint which sends an object d in D to the codomain of the above coequalizer.

An *absolute coequalizer* is a coequalizer which is preserved by any functor to any other category.

Theorem 3.2 (Corollary of Beck's theorem [Mac98]). *If a functor $U' \colon D' \to C$ is monadic, then U' creates coequalizers of those parallel pairs f, g in D' for which $U'f, U'g$ has an absolute coequalizer in C.*

Example 3.3. The forgetful functor U' from CSLat to Set is monadic. Therefore, it creates coequalizers of those parallel pairs f, g in CSLat for which $U'f, U'g$ has an absolute coequalizer in Set.

The following theorem appears essentially in the paper [Lin69]

Theorem 3.4. *Let $P = (P, \mu^P, \eta^P)$ and $P' = (P', \mu^{P'}, \eta^{P'})$ be monads on C. Let ι be a monad map from P to P'. For a P-algebra (c,p), let $e_{(c,p)} \colon P'c \to E(c,p)$ be an absolute coequalizer of $P'p$ and $\mu_c^{P'} \circ P'\iota_c$ in C.*

$$P'Pc \xrightarrow[\mu_c^{P'} \circ P'\iota_c]{\quad P'p \quad} P'c \xrightarrow{\; e_{(c,p)} \;} E(c,p)$$

Then, the functor $- \circ \iota \colon P'\text{-Alg} \to P\text{-Alg}$ has a left adjoint L, where $L(c,p)$ is the P'-algebra on $E(c,p)$ created by the forgetful functor from $P'\text{-Alg}$ to C.

Proof. Let $D = P\text{-Alg}$ and $D' = P'\text{-Alg}$ in Theorem 3.1. Let U be the forgetful functor from $P\text{-Alg}$ to C. Let U' be the forgetful functor from $P'\text{-Alg}$ to C. The composition of U and $- \circ \iota$ is natural isomorphic to U'. For a P-algebra (c,p), U' sends the parallel pair in Theorem 3.1 to

$$U'F'U\epsilon_{(c,p)} = U'F'Up = U'F'p = P'p$$

and

$$
\begin{aligned}
U'\epsilon'_{F'U(c,p)} &\circ U'F'U\epsilon_{(-\circ\iota)F'U(c,p)} \circ U'F'UF\eta'_{U(c,p)}\\
&= U'\epsilon'_{F'U(c,p)} \circ U'F'U\epsilon_{(P'c,\mu_c^{P'}\circ\iota_{P'c})} \circ U'F'UF\eta'_{U(c,p)}\\
&= \mu_c^{P'} \circ P'(\mu_c^{P'} \circ \iota_{P'c}) \circ P'P\eta_c'\\
&= \mu_c^{P'} \circ P'\mu_c^{P'} \circ P'P\eta_c^{P'} \circ P'\iota_c\\
&= \mu_c^{P'} \circ P'\iota_c \; .
\end{aligned}
$$

These pairs have an absolute coequalizer $e_{(c,p)}$. Since U' is monadic, $- \circ \iota$ has the left adjoint L by Theorem 3.2 and Theorem 3.1.   □

Theorem 3.5. *If the assumptions of Theorem 3.4 hold and for all P-algebra* (c,p)*,* $e_{(c,p)}$ *has a right inverse* $r_{(c,p)}$*,*

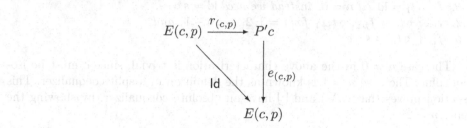

then $L(c,p)$ *is the pair of* $E(c,p)$ *and the following* P'*-structure map.*

$$P'E(c,p) \xrightarrow{P'r_{(c,p)}} P'P'c \xrightarrow{\mu_c^{P'}} P'c \xrightarrow{e_{(c,p)}} E(c,p)$$

Proof. There exist a unique object $(E(c,p), p')$ and a unique arrow $f\colon (P'c, \mu_c^{P'}) \to (E(c,p), p')$ in P'-Alg satisfying $U'f = e_{(c,p)}$ by Theorem 3.4. Since f is $e_{(c,p)}$ itself and it is a P'-algebra homomorphism, the following diagram commutes.

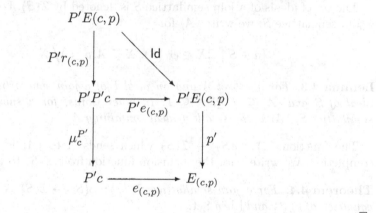

$\square$

By Theorem 3.4, the forgetful functor $- \circ \iota$ from CSLat to SLat has a left adjoint if Set has an absolute coequalizer of $\wp(\bigvee)\colon \wp(\wp_f(S)) \to \wp(S)$ and $\bigcup\colon \wp(\wp_f(S)) \to \wp(S)$ for a join semilattice $(S, \leq, \bigvee)$.

4   Ideal Completion as Absolute Coequalizer

This section shows that Set has an absolute coequalizer of $\wp(\bigvee)$ and $\bigcup$ for a join semilattice $(S, \leq, \bigvee)$.

Absolute coequalizers have a diagrammatic characterization [Par69].

Theorem 4.1. $e\colon Y \to Z$ *is an absolute coequalizer of* $f_0\colon X \to Y$ *and* $f_1\colon X \to Y$ *if and only if there exist* $s\colon Z \to Y$*, finite arrows* $t_1, \cdots, t_n\colon Y \to X$ *and a sequence of binary digits* $j_1, \cdots, j_n \in \{0,1\}$ *such that*

1. $e \circ f_0 = e \circ f_1$,
2. $e \circ s = \mathsf{Id}$,
3. $f_{j_1} \circ t_1 = \mathsf{Id}$ *(if $n = 0$, instead we need $\mathsf{Id} = s \circ e$)*,
4. $f_{1-j_i} \circ t_i = f_{j_{i+1}} \circ t_{i+1}$ *for* $i = 1, 2, \cdots, n-1$, *and*
5. $f_{1-j_n} \circ t_n = s \circ e$.

The case $n = 0$ in the above characterization is trivial, since e must be isomorphic. The case $n = 1$ is known as the definition of a split coequalizer. This section proves that $\wp(\bigvee)$ and $\bigcup$ have an absolute coequalizer, by showing the case $n = 2$.

Definition 4.2. Let S be a join semilattice. An *ideal* is a subset A of S such that

- A is closed under finite join operation $\bigvee$,
- A is closed downward under $\leq$.

Since an ideal A is closed under finite join, A must contain the least element $\bigvee \emptyset = 0$. Thus, ideals are not empty.

The set of ideals of a join semilattice S is denoted by $\mathcal{I}(S)$. For a subset A of a join semilattice S, we write $\langle A \rangle$ for

$$\{ a \in S \mid \exists X \in \wp_f(S). \ X \subseteq A, a \leq \bigvee X \} \ .$$

Lemma 4.3. *For a subset A and an ideal I of a join semilattice S, $\langle A \rangle$ is an ideal of S and $\langle A \rangle \subseteq I$ iff $A \subseteq I$. In other words, for a subset A of a join semilattice S, $\langle A \rangle$ is the smallest ideal containing A.*

The function $\langle \_ \rangle \colon \wp(S) \to \mathcal{I}(S)$ which sends A to $\langle A \rangle$ is called an ideal completion. We write r for the inclusion function from $\mathcal{I}(S)$ to $\wp(S)$.

Theorem 4.4. *For a join semilattice S, $\langle \_ \rangle \colon \wp(S) \to \mathcal{I}(S)$ is an absolute coequalizer of $\wp(\bigvee)$ and $\bigcup$ in* Set.

$$\wp(\wp_f(S)) \ \overset{\wp(\bigvee)}{\underset{\bigcup}{\rightrightarrows}} \ \wp(S) \ \overset{\langle \_ \rangle}{\longrightarrow} \ \mathcal{I}(S)$$

Proof. We define $\wp_f \colon \wp(S) \to \wp(\wp_f(S))$ and **down**$\colon \wp(S) \to \wp(\wp_f(S))$ as follows.

$$\wp_f(A) = \{ X \in \wp_f(S) \mid X \subseteq A \}$$

$$\mathbf{down}(A) = \{ \{a, b\} \mid a \in S, b \in A, a \leq b \}$$

We prove that the above diagrammatic characterization of absolute coequalizers where $f_0 = \wp(\bigvee)$, $f_1 = \bigcup$, and $e = \langle \_ \rangle$, by taking $s = r$, $n = 2$, $t_1 = \wp_f$, $t_2 = \mathbf{down} \circ \wp(\bigvee) \circ \wp_f$, $j_1 = 1$, and $j_2 = 0$. That is, we show the following equations.

1. $\langle \_ \rangle \circ \wp(\bigvee) = \langle \_ \rangle \circ \bigcup$

2. $\langle\_\rangle \circ r = \mathsf{Id}$
3. $\bigcup \circ \wp_f = \mathsf{Id}$
4. $\wp(\bigvee) \circ \wp_f = \wp(\bigvee) \circ \mathbf{down} \circ \wp(\bigvee) \circ \wp_f$
5. $\bigcup \circ \mathbf{down} \circ \wp(\bigvee) \circ \wp_f = r \circ \langle\_\rangle$

(1) Let α be an element of $\wp(\wp_f(S))$. We have $\langle\wp(\bigvee)(\alpha)\rangle \subseteq \langle\bigcup\alpha\rangle$ as follows.

$$\langle\wp(\textstyle\bigvee)(\alpha)\rangle \subseteq \langle\textstyle\bigcup\alpha\rangle$$
$$\Longleftrightarrow \wp(\textstyle\bigvee)(\alpha) \subseteq \langle\textstyle\bigcup\alpha\rangle \qquad \text{(by Lemma 4.3)}$$
$$\Longleftrightarrow \textstyle\bigvee X \in \langle\textstyle\bigcup\alpha\rangle \qquad (\forall X \in \alpha)$$
$$\Longleftarrow X \subseteq \langle\textstyle\bigcup\alpha\rangle \qquad (\forall X \in \alpha) \text{ (since an ideal is closed under finite join)}$$
$$\Longleftrightarrow \textstyle\bigcup\alpha \subseteq \langle\textstyle\bigcup\alpha\rangle$$
$$\Longleftrightarrow \langle\textstyle\bigcup\alpha\rangle \subseteq \langle\textstyle\bigcup\alpha\rangle \qquad \text{(by Lemma 4.3)}$$

Conversely, we have $\langle\bigcup\alpha\rangle \subseteq \langle\wp(\bigvee)(\alpha)\rangle$ as follows.

$$\langle\textstyle\bigcup\alpha\rangle \subseteq \langle\wp(\textstyle\bigvee)(\alpha)\rangle$$
$$\Longleftrightarrow \textstyle\bigcup\alpha \subseteq \langle\wp(\textstyle\bigvee)(\alpha)\rangle \qquad \text{(by Lemma 4.3)}$$
$$\Longleftrightarrow X \subseteq \langle\wp(\textstyle\bigvee)(\alpha)\rangle \qquad (\forall X \in \alpha)$$
$$\Longleftarrow \textstyle\bigvee X \in \langle\wp(\textstyle\bigvee)(\alpha)\rangle \qquad (\forall X \in \alpha) \text{ (since an ideal is closed downward)}$$
$$\Longleftrightarrow \wp(\textstyle\bigvee)(\alpha) \subseteq \langle\wp(\textstyle\bigvee)(\alpha)\rangle$$
$$\Longleftrightarrow \langle\wp(\textstyle\bigvee)(\alpha)\rangle \subseteq \langle\wp(\textstyle\bigvee)(\alpha)\rangle \qquad \text{(by Lemma 4.3)}$$

Therefore, for each $\alpha \in \wp(\wp_f(S))$, $\langle\wp(\bigvee)(\alpha)\rangle = \langle\bigcup\alpha\rangle$.

(2) The inclusion function $r \colon \mathcal{I}(S) \to \wp(S)$ is a right inverse of $\langle\_\rangle$, because an ideal I is the smallest ideal containing I itself.

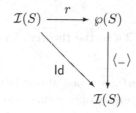

(3) The function $\wp_f$ is a right inverse of $\bigcup\colon \wp(\wp_f(S)) \to \wp(S)$, because a subset A of S is represented by the union of all finite subsets of A.

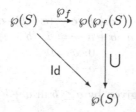

(4) The function **down** is a right inverse of $\wp(\bigvee)\colon \wp(\wp_f(S)) \to \wp(S)$, because a subset A of S satisfies $\wp(\bigvee)(\mathbf{down}(A)) = \{\bigvee\{a,b\} \mid b \in A, a \le b\} = A$.

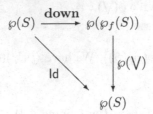

Therefore, we have $\wp(\bigvee) \circ \mathbf{down} \circ \wp(\bigvee) \circ \wp_f = \mathsf{Id} \circ \wp(\bigvee) \circ \wp_f = \wp(\bigvee) \circ \wp_f$.

(5) By the definition of $\langle\_\rangle$, we have the following equation.

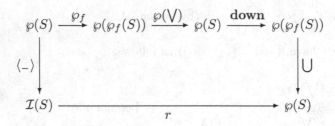

□

Example 4.5. By Theorem 4.4, Theorem 3.4, and Theorem 3.5, the forgetful functor from CSLat to SLat has a left adjoint, which sends a join semilattice S to $(\mathcal{I}(S), \subseteq, \bigvee)$ satisfying $\bigvee \alpha = \langle \bigcup \alpha \rangle$.

5   Liftable Adjunctions between Eilenberg-Moore Categories

This section extends Theorem 3.4 to the theorem for quantales and idempotent semirings.

Definition 5.1. An *idempotent semiring*, abbreviated as *I-semiring* is a tuple $(S, +, \cdot, 0, 1)$ with a set S, two binary operations $+$ and $\cdot$, and $0, 1 \in S$ satisfying the following properties:

- $(S, +, 0)$ is an idempotent commutative monoid.
- $(S, \cdot, 1)$ is a monoid.
- For all $a, b, c \in S$,

$$a \cdot c + b \cdot c = (a + b) \cdot c$$
$$a \cdot b + a \cdot c = a \cdot (b + c)$$
$$0 \cdot a = 0$$
$$a \cdot 0 = 0$$

where the *natural order* $\le$ is given by $a \le b$ iff $a + b = b$.

We often abbreviate $a \cdot b$ to ab.

The natural order $\leq$ on an I-semiring is a join semilattice, where its join operation is given by $\bigvee \emptyset = 0$ and, for a finite subset $A \subseteq S$ containing a, $\bigvee A = a + (\bigvee A \setminus \{a\})$.

Example 5.2. Let Σ be a finite set and Σ^* the set of finite words (strings) over Σ. Then, the finite power set $\wp_f(\Sigma^*)$ of Σ^* forms an I-semiring together with the union, concatenation, empty set, and the singleton set of the empty word.

IS denotes the category whose objects are I-semirings and whose arrows are homomorphisms between them.

Definition 5.3. A *quantale* S is an I-semiring satisfying the following properties: For each $A \subseteq S$ and $a \in S$,

- the least upper bound $\bigvee A$ of A exists in S,
- $(\bigvee A)a = \bigvee\{xa \mid x \in A\}$, and
- $a(\bigvee A) = \bigvee\{ax \mid x \in A\}$.

So, a quantale is a complete I-semiring or an **S**-algebra [Con71]. Homomorphisms between quantales are semiring homomorphisms preserving arbitrary joins.

Example 5.4. Let Σ be a finite set and Σ^* the set of finite words (strings) over Σ. Then, the power set $\wp(\Sigma^*)$ of Σ^* forms a quantale together with the union, concatenation, empty set, and the singleton set of the empty word.

Qt denotes the category whose objects are quantales and whose arrows are homomorphisms between them.

Remark 5.5. I-semirings need not be quantales. For example, an I-semiring $\wp_f(\Sigma^*)$ is not a quantale since it is not closed under arbitrary unions.

We recall the notion of distributive laws between two monads [Bec69]. Maps between distributive laws can be defined in the 2-categorical way [Str72], however, this paper defines them in the elementary way.

Definition 5.6 (distributive law). Let $T = (T, \mu^T, \eta^T)$ and $P = (P, \mu^P, \eta^P)$ be monads on a category C. A *distributive law* θ of P over T is a natural transformation from TP to PT satisfying the following diagrams.

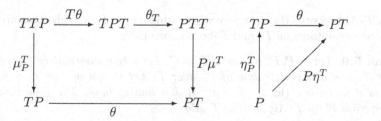

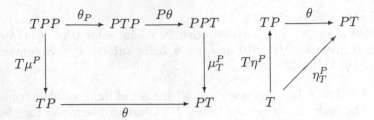

Definition 5.7. Let T, T', P, P' be monads on C. Let θ be a distributive law of P over T. Let θ' be a distributive law of P' over T'. A *morphism* $(\tau, \pi)\colon \theta \to \theta'$ of *distributive laws* consists of monad maps $\tau\colon T \to T'$ and $\pi\colon P \to P'$ satisfying the following diagram.

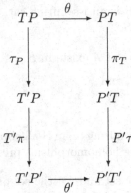

Definition 5.8. Let P and T be monads on a category C. Let θ be a distributive law of P over T. A $P \circ_\theta T$-*algebra* is a tuple (c, t, p) such that

- c is an object in C,
- The pair of c and $t\colon Tc \to c$ is a T-algebra,
- The pair of c and $p\colon Pc \to c$ is a P-algebra, and
- $p \circ Pt \circ \theta_c = t \circ Tp$.

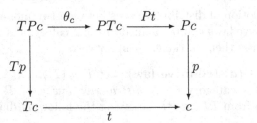

$P \circ_\theta T$-Alg denotes the category whose objects are $P \circ_\theta T$-algebras and whose arrows are simultaneous T- and P-homomorphisms.

Lemma 5.9. *Let T, P, P' be monads on C. Let θ be a distributive law of P over T. Let θ' be a distributive law of P' over T. Let ι be a monad map such that (Id, ι) is a morphism $(\mathsf{Id}, \iota)\colon \theta \to \theta'$ of distributive laws. The following G is a functor from $P' \circ_{\theta'} T$-Alg to $P \circ_\theta T$-Alg.*

- For a $P' \circ_{\theta'} T$-algebra (c, t, p'), $G(c, t, p') = (c, t, p' \circ \iota_c)$.
- For a $P' \circ_{\theta'} T$-algebra homomorphism f, $Gf = f$.

We write $- \circ \iota$ for the above functor G.

Example 5.10. Let $T = (\_)^*$ be the monad for finite sequences on Set. Then, T-Alg is equivalent to the category Mon whose objects are monoids and whose arrows are homomorphisms between them.

Let $\wp_f$ be the finite powerset monad $(\wp_f, \bigcup, \{\_\})$ on Set. Let $\wp$ be the powerset monad $(\wp, \bigcup, \{\_\})$ on Set. There exists a distributive law θ of T over $\wp_f$, and there exists a distributive law θ' of T over $\wp$ as follows.

$$\theta_X(S_1 \cdot S_2 \cdots S_n) = \{x_1 \cdot x_2 \cdots x_n \mid x_1 \in S_1, x_2 \in S_2, \ldots, x_n \in S_n\}$$

$$\theta'_X(S_1 \cdot S_2 \cdots S_n) = \{x_1 \cdot x_2 \cdots x_n \mid x_1 \in S_1, x_2 \in S_2, \ldots, x_n \in S_n\}$$

Then, $\wp_f \circ_\theta T$-Alg is equivalent to the category IS and $\wp \circ_{\theta'} T$-Alg is equivalent to the category Qt. Let ι_X be the inclusion function from $\wp_f(X)$ to $\wp(X)$. $(\text{Id}, \iota) \colon \theta \to \theta'$ is a morphism of distributive laws.

Lemma 5.11. *The forgetful functor from $P \circ_\theta T$-Alg to C is monadic.*

Example 5.12. The forgetful functor from Mon to Set, the forgetful functor from IS to Set, and the forgetful functor from Qt to Set are monadic.

Lemma 5.13. *The forgetful functor from $P \circ_\theta T$-Alg to T-Alg is monadic. The left adjoint to this forgetful functor sends a T-algebra (c, t) to $(Pc, Pt \circ \theta_c, \mu_c^P)$ and a T-homomorphism f to Pf. The unit for a T-algebra (c, t) is $\eta_c^P \colon (c, t) \to (Pc, Pt \circ \theta_c)$. The counit for a $P \circ_\theta T$-algebra (c, t, p) is $p \colon (Pc, Pt \circ \theta_c, \mu_c^P) \to (c, t, p)$.*

Example 5.14. The forgetful functor from IS to Mon and the forgetful functor from Qt to Mon are monadic.

Theorem 5.15. *Let $T = (T, \mu^T, \eta^T)$, $P = (P, \mu^P, \eta^P)$, and $P' = (P', \mu^{P'}, \eta^{P'})$ be monads on C. Let θ be a distributive law of P over T. Let θ' be a distributive law of P' over T. Let ι be a monad map such that (Id, ι) is a morphism $(\text{Id}, \iota) \colon \theta \to \theta'$ of distributive laws. For a $P \circ_\theta T$-algebra (c, t, p), let $e_{(c,p)} \colon P'c \to E(c, p)$ be an absolute coequalizer of $P'p$ and $\mu_c^{P'} \circ P'\iota_c$ in C.*

$$P'Pc \overset{P'p}{\underset{\mu_c^{P'} \circ P'\iota_c}{\rightrightarrows}} P'c \xrightarrow{e_{(c,p)}} E(c, p)$$

Then, the functor $- \circ \iota \colon P' \circ_{\theta'} T$-Alg $\to P \circ_\theta T$-Alg has a left adjoint L, where $L(c, t, p)$ is the $P' \circ_{\theta'} T$-algebra on $E(c, p)$ created by the forgetful functor from $P' \circ_{\theta'} T$-Alg to C.

Proof. Let $D = P \circ_\theta T$-Alg and $D' = P' \circ_{\theta'} T$-Alg in Theorem 3.1. Let U be the forgetful functor from $P \circ_\theta T$-Alg to T-Alg. Let U' be the forgetful functor from $P' \circ_{\theta'} T$-Alg to T-Alg. The composition of U and $- \circ \iota$ is natural isomorphic to U'. For a $P \circ_\theta T$-algebra (c, t, p), the forgetful functor U' sends the parallel pair in Theorem 3.1 to

$$U'F'U\epsilon_{(c,t,p)} = U'F'Up = U'F'p = P'p$$

and

$$
\begin{aligned}
&U'\epsilon'_{F'U(c,t,p)} \circ U'F'U\epsilon_{(-\circ\iota)F'U(c,t,p)} \circ U'F'UF\eta'_{U(c,t,p)} \\
&= U'\epsilon'_{F'U(c,t,p)} \circ U'F'U\epsilon_{(P'c,P'to\theta'_c,\mu_c^{P'}\circ\iota_{P'c})} \circ U'F'UF\eta'_{U(c,t,p)} \\
&= \mu_c^{P'} \circ P'(\mu_c^{P'} \circ \iota_{P'c}) \circ P'P\eta'_c \\
&= \mu_c^{P'} \circ P'\mu_c^{P'} \circ P'P'\eta'_c \circ P'\iota_c \\
&= \mu_c^{P'} \circ P'\iota_c .
\end{aligned}
$$

Moreover, the forgetful functor from T-Alg to C sends these pairs to the same arrows. They have an absolute coequalizer. The forgetful functor from $P' \circ_{\theta'} T$-Alg to C is monadic.

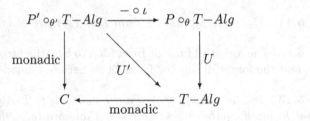

Therefore, $- \circ \iota$ has the left adjoint L by Theorem 3.1 and Theorem 3.2. □

Theorem 5.16. *If the assumptions of Theorem 5.15 hold and for all $P \circ_\theta T$-algebra (c, t, p), $e_{(c,p)}$ has a right inverse $r_{(c,p)}$,*

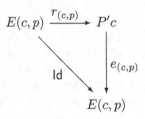

then $L(c, p)$ is the tuple of $E(c, p)$ and the following $P' \circ_{\theta'} T$-structure map.

$$TE(c,p) \xrightarrow{Tr_{(c,p)}} TP'c \xrightarrow{\theta'_c} P'Tc \xrightarrow{P't} P'c \xrightarrow{e_{(c,p)}} E(c,p)$$

$$P'E(c,p) \xrightarrow{P'r_{(c,p)}} P'P'c \xrightarrow{\mu_c^{P'}} P'c \xrightarrow{e_{(c,p)}} E(c,p)$$

Proof. There exist a unique object $(E(c,p),t',p')$ and a unique arrow $f\colon (P'c,P't\circ\theta'_c,\mu_c^{P'})\to(E(c,p),t',p')$ in $P'_\theta T$-Alg satisfying $U'f=e_{(c,p)}$ by Theorem 5.15. Since f is $e_{(c,p)}$ itself and it is simultaneous a T-homomorphism and a P'-homomorphism, both of the following diagrams commute.

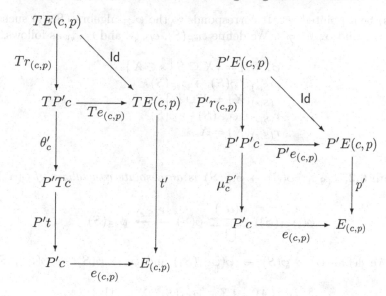

Example 5.17. By Theorem 5.15, Theorem 5.16, and Theorem 4.4, the forgetful functor from Qt to IS has a left adjoint, which sends an idempotent semiring $(S,+,\cdot,0,1)$ to $(\mathcal{I}(S),\subseteq,\bigvee_I,\cdot_I,1_I)$ satisfying

$$\bigvee_I \alpha=\langle\bigcup\alpha\rangle\ ,\quad J\cdot_I K=\langle\{a\cdot b\mid a\in J, b\in K\}\rangle\ ,\quad 1_I=\langle\{1\}\rangle\ .$$

6   Pointed Sets and Absolute Coequalizer

This section shows our third leading example, which is an adjunction between the category of complete join semilattices and the category of pointed sets.

Definition 6.1. A *pointed set* is a set equipped with a chosen element, which is called the *base point*. Homomorphisms between pointed sets are those functions that map a base point to another.

We write Set$_*$ for the category whose objects are pointed sets and whose arrows are homomorphisms between them. Let G be the functor from CSLat to Set$_*$ which maps $(S,\leq,\bigvee)$ to $(S,\perp_S)$ where $\perp_S$ is the least element of S. G has the left adjoint which sends a pointed set (X,x) to $(\{A\subseteq X\mid x\in A\},\subseteq)$.

We show that the above adjunction is also given by Theorem 3.4. Set$_*$ is equivalent to the Eilenberg-Moore category $\wp_{\leq 1}$-Alg for the submonad $\wp_{\leq 1}$ of the powerset monad $\wp$, whose endofunctor returns the set of all singleton subsets and the emptyset.

$$\wp_{\leq 1}(X) = \{A \mid A \subseteq X, |A| \leq 1\} = \{\{x\} \mid x \in X\} \cup \{\emptyset\}$$

Let (S, s) be a pointed set. It corresponds to the $\wp_{\leq 1}$-algebra (S, α_s) such that $\alpha_s(\{x\}) = x$ and $\alpha_s(\emptyset) = s$. We define $\wp_{s\in}(S)$, $e_{(S,s)}$, and $r_{(S,s)}$ as follows.

$$\wp_{s\in}(S) = \{X \subseteq S \mid s \in X\}$$
$$e_{(S,s)} \colon \wp(S) \to \wp_{s\in}(S)$$
$$e_{(S,s)}(X) = X \cup \{s\}$$
$$r_{(S,s)} \colon \wp_{s\in}(S) \to \wp(S)$$
$$r_{(S,s)}(X) = X$$

Theorem 6.2. $e_{(S,s)} \colon \wp(S) \to \wp_{s\in}(S)$ *is an absolute coequalizer of* $\wp(\alpha_s)$ *and* $\bigcup$ *in* Set.

$$\wp(\wp_{\leq 1}(S)) \underset{\bigcup}{\overset{\wp(\alpha_s)}{\rightrightarrows}} \wp(S) \xrightarrow{e_{(S,s)}} \wp_{s\in}(S)$$

Proof. We define $\wp_{\leq 1} \colon \wp(S) \to \wp(\wp_{\leq 1}(S))$ and $\wp_{=1} \colon \wp(S) \to \wp(\wp_{\leq 1}(S))$ as follows.

$$\wp_{\leq 1}(A) = \{X \in \wp_{\leq 1}(S) \mid X \subseteq A\}$$

$$\wp_{=1}(A) = \{\{x\} \mid x \in A\}$$

We prove that the diagrammatic characterization of absolute coequalizers where $f_0 = \wp(\alpha_s)$, $f_1 = \bigcup$, and $e = e_{(S,s)}$, by taking $s = r_{(S,s)}$, $n = 2$, $t_1 = \wp_{\leq 1}$, $t_2 = \wp_{=1} \circ \wp(\alpha_s) \circ \wp_{\leq 1}$, $j_1 = 1$, and $j_2 = 0$. That is, we show the following equations.

1. $e_{(S,s)} \circ \wp(\alpha_s) = e_{(S,s)} \circ \bigcup$
2. $e_{(S,s)} \circ r_{(S,s)} = \mathsf{Id}$
3. $\bigcup \circ \wp_{\leq 1} = \mathsf{Id}$
4. $\wp(\alpha_s) \circ \wp_{\leq 1} = \wp(\alpha_s) \circ \wp_{=1} \circ \wp(\alpha_s) \circ \wp_{\leq 1}$
5. $\bigcup \circ \wp_{=1} \circ \wp(\alpha_s) \circ \wp_{\leq 1} = r_{(S,s)} \circ e_{(S,s)}$

Let β be an element of $\wp(\wp_{\leq 1}(S))$. We have (1) as follows.

$$
\begin{aligned}
&e_{(S,s)}(\wp(\alpha_s)(\beta)) \\
&= \wp(\alpha_s)(\beta) \cup \{s\} \\
&= \{x \mid \{x\} \in \beta\} \cup \{s \mid \emptyset \in \beta\} \cup \{s\} \\
&= \{x \mid \{x\} \in \beta\} \cup \{s\} \\
&= (\bigcup \beta) \cup \{s\} \\
&= e_{(S,s)}(\bigcup \beta)
\end{aligned}
$$

For $X \in \wp_{s\in}(S)$, we have (2) $e_{(S,s)}(r_{(S,s)}(X)) = e_{(S,s)}(X) = X \cup \{s\} = X$, since X contains s. For $X \in \wp(S)$, we have (3) $\bigcup(\wp_{\leq 1}(X)) = \bigcup\{\{x\} \mid x \in X\} = X$.

We have $\wp(\alpha_s) \circ \wp_{=1} = \mathsf{Id}$ as follows.

$$\wp(\alpha_s)(\wp_{=1}(X))$$
$$= \wp(\alpha_s)(\{\{x\} \mid x \in X\})$$
$$= \{\alpha_s(\{x\}) \mid x \in X\}$$
$$= \{x \mid x \in X\}$$
$$= X.$$

Therefore, (4) holds. We have (5) as follows.

$$\bigcup(\wp_{=1}(\wp(\alpha_s)(\wp_{\leq 1}(X))))$$
$$= \wp(\alpha_s)(\wp_{\leq 1}(X))$$
$$= \wp(\alpha_s)(\{\{x\} \mid x \in X\} \cup \{\emptyset\})$$
$$= \{\alpha_s(\{x\}) \mid x \in X\} \cup \{\alpha_s(\emptyset)\}$$
$$= \{x \mid x \in X\} \cup \{s\}$$
$$= X \cup \{s\}$$
$$= e_{(S,s)}(X)$$
$$= r_{(S,s)}(e_{(S,s)}(X))$$

$\square$

By the above theorem, we can get an adjunction between $\wp$-Alg and $\wp_{\leq 1}$-Alg, that is, an adjunction CSLat and Set$_*$.

For the free monoid monad T, there exists a distributive law θ'' of T over $\wp_{\leq 1}$ as follows.

$$\theta''_X(S_1 \cdot S_2 \cdot \dots \cdot S_n) = \{x_1 \cdot x_2 \cdot \dots \cdot x_n \mid x_1 \in S_1, x_2 \in S_2, \dots, x_n \in S_n\}$$

If all of $S_1 \cdot S_2 \cdot \dots \cdot S_n$ are singletons, $\theta''_X(S_1 \cdot S_2 \cdot \dots \cdot S_n)$ is also a singleton. Otherwise, $\theta''_X(S_1 \cdot S_2 \cdot \dots \cdot S_n)$ is the emptyset.

By Theorem 5.15, we can lift the above adjunction to an adjunction between $\wp \circ_{\theta'} T$-Alg and $\wp_{\leq 1} \circ_{\theta''} T$-Alg.

7   Conclusion and Future Work

Our main theorem is Theorem 5.15. It provided a sufficient condition for the functor from $P' \circ_{\theta'} T$-Alg to $P \circ_\theta T$-Alg to have a left adjoint. This result includes the cases of the forgetful functor from CSLat to SLat and the forgetful functor from Qt to IS. In both cases, left adjoints are given by ideal completion. Therefore, ideal completion can give an adjunction between the category of join semilattices over T-algebras and the category of complete join semilattices over T-algebras for a general monad T satisfying certain distributive law.

Our result also includes a different example from ideals. This example is an adjunction between the category of complete join semilattices and the category of pointed sets. It can be given by taking a submonad $\wp_{\leq 1}$ of the powerset monad as P.

It is a future work to show that our main theorem includes other examples than submonads of the powerset monad. For example, an adjunction between the category of rings to the category of semirings might be given, when P is the free commutative monoid monad, P' is the free commutative group monad, and T is the free monoid monad.

Acknowledgements. We thank to the anonymous referees for their helpful comments and suggestions. This work was supported by JSPS Grant Numbers 24700017 and 25330016.

References

[BW85]   Barr, M., Wells, C.: Toposes, Triples and Theories, Grundlagen der Mathematischen Wissenschaften, vol. 278. Springer (1985)

[Bec69]   Beck, J.M.: Distributive laws. In: Seminar on Triples and Categorical Homology Theory. Lecture Notes in Mathematics, vol. 80, pp. 119–140. Springer (1969)

[Con71]   Conway, J.H.: Regular Algebra and Finite Machines. Chapman and Hall (1971)

[Lin69]   Linton, F.E.J.: Coequalizers in categories of algebras. In: Seminar on Triples and Categorical Homology Theory. Lecture Notes in Mathematics, vol. 80, pp. 75–90. Springer (1969)

[Mac98]   Mac Lane, S.: Categories for the Working Mathematician, 2nd edn. Springer (1998)

[NF12]   Nishizawa, K., Furusawa, H.: Ideal Completion of Join Semilattics over T-Algebra. Bulletin of Tottori University of Environmental Studies 9, 91–103 (2012)

[Par69]   Pare, R.: Absolute coequalizers. In: Category Theory, Homology Theory and Their Applications. Lecture Notes in Mathematics, vol. 86, pp. 132–145. Springer (1969)

[Str72]   Street, R.: The formal theory of monads. Journal of Pure and Applied Algebra 2, 149–168 (1972)

Higher-Order Arrow Categories

Michael Winter*

Department of Computer Science,
Brock University,
St. Catharines, Ontario, Canada, L2S 3A1
mwinter@brocku.ca

Abstract. Arrow and Goguen categories were introduced as a suitable
categorical and algebraic description of $\mathcal{L}$-fuzzy relations, i.e., of relations
using membership values from an arbitrary complete Heyting algebra $\mathcal{L}$
instead of truth values or elements from the unit interval $[0, 1]$. Higher-
order fuzziness focuses on sets or relations that use membership values
that are fuzzy themselves. Fuzzy membership values are functions that
assign to a each membership value a degree up to which the value is considered to be the membership degree of the element in question. In this
paper we want to extend the theory of arrow categories to higher-order
fuzziness. We will show that the arrow category of type $(n+1)$-fuzziness
is in fact the Kleisli category over the category of type n-fuzziness for a
suitable monad.

1 Introduction

The theory of relation algebras and allegories or Dedekind categories [3,9,10],
in particular, provides an algebraic framework to reason about relations. These
theories do not only cover binary relations between sets. They are also suitable
to reason about so-called $\mathcal{L}$-relations. Such a relation uses multiple degrees of
membership from a lattice $\mathcal{L}$ for every pair. Formally, such a relation R between
sets A and B is a function $R : A \times B \to \mathcal{L}$. However, not all aspects of $\mathcal{L}$-relations
can be expressed in Dedekind categories. For example, an $\mathcal{L}$-fuzzy relation can be
crisp, i.e., the membership value of every pair is either 0 (smallest element of $\mathcal{L}$)
or 1 (greatest element of $\mathcal{L}$). Even though several abstract notions of crispness in
Dedekind categories have been proposed [4,7,8], it was shown that this property
cannot be expressed in the language of allegories or Dedekind categories [15,21].
Therefore, Goguen and arrow categories were introduced adding two additional
operations to the theory of Dedekind categories. Several papers covered the
theory of those categories including applications to fuzzy controllers [5,16,17,18].

In this paper we are interested in extending the theory of arrow categories to
higher-order fuzziness. A higher-order, or type-2, fuzzy set or relation uses fuzzy
membership values, i.e., the degree of membership is itself fuzzy. This is modeled
by using functions $\mathcal{L} \to \mathcal{L}$ as the underlying lattice. Such a function provides for

---

* The author gratefully acknowledges support from the Natural Sciences and Engineering Research Council of Canada.

every degree of membership a degree up to which this is considered to be the actual membership value. Relations based on those functions also establish an arrow category, of course. We are interested in the relationship between type-1 ($\mathcal{L}$-relations) and type-2 ($\mathcal{L} \to \mathcal{L}$-relations) fuzziness. In our approach we use the extension of an object A which is an abstract version of the Cartesian product of A and the underlying lattice $\mathcal{L}$ [20]. It turns out that type-2 fuzziness can be defined as an arrow category that is isomorphic to the Kleisli category induced by a product functor based on the extension of objects. These concepts play an important role in higher-order fuzzy controllers in which the type reducer transforms type-2 sets or relations into type-1 entities.

The remainder of this paper is organized as follows. In Section 2 we will provide the required background on Dedekind and arrow categories. Section 3 recalls a construction introduced in [20], called the extension of an object. This construction will play an important role in our approach to higher-order fuzziness. In Section 4 we will show that extension induces a monad structure. It is also shown that the corresponding Kleisli category is an arrow category. This category is used in Section 5 in order to define an abstract notion of a type-2 arrow category. Finally, Section 4 provides a conclusion and an outlook to future work.

2 Dedekind and Arrow Categories

In this section we want to recall some basic notions from category, allegory theory and the theory of arrow categories. For more details we refer the reader to [3,21].

We will write $R : A \to B$ to indicate that a morphism R of a category $\mathcal{R}$ has source A and target B. We will use ; to denote composition in a category, which has to be read from left to right, i.e., $R; S$ means R first, and then S. The identity morphism on A is denoted $\mathbb{I}_A$.

Suppose $\mathcal{C}$ is a category. Then a monad on $\mathcal{C}$ is a triple (F, η, μ) consisting of a endo-functor $F : \mathcal{C} \to \mathcal{C}$ and two natural transformations $\eta : I \to F$, i.e., from the identity functor to F, and $\mu : F^2 \to F$, i.e., from the functor obtained by applying F twice to F, so that the following two diagrams commute:

$$
\begin{array}{ccc}
F(F(F(A))) & \xrightarrow{\mu_{F(A)}} & F(F(A)) \\
\downarrow{\scriptstyle F(\mu_A)} & & \downarrow{\scriptstyle \mu_A} \\
F(F(A)) & \xrightarrow{\mu_A} & F(A)
\end{array}
\qquad
\begin{array}{ccccc}
F(A) & \xrightarrow{F(\eta_A)} & F(F(A)) & \xleftarrow{\eta_{F(A)}} & F(F(A)) \\
& {\scriptstyle \mathbb{I}_{F(A)}} \searrow & \downarrow{\scriptstyle \mu_A} & \swarrow {\scriptstyle \mathbb{I}_{F(A)}} & \\
& & F(A) & &
\end{array}
$$

Monads allow one to define new categories based on the additional behavior encoded in the functor. The Kleisli category $\mathcal{C}_F$ has the same objects as $\mathcal{C}$. A morphism in $\mathcal{C}_F$ from A to B is a morphism from A to $F(B)$ in $\mathcal{C}$. η acts as an identity for the composition $R;_F S = R; F(S); \mu$.

Now we want to recall some fundamentals on Dedekind categories [9,10]. Categories of this type are called locally complete division allegories in [3].

Definition 1. *A Dedekind category $\mathcal{R}$ is a category satisfying the following:*

1. *For all objects A and B the collection $\mathcal{R}[A, B]$ is a complete Heyting algebra. Meet, join, the induced ordering, the least and the greatest element are denoted by $\sqcap, \sqcup, \sqsubseteq, \bot\!\!\!\bot_{AB}, \top\!\!\!\top_{AB}$, respectively.*
2. *There is a monotone operation $\breve{}$ (called converse) mapping a relation $Q : A \to B$ to $Q^{\smallsmile} : B \to A$ such that for all relations $Q : A \to B$ and $R : B \to C$ the following holds: $(Q; R)^{\smallsmile} = R^{\smallsmile}; Q^{\smallsmile}$ and $(Q^{\smallsmile})^{\smallsmile} = Q$.*
3. *For all relations $Q : A \to B, R : B \to C$ and $S : A \to C$ the modular law $(Q; R) \sqcap S \sqsubseteq Q; (R \sqcap (Q^{\smallsmile}; S))$ holds.*
4. *For all relations $R : B \to C$ and $S : A \to C$ there is a relation $S/R : A \to B$ (called the left residual of S and R) such that for all $X : A \to B$ the following holds: $X; R \sqsubseteq S \iff X \sqsubseteq S/R$.*

Notice that a complete Heyting algebra has an implication operation $\to$, i.e., we have $X \sqsubseteq Q \to R$ iff $X \sqcap Q \sqsubseteq R$. We will use the abbreviation $Q \leftrightarrow R = (Q \to R) \sqcap (R \to Q)$ throughout the paper.

Throughout this paper we will use some basic properties of relations such as $\bot\!\!\!\bot_{AB}^{\smallsmile} = \bot\!\!\!\bot_{BA}, \top\!\!\!\top_{AB}^{\smallsmile} = \top\!\!\!\top_{BA}, \mathbb{I}_A^{\smallsmile} = \mathbb{I}_A$, the monotonicity of all operations, and the fact that composition distributes over join from both sides without mentioning.

Notice that we have $\top\!\!\!\top_{AA}; \top\!\!\!\top_{AB} = \top\!\!\!\top_{AB}; \top\!\!\!\top_{BB} = \top\!\!\!\top_{AB}$, but the general equation $\top\!\!\!\top_{AB}; \top\!\!\!\top_{BC} = \top\!\!\!\top_{AC}$ does not necessarily hold [14]. If it does hold for all objects A, B and C, then we call the Dedekind category uniform.

An important class of relations is given by maps.

Definition 2. *Let $\mathcal{R}$ be a Dedekind category. The a relation $Q : A \to B$ is called*

1. *univalent (or partial function) iff $Q^{\smallsmile}; Q \sqsubseteq \mathbb{I}_B$,*
2. *total iff $\mathbb{I}_A \sqsubseteq Q; Q^{\smallsmile}$,*
3. *injective iff $Q^{\smallsmile}$ is univalent,*
4. *surjective iff $Q^{\smallsmile}$ is total,*
5. *a map iff Q is total and univalent.*

It is well-known that Q is total iff $Q; \top\!\!\!\top_{BC} = \top\!\!\!\top_{AC}$. We will use this and the corresponding property for surjective relations without mentioning.

In the remainder of the paper we will often use lower case letter to indicate that a relation is a map, i.e., we will use f if f is a map. In the following lemma we have summarized some important properties of mappings and univalent relations. As above a proof can be found in [2,11,12,13,14].

Lemma 1. *Let $\mathcal{R}$ be a Dedekind category. Then we have for all $Q : A \to B, R : A \to C, S, T : B \to C$ and maps $f : B \to C$*

1. *$Q; f \sqsubseteq R$ iff $Q \sqsubseteq R; f^{\smallsmile}$,*
2. *if Q is univalent, then $Q; (S \sqcap T) = Q; S \sqcap Q; T$,*
3. *if S is univalent, then $(Q \sqcap R; S^{\smallsmile}); S = Q; S \sqcap R$.*

A relator is a functor between Dedekind categories that is monotonic (with respect to $\sqsubseteq$) and preserves converse. Relators generalize functors on the subcategory of mappings to the whole Dedekind category in a natural way.

A unit 1 is an object of a Dedekind category so that $\mathbb{I}_1 = \mathbb{T}_{11}$ and $\mathbb{T}_{A1}$ is total for all objects A. A unit is an abstract version of a singleton set, and, hence, the relational version of a terminal object. In the subcategory of mappings a unit becomes a terminal object. This immediately shows that a unit is unique up to isomorphism.

The abstract version of a cartesian product is given by a relational product. Notice that a relational product is not a categorical product within the full Dedekind category. However, it is a categorical product in the subcategory of maps.

Definition 3. *The relational product of two objects A and B is an object $A \times B$ together with two relations $\pi : A \times B \to A$ and $\rho : A \times B \to B$ so that the following equations hold*

$$\pi^{\smile}; \pi \sqsubseteq \mathbb{I}_A, \qquad \rho^{\smile}; \rho \sqsubseteq \mathbb{I}_B, \qquad \pi^{\smile}; \rho = \mathbb{T}_{AB}, \qquad \pi; \pi^{\smile} \sqcap \rho; \rho^{\smile} = \mathbb{I}_{A \times B}.$$

A Dedekind category has products if the relational product for each pair of objects exists.

We will use the abbreviations $\langle Q, R \rangle = Q; \pi^{\smile} \sqcap R; \rho^{\smile}$ and $S \times T = \langle \pi; S, \rho; T \rangle$, i.e., $S \times T = \pi; S; \pi^{\smile} \sqcap \rho; T; \rho^{\smile}$. Notice that the equations

$$\langle Q, R \rangle; (S \times T) = \langle Q; S, R; T \rangle, \qquad \langle Q, R \rangle; \langle U, V \rangle^{\smile} = Q; U^{\smile} \sqcap R; V^{\smile}$$

only hold under certain assumptions. This fact is known as the sharpness (or unsharpness) problem of relational products.

Given a complete Heyting algebra $\mathcal{L}$ an $\mathcal{L}$-relation R between to sets A and B is a function $R : A \times B \to \mathcal{L}$. The values in $\mathcal{L}$ serve as degree of membership, i.e., they indicate the degree of relationship between two elements from A and B. Notice that regular binary relations between sets are a special case of $\mathcal{L}$-relations where $\mathcal{L}$ is the set $\mathcal{B} = \{\text{true, false}\}$ of truth values. The collection of all $\mathcal{L}$-relations between sets together with the standard definition of the operations constitutes a Dedekind category, normally denoted by $\mathcal{L}$-Rel. Within such a category the underlying lattice $\mathcal{L}$ of membership values can be identified by the scalar relations on an object.

Definition 4. *A relation $\alpha : A \to A$ is called a scalar on A iff $\alpha \sqsubseteq \mathbb{I}_A$ and $\mathbb{T}_{AA}; \alpha = \alpha; \mathbb{T}_{AA}$.*

The notion of scalars was introduced by Furusawa and Kawahara [7]. It is equivalent to the notion of ideal elements, i.e., relations $R : A \to B$ that satisfy $\mathbb{T}_{AA}; R; \mathbb{T}_{BB} = R$. These relations were introduced by Jónsson and Tarski [6].

The language of Dedekind categories is too weak to grasp the notion of a crisp relation [15,21]. A crisp relation is an $\mathcal{L}$-relation that only uses 0 (least element of $\mathcal{L}$) and 1 (greatest element of $\mathcal{L}$) as membership values. The Dedekind category of crisp relations can be identified with the category of regular binary relations. In order to add the notion of crispness to the theory of Dedekind categories so-called arrow category were introduced [19,21]. The standard down-arrow operation maps an $\mathcal{L}$-relation R to the greatest crisp relation included in R. Analogously, the standard up-arrow operation maps R to the least crisp relation that includes R.

Definition 5. *An arrow category $\mathcal{A}$ is a Dedekind category with $\mathbb{T}_{AB} \neq \mathbb{\bot}_{AB}$ for all objects A and B together with two operations $^\uparrow$ and $^\downarrow$ satisfying the following:*

1. $R^\uparrow, R^\downarrow : A \to B$ for all $R : A \to B$.
2. $(^\uparrow, ^\downarrow)$ is a Galois correspondence, i.e., $Q^\uparrow \sqsubseteq R$ iff $Q \sqsubseteq R^\downarrow$ for all $Q, R : A \to B$.
3. $(R^\smile ; S^\downarrow)^\uparrow = R^{\uparrow\smile} ; S^\downarrow$ for all $R : B \to A$ and $S : B \to C$.
4. If $\alpha \neq \mathbb{\bot}_{AA}$ is a non-zero scalar then $\alpha^\uparrow = \mathbb{I}_A$.
5. $(Q \sqcap R^\downarrow)^\uparrow = Q^\uparrow \sqcap R^\downarrow$ for all $Q, R : A \to B$.

A relation $R : A \to B$ of an arrow category $\mathcal{A}$ is called crisp iff $R^\uparrow = R$. The collection of crisp relations is closed under all operations of a Dedekind category, and, hence, forms a sub-Dedekind category of $\mathcal{A}$.

Arrow categories are always uniform [21]. As a consequence all projections are surjective as the computation

$$\mathbb{T}_{CA} = \mathbb{T}_{CB} ; \mathbb{T}_{BA} = \mathbb{T}_{CB} ; \rho^\smile ; \pi \sqsubseteq \mathbb{T}_{CA \times B} ; \pi$$

shows. In the remainder of the paper we will use these properties without mentioning.

The following lemma lists some further properties of relations in arrow categories that we will be using in the remainder of the paper. A proof can be found in [21].

Lemma 2. *Let $\mathcal{A}$ be an arrow category. Then we have for all $Q, Q_i : A \to B$ for $i \in I$ and $R : B \to C$*

1. $(\bigsqcap_{i \in I} Q_i)^\downarrow = \bigsqcap_{i \in I} Q_i^\downarrow$,
2. $Q^{\smile\downarrow} = Q^{\downarrow\smile}$,
3. *if R is crisp, then $(Q; R)^\uparrow = Q^\uparrow ; R$.*

3   Extension of an Object

The extension $A^\sharp$ of an object A was introduced in [20]. This construction is motivated by pairing each element of A with all membership values from $\mathcal{L}$. In addition to representing the membership values by ideals or scalars this construction also allows to obtain those values as crisp points, i.e., as crisp mappings $p : 1 \to 1^\sharp$. For further details we refer to [20].

Later we will use the extension of an object to define an arrow category of type-2 fuzziness. This will be done by defining a suitable Kleisli category. The whole approach is based on the following simple idea. A type-2 $\mathcal{L}$-relation, i.e., a $\mathcal{L} \to \mathcal{L}$-relation, between the sets A and B is a function $R : A \times B \to (\mathcal{L} \to \mathcal{L})$. It is well-known that such functions are isomorphic to functions from $A \times (B \times \mathcal{L}) \to \mathcal{L}$. Notice that the latter are $\mathcal{L}$-relations from A to $B^{\sharp}$.

Definition 6. *Let A be an object of an arrow category. An object $A^{\sharp}$ together with two relations $\eta_A, \nu_A : A \to A^{\sharp}$ is called the* extension *of A iff*

1. *η_A is crisp,*
2. *$\mathbb{T}_{AA} ; \nu_A = \nu_A$,*
3. *$\eta_A ; \breve{\eta_A} = \mathbb{I}_A$,*
4. *$\breve{\nu_A}^{\sharp} \sqcap \breve{\eta_A} ; \eta_A = \mathbb{I}_{A^{\sharp}}$,*
5. *$Q^{\sharp} ; \breve{\eta_A} = \mathbb{T}_{BA}$ for every relation $Q : B \to A$,*

where $Q^{\sharp} : B \to A^{\sharp}$ is defined by $Q^{\sharp} = ((Q ; \eta_A) \leftrightarrow (\mathbb{T}_{BA} ; \nu_A))^{\downarrow}$.

In order to explain the definition above in more detail we want to provide the concrete implementation of ν_A, η_A and the operation $(.)^{\sharp}$. The relation ν_A relates a value x to all pairs (y, d) by the degree d, and η_A relates x with (x, d) by degree 1. For example, if $A = \{A, B, C\}$ and $\mathcal{L}_3 = \{0, m, 1\}$ is the linear ordering on three elements with $0 \sqsubseteq m \sqsubseteq 1$, then we obtain $A^{\sharp} = A \times \mathcal{L}_3 = \{(A, 0), (A, m), (A, 1), (B, 0), (B, m), (B, 1), (C, 0), (C, m), (C, 1)\}$. The following matrices visualize the relations ν_A and η_A. Recall that the i-th row (resp. column) of the matrix corresponds to the i-th element of A (resp. $A^{\sharp}$). In addition, we added dividers in order to indicate the grouping within $A^{\sharp}$ with respect to one value of A:

$$\nu_A = \begin{pmatrix} 0 & m & 1 & 0 & m & 1 & 0 & m & 1 \\ 0 & m & 1 & 0 & m & 1 & 0 & m & 1 \\ 0 & m & 1 & 0 & m & 1 & 0 & m & 1 \end{pmatrix} \qquad \eta_A = \begin{pmatrix} 1 & 1 & 1 & 0 & 0 & 0 & 0 & 0 & 0 \\ 0 & 0 & 0 & 1 & 1 & 1 & 0 & 0 & 0 \\ 0 & 0 & 0 & 0 & 0 & 0 & 1 & 1 & 1 \end{pmatrix}$$

Now suppose $Q : A \to A$ is the following relation

$$Q = \begin{pmatrix} 0 & m & 1 \\ 1 & m & 0 \\ m & 0 & 0 \end{pmatrix}.$$

The relation $Q^{\sharp} : A \to A^{\sharp}$ is a crisp relation that relates $u \in A$ to a pair $(x, d) \in A^{\sharp}$ iff u and x are related in Q with degree d, i.e., we obtain $Q^{\sharp} : A \to A^{\sharp}$ as

$$Q^{\sharp} = \begin{pmatrix} 1 & 0 & 0 & 0 & 1 & 0 & 0 & 0 & 1 \\ 0 & 0 & 1 & 0 & 1 & 0 & 1 & 0 & 0 \\ 0 & 1 & 0 & 1 & 0 & 0 & 1 & 0 & 0 \end{pmatrix}.$$

For more details on those constructions and the axioms above we refer to [20].

The following theorem was shown in [20] and verifies that $A^{\sharp}$ is indeed a relational product.

Theorem 1. *Let $\mathcal{A}$ be an arrow category with extensions and unit 1. Then the extension $A^{\sharp}$ of A together with the relations $\pi := \eta_{\breve{A}}$ and $\rho := (\nu_{\breve{A}}; \mathbb{T}_{A1})^{\sharp}$ is a relational product of A and $1^{\sharp}$.*

In addition to the theorem above we recognize that both projections are crisp. This will become an important property in defining the arrow operations in the Kleisli category based on the extension of an object.

4   An Arrow Category Based on a Product Monad

It is well-known that an object B with a monoid structure leads to a monad based on the functor $P(X) = X \times B$. The monoid structure is essential to define the monad operations as functions. The neutral element of the monoid is used to define the embedding $\eta_A : A \to P(A)$, and the reduction $\mu_A : P(P(A)) \to P(A)$ uses the monoid operation in order to obtain one B element from the two elements given. If η and μ are not required to be mappings, the monoid structure is not needed, i.e., in the case of relations it is possible to define a monad without requiring any monoid structure on B. In this section we will require that $\mathcal{R}$ is a Dedekind category and L is an object of $\mathcal{R}$ so that the product $A \times L$ for every object A exists. We define an endo-relator $P : \mathcal{R} \to \mathcal{R}$ by $P(X) = X \times L$ and $P(Q) = Q \times \mathbb{I}_L$. Furthermore, we define two morphisms $\eta_A : A \to P(A)$ and $\mu_A : P(P(A)) \to P(A)$ by $\eta_A = \pi^{\smile}$ and $\mu_A = \pi \sqcap \rho; \rho^{\smile}$. Notice that μ_A is not total, and that η_A is not univalent, i.e., both relations are not mappings.

Notice that the assumption that $A \times L$ exists for all L makes the products sharp, i.e., the equations mentioned in Section 2 do hold. This is due to fact that in this case certain additional products exist. For details we refer to [14].

Lemma 3.   *1. μ_A is injective,*
2. $\pi^{\smile}; \mu_A = \mathbb{I}_{P(A)}$,
3. $\rho^{\smile}; \mu_A = \rho^{\smile}$,
4. $\mu_A = \langle \pi; \pi, \pi; \rho \sqcap \rho \rangle$,
5. $\mu_A = \mu_{P(A)}^{\smile}; P(\mu_A; \eta_A^{\smile})$,
6. $\eta_A; P(\eta_A) = \eta_A; \eta_{P(A)}$,
7. $P(\eta_A^{\smile}); \eta_A^{\smile}; \eta_A = P(\eta_A^{\smile}; \eta_A); \eta_{P(A)}^{\smile}$,
8. $P(Q); \mu_B = \pi; Q \sqcap \rho; \rho^{\smile}$ and $\mu_A^{\smile}; P(Q); \mu_B = Q \sqcap \rho; \rho^{\smile}$,
9. $P(Q) \sqcap P(R) = P(Q \sqcap R)$.

Proof.   1. The assertion follows immediately from

$$\mu_A; \mu_A^{\smile} = (\pi \sqcap \rho; \rho^{\smile}); (\pi^{\smile} \sqcap \rho; \rho^{\smile})$$
$$\sqsubseteq \pi; \pi^{\smile} \sqcap \rho; \rho^{\smile} \rho; \rho^{\smile}$$
$$\sqsubseteq \pi; \pi^{\smile} \sqcap \rho; \rho^{\smile}$$
$$= \mathbb{I}_{P(A)}$$

2. We obtain

$$\pi^{\smile}; \mu_A = \pi^{\smile}; (\pi \sqcap \rho; \rho^{\smile})$$
$$= \mathbb{I}_A \sqcap \pi^{\smile}; \rho; \rho^{\smile} \qquad \text{Lemma 1(3)}$$
$$= \mathbb{I}_A. \qquad \pi^{\smile}; \rho = \mathbb{T}_{AB} \text{ and } \rho \text{ total}$$

3. Similar to the previous case we get

$$\rho^{\smile}; \mu_A = \rho^{\smile}; (\pi \sqcap \rho; \rho^{\smile})$$
$$= \rho^{\smile}; \pi \sqcap \rho^{\smile} \qquad \text{Lemma 1(3)}$$
$$= \rho^{\smile} \qquad \rho^{\smile}; \pi = \mathbb{T}_{BA}$$

4. From the computation

$$\mu_A = \pi \sqcap \rho; \rho^{\smile}$$
$$= \pi; (\pi; \pi^{\smile} \sqcap \rho; \rho^{\smile}) \sqcap \rho; \rho^{\smile}$$
$$= \pi; \pi; \pi^{\smile} \sqcap \pi; \rho; \rho^{\smile} \sqcap \rho; \rho^{\smile} \qquad \text{Lemma 1(2)}$$
$$= \pi; \pi; \pi^{\smile} \sqcap (\pi; \rho \sqcap \rho); \rho^{\smile} \qquad \text{Lemma 1(2)}$$
$$= \langle \pi; \pi, \pi; \rho \sqcap \rho \rangle$$

we obtain the assertion.

5. We immediately compute

$$\mu_{P(A)}^{\smile}; P(\mu_A; \eta_A^{\smile}) = (\pi^{\smile} \sqcap \rho; \rho^{\smile}); ((\mu_A; \eta_A^{\smile}) \times \mathbb{I}_L)$$
$$= \langle \mathbb{I}_{P(P(A))}, \rho \rangle; ((\mu_A; \eta_A^{\smile}) \times \mathbb{I}_L)$$
$$= \langle \mu_A; \eta_A^{\smile}, \rho \rangle$$
$$= \mu_A; \pi; \pi^{\smile} \sqcap \rho; \rho^{\smile}$$
$$= \mu_A; (\pi; \pi^{\smile} \sqcap \mu_A^{\smile}; \rho; \rho^{\smile}) \qquad \text{(1) and Lemma 1(2)}$$
$$= \mu_A; (\pi; \pi^{\smile} \sqcap \rho; \rho^{\smile}) \qquad \text{(3)}$$
$$= \mu_A.$$

6. Consider the following computation

$$\eta_A; P(\eta_A) = \pi^{\smile}; (\pi; \pi^{\smile}; \pi^{\smile} \sqcap \rho; \rho^{\smile})$$
$$= \pi^{\smile}; \pi^{\smile} \sqcap \pi^{\smile}; \rho; \rho^{\smile} \qquad \text{Lemma 1(3)}$$
$$= \pi^{\smile}; \pi^{\smile} \qquad \rho \text{ total}$$
$$= \eta_A; \eta_{P(A)}.$$

7. We calculate

$$P(\eta_A^\smile); \eta_A^\smile; \eta_A = (\pi; \pi; \pi^\smile \sqcap \rho; \rho^\smile); \pi; \pi^\smile$$

$$= (\pi; \pi \sqcap \rho; \rho^\smile; \pi); \pi^\smile \qquad \text{Lemma 1(3)}$$

$$= \pi; \pi; \pi^\smile \qquad\qquad \rho \text{ total}$$

$$= \pi; \pi; \pi^\smile \sqcap \rho; \rho^\smile; \pi \qquad \rho \text{ total}$$

$$= (\pi; \pi; \pi^\smile; \pi^\smile \sqcap \rho; \rho^\smile); \pi \qquad \text{Lemma 1(3)}$$

$$= P(\eta_A^\smile; \eta_A); \eta_{P(A)}^\smile.$$

8. We only show the second assertion. The other property follows analogously. We obtain

$$\mu_A^\smile; P(Q); \mu_B = (\pi^\smile \sqcap \rho; \rho^\smile); (Q \times \mathbb{I}_L); (\pi \sqcap \rho; \rho^\smile)$$

$$= \langle \mathbb{I}_{P(A)}, \rho \rangle; (Q \times \mathbb{I}_L); \langle \mathbb{I}_{P(A)}, \rho \rangle^\smile$$

$$= Q \sqcap \rho; \rho^\smile.$$

9. We have

$$P(Q) \sqcap P(R) = \pi; Q; \pi^\smile \sqcap \pi; R; \pi^\smile \sqcap \rho; \rho^\smile$$

$$= \pi; (Q \sqcap R); \pi^\smile \sqcap \rho; \rho^\smile \qquad \text{Lemma 1(2)}$$

$$= P(Q \sqcap R).$$

This completes the proof.   □

The first theorem of this section shows that the definitions above lead to a monad, and, hence, to a Kleisli category.

Theorem 2. *Let $\mathcal{R}$ be a Dedekind category and L be an object of $\mathcal{R}$ so that the product $A \times L$ exists for every object A. Then (P, η, μ) is a monad on $\mathcal{R}$.*

Proof. First of all, we verify that η is a natural transformation by calculating

$$\eta_A; P(Q) = \pi^\smile; (Q \times \mathbb{I}_L)$$

$$= Q; \pi^\smile \qquad\qquad \text{Lemma 1(3)}$$

$$= Q; \eta_B.$$

Another computation shows that μ is also a natural transformation

$$\mu_A; P(Q) = \langle \pi; \pi, \pi; \rho \sqcap \rho \rangle; (Q \times \mathbb{I}_L) \qquad \text{Lemma 3(4)}$$

$$= \langle \pi; \pi; Q, \pi; \rho \sqcap \rho \rangle \qquad\qquad \text{Lemma 3(4)}$$

$$= \pi; \pi; Q; \pi^\smile \sqcap (\pi; \rho \sqcap \rho); \rho^\smile$$

$$= \pi; (\pi; Q; \pi^\smile \sqcap \rho; \rho^\smile) \sqcap \rho; \rho^\smile \qquad \text{Lemma 1(2) twice}$$

$$= \pi; P(Q) \sqcap \rho; \rho^\smile$$

$$= \pi; P(Q); \pi^\smile; \mu_B \sqcap \rho; \rho^\smile; \mu_B \qquad \text{Lemma 3(2,3)}$$

$$= (\pi; P(Q); \pi^\smile \sqcap \rho; \rho^\smile); \mu_B \qquad \text{Lemma 3(1) and 1(2)}$$

$$= P(P(Q)); \mu_B.$$

The first commuting diagram follows from

$$P(\mu_A); \mu_A = (\pi; \mu_A; \pi^\smile \sqcap \rho; \rho^\smile); \mu_A$$

$$= \pi; \mu_A; \pi^\smile; \mu_A \sqcap \rho; \rho^\smile; \mu_A \qquad \text{Lemma 3(1) and 1(2)}$$

$$= \pi; \mu_A \sqcap \rho; \rho^\smile \qquad \text{Lemma 3(2,3)}$$

$$= \pi; (\pi \sqcap \rho; \rho^\smile) \sqcap \rho; \rho^\smile$$

$$= \pi; \pi \sqcap (\pi; \rho \sqcap \rho); \rho^\smile \qquad \text{Lemma 1(2) twice}$$

$$= \pi; \pi; \pi^\smile; \mu_A \sqcap (\pi; \rho \sqcap \rho); \rho^\smile; \mu_A \qquad \text{Lemma 3(2,3)}$$

$$= (\pi; \pi; \pi^\smile \sqcap (\pi; \rho \sqcap \rho); \rho^\smile); \mu_A \qquad \text{Lemma 3(1) and 1(2)}$$

$$= \langle \pi; \pi, \pi; \rho \sqcap \rho \rangle; \mu_A$$

$$= \mu_{P(A)}; \mu_A. \qquad \text{Lemma 3(4)}$$

Finally, the following computations

$$P(\eta_A); \mu_A = (\pi; \pi^\smile; \pi^\smile \sqcap \rho; \rho^\smile); \mu_A$$

$$= \pi; \pi^\smile; \pi^\smile; \mu_A \sqcap \rho; \rho^\smile; \mu_A \qquad \text{Lemma 3(1) and 1(2)}$$

$$= \pi; \pi^\smile \sqcap \rho; \rho^\smile \qquad \text{Lemma 3(2,3)}$$

$$= \mathbb{I}_{P(A)}$$

$$\eta_{P(A)}; \mu_A = \pi^\smile; \mu_A$$

$$= \mathbb{I}_{P(A)} \qquad \text{Lemma 3(2)}$$

verify that the second diagram commutes as well. $\qquad\qquad\qquad \square$

Recall that the composition in the Kleisli category $\mathcal{R}_P$ is given by $Q;_P R = Q; P(R); \mu_C$ with η as identity. This Kleisli category can be made into a Dedekind category by using the meet and join operation from $\mathcal{R}$ and the following definitions of a converse and residual operation:

$$Q^\cup = \eta_B; \mu_B^\smile; P(Q^\smile), \qquad S/_P R = S/(P(R); \mu_C).$$

Theorem 3. *Let $\mathcal{R}$ be a Dedekind category and L be an object of $\mathcal{R}$ so that the product $A \times L$ exists for every object A. Then the Kleisli category $\mathcal{R}_P$ together with the operations defined above forms a Dedekind category.*

Proof. First of all the converse operation is monotonic because P is a relator. Furthermore, we have

$$(Q;_P R)^\cup$$

$$= \eta_C; \mu_C^\smile; P((Q;_P R)^\smile)$$

$$= \eta_C; \mu_C^\smile; P(\mu_C^\smile; P(R^\smile); Q^\smile)$$

$$= \eta_C; \mu_C^\smile; P(\mu_C^\smile); P(P(R^\smile)); P(Q^\smile)$$

$$= \eta_C; \mu_C^\smile; \mu_{P(C)}^\smile; P(P(R^\smile)); P(Q^\smile) \qquad \text{Monad property}$$

$$= \eta_C; \mu_C^{\smile}; P(R^{\smile}); \mu_B^{\smile}; P(Q^{\smile}) \qquad \text{μ natural transformation}$$

$$= \eta_C; \mu_C^{\smile}; P(R^{\smile}); P(\eta_B; \mu_B^{\smile}); \mu_B^{\smile}; P(Q^{\smile}) \qquad \text{Lemma 3(5)}$$

$$= \eta_C; \mu_C^{\smile}; P(R^{\smile}); P(\eta_B; \mu_B^{\smile}); P(P(Q^{\smile})); \mu_A^{\smile} \qquad \text{μ natural transformation}$$

$$= R^{\sqcup}; P(Q^{\sqcup}); \mu_A^{\smile}$$

$$= R^{\sqcup};_P Q^{\sqcup}.$$

The final property of a converse operation is shown by

$$Q^{\sqcup\sqcup} = \eta_A; \mu_A^{\smile}; P(Q^{\sqcup\smile})$$

$$= \eta_A; \mu_A^{\smile}; P(P(Q)); P(\mu_B; \eta_B^{\smile})$$

$$= \eta_A; P(Q); \mu_{P(B)}^{\smile}; P(\mu_B; \eta_B^{\smile}) \qquad \text{μ natural transformation}$$

$$= Q; \eta_B; \mu_{P(B)}^{\smile}; P(\mu_B; \eta_B^{\smile}) \qquad \text{η natural transformation}$$

$$= Q; \eta_B; \mu_B \qquad \text{Lemma 3(5)}$$

$$= Q. \qquad \text{Lemma 3(2)}$$

The next computation verifies the modular law of Dedekind categories

$$Q;_P R \sqcap S$$

$$= Q; P(R); \mu_C \sqcap S$$

$$\sqsubseteq Q; (P(R); \mu_C \sqcap Q^{\smile}; S)$$

$$= Q; (\pi; R \sqcap \rho; \rho^{\smile} \sqcap Q^{\smile}; S) \qquad \text{Lemma 3(8)}$$

$$= Q; (\pi; R \sqcap \rho; \rho^{\smile} \sqcap Q^{\smile}; S \sqcap \rho; \rho^{\smile})$$

$$= Q; (P(R); \mu_C \sqcap \mu_B^{\smile}; P(Q^{\smile}; S); \mu_C) \qquad \text{Lemma 3(8)}$$

$$= Q; (P(R); \mu_C \sqcap \mu_B^{\smile}; P(Q^{\smile}; S); \mu_{P(C)}^{\smile}; P(\mu_C); \mu_C) \qquad \text{Lemma 3(8)}$$

$$= Q; (P(R); \mu_C \sqcap \mu_B^{\smile}; \mu_{P(B)}^{\smile}; P(P(Q^{\smile}; S)); P(\mu_C); \mu_C) \qquad \text{μ nat. trans.}$$

$$= Q; (P(R); \mu_C \sqcap P(\eta_B; \mu_B^{\smile}); \mu_{P(B)}; \mu_{P(B)}^{\smile}; P(P(Q^{\smile}; S)); P(\mu_C); \mu_C)$$

$$\qquad \text{Lemma 3(5)}$$

$$\sqsubseteq Q; (P(R); \mu_C \sqcap P(\eta_B; \mu_B^{\smile}); P(Q^{\smile}; S); \mu_C); \mu_C) \qquad \text{Lemma 3(1)}$$

$$= Q; (P(R) \sqcap P(\eta_B; \mu_B^{\smile}; P(Q^{\smile}; S); \mu_C)); \mu_C \qquad \text{Lemma 3(1) \& 1(2)}$$

$$= Q; (P(R) \sqcap P(Q^{\sqcup};_P S)); \mu_C$$

$$= Q; P(R \sqcap Q^{\sqcup};_P S); \mu_C \qquad \text{Lemma 3(9)}$$

$$= Q;_P (R \sqcap Q^{\sqcup};_P S).$$

Finally, the computation

$$X;_P R \sqsubseteq S \iff X; P(R); \mu_C \sqsubseteq S$$

$$\iff X \sqsubseteq S/(P(R); \mu_C)$$

$$\iff X \sqsubseteq S/_P R$$

verifies that the residual does exist. $\qquad\qquad\qquad\qquad\qquad\qquad\qquad\qquad$ □

Our next goal is to show that $\mathcal{R}_P$ is actually an arrow category. Therefore, we define

$$Q^{\Uparrow} = (Q; \eta_B^{\smallsmile}; \eta_B)^{\uparrow}, \qquad Q^{\Downarrow} = (Q/(\eta_B^{\smallsmile}; \eta_B))^{\downarrow}.$$

In order to be able to proof the following theorem we have to require that the projections are crisp, i.e., that $\pi^{\uparrow} = \pi$ and $\rho^{\uparrow} = \rho$. In this case we will call the product crisp. Notice that this requirement is not necessarily an additional assumption [16].

Theorem 4. *Let $\mathcal{A}$ be an arrow category and L be an object of $\mathcal{A}$ so that a crisp product $A \times L$ exists for every object A. Then the Dedekind category $\mathcal{A}_P$ together with the operations defined above forms an arrow category.*

Proof. First notice that $\eta = \pi^{\smallsmile}$ is crisp. Therefore, we have

$$\begin{aligned}
Q^{\Uparrow} \sqsubseteq R &\iff (Q; \eta_B^{\smallsmile}; \eta_B)^{\uparrow} \sqsubseteq R \\
&\iff Q^{\uparrow}; \eta_B^{\smallsmile}; \eta_B \sqsubseteq R \qquad\qquad \text{Lemma 2(3)} \\
&\iff Q^{\uparrow} \sqsubseteq R/(\eta_B^{\smallsmile}; \eta_B) \\
&\iff Q \sqsubseteq (R/(\eta_B^{\smallsmile}; \eta_B))^{\downarrow} \\
&\iff Q \sqsubseteq R^{\Downarrow}.
\end{aligned}$$

In order to show that $(Q^{\cup}; _P R^{\Downarrow})^{\Uparrow} = Q^{\Uparrow\cup}; _P R^{\Downarrow}$ we show $Q^{\cup\Uparrow} = Q^{\Uparrow\cup}$ and $(Q; _P R^{\Downarrow})^{\Uparrow} = Q^{\Uparrow}; _P R^{\Downarrow}$. Notice that $\mathcal{R}_P$ is a Dedekind category so that $\mathbb{I}^{\cup} = \mathbb{I}$, i.e., we have $(*)\ \eta_A; \mu_A^{\smallsmile}; P(\eta_A^{\smallsmile}) = \eta_A$. Now, we conclude

$$\begin{aligned}
Q^{\cup\Uparrow} &= (\eta_B; \mu_B^{\smallsmile}; P(Q^{\smallsmile}); \eta_A^{\smallsmile}; \eta_A)^{\uparrow} \\
&= (\eta_B; (\eta_A; P(Q); \mu_B)^{\smallsmile}; \eta_A)^{\uparrow} \\
&= (\eta_B; Q^{\smallsmile}; \eta_A)^{\uparrow} && \eta \text{ left identity in monad} \\
&= \eta_B; Q^{\uparrow\smallsmile}; \eta_A && \text{Lemma 2(3)} \\
&= \eta_B; \eta_{P(B)}; P(Q^{\uparrow\smallsmile}) && \eta \text{ natural transformation} \\
&= \eta_B; P(\eta_B); P(Q^{\uparrow\smallsmile}) && \text{Lemma 3(6)} \\
&= \eta_B; \mu_B^{\smallsmile}; P(\eta_B^{\smallsmile}); P(\eta_B); P(Q^{\uparrow\smallsmile}) && \text{by } (*) \\
&= \eta_B; \mu_B^{\smallsmile}; P((Q^{\uparrow}; \eta_B^{\smallsmile}; \eta_B)^{\smallsmile}) \\
&= \eta_B; \mu_B^{\smallsmile}; P(Q^{\Uparrow\smallsmile}) && \text{Lemma 2(3)} \\
&= Q^{\Uparrow\cup}.
\end{aligned}$$

In order to show the second property notice that μ is crisp and that $P(Q)$ is crisp for a crisp Q because the projections are crisp. Furthermore, from

$$Q^{\Downarrow}; \eta_{\breve{B}}; \eta_B; \eta_{\breve{B}}; \eta_B = Q^{\Downarrow}; \eta_{\breve{B}}; \eta_B \qquad\qquad \eta_B; \eta_{\breve{B}} = \mathbb{I}_B$$
$$= (Q/(\eta_{\breve{B}}; \eta_B))^{\downarrow}; \eta_{\breve{B}}; \eta_B$$
$$\sqsubseteq (Q/(\eta_{\breve{B}}; \eta_B)); \eta_{\breve{B}}; \eta_B$$
$$\sqsubseteq Q$$

we conclude $Q^{\Downarrow}; \eta_{\breve{B}}; \eta_B \sqsubseteq Q/(\eta_{\breve{B}}; \eta_B)$, and, hence, $Q^{\Downarrow}; \eta_{\breve{B}}; \eta_B \sqsubseteq Q^{\Downarrow}$ since the left-hand side is crisp. This implies $(**)$ $Q^{\Downarrow}; \eta_{\breve{B}}; \eta_B = Q^{\Downarrow}$ since η_B is surjective. Now, we compute

$$
\begin{aligned}
(Q;_P R^{\Downarrow})^{\Uparrow} &= (Q; P(R^{\Downarrow}); \mu_C; \eta_{\breve{C}}; \eta_C)^{\uparrow} \\
&= Q^{\uparrow}; P(R^{\Downarrow}); \mu_C; \eta_{\breve{C}}; \eta_C && \text{see above and Lemma 2(3)} \\
&= Q^{\uparrow}; P(R^{\Downarrow}; \eta_{\breve{C}}; \eta_C); \mu_C; \eta_{\breve{C}}; \eta_C && \text{by } (**) \\
&= Q^{\uparrow}; P(R^{\Downarrow}; \eta_{\breve{C}}); \eta_{\breve{C}}; \eta_C && \text{by } (*) \\
&= Q^{\uparrow}; P(R^{\Downarrow}; \eta_{\breve{C}}; \eta_C); \eta_{\breve{P(C)}} && \text{Lemma 3(7)} \\
&= Q^{\uparrow}; P(R^{\Downarrow}); \eta_{\breve{P(C)}} && \text{by } (**) \\
&= Q^{\uparrow}; \eta_{\breve{B}}; R^{\Downarrow} && \eta \text{ natural transformation} \\
&= Q^{\uparrow}; \eta_{\breve{B}}; R^{\Downarrow}; \eta_{P(C)}; \mu_C && \text{Lemma 3(2)} \\
&= Q^{\uparrow}; \eta_{\breve{B}}; \eta_B; P(R^{\Downarrow}); \mu_C && \eta \text{ natural transformation} \\
&= (Q; \eta_{\breve{B}}; \eta_B)^{\uparrow}; P(R^{\Downarrow}); \mu_C && \text{Lemma 2(3)} \\
&= Q^{\Uparrow};_P R^{\Downarrow}
\end{aligned}
$$

Suppose $\alpha \neq \perp\!\!\!\perp_{AA}$ is a scalar in $\mathcal{A}_P$. We have to show that $\alpha^{\Uparrow} = \eta_A$ because η_A is the identity in the Kleisli category $\mathcal{A}_P$. From the fact that α is a scalar in $\mathcal{A}_P$ we get $\alpha \sqsubseteq \eta_A$ and $\alpha; P(\mathbb{T}_{AP(A)}); \mu_A = \mathbb{T}_{AP(A)}; P(\alpha); \mu_A$. In order to proceed we want to show that $\alpha; \eta_{\breve{A}}$ is a non-zero scalar in $\mathcal{A}$. First of all, we have $\alpha; \eta_{\breve{A}} \sqsubseteq \eta_A; \eta_{\breve{A}} = \mathbb{I}_A$. In addition, $\alpha; \eta_{\breve{A}} = \perp\!\!\!\perp_{AA}$ implies $\alpha \sqsubseteq \alpha; \eta_{\breve{A}}; \eta_A = \perp\!\!\!\perp_{AP(A)}$, a contradiction. Last but not least, we obtain

$$
\begin{aligned}
\alpha; \eta_{\breve{A}}; \mathbb{T}_{AA} &= \alpha; P(\mathbb{T}_{AA}); \eta_{\breve{A}} && \eta \text{ natural transformation} \\
&= \alpha; P(\mathbb{T}_{AP(A)}; \eta_A); \mu_A; \eta_{\breve{A}} && \text{by } (*) \\
&= \alpha; P(\mathbb{T}_{AP(A)}); \mu_A; \eta_{\breve{A}} && \eta_A \text{ surjective} \\
&= \mathbb{T}_{AP(A)}; P(\alpha); \mu_A; \eta_{\breve{A}} && \text{assumption} \\
&= \mathbb{T}_{AA}; \eta_A; P(\alpha); \mu_A; \eta_{\breve{A}} && \eta_A \text{ surjective} \\
&= \mathbb{T}_{AA}; \alpha; \eta_{\breve{A}}. && \eta \text{ left identity in monad}
\end{aligned}
$$

Since $\mathcal{A}$ is an arrow category we conclude $(\alpha; \eta_A^{\smile})^{\uparrow} = \mathbb{I}_A$. From this we compute

$$
\begin{aligned}
\alpha^{\Uparrow} &= (\alpha; \eta_A^{\smile}; \eta_A)^{\uparrow} \\
&= (\alpha; \eta_A^{\smile})^{\uparrow}; \eta_A && \text{Lemma 2(3)} \\
&= \eta_A && \text{see above}
\end{aligned}
$$

In order to show the remaining property we fist compute

$$
\begin{aligned}
((Q \sqcap R^{\Downarrow}); \eta_B^{\smile}; \eta_B)^{\uparrow} &\sqsubseteq (Q; \eta_B^{\smile}; \eta_B \sqcap R^{\Downarrow}; \eta_B^{\smile}; \eta_B)^{\uparrow} \\
&= (Q; \eta_B^{\smile}; \eta_B \sqcap R^{\Downarrow})^{\uparrow}, && \text{by } (**) \\
(Q; \eta_B^{\smile}; \eta_B \sqcap R^{\Downarrow})^{\uparrow} &\sqsubseteq ((Q \sqcap R^{\Downarrow}; \eta_B^{\smile}; \eta_B); \eta_B^{\smile}; \eta_B)^{\uparrow} \\
&= ((Q \sqcap R^{\Downarrow}); \eta_B^{\smile}; \eta_B)^{\uparrow} && \text{by } (**)
\end{aligned}
$$

The following two computations

$$
\begin{aligned}
(Q \sqcap R^{\Downarrow})^{\Uparrow} &= ((Q \sqcap R^{\Downarrow}); \eta_B^{\smile}; \eta_B)^{\uparrow} \\
&= (Q; \eta_B^{\smile}; \eta_B \sqcap R^{\Downarrow})^{\uparrow} && \text{see above} \\
&= (Q; \eta_B^{\smile}; \eta_B)^{\uparrow} \sqcap R^{\Downarrow} && \mathcal{A} \text{ arrow category} \\
&= Q^{\Uparrow} \sqcap R^{\Downarrow},
\end{aligned}
$$

shows the remaining property of an arrow category.   □

5   Higher-Order Arrow Categories

In this section we want to combine the results from the previous sections in order to obtain a suitable notion of a type-2 arrow category. First of all, the combination of Theorem 1 and the Theorems 2-4 implies the following corollary.

Corollary 1. *If $\mathcal{A}$ is an arrow category with extensions and a unit 1, then $\mathcal{A}_P$ is an arrow category.*

We will use this construction in order to define a type-2 arrow category over a ground (arrow) category.

Definition 7. *Let $\mathcal{A}_1$ be an arrow category with extensions and a unit 1. An arrow category $\mathcal{A}_2$ is called type-2 arrow category over $\mathcal{A}_1$ iff $\mathcal{A}_2$ is isomorphic to the Kleisli category $\mathcal{A}_{1P}$.*

It remains to show that the concrete arrow category of type-2 $\mathcal{L}$-relations is indeed a type-2 arrow category over the arrow category of $\mathcal{L}$-relations.

Theorem 5. *The arrow category $(\mathcal{L} \to \mathcal{L})$-Rel is a type-2 arrow category over the arrow category $\mathcal{L}$-Rel.*

Proof. Notice that all categories do have the same objects. Therefore, it is sufficient to provide the required isomorphism φ for relations only. We define $\varphi : ((\mathcal{L} \to \mathcal{L}) - \text{Rel})[A, B] \to (\mathcal{L} - \text{Rel})[A, B \times \mathcal{L}]$ by

$$(\varphi_{AB}(Q))(x, (y, d)) = Q(x, y)(d).$$

An easy verification shows that φ is indeed an isomorphism and maps the operations in $(\mathcal{L} \to \mathcal{L}) - \text{Rel}$ to the corresponding operation in the Kleisli category $\mathcal{L} - \text{Rel}_P$. We only show the case of the up-arrow operation here and omit the other proofs due to lack of space. First we have

$$(\varphi_{AB}(Q); \eta_{\breve{B}}; \eta_B)(x, (y, d)) = \bigsqcup_{e \in \mathcal{L}} \varphi_{AB}(Q)(x, (y, e)) = \bigsqcup_{e \in \mathcal{L}} Q(x, y)(e)$$

so that $(\varphi_{AB}(Q); \eta_{\breve{B}}; \eta_B)(x, (y, d)) \neq 0$ iff $Q(x, y) \neq \overline{0}$ follows where $\overline{0}$ is the constant function returning 0. We conclude

$$
\begin{aligned}
\varphi_{AB}(Q)^{\Uparrow}(x, (y, d)) &= \left\{ \begin{array}{ll} 1 & \text{iff } (\varphi_{AB}(Q); \eta_{\breve{B}}; \eta_B)(x, (y, d)) \neq 0 \\ 0 & \text{otherwise} \end{array} \right\} \\
&= \left\{ \begin{array}{ll} 1 & \text{iff } Q(x, y) \neq \overline{0} \\ 0 & \text{otherwise} \end{array} \right\} \\
&= Q^{\uparrow}(x, y)(d) \\
&= \varphi_{AB}(Q^{\uparrow})(x, (y, d)).
\end{aligned}
$$
$\square$

6   Conclusion and Future Work

In this paper we extended the theory of arrow and Goguen categories to higher-order fuzziness. Future work will concentrate on several aspects of this extended theory. First of all, an adaption of the framework using arrow categories for fuzzy controllers is of interest. Besides some general modifications and generalizations the operations required for the so-called type reduction step are important. We will investigate some general operations based on relational constructions such as order-based operations and relational integration. In addition, the more specific interval based operations are important in practical applications.

Another area for future research will focus on the implementation of this extended theory in an interactive proof system such as Coq in order to develop an environment that can be used for formal verification of software based on fuzzy methods.

References

1. Castillo, O., Melin, P.: Type-2 Fuzzy Logic: Theory and Applications. STUDFUZZ, vol. 223. Springer (2008)
2. Chin L.H., Tarski A.: Distributive and modular laws in the arithmetic of relation algebras. University of California Press, Berkeley and Los Angeles (1951)

3. Freyd, P., Scedrov, A.: Categories, Allegories. North-Holland (1990)
4. Furusawa, H.: Algebraic Formalizations of Fuzzy Relations and Their Representation Theorems. PhD-Thesis, Department of Informatics, Kyushu University, Japan (1998)
5. Furusawa, H., Kawahara, Y., Winter, M.: Dedekind Categories with Cutoff Operators. Fuzzy Sets and Systems 173, 1–24 (2011)
6. Jónsson, B., Tarski, A.: Boolean algebras with operators, I, II. Amer. J. Math. 73, 74, 891-939, 127–162 (1951, 1952)
7. Kawahara, Y., Furusawa, H.: Crispness and Representation Theorems in Dedekind Categories. DOI-TR 143. Kyushu University (1997)
8. Kawahara, Y., Furusawa, H.: An Algebraic Formalization of Fuzzy Relations. Fuzzy Sets and Systems 101, 125–135 (1999)
9. Olivier, J.P., Serrato, D.: Catégories de Dedekind. Morphismes dans les Catégories de Schröder. C.R. Acad. Sci. Paris 290, 939–941 (1980)
10. Olivier, J.P., Serrato, D.: Squares and Rectangles in Relational Categories - Three Cases: Semilattice, Distributive lattice and Boolean Non-unitary. Fuzzy Sets and Systems 72, 167–178 (1995)
11. Schmidt G., Ströhlein T.: Relationen und Graphen. Springer (1989); English version: Relations and Graphs. Discrete Mathematics for Computer Scientists. EATCS Monographs on Theoret. Comput. Sci. Springer (1993)
12. Schmidt, G., Hattensperger, C., Winter, M.: Heterogeneous Relation Algebras. In: Brink, C., Kahl, W., Schmidt, G. (eds.) Relational Methods in Computer Science. Advances in Computer Science. Springer, Vienna (1997)
13. Schmidt, G.: Relational Mathematics. Encyplopedia of Mathematics and its Applications 132 (2011)
14. Winter, M.: Strukturtheorie heterogener Relationenalgebren mit Anwendung auf Nichtdetermismus in Programmiersprachen. Dissertationsverlag NG Kopierladen GmbH, München (1998)
15. Winter, M.: A new Algebraic Approach to L-Fuzzy Relations Convenient to Study Crispness. INS Information Science 139, 233–252 (2001)
16. Winter, M.: Relational Constructions in Goguen Categories. In: de Swart, H. (ed.) Participants Proceedings of the 6th International Seminar on Relational Methods in Computer Science (RelMiCS), pp. 222–236. Katholieke Universiteit Brabant, Tilburg (2001)
17. Winter, M.: Derived Operations in Goguen Categories. TAC Theory and Applications of Categories 10(11), 220–247 (2002)
18. Winter, M.: Representation Theory of Goguen Categories. Fuzzy Sets and Systems 138, 85–126 (2003)
19. Winter, M.: Arrow Categories. Fuzzy Sets and Systems 160, 2893–2909 (2009)
20. Winter, M.: Membership Values in Arrow Categories. Submitted to Fuzzy Sets and Systems (October 2013)
21. Winter, M.: Goguen Categories - A Categorical Approach to L-fuzzy relations. Trends in Logic 25 (2007)

Type-2 Fuzzy Controllers in Arrow Categories

Michael Winter[1,*], Ethan Jackson[1], and Yuki Fujiwara[2]

[1] Department of Computer Science,
Brock University,
St. Catharines, Ontario, Canada, L2S 3A1
{mwinter,ej08ti}@brocku.ca
[2] Kojima Laboratory,
Kobe University,
Kobe, Hyogo, Japan
yuki.fujiwara@kojimalab.com

Abstract. Arrow categories as a suitable categorical and algebraic description of $\mathcal{L}$-fuzzy relations have been used to specify and describe fuzzy controllers in an abstract manner. The theory of arrow categories has also been extended to include higher-order fuzziness. In this paper we use this theory in order to develop an appropriate description of type-2 fuzzy controllers. An overview of the relational representation of a type-1 fuzzy controller is given before discussing the extension to a type-2 controller. We discuss how to model type reduction, an essential component of any type-2 controller. In addition, we provide a number of examples of general type reducers.

1 Introduction

Dedekind categories are a fundamental tool to reason about relations. In addition to the standard model of binary relations these categories, and some related structures such as allegories and relation algebras, also cover $\mathcal{L}$-fuzzy relations, i.e., relations in which every pair of elements is only related up to a certain degree indicated by a membership value from the complete Heyting algebra $\mathcal{L}$. Formally, an $\mathcal{L}$-fuzzy relation R (or $\mathcal{L}$-relation for short) between a set A and a set B is a function $R : A \times B \to \mathcal{L}$. It was shown that certain notions such as crispness cannot be expressed in Dedekind categories. Therefore, the theory of arrow and Goguen categories has been established as an algebraic and categorical framework to reason about these $\mathcal{L}$-fuzzy relations [19].

Type-2 fuzzy sets, introduced by Zadeh [27], use membership degrees that are themselves fuzzy sets, i.e., the lattice of membership degrees is the lattice $\mathcal{L} \to \mathcal{L}$ of functions on the lattice $\mathcal{L}$. This approach allows to describe uncertainty in determining an exact membership value for an element of a fuzzy set. For example, if x is in the type-2 $\mathcal{L}$-fuzzy set A to a degree f, then we only know for each membership degree $d \in \mathcal{L}$ that it is the degree of the membership of x in A

---

* The author gratefully acknowledges support from the Natural Sciences and Engineering Research Council of Canada.

P. Höfner et al. (Eds.): RAMiCS 2014, LNCS 8428, pp. 293–308, 2014.

up to the degree $f(d)$. Since $\mathcal{L} \to \mathcal{L}$ is also a complete Heyting algebra, type-2 $\mathcal{L}$-relations also form an arrow category. The abstract relationship between the arrow category of $\mathcal{L}$-relations and $(\mathcal{L} \to \mathcal{L})$-relations was investigated in [26]. In addition, an abstract definition of a type-2 arrow category over a ground arrow category has been provided.

The theory of arrow and Goguen categories has been studied intensively [5,19,20,21,22,24]. In addition, the theory has been used to model and specify $\mathcal{L}$-fuzzy controllers [23]. This allows formal reasoning about controllers as well as stepwise refinement of a specification towards an implementation. In this paper we want to model type-2 $\mathcal{L}$-fuzzy controllers based on the theory of type-2 arrow categories. We will discuss some differences between the approaches to type-1 vs. type-2 controllers. In particular, we will discuss type reducers and several general methods useful in order to define them.

2   Mathematical Preliminaries

In this section we want to recall some basic notions from lattice, category and allegory theory. For further details we refer to [1] and [4].

We will write $R : A \to B$ to indicate that a morphism R of a category $\mathcal{R}$ has source A and target B. We will use ; to denote composition in a category, which has to be read from left to right, i.e., $R; S$ means R first, and then S. The collection of all morphisms $R : A \to B$ is denoted by $\mathcal{R}[A, B]$. The identity morphism on A is written as $\mathbb{I}_A$.

A distributive lattice $\mathcal{L}$ is called a complete Heyting algebra iff $\mathcal{L}$ is complete and $x \sqcap \bigsqcup M = \bigsqcup_{y \in M} (x \sqcap y)$ holds for all $x \in \mathcal{L}$ and $M \subseteq \mathcal{L}$. Notice that complete Heyting algebras have relative pseudo complements, i.e., for each pair $x, y \in \mathcal{L}$ there is a greatest element $x \to y$ with $x \sqcap (x \to y) \sqsubseteq y$. By $x \leftrightarrow y$ we abbreviate the element $x \to y \sqcap y \to x$.

We will use the framework of Dedekind categories [11,12] throughout this paper as a basic theory of relations. Categories of this type are called locally complete division allegories in [4].

Definition 1. *A Dedekind category $\mathcal{R}$ is a category satisfying the following:*

1. *For all objects A and B the collection $\mathcal{R}[A, B]$ is a complete Heyting algebra. Meet, join, the induced ordering, the least and the greatest element are denoted by $\sqcap, \sqcup, \sqsubseteq, \bot\!\!\bot_{AB}, \top\!\!\top_{AB}$, respectively.*
2. *There is a monotone operation $\smile$ (called converse) mapping a relation $Q : A \to B$ to $Q^\smile : B \to A$ such that for all relations $Q : A \to B$ and $R : B \to C$ the following holds: $(Q; R)^\smile = R^\smile; Q^\smile$ and $(Q^\smile)^\smile = Q$.*
3. *For all relations $Q : A \to B, R : B \to C$ and $S : A \to C$ the modular law $(Q; R) \sqcap S \sqsubseteq Q; (R \sqcap (Q^\smile; S))$ holds.*
4. *For all relations $R : B \to C$ and $S : A \to C$ there is a relation $S/R : A \to B$ (called the left residual of S and R) such that for all $X : A \to B$ the following holds: $X; R \sqsubseteq S \iff X \sqsubseteq S/R$.*

As mentioned in the introduction, the collection of binary relations between sets as well as the collection of $\mathcal{L}$-relations between sets form a Dedekind category. In the remainder of the paper we will sometimes use a matrix representation in order to visualize examples in $\mathcal{L}$-relations. For example, if $R : A \times B \to \mathcal{L}$ with $A = \{a, b, c\}$ and $B = \{1, 2, 3, 4\}$ is a $\mathcal{L}$-relation, then the situation can be visualized by

$$R := \begin{pmatrix} 1 & k & l & 0 \\ 0 & k & m & 0 \\ 0 & 1 & l & m \end{pmatrix} \qquad \mathcal{L} :=$$

The matrix representation should be read as follows. The l in the first row of the matrix indicates that the first element of A, the element a, is in relation R to the third element of B, the element 3, to the degree l.

Based on the residual operation it is possible to define a right residual by $R\backslash S = (S/R)^{\smile}$. This operation is characterized by

$$R; X \sqsubseteq S \iff X \sqsubseteq R\backslash S.$$

A symmetric version of the residuals is given by the symmetric quotient $\mathrm{syQ}(Q, R)$ of two relations. This construction is defined by

$$\mathrm{syQ}(Q, R) = (Q\backslash R) \sqcap (Q^{\smile}/R^{\smile}).$$

An order relation $E : A \to A$ is a relation that is reflexive, transitive, and antisymmetric, i.e., it satisfies $\mathbb{I}_A \sqsubseteq E$, $E; E \sqsubseteq E$, and $E \sqcap E^{\smile} \sqsubseteq \mathbb{I}_A$. For a given relation $X : B \to A$ it is possible to compute the least upper bound of all elements related to one $b \in B$ using the following construction

$$\mathrm{lub}_E(X) = (X^{\smile}\backslash E) \sqcap ((X^{\smile}\backslash E)^{\smile}\backslash E^{\smile}).$$

The greatest lower bounds are computed by reversing the order, i.e., $\mathrm{glb}_E(X) = \mathrm{lub}_{E^{\smile}}(X)$. For more details on these constructions we refer to [14,15,17,18].

Another important class of relations is given by maps.

Definition 2. *Let $\mathcal{R}$ be a Dedekind category. Then a relation $Q : A \to B$ is called*

1. *univalent (or a partial function) iff $Q^{\smile}; Q \sqsubseteq \mathbb{I}_B$,*
2. *total iff $\mathbb{I}_A \sqsubseteq Q; Q^{\smile}$,*
3. *injective iff $Q^{\smile}$ is univalent,*
4. *surjective iff $Q^{\smile}$ is total,*
5. *a map iff Q is total and univalent.*

A relator is a functor between Dedekind categories that is monotonic (with respect to $\sqsubseteq$) and preserves converses. Relators generalize functors on the subcategory of mappings to the whole Dedekind category in a natural way.

The relational version of a terminal object is a unit. A unit 1 is an object of a Dedekind category so that $\mathbb{I}_1 = \mathbb{T}_{11}$ and $\mathbb{T}_{A1}$ is total for all objects A. Notice that a unit is a terminal object in the subcategory of mappings.

The abstract version of a Cartesian product is given by a relational product. Similar to a unit a relational product is the relational version of the well-known categorical construction of a product.

Definition 3. *The relational product of two objects A and B is an object $A \times B$ together with two relations $\pi : A \times B \to A$ and $\rho : A \times B \to B$ so that the following equations hold*

$$\pi^{\smile};\pi \sqsubseteq \mathbb{I}_A, \quad \rho^{\smile};\rho \sqsubseteq \mathbb{I}_B, \quad \pi^{\smile};\rho = \mathbb{T}_{AB}, \quad \pi;\pi^{\smile} \sqcap \rho;\rho^{\smile} = \mathbb{I}_{A \times B}.$$

Another important construction is based on forming the disjoint union of sets.

Definition 4. *Let A and B be objects of a Dedekind category. An object $A + B$ together with two relations $\iota : A \to A + B$ and $\kappa : B \to A + B$ is called a relational sum of A and B iff*

$$\iota;\iota^{\smile} = \mathbb{I}_A, \quad \kappa;\kappa^{\smile} = \mathbb{I}_B, \quad \iota;\kappa^{\smile} = {\perp\!\!\!\perp}_{AB}, \quad \iota^{\smile};\iota \sqcup \kappa^{\smile};\kappa = \mathbb{I}_{A+B}.$$

Last but not least, a relational power is an abstract version of powersets, i.e., the set of all subsets of a set.

Definition 5. *Let A be an object of a Dedekind category. An object $\mathcal{P}(A)$ together with a relation $\varepsilon : A \to \mathcal{P}(A)$ is called a relational power of A iff*

$$\mathrm{syQ}(\varepsilon, \varepsilon) \sqsubseteq \mathbb{I}_{\mathcal{P}(A)}, \quad \mathrm{syQ}(R^{\smile}, \varepsilon) \text{ is total for every } R : B \to A.$$

Notice that $\mathrm{syQ}(R^{\smile}, \varepsilon)$ is, in fact, a map that maps x to the set of elements that x is related to in R.

In a Dedekind category one can identify the underlying lattice $\mathcal{L}$ of membership values by the scalar relations on an object.

Definition 6. *A relation $\alpha : A \to A$ is called a scalar on A iff $\alpha \sqsubseteq \mathbb{I}_A$ and $\mathbb{T}_{AA};\alpha = \alpha;\mathbb{T}_{AA}$.*

The notion of scalars was introduced by Furusawa and Kawahara [9] and is equivalent to the notion of ideals, i.e., relations $R : A \to B$ that satisfy $\mathbb{T}_{AA};R;\mathbb{T}_{BB} = R$, which were introduced by Jónsson and Tarski [8].

The next definition introduces arrow categories, i.e., the basic theory for $\mathcal{L}$-relations.

Definition 7. *An arrow category $\mathcal{A}$ is a Dedekind category with $\mathbb{T}_{AB} \neq {\perp\!\!\!\perp}_{AB}$ for all A, B and two operations $\uparrow$ and $\downarrow$ satisfying:*

1. *$R^{\uparrow}, R^{\downarrow} : A \to B$ for all $R : A \to B$*
2. *$(\uparrow, \downarrow)$ is a Galois correspondence, i.e., we have $Q^{\uparrow} \sqsubseteq R$ iff $Q \sqsubseteq R^{\downarrow}$ for all $Q, R : A \to B$.*

3. $(R^{\smile}; S^{\downarrow})^{\uparrow} = R^{\uparrow\smile}; S^{\downarrow}$ for all $R : B \to A$ and $S : B \to C$
4. $(Q \sqcap R^{\downarrow})^{\uparrow} = Q^{\uparrow} \sqcap R^{\downarrow}$ for all $Q, R : A \to B$
5. If $\alpha_A \neq \sqcup\!\sqcup_{AA}$ is a non-zero scalar then $\alpha_A^{\uparrow} = \mathbb{I}_A$.

A relation that satisfies $R^{\uparrow} = R$, or equivalently $R^{\downarrow} = R$, is called crisp. Notice that the complete Heyting algebra of scalar relations on each object are isomorphic.

If α is scalar (cf. Definition 6), then the relation $(\alpha \backslash R)^{\downarrow}$ is called the α-cut of R. For $\mathcal{L}$-relations this construction can be characterized as follows. The elements x and y are in the (crisp) relation $(\alpha \backslash R)^{\downarrow}$ iff x and y are related in R by a degree greater or equal to α.

In fuzzy theory t-norms and t-conorms are essential for defining new operations for fuzzy sets or relations. The corresponding notion for $\mathcal{L}$-fuzzy relations is given by complete lattice-ordered semigroups introduced in [6]. We refer to that paper for more details on those operations. For the current paper we only need to know how to define new composition, join, and meet operations for relations based on such a semigroup operation. These operations play an important role in any implementation of a fuzzy controller.

Definition 8. Let Q, R be relations, $\otimes \in \{\sqcap, \sqcup, ;\}$ such that $Q \otimes R$ is defined, and $*$ the operation of a complete lattice-ordered semigroup on the set of scalar relations. Then we define

$$Q \otimes_* R := \bigsqcup_{\alpha, \beta \ scalars} (\alpha * \beta); ((\alpha \backslash Q)^{\downarrow} \otimes (\beta \backslash R)^{\downarrow}).$$

Notice that the abstract definition above corresponds in the case of $\mathcal{L}$-relations to the well-known definitions. In particular, we have for $\mathcal{L}$-relations $Q, R : A \to B$ and $S : B \to C$ the following property

$$(Q \sqcap_* R)(x, y) = Q(x, y) * R(x, y),$$

$$(Q ;_* S)(x, z) = \bigsqcup_{y \in B} Q(x, y) * S(y, z).$$

A type-2 $\mathcal{L}$-relation Q between A and B uses functions from $\mathcal{L}$ to $\mathcal{L}$ as membership values, i.e., is a function $Q : A \times B \to (\mathcal{L} \to \mathcal{L})$. Such a relation corresponds to an $\mathcal{L}$-relation $R : A \times (B \times \mathcal{L}) \to \mathcal{L}$ by $Q(x, y)(d_1) = d_2$ iff $R(x, (y, d_1)) = d_2$. This correspondence motivates the abstract treatment of higher-order fuzziness in the context of arrow categories. We need an abstract version of product $B \times \mathcal{L}$. This is provided by the notion of an extension $B^{\sharp}$ of an object B.

Definition 9. Let A be an object of an arrow category. An object $A^{\sharp}$ together with two relations $\eta_A, \nu_A : A \to A^{\sharp}$ is called the extension of A iff

1. η_A is crisp,
2. $\mathbb{T}_{AA}; \nu_A = \nu_A$,
3. $\eta_A; \eta_A^{\smile} = \mathbb{I}_A$,

4. $\nu_A^{\sharp}{}^{\smile} \sqcap \eta_A{}^{\smile}; \eta_A = \mathbb{I}_{A^{\sharp}}$,

5. $Q^{\sharp}; \eta_A{}^{\smile} = \mathbb{T}_{BA}$ *for every relation* $Q : B \to A$,

where $Q^{\sharp} = ((Q; \eta_A) \leftrightarrow (\mathbb{T}_{BA}; \nu_A))^{\downarrow}$.

The concrete implementation of ν_A, η_A and the operation $Q^{\sharp}$ can be explained as follows. The relation ν_A relates a value x to all pairs (y, d) by the degree d, and η_A relates x with all (x, d) by degree 1. The relation $Q^{\sharp}$ is a crisp relation that relates $u \in B$ to a pair $(x, d) \in A^{\sharp}$ iff u and x are related in Q with degree d.

In addition to the explanation above we want to provide an example that illustrates the definition of an extension. Consider the sets $A = \{\text{Joe}, \text{Peter}, \text{Jim}\}$ and $B = \{\text{Chrystler}, \text{BMW}, \text{VW}, \text{Hyundai}\}$, the complete Heyting algebra $\mathcal{L}_3$ (see below) and the relation $Q : A \to B$, read as 'likes', given by the following:

$$
\begin{array}{c}
1 \\
| \\
m \\
| \\
0
\end{array}
\qquad\qquad
Q = \begin{pmatrix} 0 & m & 1 & 0 \\ 1 & m & 0 & m \\ m & 0 & 0 & 1 \end{pmatrix}
$$

According to the first three entries of the first row of the matrix above, Joe does not like Chrystler (degree 0), he likes BMW somewhat (degree m), and he definitely likes VW (degree 1). The extension $B^{\sharp}$ of B is the set $B^{\sharp} = B \times \mathcal{L}_3 = \{(\text{Chrystler}, 0), (\text{Chrystler.}, m), (\text{Chrystler}, 1), (\text{BMW}, 0), \ldots, (\text{Hyundai}, 1)\}$. The relations $\nu_B, \eta_B : B \to B^{\sharp}$ and the relation $Q^{\sharp}$ are displayed below. We added dividers in order to indicate the grouping within $Q^{\sharp}$ with respect to one value of B:

$$
\nu_B = \begin{pmatrix} 0 & m & 1 & 0 & m & 1 & 0 & m & 1 & 0 & m & 1 \\ 0 & m & 1 & 0 & m & 1 & 0 & m & 1 & 0 & m & 1 \\ 0 & m & 1 & 0 & m & 1 & 0 & m & 1 & 0 & m & 1 \\ 0 & m & 1 & 0 & m & 1 & 0 & m & 1 & 0 & m & 1 \end{pmatrix}
\qquad
\eta_B = \begin{pmatrix} 1 & 1 & 1 & 0 & 0 & 0 & 0 & 0 & 0 & 0 & 0 & 0 \\ 0 & 0 & 0 & 1 & 1 & 1 & 0 & 0 & 0 & 0 & 0 & 0 \\ 0 & 0 & 0 & 0 & 0 & 0 & 1 & 1 & 1 & 0 & 0 & 0 \\ 0 & 0 & 0 & 0 & 0 & 0 & 0 & 0 & 0 & 1 & 1 & 1 \end{pmatrix}
$$

$$
Q^{\sharp} = \begin{pmatrix} 1 & 0 & 0 & 0 & 1 & 0 & 0 & 0 & 1 & 1 & 0 & 0 \\ 0 & 0 & 1 & 0 & 1 & 0 & 1 & 0 & 0 & 0 & 1 & 0 \\ 0 & 1 & 0 & 1 & 0 & 0 & 1 & 0 & 0 & 0 & 0 & 1 \end{pmatrix}
$$

For more details on those constructions and the axioms above we refer to [25].

The following theorem was shown in [25] and verifies that $A^{\sharp}$ is indeed a relational product.

Theorem 1. *Let* $\mathcal{A}$ *be an arrow category with extensions and unit* 1. *Then the extension* $A^{\sharp}$ *of* A *together with the relations* $\pi := \eta_A{}^{\smile}$ *and* $\rho := (\nu_A{}^{\smile}; \mathbb{T}_{A1})^{\sharp}$ *is a relational product of* A *and* $1^{\sharp}$.

The object $1^{\sharp}$ is another way of representing the membership values within an arrow category. It has been shown in [25] that $\nu_1 \backslash \nu_1$ is an order relation on $1^{\sharp}$ that internalizes the order of membership values.

Based on the extension it is possible to define a monad structure on $\mathcal{A}$. We are interested in the Kleisli category induced by this monad. Recall that the

Kleisli category $\mathcal{C}_F$ induced by a monad (F, η, μ) over $\mathcal{C}$ has the same objects as $\mathcal{C}$, the morphisms in $\mathcal{C}_F$ from A to B are the morphisms of $\mathcal{C}$ from A to $F(B)$, and composition $;_F$ and identities are given by $f;_F g = f; F(g); \mu$ and η. The following result was shown in [26].

Theorem 2. *Let A be an arrow category with extensions with a unit 1. Furthermore, let $\pi : A^\sharp \to A$ and $\rho : A^\sharp \to 1^\sharp$ be the projections from Theorem 1. Then the endo-relator P defined by $P(A) = A^\sharp$ and $P(Q) = \pi; Q; \pi^\smile \sqcap \rho; \rho^\smile$ together with the natural transformations $\eta_A : A \to A^\sharp$ and $\mu_A : A^{\sharp\sharp} \to A^\sharp$ defined by $\eta_A := \pi^\smile$ and $\mu_A := \pi \sqcap \rho; \rho^\smile$ forms a monad. Furthermore, the Kleisli category A_P with the lattice structure inherited from A and the operations*

$$Q^\cup := \eta_B; \mu_{\widetilde{B}}; P(Q^\smile), \qquad S/_P R := S/(P(R); \mu_C),$$

$$Q^\Downarrow := (Q; \eta_B; \eta_{\widetilde{B}})^\downarrow, \qquad Q^\Uparrow := (Q/(\eta_{\widetilde{B}}; \eta_B))^\downarrow,$$

forms an arrow category.

The previous theorem is the basis for the following definition.

Definition 10. *Let A_1 be an arrow category with extensions. An arrow category A_2 is called type-2 arrow category over A_1 iff A_2 is isomorphic to the Kleisli category A_{1P}.*

Due to the definition above the collections of objects in A_1 and A_2 are isomorphic. In the remainder of the paper we will identify them and assume that A_1 and A_2 have the same objects. We will write $R : A \to_i B$ for $i = 1, 2$ in order to indicate from which category R is.

In order to explain the previous definition we want to illustrate the relationship between $(\mathcal{L} \to \mathcal{L})$-relations to the Kleisli category based on the extension by an example. Consider the following type-2 relation where $\mathcal{L} = \{0, m, 1\}$ with $0 \sqsubseteq m \sqsubseteq 1$:

$$R : A \to_2 B = \left(\begin{array}{ccc} \begin{array}{c|c} x & f(x) \\ \hline 0 & 1 \\ m & 0 \\ 1 & 0 \end{array} & \begin{array}{c|c} x & f(x) \\ \hline 0 & 0 \\ m & 1 \\ 1 & m \end{array} & \begin{array}{c|c} x & f(x) \\ \hline 0 & 0 \\ m & m \\ 1 & 1 \end{array} \\ \hline \begin{array}{c|c} x & f(x) \\ \hline 0 & 0 \\ m & 0 \\ 1 & 1 \end{array} & \begin{array}{c|c} x & f(x) \\ \hline 0 & m \\ m & m \\ 1 & 0 \end{array} & \begin{array}{c|c} x & f(x) \\ \hline 0 & 0 \\ m & m \\ 1 & m \end{array} \end{array} \right).$$

This relation corresponds to the following type-1 relation between A and $B^\sharp$

$$\tilde{R} : A \to_1 B^\sharp = \left(\begin{array}{ccc|ccc|ccc} 1 & 0 & 0 & 0 & 1 & m & 0 & m & 1 \\ 0 & 0 & 1 & m & m & 0 & 0 & m & m \end{array} \right).$$

Notice that we have grouped the columns in the matrix above that belong to the same value in B (but different values in $\mathcal{L}$). In the remainder of the paper we use the notion $\tilde{R}$ to indicate the relation in A_1 that is isomorphic to R as a relation in A_{1P}.

3 $\mathcal{L}$-Fuzzy Controllers

In this section we want to recall the Mamdani approach to fuzzy controllers. We will concentrate on the difference of type-1 vs. type-2 controllers in particular. Following a general overview we show how to model each component of a controller in the theory of arrow categories. Notice that type-1 controllers were already discussed in [23].

3.1 The Mamdani Approach to Fuzzy Controllers

In the Mamdani approach a fuzzy controller consists of several components [10]. In the case of a type-1 controller the general structure is visualized in Figure 1. Notice that fuzzy values are only used inside the controller. The communication with the actual process is based on crisp data.

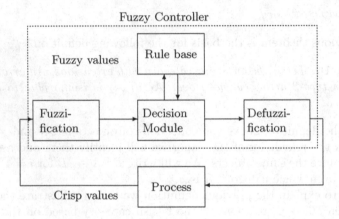

Fig. 1. Components of a type-1 fuzzy controller

The fuzzification component transforms crisp input into fuzzy values that can be processed by the controller. The rule base hosts the rules of the system. The decision module selects appropriate rules from the rule base and applies them to the fuzzified input. Finally, the defuzzifier converts the fuzzy result into a crisp value that can be used as input for the process.

A type-2 fuzzy controller adds one additional component to the model above (cf. [2]). In addition, within a type-2 controller type-1 and type-2 fuzzy sets and values are used. The structure of such a type-2 controller, including the information about what kind of values are being used, is summarized in Figure 2.

The type reducer is responsible to transform type-2 fuzzy values coming from the decision module into type-1 fuzzy values that can be passed to the defuzzifier.

The individual components of the controller will be discussed with more detail in the following sections.

Type-2 Fuzzy Controller

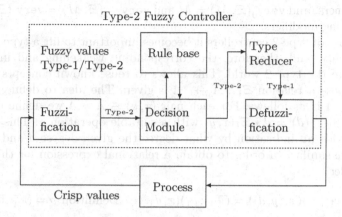

Fig. 2. Components of a type-2 fuzzy controller

3.2 Linguistic Entities and Variables

A basic concept in any fuzzy controller is the use of linguistic entities. A linguistic entity is an abstract notion usually phrased in natural language that represents values like "extremely high speed", "hot water", "very heavy rain", and so on. It is important to notice that linguistic entities are often constructed from a basic notion by applying a so-called linguistic modifier. For example, "very hot" is based on the notion of "hot" which is then modified by the intensifying modifier "very".

Variables ranging over linguistic entities are called linguistic variables. Linguistic variables and entities are normally used in the rules of the controller. Within the mathematical model of a controller linguistic entities are interpreted as suitable $\mathcal{L}$-fuzzy sets, and linguistic variables as variables over those sets.

As usual in the theory of relations, we describe a subset of A by a relation $M : 1 \rightarrow A$, where 1 is the unit of the corresponding relational category. A linguistic modifier now becomes an operation on relations. In this paper we want to consider only one kind of modifiers. These modifiers originate from relations that model approximate equalities. Such a relation Ξ indicates whether two elements are approximately equal. Therefore, Ξ is reflexive $\mathbb{I}_A \sqsubseteq \Xi$ and symmetric $\Xi^{\smile} \sqsubseteq \Xi$. However, Ξ is usually not transitive because this usually leads to inconsistencies [3]. Given such an approximate equality and a fuzzy set $M : 1 \rightarrow A$, the operation $M ;_{*} \Xi$ for some suitable lattice-ordered semigroup operation $*$ weakens the notion represented by M. On the other hand, the operation $M \rhd_{*} \Xi$, where $\rhd_{*}$ is the residual operation with respect to $;_{*}$, intensifies the notion represented by M. For example, if Ξ is an approximate equality on temperature degrees, then $M ;_{*} \Xi$ has a wider range than M, and $M \rhd_{*} \Xi$ has a smaller range than M.

A weakening or intensifying modifier based on Ξ can be applied multiple times to one fuzzy set to get even stronger (or weaker) results. We define $\mathtt{roughly}_0(\Xi, M) = M$ and $\mathtt{roughly}_{i+1}(\Xi, M) = \mathtt{roughly}_i(\Xi, M) ;_{*} \Xi$ as weakening modifiers, and $\mathtt{very}_0(\Xi, M) = M$ and $\mathtt{very}_{i+1}(\Xi, M) = \mathtt{very}_i(\Xi, M) \triangleright_* \Xi$ as intensifying modifiers.

In the case of type-2 controllers it becomes important to lift a type-1 approximate equality, and, therefore, the corresponding weakening and intensifying modifiers, to the type-2 world. This allows to reuse known concepts and relations. Suppose a relation $\Xi_1 : A \to_1 A$ is given. The idea to define a relation $\Xi_2 : A \to_2 A$ is as follows. For each pair $(x, y) \in A \times A$ we define a function $f : \mathcal{L} \to \mathcal{L}$ by $f(d) = d * \Xi_1(x, y)$ with a suitable operation $*$. This operation should provide an indication by what degree the given degree d and the value $\Xi_1(x, y)$ are similar. In order to obtain a relational expression for this idea we first consider

$$
\begin{aligned}
(\mathbb{T}_{A1}; \nu_1; \rho^{\smile})(x, (y, d)) &= (\mathbb{T}_{A1}; \nu_1)(x, d) &&\text{Definition } \rho = (\nu_A^{\smile}; \mathbb{T}_{A1})^{\sharp} \\
&= \nu_1(*, d) \\
&= d. &&\text{Definition } \nu_1
\end{aligned}
$$

Now, we combine the part above with Ξ_1 by an $*$-based operation and obtain

$$
\begin{aligned}
(\mathbb{T}_{A1}; \nu_1; \rho^{\smile} \sqcap_* \Xi_1; \pi^{\smile})(x, (y, d)) &= (\mathbb{T}_{A1}; \nu_1; \rho^{\smile})(x, (y, d)) * (\Xi_1; \pi^{\smile})(x, (y, d)) \\
&= d * \Xi_1(x, y).
\end{aligned}
$$

Overall for the given relation $\Xi_1 : A \to_1 A$ we obtain a relation $\Xi_2 : A \to_2 A$ so that $\tilde{\Xi}_2 = \Xi_1; \pi^{\smile} \sqcap_* \mathbb{T}_{A1}; \nu_1; \rho^{\smile}$. Notice that it depends on the operation $*$ whether Ξ_2 is also an approximate equality. For example, in order to obtain a reflexive relation we need that $1 * x = 1$. A suitable choice is $\to$, i.e., the implication operation of the lattice.

3.3   Fuzzification

Fuzzification is the process of changing crisp values x to an $\mathcal{L}$-fuzzy set $F(x)$. In our approach input values from A are interpreted by points, i.e., by crisp functions $x : 1 \to A$. The operation F often depends on the actual application. Occasionally parameters of F or F itself is constantly modified in order to validate the controller. However, a typical choice of F is simply a weakening operation, i.e., $F(x) = \mathtt{very}_1(x)$ for suitable choices of $*$ and Ξ. Notice that our discussion in the previous section allows to obtain such a weakening operation on the type-2 level from a given approximate equality on the type-1 level.

3.4   Rule Base

In this paper fuzzy controllers use rules of the form

$$\textbf{if } x \textbf{ is } M, \textbf{ then } y = N,$$

where x and y are linguistic variables considered as the input resp. as the output and M and N are fuzzy sets representing linguistic entities. The collection of all linguistic entities used in the if-parts of the rules can be considered as the input set and the collection of linguistic entities used in the then-part of the rules as the output set of the rule base. Since every linguistic entity is modeled by a copy of the unit 1 and a rule base is finite, the input as well as the output of the rule base are given by a suitable number of copies of the unit $1 + \ldots + 1$. The rule base itself becomes a crisp relation between those objects. For example the rules of the form

$$\textbf{if } x \textbf{ is } M_1, \textbf{ then } y = N_1,$$
$$\textbf{if } x \textbf{ is } M_2, \textbf{ then } y = N_1,$$
$$\textbf{if } x \textbf{ is } M_3, \textbf{ then } y = N_2$$

then the rule base is modeled by a relation $R : 1 + 1 + 1 \to 1 + 1$ defined by

$$R := \iota_1{}^{\smile}; \iota_1' \sqcup \iota_2{}^{\smile}; \iota_1' \sqcup \iota_3{}^{\smile}; \iota_2' = \begin{pmatrix} 1 & 0 \\ 1 & 0 \\ 0 & 1 \end{pmatrix}$$

where $\iota_1, \iota_2, \iota_3$ resp. ι_1', ι_2' are the crisp injections from 1 to $1+1+1$ resp. $1+1$. In general, if $R(i)$ denotes the set of indices of output linguistic entities related by the rule base to the input linguistic entity i, the relation R is of the form $R = \bigsqcup_{i \in I \text{ and } j \in R(i)} \iota_i{}^{\smile}; \iota_j'$. Recall that each linguistic entity is modeled by a relation.

In particular, each input linguistic entity is modeled by a relation $Q_i : 1 \to A$ ($i \in I$), and each output linguistic entity is modeled by a relation $S_j : 1 \to B$ ($j \in J$). If we combine these relation with the relation R we obtain the core $T : A \to B$ as

$$T := (\bigsqcup_{i \in I} Q_i{}^{\smile}; \iota_i); R; (\bigsqcup_{j \in J} \iota_j'; S_j).$$

The core T is visualized in Figure 3. Notice that the composition operations in the expression above can be generalized to $*$-based operations. However, replacing one of the two is sufficient because replacing both compositions by a $*$-based operation is redundant [23].

3.5 Decision Module

The decision module describes which rules and how they are applied to the actual input of the controller. In our model this corresponds to the question how to combine of the fuzzified input $F(x)$ with the core of the controller T. Usually, an optimistic view is taken towards computing the degree of activation for a rule [23]. This corresponds to a $*$-based composition operation and leads to the expression $U(x) := F(x);_* T$. A pessimistic view would correspond to a residual operation based on on $*$. However, this approach is barely taken in actual implementations of fuzzy controllers [7].

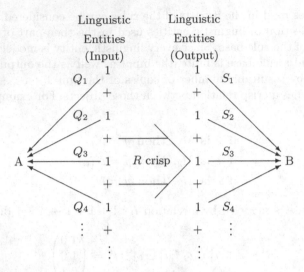

Fig. 3. Core of a fuzzy controller

3.6   Defuzzification

Defuzzification is the process by which fuzzy output from the decision module is converted back to a crisp value. In our model this requires an operation that maps an $\mathcal{L}$-fuzzy relation $U(x) : 1 \to B$ to a crisp relation $\mathcal{D}(U(x)) : 1 \to B$. There are several different operations used in concrete implementations of fuzzy controllers. A lot of them are specific for the unit interval and cannot be generalized to arbitrary lattices. However, there are also some general methods available. In this paper we only want to provide one example. For further operations we refer to [23]. Given an operation Θ, mapping each crisp $x : 1 \to A$ to a scalar $\Theta(x)$ on B we define

$$\mathcal{D}(U(x)) := (\Theta(x)\backslash U(x))^{\downarrow}.$$

An obvious choice for the operation Θ is $\Theta(x) = x; S_A$ where $S_A \sqsubseteq \mathbb{I}_A$ is a partial identity.

3.7   Type Reduction

The type reducer in a type-2 controller has to map the type-2 relation $U(x) : 1 \to_2 B$ to a type-1 relation $\mathcal{R}(U(x)) : 1 \to_1 B$. In this section we discuss several possible operations. In order to do so we will assume that $R : A \to_2 B$ is a type-2 relation and that $\tilde{R} : A \to_1 B^{\sharp}$ is the corresponding type-1 relation as discussed in Section 2. We want to focus on general operations that do not make any assumptions on the underlying lattice $\mathcal{L}$. If we, for example, assume that $\mathcal{L}$ is linear, then more specific operations become available.

Forming Suprema: A very simple operation is $\mathcal{R}(R) = \tilde{R}; \eta_B^\smile$, i.e., is to compose $\tilde{R}$ with $\eta_B^\smile$. In the concrete case we obtain

$$(\tilde{R}; \eta_B^\smile)(x, z) = \bigsqcup_{(y,d) \in B^\sharp} \tilde{R}(x, (y, d)) \sqcap \eta_B(z, (y, d))$$

$$= \bigsqcup_{d \in \mathcal{L}} \tilde{R}(x, (z, d)) \qquad\qquad \text{Definition } \eta_B$$

$$= \bigsqcup_{d \in \mathcal{L}} R(x, z)(d),$$

i.e., this construction uses the supremum of all membership values in the function $R(x, z)$. This simple operation can be generalized by using $\pi; \pi^\smile \sqcap \rho; S_{1\sharp}; \rho^\smile :$ $B^\sharp \to_1 B^\sharp$ with $S_{1\sharp} \sqsubseteq \mathbb{I}_{1\sharp}$ as a partial identity. In this case we define $\mathcal{R}(R) =$ $\tilde{R}; (\pi; \pi^\smile \sqcap \rho; S_{1\sharp}; \rho^\smile); \eta_B^\smile$. For $S_{1\sharp} = \mathbb{I}_{1\sharp}$ we obviously get the original operation. If $S_{1\sharp} \neq \mathbb{I}_{1\sharp}$ is a crisp relation, then we exclude certain elements, i.e., we form the supremum on selected elements only. A special case of this situation is given when $S_{1\sharp}$ is an atom. Then type reduction selects the membership degree of the elements represented by $S_{1\sharp}$. Last but not least, if $S_{1\sharp}$ is not crisp, then we obtain a weighted supremum.

Relational Integration: Relational integration was introduced in [16]. This construction generalizes the well-known Choquet and Sugeno integrals to non-linear orderings in an algebraic fashion. These integrals have been used in many applications. In terms of an aggregation function they behave nicely, i.e., they are continuous, non-decreasing, and stable under certain interval preserving transformations. In addition, if the underlying measure (or belief function) is additive, they reduce to a weighted arithmetic mean. The relational integral is formed for a map $X : C \to \mathcal{L}$ that assigns to each criteria in C a value from the lattice $\mathcal{L}$. Forming such an integral is then based on a belief mapping (or relational measure) $\mu : \mathcal{P}(C) \to \mathcal{L}$, i.e., a mapping that is monotonic with respect to the subset order $\varepsilon \backslash \varepsilon$ and the order E of $\mathcal{L}$ and that preserves least and greatest element. The situation is visualized in the following diagram:

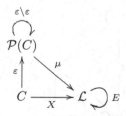

The integral $(R) \int X \circ \mu : 1 \to \mathcal{L}$ is now defined by:

$$(R) \int X \circ \mu = \mathrm{lub}_E(\mathbb{T}_{1C}; \mathrm{glb}_E(X \sqcup \mathrm{syQ}(X; E^\smile; X^\smile, \varepsilon); \mu)).$$

The integral can be explained as follows. For a given element of C we form the set of all those criteria with a higher (or equal) valuation (by X). The resulting set is measured by μ. Now we take the meet of this value and the valuation of the element itself. Finally, the least upper bound over all elements is taken. For more details about relational measures and integrals we refer to [13,16,17].

The idea for applying relational integration for type reduction is as follows. A map $f : \mathcal{L} \to \mathcal{L}$ can be seen as a valuation of the elements in $\mathcal{L}$ by itself. The integral now aggregates these valuations into a single value. Unfortunately, we cannot apply the relational integral directly because it is defined for a single map X only. In our situation we encounter multiple maps simultaneously that we have to treat together but that we want to handle separately. In order to explain and visualize the situation even better we will identify the isomorphic objects $A^{\sharp}$ and $A \times 1^{\sharp}$ in this section. Now, consider the relation $\tilde{R}^{\sharp} : A \to_1 (B \times 1^{\sharp}) \times 1^{\sharp}$. We can transform this relation into a map $Y : (A \times B) \times 1^{\sharp} \to 1^{\sharp}$ that is a valuation of elements in $1^{\sharp}$ by elements from $1^{\sharp}$ for each pair in $A \times B$. Y is defined by

$$Y := (\pi; \pi; \tilde{R}^{\sharp} \sqcap (\pi; \rho; \pi^{\smile} \sqcap \rho; \rho^{\smile}); \pi^{\smile}); \rho.$$

The situation can be visualized by the following diagram:

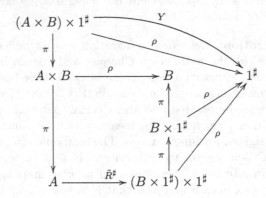

We need to compute the relational integral for each pair separately, i.e., we have to define $(R) \int_A X \circ \mu$ in the general situation below:

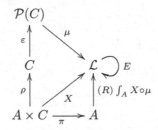

We define

$$(R) \int_A X \circ \mu = \text{lub}_E(\pi^{\smile}; \text{glb}_E(X \sqcup \text{syQ}(\pi; \pi^{\smile} \sqcap X; E^{\smile}; X^{\smile}, \varepsilon); \mu)).$$

Notice that if $A = 1$, i.e., $A \times C$ is isomorphic to C, we obtain the original definition since $\pi = \pi_{C1}$ in this case.

Given a relational measure $\mu : \mathcal{P}(1^\sharp) \to 1^\sharp$ we define a type reducer by

$$\mathcal{R}(R) = (R) \int_A ((\pi; \pi; \tilde{R}^\sharp \sqcap (\pi; \rho; \pi^\smile \sqcap \rho; \rho^\smile); \pi^\smile); \rho) \circ \mu.$$

4 Conclusion and Future Work

In this paper we have shown how to model type-1 and type-2 $\mathcal{L}$-fuzzy controllers in the framework of arrow categories. In addition to the general approach we have investigated several operations suitable for lifting relations from type-1 to type-2 and for type reduction.

Future work will concentrate on an implementation of the approach in an interactive theorem prover such as Coq or Isabelle. Such an implementation would allow to handle the specification and the implementation of a controller within one system. Then it becomes possible to formally proof properties of the controller within a formal system that supports semi-automatic reasoning.

References

1. Birkhoff, G.: Lattice Theory, 3rd edn., vol. XXV. American Mathematical Society Colloquium Publications (1940)
2. Castillo, O., Melin, P.: Type-2 Fuzzy Logic: Theory and Applications. STUDFUZZ, vol. 223. Springer (2008)
3. De Cock, M., Radzikowska, A.M., Kerre, E.E.: Modelling Linguistic Modifiers Using Fuzzy-Rough Structures. In: Proceedings of IPMU 2000, vol. III, pp. 1735–1742 (2000)
4. Freyd, P., Scedrov, A.: Categories, Allegories. North-Holland (1990)
5. Furusawa, H., Kawahara, Y., Winter, M.: Dedekind Categories with Cutoff Operators. Fuzzy Sets and Systems 173, 1–24 (2011)
6. Goguen, J.A.: L-fuzzy sets. J. Math. Anal. Appl. 18, 145–157 (1967)
7. Gottwald, S.: Fuzzy Sets and Fuzzy Logic. Foundations of Application – from a Mathematical Point of View. Vieweg (1993)
8. Jónsson, B., Tarski, A.: Boolean algebras with operators, I, II. Amer. J. Math. 73, 74, 891–939, 127-162 (1951, 1952)
9. Kawahara, Y., Furusawa, H.: Crispness and Representation Theorems in Dedekind Categories. DOI-TR 143. Kyushu University (1997)
10. Mamdani, E.H., Gaines, B.R.: Fuzzy Reasoning and its Application. Academic Press, London (1987)
11. Olivier, J.P., Serrato, D.: Catégories de Dedekind. Morphismes dans les Catégories de Schröder. C.R. Acad. Sci. Paris 290, 939–941 (1980)
12. Olivier, J.P., Serrato, D.: Squares and Rectangles in Relational Categories - Three Cases: Semilattice, Distributive lattice and Boolean Non-unitary. Fuzzy Sets and Systems 72, 167–178 (1995)
13. Schmidt, G., Berghammer, R.: Relational measures and integration in preference modeling. JLAP 76, 112–129 (2008)

14. Schmidt, G., Hattensperger, C., Winter, M.: Heterogeneous Relation Algebras. In: Brink, C., Kahl, W., Schmidt, G. (eds.) Relational Methods in Computer Science. Springer, Vienna (1997)
15. Schmidt, G., Ströhlein, T.: Relationen und Graphen. Springer (1989); English version: Relations and Graphs. Discrete Mathematics for Computer Scientists. EATCS Monographs on Theoret. Comput. Sci., Springer (1993)
16. Schmidt, G.: Relational Measures and Integration. In: Schmidt, R.A. (ed.) RelMiCS/AKA 2006. LNCS, vol. 4136, pp. 343–357. Springer, Heidelberg (2006)
17. Schmidt, G.: Relational Mathematics. Encyclopedia of Mathematics and its Applications 132 (2011)
18. Winter, M.: Strukturtheorie heterogener Relationenalgebren mit Anwendung auf Nichtdetermismus in Programmiersprachen. Dissertationsverlag NG Kopierladen GmbH, München (1998)
19. Winter, M.: A new Algebraic Approach to L-Fuzzy Relations Convenient to Study Crispness. INS Information Science 139, 233–252 (2001)
20. Winter, M.: Relational Constructions in Goguen Categories. In: de Swart, H. (ed.) Participants Proceedings of the 6th International Seminar on Relational Methods in Computer Science (RelMiCS), pp. 222–236. Katholieke Universiteit Brabant, Tilburg (2001)
21. Winter, M.: Derived Operations in Goguen Categories. TAC Theory and Applications of Categories 10(11), 220–247 (2002)
22. Winter, M.: Representation Theory of Goguen Categories. Fuzzy Sets and Systems 138, 85–126 (2003)
23. Winter, M.: Goguen Categories - A Categorical Approach to L-fuzzy relations. Trends in Logic 25 (2007)
24. Winter, M.: Arrow Categories. Fuzzy Sets and Systems 160, 2893–2909 (2009)
25. Winter, M.: Membership Values in Arrow Categories. Submitted to Fuzzy Sets and Systems (October 2013)
26. Winter, M.: Higher-order arrow categories. In: Höfner, P., Jipsen, P., Kahl, W., Müller, M.E. (eds.) RAMiCS 2014. LNCS, vol. 8428, pp. 277–292. Springer, Heidelberg (2014)
27. Zadeh, L.A.: The concept of a linguistic variable and its application to approximate reasoning - I. Information Sciences 8, 199–249 (1975)

Relation Algebra and RelView Applied to Approval Voting

Rudolf Berghammer, Nikita Danilenko, and Henning Schnoor

Institut für Informatik
Christian-Albrechts-Universität Kiel
Olshausenstraße 40, 24098 Kiel, Germany
{rub,nda,schnoor}@informatik.uni-kiel.de

Abstract. In this paper we demonstrate how relation algebra and a BDD-based tool can be combined to solve computational problems of voting systems. We concentrate on approval voting and model this kind of voting within relation algebra. Based on this, we then formally develop relation-algebraic specifications of two important control problems from their logical specifications. They can be transformed immediately into the programming language of the BDD-based Computer Algebra system RelView. Therefore, this tool can be used to solve the problems and to visualize the computed results. The entire approach is extremely formal but also very flexible. In combination with RelView it is especially appropriate for prototyping and experimentation, and as such very instructive for educational purposes.

1 Introduction

For many years relation algebra in the sense of [14,15] has been used widely by mathematicians and computer scientists as conceptual and methodological base of their work. A first reason of its importance is that many fundamental objects of discrete mathematics and computer science are relations (like orders), can be seen as relations (like directed graphs), or can be easily modeled via relations (like simple games). Another advantage of relation algebra is that it allows very concise and exact problem specifications and, especially if combined with predicate logic, extremely formal and precise calculations. This drastically reduces the danger of making mistakes. A third reason for the use of relation algebra is that it has a fixed and surprisingly small set of operations all of which can be implemented efficiently and with reasonable effort on finite carrier sets. Thus, a specific purpose Computer Algebra system for relation algebra can be implemented with reasonable effort, too. At Kiel University we have developed such a system, called RelView (see [3,17]). It uses reduced ordered binary decision diagrams (ROBDDs) to implement relations very efficiently, provides an own programming language with a lot of pre-defined operations, and has visualization and animation facilities which are not easily found in other software tools and which are most helpful e.g., for prototypic computations, experimenting with difficult concepts, and for understanding and testing specifications and programs.

P. Höfner et al. (Eds.): RAMiCS 2014, LNCS 8428, pp. 309–326, 2014.
© Springer International Publishing Switzerland 2014

Originally voting systems and related notions have been introduced in political science and economics to investigate and model problems and phenomena that may arise during the aggregation of a collective preference relation from the preferences of a group of individuals. More recently they have also been recognized as important in various areas of computer science like multi-agent systems, planning, similarity search and the design of ranking algorithms. For more references we refer to the overview papers [6,7].

A central notion of voting systems is that of a voting rule, which specifies the preference aggregation. There is a list of such rules and it is known that different rules may lead to different results; see [4,6,13] for example. An important problem is the susceptibility of voting systems to manipulations. As in the case of preference aggregation there are also several possibilities to manipulate. Inspired by the seminal papers [1,2], in recent years computer scientists have investigated the hardness of manipulations using methods of complexity theory. For some rules and types of manipulation it has been shown that finding a beneficial manipulation is NP-hard, for other rules and types of manipulation, however, efficient algorithms exist to compute a beneficial manipulation. See e.g., [1,2,9], the overview paper [16], and the references of Section 3.2 of [6].

In this paper we apply relation algebra and the RELVIEW tool to voting systems. We concentrate on a specific rule known as approval voting. Here the individual preferences are the sets of alternatives the single voters approve and then collectively a is (weakly) preferred to b if the number of approvals of a is equal or greater than the number of approvals of b. Furthermore, we only consider two types of manipulation, known as constructive control by deleting and adding voters, respectively. Here the assumption is that the authority conducting the election (usually called the chair) knows all individual preferences and tries to achieve a specific result by a strategic manipulation of the set of voters. The chair's knowledge of all individual preferences and the ability to delete or add voters by 'dirty tricks' are worst-case assumptions that are not entirely unreasonable in some settings, e.g., in case of governing boards, commissions of political institutions, and meetings of members of a club.

First, we show how approval voting can be modeled by means of relation algebra. Based on this, we then formally develop relation-algebraic specifications of the above control problems from their specifications as formulae in predicate logic. They can be transformed immediately into the programming language of RELVIEW and, hence, the tool can be used to compute solutions and to visualize them. Because RELVIEW has a very efficient ROBDD-implementation of relations, the tool is able to deal with non-trivial examples as those mentioned at the end of the preceding paragraph.

2 Relation-Algebraic Preliminaries

In this section we recall some preliminaries of (typed, heterogeneous) relation algebra. For more details, see [14,15], for example.

Given two sets X and Y, we write $R : X \leftrightarrow Y$ if R is a (typed, binary) relation with source X and target Y, i.e., a subset of the direct product $X \times Y$. If the two sets X and Y of R's *type* $X \leftrightarrow Y$ are finite, then we may consider R as a Boolean matrix. Since a Boolean matrix interpretation is well suited for many purposes and also used by the RELVIEW tool as the main possibility to visualize relations, in the present paper we frequently use matrix terminology and notation. In particular, we speak about the entries, rows and columns of a relation/matrix and write $R_{x,y}$ instead of $(x, y) \in R$ or $x\, R\, y$. We assume the reader to be familiar with the basic operations on relations, viz. R^{T} (*transposition*), $\overline{R}$ (*complement*), $R \cup S$ (*union*), $R \cap S$ (*intersection*) and $R; S$ (*composition*), the predicate $R \subseteq S$ (*inclusion*) and the special relations O (*empty relation*), L (*universal relation*) and I (*identity relation*). In case of O, L and I we overload the symbols, i.e., avoid the binding of types to them.

For $R : X \leftrightarrow Y$ and $S : X \leftrightarrow Z$, their *symmetric quotient* $\mathrm{syq}(R, S) : Y \leftrightarrow Z$ is defined by $\mathrm{syq}(R, S) = \overline{R^{\mathsf{T}}; \overline{S}} \cap \overline{\overline{R}^{\mathsf{T}}; S}$. From this definition we get

$$\mathrm{syq}(R, S)_{y,z} \iff \forall x \in X : R_{x,y} \leftrightarrow S_{x,z}, \tag{1}$$

for all $y \in Y$ and $z \in Z$. This logical (usually called *pointwise*) specification of symmetric quotients will be used several times in Section 3 to Section 5.

In the remainder of this section we introduce some additional relation-algebraic constructs concerning the modeling of direct products and sets. They will also play an important role in Sections 3 to 5.

To model direct products $X \times Y$ of sets X and Y relation-algebraically, the *projection relations* $\pi : X \times Y \leftrightarrow X$ and $\rho : X \times Y \leftrightarrow Y$ are the convenient means. They are the relational variants of the well-known projection functions and, hence, fulfill for all $(x, y) \in X \times Y$, $z \in X$, and $u \in Y$ the following equivalences.

$$\pi_{(x,y),z} \iff x = z \qquad\qquad \rho_{(x,y),u} \iff y = u \tag{2}$$

The projection relations enable us to specify the well-known pairing operation of functional programming relation-algebraically. The *pairing* (or *fork*) of the relations $R : Z \leftrightarrow X$ and $S : Z \leftrightarrow Y$ is defined as relation $[R, S] = R; \pi^{\mathsf{T}} \cap S; \rho^{\mathsf{T}} : Z \leftrightarrow X \times Y$, where π and ρ are as above. From the properties (2) and this definition we get that for all elements $z \in Z$ and pairs $(x, y) \in X \times Y$ it holds

$$[R, S]_{z,(x,y)} \iff R_{z,x} \wedge S_{z,y}. \tag{3}$$

Having modeled direct products via the projection relations and introduced the relation-algebraic version of pairing, we now show how to model sets and important set-theoretic constructions within relation algebra.

Vectors are a first well-known means to model sets. For the purpose of this paper it suffices to define them as relations $r : X \leftrightarrow \mathbf{1}$ (we prefer lower case letters in this context) with a specific singleton set $\mathbf{1} = \{\bot\}$ as target. Then relational vectors can be considered as Boolean column vectors. To be consonant with the usual notation, we always omit the second subscript, i.e., write r_x instead of $r_{x,\bot}$. Then the vector r *describes the subset* Y of X if r_x and $x \in Y$ are equivalent, for all $x \in X$. A *point* $p : X \leftrightarrow \mathbf{1}$ is a specific vector with precisely one 1-entry. Consequently, it describes a singleton subset $\{x\}$ of X and we then say that it *describes the element* x of X. If $r : X \leftrightarrow \mathbf{1}$ is a vector and Y the subset of X it describes, then $inj(v) : Y \leftrightarrow X$ denotes the *embedding relation* of Y into X. In Boolean matrix terminology this means that the relation $inj(v)$ is obtained from the identity relation $\mathsf{I} : X \leftrightarrow X$ by deleting all rows which do not correspond to an element of Y, and pointwise this means that for all $y \in Y$ and $x \in X$ it holds $inj(r)_{y,x}$ iff $y = x$. In conjunction with powersets 2^X we will use *membership relations* $\mathsf{M} : X \leftrightarrow 2^X$ and *size comparison relations* $\mathsf{S} : 2^X \leftrightarrow 2^X$. Pointwise they are defined for all $x \in X$ and $Y, Z \in 2^X$ as follows.

$$\mathsf{M}_{x,Y} \iff x \in Y \qquad\qquad \mathsf{S}_{Y,Z} \iff |Y| \leq |Z| \qquad\qquad (4)$$

A combination of M with embedding relations allows a *column-wise enumeration* of arbitrary subsets $\mathfrak{S}$ of 2^X. Namely, if the vector $r : 2^X \leftrightarrow \mathbf{1}$ describes the subset $\mathfrak{S}$ in the sense defined above and we define $S = \mathsf{M}; inj(r)^{\mathsf{T}} : X \leftrightarrow \mathfrak{S}$, then for all $x \in X$ and $Y \in \mathfrak{S}$ it holds $S_{x,Y}$ iff $x \in Y$. In the Boolean matrix model this means that the sets of $\mathfrak{S}$ are precisely described by the columns of S, if the columns are considered as vectors of type $X \leftrightarrow \mathbf{1}$. Or, in other words, $\mathsf{M}; inj(r)^{\mathsf{T}}$ deletes all columns from M which correspond to 0-entries of r. Finally, we will use an operation *filter* that takes a vector $v : 2^X \leftrightarrow \mathbf{1}$ and a natural number k as inputs and yields a vector $filter(v, k) : 2^X \leftrightarrow \mathbf{1}$ such that

$$filter(v, k)_Y \iff v_Y \wedge |Y| = k, \qquad\qquad (5)$$

for all $Y \in 2^X$. Hence, if the vector $v : 2^X \leftrightarrow \mathbf{1}$ describes the subset $\mathfrak{S}$ of 2^X, then the vector $filter(v, k)$ describes the subset $\{Y \in \mathfrak{S} \mid |Y| = k\}$.

As from now we assume all constructions of this section to be at hand. Each of them is available in the programming language of the RELVIEW tool via a pre-defined operation (the only exception is *filter*, but it easily can be realized via a pre-defined operation *cardfilter* with $cardfilter(v, k)_Y$ iff v_Y and $|Y| < k$, for all $Y \in 2^X$) and the implementing ROBDDs are comparatively small. E.g., a size-comparison relation $\mathsf{S} : 2^X \leftrightarrow 2^X$ can be implemented by a ROBDD with $O(|X|^2)$ nodes and for a membership relation $\mathsf{M} : X \leftrightarrow 2^X$ even $O(|X|)$ nodes suffice. For details, see [11,12].

3   A Relation-Algebraic Model of Approval Voting

In Computational Social Choice a voting system consists of a finite and non-empty set N of voters (players, agents, individuals), a finite and non-empty set A of alternatives (candidates), the individual preferences (choices, wishes) of the voters, and a voting rule that aggregates the winners from the individual preferences. A well-known voting rule is *approval voting*, see e.g., [5]. Here each voter may approve (that is, vote for) as many alternatives as he wishes and then the alternatives with the most approvals are defined as the winners of the election. The rule is frequently favored if multiple alternatives may be elected and used, for instance, by the Mathematical Association of America. At this point we should also mention that during a workshop, held in France from July

30 to August 2, 2010, the participants (all experts in voting theory) elected the best voting rule. First, they agreed to use approval voting as voting rule for the election. Then among the 18 nominated voting rules approval voting won the election. See [10] for details.

If A_i denotes the set of alternatives the voter $i \in N$ approves, then a voting system with approval voting as voting rule, in the remainder of the paper called an *approval election*, can be specified formally as a triple $(N, A, (A_i)_{i \in N})$. The comparison of the alternatives with regard to the number of their approvals leads to its *dominance relation* $D : A \leftrightarrow A$, that is defined pointwise by

$$D_{a,b} \iff |\{i \in N \mid a \in A_i\}| \geq |\{i \in N \mid b \in A_i\}|, \tag{6}$$

for all $a, b \in A$. In case of $D_{a,b}$ we say that a *dominates* b in $(N, A, (A_i)_{i \in N})$. From the dominance relation, finally, the set of winners of the approval election is defined as $\{a \in A \mid \forall b \in A : D_{a,b}\}$. To treat approval elections with relation-algebraic means, we first have to model them accordingly.

Definition 3.1. *The relation* $P : N \leftrightarrow A$ *is a relational model of the approval election* $(N, A, (A_i)_{i \in N})$ *iff* $P_{i,a}$ *is equivalent to* $a \in A_i$, *for all* $i \in N$ *and* $a \in A$.

Hence, if P is a relational model of the approval election $(N, A, (A_i)_{i \in N})$, then the relationship $P_{i,a}$ holds iff voter i approves alternative a, for all $i \in N$ and $a \in A$. Next, we present a small example for a relational model that will be used during the next sections, too.

Example 3.1. We consider the following relation P in its RELVIEW representation as a labeled 12×8 Boolean matrix, that is, as a matrix with 12 rows and 8 columns and 1 and 0 as entries only. From the labels of the rows and the columns, respectively, we see that the voters are the natural numbers from 1 to 12 and the alternatives are the 8 letters from a to h.

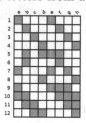

In a RELVIEW matrix a black square means a 1-entry and a white square means a 0-entry so that, for example, voter 1 approves the two alternatives a and e and voter 2 approves the two alternatives b and f. I.e., we have $A_1 = \{a, e\}$ and $A_2 = \{b, f\}$.

In the remainder of the paper we always denote in case of an approval election $(N, A, (A_i)_{i \in N})$ its relational model by the letter P and its dominance relation by the letter D. As a first result we show how the dominance relation can be obtained from the relational model via symmetric quotients, the membership relation $M : N \leftrightarrow 2^N$, and the size comparison relations $S : 2^N \leftrightarrow 2^N$.

Theorem 3.1. *For each approval election $(N, A, (A_i)_{i \in N})$ and its dominance relation $D : A \leftrightarrow A$ we have*

$$D = syq(P, \mathsf{M}); \mathsf{S}^\mathsf{T}; syq(P, \mathsf{M})^\mathsf{T}. \tag{7}$$

Proof. In the first step we use property (1) and the first property of (4) to prove for all alternatives $c \in A$ and sets $Z \in 2^N$ the following equivalence.

$$
\begin{aligned}
syq(P, \mathsf{M})_{c,Z} &\iff \forall i \in N : P_{i,c} \leftrightarrow \mathsf{M}_{i,Z} \\
&\iff \forall i \in N : P_{i,c} \leftrightarrow i \in Z \\
&\iff \{i \in N \mid P_{i,c}\} = Z
\end{aligned}
$$

Now, let $a, b \in A$ be arbitrary alternatives. Then we can calculate as follows, where the first step uses the pointwise specification (6) of D, the second step applies that P is a relational model of $(N, A, (A_i)_{i \in N})$, and in the fourth step the above equivalence and the second property of (4) are used.

$$
\begin{aligned}
D_{a,b} &\iff |\{i \in N \mid a \in A_i\}| \geq |\{i \in N \mid b \in A_i\}| \\
&\iff |\{i \in N \mid P_{i,a}\}| \geq |\{i \in N \mid P_{i,b}\}| \\
&\iff \exists X, Y \in 2^N : \{i \in N \mid P_{i,a}\} = X \wedge \{i \in N \mid P_{i,b}\} = Y \wedge |X| \geq |Y| \\
&\iff \exists X \in 2^N : syq(P, \mathsf{M})_{a,X} \wedge \exists Y \in 2^N : \mathsf{S}^\mathsf{T}_{X,Y} \wedge syq(P, \mathsf{M})^\mathsf{T}_{Y,b} \\
&\iff \exists X \in 2^N : syq(P, \mathsf{M})_{a,X} \wedge (\mathsf{S}^\mathsf{T}; syq(P, \mathsf{M})^\mathsf{T})_{X,b} \\
&\iff (syq(P, \mathsf{M}); \mathsf{S}^\mathsf{T}; syq(P, \mathsf{M})^\mathsf{T})_{a,b}
\end{aligned}
$$

Finally, the definition of the equality of relations shows the desired result. □

It remains to specify the winners of an approval election relation-algebraically. How this can be done by means of a vector is shown next.

Theorem 3.2. *Based on the dominance relation $D : A \leftrightarrow A$ and the universal vector $\mathsf{L} : A \leftrightarrow \mathbf{1}$, the following vector $win : A \leftrightarrow \mathbf{1}$ describes the set of winners of an approval election $(N, A, (A_i)_{i \in N})$ as a subset of A.*

$$win = \overline{\overline{D}; \mathsf{L}} \tag{8}$$

Proof. We take an arbitrary alternative $a \in A$ and proceed as follows.

$$win_a \iff \overline{\overline{D}; \mathsf{L}}_a \iff \neg \exists b \in A : \overline{D}_{a,b} \wedge \mathsf{L}_b \iff \forall b \in A : D_{a,b}$$

As a consequence, the vector win describes the subset $\{a \in A \mid \forall b \in A : D_{a,b}\}$ of A as claimed. □

The relation-algebraic specifications (7) and (8) of Theorem 3.1 and Theorem 3.2, respectively, can be translated immediately into the programming language of RELVIEW. To give an impression of what the resulting code looks like, in the following we present a relational program `DomRel` for the computation of D and a relational function `winVec` for the computation of win. In `DomRel` the application of the pre-defined operation On1 yields the empty vector $O : N \leftrightarrow \mathbf{1}$. An argument of such a type is necessary for the two pre-defined operations epsi and cardrel to compute the relations $M : N \leftrightarrow 2^N$ and $S : 2^N \leftrightarrow 2^N$, respectively. The pre-defined operation dom used in winVec realizes the composition of its argument with an appropriate universal vector, i.e., yields a vector description of the domain of its argument.

```
DomRel(P)
  DECL M, S
  BEG   M = epsi(On1(P));
        S = cardrel(On1(P))
        RETURN syq(P,M) * S^ * syq(P,M)^
  END.

winVec(P) = -dom(-DomRel(P)).
```

After the headline with the name and the formal parameter, in its declaration part the program DomRel introduces two variables for relations, viz. M and S. In the subsequent body, first $M : N \leftrightarrow 2^N$ is computed and assigned to M and then $S : 2^N \leftrightarrow 2^N$ is computed and assigned to S. The final RETURN-clause of the program is nothing else than the formulation of $syq(P, \mathsf{M}); \mathsf{S}^\mathsf{T}; syq(P, \mathsf{M})^\mathsf{T}$ in RELVIEW-syntax, with *, ^, and - as symbols for the three relation-algebraic operations of composition, transposition, and complementation, respectively. In the same way, -dom(-DomRel(P)) is the RELVIEW version of $\overline{DomRel(P)}; \mathsf{L}$.

Example 3.2. If we apply DomRel and winVec to the relational model of approval election introduced in Example 3.1, then the tool depicts the results D and *win* as shown in the following two pictures.

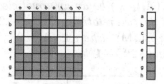

Hence, in case of our running example, the picture on the right shows that the alternatives b, f, g and h are the winners. This fact also can be obtained from the dominance relation on the left, since an alternative is a winner iff it dominates each alternative, i.e., the corresponding row of D consists of 1-entries only.

4   Relation-Algebraic Solutions of Control Problems

There are several possibilities to manipulate elections. In the already mentioned paper [2] the hardness of so-called *constructive* manipulations is investigated. By such manipulations the chair of an election tries to make a specific alternative a^* a winner, where it is assumed that he knows all the individual preferences. If the chair's manipulations shall prevent a^* from winning, then they are called *destructive*. The complexity of certain destructive manipulations have been investigated in [9] for the first time. Manipulations that delete or add voters and alternatives, respectively, are called *control manipulations*. In case of approval voting the following facts hold for them (see again [2,9]): Constructive control by deleting or adding voters is computationally hard, contrary to the destructive control by deleting or adding voters. For the latter there exist efficient algorithms. In case of destructive control by deleting alternatives it is even impossible for the chair to reach his goal and the same is true in case of constructive control by adding alternatives. One says that approval voting is immune to these types of control. For the remaining two types (constructive control by deleting alternatives and destructive control by adding alternatives) there exist straight-forward efficient manipulation algorithms.

In the following we combine relation algebra and RELVIEW to compute solutions for the two computationally hard control problems for approval voting. We start with the constructive control by deleting voters. Usually, the task is formulated as a minimization-problem: Given an approval election $(N, A, (A_i)_{i \in N})$ and a specific alternative a^*, determine a minimum set (w.r.t. size) of voters Y such that the removal of Y from N makes a^* a winner. To solve this task, we reformulate it as a maximization-problem and ask for a maximum set (w.r.t. size) of voters X such that a^* wins in the approval election $(X, A, (A_i)_{i \in X})$, i.e., subject to the condition that only voters from X are allowed to vote. It is obvious that from X then a desired Y is obtained via $Y = N \setminus X$.

As a first step towards a solution of the new task we relation-algebraically specify a vector that describes all candidate sets, that is, neglect the condition that X has to be a maximum set. In the corresponding Theorem 4.1 besides $M : N \leftrightarrow 2^N$ and $S : 2^N \leftrightarrow 2^N$ the first projection relation $\pi : 2^N \times A \leftrightarrow 2^N$ of the direct product $2^N \times A$ is used.

Theorem 4.1. *Let $(N, A, (A_i)_{i \in N})$ be an approval election, $a^* \in A$ be described by the point $p : A \leftrightarrow \mathbf{1}$, and the two relations $E : 2^N \leftrightarrow 2^N$ and $F : 2^N \leftrightarrow 2^N \times A$ be defined by $E = syq(\mathsf{M} \cap P; p; \mathsf{L}, \mathsf{M})$, where $\mathsf{L} : \mathbf{1} \leftrightarrow 2^N$, and $F = syq(\mathsf{M}, [\mathsf{M}, P])$. If we define the vector $dcand : 2^N \leftrightarrow \mathbf{1}$ by*

$$dcand = \overline{(E; (\mathsf{S} \cap \overline{\mathsf{S}}^\mathsf{T}); F \cap \pi^\mathsf{T}); \mathsf{L}}, \qquad (9)$$

where $\mathsf{L} : 2^N \times A \leftrightarrow \mathbf{1}$, then we have for all $X \in 2^N$ that $dcand_X$ holds iff the alternative a^* wins in the approval election $(X, A, (A_i)_{i \in X})$.

Proof. Let an arbitrary set $X \in 2^N$ be given. First, we show for all $i \in N$ the following property, where the assumption that $a^* \in A$ is described by $p : A \leftrightarrow \mathbf{1}$ is used in the second step.

$$(P; p)_i \iff \exists a \in A : P_{i,a} \wedge p_a \iff \exists a \in A : P_{i,a} \wedge a = a^* \iff P_{i,a^*}$$

Based on this equivalence, property (1), the first property of (4), and that the relation P is a relational model of $(N, A, (A_i)_{i \in N})$, in the next part we prove for all sets $Y \in 2^N$ the subsequent equivalence.

$$E_{X,Y} \iff syq(\mathsf{M} \cap P; p; \mathsf{L}, \mathsf{M})_{X,Y}$$
$$\iff \forall i \in N : (\mathsf{M} \cap P; p; \mathsf{L})_{i,X} \leftrightarrow \mathsf{M}_{i,Y}$$
$$\iff \forall i \in N : (\mathsf{M}_{i,X} \wedge (P; p; \mathsf{L})_{i,X}) \leftrightarrow \mathsf{M}_{i,Y}$$
$$\iff \forall i \in N : (i \in X \wedge (P; p)_i) \leftrightarrow i \in Y$$
$$\iff \forall i \in N : (i \in X \wedge P_{i,a^*}) \leftrightarrow i \in Y$$
$$\iff \{i \in X \mid P_{i,a^*}\} = Y$$
$$\iff \{i \in X \mid a^* \in A_i\} = Y$$

Third, we additionally use property (3) and treat the relation F in a similar way, i.e., calculate for all sets $Z \in 2^N$ and alternatives $b \in A$ as follows.

$$F_{Z,(X,b)} \iff (syq(\mathsf{M}, [\mathsf{M}, P]))_{Z,(X,b)}$$
$$\iff \forall i \in N : \mathsf{M}_{i,Z} \leftrightarrow [\mathsf{M}, P]_{i,(X,b)}$$
$$\iff \forall i \in N : i \in Z \leftrightarrow (\mathsf{M}_{i,X} \wedge P_{i,b})$$
$$\iff \forall i \in N : i \in Z \leftrightarrow (i \in X \wedge P_{i,b})$$
$$\iff Z = \{i \in X \mid P_{i,b}\}$$
$$\iff Z = \{i \in X \mid b \in A_i\}$$

Now we combine the last two results and obtain for all alternatives $b \in A$ the following equivalence, where the second property of (4) implies that the relationship $(S \cap \overline{S}^{\mathsf{T}})_{Y,Z}$ holds iff $|Y| < |Z|$.

$$(E; (S \cap \overline{S}^{\mathsf{T}}); F)_{X,(X,b)}$$
$$\iff \exists Y, Z \in 2^N : E_{X,Y} \wedge (S \cap \overline{S}^{\mathsf{T}})_{Y,Z} \wedge F_{Z,(X,b)}$$
$$\iff \exists Y, Z \in 2^N : E_{X,Y} \wedge |Y| < |Z| \wedge F_{Z,(X,b)}$$
$$\iff \exists Y, Z \in 2^N : Y = \{i \in X \mid a^* \in A_i\} \wedge Z = \{i \in X \mid b \in A_i\} \wedge |Y| < |Z|$$
$$\iff |\{i \in X \mid a^* \in A_i\}| < |\{i \in X \mid b \in A_i\}|$$

Finally, we apply this result in combination with the first property of (2) and calculate for the assumed set $X \in 2^N$ as follows.

$$dcand_X \iff \overline{(E; (S \cap \overline{S}^{\mathsf{T}}); F \cap \pi^{\mathsf{T}}); \mathsf{L}}_X$$
$$\iff \neg \exists U \in 2^N, b \in A : (E; (S \cap \overline{S}^{\mathsf{T}}); F \cap \pi^{\mathsf{T}})_{X,(U,b)} \wedge \mathsf{L}_{(U,b)}$$
$$\iff \neg \exists U \in 2^N, b \in A : (E; (S \cap \overline{S}^{\mathsf{T}}); F)_{X,(U,b)} \wedge \pi_{(U,b),X}$$
$$\iff \neg \exists U \in 2^N, b \in A : (E; (S \cap \overline{S}^{\mathsf{T}}); F)_{X,(U,b)} \wedge U = X$$
$$\iff \neg \exists b \in A : (E; (S \cap \overline{S}^{\mathsf{T}}); F)_{X,(X,b)}$$
$$\iff \neg \exists b \in A : |\{i \in X \mid a^* \in A_i\}| < |\{i \in X \mid b \in A_i\}|$$
$$\iff \forall b \in A : |\{i \in X \mid a^* \in A_i\}| \geq |\{i \in X \mid b \in A_i\}|$$

The last formula of this calculation states that the alternative a^* wins in the approval election $(X, A, (A_i)_{i \in X})$ and this concludes the proof. $\qquad \square$

In the second step towards a solution of our task we take the vector *dcand* of Theorem 4.1 and specify with its help a further vector *dsol* that describes the set of maximum sets of the set *dcand* described as a subset of 2^N. Doing so, S denotes again the corresponding size-comparison relation.

Theorem 4.2. *Given the vector* $dcand : 2^N \leftrightarrow \mathbf{1}$ *of the relation-algebraic specification (9), we specify a vector* $dsol : 2^N \leftrightarrow \mathbf{1}$ *as follows.*

$$dsol = dcand \cap \overline{\overline{\mathsf{S}}^\mathsf{T} ; dcand} \tag{10}$$

Then we have for all $X \in 2^N$ *that* $dsol_X$ *holds iff* X *is a maximum set such that the alternative* $a^* \in A$ *wins in the approval election* $(X, A, (A_i)_{i \in X})$.

Proof. We assume an arbitrary set $X \in 2^N$ and then calculate as given below, using the second property of (4).

$$
\begin{aligned}
dsol_X &\iff (dcand \cap \overline{\overline{\mathsf{S}}^\mathsf{T} ; dcand})_X \\
&\iff dcand_X \wedge \overline{\overline{\mathsf{S}}^\mathsf{T} ; dcand}\,_X \\
&\iff dcand_X \wedge \neg \exists\, Y \in 2^N : \overline{\mathsf{S}}_{Y,X} \wedge dcand_Y \\
&\iff dcand_X \wedge \forall\, Y \in 2^N : dcand_Y \to \mathsf{S}_{Y,X} \\
&\iff dcand_X \wedge \forall\, Y \in 2^N : dcand_Y \to |Y| \leq |X|
\end{aligned}
$$

Now the claim follows from Theorem 4.1.   □

Hence, the vector *dsol* describes the set $\mathfrak{S}$ of all solutions of our control problem as a subset of 2^N. It is rather troublesome to get from the vector *dsol* the solutions of the control problem as 'concrete' sets or relational vectors. We have to compare each 1-entry of *dsol* with the corresponding column of the membership relation $\mathsf{M} : N \leftrightarrow 2^N$. It is much simpler to apply the technique of Section 2 and to enumerate the set $\mathfrak{S}$ column-wise via the relation $\mathsf{M}; inj(dsol)^\mathsf{T} : N \leftrightarrow \mathfrak{S}$. As we will demonstrate in a moment, such a representation also allows to identify immediately the minimum sets of voters the original version of the control problem asks for.

The fundamental laws of relation algebra yield $\overline{\overline{\mathsf{S}}^\mathsf{T} ; dcand} = \overline{\overline{dcand^\mathsf{T} ; \overline{\mathsf{S}}}}^\mathsf{T}$ We used the right-hand expression of this equation to introduce the vector *dsol* in the RELVIEW program that follows from the relation-algebraic specification (10). The reason for the new definition of *dsol*, instead of the original one, is caused by the implementation of relations in RELVIEW. Namely, a ROBDD-implementation of relations implies that, compared with a simple Boolean matrix implementation, transposition may become more costly. This is due to the fact that it requires to exchange the variables for the encoding of the elements of the source with the variables for the encoding of the elements of the target. But in case of $\mathbf{1}$ as source or target transposition only means to exchange source and target (in RELVIEW: two numbers), the ROBDD of the relation remains unchanged. See [12] for details.

Example 4.1. Using the just mentioned RELVIEW program, we have solved the constructive control problem via the removal of voters for the relational model P of our running example and each of the 8 alternatives. The following series of RELVIEW pictures shows the corresponding column-wise enumerations of the sets of solutions, where the first relation corresponds to the case $a^* = a$, the second one to the case $a^* = b$, the third one to the case $a^* = c$, and so on.

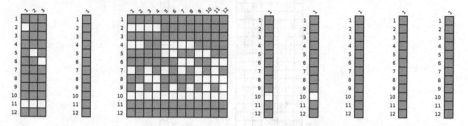

The universal vectors at position 2, 6, 7 and 8 show that the alternatives b, f, g and h win if all voters are allowed to vote, i.e., no voter has to be deleted to ensure win. From the remaining two vectors at position 4 and 5 we obtain that precisely the removal of voter 10 makes the alternatives d and e winning. To ensure that the alternative a wins, at least two voters have to be deleted. Which pairs suffice is depicted by the three columns of the Boolean matrix at position 1, viz. $2, 11$ and $5, 11$ and $6, 11$. Finally, from the Boolean matrix at position 3 we see that at least four voters have to be deleted if the alternative c is intended as a winner and there are 12 possibilities to reach this goal with deleting exactly four voters.

In the remainder of the section we consider the constructive control by adding voters. Here besides the original approval election and the specific alternative a^* a second approval election is given such that the sets of alternatives coincide, the sets of voters are disjoint, and the chair also knows all new individual preferences. The task is then to determine a minimum set of new voters such that their addition to the original ones makes a^* a winner.

A relation-algebraic solution of this control problem is possible using Theorem 4.1. To be consonant with the notation used in it, we assume now that N denotes the set of all voters, i.e., $N = N_o \cup N_a$ with N_o as the set of original voters and N_a as the set of additional (new) voters, and that $P : N \leftrightarrow A$ is again the relational model of the entire approval election. In terms of relation algebra P is the *relational sum* (in the sense of [15]) $P_o + P_a$ of the relational models $P_o : N_o \leftrightarrow A$ of the original election and $P_a : N_a \leftrightarrow A$ of the additional election. Pointwise this means that P behaves like P_o for all pairs with first component from N_o and like P_a for all pairs with first component from N_a. In RELVIEW the Boolean matrix of $P_o + P_a$ is formed by putting the matrices of P_o and P_a one upon another.

Example 4.2. We consider again our running example, that is, the relational model of Example 3.1. As additional voters we assume the natural numbers from 13 to 20. The 20×8 RELVIEW matrix below shows the relational model of the entire election as an extension of the relational model of Example 3.1. As new rows we have those labeled by 13 to 20. The 20×1 RELVIEW vector right of the matrix describes the subset $\{1, 2, \ldots, 12\}$ of original voters.

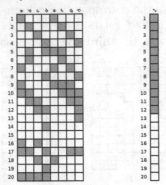

For example, from row 13 of the Boolean matrix we see that the new voter 13 approves only alternative h and from row 15 we see that the new voter 15 approves no alternative.

How to solve the constructive control problem by adding voters is shown now. In the corresponding Theorem 4.3 we use the membership relation $\mathsf{M} : N \leftrightarrow 2^N$ and the size-comparison relation $\mathsf{S} : 2^N \leftrightarrow 2^N$.

Theorem 4.3. *Assume that the vector* $ori : N \leftrightarrow \mathbf{1}$ *describes the set* N_o *of the original voters as a subset of* N, *the point* $p : A \leftrightarrow \mathbf{1}$ *describes* $a^* \in A$, *and* $dcand : 2^N \leftrightarrow \mathbf{1}$ *is the vector of the relation-algebraic specification (9). If we define a vector* $asol : 2^N \leftrightarrow \mathbf{1}$ *by*

$$asol = acand \cap \overline{\overline{\mathsf{S}}^{\mathsf{T}} ; acand}, \tag{11}$$

where $acand = \overline{\overline{\mathsf{M}}^{\mathsf{T}} ; ori} \cap dcand : 2^N \leftrightarrow \mathbf{1}$, *then we have for all* $X \in 2^N$ *that* $asol_X$ *holds iff* X *is a minimum set that contains* N_o *and the alternative* a^* *wins in the approval election* $(X, A, (A_i)_{i \in X})$.

Proof. Let an arbitrary set $X \in 2^N$ be given. Using that the vector ori describes the subset N_o of N and the first property of (4), we get the following result.

$$\overline{\overline{\mathsf{M}}^{\mathsf{T}} ; ori}_X \iff \neg \exists\, i \in N : \overline{\mathsf{M}}^{\mathsf{T}}_{X,i} \wedge ori_i$$
$$\iff \neg \exists\, i \in N : \overline{\mathsf{M}}_{i,X} \wedge ori_i$$
$$\iff \forall\, i \in N : i \notin X \to i \notin N_o$$
$$\iff \forall\, i \in N : i \in N_o \to i \in X$$
$$\iff N_o \subseteq X$$

In combination with Theorem 4.1 the above equivalence implies that $acand_X$ holds iff X is a set that contains N_o as subset and, furthermore, a^* wins in $(X, A, (A_i)_{i \in X})$. That $asol_X$ holds iff X is a minimum set with these two properties can be shown as in the proof of Theorem 4.2. □

$$\text{Besides } asol = acand \cap \overline{\overline{acand^{\mathsf{T}};\mathsf{S}}}^{\mathsf{T}},$$

we used $acand = \overline{\overline{ori^{\mathsf{T}};\mathsf{M}}}^{\mathsf{T}} \cap dcand$ as definition of the vector $acand$ in the RELVIEW program following from the relation-algebraic specification (11) to avoid the costly transpositions of the size comparison relation S and the membership relation M. In the following example we demonstrate what this RELVIEW program yields as results if the relation P and the vector ori of Example 4.2 are taken as inputs. To enhance readability, we apply the technique of Section 2 again and enumerate the solutions column-wise. Furthermore, we intersect each such enumeration relation $S : N \leftrightarrow \mathfrak{S}$ with the relation $ori; \mathsf{L} : N \leftrightarrow \mathfrak{S}$, where $\mathsf{L} : \mathbf{1} \leftrightarrow \mathfrak{S}$, since $S \cap ori; \mathsf{L}$ column-wise enumerates the solutions of the original control problem.

Example 4.3. Again we have solved the constructive control problem via the addition of voters for each of the eight alternatives. The following eight REL-VIEW pictures show the corresponding column-wise enumerations of the sets of voters which have to be added to ensure win.

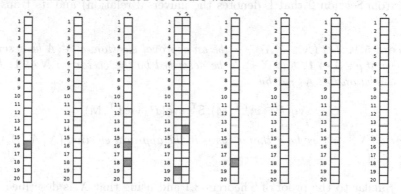

The order of the pictures is as in Example 4.2. Hence, the empty vectors at positions 2, 6, 7 and 8 show that the alternatives b, f, g and h win without any additional voter (see again Example 3.2). To ensure that one of the alternatives a, d or e wins, at least one least one voter in has to be added in each case. The vectors at positions 1 and 5 show that in case of the alternatives a and e the only possibilities are 16 and 18, respectively, and the matrix at position 4 shows that in case of the alternative d there are two possibilities, viz. 19 and 14. Finally, from the vector at position 3 we get that at least two voters have to be added to ensure that the alternative c wins and the only possibility is to add 16 and 18.

Each approval election possesses winners. Furthermore, for each constructive control problem by deleting voters there exist solutions. The extreme situation is that all voters have to be deleted to ensure that the specific alternative a^* wins. Then the dominance relation equals the universal relation $\mathsf{L} : A \leftrightarrow A$ and, as a consequence, all alternatives win. In contrast with constructive control by deleting voters, in case of constructive control by adding voters it may be impossible to make a^* winning by adding new voters from the given set N_a (for example, this happens in the extreme case when a^* does not win the original election and N_a is the empty set). Furthermore, it should be mentioned that the literature frequently also considers the so-called *single winner* variant of approval election. Here $a \in A$ dominates $b \in A$ if $|\{i \in N \mid a \in A_i\}| > |\{i \in N \mid b \in A_i\}|$ and a^* wins if it dominates all other alternatives. In this approach it may happen that no winners exist and both control problems we have considered are unsolvable for some instances.

5   Alternative Approaches

In settings as mentioned at the beginning of the paper, a chair is usually able to hide his manipulation only if the changes of the set of voters are not too large. Assuming this, we next present an alternative relation-algebraic approach to solve our control problems, where we restrict ourselves to the removal of voters. The approach is based on the following generalization of Theorem 3.1 that, besides the relations $\mathsf{M} : N \leftrightarrow 2^N$ and $\mathsf{S} : 2^N \leftrightarrow 2^N$, uses the vector $\mathsf{L} : A \leftrightarrow \mathbf{1}$ (recall from Section 2 that L denotes the universal relation) and its transpose $\mathsf{L}^\mathsf{T} : \mathbf{1} \leftrightarrow A$.

Theorem 5.1. *Let* $(N, A, (A_i)_{i \in N})$ *be an approval election,* $a^* \in A$ *be described by the point* $p : A \leftrightarrow \mathbf{1}$, *and* $X \subseteq N$ *be described by the vector* $v : N \leftrightarrow \mathbf{1}$. *If we define a relation* $C : A \leftrightarrow A$ *by*

$$C = syq(P \cap v; \mathsf{L}^\mathsf{T}, \mathsf{M}); \mathsf{S}^\mathsf{T}; syq(P \cap v; \mathsf{L}^\mathsf{T}, \mathsf{M})^\mathsf{T}, \tag{12}$$

then for all $X \in 2^N$ *we have that* a^* *wins in the approval election* $(X, A, (A_i)_{i \in X})$ *iff* $p \subseteq \overline{C}; \mathsf{L}$.

Proof. Similar to the proof of Theorem 3.1 and using that X is described by v as a subset of N, the following result can be shown for all $b, c \in A$.

$$C_{b,c} \iff |\{i \in X \mid b \in A_i\}| \geq |\{i \in X \mid c \in A_i\}|$$

Hence, C is the dominance relation of the approval election $(X, A, (A_i)_{i \in X})$. Now the claim follows from the calculation

$$
\begin{aligned}
p \subseteq \overline{C}; \mathsf{L} &\iff \forall b \in A : p_b \to \overline{C; \mathsf{L}}_b \\
&\iff \forall b \in A : b = a^* \to \neg \exists c \in A : \overline{C}_{b,c} \wedge \mathsf{L}_c \\
&\iff \forall b \in A : b = a^* \to \forall c \in A : C_{b,c} \\
&\iff \forall c \in A : C_{a^*,c}
\end{aligned}
$$

using that the point p describes a^* as an element of A.   □

Assume that the point $q : 2^N \leftrightarrow \mathbf{1}$ *describes a set of voters as an element of* 2^N, where $(N, A, (A_i)_{i \in N})$ is as above. Then the same set is *described as a subset of* N by the vector $\mathsf{M}; q : N \leftrightarrow \mathbf{1}$, with $\mathsf{M} : N \leftrightarrow 2^N$ as membership relation. See e.g., [15]. For a given $k \in \mathbb{N}$ with $0 \leq k \leq |N|$ this relationship allows immediately to test whether there exists a set X of k voters that makes a^*

winning. We only have to run through all $\binom{|N|}{k}$ points $q : 2^N \leftrightarrow \mathbf{1}$ with the property $q \subseteq \mathit{filter}(\mathsf{L}, k)$, where the operation *filter* is as in specification (5) and $\mathsf{L} : 2^N \leftrightarrow \mathbf{1}$, to compute in each case the relation C of the relation-algebraic specification (12) with v given as $v = \mathsf{M}; q$, and to test whether $p \subseteq \overline{C}; \mathsf{L}$ holds. The vector v^* of the first successful test describes a set X we are looking for. If we start the tests with $k = |N|$ and decrease k by 1 as long as the tests fail, then v^* even describes a maximum set X such that a^* wins, i.e., a solution of the control problem we presently consider. The just sketched approach can be easily formulated in the programming language of RELVIEW using while-loops and pre-defined operations such as *point* for the selection of a point $\mathit{point}(r) : X \leftrightarrow \mathbf{1}$ that is contained in a non-empty vector $r : X \leftrightarrow \mathbf{1}$.

Example 5.1. We consider the approval election $(N, A, (A_i)_{i \in N})$ the relational model of which is given by the 20×8 RELVIEW matrix of Example 4.2. I.e., we have $N = \{1, \ldots, 20\}$ as set of voters, $A = \{a, \ldots, h\}$ as set of alternatives, and $A_1 = \{a, e\}$, $A_2 = \{b, f\}$, $\ldots$, $A_{20} = \{a, b, c, d\}$ as the sets of alternatives the single voters approve.

RELVIEW computed that alternative d wins, the alternatives a, b and h win if one voter is deleted, the alternatives c and g win if two voters are deleted, and the alternatives e and f win if three voters are deleted. To demonstrate the enormous improvement of efficiency for cases where only a few voters have to be deleted, we present the running times for alternative f. They have been obtained with the newest version of RELVIEW (version 8.1, released September 2012 and available free of charge via the Web site [17]) on an ordinary desktop PC with CPU AMD Phenom II X4 810, 2.6 GHz and 16 GB RAM and running Linux.

With the approach of Section 4 it took 77.41 sec. to compute that f wins precisely if one of the triples $4, 17, 20$ or $8, 17, 20$ or $12, 17, 20$ is deleted, whereas with the approach of this section the first solution $4, 17, 20$ was found within 0.26 sec. only.

In this approach we basically use brute force on the powerset of the set of voters to check whether the removal of a certain set of voters ensures that the alternative a^* wins. Now let us assume, that the powerset of the voters is given as a list, such that the elements are sorted increasingly according to their cardinality. We can then traverse this list from left to right until we find some set P, such that a^* wins if all voters from P are removed. Since the list is sorted increasingly w.r.t. cardinality we also have that the first set we find is a minimal (w.r.t. cardinality) set as well, since any set with strictly smaller cardinality would have been found earlier in the traversal.

This implementation idea fits well into the lazy functional language Haskell. We used Haskell to implement this search based upon the following two components. The outer component is a function that takes an alternative and a set of voters and checks whether the removal of the set of voters results in the alternative being amongst the winners. This function is then applied to every element of the "powerlist" and the search is stopped as soon as such a set is found. The test is based upon a multiplication of a matrix with a vector which in this case results in complexity of $\mathcal{O}(|A| \cdot |N|)$. (We use a naïve implementation of the multiplication function, since our test cases are rather small and we want to check performance on a prototypical level only.) The inner component uses a technique from functional programming that is commonly used in the context of memoization. The actual implementation is inspired by the tabulated computation of binomial coefficients [8]. Altogether the functional implementation is a rather short program with a very good space behavior. It consumes less than 50 kB of memory for test cases with 40 voters.

Example 5.2. To give an impression on the running times of our Haskell program, we consider the set $N = \{1, \ldots, 40\}$ of 40 voters and the set $A = \{a^*, b\}$ of two alternatives. Furthermore, for a given set I of voters we define the set $A_i = \{b\}$ for all $i \in I$ and the set $A_i = \emptyset$ for all $i \in N \setminus I$. Clearly, the minimal set of voters that has to be removed for a^* to win is precisely the set I, since removing I from N results in no votes at all and thus every alternative gets exactly the same number of votes. Obviously, the set described above can be computed in linear time (w.r.t. voters): since no voter has voted for alternative a^*, we need to remove every voter who did vote and determining, whether a voter has voted or not, can be computed in constant time. (We can assume that the votes of a voter are stored in a list so that we merely need to check a list for being empty.) Nevertheless we use the brute force approach since it allows us to compare average and worst case behaviors of different implementations in a simple fashion.

We have experimented with randomly generated sets I of specified cardinalities k using again the PC mentioned in Example 5.1. Doing so, we ran the program on random cases and on cases where the last possible set of a given cardinality is the first that results in a^* winning. The next table summarizes our measurements (with the running times given in seconds).

k	1	2	3	4	5	6	7	8	9	10
random	0.01	0.01	0.01	0.1	0.66	4.24	23.79	109.76	433.21	1315.75
worst	0.01	0.02	0.04	0.32	2.03	11.61	60.38	248.68	948.64	3055.69

Clearly, removing a small number of voters comes with negligible running times of only a couple of seconds, whereas removing 10 of 40 voters can take about 51 minutes already.

We also implemented the brute force solution as a C-program. To enumerate all subsets of the set of voters, we used a very simple method, namely a for-loop using an integer counter, where the bits in the binary representation are interpreted as indicators whether a certain voter is an element of the currently studied subset or not. In an election with two alternatives and randomly distributed votes we obtained the following running times (in seconds), depending on the number n of voters:

n	20	25	26	27	28	29	30	31	32	33	34	35
time	0.2	8	16	34	50	144	293	590	1196	2445	5093	10359

The running time shows the expected exponential behaviour, where the running time is roughly doubled for each additional voter.

6 Conclusion

In this paper we have used relation algebra to model approval voting and to develop solutions for two hard control problems. All relation-algebraic specifications are algorithmic and can be evaluated by the ROBDD-based tool REL-VIEW after a straightforward translation into the tool's programming language. To demonstrate the visualization facilities of RELVIEW, we have used a small running example. We also have compared its computational power with that of conventional programming languages.

The correctness of all relation-algebraic specifications and, hence, also of the corresponding RELVIEW programs is guaranteed by the highly formal calculations. In the paper we have used the prevalent mathematical theorem-proof-style to emphasize the results and to enhance readability. But, in fact, we have obtained the results by developing them formally from the original logical specifications. We regard this goal-oriented development of algorithms from logical specifications, that are correct by construction, as the first main advantage of our approach. As its second main advantage we regard its computer-support by means of an appropriate tool. All results we have developed are expressed by very short and concise RELVIEW programs. Thus, these are easy to alter in case of slightly changed specifications, e.g., if to win means to be a single winner. Combining this with RELVIEW's possibilities for visualization and animation allows to experiment with established as well as new concepts while avoiding unnecessary overhead. This makes RELVIEW very useful for scientific research, since nowadays in computer science systematic experiments are accepted as a means for obtaining new insights and results. In this regard the very efficient ROBDD-implementation of relations is of immense help since it also allows to experiment with non-trivial examples.

Acknowledgment. We thank the unknown referees for their comments and suggestions. They helped to improve the paper.

References

1. Bartholdi III, J.J., Tovey, C.A., Trick, M.A.: The computational difficulty of manipulating an election. Social Choice and Welfare 6, 227–241 (1989)
2. Bartholdi III, J.J., Tovey, C.A., Trick, M.A.: How hard is it to control an election? Mathematical and Computer Modeling 16, 27–40 (1992)
3. Berghammer, R., Neumann, F.: RELVIEW – An OBDD-based Computer Algebra system for relations. In: Ganzha, V.G., Mayr, E.W., Vorozhtsov, E.V. (eds.) CASC 2005. LNCS, vol. 3718, pp. 40–51. Springer, Heidelberg (2005)

4. Brams, S.J., Fishburn, P.C.: Voting procedures. In: Arrow, K., Sen, A., Suzumara, K. (eds.) Handbook of Social Choice and Welfare, vol. 1, pp. 173–236. North-Holland (2002)
5. Brams, S.J., Fishburn, P.C.: Approval voting, 2nd edn. Springer (2007)
6. Brandt, F., Conitzer, V., Endriss, U.: Computational social choice. In: Weiss, G. (ed.) Multiagent Systems, 2nd edn., pp. 213–283. MIT Press (2013)
7. Chevaleyre, Y., Endriss, U., Lang, J., Maudet, N.: A short introduction to computational social choice. In: van Leeuwen, J., Italiano, G.F., van der Hoek, W., Meinel, C., Sack, H., Plášil, F. (eds.) SOFSEM 2007. LNCS, vol. 4362, pp. 51–69. Springer, Heidelberg (2007)
8. Fischer S.: Tabulated binomial coefficients, http://www-ps.informatik.uni-kiel.de/~sebf/haskell/tabulated-binomial-coefficients.lhs.html
9. Hemaspaandra, E., Hemaspaandra, L., Rothe, J.: Anyone but him: The complexity of precluding an alternative. Artificial Intelligence 171, 255–285 (2007)
10. Laslier, J.-F.: And the loser is . . . plurality voting. In: Felsenthal, D.S., Machover, M. (eds.) Electoral Systems, Studies in Choice and Welfare, pp. 327–351. Springer (2012)
11. Leoniuk, B.: ROBDD-based implementation of relational algebra with applications. Dissertation, Universität Kiel (2001) (in German)
12. Milanese, U.: On the implementation of a ROBDD-based tool for the manipulation and visualization of relations. Dissertation, Universität Kiel (2003) (in German)
13. Nurmi, H.: On the difficulty of making social choices. Theory and Decision 38, 99–119 (1998)
14. Schmidt, G., Ströhlein, T.: Relations and graphs, Discrete mathematics for computer scientists. EATCS Monographs on Theoretical Computer Science. Springer (1993)
15. Schmidt, G.: Relational mathematics. Encyclopedia of Mathematics and its Applications, vol. 132. Cambridge University Press (2010)
16. Walsh, T.: Is computational complexity a barrier to manipulation? Annals of Mathematics and Artificial Intelligence 62, 7–26 (2011)
17. http://www.informatik.uni-kiel.de/~progsys/relview

Relational Lattices

Tadeusz Litak[1,*], Szabolcs Mikulás[2], and Jan Hidders[3]

[1] Informatik 8, Friedrich-Alexander-Universität Erlangen-Nürnberg
Martensstraße 3, 91058 Erlangen, Germany
tadeusz.litak@gmail.com
[2] School of Computer Science and Information Systems,
Birkbeck, University of London, WC1E 7HX London, UK
szabolcs@dcs.bbk.ac.uk
[3] Delft University of Technology, Elektrotechn., Wisk. and Inform.,
Mekelweg 4, 2628CD Delft, The Netherlands
A.J.H.Hidders@tudelft.nl

Abstract. *Relational lattices* are obtained by interpreting lattice connectives as *natural join* and *inner union* between database relations. Our study of their equational theory reveals that the variety generated by relational lattices has not been discussed in the existing literature. Furthermore, we show that addition of just the *header constant* to the lattice signature leads to undecidability of the quasiequational theory. Nevertheless, we also demonstrate that relational lattices are not as intangible as one may fear: for example, they do form a pseudoelementary class. We also apply the tools of Formal Concept Analysis and investigate the structure of relational lattices via their standard contexts.

Keywords: relational lattices, relational algebra, database theory, algebraic logic, lattice theory, cylindric algebras, Formal Concept Analysis, standard context, incidence relation, arrow relations.

1 Introduction

We study a class of lattices with a natural database interpretation [Tro, ST06, Tro05]. It does not seem to have attracted the attention of algebraists, even those investigating the connections between algebraic logic and relational databases (see, e.g., [IL84] or [DM01]).

The connective *natural join* (which we will interpret as lattice meet!) is one of the basic operations of Codd's *(named) relational algebra* [AHV95, Cod70].

---

* We would like to thank *Vadim Tropashko* and *Marshall Spight* for introducing the subject to the third author (who in turn introduced it to the other two) and discussing it in the usenet group comp.databases.theory, *Maarten Marx, Balder ten Cate, Jan Paredaens* for additional discussions and general support in an early phase of our cooperation and the *referees* for the comments. The first author would also like to acknowledge: a) *Peter Jipsen* for discussions in September 2013 at the Chapman University leading to recovery, rewrite and extension of the material (in particular for Sec. 5) and b) suggestions by participants of: TACL'09, ALCOP 2010 and the Birmingham TCS seminar (in particular for Sec. 2.1 and 6).

P. Höfner et al. (Eds.): RAMiCS 2014, LNCS 8428, pp. 327–343, 2014.
© Springer International Publishing Switzerland 2014

Incidentally, it is also one of its few genuine algebraic operations—i.e., defined for all arguments. Codd's "algebra", from a mathematical point of view, is only a *partial algebra*: some operations are defined only between relations with suitable headers, e.g., the (set) union or the difference operator. Apart from the issues of mathematical elegance and generality, this partial nature of operations has also unpleasant practical consequences. For example, queries which do not observe constraints on headers can *crash* [VdBVGV07].

It turns out, however, that it is possible to generalize the union operation to *inner union* defined on all elements of the algebra and lattice-dual to natural join. This approach appears more natural and has several advantages over the embedding of relational "algebras" in cylindric algebras proposed in [IL84]. For example, we avoid an artificial uniformization of headers and hence queries formed with the use of proposed connectives enjoy the *domain independence property* (see, e.g., [AHV95, Ch. 5] for a discussion of its importance in databases).

We focus here on the (quasi)equational theory of natural join and inner union. Apart from an obvious mathematical interest, Birkhoff-style equational inference is the basis for certain query optimization techniques where algebraic expressions represent query evaluation plans and are rewritten by the optimizer into equivalent but more efficient expressions. As for *quasiequations*, i.e., definite Horn clauses over equalities, reasoning over many database constraints such as key constraints and foreign keys can be reduced to quasiequational reasoning. Note that an optimizer can consider more equivalent alternatives for the original expression if it can take the specified database constraints into account.

Strikingly, it turned out that relational lattices does not seem to fit anywhere into the rather well-investigated landscape of equational theories of lattices [JR92, JR98]. Nevertheless, there were some indications that the considered choice of connectives may lead to positive results concerning decidability/axiomatizability even for quasiequational theories. There is an elegant procedure known as *the chase* [AHV95, Ch. 8] applicable for certain classes of queries and database constraints similar to those that can be expressed with the natural join and inner union.

To our surprise, however, it turned out that when it comes to decidability, relational lattices seem to have a lot in common with other "untamed" structures from algebraic logic such as Tarski's relation algebras or cylindric algebras. As soon as an additional *header constant* H is added to the language, one can encode the word problem for semigroups in the quasiequational theory using a technique introduced by Maddux [Mad80]. This means that decidability of query equivalence under constraints for restricted positive database languages does not translate into decidability of corresponding quasiequational theories. However, our Theorem 4.7 and Corollary 4.8 do not rule out possible finite axiomatization results (except for quasiequational theory of *finite* structures) or decidability of equational theory.[1] And with H removed, i.e., in the pure lattice signature, the picture is completely open. Of course, such a language would be rather weak from a database point of view, but natural for an algebraist.

---

[1] Note, however, that an extension of our signature to a language with EDPC or a discriminator term would result in an undecidable *equational* theory.

We also obtained a number of positive results. First of all, concrete relational lattices are pseudoelementary and hence their closure under subalgebras and products is a quasivariety—Theorem 4.1 and Corollary 4.3. The proof yields an encoding into a sufficiently rich (many-sorted) first-order theory with finitely many axioms. This opens up the possibility of using generic proof assistants like Isabelle or Coq in future investigations—so far, we have only used Prover9/Mace4 to study interderivability of interesting (quasi)equations.[2] We have also used the tools of Formal Concept Analysis (Theorem 5.3) to investigate the dual structure of full concrete relational lattices and establish, e.g., their subdirect irreducibility (Corollary 5.4). Theorem 5.3 is likely to have further applications—see the discussion of Problem 6.1.

The structure of the paper is as follows. In Section 2, we provide basic definitions, establish that relational lattices are indeed lattices and note in passing a potential connection with category theory in Section 2.1. Section 3 reports our findings about the (quasi)equational theory of relational lattices: the failure of most standard properties such as weakening of distributivity (Theorem 3.2), those surprising equations and properties that still hold (Theorem 3.4) and dependencies between them (Theorem 3.5). In Section 4, we focus on quasiequations and prove some of most interesting results discussed above, both positive (Theorem 4.1 and Corollaries 4.2–4.4) and negative ones (Theorem 4.7 and Corollaries 4.8–4.9). Section 5 analyzes *standard contexts*, *incidence* and *arrow relations* [GW96] of relational lattices. Section 6 concludes and discusses future work, in particular possible extensions of the signature in Section 6.1.

2   Basic Definitions

Let $\mathcal{A}$ be a set of *attribute names* and $\mathcal{D}$ be a set of *domain values*. For $H \subseteq \mathcal{A}$, a *H-sequence from* $\mathcal{D}$ or an *H-tuple over* $\mathcal{D}$ is a function $x : H \to \mathcal{D}$, i.e., an element of $^H\mathcal{D}$. H is called the *header* of x and denoted as $h(x)$. The *restriction of x to H'* is denoted as $x[H']$ and defined as $x[H'] := \{(a,v) \in x \mid a \in H'\}$, in particular $x[H'] = \emptyset$ if $H' \cap h(x) = \emptyset$. We generalize this to the *projection of a set of H-sequences X to a header H'* which is $X[H'] := \{x[H'] \mid x \in X\}$. A *relation* is a pair $r = (H_r, B_r)$, where $H_r \subseteq \mathcal{A}$ is the *header* of r and $B_r \subseteq {}^{H_r}\mathcal{D}$ the *body* of r. The collections of all relations over $\mathcal{D}$ whose headers are contained in $\mathcal{A}$ will be denoted as $R(\mathcal{D}, \mathcal{A})$. For the relations r, s, we define the *natural join* $r \bowtie s$, and *inner union* $r \oplus s$:

$$r \bowtie s := (H_r \cup H_s, \{x \in {}^{H_r \cup H_s}\mathcal{D} \mid x[H_r] \in B_r \text{ and } x[H_s] \in B_s\})$$

$$r \oplus s := (H_r \cap H_s, \{x \in {}^{H_r \cap H_s}\mathcal{D} \mid x \in B_r[H_s] \text{ or } x \in B_s[H_r]\})$$

In our notation, $\bowtie$ always binds stronger than $\oplus$. The *header constant* $\mathsf{H} := (\emptyset, \emptyset)$ plays a special role: for any r, $(H_r, B_r) \bowtie \mathsf{H} = (H_r, \emptyset)$ and hence r_1 and r_2 have the same headers iff $\mathsf{H} \bowtie r_1 = \mathsf{H} \bowtie r_2$. Note also that the projection of r_1 to H_{r_2} can be defined as $r_1 \oplus (\mathsf{H} \bowtie r_2)$. In fact, we can identify $\mathsf{H} \bowtie r$ and H_r. We denote $(R(\mathcal{D}, \mathcal{A}), \bowtie, \oplus, \mathsf{H})$ as $\mathfrak{R}^H(\mathcal{D}, \mathcal{A})$, with $\mathcal{L}_{\mathsf{H}}$ denoting the corresponding algebraic signature. $\mathfrak{R}(\mathcal{D}, \mathcal{A})$ is its reduct to the signature $\mathcal{L} := \{\bowtie, \oplus\}$.

---

[2] It is worth mentioning that the database inventor of relational lattices has in the meantime developed a dedicated tool [Tro].

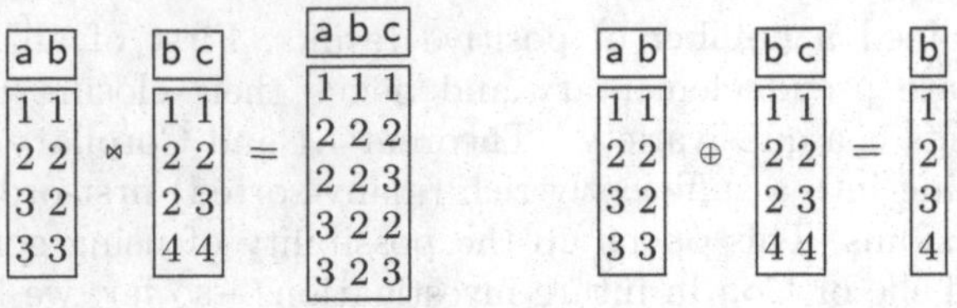

$$
\begin{array}{|cc|}\hline a & b \\\hline 1 & 1 \\ 2 & 2 \\ 3 & 2 \\ 3 & 3 \\\hline\end{array}
\;\bowtie\;
\begin{array}{|cc|}\hline b & c \\\hline 1 & 1 \\ 2 & 2 \\ 2 & 3 \\ 4 & 4 \\\hline\end{array}
\;=\;
\begin{array}{|ccc|}\hline a & b & c \\\hline 1 & 1 & 1 \\ 2 & 2 & 2 \\ 2 & 2 & 3 \\ 3 & 2 & 2 \\ 3 & 2 & 3 \\\hline\end{array}
\qquad
\begin{array}{|cc|}\hline a & b \\\hline 1 & 1 \\ 2 & 2 \\ 3 & 2 \\ 3 & 3 \\\hline\end{array}
\;\oplus\;
\begin{array}{|cc|}\hline b & c \\\hline 1 & 1 \\ 2 & 2 \\ 2 & 3 \\ 4 & 4 \\\hline\end{array}
\;=\;
\begin{array}{|c|}\hline b \\\hline 1 \\ 2 \\ 3 \\ 4 \\\hline\end{array}
$$

Fig. 1. Natural join and inner union. In this example, $\mathcal{A} = \{a, b, c\}$, $D = \{1, 2, 3, 4\}$.

Lemma 2.1. *For any D and $\mathcal{A}$, $\mathfrak{R}(D, \mathcal{A})$ is a lattice.*

Proof. This result is due to Tropashko [Tro, ST06, Tro05], but let us provide an alternative proof. Define $Dom := \mathcal{A} \cup {}^{\mathcal{A}}D$ and for any $X \subseteq Dom$ set

$$Cl(X) := X \cup \{x \in {}^{\mathcal{A}}D \mid \exists y \in (X \cap {}^{\mathcal{A}}D).\, x[\mathcal{A} - X] = y[\mathcal{A} - X]\}.$$

In other words, $Cl(X)$ is the sum of $X \cap \mathcal{A}$ (the set of attributes contained in X) with the cylindrification of $X \cap {}^{\mathcal{A}}D$ along the axes in $X \cap \mathcal{A}$. It is straightforward to verify Cl is a closure operator and hence Cl-closed sets form a lattice, with the order being obviously $\subseteq$ inherited from the powerset of Dom. It remains to observe $\mathfrak{R}(D, \mathcal{A})$ is isomorphic to this lattice and the isomorphism is given by

$$(H, B) \mapsto (\mathcal{A} - H) \cup \{x \in {}^{\mathcal{A}}D \mid x[H] \in B\}. \qquad \square$$

We call $\mathfrak{R}(D, \mathcal{A})$ the *(full) relational lattice over $(D,\mathcal{A})$*. We also use the alternative name *Tropashko lattices* to honor the inventor of these structures. The lattice order given by $\bowtie$ and $\oplus$ is

$$(H_r, B_r) \sqsubseteq (H_s, B_s) \text{ iff } H_s \subseteq H_r \text{ and } B_r[H_s] \subseteq B_s.$$

For classes of algebras, we use $\mathbb{H}, \mathbb{S}, \mathbb{P}$ to denote closures under, respectively, homomorphisms, (isomorphic copies of) subalgebras and products. Let

$$\mathcal{R}^{\mathsf{H}}_{\mathrm{fin}} := \mathbb{S}\{\mathfrak{R}^{\mathsf{H}}(D, \mathcal{A}) \mid D, \mathcal{A} \text{ finite}\}, \quad \mathcal{R}^{\mathsf{H}}_{\mathrm{unr}} := \mathbb{S}\{\mathfrak{R}^{\mathsf{H}}(D, \mathcal{A}) \mid D, \mathcal{A} \text{ unrestricted}\}$$

and let $\mathcal{R}_{\mathrm{fin}}$ and $\mathcal{R}_{\mathrm{unr}}$ denote the lattice reducts of respective classes.

2.1 Relational Lattice as the Grothendieck Construction

Given D and $\mathcal{A}$, a category theorist may note that

$$F^{\mathcal{A}}_{D} : \quad \mathcal{P}^{\supseteq}(\mathcal{A}) \ni H \longrightarrow \mathcal{P}({}^{H}D) \in \mathbf{Cat}$$

$$F^{\mathcal{A}}_{D}(H \supseteq H') := \quad ({}^{H}D \supseteq B \mapsto B[H'] \subseteq {}^{H'}D)$$

defines a *quasifunctor* assigning to an element of the powerset $\mathcal{P}^{\supseteq}(\mathcal{A})$ (considered as a poset with reverse inclusion order) the poset $\mathcal{P}({}^{H}D)$ considered as a small category. Then one readily notes that $\mathfrak{R}(D, \mathcal{A})$ is an instance of what is known as the (covariant) *Grothendieck construction/completion*[3] of $F_{\mathcal{D}}^{\mathcal{A}}$ [Jac99, Definition 1.10.1] denoted as $\int^{\mathcal{P}^{\supseteq}(\mathcal{A})} F_{\mathcal{D}}^{\mathcal{A}}$. As such considerations are irrelevant for the rest of our paper, for the time being we just note this category-theoretical connection as a curiosity, but it might lead to an interesting future study.

3   Towards the Equational Theory of Relational Lattices

Let us begin the section with an open

Problem 3.1. Are $\mathbb{SP}(\mathcal{R}_{unr}^{H}) = \mathbb{HSP}(\mathcal{R}_{unr}^{H})$ and $\mathbb{SP}(\mathcal{R}_{unr}) = \mathbb{HSP}(\mathcal{R}_{unr})$?

If the answer is "no", it would mean that relational lattices should be considered a quasiequational rather than equational class (cf. Corollary 4.3 below). Note also that the decidability of equational theories seems of less importance from a database point of view than decidability of quasiequational theories. Nevertheless, relating to already investigated varieties of lattices seems a good first step. It turns out that weak forms of distributivity and similar properties (see [JR92, JR98, Ste99]) tend to fail dramatically:

Theorem 3.2. $\mathcal{R}_{fin}$ *(and hence* $\mathcal{R}_{unr}$*) does not have any of the following properties (see the above references or the proof below for definitions):*

1. *upper- and lower-semidistributivity,*
2. *almost distributivity and neardistributivity,*
3. *upper- or lower-semimodularity (and hence also modularity),*
4. *local distributivity/local modularity,*
5. *the Jordan–Dedekind chain condition,*
6. *supersolvability.*

Proof. For most clauses, it is enough to observe that $\mathfrak{R}(\{0,1\}, \{0\}))$ is isomorphic to L_4, one of the covers of the non-modular lattice N_5 in [McK72] (see also [JR98]): a routine counterexample in such cases. In more detail:
Clause 1:   Recall that *semidistributivity* is the property:
  $a \oplus b = a \oplus c$ implies $a \oplus b = a \oplus (b \times c)$.
  Now take a to be H and b and c to be the atoms with the header $\{0\}$.
Clause 2:   This is a corollary of Clause 1, see [JR92, Th 4.2 and Sec 4.3].
Clause 3:   Recall that *semimodularity* is the property:
  if $a \times b$ covers a and b, then $a \oplus b$ is covered by a and b.
  Again, take a to be H and b to be either of the atoms with the header $\{0\}$.
Clause 4:   This is a corollary of Clause 3, see [Mae74].
Clause 5:   Recall that *the Jordan-Dedekind chain condition* is the property that the cardinalities of two maximal chains between common end points are equal. This obviously fails in N_5.

---

[3] Note that to preserve the lattice structure of $\mathfrak{R}(\mathcal{D}, \mathcal{A})$ we *cannot* consider $F_{\mathcal{D}}^{\mathcal{A}}$ as a functor into **Set**, which would yield a special case of the Grothendieck construction known as the *category of elements*. Note also that we chose the covariant definition on $\mathcal{P}^{\supseteq}(\mathcal{A})$ rather than the contravariant definition on $\mathcal{P}(\mathcal{A})$ to ensure the order $\sqsubseteq$ does not get reversed inside each slice $\mathcal{P}(^{H}\mathcal{D})$.

Clause 6:   Recall that for finite lattices, *supersolvability* [Sta72] boils down to the existence of a maximal chain generating a distributive lattice with any other chain. Again, this fails in N_5.   □

Remark 3.3. Theorem 3.2 has an additional consequence regarding the notion called rather misleadingly *boundedness* in some references (see e.g., [JR92, p. 27]): being an image of a freely generated lattice by a *bounded morphism*. We use the term *McKenzie-bounded*, as McKenzie showed that for finite subdirectly irreducible lattices, this property amounts to splitting the lattice of varieties of lattices [JR92, Theorem 2.25]. Finite Tropashko lattices are subdirectly irreducible (Corollary 5.4 below) but Clause 1 of Theorem 3.2 entails they are not McKenzie-bounded by [JR92, Lemma 2.30].

Nevertheless, Tropashko lattices do not generate the variety of all lattices. The results of our investigations so far on valid (quasi)equations are summarized by the following theorems:

Theorem 3.4. *Axioms of $\underline{R}^{\mathsf{H}}$ in Table 1 are valid in $\mathcal{R}_{\mathrm{unr}}^{\mathsf{H}}$ (and consequently in $\mathcal{R}_{\mathrm{fin}}^{\mathsf{H}}$). Similarly, axioms of $\underline{R}$ are valid in $\mathcal{R}_{\mathrm{unr}}$ (and consequently $\mathcal{R}_{\mathrm{fin}}$).*

Table 1. (Quasi)equations Valid in Tropashko Lattices

Class $\underline{R}^{\mathsf{H}}$ in the signature $\mathcal{L}_{\mathsf{H}}$:

all lattice axioms

AxRH1 $\mathsf{H} \ltimes x \ltimes (y \oplus z) \oplus y \ltimes z = (\mathsf{H} \ltimes x \ltimes y \oplus z) \ltimes (\mathsf{H} \ltimes x \ltimes z \oplus y)$
AxRH2 $x \ltimes (y \oplus z) = x \ltimes (z \oplus \mathsf{H} \ltimes y) \oplus x \ltimes (y \oplus \mathsf{H} \ltimes z)$
AxRL1 $x \ltimes y \oplus x \ltimes z = x \ltimes (y \ltimes (x \oplus z) \oplus z \ltimes (x \oplus y))$

Class $\underline{R}$ in the signature $\mathcal{L}$ (without H):

all lattice axioms **together with AxRL1** and

AxRL2 $t \ltimes ((x \oplus y) \ltimes (x \oplus z) \oplus (u \oplus w) \ltimes (u \oplus v)) =$
             $= t \ltimes ((x \oplus y) \ltimes (x \oplus z) \oplus u \oplus w \ltimes v) \oplus t \ltimes ((u \oplus w) \ltimes (u \oplus v) \oplus x \oplus y \ltimes z)$
             (in $\mathcal{L}_{\mathsf{H}}$, AxRL2 is derivable from AxRH1 and AxRH2 above)

Additional (quasi)equations derivable in $\underline{R}^{\mathsf{H}}$ and $\underline{R}$:

Qu1	$x \oplus y = x \oplus z$	$\Rightarrow$ $x \ltimes (y \oplus z) = x \ltimes y \oplus x \ltimes z.$
Qu2	$\mathsf{H} \ltimes (x \oplus y) = \mathsf{H} \ltimes (x \oplus z)$	$\Rightarrow$ $x \ltimes (y \oplus z) = x \ltimes y \oplus x \ltimes z.$
Eq1	$\mathsf{H} \ltimes x \ltimes (y \oplus z)$	$= \mathsf{H} \ltimes x \ltimes y \oplus \mathsf{H} \ltimes x \ltimes z$
Der1	$\mathsf{H} \ltimes x \oplus x \ltimes y$	$= x \ltimes (y \oplus \mathsf{H} \ltimes x)$

Theorem 3.5. *Assuming all lattice axioms, the following statements hold:*

1. Axioms of $\underline{R}$ are mutually independent.

2. *Each of the axioms of $\underline{R}^H$ is independent from the remaining ones, with a possible exceptions of AxRL1.*
3. *[PMV07] AxRL1 forces Qu1.*
4. *Qu2 together with Eq1 imply AxRL2.*
5. *Eq1 is implied by AxRH1. The converse implication does not hold even in presence of AxRL1.*
6. *AxRH1 and AxRH2 jointly imply Qu2, although each of the two equations separately is too weak to entail Qu2. In the converse direction, Qu2 implies AxRH2 but not AxRH1.*
7. *AxRH1 implies Der1.*

Proof. Clause 1:   The example showing that the validity of AxRL2 does not imply the validity of AxRL1 is the non-distributive diamond lattice M_3, while the reverse implication can be disproved with an eight-element model:

Clause 2:   Counterexamples can be obtained by appropriate choices of the interpretation of H in the pentagon lattice.
Clause 4:   Direct computation.
Clause 5:   The first part has been proved with the help of Prover9 (66 lines of proof). The counterexample for the converse is obtained by choosing H to be the top element of the pentagon lattice.
Clause 6:   Prover9 was able to prove the first statement both in presence and in absence of AxRL1, although there was a significant difference in the length of both proofs (38 lines vs. 195 lines). The implication from Qu2 to AxRH2 is straightforward. All the necessary counterexamples can be found by appropriate choices of the interpretation of H in the pentagon lattice.
Clause 7:   Substitue x for z and use the absorption law.   □

AxRL1 comes from [PMV07] as an example of an equation which forces *the Huntington property* (distributivity under unique complementation). Qu1 is a form of weak distributivity, denoted as $CD_\vee$ in [PMV07] and $WD_\wedge$ in [JR98].

Problem 3.6. Are the equational theories of $\mathcal{R}_{unr}^H$ and $\mathcal{R}_{fin}^H$ equal?

Problem 3.7. Is the equational theory of $\mathcal{R}_{unr}^H$ ($\mathcal{R}_{unr}$) equal to $\underline{R}^H$ ($\underline{R}$, respectively)? If not, is it finitely axiomatizable at all?

If the answer to the last question is in the negative, one can perhaps attempt a rainbow-style argument from algebraic logic [HH02].

4   Relational Lattices as a Quasiequational Class

In the introduction, we discussed why an axiomatization of valid *quasi*equations is desirable from a DB point of view. There is also an algebraic reason: the class of representable Tropashko lattices (i.e., the $\mathbb{SP}$-closure of concrete ones) is a *quasi*variety. This is a corollary of a more powerful result:

Theorem 4.1. $\mathcal{R}_{\mathrm{unr}}^{\mathsf{H}}$ *and* $\mathcal{R}_{\mathrm{unr}}$ *are pseudoelementary classes.*

Proof. (sketch) Assume a language with sorts A, F, D and R. The connectives of $\mathcal{L}_{\mathsf{H}}$ live in R, we also have a relation symbol $inR : (F \cup A) \times R$ and a function symbol $assign : (F \times A) \mapsto D$. The interpretation is suggested by the closure system used in the proof of Lemma 2.1. That is, A denotes $\mathcal{A}$, F denotes $^{\mathcal{A}}\mathcal{D}$, D denotes $\mathcal{D}$ and R denotes the family of Cl-closed subsets of Dom. Moreover, $assign(f, a)$ denotes the value of the $\mathcal{A}$-sequence denoted by f on the attribute a and $inR(x, r)$—the membership of an attribute/sequence in the closed subset of Dom denoted by r. One needs to postulate the following axioms: "F and R are extensional" (the first via injectivity of $assign$, the second via axioms on inR); "each element of R is Cl-closed"; "$\ltimes$ and $\oplus$ are genuine infimum/supremum on R". For $\mathcal{R}_{\mathrm{unr}}^{\mathsf{H}}$, we add an axiom "$inR$ assigns no elements of F and all elements of A (the latter means all attributes are *irrelevant* for the element under consideration!) to H". $\qquad\square$

Corollary 4.2. $\mathcal{R}_{\mathrm{unr}}^{\mathsf{H}}$ *and* $\mathcal{R}_{\mathrm{unr}}$ *are closed under ultraproducts.*

Corollary 4.3. *The* $\mathbb{SP}$-*closures of* $\mathcal{R}_{\mathrm{unr}}^{\mathsf{H}}$ *and* $\mathcal{R}_{\mathrm{unr}}$ *are quasiequational classes.*

Corollary 4.4. *The quasiequational, universal and elementary theories of* $\mathcal{R}_{\mathrm{unr}}^{\mathsf{H}}$ *and* $\mathcal{R}_{\mathrm{unr}}$ *are recursively enumerable.*

Proof. The proof of Theorem 4.1 uses finitely many axioms. $\qquad\square$

Note that postulating that headers are *finite* subsets of $\mathcal{A}$ would break the proof of Theorem 4.1: such conditions are not first-order. However, concrete database instances always belong to $\mathcal{R}_{\mathrm{fin}}^{\mathsf{H}}$ and we will show now that the decidability status of the quasiequational theory of $\mathcal{R}_{\mathrm{unr}}^{\mathsf{H}}$ and $\mathcal{R}_{\mathrm{fin}}^{\mathsf{H}}$ is the same. Moreover, an undecidability result also obtains for the corresponding abstract class, much like for relation algebras and cylindric algebras—in fact, we build on a proof of Maddux [Mad80] for CA_3—and we *do not even need all the axioms* of $\underline{R}^{\mathsf{H}}$ to show this! Let $\underline{RH1}$ be the variety of $\mathcal{L}_{\mathsf{H}}$-algebras axiomatized by the lattice axioms and AxRH1. Let us list some basic observations:

Proposition 4.5.

1. $\mathcal{R}_{\mathrm{fin}}^{\mathsf{H}} \subset \mathcal{R}_{\mathrm{unr}}^{\mathsf{H}} \subset \mathbb{SP}(\mathcal{R}_{\mathrm{unr}}^{\mathsf{H}}) \subseteq \underline{R}^{\mathsf{H}} \subset \underline{RH1}$.
2. *Der1 holds in* $\underline{RH1}$.
3. *AxRH1 holds whenever* H *is interpreted as the bottom of a bounded lattice.*
4. *AxRH1 holds for an arbitrary choice of* H *in a distributive lattice.*

Proof. Clause 2 holds by clause 7 of Theorem 3.5. The remaining ones are straightforward to verify. $\qquad\square$

Note, e.g., that interpreting H as $\perp$ in AxRH2 would only work if the lattice is distributive, so Clause 3 would not hold in general for AxRH2. In order to state our undecidability result, we need first

Definition 4.6. *Let $\bar{e} = (u_0, u_1, u_2, e_0, e_1)$ be an arbitrary 5-tuple of variables. We abbreviate $u_0 \bowtie u_1 \bowtie u_2$ as u. For arbitrary L-terms s, t define*

$$\mathbf{c}_0^{\bar{e}} \langle t \rangle := u \bowtie (\mathsf{H} \bowtie u_1 \bowtie u_2 \oplus u \bowtie t),$$

$$\mathbf{c}_1^{\bar{e}} \langle t \rangle := u \bowtie (\mathsf{H} \bowtie u_0 \bowtie u_2 \oplus u \bowtie t),$$

$$\mathbf{c}_2^{\bar{e}} \langle t \rangle := u \bowtie (\mathsf{H} \bowtie u_0 \bowtie u_1 \oplus u \bowtie t),$$

$$s \circ^{\bar{e}} t := \mathbf{c}_2^{\bar{e}} \left\langle \mathbf{c}_1^{\bar{e}} \left\langle e_0 \bowtie \mathbf{c}_2^{\bar{e}} \langle s \rangle \right\rangle \bowtie \mathbf{c}_0^{\bar{e}} \left\langle e_1 \bowtie \mathbf{c}_2^{\bar{e}} \langle s \rangle \right\rangle \right\rangle.$$

Let $T_n(x_1, \ldots, x_n)$ be the collection of all semigroup terms in n variables. Whenever $\bar{e} = (x_{n+1}, \ldots, x_{n+5})$ define the translation $\tau^{\bar{e}}$ of semigroup terms as follows: $\tau^{\bar{e}}(x_i) := x_i$ for $i \leq n$ and $\tau^{\bar{e}}(s \circ t) := s \circ^{\bar{e}} t$ for any $s, t \in T_n(x_1, \ldots, x_n)$.

Whenever $\bar{e}$ is clear from the context, we will drop it to ensure readability. Now we can formulate

Theorem 4.7. *For any $p_0, \ldots, p_m, r_0, \ldots, r_m, s, t \in T_n(x_1, \ldots, x_n)$, the following conditions are equivalent :*

(I) The quasiequation

(Qu3) $\forall x_1, \ldots, x_n. (p_0 = r_0 \& \ldots \& p_m = r_m \Rightarrow s = t)$

holds in all semigroups (finite semigroups).

(II) For $\bar{e} = (x_{n+1}, \ldots, x_{n+5})$ as in Definition 4.6, the quasiequation

$$\forall x_0, x_1, \ldots, x_{n+5}. \, (\tau^{\bar{e}}(p_0) = \tau^{\bar{e}}(r_0) \& \ldots \tau^{\bar{e}}(p_m) = \tau^{\bar{e}}(r_m) \, \&$$

(Qu4)
$$\& \, x_{n+4} = \mathbf{c}_0^{\bar{e}} \langle x_{n+4} \rangle \, \& \, x_{n+5} = \mathbf{c}_1^{\bar{e}} \langle x_{n+5} \rangle) \Rightarrow$$
$$\Rightarrow \tau^{\bar{e}}(s) \circ^{\bar{e}} \mathbf{c}_1^{\bar{e}} \langle x_0 \rangle = \tau^{\bar{e}}(t) \circ^{\bar{e}} \mathbf{c}_1^{\bar{e}} \langle x_0 \rangle))$$

holds in every member of $\mathcal{R}_{\mathrm{unr}}^{\mathsf{H}}$ (every member of $\mathcal{R}_{\mathrm{fin}}^{\mathsf{H}}$).

(III) Qu4 above holds in every member of $\underline{RH1}$ (finite member of $\underline{RH1}$).

Proof. (I) $\Rightarrow$ (III). By contraposition:

Take any $\mathfrak{A} \in \underline{RH1}$ and arbitrarily chosen elements $u_0, u_1, u_2 \in \mathfrak{A}$. In order to use Maddux's technique, we have to prove that for any $a, b \in \mathfrak{A}$ and $k, l < 3$

(b) $\mathbf{c}_k \langle \mathbf{c}_k \langle a \rangle \rangle = \mathbf{c}_k \langle a \rangle$,
(c) $\mathbf{c}_k \langle a \bowtie \mathbf{c}_k \langle b \rangle \rangle = \mathbf{c}_k \langle a \rangle \bowtie \mathbf{c}_k \langle b \rangle$,
(d) $\mathbf{c}_k \langle \mathbf{c}_l \langle a \rangle \rangle = \mathbf{c}_l \langle \mathbf{c}_k \langle a \rangle \rangle$

(we deliberately keep the same labels as in the quoted paper), where $\mathbf{c}_k \langle a \rangle$ is defined in the same way as in Definition 4.6 above. We will denote by $u_{\hat{k}}$ the product of u_i's such that $i \in \{0, 1, 2\} - \{k\}$. For example, $u_{\hat{0}} = u_1 \bowtie u_2$.

For (b):

$$
\begin{aligned}
L &= u \bowtie (\mathsf{H} \bowtie u_{\hat{k}} \oplus u \bowtie (\mathsf{H} \bowtie u_{\hat{k}} \oplus u \bowtie a)) \\
&= u \bowtie (\mathsf{H} \bowtie u_{\hat{k}} \bowtie (u \oplus \mathsf{H} \bowtie u_{\hat{k}} \oplus u \bowtie a) \oplus u \bowtie (\mathsf{H} \bowtie u_{\hat{k}} \oplus u \bowtie a)) && \text{by lattice laws} \\
&= u \bowtie (\mathsf{H} \bowtie u_{\hat{k}} \bowtie u \oplus \mathsf{H} \bowtie u_{\hat{k}} \oplus u \bowtie a) \bowtie (\mathsf{H} \bowtie u_{\hat{k}} \bowtie (\mathsf{H} \bowtie u_{\hat{k}} \oplus u \bowtie a) \oplus u) && \text{by AxRH1} \\
&= u \bowtie (\mathsf{H} \bowtie u_{\hat{k}} \oplus u \bowtie a) \bowtie (\mathsf{H} \bowtie u_{\hat{k}} \oplus u) && \text{by lattice laws} \\
&= u \bowtie (\mathsf{H} \bowtie u_{\hat{k}} \oplus u \bowtie a) && \text{by lattice laws} \\
&= R.
\end{aligned}
$$

(c) is proved using a similar trick:

$$
\begin{aligned}
L &= u{\scriptstyle\times}(\mathsf{H}{\scriptstyle\times}u_{\hat{k}} \oplus u{\scriptstyle\times}a{\scriptstyle\times}(\mathsf{H}{\scriptstyle\times}u_{\hat{k}} \oplus u{\scriptstyle\times}b)) \\
&= u{\scriptstyle\times}(\mathsf{H}{\scriptstyle\times}u_{\hat{k}}{\scriptstyle\times}(u{\scriptstyle\times}a \oplus \mathsf{H}{\scriptstyle\times}u_{\hat{k}} \oplus u{\scriptstyle\times}b) \oplus u{\scriptstyle\times}a{\scriptstyle\times}(\mathsf{H}{\scriptstyle\times}u_{\hat{k}} \oplus u{\scriptstyle\times}b)) && \text{by lattice laws} \\
&= u{\scriptstyle\times}(\mathsf{H}{\scriptstyle\times}u_{\hat{k}}{\scriptstyle\times}u{\scriptstyle\times}a \oplus \mathsf{H}{\scriptstyle\times}u_{\hat{k}} \oplus u{\scriptstyle\times}b){\scriptstyle\times}(\mathsf{H}{\scriptstyle\times}u_{\hat{k}}{\scriptstyle\times}(\mathsf{H}{\scriptstyle\times}u_{\hat{k}} \oplus u{\scriptstyle\times}b) \oplus u{\scriptstyle\times}a) && \text{by AxRH1} \\
&= u{\scriptstyle\times}(\mathsf{H}{\scriptstyle\times}u_{\hat{k}} \oplus u{\scriptstyle\times}b){\scriptstyle\times}(\mathsf{H}{\scriptstyle\times}u_{\hat{k}} \oplus u{\scriptstyle\times}a) && \text{by lattice laws} \\
&= R.
\end{aligned}
$$

(d) is obviously true for $k = l$, hence we can restrict attention to $k \neq l$. Let j be the remaining element of $\{0, 1, 2\}$. Thus,

$$
\begin{aligned}
L &= u{\scriptstyle\times}(\mathsf{H}{\scriptstyle\times}u_l{\scriptstyle\times}u_j \oplus u{\scriptstyle\times}(\mathsf{H}{\scriptstyle\times}u_k{\scriptstyle\times}u_j \oplus u{\scriptstyle\times}a)) \\
&= u{\scriptstyle\times}(\mathsf{H}{\scriptstyle\times}u_l{\scriptstyle\times}u_j \oplus u_l{\scriptstyle\times}(\mathsf{H}{\scriptstyle\times}u_k{\scriptstyle\times}u_j \oplus u{\scriptstyle\times}a)) && \text{by Der1} \\
&= u{\scriptstyle\times}(\mathsf{H}{\scriptstyle\times}u_l{\scriptstyle\times}u_j{\scriptstyle\times}(u_l \oplus \mathsf{H}{\scriptstyle\times}u_k{\scriptstyle\times}u_j \oplus u{\scriptstyle\times}a) \oplus u_l{\scriptstyle\times}(\mathsf{H}{\scriptstyle\times}u_k{\scriptstyle\times}u_j \oplus u{\scriptstyle\times}a)) && \text{by lattice laws} \\
&= u{\scriptstyle\times}(\mathsf{H}{\scriptstyle\times}u_l{\scriptstyle\times}u_j \oplus \mathsf{H}{\scriptstyle\times}u_k{\scriptstyle\times}u_j \oplus u{\scriptstyle\times}a){\scriptstyle\times}(\mathsf{H}{\scriptstyle\times}u_l{\scriptstyle\times}u_j{\scriptstyle\times}(\mathsf{H}{\scriptstyle\times}u_k{\scriptstyle\times}u_j \oplus u{\scriptstyle\times}a) \oplus u_l) && \text{by AxRH1} \\
&= u{\scriptstyle\times}(\mathsf{H}{\scriptstyle\times}u_l{\scriptstyle\times}u_j \oplus \mathsf{H}{\scriptstyle\times}u_k{\scriptstyle\times}u_j \oplus u{\scriptstyle\times}a){\scriptstyle\times}u_l && \text{by lattice laws} \\
&= u{\scriptstyle\times}(\mathsf{H}{\scriptstyle\times}u_l{\scriptstyle\times}u_j \oplus \mathsf{H}{\scriptstyle\times}u_k{\scriptstyle\times}u_j \oplus u{\scriptstyle\times}a) && \text{by lattice laws}
\end{aligned}
$$

and in the last term, u_l and u_k may be permuted by commutativity. We then obtain the right side of the equation via an analogous sequence of transformations in the reverse direction, with the roles of u_k and u_l replaced.

The rest of the proof mimics the one in [Mad80]. In some detail: assume there is $\bar{e} = (u_0, u_1, u_2, e_0, e_1) \in \mathfrak{A}$ such that

(a) $\mathbf{c}_0^{\bar{e}} \langle e_0 \rangle = e_0$, $\mathbf{c}_1^{\bar{e}} \langle e_1 \rangle = e_1$

holds. Using (a)–(d) we prove that for every $a, b \in \mathfrak{A}$ the following hold:

(i) $\mathbf{c}_1^{\bar{e}} \langle a \circ^{\bar{e}} b \rangle = a \circ^{\bar{e}} \mathbf{c}_1^{\bar{e}} \langle b \rangle$,
(ii) $a \circ^{\bar{e}} \mathbf{c}_1^{\bar{e}} \langle b \rangle = \mathbf{c}_1^{\bar{e}} \langle \mathbf{c}_2^{\bar{e}} \langle a \rangle {\scriptstyle\times} \mathbf{c}_0^{\bar{e}} \langle \mathbf{c}_2^{\bar{e}} \langle e_0 {\scriptstyle\times} e_1 {\scriptstyle\times} \mathbf{c}_2^{\bar{e}} \langle \mathbf{c}_1^{\bar{e}} \langle b \rangle \rangle \rangle \rangle \rangle$,
(iii) $(a \circ^{\bar{e}} b) \circ^{\bar{e}} \mathbf{c}_1^{\bar{e}} \langle c \rangle = a \circ^{\bar{e}} (b \circ^{\bar{e}} \mathbf{c}_1^{\bar{e}} \langle c \rangle)$,
(iv) $((a \circ^{\bar{e}} b) \circ^{\bar{e}} c) \circ^{\bar{e}} \mathbf{c}_1^{\bar{e}} \langle d \rangle = (a \circ^{\bar{e}} (b \circ^{\bar{e}} c)) \circ^{\bar{e}} \mathbf{c}_1^{\bar{e}} \langle d \rangle$.

Now pick $\mathfrak{A}$ witnessing the failure of Qu4 together with $\bar{e} = (u_0, u_1, u_2, e_0, e_1)$ such that elements of $\bar{e}$ interpret variables $(x_{n+1}, \ldots, x_{n+5})$ in Qu4. This means (a) is satisfied, hence (i)–(iv) hold for every element of $\mathfrak{A}$. We define an equivalence relation $\equiv$ on $\mathfrak{A}$:

$a \equiv b$ iff for all $c \in \mathfrak{A}$, $a \circ^{\bar{e}} \mathbf{c}_1^{\bar{e}} \langle c \rangle = b \circ^{\bar{e}} \mathbf{c}_1^{\bar{e}} \langle c \rangle$.

We take $\circ^{\bar{e}}$ to be the semigroup operation on $\mathfrak{A}/ \equiv$. Following [Mad80], we use (i)–(iv) to prove that this operation is well-defined (i.e., independent of the choice of representatives) and satisfies semigroup axioms. It follows from the assumptions that the semigroup thus defined fails Qu3.

(III) $\Rightarrow$ (II). Immediate.

(II) $\Rightarrow$ (I). In analogy to [Mad80], given a semigroup $\mathfrak{B} = (B, \circ, \mathsf{u})$ failing Qu3 and a valuation v witnessing this failure, consider $\mathfrak{R}(B, \{0, 1, 2\})$ with a valuation w defined as follows:

$$w(x_0) := (\{0, 1, 2\}, \{\{(0, v(r)), (1, a), (2, b)\} \mid a, b \in \mathfrak{B}\}),$$
$$w(x_i) := (\{0, 1, 2\}, \{\{(0, a), (1, a \circ v(x_i)), (2, b)\} \mid a, b \in \mathfrak{B}\}), \qquad i \leq n,$$
$$w(x_{n+i}) := (\{i\}, \{\{(i, b)\} \mid b \in \mathfrak{B}\}), \qquad\qquad (0 < i \leq 3),$$
$$w(x_{n+4}) := (\{0, 1, 2\}, \{\{(0, a), (1, b), (2, b)\} \mid a, b \in \mathfrak{B}\}),$$
$$w(x_{n+5}) := (\{0, 1, 2\}, \{\{(0, b), (1, a), (2, b)\} \mid a, b \in \mathfrak{B}\}).$$

It is proved by induction that

$$w(\tau^{\overline{e}}(t)) = (\{0, 1, 2\}, \{\{(0, a), (1, a \circ v(t)), (2, b)\} \mid a, b \in \mathfrak{B}\})$$

(where $e = (x_{n+1}, \ldots, x_{n+5})$) for every $t \in T(x_1, \ldots x_n)$ and also

$$w(\tau^{\overline{e}}(s) \circ^{\overline{e}} \mathbf{c}_1^{\overline{e}} \langle x_0 \rangle) = (\{0, 1, 2\}, \{\{(0, a), (1, b), (2, c)\} \mid a, b, c \in \mathfrak{B}, v(r) \circ a = v(s)\}),$$
$$w(\tau^{\overline{e}}(r) \circ^{\overline{e}} \mathbf{c}_1^{\overline{e}} \langle x_0 \rangle) = (\{0, 1, 2\}, \{\{(0, a), (1, b), (2, c)\} \mid a, b, c \in \mathfrak{B}, v(r) \circ a = v(r)\}).$$

Any tuple whose value for attribute 0 is u belongs to the first relation, but not to the second. Thus w is a valuation refuting Qu4. $\qquad\square$

Corollary 4.8. *The quasiequational theory of any class of algebras between $\mathcal{R}_{\text{fin}}^{\mathsf{H}}$ and $\underline{RH}1$ is undecidable.*

Proof. Follows from Theorem 4.7 and theorems of Gurevič [Gur66, GL84] and Post [Pos47] (for finite and arbitrary semigroups, respectively). $\qquad\square$

Corollary 4.9. *The quasiequational theory of $\mathcal{R}_{\text{fin}}^{\mathsf{H}}$ is not finitely axiomatizable.*

Proof. Follows from Theorem 4.7 and the Harrop criterion [Har58]. $\qquad\square$

Problem 4.10. Are the quasiequational theories of $\mathcal{R}_{\text{unr}}$ and $\mathcal{R}_{\text{fin}}$ (i.e., of lattice reducts) decidable?

5 The Concept Structure of Tropashko Lattices

Given a finite lattice $\mathcal{L}$ with $\mathfrak{J}(\mathcal{L})$ and $\mathfrak{M}(\mathcal{L})$ being the sets of its, respectively, join- and meet-irreducibles, let us follow Formal Concept Analysis [GW96] and investigate the structure of $\mathcal{L}$ via its *standard context* $\text{con}(\mathcal{L}) := (\mathfrak{J}(\mathcal{L}), \mathfrak{M}(\mathcal{L}), \mathsf{I}_{\leq})$, where $\mathsf{I}_{\leq} := \leq \cap (\mathfrak{J}(\mathcal{L}) \times \mathfrak{M}(\mathcal{L}))$. Set

$$g \swarrow m : \quad g \text{ is } \leq\text{-minimal in } \{h \in \mathfrak{J}(\mathcal{L}) \mid \text{not } h \mathsf{I}_{\leq} m\},$$
$$g \nearrow m : \quad m \text{ is } \leq\text{-maximal in } \{n \in \mathfrak{M}(\mathcal{L}) \mid \text{not } g \mathsf{I}_{\leq} n\},$$
$$g \nearrow\!\!\!\!\!\swarrow m : \quad g \swarrow m \,\&\, g \nearrow m.$$

Let also $\swarrow\!\!\!\!\swarrow$ be the smallest relation containing $\swarrow$ and satisfying the condition

$$g \not\nearrow m, \ h \nearrow m \text{ and } h \nwarrow n \text{ imply } g \not\nwarrow n;$$

in a more compact notation, $\not\nwarrow \circ \nearrow \circ \nwarrow \subseteq \not\nwarrow$. We have the following

Proposition 5.1. *[GW96, Theorem 17] A finite lattice is*

- *subdirectly irreducible iff there is $m \in \mathfrak{M}(\mathcal{L})$ such that $\not\nwarrow \supseteq \mathfrak{J}(\mathcal{L}) \times \{m\}$,*
- *simple iff $\not\nwarrow = \mathfrak{J}(\mathcal{L}) \times \mathfrak{M}(\mathcal{L})$.*

Let us describe $\mathfrak{J}(\mathfrak{R}(\mathcal{D}, \mathcal{A}))$ and $\mathfrak{M}(\mathfrak{R}(\mathcal{D}, \mathcal{A}))$ for finite $\mathcal{D}$ and $\mathcal{A}$. Set

$$
\begin{aligned}
\mathcal{ADom}_{\mathcal{D},\mathcal{A}} &:= \{\mathsf{adom}(x) \mid x \in {}^{\mathcal{A}}\mathcal{D}\} && \text{where } \mathsf{adom}(x) && := (\mathcal{A}, \{x\}), \\
\mathcal{AAtt}_{\mathcal{D},\mathcal{A}} &:= \{\mathsf{aatt}(a) \mid a \in \mathcal{A}\} && \text{where } \mathsf{aatt}(a) && := (\mathcal{A} - \{a\}, \emptyset), \\
\mathcal{CoDom}_{\mathcal{D},H} &:= \{\mathsf{codom}^H(x) \mid x \in {}^{H}\mathcal{D}\} && \text{where } \mathsf{codom}^H(x) && := (H, {}^{H}\mathcal{D} - \{x\}), \\
\mathcal{CoAtt}_{\mathcal{D},\mathcal{A}} &:= \{\mathsf{coatt}(a) \mid a \in \mathcal{A}\} && \text{where } \mathsf{coatt}(a) && := (\{a\}, {}^{\{a\}}\mathcal{D}),
\end{aligned}
$$

$$
\begin{aligned}
\mathcal{J}_{\mathcal{D},\mathcal{A}} &:= & \mathcal{ADom}_{\mathcal{D},\mathcal{A}} \cup \mathcal{AAtt}_{\mathcal{D},\mathcal{A}}, \\
\mathcal{M}_{\mathcal{D},\mathcal{A}} &:= & \mathcal{CoAtt}_{\mathcal{D},\mathcal{A}} \cup \bigcup_{H \subseteq \mathcal{A}} \mathcal{CoDom}_{\mathcal{D},H}.
\end{aligned}
$$

It is worth noting that $\mathfrak{R}(\mathcal{D}, \mathcal{A})$ naturally divides into what we may call *boolean H-slices*—i.e., the powerset algebras of ${}^{H}\mathcal{D}$ for each $H \subseteq \mathcal{A}$. Furthermore, the projection mapping from H-slice to H'-slice where $H' \subseteq H$ is a join-homomorphism. Lastly, note that the bottom elements of H-slices—i.e., elements of the form $(H, \emptyset)$—and top elements of the form $(H, {}^{H}\mathcal{D})$ form two additional boolean slices, which we may call the *lower attribute slice* and *the upper attribute slice*, respectively. Both are obviously isomorphic copies of the powerset algebra of $\mathcal{A}$. The intention of our definition should be clear then:

- The join-irreducibles are only the atoms of the $\mathcal{A}$-slice (i.e., the slice with the longest tuples) plus the atoms of the lower attribute slice.
- The meet-irreducibles are much richer: they consists of the coatoms of *all H-slices* (note $\mathcal{M}_{\mathcal{D},\mathcal{A}}$ includes H as the sole element of $\mathcal{CoDom}_{\mathcal{D},\emptyset}$) plus all coatoms of the *upper* attribute slice.

Let us formalize these two itemized points as

Theorem 5.2. *For any finite $\mathcal{A}$ and $\mathcal{D}$ such that $|\mathcal{D}| \geq 2$, we have*

$$
\begin{aligned}
\mathcal{J}_{\mathcal{D},\mathcal{A}} &= \mathfrak{J}(\mathfrak{R}(\mathcal{D}, \mathcal{A})), && \text{(join-irreducibles)} \\
\mathcal{M}_{\mathcal{D},\mathcal{A}} &= \mathfrak{M}(\mathfrak{R}(\mathcal{D}, \mathcal{A})). && \text{(meet-irreducibles)}
\end{aligned}
$$

Proof. (join-irreducibles): To prove the $\subseteq$-direction, simply observe that the elements of $\mathcal{J}_{\mathcal{D},\mathcal{A}}$ are exactly the atoms of $\mathfrak{R}(\mathcal{D}, \mathcal{A})$. For the converse, note that

- every element in a H-slice is a join of the atoms of this slice, as each H-slice has a boolean structure and in the boolean case atomic = atomistic,
- the header elements $(H, \emptyset)$ are joins of elements of $\mathcal{AAtt}_{\mathcal{D},\mathcal{A}}$,
- the atoms of H-slices are joins of header elements with elements of $\mathcal{AAtt}_{\mathcal{D},\mathcal{A}}$. Hence, no element of $\mathfrak{R}(\mathcal{D}, \mathcal{A})$ outside $\mathcal{AAtt}_{\mathcal{D},\mathcal{A}}$ can be join-irreducible.

(meet-irreducibles): This time, the $\supseteq$-direction is easier to show: $\mathcal{M}_{\mathcal{D},\mathcal{A}}$ includes the coatoms of the H-slices and the upper attribute slices. Hence, the basic properties of finite boolean algebras imply all meet-irreducibles must be contained in $\mathcal{M}_{\mathcal{D},\mathcal{A}}$: every element of $\mathfrak{R}(\mathcal{D},\mathcal{A})$ can be obtained as an intersection of elements of $\mathcal{M}_{\mathcal{D},\mathcal{A}}$. For the $\subseteq$-direction, it is clear that elements of $CoAtt_{\mathcal{D},\mathcal{A}}$ are meet-irreducible, as they are coatoms of the whole $\mathfrak{R}(\mathcal{D},\mathcal{A})$. This also applies to $H \in CoDom_{\mathcal{D},\emptyset}$. Now take $\mathrm{codom}^H(x) = (H, {}^H\mathcal{D} - \{x\})$ for a non-empty $H = \{1,\ldots,h\}$ and $x = (x_1,\ldots x_h) \in {}^H\mathcal{D}$ and assume $\mathrm{codom}^H(x) = r \bowtie s$ for $r, s \neq \mathrm{codom}^H(x)$. That is, $H = H_r \cup H_s$ and

$${}^H\mathcal{D} - \{x\} = \{y \in {}^{H_r \cup H_s}\mathcal{D} \mid y[H_r] \in B_r \quad \text{and} \quad y[H_s] \in B_s\}.$$

Note that wlog $H_r \subsetneq H$ and $r \subseteq \mathrm{codom}^{H_r}(z)$ for some $z \in {}^{H_r}\mathcal{D}$; otherwise, if both r and s were top elements of their respective slices, their meet would be $(H, {}^H\mathcal{D})$. Thus ${}^H\mathcal{D} - \{x\} \subseteq \{y \in {}^H\mathcal{D} \mid y[H_r] \neq z\}$ and by contraposition

$$\{y \in {}^H\mathcal{D} \mid y[H_r] = z\} \subseteq \{x\}. \tag{1}$$

This means that $z = x[H_r]$. But now take any $i \in H - H_r$, pick any $d \neq x_i$ (here is where we use the assumption that $|\mathcal{D}| \geq 2$) and set

$$x' := (x_1, \ldots, x_{i-1}, d, x_{i+1}, \ldots, x_h).$$

Clearly, $x'[H_r] = x[H_r] = z$, contradicting (1).   □

Theorem 5.3. *Assume $\mathcal{D}, \mathcal{A}$ are finite sets such that $|\mathcal{D}| \geq 2$ and $\mathcal{A} \neq \emptyset$. Then $l_\leq$, $\swarrow$, $\nearrow$ and $\mathcal{U}$ look for $\mathfrak{R}(\mathcal{D},\mathcal{A})$ as follows:*

$r =$	$\mathrm{adom}(x)$	$\mathrm{aatt}(a)$	$\mathrm{adom}(x)$	$\mathrm{aatt}(a)$
$s =$	$\mathrm{coatt}(a)$	$\mathrm{coatt}(b)$	$\mathrm{codom}^H(y)$	$\mathrm{codom}^H(y)$
$r \, l_\leq s$	*always*	$u \neq b$	$x[H] \neq y$	$u \not\subseteq H$
$r \swarrow s$	*never*	$a = b$	$x[H] = y$	$a \in H$
$r \nearrow s$	*never*	$a = b$	$x[H] = y$	*never*
$r \, \mathcal{U} \, s$	*never*	$a = b$	*always*	*always*

Proof (Sketch).

For the $l_\leq$-row: this is just spelling out the definition of $\leq$ on $\mathfrak{R}(\mathcal{D},\mathcal{A})$ as restricted to $\mathcal{J}_{\mathcal{D},\mathcal{A}} \times \mathcal{M}_{\mathcal{D},\mathcal{A}}$.

For the $\swarrow$-row: the set of join-irreducibles consists of only of the atoms of the whole lattice, hence $\swarrow$ is just the complement of $\leq$.

This observation already yields $\nearrow \subseteq \swarrow$ and $\swarrow = \nearrow$. The last missing piece of information to define $\nearrow$ is provided by the analysis of restriction of $\leq$ to $\mathcal{M}_{\mathcal{D},\mathcal{A}} \times \mathcal{M}_{\mathcal{D},\mathcal{A}}$:

for
$$\begin{array}{lll}
r = \mathrm{coatt}(a), & s = \mathrm{coatt}(b), & & \text{never,} \\
r = \mathrm{coatt}(a), & s = \mathrm{codom}^H(x), & r \leq s \ \text{iff} & \text{never,} \\
r = \mathrm{codom}^H(x), & s = \mathrm{coatt}(a), & & a \in H, \\
r = \mathrm{codom}^H(x), & s = \mathrm{codom}^H(y), & & \text{never.}
\end{array}$$

Finally, for $\not\Longleftarrow$ we need to observe that composing $\not\swarrow$ with $\nearrow \circ \not\swarrow$ does not allow to reach any new elements of $CoAtt_{\mathcal{D},\mathcal{A}}$. As for elements of $\mathcal{M}_{\mathcal{D},\mathcal{A}}$ of the form $\mathsf{codom}^H(y)$, note that

$$\exists h.(h \nearrow \mathsf{coatt}(a) \,\&\, h \not\swarrow \mathsf{codom}^H(y)) \text{ if } a \in H, \tag{2}$$

$$\exists h.(h \nearrow \mathsf{codom}^{H_x}(x) \,\&\, h \not\swarrow \mathsf{codom}^{H_y}(y)) \text{ if } x[H_x \cap H_y] = y[H_x \cap H_y]. \tag{3}$$

Furthermore, we have that

- for any $x \in {}^{\mathcal{A}}\mathcal{D}$ and any $H \subseteq \mathcal{A}$, $\mathsf{adom}(x) \not\swarrow \mathsf{codom}^H(x[H])$,
- for any $a \in \mathcal{A}$ and any $x \in {}^{\mathcal{A}}\mathcal{D}$, $\mathsf{aatt}(a) \not\swarrow \mathsf{codom}^{\mathcal{D}}(x)$.

Using (3), we obtain then that $\mathcal{J}_{\mathcal{D},\mathcal{A}} \times \{H\} \subseteq \not\Longleftarrow$ and using (3) again—that $\mathcal{J}_{\mathcal{D},\mathcal{A}} \times \{\mathsf{codom}^H(y)\} \subseteq \not\Longleftarrow$ for any $y \in {}^{\mathcal{A}}\mathcal{D}$ and any $H \subseteq \mathcal{A}$. □

Corollary 5.4. *If $\mathcal{D}, \mathcal{A}$ are finite sets such that $|\mathcal{D}| \geq 2$ and $\mathcal{A} \neq \emptyset$, then $\mathfrak{R}(\mathcal{D}, \mathcal{A})$ is subdirectly irreducible but not simple.*

Proof. Follows immediately from Proposition 5.1 and Theorem 5.3. □

6 Conclusions and Future Work

6.1 Possible Extensions of the Signature

Clearly, it is possible to define more operations on $\mathcal{R}_{\mathrm{unr}}^H$ than those present in $\mathcal{L}_H$. Thus, our first proposal for future study, regardless of the negative result in Corollary 4.8, is a systematic investigation of extensions of the signature. Let us discuss several natural ones; see also [ST06, Tro].

The top element $\top := (\emptyset, \{\emptyset\})$. Its inclusion in the signature would be harmless, but at the same time does not appear to improve expressivity a lot.

The bottom element $\bot := (\mathcal{A}, \emptyset)$. Whenever $\mathcal{A}$ is infinite, including $\bot$ in the signature would exclude subalgebras consisting of relations with finite headers— i.e., exactly those arising from concrete database instances. Another undesirable feature is that the interpretation of $\bot$ depends on $\mathcal{A}$, i.e., the collection of all possible attributes, which is not explicitly supplied by a query expression.

The full relation $\mathsf{U} := (\mathcal{A}, {}^{\mathcal{A}}\mathcal{D})$ [Tro, ST06]. Its inclusion would destroy the *domain independence property (d.i.p.)* [AHV95, Ch. 5] mentioned above. Note that for non-empty $\mathcal{A}$ and $\mathcal{D}$, U is a complement of H.

Attribute constants $\underline{a} := (\{a\}, \emptyset)$, for $a \in \mathcal{A}$. We touch upon an important difference between our setting and that of both *named SPJR algebra* and *unnamed SPC algebra* in [AHV95, Ch. 4], which are *typed*: expressions come with an explicit information about their headers (*arities* in the unnamed case). Our expressions are untyped *query schemes*. On the one hand, $\mathcal{L}_H$ allows, e.g., *projection of r to the header of s*: $r \oplus (s \ltimes H)$, which does not correspond to any *single* SPJR expression. On the other hand, only with attribute constants we can write the SPJR *projection of r to a <u>concrete</u> header* $\{a_1, \ldots, a_n\}$: $\pi_{a_1,\ldots,a_n}(r) := r \oplus \underline{a_1} \ltimes \ldots \ltimes \underline{a_n}$.

Unary singleton constants $(\underline{a} : d) := (\{a\}, \{(a : d)\})$, for $a \in \mathcal{A}$, $d \in \mathcal{D}$. These are among the *base SPJR queries* [AHV95, p. 58]. Note they add more expressivity than attribute constants: whenever the signature includes $(\underline{a} : d)$ for some $d \in \mathcal{D}$, we have $\underline{a} = (\underline{a} : d) \ast H$. They also allow to define $\top$ as $\top = (\underline{a} : d) \oplus H$ and, more importantly, the SPJR *constant-based selection queries* $\sigma_{\underline{a}=d}(r) := r \ast (\underline{a} : d)$.

The equality constant $\Delta := (\mathcal{A}, \{x \in {}^{\mathcal{A}}\mathcal{D} \mid \forall a, a'. x(a) = x(a')\})$. With it, we can express the *equality-based selection queries*: $\sigma_{\underline{a}=\underline{b}}(r) := r \ast (\Delta \oplus \underline{a} \ast \underline{b})$. But the interpretation of Δ violates d.i.p., hence we prefer *the inner equality operator*:

$$\bar{\bar{r}} := (H_r, \{x \in {}^{H_r}\mathcal{D} \mid \exists x' \in r. \exists a' \in H_r. \forall a \in H_r. x(a) = x'(a')\}),$$

which also allows to define $\sigma_{\underline{a}=\underline{b}}(r)$ as $r \ast (\bar{\bar{r}} \oplus \underline{a} \ast \underline{b})$.

The header-narrowing operator $r \pitchfork s := (H_r - H_s, \{x[H_r - H_s] \mid x \in H_r\})$. This one is perhaps more surprising, but now we can define the *attribute renaming* operators [AHV95, p. 58] as $\rho_{\underline{a} \mapsto \underline{b}}(r) := (r \ast \overline{(r \oplus \underline{a})} \ast (\underline{b} : d)) \pitchfork \underline{a}$, where $d \in \mathcal{D}$ is arbitrary. Instead of using $\pitchfork$, one could add constants for elements aatt(a) introduced in Section 5, but this would lead to the same criticism as $\bot$ above: indeed, such constants would make $\bot$ definable as $\bot = \mathsf{aatt}(a) \ast \underline{a}$.

Overall, one notices that just to express the operators discussed in [AHV95, Ch. 4], it would be sufficient to add special constants, but more care is needed in order to preserve the d.i.p. and similar relativization/finiteness properties.

The difference operator $r - s := (H_r, \{x \in B_r \mid x \notin B_s\})$. This is a very natural extension from the DB point of view [AHV95, Ch. 5], which leads us beyond the SPJRU setting towards the question of *relational completeness* [Cod70]. Here again we break with the partial character of Codd's original operator. Another option would be $(H_{r \cap s}, \{x \in B_r[H_s] \mid x \notin B_s[H_r]\})$, but this one can be defined with the difference operator proposed here as $(r \oplus s) - (s \oplus (r \ast H))$.

6.2 Summary and Other Directions for Future Research

We have seen that relational lattices form an interesting class with rather surprising properties. Unlike Codd's relational algebra, all operations are total and in contrast to the encoding of relational algebras in cylindric algebras, the domain independence property obtains automatically. We believe that with the extensions of the language proposed in Section 6.1, one can ultimately obtain most natural algebraic treatment of SPRJ(U) operators and relational query languages. Besides, given how well investigated the lattice of varieties of lattices is in general [JR92], it is intriguing to discover a class of lattices with a natural CS motivation which does not seem to fit anywhere in the existing picture.

To save space and reader's patience, we are not going to recall again all the conjectures and open questions posed above, but without settling them we cannot claim to have grasped how relational lattices behave as an algebraic class. None of them seems trivial, even with the rich supply of algebraic logic tools

342   T. Litak, S. Mikulás, and J. Hidders

available in existing literature. A reference not mentioned so far and yet potentially relevant is [Cra74]. An interesting feature of Craig's setting from our point of view is that it allows tuples of varying arity.

We would also like to mention the natural question of *representability*:

Problem 6.1 (Hirsch). Given a finite algebra in the signature $\mathcal{L}_H$ ($\mathcal{L}$), is it decidable whether it belongs to $\mathbb{SP}(\mathcal{R}_{unr}^H)$, $\mathbb{SP}(\mathcal{R}_{fin}^H)$ ($\mathbb{SP}(\mathcal{R}_{unr})$, $\mathbb{SP}(\mathcal{R}_{fin})$)?

We believe that the analysis of the concept structure of finite relational lattices in Section 5 may lead to an algorithm recognizing whether the concept lattice of a given context belongs to $\mathbb{SP}(\mathcal{R}_{fin}^H)$ (or $\mathbb{SP}(\mathcal{R}_{fin})$). It also opens the door to a systematic investigation of a research problem suggested by Yde Venema: *duality theory of relational lattices*. See also Section 2.1 above for another category-theoretical connection.

References

[AHV95]   Abiteboul, S., Hull, R., Vianu, V.: Foundations of Databases. Addison-Wesley (1995)
[Cod70]   Codd, E.F.: A Relational Model of Data for Large Shared Data Banks. Commun. ACM 13, 377–387 (1970)
[Cra74]   Craig, W.: Logic in Algebraic Form. Three Languages and Theories. Studies in Logic and the Foundations of Mathematics, p. 72. North Holland (1974)
[DM01]   Düntsch, I., Mikulás, S.: Cylindric structures and dependencies in relational databases. Theor. Comput. Sci. 269, 451–468 (2001)
[GW96]   Ganter, B., Wille, R.: Applied Lattice Theory: Formal Concept Analysis. In: Grätzer, G. (ed.) General Lattice Theory, 2nd edn., Birkhäuser (1996)
[Gur66]   Gurevich, Y.: The word problem for certain classes of semigroups. Algebra and Logic 5, 25–35 (1966)
[GL84]   Gurevich, Y., Lewis, H.R.: The Word Problem for Cancellation Semigroups with Zero. The Journal of Symbolic Logic 49, 184–191 (1984)
[Har58]   Harrop, R.: On the existence of finite models and decision procedures for propositional calculi. Mathematical Proceedings of the Cambridge Philosophical Society 54, 1–13 (1958)
[HH02]   Hirsch, R., Hodkinson, I.: Relation Algebras by Games. Studies in Logic and the Foundations of Mathematics, vol. 147. Elsevier (2002)
[IL84]   Imieliński, T., Lipski, W.: The Relational Model of Data and Cylindric Algebras. J. Comput. Syst. Sci. 28, 80–102 (1984)
[Jac99]   Jacobs, B.: Categorical Logic and Type Theory. Studies in Logic and the Foundations of Mathematics, vol. 141. North Holland, Amsterdam (1999)
[JR92]   Jipsen, P., Rose, H.: Varieties of Lattices. Lecture Notes in Mathematics, vol. 1533. Springer (1992)
[JR98]   Jipsen, P., Rose, H.: Varieties of Lattices. In: Grätzer, G. (ed.) General Lattice Theory, pp. 555–574. Birkhäuser (1998); Appendix F to the second edition
[Mad80]   Maddux, R.: The Equational Theory of CA_3 is Undecidable. The Journal of Symbolic Logic 45, 311–316 (1980)

[Mae74]   Maeda, S.: Locally Modular Lattices and Locally Distributive Lattices.
            Proceedings of the American Mathematical Society 44, 237–243 (1974)
[McK72]   McKenzie, R.: Equational bases and non-modular lattice varieties.
            Trans. Amer. Math. Soc. 174, 1–43 (1972)
[PMV07]   Padmanabhan, R., McCune, W., Veroff, R.: Lattice Laws Forcing Dis-
            tributivity Under Unique Complementation. Houston Journal of Math-
            ematics 33, 391–401 (2007)
[Pos47]   Post, E.L.: Recursive Unsolvability of a Problem of Thue. The Journal
            of Symbolic Logic 12, 1–11 (1947)
[ST06]   Spight, M., Tropashko, V.: First Steps in Relational Lattice (2006),
            http://arxiv.org/abs/cs/0603044
[Sta72]   Stanley, R.P.: Supersolvable lattices. Algebra Universalis 2, 197–217
            (1972)
[Ste99]   Stern, M.: Semimodular Lattices. Encyclopedia of Mathematics and its
            Applications, vol. 73. Cambridge University Press (1999)
[Tro]   Tropashko, V.: The website of QBQL: Prototype of relational lattice
            system, https://code.google.com/p/qbql/
[Tro05]   Tropashko, V.: Relational Algebra as non-Distributive Lattice (2005),
            http://arxiv.org/abs/cs/0501053
[VdBVGV07] Van den Bussche, J., Van Gucht, D., Vansummeren, S.: A crash course on
            database queries. In: PODS 2007: Proceedings of the Twenty-Sixth ACM
            SIGMOD-SIGACT-SIGART Symposium on Principles of Database Sys-
            tems, pp. 143–154. ACM, New York (2007)

Towards Finding Maximal Subrelations
with Desired Properties

Martin Eric Müller[1]

University Augsburg, Dept. Computer Science
m.e.mueller@acm.org

Abstract. As soon as data is noisy, knowledge as it is represented in an information system becomes unreliable. Features in databases induce equivalence relations—but knowledge discovery takes the other way round: given a relation, what could be a suitable functional description? But the relations we work on are noisy again. If we expect to record data for learning a classification of objects then it can well be the real data does not create a reflexive, symmetric and transitive relation although we know it should be. The usual approach taken here is to build the closure in order to ensure desired properties. This, however, leads to overgeneralisation rather quickly.

In this paper we present our first steps towards finding maximal subrelations that satisfy the desired properties. This includes a discussion of different properties and their representations, several simple measures on relations that help us comparing them and a few distance measures that we expect to be useful when searching for maximal subrelations.

1 Motivation

Information systems are used to represent *data* and their (observed) properties in some (informative) systematic way:

$$\mathcal{I} = \langle U, \mathbb{F}, \mathbb{V} \rangle, \text{ with } \mathbb{F} = \{f_j : U \to V_j \ : \ j \in \mathbf{n}\}, \mathbb{V} = \{V_j : j \in \mathbf{n}\} \qquad (1)$$

where $\mathbb{F}$ is the set of all total *features* describing the elements of $U = \{x_i : i \in \mathbf{m}\}$. A feature f_j is called an *attribute*, iff $|\text{cod}(f_j)| = 2$; sets of attributes are also known as *contexts*, [GW99]. Every $f_j \in \mathbb{F}$ induces a canonical equivalence relation R_j. For $\mathbf{F} \subseteq \mathbb{F}$ the set of induced equivalence relations are denoted by $\mathbf{R} \subseteq \mathbb{R}$. A *concept* is a subset $s \subseteq U$ which is defined in terms of basic set operations on the equivalence classes of induced by relations $R_j \in \mathbf{R}$. They can be expressed in conjunctive normal form

$$s = \bigcap_{D \in P} \bigcup_{\langle i,j \rangle \in D} [x_i]_{R_j}, \quad \text{for some } P \subseteq \mathbf{2^{m \times n}} \qquad (2)$$

which is an expression only parametrised by relation identifiers and domain element identifiers serving as class representatives. The second approach to concepts is that all elements belonging to a certain concept are *indiscernible* with respect

P. Höfner et al. (Eds.): RAMiCS 2014, LNCS 8428, pp. 344–361, 2014.

to a certain set of *knowledge*. Arguing in reverse direction, we say that we can (uniquely) identify x if we are able to tell the difference between x and any other element. Therefore, given the question whether an x is-a s is the same as the question whether our knowledge suffices to prove that x is indiscernible from any object of which we know it belongs to s.

The problem is that real data is *noisy* and the information system unintentionally carries some misinformation.

Once we are given an information system $\mathcal{I}$, hypotheses can be induced in many ways (c.f.[Mül12]), but there arise several problems when all we have are *noisy relations of similarities of domain objects*. The reverse construction of $\mathbb{F}$ for such a set of relations on $\mathbf{F} \subseteq 2^{U \times U}$ will, in the general case, fail, for these relations are not neccessarily total or functional or simply do not exhibit any desired relation property. One way to force such properties is by building closures, which on the other hand, may result in huge information loss: In the worst case one might yield for some closure operation $\otimes$ that $R^{\otimes} = \mathbb{T}$ and, thus, $U/R^{\otimes} = \{[x]\}$.[1] One consequence expressed in terms of rough set data analysis is that for all $s \subset U$ and $\mathbf{R}' := \mathbf{R} \cup \{R^{\otimes}\}$,

$$[\![\mathbf{R}']\!]s = \emptyset \text{ and } \langle\!\langle\mathbf{R}'\rangle\!\rangle s = U.$$

where $[\![\cdot]\!]$ and $\langle\!\langle\cdot\rangle\!\rangle$ are the lower and upper approximations (see [Paw84, Pol02, Mül12]). Then, no object x could be discerned from any (other) object y.

The problem at hand can be described as follows: Given a relation R that we want to satisfy a certain property. Then, its closure $R^{\otimes}$ is the smallest superset satisfying this property. The largest *subset* of R satisfying the desired property need not be defined uniquely.

In this paper, we shall discover how to approach the required properties for finding suitable classes on a given set of data.

2   Interesting Relation Properties

One can identify equivalence relations in matrix notation by simultaneously rearranging rows and columns such that the matrix becomes a *square diagonal* matrix.[2] What sounds simple in theory, becomes cumbersome in practice: How can one efficiently prove or disprove whether a relation matrix can be rearranged into a desired normal form to show it satisfies certain properties? There exist several, weaker versions of equivalence that one might hope are easier to identify.

Notational Conventions. In the following, $\langle \text{Reln}(U), \subseteq, \cap, \cup, \mathbb{T}, \perp\!\!\!\perp \rangle$ shall denote the complete lattice of binary relations on a finite base set U with $\mathbb{T} = U \times U$ and $\perp\!\!\!\perp = \emptyset$. $\mathit{1}$ denotes the identity relation and $\mathit{1}_n := \mathit{1} \cup \{\langle x_i, x_j \rangle : |i - j| < n\}$ the *band* of radius n around $\mathit{1}$. $x_{\bullet}R$ and $R_{\bullet}y$ denote the image and preimage of x and y under R. Closure operations are written post-superscript ($R^{\otimes}$) their

---

[1] Where $\mathbb{T} := U \times U$ and U/R is the quotient (i.e. partition).
[2] Matrix normal forms and associated relation properties are extensively dealt with in [Sch11].

"duals" with a pre-subscript ($_\otimes R$). We define the *(principal)* up- or *downsets*[3] of R by

$$R\!\uparrow := \{P \in \mathrm{Reln}(U) : R \subseteq P\} \text{ and } R\!\downarrow := \{P \in \mathrm{Reln}(U) : P \subseteq R\}. \qquad (3)$$

$\overline{R}$, $P \,\mathring{,}\, Q$, $R^\smile$, R^* and $R^{\mathsf{d}} := \overline{R}^\smile$ denote complementation, composition, converse, the reflexive transitive closure and duality resp.

2.1   Biorders (Ferrers relations)

A binary relation on U is called a *biorder* (or *Ferrers relation*), iff:

$$R \,\mathring{,}\, R^{\mathsf{d}} \,\mathring{,}\, R \subseteq R. \qquad (4)$$

Pointwise, we get $uRx \wedge vRy \wedge \neg uRy \implies vRx$ (see, e.g.[GW99]) which intuitively implies that for any two elements having at least one common image, one image set is a subset of the other, which together means that all elements' image sets form a sequence of subsets:

$$\forall x, y \in U : \ R_\bullet x \neq \emptyset \wedge R_\bullet y \neq \emptyset \implies x_\bullet R \subseteq y_\bullet R \vee y_\bullet R \subseteq x_\bullet R \qquad (5)$$

In matrix notation it means that if wRx and yRz then also wRz or yRx. Visually speaking, there is a representation of R in echelon normal form.[4] This is equivalent to stating that there exists an ordering of the domain elements such that their images form a monotone sequence of sub- (or super-) sets:

$$R \text{ is a biorder} \iff \exists f : \forall x, y \in U : f(x) \leq f(y) \longrightarrow x_\bullet R \subseteq y_\bullet R \qquad (6)$$
$$\implies \exists f : \forall x, y \in U : f(x) \leq f(y) \longrightarrow |x_\bullet R| \leq |y_\bullet R|, \qquad (7)$$

where f is an enumeration (injective function $f : U \to \mathbb{N}_0$). The former statement is proved in the appendix, the latter one is used for the implementation of an (efficient) test procedure.[5] By $\mathrm{FerR}(U)$ we denote the set of all biorders on U. The definition of a proper closure operator appears straightforward:

$$R^{\updownarrow} := \text{the smallest relation } Q \supseteq R \text{ s.t. } Q \text{ is a biorder} \qquad (8)$$
$$= R \,\mathring{,}\, (\, R^{\mathsf{d}} \,\mathring{,}\, R\,)^* , \qquad (9)$$

---

[3] See [Bly05] or [GW99] for principal filters and ideals.

[4] A matrix is in echelon form, if there is a monotone descending slope of **1**-entries and all entries above are **0**s. To be precise, we speak of *column echelon* normal forms since we want the **1**-entries to be aligned in the lower left corner of the matrix.

[5] After writing this paragraph and the corresponding proofs in the appendix, the author found that this property corresponds to the notion of *block transitivity* as defined through *fringes* in [Sch11] or *order shaped* relations as in [Win04]. There is, however no proof directly showing the equivalence of a relation being a biorder and having a sequent subset matrix representation from which one could infer an algorithm in a natural way. See also [CDF03].

but the $*$-operation causes serious problems since $\mathrm{FerR}(U)$ is not closed under $\cup$ or $\cap$: Therefore, $R^{\updownarrow}$ is a uniquely defined biorder containing R (by equation (9)), but it is not necessarily a *minimal* biorder containing R as stated in (8).[6] To verify a relation is a biorder one has to show that $R = R^{\updownarrow}$. It is, iff $\forall x \in U : x_\bullet R = x_\bullet R^{\updownarrow} = x_\bullet(R_9^\circ(R^{\mathrm{d}}{}_9^\circ R)^*) = x_\bullet(R_9^\circ(R^{\mathrm{d}}{}_9^\circ R))$. A naïve implementation would require the computation of R^{d} plus $\mathcal{O}(n^3)$ tests for equality.[7] An $\mathcal{O}(n^2)$-algorithm for checking relations being biorders is given in the appendix.

2.2 Difunctionality

$R \in \mathrm{Reln}(U)$ is called *difunctional*, iff

$$R_9^\circ R^\smile{}_9^\circ R \subseteq R,$$

which means that any two elements $x, y \in U$ sharing a common element in their images, have equal image sets:

$$R \text{ is difunctional} \iff \forall x, y \in U : x_\bullet R \cap y_\bullet R \neq \emptyset \longrightarrow x_\bullet R = y_\bullet R. \quad (10)$$

The only relations that are difunctional and biorders are $\top, \bot \in \mathrm{Reln}(U)$. The set $\mathrm{DifR}(U)$ of difunctional relations on U forms a poset $\langle \mathrm{DifR}(U), \subseteq \rangle$. It is closed under $\cap$ but not under $\cup$ which can easily be proved using the matrix notation supposing that a relation is difunctional if and only if its incidence matrix can be represented in *block form*[8] (this is an immediate consequence of the proof of equation (10) in the appendix). $\mathrm{DifR}(U)$ is not closed under complementation (simple counterexample or by contradiction and deMorgan of a $\cap$-expression). The *difunctional closure* is defined as

$$R^\dagger := R_9^\circ(R^\smile{}_9^\circ R)^*. \quad (11)$$

A proof is given in [Sch11]. Again, it does not suffice to add the condition of difunctionality to the relation since we may have only "pointwise connected" images (for example 1_n), but in contrast to biorder closures, the difunctional closure of a relation only requires mutual "blind copies" of images without the need for checking set inclusion.

2.3 Transitivity

The *transitive closure* of a relation R is $R^+ := \bigcup_{i \in \mathbb{N}} R^i$. The reflexive, transitive closure is $R^* = R^+ \cup 1 = \bigcup_{i \in \mathbb{N}_0} R^i$. We denote the set of transitive relations

---

[6] As an example, see R as in section 4.1. For $R = \left(\begin{smallmatrix}100\\101\\110\end{smallmatrix}\right)$, both $P_1 = \left(\begin{smallmatrix}100\\101\\110\end{smallmatrix}\right)$ and $P_2 = \left(\begin{smallmatrix}111\\111\\110\end{smallmatrix}\right)$ are minimal biorders containing R, whereas $R^{\updownarrow} = \left(\begin{smallmatrix}100\\111\\111\end{smallmatrix}\right)$ is even bigger. But in contrast to the minimal relations it is unique in the sense that there are no other biorders of the *same cardinality* containing R.

[7] Cubic runtime follows from a three-fold iteration similar to the naïve algorithm for the difunctional closure.

[8] A matrix that can be partitioned into non-overlapping rectangular regions with all values in such a region being either **1** or **0**.

on U by $\mathrm{TrnR}(U)$. $\langle \mathrm{TrnR}(U), \subseteq, \cap, \bot\!\!\!\bot, \top\!\!\!\top \rangle$ forms a complete meet-semilattice. Transitivity is an important property that is hard to come by: There is no efficiently computable formula known to determine the number of transitive relations on a domain of n elements (a few asymptotic results can be found in [Pfe04, Kla04]), and the only well-known algorithm for computing the transitive closure (the Roy-Warshall(-Floyd) algorithm, see appendix) is of cubic runtime complexity. Recent algorithms have a worst case behaviour of less than $\mathcal{O}(n^{2.4})$ but require special normal forms or additional relation properties.

2.4 Equivalence

Let $\mathrm{EquR}(U)$ be the set of all equivalence relations on U. Then, $\langle \mathrm{EquR}(U), \subseteq \rangle$ forms a $\cap$-semilattice $\langle \mathrm{EquR}(U), \subseteq, \cap, 1_U, \top\!\!\!\top \rangle$. It does not form a complete lattice in the sense that it is not closed under $\cup$ meaning that the supremum is not defined. By defining

$$x P \sqcup R y := \exists n \in \mathbb{N}_0 \exists Q_0, \ldots, Q_{n-1} \in \{P, R\} : x\, Q_0 \mathbin{\substack{\circ\\\circ}} \cdots \mathbin{\substack{\circ\\\circ}} Q_{n-1} \qquad (12)$$

we find $\langle \mathrm{EquR}(U), \subseteq, \cap, \sqcup, \bot\!\!\!\bot, \top\!\!\!\top \rangle$ with $\bot\!\!\!\bot = \mathit{1}$ to be a complete lattice.[9] The *equivalence closure* of an arbitrary binary relation $R \in \mathrm{Reln}(U)$ is defined as follows:

$$R^{\equiv} := \text{ the smallest relation } P \in \mathrm{EquR}(U) \text{ with } R \subseteq P. \qquad (13)$$

Equivalence relations can be visualised in block-diagonal[10] form by rearranging rows and columns simultaneously (since $R\breve{\ } \subset R$). Therefore, relations can be checked in $\mathcal{O}(n^2)$ for being equivalent by reordering through sorting and checking class equality in at most $\mathcal{O}(n^2)$. Having reflexivity and difunctionality we represent the *equivalence closure* as

$$R^{\equiv} := R^{\mathit{1}\,\dagger}. \qquad (14)$$

For a proof, see the appendix. The most important properties that we have reported so far are summarised in table 1.

3   Relation Footprints

Let $R \in \mathrm{Reln}(U)$. Up- and downsets can be restricted to subsets of $\mathrm{Reln}(U)$ that are defined by the according relation properties:

$$R{\uparrow}_{\{\leftrightarrow, \mathit{1}, \dagger, \updownarrow, +, \equiv\}} := \{P \in \mathrm{SymR}, \mathrm{RefR}, \mathrm{FerR}, \mathrm{TrnR}, \mathrm{EquR}(U) : R \subseteq P\} \qquad (15)$$

---

[9] See the definition of $\bigvee$ on $\mathfrak{E}(U)$ in [GW99].

[10] A partition of a square matrix into squares with all diagonal elements being a diagonal element of a square matrix, too. All the squares intersecting the diagonal are **1**s, all others are **0**. Block diagonal matrices are subsets of band matrices (n-diagonal), [RZ11].

Table 1. Summary of closure properties and the poset of relation types

	closed				
	$\cup$	$\cap$	$^{-}$	$\S$	$\perp\!\!\!\perp$
RefR	+	+	−	+	*1*
SymR	+	+	+	+	*1*
FerR	−	−	+	+	$\emptyset$
DifR	−	+	−	+	$\emptyset$
TrnR	−	−	+	+	*1*
EquR	−	+	−	+	*1*

$$R \in p\,\text{EquR}(U) :\Longleftrightarrow R\S R \subseteq R \wedge R^{\smile} \subseteq R.$$

$\subseteq_{\text{Reln}}$-extrema of restricted up- or downsets do *not* always coincide with closure operations (or their dual concepts). Therefore, we define

$$\otimes R := \max R\!\downarrow_\otimes \text{ for } \otimes \in \{\leftrightarrow, 1, \dagger, \updownarrow, +, \equiv\}, \tag{16}$$

as the set of $\subseteq_\otimes$-maximal relations. It is kind of an interior operator corresponding to the dual of the closure operator $R^\otimes$ and delivers a not neccessarily singleton set of solutions. As an example, consider $U = \{a, b, c\}$ and $\otimes = +$ on $\text{Reln}(U) \ni R \notin \text{TrnR}(U)$ (since $b\overline{R}d$):

$$R = \begin{pmatrix} 1 & 1 & 1 \\ 0 & 1 & 0 \\ 0 & 0 & 1 \end{pmatrix}, \; {}_+R = \left\{ \begin{pmatrix} 1 & 1 & 1 \\ 0 & 1 & 0 \\ 0 & 0 & 0 \end{pmatrix} \begin{pmatrix} 1 & 1 & 1 \\ 0 & 0 & 0 \\ 0 & 0 & 1 \end{pmatrix} \right\}. \tag{17}$$

To find a subrelation with desired properties, one has to search for it, and to search efficiently, one needs heuristics guiding the way. We will now discuss a few methods to determine relation "footprints".[11]

3.1 Signatures and Permutations

For $U = \{x_i : i \in \mathbf{n}\}$, a matrix representation

$$
\begin{array}{c|ccc}
 & x_0 & \cdots & x_{n-1} \\
\hline
x_0 & & & \\
\vdots & & \ddots & \\
x_{n-1} & & &
\end{array}
= \begin{pmatrix} x_0 R x_0 & \cdots & x_0 R x_{n-1} \\ \vdots & \ddots & \vdots \\ x_{n-1} R x_0 & \cdots & x_{n-1} R x_{n-1} \end{pmatrix}
$$

or $R[i][j] = (x_{ij})_{i,j\in\mathbf{n}}$ with $x_{ij} = \mathbf{1} :\Longleftrightarrow x_i R x_j$ of a relation $R \in \text{Reln}(U)$ is said to be of a *signature* $\sigma_R := \langle \sigma_R^X, \sigma_R^Y \rangle := \langle 0 \cdots (n-1), 0 \cdots (n-1) \rangle$. Each part of the signature corresponds to a bijective indexing function $idx : \mathbf{n} \to U$ such that the i-th component (i.e. row or column) defines to which element

---

[11] They are called footprints but not "fingerprints" because they are not unique.

$x_{idx(i)} \in U$ it refers.[12] Reordering rows or columns simply means to rearrange the index strings using a permutation.[13] Two relations P and Q are said to be *weakly shape equivalent*, iff there are permutations of their signatures such that the matrices (disregarding the signatures) are equal:

$$P \simeq Q :\Longleftrightarrow \exists \pi_X, \pi_Y : P[x_i][y_j] = Q[x_{\pi_X(i)}][y_{\pi_Y(j)}] \tag{18}$$

P and Q are called *strongly shape equivalent* $(P \overset{!}{\simeq} Q)$, iff $P \simeq Q$ and $\pi_X = \pi_Y$.[14] Given a matrix R of signature $\sigma_R = \langle i_0 \cdots i_{n-1}, j_0 \cdots j_{n-1} \rangle$, we denote by

$$\vec{x}_k := \langle x_{i_k} R x_{j_0}, \ldots, x_{i_k} R x_{j_{n-1}} \rangle_{\mathtt{bin}} \tag{19}$$

$$\overleftarrow{y}_l := \langle y_{j_{n-1}} R y_{i_l}, \ldots, y_{j_0} R y_{i_l} \rangle_{\mathtt{bin}} \tag{20}$$

the fixed length n-bit vectors (and, hence natural numbers) that we get when reading the k-th row from left to right and the l-th column bottom up (!)[15], respectively. The cardinality of $|\vec{x}_k| := |x_{k\bullet}R|$ corresponds to the number of **1**s in the row; the same holds for $|\overleftarrow{x}_l| := |Rx_{\bullet l}|$.

Vectors as Footprints. The cardinality of a relation $|R|$ is simply the sum of all vector cardinalities $\sum_{i \in \mathbf{n}} |\vec{x}_i|$. The *density* of a matrix can be measured by the ratio of the relation's cardinality and relation size, $|R|/n^2$. Its *relative reflexivity* is the size of the subidentity contained in R compared to n, $|R \cap \mathbf{1}|/n = (\sum_{i \in \mathbf{n}} x_{ii})n$.

Recall the definition of R from equation (17). There we have $\sigma_R = \langle abc, abc \rangle$. such that for $\pi_X(\sigma_R^X) = \{\langle a,c \rangle, \langle b,a \rangle, \langle c,b \rangle\}$ and $\pi_Y(\sigma_R^Y) = \{\langle c,a \rangle, \langle b,c \rangle, \langle a,b \rangle\}$ we obtain a new signature $\sigma_R' = \langle \pi_Y(\sigma_R^X), \pi_Y(\sigma_R^X) \rangle = \langle cab, bca \rangle$. The result of the reordering is:

R_σ	a b c	$R_{\sigma'}$	b c a	$R_{\sigma''}$	c b a	$R_{\sigma'''}$	b c a
a	1 1 1	c	0 1 0	b	0 1 0	b	1 0 0
b	0 1 0	a	1 1 1	c	1 0 0	c	0 1 0
c	0 0 1	b	1 0 0	a	1 1 1	a	1 1 1

Both matrices represent the same relation in just another permutation of rows and columns. Hence, the scores (cardinality of (pre-) images) remain the same.

---

[12] This is equivalent to saying that σ_R^X and σ_R^Y are permutations of $\mathbf{n}$, [Bón12].

[13] Here, the word "permutation" means a bijective mapping from one permutation of $\mathbf{n}$ onto another. We use the same word for both assuming that the context disambiguates.

[14] Using $\simeq$ for (strong) shape equivalence originates in the isomorphism of permutation groups. — Also, permutations are *renamings* of domain elements. Hence, by some abuse of notation and for later purposes, we also define renamings as: $\exists \rho : Q = \rho(P) :\Longleftrightarrow \exists \pi_{1,2} : xPy \longleftrightarrow \pi_1(x)Q\pi_2(y)$ such that $P \simeq Q$ iff there is some ρ such that $\rho(P) = Q$. This is a very informal way of writing, because the renaming function does not operate on the relation but rather on the set of possible sequences of indices for enumerating the objects of the domain.

[15] Reading the l-th column top down yields mirror($\overleftarrow{y}_l$). The direction of reading is interchangeable.

The bit-vector reading base 10 results in quite *different* values:

$$
\begin{array}{c|cccc}
 & \sigma & \sigma' & \sigma'' & \sigma''' \\
\hline
\vec{x}_0 & 111_{\text{bin}} = 7_{\text{dec}} & 001_{\text{bin}} = 1_{\text{dec}} & 010_{\text{bin}} = 2_{\text{dec}} & 100_{\text{bin}} = 4_{\text{dec}} \\
\vec{x}_1 & 010_{\text{bin}} = 2_{\text{dec}} & 011_{\text{bin}} = 3_{\text{dec}} & 100_{\text{bin}} = 4_{\text{dec}} & 010_{\text{bin}} = 2_{\text{dec}} \\
\vec{x}_2 & 001_{\text{bin}} = 1_{\text{dec}} & 101_{\text{bin}} = 5_{\text{dec}} & 111_{\text{bin}} = 7_{\text{dec}} & 111_{\text{bin}} = 7_{\text{dec}} \\
\hline
\overline{x}_0 & 010_{\text{bin}} = 2_{\text{dec}} & 110_{\text{bin}} = 6_{\text{dec}} & 110_{\text{bin}} = 6_{\text{dec}} & 101_{\text{bin}} = 5_{\text{dec}} \\
\overline{x}_1 & 111_{\text{bin}} = 7_{\text{dec}} & 011_{\text{bin}} = 3_{\text{dec}} & 101_{\text{bin}} = 5_{\text{dec}} & 110_{\text{bin}} = 6_{\text{dec}} \\
\overline{x}_2 & 100_{\text{bin}} = 4_{\text{dec}} & 010_{\text{bin}} = 2_{\text{dec}} & 100_{\text{bin}} = 4_{\text{dec}} & 100_{\text{bin}} = 4_{\text{dec}}
\end{array}
\tag{21}
$$

Note that for σ'', we have an increasing sequence of values for $\vec{x}$ and a decreasing sequence for $\overline{x}$.

Vector Weights. The same cardinality of images and preimages do not imply equality of the sets, but we can infer quite useful information if cardinalities do *not* coincide. If for two elements the image cardinalities are different, the images are unequal. In the case of biorders we know that the smaller one must be a subset of the larger one and in the case of difunctionality the intersection of both is empty.

In order to distinguish different vectors with the same number of 1s in them, we define:

$$
\|\vec{x}_i\| := \sum_{j \in \mathbf{n}} (n - j) * x_i R x_j.
\tag{22}
$$

This way, the score for a vector increases the farther to the left the 1s are located within $\vec{x}_i$ and unequal images of same size are assigned different values. It is easy to see that $x_i \bullet R \subseteq x_j \bullet R$ implies $\|\vec{x}_i\| \leq \|\vec{x}_j\|$.

Run-Length Encoding of Matrices. Using *run-length-encoding* (RLE) of $\vec{x}_i$ and $\overline{x}_i$, we have a very simple measure for both the homogeneity of single rows or columns and the distance between two such vectors:[16]

$$
\begin{aligned}
\text{alt}(\vec{x}) &= \text{alt}(x_0 \cdots x_{n-1}) \\
&:= \langle 0 \rangle \circ \langle i : x_i \neq x_{i-1}, 1 \leq i \leq n - 1 \rangle
\end{aligned}
\tag{23}
$$

$$
\text{rle}(\vec{x}) := [\text{alt}(\vec{x})[i+1] - \text{alt}(\vec{x})[i] : 0 \leq i < \ell(\text{alt}(\vec{x}))]
\tag{24}
$$

Two vectors $\vec{x}$ and $\vec{y}$ are equal, iff $\text{rle}(\vec{x}) = \text{rle}(\vec{y})$ and $x_0 = y_0$. Hence, we agree to add a prefix **1** or **0** indicating the first symbol of the encoded string: $1[4,1,3]$ encodes $\langle 11110111 \rangle$. When interpreted as binary strings with bitwise complementation, $\vec{x} = \overline{\vec{y}}$, iff $\text{rle}(\vec{x}) = \text{rle}(\vec{y})$ and $x_0 = \overline{y_0}$. We define an alphanumeric ordering on the (possibly different length) RLE strings by:

$$
\vec{x} \underset{\text{rle}}{\leq} \vec{y} :\Longleftrightarrow x_0 < y_o \vee x_1 \cdots x_{k-1} \underset{\text{rle}}{\leq} y_1 \cdots y_{l-1}.
\tag{25}
$$

---

[16] To avoid confusion, we denote sequences or strings enclosed in $\langle \cdots \rangle$ without separating commas. Sequences can be concatenated using the $\circ$ operator and can be described intensionally like sets: $s = \langle s_0 \cdots s_{n-1} \rangle = \langle i : i \in \mathbf{n} \rangle$. Sequence elements can be accessed similar to array elements $s[i] = \langle s_0 \cdots s_{n-1} \rangle [i] = s_i$.

The shorter a RLE-code, the more homogenous a vector. The ordering as defined in equation (25) can also be computed arithmetically by adding a prefix of values n such that all RLE codes have the same length. From the point of view of compression this is not a wise thing to do, but then

$$\vec{x} \underset{\text{rle}}{\leq} \vec{y} \iff x'_{\mathrm{n}} \leq y'_{\mathrm{n}}$$

where x'_{n} and y'_{n} are the integers base n after adding the prefixes.

Structural Footprints. We use the following formulae to compute a heuristic measure for the "staircaseness" of a given matrix:

$$sc_X(\sigma_R^X) := \sum_{i \in \mathbf{n}} (n - (i+1))(\vec{x}_{i+1} - \vec{x}_i) \tag{26}$$

$$sc_Y(\sigma_R^Y) := \sum_{i \in \mathbf{n}} (i+1)\,(\vec{x}_{i+1} - \vec{x}_i) \tag{27}$$

Example. The heuristic measures for the example above deliver the following values:

	σ	σ'	σ''	σ'''
$sc_X(\sigma_R^X)$	−11	6	7	1
$sc_Y(\sigma_R^Y)$	−1	−5	−3	−3
$h_{echelon}$	−12	1	4	−2

A simple summation ranks σ'' as the most promising echelon candidate. The idea behind this heuristics is to get a rough picture of a given relation R so as to bring it into a form that allows for generating a good candidate P for $P \in \,{}_{\updownarrow}R$.

3.2   Comparing Relation Matrices

Now that we have a toolset of measures, we collect a few heuristics to guide our search for maximal subrelations satisfying a desired property.

Distance by Difference: A Hamming-Like Dissimilarity. Computing the distance for two relations on a set of n elements is, in general, in $\mathcal{O}(n^2)$. The simplest measure for the distance between two relations is the number of matrix entries in which the two relations disagree:

$$\mathrm{dist}^{\vee}(P,Q) := \sum_{i,j \in \mathbf{n}, i \neq j} x_i P x_j \dot{\vee} x_i Q x_j \qquad |\colon \text{ Defn. as, e.g. by [GPR08]} \tag{28}$$

$$= |(P \cap \overline{Q}) \cup (\overline{P} \cap Q)| - |\mathbb{1} \cap ((P \cap \overline{Q}) \cup (\overline{P} \cap Q))| \tag{29}$$

$$\leq |(P \cap \overline{Q}) \cup (\overline{P} \cap Q)| \tag{30}$$

where $\dot{\vee}$ denotes exclusive disjunction. This measure disregards differences in subidentities contained in the relations. It satisfies the separation axiom (for in line (29), the minuend is always a subset), it is symmetric (due to $\dot{\vee}$), identical arguments deliver zero distance, and it also satisfies the triangle inequation. [GPR08] silently presuppose subadditivity, but since this not as trivial as it seems at first sight, we provide a proof in the appendix.

Based on $\text{dist}^{\vee}$, one can also define asymmetric and weaker distance measures:

$$\text{dist}^{\rightarrow}(P,Q) := |(\overline{P} \cup Q)| - n \text{ and } \text{dist}^{\leftarrow}(P,Q) := |(\overline{P} \cup Q)| - n \qquad (31)$$

are both lower bounds for $\text{dist}^{\vee}$. If $\text{dist}^{\rightarrow}(P,Q) = \text{dist}^{\leftarrow}(P,Q)$ then $\text{dist}^{\vee}(P,Q) = n^2 - \text{dist}^{\leftarrow}(P-Q) + n = n^2 - \text{dist}^{\rightarrow}(P-Q) + n$. Of course, all these measures are invariant against replacing P and Q by weakly shape equivalent versions P' and Q' resulting from the *same* renaming: $P' = \rho(P)$ and $Q' = \rho(Q)$. The time it takes to compute any of these distance values takes $n^2 - n = n(n-1)$.

Distance by Integer Values. Let there be two vectors $\vec{x}^P \neq \vec{x}^Q$ describing $x_\bullet P$ and $x_\bullet Q$ respecitvely. We define the *(squared) Signed Encoding Length distance*:

$$\text{dist}_{\text{SEL}} := (-1) * (x_0^P \neq x_0^Q) * (\ell(\text{rle}(\vec{x}^P)) - \ell(\text{rle}(\vec{x}^Q)))^2. \qquad (32)$$

This measure only reveils information about the relative homogeneity of the vectors, but not at all any information about the number of entries where the relations disagree. A more informative way of talking about the difference between relations has been described in the previous section and we now shall develop a method to extract $\text{dist}^{\vee}$ from the RLE encodings. Two vectors are *different*, if at at least one position RLE values unequal (and thus, lose alignment). Suppose the two RLE codes have a common prefix and differ for the first time in their k-th arguments. Let, without loss of generality, $\text{rle}(\vec{x}^P)[k] < \text{rle}(\vec{x}^Q)[k]$. We then know that the two vectors differ in at least one position and we also know that

$$x_l^P \neq x_l^Q \text{ where } l = \sum_{0 \leq i < k} \text{rle}(\vec{x}^P)[i] + \min\left\{\text{rle}(\vec{x}^P)[k], \text{rle}(\vec{x}^Q)[k]\right\}.$$

Then, we also know that the following

$$\max\left\{\text{rle}(\vec{x}^P)[k], \text{rle}(\vec{x}^Q)[k]\right\} - \min\left\{\text{rle}(\vec{x}^P)[k], \text{rle}(\vec{x}^Q)[k]\right\}$$

have different values, too:

	3	2	2	3
$\vec{x}_k^P$	0 0 0	1 1	0 0	1 1 1
$\vec{x}_k^Q$	0 0 1	1 1	1 1	1 1 0
	2		7		1

Hence, we can compute the number of places in which both vectors disagree as follows:

i	j	$\mathrm{rle}(x_k^P)[i]$	$\mathrm{rle}(x_k^Q)[j]$	Counter
0	0	3	2	0
0'	1	$3-2=1$	7	$0+1$
1	1'	2	$7-1=6$	
2	1'	2	$6-2=4$	$1+2$
3	1'	3	$4-2=2$	
3'	2	$3-2=1$	1	$3+1$
				4

As one can see, the bit difference counter increases every second row starting at the first position of disagreement because it always takes two toggles to retain the original bit value. We then successively substract the minimum of the remaining block lengths from the larger one and skip an addition step if the longer block continues even farther then the current shorter one.

Since both encodings are aligned at the end of the row, we can concatenate *all* row vector codes to one RLE-encoding of the entire matrices for each P and Q and, hence, compute $\mathrm{dist}^\vee$ in a time linear in the matrix *RLE code length* rather than in matrix size n^2 (in the example above it took us 5 steps while the vector length is 10). This motivates the idea of sorting matrices for another reason than just their visual appearance: The better it is ordered, the longer the average sequences or *runs* and, therefore, the shorter the code.

4   Finding Subrelations with Desired Properties

Sorting helps to bring matrices into a form that is "closer" to block matrices, block diagonal matrices or echelon matrices.[17] Once matrices are sorted with the intention to show a certain property one can identify entries that violate the restrictions.

4.1   Biorders

Recall the table in equation (21). The visual appearance of $R_{\sigma''}$ comes close to an echelon form and it is easy to see that by removing bRb, the relation becomes a biorder. Alternatively, removing cRc also yields a biorder and both are of the same cardinality and coincide with the set $_+R$. Similarly, For example, the following matrix defines a non-biorder R with the $\mathbf{0}$ denoting a "missing" entry and $\mathbf{1}'$ and $\mathbf{1}''$ as optional entries that need to be deleted to make R a biorder:

$$R = \begin{pmatrix} \mathbf{1} & 0 & 0 \\ \mathbf{1} & \mathbf{1}' & \underline{\mathbf{0}} \\ \mathbf{1} & \underline{\mathbf{0}} & \mathbf{1}'' \end{pmatrix}, \; _\updownarrow R = \left\{ \begin{pmatrix} \mathbf{1} & 0 & 0 \\ \mathbf{1} & 0 & 0 \\ \mathbf{1} & 0 & \mathbf{1} \end{pmatrix} \begin{pmatrix} \mathbf{1} & 0 & 0 \\ \mathbf{1} & \mathbf{1} & 0 \\ \mathbf{1} & 0 & 0 \end{pmatrix} \right\} = \{P_1, P_2\}.$$

---

[17] Another idea that comes to mind when permuting matrices so as to "cluster" **1**s is that of memory efficient binary representations of decimals, [TCC75, Cow02], and block-wise binary string multiplication, [Boo51].

We proceed as follows: After sorting the rows by increasing $|\vec{x}|$, the density increases the closer we come to the bottom. Next, the columns are sorted with respect to $\|\vec{y}\|$ or $\vec{y}_{dec}$. As an example,

$$R = \begin{pmatrix} 1\,0\,0\,0 \\ 0\,1\,1\,1 \\ 0\,0\,1\,0 \\ 1\,0\,1\,1 \end{pmatrix} \quad \text{becomes} \quad \rho(R) = \begin{pmatrix} 1\,0\,0\,0 \\ 0\,1\,0\,0 \\ 0\,1\,1\,1 \\ 1\,1\,1\,0 \end{pmatrix}$$

where $\sigma_R = \langle abcd, abcd \rangle$ and $\sigma'_R = \langle acbd, acdb \rangle$. The RLE-encodings are:

Rows	$1[1,3]$	$0[1,1,2]$	$0[1,3]$	$1[3.1]$
Columns	$1[1,2,1]$	$1[3,1]$	$1[2,2]$	$0[1,1,2]$

It shows immediately that there are proper subsets ($1[1,3]$ and $1[3,1]$), intersections ($0[1,3]$ and $1[3,1]$) and even complementary rows ($1[1,3]$ and $0[1,3]$). From all those columns where $\vec{y}$ has a prefix of 1's the leftmost is the one with an RLE-code greater than two, therefore it contains a "gap" inside the echelon and should be deleted. A test verifies that $_{\uparrow}R = \{R - \{\langle a, a \rangle, \langle a, d \rangle\}\}$. The same method also delivers P_1 and P_2 for the introducing example—depending on whether we focus on the RLE-codes of the rows or the columns when deciding which entries we delete.

4.2 Difunctionality

Difunctional relations can be displayed in a special block matrix form with maximal three blocks per row or column (see section 3.1). This means there are a whole lot of permutations of signatures against which the matrix shape is invariant. But the good news is that any difunctional relation R can be reordered such that its row and colum vectors all have of length 3 and less and that, for every block the codes are equal. Where they are not, they must not intersect with any other code which means that no code segment of $\mathrm{rle}(\vec{x})$ may cover any point in $\mathrm{alt}(\vec{y})$. This overlap condition can be formalised as:

$$x_0 = y_0 \longrightarrow \mathrm{alt}(\vec{x}) = \mathrm{alt}(\vec{y}).$$

The simplest way to check for overlaps between two rows is by joining the sequence of alteration indices and delete *all* those entries where the rows at a certain point have different values. Difunctionality is a strong property as it requires equality or disjointness of image sets. Hence there are not many options when encountering an object that is and is not element of intersecting image sets: it has to be deleted.

Surprisingly, the first idea that comes to mind appears to be the best, too: Just as in biorder reduction, we simply sort the rows by $|\vec{x}|$ and then do the same on the columns $\vec{y}$. Now the matrix is roughly ordered by density. If $\vec{x}_i \& \vec{x}_j = \vec{x}_i = \vec{x}_j$, we carry on. If $\vec{x}_i \& \vec{x}_j \neq 0^n$, we have a nonempty intersection on the image sets. Hence, we remove from x_i all those 1-entries, that are 0 in $\vec{x}_j$:

$x_i := x_j := x_i \,\&\, x_j$. Forcing difunctionality is quite easy: all we need is the conjunction $\&$ or intersection of image sets. At the same time this suggests that for very more or less noisy relations R their interior $_+R$ is nearly $\perp\!\!\!\perp$.

4.3 Transitivity

Consider the relation

$$R = \begin{pmatrix} 1\,1\,0\,1 \\ 0\,0\,1\,0 \\ 0\,1\,1\,1 \\ 0\,1\,0\,0 \end{pmatrix}$$

It is not transitive, since bRc and $c_\bullet R = \{b, c, d\}$ but neither bRb nor bRd. A closer inspection shows that in fact

$$_+R = \{R - \{\langle b, c \rangle\}, R - \{\langle d, b \rangle\}\}.$$

One could say that b *witnesses a gap at* c and that b *points to gaps* in d and b. The common thing of the two relations in $_+R$ is that the removed tuples always involved the domain element b.

In other words, we should try deleting those entries first, which then also resolve violations of the transitivity principle in interaction with other elements. Finding a suitable, quick algorithm is not easy but [FR95] showed that

$$R \cap R^\lhd \subseteq R \text{ and } R \cap R^\rhd R \text{ are transitive} \tag{33}$$

for any $R \in \mathrm{Reln}(U)$ where $R^\lhd$ and $R^\rhd$ are so-called *left* and *right covering relations* or *traces*:

$$\begin{aligned} xR^\lhd y :&\Longleftrightarrow \bigwedge_{z \in U} zRx \longrightarrow zRy \qquad |\text{: Left trace/covering relation} \\ xR^\rhd y :&\Longleftrightarrow \bigwedge_{z \in U} yRz \longrightarrow xRz \qquad |\text{: Right trace/covering relation} \end{aligned} \tag{34}$$

It even follows that transitive relations $R \in \mathrm{TrnR}(U)$ can be defined by intersections of transitive biorders $P_i \in \mathrm{TrnR}(U) \cap \mathrm{FerR}(U)$. This conclusion does not sound very exciting since from section 2.1 and table 1 we already know that $\mathrm{FerR}(U)$ is not closed under $\cap$. But with a (relatively) efficient algorithm to solve the problem of finding biorder subrelations and weakly shape equivalent versions thereof we might find good candidates for maximal transitive subrelations, too.

[GPR08] present algorithms for defining so-called exterior and interior approximations as well as approximate fittings corresponding to R^+, $_+R$ and a mixed form that allows for both adding and deleting elements from R. The algorithm is based on the distance measure presented in section 3.2. Similar to the concepts of interior approximations and maximal transitive subsets are *transitive reductions*: P is called a transitive reduction of R iff P is a *minimal* subset of R such that $P^+ = R^+$. Reductions have been studied intensively, especially in the context of graph theory, and it has been shown that the time complexity of an according algorithm lies in between $\mathcal{O}(n^2) \subseteq \mathcal{O}(n^3)$.

Bringing together the ideas of Fodor/Roubens and how to identify witnesses and gaps efficiently using suitable heuristics and probabilistic approaches (see [Bón12]) to our minimisation problem are subject to current work.

4.4 Equivalence

The final goal of our work is to find *equivalence* relations, where the hard part is transitivity. On the other hand, we have shown that reflexive difunctional relations are equivalence relations. Here, reflexivity plays a minor, but crucial role: If $1 \not\subseteq R$, then there can't be a subset of R which is an equivalence relation. If, on the other hand, we *postulate*[18] $1 \subseteq R$, the problem boils down to finding difunctional subrelations.

5 Conclusion

In this article we described the problems of search in a sublattice of $\mathrm{Reln}(U)$ that is upper bounded by a given relation R and that is lower bounded by a relation $\underline{R}$ which is the infimum of $_\otimes R$.

The relations that are in focus of our work are biorders, difunctional relations and transitive relations. The substructures of $\mathrm{FerR}(U)$, $\mathrm{DifR}(U)$ and $\mathrm{TrnR}(U)$ were described in section 2; including a few considerations about equivalences between relation properties and their corresponding representations from the point of view of implementing efficient algorithms.

In the further course of developing according algorithms, we focused on measures to speed up finding elements in $_\otimes R$. We introduced the idea of using lossless RLE-compressed representations of vectors which result in memory efficiency gain, a speedup (modulo the compression procedure itself) and a heuristic measure for the distance of relations in order to overcome shortcomings of the $\mathrm{dist}^\vee$-measure. Another interesting perspective is to introduce weights or a preference structure describing which parts of a relation may be altered and which should be left as they are. One of the many tasks ahead is a deeper investigation of the product of such a preference relation together with an ordering of the relation matrix as described in section 3.1.

In the first part many results on the algorithmic complexity of algorithms for testing relation properties were reported (most of them already well-known) and we presented one with a lower bound pushed from cubic down to square. Still it remains to find more precise runtime approximations; many of the algorithms can be optimized by taking into account whether a matrix is sparse or dense. Especially with respect to the numerical footprints of relations (see sections 3.1

---

[18] This does not hurt in the context of information systems since we deal with equivalence relations induced by functions. So if an element has none or several different values for one and the same feature, we know there is something wrong in our underlying data collection.

and 3.2) there is still a lot of work to be done; we couldn't find a satisfying answer to the question of how to identify witnesses and the use of traces in the discussion of finding elements in $_+R$ in section 4.3.

In this article we ignored the space complexity of the presented algorithms; an issue that is of crucial importance in the context of "big data". On the other hand, it offers further crosslinks between the combinatoric problems of permutations, relation properties and matrix representation (for example, sky-line matrices or map-and-reduce methods are space-preserving representation and time-saving processing techniques that come into play there).

In the course of this research we also hope to discover new heuristics for a better guidance when searching Reln(U) for subrelations of desired properties—this issue has been pointed out by [FR95] already, but there is much research in the area of combinatorics of permutations and matrices ([RZ11, Zha99, Bón12] as well as in optimisation (e.g. [Nik70]) that has not been considered in the context of relational data analysis.

Acknowledgements. The author wishes to thank Bernhard Möller and an anonymous reviewer for valuable comments. Rudolf Berghammer provided a valuable comment on related work. Sebastian Pospiech implemented an algorithm for $_+R$ for the RELA-X project.

References

[Bly05]   Blyth, T.S.: Lattices and ordered algebraic structures. Springer (2005)
[Bón12]   Bóna, M.: Combinatorics of Permutations, 2nd edn. Chapman and Hall/CRC (2012)
[Boo51]   Booth, A.D.: A Signed Binary Multiplication Technique. Quarterly Journal of Mechanics and Applied Mathematics, 236–240 (1951)
[CDF03]   Christophe, J., Doignon, J.-P., Fiorini, S.: Counting Biorders. Journal of Integer Sequences 6 (2003)
[Cow02]   Cowlishaw, M.: Densely packed decimal encoding. IEE Proceedings – Computers and Digital Techniques 149 (2002)
[FR95]   Fodor, J.C., Roubens, M.: Structure of transitive valued binary relations. Mathematical Social Sciences (1995)
[GW99]   Ganter, B., Wille, R.: Formal Concept Analysis. Springer (1999)
[GPR08]   Gonzalez-Pachon, J., Romero, C.: A method fir obtaining transitive approximations of a binary relation. Annals of Operations Research (2008)
[Kla04]   Klaška, J.: Transitivity and partial order. Mathematica Bohemnia 122 (2004)
[Mül12]   Müller, M.E.: Relational Knowledge Discovery. Cambridge University Press (2012)
[Nik70]   Nikaido, H.: Introduction to sets and mappings in modern economics. North-Holland (1970)
[Paw84]   Pawlak, Z.: On rough sets. Bulletin of the EATCS 24, 94–184 (1984)
[Pfe04]   Pfeiffer, G.: Counting transitive relations. Journal of integer sequences 7 (2004)
[Pol02]   Polkowski, L.: Rough Sets - Mathematical Foundations. Advances in Soft Computing. Physica (2002)

[RZ11]   Rudolf Zurmühl, S.F.: Matrizen und ihre Anwendungen. Springer Orig. Erstausgabe 1950 (2011)

[Sch11]   Schmidt, G.: Relational Mathematics, Encyclopedia of Mathematics and its Applications, vol. 132. Cambridge University Press (2011)

[TCC75]   Tien Chi Chen, I. T.H.: Storage-efficient representation of decimal data. Communications of the ACM 18, 49–52 (1975)

[Win04]   Winter, M.: Decomposing relations into orderings. In: Berghammer, R., Möller, B., Struth, G. (eds.) RelMiCS/Kleene-Algebra Ws 2003. LNCS, vol. 3051, pp. 265–277. Springer, Heidelberg (2004)

[Zha99]   Zhang, F.: Matrix Theory: Basic Results and Techniques. Springer (1999)

6   Appendix: Proofs

Theorem: Biorders are sequences of subsets of images (equation 6).
This theorem is also stated in [CDF03] but remains unproved. In order to show equation (6) we first prove the reverse direction.

"$\Longleftarrow$" by contradiction: Assume some f satisfying $\forall x, y : f(x) \leq f(y) \longrightarrow x_\bullet R \subseteq y_\bullet R$ is given. We further assume that R is not a biorder:

$$R \text{ is not a biorder } \Longleftrightarrow R\, \overset{\circ}{,}\, R^d\, \overset{\circ}{,}\, R \not\subseteq R \tag{35}$$

Then we need to have $a, b, c, d \in U$ such that

with $aRb, cRd, c\overline{R}b, a\overline{R}d$, eqn.(35) yields: $aRb\overline{R}cRd$ but $a\overline{R}d$. (36)

We examine a and c:

1. If $f(a) \leq f(c)$, then $a_\bullet R \subseteq c_\bullet R$. By construction, there exists b with $b \in a_\bullet R \subseteq c_\bullet R \not\ni b$, hence a contradiction.
2. If $f(c) \leq f(a)$, then $c_\bullet R \subseteq a_\bullet R$. By construction, there exists d with $d \in c_\bullet R \subseteq a_\bullet R \not\ni d$, hence a contradiction.

The assumptions always lead to a contradiction, hence we have shown that the existence of a suitable enumeration of domain elements implies that R is a biorder.

"$\Longrightarrow$": We need to show that given a biorder, we can find an enumeration that orders all domain elements in a way such that the sequence of their image sets is $\subseteq$-isotone. We choose as enumeration a function f whose inverse $f^{-1} : \mathbb{N}_0 \to U$ delivers for the first k_0 numbers the objects whose image sets are empty, then k_1 objects whose image sets are singletons and so forth until it lists k_n objects with image sets U. Then, clearly, $f(x) \leq f(y) \Longrightarrow |x_\bullet R| \leq |y_\bullet R|$. We now have to show that $f(x) \leq f(y) \longrightarrow x_\bullet R \subseteq y_\bullet R$ given that R is a biorder:

1. Empty image sets:
   From $f(x) = 0 = |x_\bullet R|$ follows the $x_\bullet R = \emptyset$. Hence we have $0 \leq f(y) \longrightarrow \emptyset \subseteq y_\bullet R$ which is true. $\square$

2. Let $f(x) \neq 0 \neq f(y)$, i.e. both $x_\bullet R$ and $y_\bullet R$ are nonempty. Assume wlog. that $f(x) \leq f(y)$. We need to show that yRz for all $z \in x_\bullet R$. Since $f(y) \geq f(z) > 0$ we know that $y_\bullet R \neq \emptyset$, i.e. there exists some $a \in y_\bullet R$. The interesting case is where $a \notin x_\bullet R$: Then we have aRy and aR^dx. If we now assume that aRz, we have

$$yRaR^dxRz$$

and, since R is a biorder, also yRz, which completes the proof. $\square$

Theorem: $R^\equiv = R^{1^\dagger}$, (equation 14). We first show $R^\equiv \subseteq R^{1^\dagger}$ by contradiction. Let $R = R^\equiv \in \mathrm{EquR}(U)$. Then, by symmetry and twice transitivity it follows that $R \mathbin{\mathring{,}} R^\smile \mathbin{\mathring{,}} R \subseteq R \mathbin{\mathring{,}} R \mathbin{\mathring{,}} R \subseteq R \mathbin{\mathring{,}} R \subseteq R \subseteq R^{1^\dagger}$ which would contradict $R \mathbin{\mathring{,}} R^\smile \mathbin{\mathring{,}} R \not\subseteq R^{1^\dagger}$. The reverse direction requires to show $1 \subseteq R^{1^\dagger}$, $(R^{1^\dagger})^\smile \subseteq R^{1^\dagger}$, and $R^{1^\dagger} \mathbin{\mathring{,}} R^{1^\dagger} \subseteq R^{1^\dagger}$. We have

$$
\begin{aligned}
R^{1^\dagger} &= R^1 \mathbin{\mathring{,}} (R^{1^\smile} \mathbin{\mathring{,}} R^1)^* &&|: \text{Closure }^* \\
&\supseteq R^1 \mathbin{\mathring{,}} (R^{1^\smile} \mathbin{\mathring{,}} R^1) &&|: \text{Force surjectivity} \\
&\supseteq R^1 \mathbin{\mathring{,}} R^1 &&|: \text{Reflexive closure} \\
&\supseteq 1 \mathbin{\mathring{,}} 1 &&|: \mathbin{\mathring{,}}\text{-neutrality of } 1 \\
&= 1
\end{aligned}
\tag{37}
$$

A few steps yield $(R^{1^\dagger})^\smile = ((R^1)^\smile \mathbin{\mathring{,}} R^1)^* \mathbin{\mathring{,}} (R^1)^\smile$ and $(((R^1)^\smile \mathbin{\mathring{,}} R^1)^*) \mathbin{\mathring{,}} (R^1)^\smile = R^{1^\dagger}$ such that symmetry follows by:

$$
\begin{aligned}
((R^1)^\smile \mathbin{\mathring{,}} R^1)^* &\stackrel{\text{renaming}}{=} (P^\smile \mathbin{\mathring{,}} P)^* &&|: \text{Since } 1 \subseteq P \\
&\subseteq (P^\smile \mathbin{\mathring{,}} P)^* \mathbin{\mathring{,}} P &&|: \text{Again, } 1 \subseteq P \\
&\subseteq P \mathbin{\mathring{,}} (P^\smile \mathbin{\mathring{,}} P)^* \mathbin{\mathring{,}} P &&|: \text{Composition and converse} \\
&\subseteq P \mathbin{\mathring{,}} P^\smile \mathbin{\mathring{,}} ((P^\smile \mathbin{\mathring{,}} P)^*)^\smile &&|: \text{Drop surjectivity condition} \\
&\subseteq ((P^\smile \mathbin{\mathring{,}} P)^*)^\smile \\
&\stackrel{\text{renaming}}{=} (((R^1)^\smile \mathbin{\mathring{,}} R^1)^*)^\smile.
\end{aligned}
\tag{38}
$$

It remains to show that $R^{1^\dagger}$ is transitive.

$$
\begin{aligned}
R^{1^\dagger} \mathbin{\mathring{,}} R^{1^\dagger} &\stackrel{\text{renaming}}{=} P^\dagger \mathbin{\mathring{,}} P^\dagger &&|: \text{Defn. }^\dagger, \text{assoc.} \\
&= P \mathbin{\mathring{,}} (P^\smile \mathbin{\mathring{,}} P)^* \mathbin{\mathring{,}} P \mathbin{\mathring{,}} (P^\smile \mathbin{\mathring{,}} P)^* &&|: \text{Renaming} \\
&= P \mathbin{\mathring{,}} Q \mathbin{\mathring{,}} P \mathbin{\mathring{,}} Q &&|: 1 \text{ neutral wrt } \mathbin{\mathring{,}} \\
&= P \mathbin{\mathring{,}} Q \mathbin{\mathring{,}} 1 \mathbin{\mathring{,}} P \mathbin{\mathring{,}} Q &&|: 1 \subseteq P = R^1 = R \cup 1 \\
&\subseteq P \mathbin{\mathring{,}} Q \mathbin{\mathring{,}} P \mathbin{\mathring{,}} P \mathbin{\mathring{,}} Q &&|: {}^\smile \text{ over } \mathbin{\mathring{,}}, \text{symmetry} \\
&\subseteq P \mathbin{\mathring{,}} Q \mathbin{\mathring{,}} P^\smile \mathbin{\mathring{,}} P \mathbin{\mathring{,}} Q &&|: \text{Surjectivity} \\
&\subseteq P \mathbin{\mathring{,}} Q^* \mathbin{\mathring{,}} Q^* &&|: {}^* \\
&= P \mathbin{\mathring{,}} Q^* = P \mathbin{\mathring{,}} (P^\smile \mathbin{\mathring{,}} P)^* &&|: \text{Renaming} \\
&= R^1 \mathbin{\mathring{,}} (R^{1^\smile} \mathbin{\mathring{,}} R^1)^* = R^{1^\dagger} &&\square.
\end{aligned}
\tag{39}
$$

Theorem: In difunctional relations any pair x, y have either disjoint or equal images, (equation 10). The property to be shown is expressed by equation (10):

$$
\begin{aligned}
&R \,\mathring{,}\, R^{\smile} \mathring{,}\, R \subseteq R && |\text{: pointwise} \\
\Longleftrightarrow\ & \forall a,b,c,d: aRb \wedge cRb \wedge cRd \longrightarrow aRd && |\text{: Defn. } \longrightarrow \\
\Longleftrightarrow\ & \forall a,b,c,d: a\overline{R}b \vee c\overline{R}b \vee c\overline{R}d \vee aRd && |\text{: deMorgan, assoc.} \\
\Longleftrightarrow\ & \forall a,b,c,d: \overline{a\overline{R}b \wedge c\overline{R}b} \vee (c\overline{R}d \vee aRd) && |\text{: de Morgan} \\
\Longleftrightarrow\ & \forall a,b,c,d: (aRb \wedge cRb) \wedge \overline{c\overline{R}d \vee aRd} && |\text{: Quantifier} \\
\Longleftrightarrow\ & \neg\exists a,b,c,d: (aRb \wedge cRb) \wedge \overline{c\overline{R}d \vee aRd} && |\text{: deMorgan, Assoc.} \\
\Longleftrightarrow\ & \neg\exists a,b,c,d: b \in a_{\bullet}R \wedge b \in c_{\bullet}R \wedge d \in c_{\bullet}R \wedge d \notin a_{\bullet}R && |\text{: Commutativity} \\
\Longleftrightarrow\ & \neg\exists a,b,c,d: b \in c_{\bullet}R \wedge d \in c_{\bullet}R \wedge b \in a_{\bullet}R \wedge d \notin a_{\bullet}R &&
\end{aligned}
$$

There are no combinations of elements such that they share one image and, at the same time differ in their images. Hence, images are disjoint or equal. $\square$

Theorem: If P and Q are in $\uparrow R$, then $P \simeq Q$ (section 4.1). More or less trivial. We can presuppose $P \neq Q$ (trivial) and $P \not\subseteq Q$ (contradicts $P, Q \in \uparrow R$). Hence, $P \| Q$. $P \simeq Q$ is equivalent to stating there is a renaming $\rho: P \mapsto Q$ with $\rho(\langle x, y\rangle) := \langle \pi_1(x), \pi_2(y)\rangle$ inheriting bijectivity from $\pi_{1,2}$ which is preserved by building the product. If xPy and xQy, then $\pi_1(x) = x, \pi_2(y) = y, \rho(\langle x, y\rangle) = \langle x, y\rangle$. Therefore, we suppose that $x\overline{P}y$ and xQy and show that assuming $Q \overset{(*)}{=} \rho(P)$ leads to a contradiction: The equivalence

$$
xQy \overset{(*)}{\Longleftrightarrow} x\rho(P)y \overset{\text{def}}{\Longleftrightarrow} \pi_1(x)Q\pi_2(y) \overset{(*)}{\Longleftrightarrow} \pi_1(x)\rho(P)\pi_2(y) \overset{\text{def}}{\Longleftrightarrow} xPy \tag{40}
$$

contradicts $x\overline{P}y$. $\square$

The result is that we can find a *single* matrix representation for all $P_i \in \uparrow R$ and the definitions of all the different P_i can be derived from this matrix by application of permutations π_1^i and π_2^i.

Theorem: dist$^{\vee}$ is subadditive (section 3.2). Let there be $P, Q, R \in \text{Reln}(U)$. We compute:

$$
\begin{aligned}
& \text{dist}^{\vee}(P, Q) + \text{dist}^{\vee}(Q, R) \\
=\ & \sum_{i,j \in \mathbf{n}, i \neq j} x_i P x_j \vee x_i Q x_j + \sum_{i,j \in \mathbf{n}, i \neq j} x_i R x_j \vee x_i Q x_j \\
=\ & \sum_{i,j \in \mathbf{n}, i \neq j} ((x_i P x_j \vee x_i Q x_j) + (x_i R x_j \vee x_i Q x_j)) \\
& |\text{: By a lengthy set theory/truthtable argument :}| \\
\geq\ & \sum_{i,j \in \mathbf{n}, i \neq j} ((x_i P x_j \vee x_i R x_j)) = \text{dist}^{\vee}(P, R) \qquad \square.
\end{aligned}
$$

Complete Solution of a Constrained Tropical Optimization Problem with Application to Location Analysis

Nikolai Krivulin*

Saint Petersburg State University, Faculty of Mathematics and Mechanics
Universitetsky Ave. 28, 198504 Saint Petersburg, Russia
nkk<at>math.spbu.ru

Abstract. We present a multidimensional optimization problem that is formulated and solved in the tropical mathematics setting. The problem consists of minimizing a nonlinear objective function defined on vectors over an idempotent semifield by means of a conjugate transposition operator, subject to constraints in the form of linear vector inequalities. A complete direct solution to the problem under fairly general assumptions is given in a compact vector form suitable for both further analysis and practical implementation. We apply the result to solve a multidimensional minimax single facility location problem with Chebyshev distance and with inequality constraints imposed on the feasible location area.

Keywords: idempotent semifield, tropical mathematics, minimax optimization problem, single facility location problem, Chebyshev distance.

1 Introduction

Tropical (idempotent) mathematics encompasses various aspects of the theory and applications of semirings with idempotent addition and has its origin in a few pioneering works by Pandit [Pan61], Cuninghame-Green [CG62], Giffler [Gif63], Vorob'ev [Vor63] and Romanovskiĭ [Rom64]. At the present time, the literature on the topic contains several monographs, including those by Carré [Car79], Cuninghame-Green [CG79], U. Zimmermann [Zim81], Baccelli et al. [BCOQ93], Kolokoltsov and Maslov [KM97], Golan [Gol03], Heidergott, Olsder and van der Woude [HOvdW06], Gondran and Minoux [GM08], and Butkovič [But10]; as well as a rich variety of contributed papers.

Optimization problems that are formulated and solved in the tropical mathematics setting come from various application fields and form a noteworthy research domain within the research area. Certain optimization problems have appeared in the early paper [CG62], and then the problems were investigated in many works, including [CG79, Zim81, GM08, But10].

---
* This work was supported in part by the Russian Foundation for Humanities (grant No. 13-02-00338).

P. Höfner et al. (Eds.): RAMiCS 2014, LNCS 8428, pp. 362–378, 2014.
© Springer International Publishing Switzerland 2014

Tropical mathematics provides a useful framework for solving optimization problems in location analysis. Specifically, a solution in terms of tropical mathematics has been proposed by Cuninghame-Green [CG91, CG94] to solve single facility location problems defined on graphs. A different but related approach to location problems on graphs and networks has been developed by K. Zimmermann [Zim92], Hudec and K. Zimmermann [HZ93, HZ99], Tharwat and K. Zimmermann [TZ10] on the basis of the concept of max-separable functions.

Multidimensional minimax location problems with Chebyshev distance arise in various applications, including the location of emergency service facility in urban planning and the location of a component on a chip in electronic circuit manufacturing (see, e.g., Hansen, Peeters and Thisse [HPT80, HPT81]). The two-dimensional problems on the plane without constraints can be solved directly on the basis of geometric arguments, as demonstrated by Sule [Sul01] and Moradi and Bidkhori [MB09]. The solution of the multidimensional constrained problems is less trivial and requires different approaches. These problems can be solved, for instance, by using standard linear programming techniques which, however, generally offer iterative procedures and do not guarantee direct solutions.

A strict tropical mathematics approach to solve both unconstrained and constrained minimax location problems with Chebyshev distance was developed by Krivulin [Kri11, Kri12], and Krivulin and K. Zimmermann [KZ13]. The main result of [Kri11] is a direct solution to the unconstrained problem obtained by using the spectral properties of matrices in idempotent algebra. The application of another technique in [Kri12, KZ13], which is based on the derivation of sharp bounds on the objective function, shows that the solution in [Kri11] is complete.

In this paper, a new minimax Chebyshev location problem with an extended set of constraints is taken to both motivate and illustrate the development of the solution to a new general tropical optimization problem. The problem is to minimize a nonlinear objective function defined on vectors over a general idempotent semifield by means of a conjugate transposition operator. The problem involves constraints imposed on the solution set in the form of linear vector inequalities given by a matrix, and two-sided boundary constraints.

To solve the problem, we use the approach, which is proposed in [Kri13, Kri14] and combines the derivation of a sharp bound on the objective function with the solution of linear inequalities. The approach is based on the introduction of an auxiliary variable as a parameter, and the reduction of the optimization problem to the solution of a parametrized system of linear inequalities. Under fairly general assumptions, we obtain a complete direct solution to the problem and represent the solution in a compact vector form. The obtained result is then applied to solve the Chebyshev location problem, which motivated this study.

The paper is organized as follows. In Section 2, we offer an introduction to idempotent algebra to provide a formal framework for the study in the rest of the paper. Section 3 offers the preliminary results on the solution of linear inequalities, which form a basis for later proofs. The main result is included in Section 4, which starts with a discussion of previously solved problems. Furthermore, we describe the problem under study, present a complete direct solution to the problem, consider particular cases, and give illustrative examples. Finally, application of the results to location analysis is discussed in Section 5.

2 Preliminary Definitions and Notation

We start with a short, concise introduction to the key definitions, notation, and preliminary results in idempotent algebra, which is to provide a proper context for solving tropical optimization problems in the subsequent sections. The introduction is mainly based on the notation and results suggested in [Kri06, Kri09b, Kri12, Kri13], which offer strong possibilities for deriving direct solutions in a compact form. Further details on both introductory and advanced levels are available in various works published on the topic, including [CG79, Car79, Zim81, BCOQ93, KM97, Gol03, HOvdW06, ABG07, GM08, But10].

2.1 Idempotent Semifield

An idempotent semifield is an algebraic system $(\mathbb{X}, \oplus, \otimes, \mathbb{0}, \mathbb{1})$, where $\mathbb{X}$ is a non-empty carrier set, $\oplus$ and $\otimes$ are binary operations, called addition and multiplication, $\mathbb{0}$ and $\mathbb{1}$ are distinct elements, called zero and one; such that $(\mathbb{X}, \oplus, \mathbb{0})$ is a commutative idempotent monoid, $(\mathbb{X}, \otimes, \mathbb{1})$ is an abelian group, multiplication distributes over addition, and $\mathbb{0}$ is absorbing for multiplication.

In the semifield, addition is idempotent, which means the equality $x \oplus x = x$ is valid for each $x \in \mathbb{X}$. The addition induces a partial order relation such that $x \leq y$ if and only if $x \oplus y = y$ for $x, y \in \mathbb{X}$. Note that $\mathbb{0}$ is the least element in terms of this order, and so the inequality $x \neq \mathbb{0}$ implies $x > \mathbb{0}$.

Furthermore, with respect to this partial order, addition exhibits an extremal property in the form of the inequalities $x \oplus y \geq x$ and $x \oplus y \geq y$. Both addition and multiplication are monotone in each argument, which implies that the inequalities $x \leq y$ and $u \leq v$ result in the inequalities $x \oplus u \leq y \oplus v$ and $x \otimes u \leq y \otimes v$. These properties lead, in particular, to the equivalence of the inequality $x \oplus y \leq z$ with the two simultaneous inequalities $x \leq z$ and $y \leq z$.

Multiplication is invertible to allow every non-zero $x \in \mathbb{X}$ to have an inverse x^{-1} such that $x^{-1} \otimes x = \mathbb{1}$. The multiplicative inversion is antitone in the sense that if $x \leq y$ then $x^{-1} \geq y^{-1}$ for all non-zero x and y.

The integer power indicates iterated product defined, for each non-zero $x \neq \mathbb{0}$ and integer $p \geq 1$, as $x^p = x^{p-1} \otimes x$, $x^{-p} = (x^{-1})^p$, $x^0 = \mathbb{1}$ and $\mathbb{0}^p = \mathbb{0}$. We suppose the rational exponents can be defined as well, and take the semifield to be algebraically closed (radicable).

In what follows, the multiplication sign $\otimes$ will be omitted to save writing.

Typical examples of the idempotent semifield under consideration include $\mathbb{R}_{\max,+} = (\mathbb{R} \cup \{-\infty\}, \max, +, -\infty, 0)$, $\mathbb{R}_{\min,+} = (\mathbb{R} \cup \{+\infty\}, \min, +, +\infty, 0)$, $\mathbb{R}_{\max,\times} = (\mathbb{R}_+ \cup \{0\}, \max, \times, 0, 1)$, and $\mathbb{R}_{\min,\times} = (\mathbb{R}_+ \cup \{+\infty\}, \min, \times, +\infty, 1)$, where $\mathbb{R}$ denotes the set of real numbers and $\mathbb{R}_+ = \{x \in \mathbb{R} | x > 0\}$.

Specifically, the semifield $\mathbb{R}_{\max,+}$ is equipped with the maximum operator in the role of addition, and arithmetic addition as multiplication. Zero and one are defined as $-\infty$ and 0, respectively. For each $x \in \mathbb{R}$, there exists the inverse x^{-1}, which is equal to $-x$ in ordinary notation. The power x^y can be defined for all $x, y \in \mathbb{R}$ (and thus for rational y) to coincide with the arithmetic product xy. The partial order induced by addition agrees with the usual linear order on $\mathbb{R}$.

2.2   Matrix and Vector Algebra

Consider matrices over the idempotent semifield and denote the set of matrices with m rows and n columns by $\mathbb{X}^{m \times n}$. A matrix with all zero entries is the zero matrix. A matrix is column- (row-) regular if it has no zero columns (rows).

Addition, multiplication, and scalar multiplication of matrices follow the usual rules. For any matrices $A = (a_{ij}) \in \mathbb{X}^{m \times n}$, $B = (b_{ij}) \in \mathbb{X}^{m \times n}$ and $C = (c_{ij}) \in \mathbb{X}^{n \times l}$, and a scalar $x \in \mathbb{X}$, these operations are performed according to the entry-wise formulas

$$\{A \oplus B\}_{ij} = a_{ij} \oplus b_{ij}, \qquad \{AC\}_{ij} = \bigoplus_{k=1}^{n} a_{ik} c_{kj}, \qquad \{xA\}_{ij} = x a_{ij}.$$

The extremal property of the scalar addition extends to the matrix addition, which implies the entry-wise inequalities $A \oplus B \geq A$ and $A \oplus B \geq B$. All matrix operations are entry-wise monotone in each argument. The inequality $A \oplus B \leq C$ is equivalent to the two inequalities $A \leq C$ and $B \leq C$.

Furthermore, we concentrate on square matrices of order n in the set $\mathbb{X}^{n \times n}$. A matrix that has the diagonal entries set to $\mathbb{1}$, and the off-diagonal entries to $\mathbb{0}$ is the identity matrix, which is denoted by I.

The integer power of a square matrix A is routinely defined as $A^0 = I$ and $A^p = A^{p-1} A = A A^{p-1}$ for all $p \geq 1$.

The trace of a matrix $A = (a_{ij})$ is given by

$$\operatorname{tr} A = a_{11} \oplus \cdots \oplus a_{nn}.$$

A matrix that consists of one column (row) is a column (row) vector. In the following, all vectors are regarded as column vectors, unless otherwise specified. The set of column vectors of length n is denoted by $\mathbb{X}^n$. A vector with all zero elements is the zero vector. A vector is called regular if it has no zero components.

Let $x = (x_i)$ be a non-zero vector. The multiplicative conjugate transpose of x is a row vector $x^- = (x_i^-)$, where $x_i^- = x_i^{-1}$ if $x_i > 0$, and $x_i^- = \mathbb{0}$ otherwise.

It follows from the antitone property of the inverse operation that, for regular vectors x and y, the inequality $x \leq y$ implies that $x^- \geq y^-$ and vice versa.

The conjugate transposition exhibits the following properties, which are easy to verify. First, note that $x^- x = \mathbb{1}$ for each non-zero vector x.

Suppose that $x, y \in \mathbb{X}^n$ are regular vectors. Then, the matrix inequality $xy^- \geq (x^- y)^{-1} I$ holds entry-wise, and becomes $xx^- \geq I$ if $y = x$.

Finally, for any regular vector $x \in \mathbb{X}^n$, if a matrix $A \in \mathbb{X}^{n \times n}$ is row-regular, then Ax is a regular vector. If A is column-regular, then $x^- A$ is regular.

3   Solutions to Linear Inequalities

We now present solutions to linear vector inequalities, which form the basis for later investigation of constrained optimization problems. These solutions are often obtained as consequences to the solution of the corresponding equations, and are known under diverse assumptions, at different levels of generality, and in various forms (see, e.g., [Car79, CG79, Zim81, BCOQ93, ABG07, But10]).

In this section we follow the results in [Kri06, Kri09b, Kri09a, Kri13, KZ13], which offer a framework to represent the solutions in a compact vector form.

Suppose that, given a matrix $A \in \mathbb{X}^{m \times n}$ and a regular vector $d \in \mathbb{X}^m$, the problem is to find all regular vectors $x \in \mathbb{X}^n$ that satisfy the inequality

$$Ax \leq d. \tag{1}$$

The next result offers a solution obtained as a consequence of the solution to the corresponding equation [Kri09b, Kri09a], and by independent proof [KZ13].

Lemma 3.1. *For every column-regular matrix A and regular vector d, all regular solutions to inequality* (1) *are given by*

$$x \leq (d^- A)^-.$$

Furthermore, we consider the following problem: given a matrix $A \in \mathbb{X}^{n \times n}$ and a vector $b \in \mathbb{X}^n$, find all regular vectors $x \in \mathbb{X}^n$ that satisfy the inequality

$$Ax \oplus b \leq x. \tag{2}$$

To describe a complete solution to the problem, we define a function that maps every matrix $A \in \mathbb{X}^{n \times n}$ to a scalar given by

$$\mathrm{Tr}(A) = \mathrm{tr}\, A \oplus \cdots \oplus \mathrm{tr}\, A^n.$$

We also employ the asterate operator (also known as the Kleene star), which takes A to the matrix

$$A^* = I \oplus A \oplus \cdots \oplus A^{n-1}.$$

Note that the asterate possesses a useful property established by Carré [Car71]. The property states that each matrix A with $\mathrm{Tr}(A) \leq \mathbb{1}$ satisfies the entry-wise inequality $A^k \leq A^*$ for all integer $k \geq 0$. Specifically, this property makes the equality $A^* A^* = A^*$ valid provided that $\mathrm{Tr}(A) \leq \mathbb{1}$.

A direct solution to inequality (2) is given as follows [Kri06, Kri09b, Kri13].

Theorem 3.2. *For every matrix A and vector b, the following statements hold:*

1. *If $\mathrm{Tr}(A) \leq \mathbb{1}$, then all regular solutions to* (2) *are given by $x = A^* u$, where u is any regular vector such that $u \geq b$.*
2. *If $\mathrm{Tr}(A) > \mathbb{1}$, then there is no regular solution.*

4 Optimization Problems

This section is concerned with deriving complete direct solutions to multidimensional constrained optimization problems. The problems consist in minimizing a nonlinear objective function subject to both linear inequality constraints with a matrix and simple boundary constraints. We start with a short overview of the previous results, which provide solutions to problems with reduced sets of constraints. Furthermore, a complete solution to a general problem that involves both constraints is obtained under fairly general assumptions. Two special cases of the solution are discussed which improve the previous results. Finally, we present illustrative examples of two-dimensional optimization problems.

4.1 Previous Results

We start with an unconstrained problem that is examined in [Kri11] by applying extremal properties of tropical eigenvalues. Given vectors $p, q \in \mathbb{X}^n$, the problem is to find regular vectors $x \in \mathbb{X}^n$ that

$$\text{minimize} \quad x^- p \oplus q^- x. \tag{3}$$

The problem is reduced to the solving of the eigenvalue-eigenvector problem for a certain matrix. The solution is given by the next statement.

Lemma 4.1. *Let p and q be regular vectors, and*

$$\theta = (q^- p)^{1/2}.$$

Then, the minimum value in problem (3) is equal to θ and attained at each vector x such that

$$\theta^{-1} p \leq x \leq \theta q.$$

A different approach based on the solutions to linear inequalities is used in [Kri12, KZ13] to show that the above solution of problem (3) is complete. Moreover, the approach is applied to solve constrained versions of the problem. Specifically, the following problem is considered: given a matrix $B \in \mathbb{X}^{n \times n}$, find regular vectors x that

$$\begin{aligned} \text{minimize} \quad & x^- p \oplus q^- x, \\ \text{subject to} \quad & Bx \leq x. \end{aligned} \tag{4}$$

The solution, which is given in [Kri12] under some restrictive assumptions on the matrix B, can readily be extended to arbitrary matrices by using the result of Theorem 3.2, and then written in the following form.

Theorem 4.2. *Let B be a matrix with $\mathrm{Tr}(B) \leq \mathbb{1}$, p and q regular vectors, and*

$$\theta = ((B^*(q^- B^*)^-)^- p)^{1/2}. \tag{5}$$

Then, the minimum value in problem (4) is equal to θ and attained at

$$x = \theta B^*(q^- B^*)^-.$$

Note that the theorem offers a particular solution to the problem rather than provides a complete solution.

Furthermore, given vectors $g, h \in \mathbb{X}^n$, consider a problem with two-sided boundary constraints to find regular vectors x that

$$\text{minimize} \quad x^- p \oplus q^- x,$$
$$\text{subject to} \quad g \le x \le h. \tag{6}$$

The complete solution obtained in [KZ13] is as follows.

Theorem 4.3. *Let p, q, g, and h be regular vectors such that $g \le h$, and*

$$\theta = (q^- p)^{1/2} \oplus h^- p \oplus q^- g.$$

Then, the minimum in problem (6) is equal to θ and all regular solutions of the problem are given by the condition

$$g \oplus \theta^{-1} p \le x \le (h^- \oplus \theta^{-1} q^-)^-.$$

Below, we examine a new general problem, which combines the constraints in problems (4) and (6), and includes both these problems as special cases.

4.2   New Optimization Problem with Combined Constraints

We now are in a position to formulate and solve a new constrained optimization problem. The solution follows the approach developed in [Kri13, Kri14], which is based on the introduction of an auxiliary variable and the reduction of the problem to the solution of a parametrized system of linear inequalities, where the new variable plays the role of a parameter. The existence condition for the solution of the system is used to evaluate the parameter, whereas the complete solution to the system is taken as the solution to the optimization problem.

Given vectors $p, q, g, h \in \mathbb{X}^n$, and a matrix $B \in \mathbb{X}^{n \times n}$, consider the problem to find all regular vectors $x \in \mathbb{X}^n$ that

$$\text{minimize} \quad x^- p \oplus q^- x,$$
$$\text{subject to} \quad Bx \oplus g \le x, \tag{7}$$
$$x \le h.$$

The constraints in the problem can also be written in the equivalent form

$$Bx \le x,$$
$$g \le x \le h.$$

The next statement gives a complete direct solution to the problem.

Theorem 4.4. *Let B be a matrix with $\operatorname{Tr}(B) \le \mathbb{1}$, p be a non-zero vector, q and h regular vectors, and g a vector such that $h^- B^* g \le \mathbb{1}$. Define a scalar*

$$\theta = (q^- B^* p)^{1/2} \oplus h^- B^* p \oplus q^- B^* g. \tag{8}$$

Then, the minimum value in problem (7) is equal to θ and all regular solutions of the problem are given by

$$x = B^* u,$$

where u is any regular vector such that

$$g \oplus \theta^{-1} p \le u \le ((h^- \oplus \theta^{-1} q^-) B^*)^-. \tag{9}$$

Proof. Suppose that θ is the minimum of the objective function in problem (7) over all regular x, and note that $\theta \ge (q^- B^* p)^{1/2} \ge (q^- p)^{1/2} > 0$. Then, all solutions to the problem are given by the system

$$x^- p \oplus q^- x = \theta,$$
$$B x \oplus g \le x,$$
$$x \le h.$$

Since θ is the minimum of the objective function, the solution set remains unchanged if we replace the first equation by the inequality $x^- p \oplus q^- x \le \theta$ and then substitute this inequality with equivalent two inequalities as follows

$$x^- p \le \theta,$$
$$q^- x \le \theta,$$
$$B x \oplus g \le x,$$
$$x \le h.$$

After the application of Lemma 3.1 to the first two inequalities, the system becomes

$$\theta^{-1} p \le x,$$
$$x \le \theta q,$$
$$B x \oplus g \le x,$$
$$x \le h.$$

We now combine the inequalities in the system as follows. The first and third inequalities are equivalent to the inequality $B x \oplus g \oplus \theta^{-1} p \le x$.

The other two inequalities are replaced by $x^- \ge \theta^{-1} q^-$ and $x^- \ge h^-$, which are equivalent to $x^- \ge h^- \oplus \theta^{-1} q^-$, and thus to $x \le (h^- \oplus \theta^{-1} q^-)^-$.

After the rearrangement of the system, we arrive at the double inequality

$$B x \oplus g \oplus \theta^{-1} p \le x \le (h^- \oplus \theta^{-1} q^-)^-.$$

The solution of the left inequality by using Theorem 3.2 gives the result

$$x = B^* u, \qquad u \ge g \oplus \theta^{-1} p.$$

Substitution of this solution into the right inequality yields the inequality

$$B^* u \le (h^- \oplus \theta^{-1} q^-)^-,$$

which, by Lemma 3.1, has the solution

$$\boldsymbol{u} \leq ((\boldsymbol{h}^- \oplus \theta^{-1}\boldsymbol{q}^-)\boldsymbol{B}^*)^-.$$

By coupling both lower and upper bounds on $\boldsymbol{u}$, we arrive at the solution in the form of (9). The solution set defined by (9) is non-empty if and only if

$$\boldsymbol{g} \oplus \theta^{-1}\boldsymbol{p} \leq ((\boldsymbol{h}^- \oplus \theta^{-1}\boldsymbol{q}^-)\boldsymbol{B}^*)^-.$$

The left multiplication of this inequality by $(\boldsymbol{h}^- \oplus \theta^{-1}\boldsymbol{q}^-)\boldsymbol{B}^*$ and application of one property of conjugate transposition lead to

$$(\boldsymbol{h}^- \oplus \theta^{-1}\boldsymbol{q}^-)\boldsymbol{B}^*(\boldsymbol{g} \oplus \theta^{-1}\boldsymbol{p}) \leq (\boldsymbol{h}^- \oplus \theta^{-1}\boldsymbol{q}^-)\boldsymbol{B}^*((\boldsymbol{h}^- \oplus \theta^{-1}\boldsymbol{q}^-)\boldsymbol{B}^*)^- = \mathbb{1},$$

which results in the new inequality

$$(\boldsymbol{h}^- \oplus \theta^{-1}\boldsymbol{q}^-)\boldsymbol{B}^*(\boldsymbol{g} \oplus \theta^{-1}\boldsymbol{p}) \leq \mathbb{1}.$$

Since the left multiplication of the latter inequality by $((\boldsymbol{h}^- \oplus \theta^{-1}\boldsymbol{q}^-)\boldsymbol{B}^*)^-$ and the other property of conjugate transposition give the former inequality, both inequalities are equivalent. The obtained inequality can further be rewritten as

$$\theta^{-2}\boldsymbol{q}^-\boldsymbol{B}^*\boldsymbol{p} \oplus \theta^{-1}(\boldsymbol{h}^-\boldsymbol{B}^*\boldsymbol{p} \oplus \boldsymbol{q}^-\boldsymbol{B}^*\boldsymbol{g}) \oplus \boldsymbol{h}^-\boldsymbol{B}^*\boldsymbol{g} \leq \mathbb{1},$$

and then represented by the equivalent system

$$\theta^{-2}\boldsymbol{q}^-\boldsymbol{B}^*\boldsymbol{p} \leq \mathbb{1},$$
$$\theta^{-1}(\boldsymbol{h}^-\boldsymbol{B}^*\boldsymbol{p} \oplus \boldsymbol{q}^-\boldsymbol{B}^*\boldsymbol{g}) \leq \mathbb{1},$$
$$\boldsymbol{h}^-\boldsymbol{B}^*\boldsymbol{g} \leq \mathbb{1}.$$

Note that the third inequality in the system is valid by the condition of the theorem. After rearrangement of terms, the first two inequalities become

$$\theta \geq (\boldsymbol{q}^-\boldsymbol{B}^*\boldsymbol{p})^{1/2},$$
$$\theta \geq \boldsymbol{h}^-\boldsymbol{B}^*\boldsymbol{p} \oplus \boldsymbol{q}^-\boldsymbol{B}^*\boldsymbol{g},$$

and then finally lead to one inequality

$$\theta \geq (\boldsymbol{q}^-\boldsymbol{B}^*\boldsymbol{p})^{1/2} \oplus \boldsymbol{h}^-\boldsymbol{B}^*\boldsymbol{p} \oplus \boldsymbol{q}^-\boldsymbol{B}^*\boldsymbol{g}.$$

Since θ is assumed to be the minimum value of the objective function, the last inequality has to be satisfied as an equality, which gives (8).   □

4.3   Particular Cases

We now examine particular cases, in which the feasible solution set is defined either by a linear inequality with a matrix or by two-sided boundary constraints.

First, we offer a new complete solution to problem (4), which does not have the boundary constraints. A slight modification to the proof of Theorem 4.4 yields the solution in the following form.

Corollary 4.5. *Let B be a matrix with $\mathrm{Tr}(B) \leq 1$, p be a non-zero vector, and q a regular vector. Define a scalar*

$$\theta = (q^- B^* p)^{1/2}. \tag{10}$$

Then, the minimum in (4) is θ and all regular solutions are given by

$$x = B^* u, \qquad \theta^{-1} p \leq u \leq \theta (q^- B^*)^-.$$

Although the expression at (10) offers the minimum in a different and more compact form than that at (5), both representations prove to be equivalent.

To verify that these representations coincide, we first note that $B^* B^* = B^*$ and then apply the properties of conjugate transposition to write

$$B^*(q^- B^*)^- = B^*(q^- B^* B^*)^- \leq (q^- B^*)^- q^- B^* B^* (q^- B^* B^*)^- = (q^- B^*)^-,$$

which implies that the inequality $B^*(q^- B^*)^- \leq (q^- B^*)^-$ holds.

Since $B^* \geq I$, the opposite inequality $B^*(q^- B^*)^- \geq (q^- B^*)^-$ is valid as well. Both inequalities result in the equality $B^*(q^- B^*)^- = (q^- B^*)^-$, and thus in the equality $(B^*(q^- B^*)^-)^- = q^- B^*$. Finally, the right multiplication by p and extraction of square roots lead to the desired result.

Furthermore, we put B to be the zero matrix in (7) and so arrive at problem (6), which can be completely solved through a direct consequence of Theorem 4.4. Clearly, the new solution of (6) coincides with that given by Theorem 4.3, and even involves somewhat less assumptions on the vectors under consideration.

4.4   Numerical Examples and Graphical Illustration

To illustrate the results obtained above, we present examples of two-dimensional problems in the setting of the idempotent semifield $\mathbb{R}_{\max,+}$ and provide geometric interpretation on the plane with a Cartesian coordinate system.

Consider problem (7) formulated in terms of $\mathbb{R}_{\max,+}$ under the assumptions that

$$p = \begin{pmatrix} 3 \\ 14 \end{pmatrix}, \quad q = \begin{pmatrix} -12 \\ -4 \end{pmatrix}, \quad g = \begin{pmatrix} 2 \\ -8 \end{pmatrix}, \quad h = \begin{pmatrix} 6 \\ 8 \end{pmatrix}, \quad B = \begin{pmatrix} 0 & -4 \\ -8 & -6 \end{pmatrix}.$$

Prior to solving the general problem, we examine several special cases.

We start with problem (3) without constraints, which has a complete solution given by a consequence of Theorem 4.4 (see also Lemma 4.1). According to this result, the minimum in the unconstrained problem is given by

$$\theta_1 = (q^- p)^{1/2} = 9,$$

and attained if and only if the vector x satisfies the conditions

$$x_1' \leq x \leq x_1'', \qquad x_1' = \theta_1^{-1} p = \begin{pmatrix} -6 \\ 5 \end{pmatrix}, \qquad x_1'' = \theta_1 q = \begin{pmatrix} -3 \\ 5 \end{pmatrix}.$$

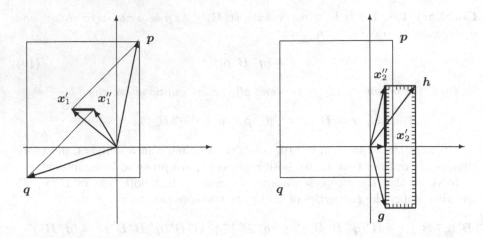

Fig. 1. Solutions to problems without constraints (left) and with two-sided boundary constraints (right)

A graphical illustration of the result is given in Fig. 1 (left), where the solutions form a horizontal segment between the ends of the vectors x_1' and x_1''.

Furthermore, we consider the problem in the form (6) with two-sided boundary constraints $g \leq x \leq h$. It follows from Theorem 4.3 (or as another consequence of Theorem 4.4) that the minimum in the problem is calculated as

$$\theta_2 = (q^- p)^{1/2} \oplus h^- p \oplus q^- g = 14.$$

The solution set consists of those vectors x that satisfy the double inequality

$$x_2' \leq x \leq x_2'', \qquad x_2' = g \oplus \theta_2^{-1} p = \begin{pmatrix} 2 \\ 0 \end{pmatrix}, \qquad x_2'' = (h^- \oplus \theta_2^{-1} q^-)^- = \begin{pmatrix} 2 \\ 8 \end{pmatrix}.$$

The solutions of the problem are indicated on Fig. 1 (right) by a thick vertical segment on the left side of the rectangle that represents the feasible set.

We now examine problem (4) with the linear inequality constraints $Bx \leq x$. We calculate

$$B^* = I \oplus B = \begin{pmatrix} 0 & -4 \\ -8 & 0 \end{pmatrix}, \qquad q^- B^* = \begin{pmatrix} 12 & 8 \end{pmatrix}.$$

The application of Corollary 4.5 gives the minimum value

$$\theta_3 = (q^- B^* p)^{1/2} = 11,$$

which is attained if and only if $x = B^* u$ for all u such that

$$u_3' \leq u \leq u_3'', \qquad u_3' = \theta^{-1} p = \begin{pmatrix} -8 \\ 3 \end{pmatrix}, \qquad u_3'' = \theta(q^- B^*)^- = \begin{pmatrix} -1 \\ 3 \end{pmatrix}.$$

After multiplication of B^* by both bounds on u, we conclude that the problem has the unique solution

$$x_3 = B^*u_3' = B^*u_3'' = \begin{pmatrix} -1 \\ 3 \end{pmatrix}.$$

Figure 2 (left) shows the solution point located on the upper side of the strip, which represents the solution of the inequality $Bx \leq x$. The columns of the matrices $B = (b_1, b_2)$ and $B^* = (b_1^*, b_2^*)$ are also included.

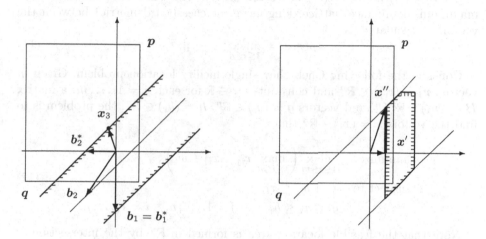

Fig. 2. Solutions to problems with linear inequality constraints (left) and with both linear inequality and two-sided boundary constraints (right)

Finally, we consider general problem (7). To solve the problem, we calculate

$$B^*p = \begin{pmatrix} 10 \\ 14 \end{pmatrix}, \qquad h^- B^*p = 6, \qquad q^- B^*g = 14.$$

It follows from Theorem 4.4 that the minimum in the problem is given by

$$\theta = (q^- B^*p)^{1/2} \oplus h^- B^*p \oplus q^- B^*g = 14.$$

This minimum is attained only at $x = B^*u$, where u is any vector such that

$$u' \leq u \leq u'', \qquad u' = g \oplus \theta^{-1}p = \begin{pmatrix} 2 \\ 0 \end{pmatrix}, \qquad u'' = ((h^- \oplus \theta^{-1}q^-)B^*)^- = \begin{pmatrix} 2 \\ 6 \end{pmatrix}.$$

Turning to the solution of the problem, we arrive at the set of vectors x that satisfy the conditions

$$x' \leq x \leq x'', \qquad x' = B^*u' = \begin{pmatrix} 2 \\ 0 \end{pmatrix}, \qquad x'' = B^*u'' = \begin{pmatrix} 2 \\ 6 \end{pmatrix}.$$

The solution is shown on Fig. 2 (right) by the thick vertical segment on the left side of the polygon which describes the feasible set.

5   Application to Location Analysis

In this section, we apply the above results to solve minimax single facility location problems, which are often called the Rawls problems [HPT80, HPT81], but also known as Messenger Boy problems [EH72] and 1-center problems [Dre11]. We consider a new constrained problem on a multidimensional space with Chebyshev distance. A complete direct solution is obtained which extends the results in [Kri11, Kri12, KZ13] by taking into account a more general system of constraints.

Let $r = (r_i)$ and $s = (s_i)$ be vectors in $\mathbb{R}^n$. The Chebyshev distance (L_∞, maximum, dominance, lattice, king-move, or chessboard metric) between the vectors is calculated as

$$\rho(r, s) = \max_{1 \le i \le n} |r_i - s_i|. \tag{11}$$

Consider the following Chebyshev single facility location problem. Given m vectors $r_j = (r_{ij}) \in \mathbb{R}^n$ and constants $w_j \in \mathbb{R}$ for each $j = 1, \ldots, m$, a matrix $B = (b_{ij}) \in \mathbb{R}^{n \times n}$, and vectors $g = (g_i) \in \mathbb{R}^n$, $h = (h_i) \in \mathbb{R}^n$, the problem is to find the vectors $x = (x_i) \in \mathbb{R}^n$ that

$$
\begin{aligned}
&\text{minimize} \quad \max_{1 \le j \le m} \left(\max_{1 \le i \le n} |r_{ij} - x_i| + w_i \right), \\
&\text{subject to} \quad x_j + b_{ij} \le x_i, \\
&\qquad\qquad g_i \le x_i \le h_i, \qquad j = 1, \ldots, n, \quad i = 1, \ldots, n.
\end{aligned} \tag{12}
$$

Note that the feasible location area is formed in $\mathbb{R}^n$ by the intersection of the hyper-rectangle defined by the boundary constraints with closed half-spaces given by the other inequalities.

To solve the problem, we represent it in terms of the semifield $\mathbb{R}_{\max,+}$. First, we put (11) in the equivalent form

$$\rho(r, s) = \bigoplus_{i=1}^{n} (s_i^{-1} r_i \oplus r_i^{-1} s_i) = s^- r \oplus r^- s.$$

Furthermore, we define the vectors

$$p = w_1 r_1 \oplus \cdots \oplus w_m r_m, \qquad q^- = w_1 r_1^- \oplus \cdots \oplus w_m r_m^-.$$

The objective function in problem (12) becomes

$$\bigoplus_{i=1}^{m} w_i \rho(r_i, x) = \bigoplus_{i=1}^{m} w_i (x^- r_i \oplus r_i^- x) = x^- p \oplus q^- x.$$

We now combine the constraints $x_j + b_{ij} \le x_i$ for all $j = 1, \ldots, n$ into one inequality for each i, and write the obtained inequalities in terms of $\mathbb{R}_{\max,+}$ as

$$b_{i1} x_1 \oplus \cdots \oplus b_{in} x_n \le x_i, \qquad i = 1, \ldots, n.$$

After rewriting the above inequalities and the boundary constraints in matrix-vector form, we obtain the problem in the form (7), where all given vectors have real components. Since these vectors are clearly regular in the sense of $\mathbb{R}_{\max,+}$, they satisfy the conditions of Theorem 4.4, which completely solves the problem.

As an illustration, consider the two-dimensional problem with given points

$$r_1 = \begin{pmatrix} -7 \\ 12 \end{pmatrix}, \quad r_2 = \begin{pmatrix} 2 \\ 10 \end{pmatrix}, \quad r_3 = \begin{pmatrix} -10 \\ 3 \end{pmatrix}, \quad r_4 = \begin{pmatrix} -4 \\ 4 \end{pmatrix}, \quad r_5 = \begin{pmatrix} -4 \\ -3 \end{pmatrix},$$

and constants $w_1 = w_3 = 2$, $w_2 = w_4 = w_5 = 1$. For the sake of simplicity, we take the same matrix B and vectors g, h as in the examples considered above.

To reduce the location problem to problem (7), we first calculate the vectors

$$p = \begin{pmatrix} 3 \\ 14 \end{pmatrix}, \quad q = \begin{pmatrix} -12 \\ -4 \end{pmatrix}.$$

These vectors define two opposite corners of the minimum rectangle which encloses all points $w_i r_i$ and $w_i^{-1} r_i$. The rectangle is depicted in Fig. 3 (left).

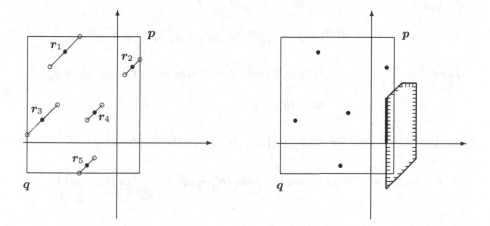

Fig. 3. Minimum enclosing rectangle (left) and solution of location problem (right)

Note that the reduced problem coincides with that examined as an example in the previous section, and thus admits the same solution. We show the solution as a thick vertical segment and the given points as black dots in Fig. 3 (right).

To conclude this section, we write the solution given by Theorem 4.4 to problem (12) in the usual form.

We first represent the entries of the matrix $B^* = (b_{ij}^*)$ in terms of ordinary operations. It follows from the definition of the asterate operator that

$$b_{ij}^* = \begin{cases} \beta_{ij}, & \text{if } i \neq j; \\ \max(\beta_{ij}, 0), & \text{if } i = j; \end{cases}$$

where the numbers β_{ij} are calculated as

$$\beta_{ij} = \max_{1 \leq k \leq n-1} \max_{\substack{1 \leq i_1, \ldots, i_{k-1} \leq n \\ i_0 = i, i_k = j}} (b_{i_0 i_1} + \cdots + b_{i_{k-1} i_k}).$$

Furthermore, we replace the operations of tropical mathematics by arithmetic operations in the rest of the statement of Theorem 4.4. By adding definitions for the vectors p and q, we obtain the following statement.

Theorem 5.1. *Let B be a matrix, and g and h be vectors such that*

$$\max_{\substack{1 \leq i,k \leq n \\ }} \max_{\substack{1 \leq i_1,\dots,i_{k-1} \leq n \\ i_0 = i_k = i}} (b_{i_0 i_1} + \cdots + b_{i_{k-1} i_k}) \leq 0,$$

$$\max_{1 \leq i,j \leq n} (b^*_{ij} - h_i + g_j) \leq 0.$$

Define vectors $p = (p_i)$ and $q = (q_i)$ with elements

$$p_i = \max_{1 \leq j \leq m} (r_{ij} + w_j), \qquad q_i = \min_{1 \leq j \leq m} (r_{ij} - w_j), \qquad i = 1,\dots,n;$$

and a scalar

$$\theta = \max_{1 \leq i,j \leq n} \left((b^*_{ij} - q_i + p_j)/2, b^*_{ij} - h_i + p_j, b^*_{ij} - q_i + g_j \right).$$

Then, the minimum in (12) is θ and all solutions $x = (x_i)$ are given by

$$x_i = \max_{1 \leq j \leq n} (b^*_{ij} + u_j), \qquad i = 1,\dots,n;$$

where the numbers u_j for each $j = 1,\dots,n$ satisfy the condition

$$\max(g_j, p_j - \theta) \leq u_j \leq \min \left(- \max_{1 \leq i \leq n} (b^*_{ij} - h_i), \theta - \max_{1 \leq i \leq n} (b^*_{ij} - q_i) \right).$$

6   Conclusions

The paper was concerned with a new multidimensional tropical optimization problem with a nonlinear objective function and inequality constraints. A complete solution was obtained based on the technique, which reduces the problem to the solution of a linear inequality with a parametrized matrix. The solution is given in a closed form in terms of simple vector operations, which offers low computational complexity and provides for efficient software implementation.

Possible directions of future research include the further extension of the problem to account for new types of objective functions and constraints. The development of new real-world applications of the results is also of interest.

Acknowledgments. The author is very grateful to the three reviewers for their careful reading of a previous draft of this paper. He thanks the reviewers for their valuable comments and illuminating suggestions that have been incorporated in the final version.

References

[ABG07]   Akian, M., Bapat, R., Gaubert, S.: Max-plus algebra. In: Hogben, L. (ed.) Handbook of Linear Algebra. Discrete Mathematics and its Applications, pp. 25-1–25-17. Taylor and Francis, Boca Raton (2007)

[BCOQ93]   Baccelli, F.L., Cohen, G., Olsder, G.J., Quadrat, J.-P.: Synchronization and Linearity: An Algebra for Discrete Event Systems. Wiley Series in Probability and Statistics. Wiley, Chichester (1993)

[But10]   Butkovič, P.: Max-linear Systems: Theory and Algorithms. Springer Monographs in Mathematics. Springer, London (2010)

[Car71]   Carré, B.A.: An algebra for network routing problems. IMA J. Appl. Math. 7, 273–294 (1971)

[Car79]   Carré, B.: Graphs and Networks. Oxford Applied Mathematics and Computing Science Series. Clarendon Press, Oxford (1979)

[CG62]   Cuninghame-Green, R.A.: Describing industrial processes with interference and approximating their steady-state behaviour. Oper. Res. Quart. 13, 95–100 (1962)

[CG91]   Cuninghame-Green, R.A.: Minimax algebra and applications. Fuzzy Sets and Systems 41, 251–267 (1991)

[CG94]   Cuninghame-Green, R.A.: Minimax algebra and applications. In: Hawkes, P.W. (ed.) Advances in Imaging and Electron Physics, vol. 90, pp. 1–121. Academic Press, San Diego (1994)

[CG79]   Cuninghame-Green, R.: Minimax Algebra. Lecture Notes in Economics and Mathematical Systems, vol. 166. Springer, Berlin (1979)

[Dre11]   Drezner, Z.: Continuous Center Problems. In: Eiselt, H.A., Marianov, V. (eds.) Foundations of Location Analysis. International Series in Operations Research and Management Science, vol. 155, pp. 63–78. Springer, New York (2011)

[EH72]   Elzinga, J., Hearn, D.W.: Geometrical solutions for some minimax location problems. Transport. Sci. 6, 379–394 (1972)

[Gif63]   Giffler, B.: Scheduling general production systems using schedule algebra. Naval Res. Logist. Quart. 10, 237–255 (1963)

[Gol03]   Golan, J.S.: Semirings and Affine Equations Over Them: Theory and Applications. Mathematics and its Applications, vol. 556. Springer, New York (2003)

[GM08]   Gondran, M., Minoux, M.: Graphs, Dioids and Semirings: New Models and Algorithms. Operations Research/Computer Science Interfaces, vol. 41. Springer, New York (2008)

[HPT80]   Hansen, P., Peeters, D., Thisse, J.-F.: Location of public services: A selective method-oriented survey. Annals of Public and Cooperative Economics 51, 9–51 (1980)

[HPT81]   Hansen, P., Peeters, D., Thisse, J.-F.: Constrained location and the Weber-Rawls problem. In: Hansen, P. (ed.) Annals of Discrete Mathematics (11) Studies on Graphs and Discrete Programming. North-Holland Mathematics Studies, vol. 59, pp. 147–166. North-Holland (1981)

[HOvdW06] Heidergott, B., Olsder, G.J., van der Woude, J.: Max-plus at Work: Modeling and Analysis of Synchronized Systems. Princeton Series in Applied Mathematics. Princeton University Press, Princeton (2006)

[HZ99]   Hudec, O., Zimmermann, K.: Biobjective center – balance graph location model. Optimization 45, 107–115 (1999)

[HZ93]   Hudec, O., Zimmermann, K.: A service points location problem with Min-Max distance optimality criterion. Acta Univ. Carolin. Math. Phys. 34, 105–112 (1993)

[KM97]   Kolokoltsov, V.N., Maslov, V.P.: Idempotent Analysis and Its Applications, Mathematics and its Applications, vol. 401. Kluwer Academic Publishers, Dordrecht (1997)

[Kri06]   Krivulin, N.K.: Solution of generalized linear vector equations in idempotent algebra. Vestnik St. Petersburg Univ. Math. 39, 16–26 (2006)

[Kri09a]   Krivulin, N.K.: Methods of Idempotent Algebra for Problems in Modeling and Analysis of Complex Systems. Saint Petersburg University Press, St. Petersburg (2009) (in Russian)

[Kri09b]   Krivulin, N.K.: On solution of a class of linear vector equations in idempotent algebra. Vestnik St. Petersburg University. Applied Mathematics, Informatics, Control Processes 10, 64–77 (2009) (in Russian)

[Kri11]   Krivulin, N.: An algebraic approach to multidimensional minimax location problems with Chebyshev distance. WSEAS Trans. Math. 10, 191–200 (2011)

[Kri12]   Krivulin, N.: A new algebraic solution to multidimensional minimax location problems with Chebyshev distance. WSEAS Trans. Math. 11, 605–614 (2012)

[Kri13]   Krivulin, N.: A multidimensional tropical optimization problem with nonlinear objective function and linear constraints. Optimization (2013)

[KZ13]   Krivulin, N., Zimmermann, K.: Direct solutions to tropical optimization problems with nonlinear objective functions and boundary constraints. In: Biolek, D., Walter, H., Utu, I., von Lucken, C. (eds.) Mathematical Methods and Optimization Techniques in Engineering, pp. 86–91. WSEAS Press (2013)

[Kri14]   Krivulin, N.: A constrained tropical optimization problem: complete solution and application example. In: Litvinov, G.L., Sergeev, S.N. (eds.) Tropical and Idempotent Mathematics and Applications, Contemp. Math. American Mathematical Society, Providence (2014)

[MB09]   Moradi, E., Bidkhori, M.: Single facility location problem. In: Farahani, R.Z., Hekmatfar, M. (eds.) Facility Location, Contributions to Management Science, pp. 37–68. Physica (2009)

[Pan61]   Pandit, S.N.N.: A new matrix calculus. J. SIAM 9, 632–639 (1961)

[Rom64]   Romanovskiĭ, I. V.: Asymptotic behavior of dynamic programming processes with a continuous set of states. Soviet Math. Dokl. 5, 1684–1687 (1964)

[Sul01]   Sule, D.R.: Logistics of facility location and allocation. Marcel Dekker Ltd., New York (2001)

[TZ10]   Tharwat, A., Zimmermann, K.: One class of separable optimization problems: Solution method, application. Optimization 59, 619–625 (2010)

[Vor63]   Vorob'ev, N.N.: The extremal matrix algebra. Soviet Math. Dokl. 4, 1220–1223 (1963)

[Zim92]   Zimmermann, K.: Optimization problems with unimodal functions in max-separable constraints. Optimization 24, 31–41 (1992)

[Zim81]   Zimmermann, U.: Linear and Combinatorial Optimization in Ordered Algebraic Structures. Annals of Discrete Mathematics, vol. 10. Elsevier, Amsterdam (1981)

Refinements of the RCC25 Composition Table

Manas Ghosh and Michael Winter

Department of Computer Science
Brock University
St. Catharines, Ontario, Canada, L2S 3A1
{mg11yq,mwinter}@brocku.ca

Abstract. Boolean Contact Algebras (BCAs) are an appropriate algebraic approach to mereotopological structures. They are Boolean algebras equipped with a binary contact relation C indicating whether two regions are considered to be in contact or not. It has been shown that BCAs with some additional properties are equivalent to the well-known Region Connection Calculus (RCC) of Randell et al. In this paper we show that the contact relation of a BCA gives rise to at least 35 atomic relationships between regions in any model of RCC. In addition, we provide a composition table of the corresponding relation algebra up to 31 atoms. This improves previous results that distinguished only 25 atomic relationships.

1 Introduction

Qualitative spatial reasoning (QSR) is an important area of Artificial Intelligence (AI) which is concerned with the qualitative aspects of representing and reasoning about spatial entities. Non-numerical relationships among spatial objects can be expressed through QSR. The majority of work in QSR has focused on single aspects of space. Probably the most important aspect is the underlying topology, i.e., the spatial relationship between regions. Relation algebras (RAs) have been shown to be a very convenient tool within spatial reasoning. One reason is that a large part of contemporary spatial reasoning is based on the investigation of "part of" relations and "contact" relations in various domains [7, 21, 22, 37]. Relation algebras were introduced into spatial reasoning in [22] with additional results published in [20, 23]. We would like to refer the reader to these papers for additional background and motivation.

As a subarea of QSR, mereotopology combines mereology, topology and algebraic reasoning. Many possible theories have been proposed for mereotopology [4, 6, 30, 31]. The most prominent theory is the region connection calculus (RCC) [7], which is based on Clarke's theory [8]. Randell in [32, 33] first proposed RCC as a logical framework for mereotopology. It was shown in [36] that models of the RCC are isomorphic to Boolean connection algebras [35]. A slightly more general approach is based on Boolean contact algebras (BCAs) which are Boolean algebras equipped with a binary contact relation C. In fact, Boolean Connection Algebras, and, hence, RCC models, are extensional and connected BCAs.

P. Höfner et al. (Eds.): RAMiCS 2014, LNCS 8428, pp. 379–394, 2014.
© Springer International Publishing Switzerland 2014

As lattices, and Boolean algebras in particular, are well-known mathematical structures, this led towards an intensive study of the properties of the RCC including several topological representation theorems [11, 13–15, 19].

The contact relation of a BCA allows to define several atomic relationships between regions. These atomic relations form a jointly exhaustive and pairwise disjoint (JEPD) set of relations, i.e., each pair of regions is in exactly one of those atomic relationships. This approach to spatial reasoning closely mirrors Allen's interval algebra [1, 2] and the corresponding interval based approach to temporal reasoning. The JEPD set of topological relations known as RCC8 has been identified to be of particular importance in the RCC theory. RCC8 consists of the relations "x is disconnected from y" (DC), "x is externally connected to y" (EC), "x partially overlaps y" (PO), "x is equal to y" (EQ), "x is a tangential proper part of y" (TPP), "x is a non-tangential proper part of y" ($NTPP$), and the converse of the latter two relations. A relation algebra was developed based on these eight atomic relations. This kind of categorization of topological relations was independently given by Egenhofer [19] in the context of geographical information systems (GIS). In order to further study contact relations, Düntsch [12, 17, 18] studied RCC8 and other relation algebras derived from mereotopology. Furthermore, he explored their expressive power with respect to topological domains.

It has been shown in [16–18] that it is possible to refine the eight atomic relations of RCC8. As a result several new algebras with up to 25 atoms were generated. New relations were obtained by splitting certain atoms from a previous algebra into two new relations. A new composition table was generated by removing certain elements after splitting in the composition table for one of the new atoms according to its definition. Splitting atoms in relation algebras is based on a general method introduced in [34]. This method generalized a previous method [3] that is not suitable in the BCA context. The reason is that this splitting procedure requires a condition to hold that is violated by all the RCC tables starting with RCC11.

In this paper we will refine the RCC25 algebra by providing 3 additional algebras with 27, 29 and 31 atoms. In addition, we will show that further atoms will split. However, the generation of algebras beyond RCC31 will require further study on regions with holes in BCAs.

2   Mathematical Preliminaries

In this section we recall some basic notions and properties. However, we will assume that the reader is familiar with the basic notions from Boolean algebras and lattice theory. For any notion used but defined here we refer to [5].

A subset $R \subseteq A \times A$ is called a binary relation on A. We will use the notation xRy instead of $(x, y) \in R$. We will also write $xRySz$ as an abbreviation for xRy and ySz.

2.1 Boolean Contact Algebra

De Laguna (1922) and Whitehead (1929) [10, 38] first used contact relations in their research. They used regions instead of points as the basic entity of geometry. Whitehead [38] has defined that two regular closed sets are in contact, if they have a non-empty intersection. Notice that the regular closed sets of a topological space form a Boolean algebra in which the meet of two regions is not their set intersection. This notion of contact is a reflexive and symmetric relation C among non-empty regions, satisfying an additional extensionality axiom. Leśniewski's classical mereology was generalized by Clarke [8] by taking a contact relation C as the basic structural element. Clarke also proposed additional axioms such as compatibility and summation. As a result he obtained complete Boolean algebras without a least element together with Whitehead's connection relation C. This observation led to the notion of BCAs.

Definition 1. *Let $\mathfrak{B} = (B, +, \cdot, {}^-, 0, 1)$ be a Boolean algebra, and $C \subseteq B \times B$ a binary relation on B. Then we consider the following properties of C:*

C_0. $0\overline{C}x$, *i.e., not $0Cx$*   *(Null disconnectedness)*
C_1. *if $x \neq 0$, then xCx*   *(Reflexivity)*
C_2. *if xCy, then yCx*   *(Symmetry)*
C_3. *if xCy and $y \leq z$, then xCz*   *(Compatibility)*
C_4. *if $xC(y + z)$, then xCy or xCz*   *(Summation)*
C_5. *if $C(x) = C(y)$, then $x = y$*   *(Extensionality)*
C_6. *if xCz or $yC\overline{z}$ for all $z \in B$, then xCy*   *(Interpolation)*
C_7. *if $x \neq 0$ and $x \neq 1$, then $xC\overline{x}$*   *(Connection)*

C *is called a contact relation if it satisfies C_0-C_4. In this case the pair $\langle \mathfrak{B}, C \rangle$ is called a Boolean contact algebra. If C satisfies C_5 in addition, it is called an extensional contact relation. A Boolean contact algebra is called connected if C also satisfies C_7.*

2.2 Relation Algebras

We will only recall the concepts from the theory of relation algebras that will be used in this paper. For further details and some of the basic algebraic properties that will be used throughout this paper we refer to [28].

Definition 2. *A structure $\mathfrak{A} = \langle A, +, \cdot, {}^-, 0, 1, {}^\smile, ;, 1' \rangle$ is called a relation algebra (RA) iff it satisfies the following:*

R1 $\langle A, +, \cdot, {}^-, 0, 1 \rangle$ *is a Boolean algebra.*
R2 $\langle A, ;, 1' \rangle$ *is a monoid.*
R3 *For all $x, y, z \in A$ the following formulas are equivalent:*

$$x; y \cdot z = 0 \iff \breve{x}; z \cdot y = 0 \iff z; \breve{y} \cdot x = 0.$$

We say that $\mathfrak{A}$ is a non-associative relation algebra (NA) if $\mathfrak{A}$ is a structure satisfying all of the axioms above except associativity of the composition operation $;$, i.e., **R2** is weakened by only requiring that $1'$ is a neutral element for composition. We will denote the set of atoms of a relation algebra $\mathfrak{B}$ by $At\mathfrak{B}$ and the set of all bijections, i.e., relations f that satisfy $\breve{f}; f = 1'$ and $f; \breve{f} = 1'$, by $Bij\mathfrak{B}$.

An integral relation algebra is a relation algebra in which the composition of any two nonzero elements is not equal to zero. This property is equivalent to condition that the identity $1'$ is an atom.

Oriented triangles can be used to visualize **R3** and its immediate consequence the so-called *cycle law* [9]. It states that the following properties are equivalent:

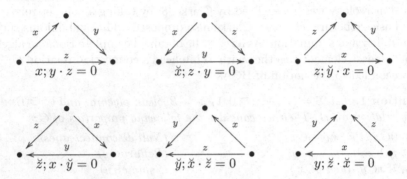

It is possible to recover a complete and atomic relation algebra from a suitable structure based on its atoms. Atom structures are very useful for storing and handling relation algebras on a computer because they take less storage space than the entire algebra. For this reason several implementations such as [26] are actually based on these structures. We consider a relational structure $\mathfrak{G} = \langle U, D, f, I \rangle$, where D is a ternary relation on U, f is a unary function $f : U \to U$, and I is a subset of U. It is possible to construct an algebra of relational type as follows.

Definition 3. *Given a relational structure* $\mathfrak{G} = \langle U, D, f, I \rangle$*, the complex algebra* $\mathfrak{Cm}\mathfrak{G} = \langle \mathcal{P}(U), \cup, \cap, ^-, \emptyset, U, \breve{\ }, ;, I \rangle$ *is based on the set-theoretic operations and constants* $\cup, \cap, ^-$ *and* $\emptyset$ *and the following operations:*

$$X; Y = \{z \in U : \exists x \in X, y \in Y (x, y, z) \in D\} \text{ and } \breve{X} = \{f(x) : x \in X\}.$$

Conversely, every relation algebra defines an atom structure on its set of atoms.

Definition 4. *An atom structure* $\mathfrak{At}\mathfrak{A} = \langle At\mathfrak{A}, D(\mathfrak{A}), f, I(\mathfrak{A}) \rangle$ *of a non-associative relation algebra* $\mathfrak{A}$ *consists of the set* $At\mathfrak{A}$ *of atoms, the set* $I(\mathfrak{A}) = \{x \in At\mathfrak{A} : x \leq 1'\}$*, the unary function* $f(x) = \breve{x}$*, and the ternary relation* $D(\mathfrak{A}) = \{(x, y, z) : x, y, z \in At\mathfrak{A}, x; y \geq z\}$*.*

The following theorem relates atom structures, relation algebras, and complex algebras. A proof can be found in [27].

Theorem 1. *Let* $\mathfrak{G} = \langle U, D, f, I \rangle$ *be a relational structure.*

1. *The following three conditions are equivalent:*
    - *i.* $\mathfrak{G}$ *is an atom structure of some complete atomic NA.*
    - *ii.* $\mathfrak{Cm}\mathfrak{G}$ *is a NA.*
    - *iii.* $\mathfrak{G}$ *satisfies condition (a) and (b):*
        - *(a) If* $(x, y, z) \in D$, *then* $(f(x), z, y) \in D$ *and* $(z, f(y), x) \in D$.
        - *(b) For all* $x, y \in U$ *we have* $x = y$ *iff there is some* $w \in I$ *such that* $(x, w, y) \in D$.
2. $\mathfrak{Cm}\mathfrak{G}$ *is a relation algebra iff* $\mathfrak{Cm}\mathfrak{G}$ *is a NA and it satisfies condition (c):*
    - *(c) For all* $x, v, w, x, y, z \in U$, *if* $(v, w, x) \in D$ *and* $(x, y, z) \in D$, *then there is some* $u \in U$ *such that* $(w, y, u) \in D$ *and* $(v, u, z) \in D$.

Condition *(a)* of the theorem above is closely related to the cycle law. This property can alternatively be formulated by using cycles. For three elements x, y, z we use the cycle $\langle x, y, z \rangle$ to denote the following set of up to six triples:

$$\langle x, y, z \rangle = \{(x, y, z), (\breve{x}, z, y), (y, \breve{z}, \breve{x}), (\breve{y}, \breve{x}, \breve{z}), (\breve{z}, x, \breve{y}), (z, \breve{y}, x)\}.$$

Using cycles we can reformulate Condition *(a)* as: if $(x, y, z) \in D$, then $\langle x, y, z \rangle \subseteq D$. In order to guarantee that the structures we consider are NAs, i.e., Condition *(a)* is satisfied, we will represent D by a set of cycles.

Condition *(c)* of the previous theorem can be visualized by the following diagram, where the u (dotted arrow) is required to exist:

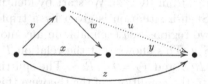

Notice that for integral relation algebras the set I of the underlying atom structure becomes a singleton.

Henkin et al. [25] introduced the method of splitting in cylindric algebra theory. He showed how to obtain non-representable cylindric algebras from representable ones. Later on H. Andréka et al. [3] formulated the way of splitting atoms in relation algebras. However, those methods cannot be applied if the algebra in question contains a bijection, and, hence, cannot be applied to any algebra starting with RCC11. In [34] a new method for splitting atoms was introduced that allows splitting of atoms in presence of bijections.

Definition 5. *Let* $\mathfrak{A}$ *and* $\mathfrak{B}$ *be atomic integral RAs. We say that* $\mathfrak{A}$ *is an extension of* $\mathfrak{B}$ *if the following conditions are satisfied:*

1. $\mathfrak{A} \supseteq \mathfrak{B}$, *i.e.,* $\mathfrak{B}$ *is a subalgebra of* $\mathfrak{A}$,
2. *every atom* $x \in \mathfrak{A}$ *is contained* $x \leq c(x)$ *in an atom* $c(x) \in \mathfrak{B}$, *called the cover of* x.

If η and θ are functions mapping elements of $At\mathfrak{B}$ to cardinals, we say that $\mathfrak{A}$ is an extension of $\mathfrak{B}$ along η and θ if $\mathfrak{A}$ is an extension $\mathfrak{B}$ and for all $x \in At\mathfrak{B}$ we have

$$\eta(x) = |\{y \in At\mathfrak{A} : y \leq x, y \neq \breve{y}\}|,$$
$$\theta(x) = |\{y \in At\mathfrak{A} : y \leq x, y = \breve{y}\}|$$

The function η and θ of the previous definition indicate into how many symmetric atoms $\theta(x)$ and how many non-symmetric atoms $\eta(x)$ the given atom x of $\mathfrak{B}$ will split into in $\mathfrak{A}$. The following theorem characterizes situations in which splitting of atoms is possible. A proof can be found in [34].

Theorem 2. *Let $\mathfrak{B}$ be a complete atomic integral RA and let η, θ be the functions mapping elements of $At\mathfrak{B}$ to cardinals, and let $\alpha(x) = \theta(x) + \eta(x)$. Then there is a complete atomic integral RA $\mathfrak{A}$ that is an extension of $\mathfrak{B}$ along η and θ if the following conditions hold for all $x, y \in At\mathfrak{B}$:*

(a) $\alpha(x) \geq 1$.
(b) $\eta(x) = \eta(\breve{x})$.
(c) $x \in Bij\mathfrak{B}$ implies $\alpha(x) = 1$.
(d) $x = \breve{x}$ implies $\eta(x)$ is even.
(e) $x \neq \breve{x}$ implies $\theta(x) = 0$.
(f) $y \in Bij\mathfrak{B}$ implies $\alpha(x; y) = \alpha(x)$.
(g) $y \in Bij\mathfrak{B}$, $x = \breve{x}$ and $\eta(x) > 0$ implies $x; y = (x; y)^{\smile}$ and $\theta(x) = \theta(x; y)$.
(h) $\alpha(x) > 1$, $y; x \neq 0$ and $y \notin Bij\mathfrak{B}$ implies $y \leq y; (x; x\breve{x} \cap 0')$.

Before we continue we want to explain how we will obtain new RCC tables, i.e., atom structures, from RCC25. We start by identifying a so-called non-extensional situation. Such a situation is given by a triple $(R, S, T) \in$ RCC25 for which we provide two regions x, y (definable in any model) so that xTy and $x\overline{R; S}y$, i.e., not $x(R; S)y$. This shows that the relation T can be split into the two relations $T_1 = T \cap R; S$ and $T_2 = T \cap \overline{R; S}$. The splitting of T is done using Theorem 2. According to the conditions of the theorem this implies that further atoms have to be split as well (Conditions *(f)* and *(g)*). These atoms are related to T by composing T with any bijection from the left and/or the right. After the splitting process we remove the cycle $\langle R, S, T_2 \rangle$ and all cycles necessary to maintain the bijections. Now, Condition *(c)* of Theorem 1 will be checked. If it is satisfied, we have obtained the new table. If not, the specific situation has to be resolved by removing one of the two triples in question. In order to decide which of the triples be removed we have to keep the triple for which an example in any model exists. After removing the cycle of one of the triples we start over. It is possible that this process ends in a situation where both triples must be kept. This is a clear indication that splitting T alone is not possible. We will encounter this situation in Section 6.

2.3 RCC Relations

In this paper we will use several relations derived from the basic contact relation C of a extensional and connected BCA. In the table below we list the most convenient among the equivalent conditions defining each of the relations. For further explanation and motivation of those relations we refer to [16–19].

$xDCy$ iff not xCy

$xECy$ iff xCy and $x \cdot y = 0$

$xECNy$ iff $xECy$ and $x + y \neq 1$

$xECDy$ iff $x = \overline{y}$

xOy iff $x \cdot y \neq 0$

$xPOy$ iff xOy and $x \not\leq y$ and $y \not\leq x$

$xPONy$ iff $xPOy$ and $x + y \neq 1$

$xPODy$ iff $xPOy$ and $x + y = 1$

$xTPPy$ iff $x < y$ and there is a z with $xECz$ and $yECz$

$xTPPAy$ iff $xTPPy$ and $x(ECN;TPP)y$

$xTPPBy$ iff $xTPPy$ and not $x(ECN;TPP)y$

$xNTPPy$ iff $x < y$ and not $xTPPy$

$xECNAy$ iff $xECNy$ and $x(TPP;TPP^\smile)y$

$xECNBy$ iff $xECNy$ and not $x(TPP;TPP^\smile)y$

In the following table we listed some sub-relations of PON and their defining properties. A "+" sign in the table means that two regions that are in the relationship indicated by the row also satisfy the property indicated by the column. A "-" sign obviously indicates the opposite.

	$ECN;TPP$	$TPP^\smile;ECN$	$TPP;TPP^\smile$	$TPP^\smile;TPP$
$PONXA1$	+	+	+	+
$PONXA2$	+	+	+	-
$PONXB1$	+	+	-	+
$PONXB2$	+	+	-	-
$PONYA1$	-	+	+	+
$PONYA2$	-	+	+	-
$PONYA1^\smile$	+	-	+	+
$PONYA2^\smile$	+	-	+	-
$PONYB$	-	+	-	+
$PONYB^\smile$	+	-	-	+
$PONZ$	-	-	+	+

The next lemma summarizes some important properties of these relations that we will need in the remainder of the paper. A proof can be found in [16].

Lemma 1. *In any extensional and connected BCA we have the following:*

1. $1' \in NTPP^\smile;NTPP$, *i.e. for all z there is some x with $xNTPPz$.*
2. $ECN = TPP;ECD$, *i.e., $xECNz$ iff $xTPP\overline{z}$.*
3. $xNTPPz$ and $yNTPPz$ iff $(x+y)NTPPz$.
4. $xNTPPy$ and $xNTPPz$ iff $xNTPPy \cdot z$.
5. $ECD;DC = NTPP^\smile$, *i.e $\overline{x}DCz$ iff $zNTPPx$.*
6. $x(ECN;TPP)z$ iff $xECN(\overline{x} \cdot z)TPPz$.
7. $xNTPPy$ and $y \leq z$ implies $xNTPPz$.

In [16] the RCC15 and RCC25 composition table were presented. These tables have also been computed by splitting certain atoms (in RCC11) and removing certain cycles. In fact the cycle $\langle TPPA, TPPA, TPPB \rangle$ was removed in addition to the cycles required by the definitions of the new relations. A proof that this cycle should be removed, i.e., a proof that this situation is impossible in all models, was not given. We provide the proof in the following lemma.

Lemma 2. *In every extensional and connected BCA we have*

$$TPPA; TPPA \cap TPP \subseteq ECN; TPP.$$

Proof. To prove this lemma, assume $x(TPPA; TPPA \cap TPP)z$. Then there is a y so that $xTPPAyTPPAz$, i.e., we have (1) $xTPPy$ (2) $yTPPz$ (3) $xTPPz$ (4) $xECN(\overline{x} \cdot y)TPPy$ (5) $yECN(\overline{y} \cdot z)TPPz$ by Lemma 1(6). We want to show (a) $xECN(\overline{x} \cdot z)$ and (b) $(\overline{x} \cdot z)TPPz$. To prove (a) at first we show $xTPP(x + \overline{z})$ which is equivalent to (a) by Lemma 1(2). We have $x \leq x + \overline{z}$. If $x = x + \overline{z}$, then $\overline{z} \leq x$, and hence $\overline{x} \leq z$. But this implies $z = 1$, and, hence, $xTPP1$ by (3), which is a contradiction to $xNTPP1$ for all x. Therefore we have $x < (x + \overline{z})$. Now assume $xNTPP(x + \overline{z})$. From the computation

$$
\begin{aligned}
(x + \overline{y}) \cdot (y + \overline{z}) &= x \cdot (y + \overline{z}) + \overline{y} \cdot (y + \overline{z}) \\
&= x \cdot y + x \cdot \overline{z} + \overline{y} \cdot y + \overline{y} \cdot \overline{z} \\
&= x + 0 + 0 + \overline{z} \qquad \text{by (1),(2) and (3)} \\
&= x + \overline{z}
\end{aligned}
$$

we obtain $xNTPP(x + \overline{y}) \cdot (y + \overline{z})$, which implies by Lemma 1(4) $xNTPP(x + \overline{y})$ and $xNTPP(y + \overline{z})$. The first property is equivalent to $xECD(\overline{x} \cdot y)$ by Lemma 1(5), a contradiction to (3). Therefore we have $xTPP(x + \overline{z})$. Now, in order to prove (b) we already have $\overline{x} \cdot z \leq z$. If $\overline{x} \cdot z = z$, then $z \leq \overline{x}$, and, hence, $x \leq z \cdot \overline{z} = 0$ by (3). But this is a contradiction to (3) since $0NTTPz$ for all z. We conclude $(\overline{x} \cdot z) < z$. Now, assume $(\overline{x} \cdot z)NTPPz$. From the computation

$$
\begin{aligned}
\overline{x} \cdot y + \overline{y} \cdot z &= \overline{x} \cdot y + \overline{x} \cdot (x + \overline{y}) \cdot z \\
&= \overline{x} \cdot (y + (x + \overline{y}) \cdot z) \\
&= \overline{x} \cdot (y \cdot z + (x + \overline{y}) \cdot z) \\
&= \overline{x} \cdot (y + x + \overline{y}) \cdot z \\
&= \overline{x} \cdot z
\end{aligned}
$$

we get $(\overline{x} \cdot y + \overline{y} \cdot z)NTPPz$, and, hence, $\overline{x} \cdot yNTPPz$ and $\overline{y} \cdot zNTPPz$ by Lemma 1(3). The second property is a contradiction to (5). So we conclude that $(\overline{x} \cdot z)TPPz$. All these facts prove the lemma.   □

3   From RCC25 to RCC27

Our investigation starts with the relation algebra RCC25. Due to lack of space we cannot provide its composition table here. We refer to [16] for the details.

Consider the diagram given in Figure 1 where we define $x = a+c$ and $z = \overline{a}\cdot s$. It shows a general $PONXB2$ situation between x and z, i.e., a situation that appears in any non-empty model. In order to verify that such a situation indeed occurs in any non-empty model we start with any region s within the model. Since $NTTP$ is surjective [16] we obtain a with $aNTTPs$. Then we choose c with $cNTPP(\overline{a}\cdot s)$.

It is easy to see that $xNTPP(x+z)$, but $z\overline{NTPP}(x+z)$. So this fact implies that $PONXB2$ can be split into two parts that we will call $PONXB2H$ and $PONXB2H^\smile$:

$$xPONXB2Hz = xPONXB2z \cap xNTPP(x+z),$$
$$xPONXB2H^\smile z = xPONXB2z \cap x\overline{NTPP}(x+z).$$

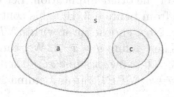

Fig. 1. $(a+c)PONXB2(\overline{a}\cdot s)$

Later we will provide a justification for the name $PONXB2H^\smile$ by showing that $PONXB2H^\smile$ is indeed the converse of $PONXB2H$. The definition of $PONXB2H$ is based on the condition $xNTPP(x+z)$. This property cannot be used to remove triples because it is not based on the composition of atomic relations. Instead it uses the algebraic operation $+$. In the following we want to show that $PONXB2H$ can be written in suitable way. Notice that a similar computation was already shown in [29].

Lemma 3. *We have* $x(ECN;\overline{O})z \Leftrightarrow xTPP(x+z)$.

Proof. First we want to prove the implication $\Rightarrow$. For this purpose we have to find a region y with $xECNy\overline{O}z$, i.e., y has to satisfy (1) $x\cdot y = 0$ (2) xCy (3) $x+y \neq 1$ (4) $z\cdot y = 0$. Now choose $y = \overline{x}\cdot\overline{z}$. The properties (1) and (4) follow immediately from the definition of y. From the assumption we conclude that $x\overline{NTPP}(x+z)$, which is equivalent to (2) by Lemma 1(5). In order to show (3) assume that $x+y = 1$. Then we have

$$1 = x + (\overline{x}\cdot\overline{z})$$
$$= (x+\overline{x})\cdot(x+\overline{z})$$
$$= (x+\overline{z}).$$

This implies $z \leq x$, and, hence, $x+z = x$. But this is a contradiction to the assumption. For the other implication assume $xNTPP(x+z)$. From the assumption we obtain a y with (1)-(4) as listed above. First, we want to show that $xTPP\overline{y}$. From (1) we get $x \leq \overline{y}$. If $x = \overline{y}$, then $xECDy$, a contradiction to the assumption $xECNy$. Since xCy we conclude $xTPP\overline{y}$. On the other hand (1) and (4) show that $x + z \leq \overline{y}$. By our assumption $xNTPP(x+z)$ and Lemma 1(7) we get $xNTPP\overline{y}$, a contradiction.   □

Using the previous lemma we are now able to provide a condition equivalent to $xNTPP(x+y)$ based on the relational operations only.

Lemma 4. *We have* $xNTPP(x+z) \Leftrightarrow x(ECN \setminus O)z$.

Proof. First consider direction $\Rightarrow$. From $xNTPP(x+z)$ we get $x\overline{TPP}(x+z)$ since $x \leq x+z$. This is equivalent to $x\overline{ECN;\overline{O}}(x+z)$ by Lemma 3, and, hence, we have $x(ECN \setminus O)z$. Now consider the other implication. Let us assume $xTPP(x+z)$. Then we get $x(ECN;\overline{O})z$ from Lemma 3. But this contradicts with $(ECN \setminus O)$. If, $x = x + z$ then $z \leq x$. Now choose an a with $aNTPP\overline{x}$, which is possible because of Lemma 1(1). Then define $y = \overline{x} \cdot \overline{a}$. We want to show that $xECNy$. First, we have $x \cdot y \leq x \cdot \overline{x} = 0$. Now, assume $x\overline{C}y$ then $yNTPP\overline{x}$ from Lemma 1(5) and hence we have $(y+a)NTPP\overline{x}$ from Lemma 1(3). Then we have

$$
\begin{aligned}
y + a &= (\overline{x} \cdot \overline{a}) + a \\
&= (\overline{x} + a) \cdot (\overline{a} + a) \\
&= (\overline{x} + a) \\
&= \overline{x} \qquad\qquad\qquad \text{since } a \leq \overline{x}.
\end{aligned}
$$

This implies $\overline{x}NTPP\overline{x}$. But that is contradiction to the assumption. This fact implies xCy. If $x + y = 1$ then $1 = x + (\overline{x} \cdot \overline{a})$, which is equivalent to $x + \overline{a}$ and that implies $a \leq x$. So we have $a \leq x + \overline{x} = 1$ and hence $aNTPP\overline{x}$ i.e $xECNy$. Now $y.z$ is equivalent to $\overline{x} \cdot \overline{a} \cdot z$ and that is 0. Which implies $z \leq x$. So, we can conclude that $y\overline{O}z$.   □

The previous two lemmas show that $PONXB2H = PONXB2 \cap (ECN \setminus O)$. This formula can be used for our method of splitting. But before we proceed with this procedure, we want to show that $PONXB2H^{\smile}$ is indeed the converse of $PONXB2H$.

Lemma 5. *We have* $(ECN \setminus O) \cap (ECN \setminus O)^{\smile} = \emptyset$.

Proof. From Lemma 4 $x(ECN \setminus O)z$ is equivalent to $xNTPP(x+z)$ and $z(ECN \setminus O)x$ is equivalent to $zNTPP(x+z)$. The latter two imply $(x+z)NTPP(x+z)$ by Lemma 1(3), which is a contradiction.   □

The next lemma shows that $(ECN \setminus O)^{\smile}$ is the complement of $ECN \setminus O$ within a $\overline{TPP;TPP^{\smile}}$ context.

Lemma 6. *We have* $\overline{(ECN \setminus O)} \cap \overline{(ECN \setminus O)^{\smile}} \subseteq TPP;TPP^{\smile}$.

Proof. Suppose we have $x\overline{(ECN \setminus O)}z$ and $x\overline{(ECN \setminus O)}^{\smile}z$. Then Lemma 4 implies $x\overline{NTPP}(x + z)$ and $z\overline{NTPP}(x + z)$. But $x\overline{NTPP}(x + z)$ is equivalent to $xTPP(x + z)$ and $z\overline{NTPP}(x + z)$ equivalent to $zTPP(x + z)$ since $x < (x + z)$ and $z < (x + z)$. □

Since $PONXB2$ is a subset of $\overline{TPP;TPP^{\smile}}$ we obtain the desired result.

Lemma 7. *We have* $PONXB2H^{\smile} = PONXB2 \cap (ECN \setminus O)^{\smile}$.

Proof. We compute

$$PONXB2H^{\smile} = PONXB2 \cap \overline{ECN \setminus O}$$
$$= PONXB2 \cap \overline{ECN \setminus O} \cap \overline{TPP;TPP^{\smile}}$$
$$= PONXB2 \cap (ECN \setminus O)^{\smile},$$

where the last line follows from the previous two lemmas. □

The final step before we start the splitting process is to express the definition of $PONXB2H$ in terms that can be used to find appropriate cycles of RCC25 which should be removed. Notice that we will use the names of new relations in the Lemmas 8-10 as an abbreviation for their definition. Those relations are not in RCC25.

Lemma 8. *In RCC25 we have* $PONXB2H = PONXB2 \cap \overline{ECN;DC}$.

Proof. Consider the following computation

$$PONXB2H = PONXB2 \cap (ECN \setminus O)$$
$$= PONXB2 \cap \overline{ECN;\overline{O}}$$
$$= PONXB2 \cap \overline{ECN;(DC \cup ECD \cup ECN)}$$
$$= PONXB2 \cap \overline{ECN;DC \cup ECN;ECD \cup ECN;ECN}$$
$$= PONXB2 \cap \overline{ECN;DC} \cap \overline{ECN;ECD} \cap \overline{ECN;ECN}$$
$$= PONXB2 \cap \overline{ECN;DC} \cap \overline{TPP} \cap \overline{ECN;ECN}$$
$$= PONXB2 \cap \overline{ECN;DC},$$

where the last line follows from $PONXB2 \neq TPP$ and the fact $PONXB2$ is not in $ECN;ECN$. □

RCC25 contains one bijection; the relation ECD. From the composition table of RCC25 we obtain $PONXB2;ECD = PONZ$ and $ECD;PONXB2 = PONZ$. This shows that we need to split $PONZ$ as well. We define

$$PONZH = PONXB2H;ECD, \quad PONZH^{\smile} = PONXB2H^{\smile};ECD.$$

As above, the next lemma indicates precisely which cycles have to be removed. The proof is similar to the proof of Lemma 8, and, therefore, omitted.

Lemma 9. *In RCC25 we have* $PONZH = PONZ \cap \overline{ECN;NTPP}$.

After splitting $PONXB2$ and $PONZ$ in RCC25, we obtain an algebra with 27 atoms, which we will call RCC27. During the splitting process we will remove initially the cycles listed below plus the cycles obtained from them by composing the relation in a cycle from the left and/or right with the bijection ECD appropriately. The first two cycles are removed in order to make sure that ECD is a bijection in the new algebra as well. The other cycles are removed considering the definition of $PONXB2H$, $PONXB2H^\smile$, $PONZH$ and $PONZH^\smile$.

1. $\langle PONXB2H, ECD, PONZH^\smile \rangle$
2. $\langle PONXB2H^\smile, ECD, PONZH \rangle$
3. $\langle ECNA, DC, PONXB2H \rangle$
4. $\langle ECNB, DC, PONXB2H \rangle$
5. $\langle DC, ECNA, PONXB2H^\smile \rangle$
6. $\langle DC, ECNB, PONXB2H^\smile \rangle$
7. $\langle ECNA, NTPP, PONZH \rangle$
8. $\langle ECNB, NTPP, PONZH \rangle$
9. $\langle NTPP^\smile, ECNA, PONZH^\smile \rangle$
10. $\langle NTPP^\smile, ECNB, PONZH^\smile \rangle$

As described in Section 2.2 we continue by checking the associativity of RCC27 based on Theorem 1. After removing additional cycles we have obtained the set of cycles that define the composition table of RCC27. This list can be found in [24].

4   From RCC25 to RCC29

Similar to the previous section consider the diagram given in Figure 2 where we define $x = a + c$ and $z = \bar{a} \cdot (s + b)$. Again, the fact that $NTPP$ is surjective guarantees that such a situation is available in any non-empty model.

It is easy to see that $xNTPP(x + z)$, but $zNTPP(x + z)$ does not hold. So this fact implies that $PONXB1$ can also be split into two parts that we will call $PONXB1H$ and $PONXB1H^\smile$:

$$xPONXB1Hz = xPONXB1z \cap xNTPP(x + z),$$
$$xPONXB1H^\smile z = xPONXB1z \cap x\overline{NTPP}(x + z).$$

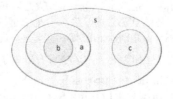

Fig. 2. $(a + c)PONXB1(\bar{a} \cdot (s + b))$

The same reasoning as in the previous section leads to the following equations for the two relations

$$PONXB1H = PONXB1 \cap (ECN \setminus O),$$
$$PONXB1H^{\smile} = PONXB1 \cap (ECN \setminus O)^{\smile}.$$

From the composition table we obtain that $PONXB1; ECD = PONYA1$, $ECD; PONXB1 = PONYA1^{\smile}$ and $ECD; PONXB1; ECD = PONXA2$ in RCC25. This shows that we need to split the relation $PONYA1$, $PONYA1^{\smile}$ and $PONXA2$ as well. We define

$$PONYA1H = PONXB1H; ECD,$$
$$PONYA1H^{\smile} = ECD; PONXB1H^{\smile},$$
$$PONYA1tH = PONXB1H^{\smile}; ECD,$$
$$PONYA1tH^{\smile} = ECD; PONXB1H,$$
$$PONXA2H = ECD; PONXB1H^{\smile}; ECD,$$
$$PONXA2H^{\smile} = ECD; PONXB1H; ECD.$$

The next lemma will tell us which cycles have to be removed concretely. The proof is very similar to Lemma 8, and, therefore, omitted.

Lemma 10. *In RCC25 we have:*

1. $PONXB1H = PONXB1 \cap \overline{ECN; DC}$,
2. $PONYA1H = PONYA1 \cap \overline{ECN; NTPP}$,
3. $PONYA1tH = PONYA1 \cap \overline{DC; TPP}$,
4. $PONXA2H = PONXA2 \cap \overline{NTPP^{\smile}; TPP}$.

As in the previous section the cycles of the RCC29 composition table are obtained by splitting the atoms above, and then removing the appropriate cycles. The result of this process can be found in [24].

5   Beyond RCC29

In this section we sketch two further refinements of the RCC25 composition table.

5.1   Generating RCC31

The algebra RCC31 is obtained by combining the splittings that were used to generate RCC27 and RCC29. Overall we are splitting the atoms $PONXA2$, $PONXB1$, $PONXB2$, $PONYA1$, $PONYA1^{\smile}$ and $PONZ$. This was done splitting the two remaining atoms in the algebra RCC29. An alternative approach could have been to split the remaining four atoms in RCC27, of course. It is worth mentioning that during the associativity test for RCC31, only one additional cycle had to be removed. Again, the resulting set of cycles can be found in [24].

5.2 Splitting ECNB

Mormann [29] introduced the concept of a hole relation. He defined the restricted hole relation by $H = ECN \cap (EC \setminus O)$. He also showed that H splits $ECNB$. Following the same procedure as in Section 3 and 4 we recognize that we have to split $TPPB$, $TPPB^{\smile}$ and $PODYB$ as well. We get the following definitions (the remaining relations are just converses of the ones listed below):

$$H = ECNB \cap \overline{ECN; DC},$$
$$TPPB1 = TPPB \cap \overline{DC; TPP},$$
$$TPPB2 = TPPB \cap \overline{ECN; NTPP},$$
$$PODYBH = PODYB \cap \overline{TPP^{\smile}; NTPP}.$$

After splitting the atoms we removed the appropriate cycles similar to the previous sections. This time we were not able to obtain a relation algebra in the iterative removal procedure based on Theorem 1 (Condition (c)). We ended in a situation were both cycles could not be removed. For an intensive study of the situation and further details we refer to [24]. This situation is similar to the RCC10 composition table [17]. The RCC10 composition table is not associative, and, hence, not a relation algebra. Splitting an additional atom in RCC10 led to RCC11 which is associative. As already mentioned in Section 2.2 it should be possible to obtain a relation algebra based on splitting $ECNB$ by splitting at least one atom in addition to the four identified here. So far we were not able to find that atom and the condition defining its sub-relations. Therefore, we leave this problem for future investigation.

6 Conclusion and Future Work

In this paper we have presented three refinements of the RCC25 relation algebra. These refinements were obtained by splitting certain atoms of the RCC25 algebra. In addition, we have identified another atom that splits but we weren't able to produce the corresponding algebra. A successful attempt to split $ECNB$ requires to identify another atom that splits so that the resulting composition table is associative. This work will be part of future research.

In addition to $ECNB$ we recognized that there are two $TPPA$ situations that indicate that $TPPA$ splits as well. The corresponding definitions of the sub-relations of $TPPA$ are:

$$TPPA1 = TPPA \cap TPPA; TPPB2, \quad TPPA2 = TPPA \cap \overline{TPPA; TPPB2}.$$

Notice that the definitions above use $TPPB2$, a relation that is generated while splitting $ECNB$. In order to generate a relation algebra that contains $TPPA1$ and $TPPA2$ as atoms we need to split $ECNB$ successfully first.

References

1. Allen, J.F.: Maintaining knowledge about temporal intervals. Communications of the ACM 26(11), 832–843 (1983)
2. Allen, J.F., Hayes, P.J.: A common sense theory of time. In: Proceedings 9th IJCAI, Los Angeles, pp. 528–531 (1985)
3. Andréka, H., Maddux, R.D., Nemeti, I.: Splitting in relation algebras. Proceedings of the AMS 111(4), 1085–1093 (1991)
4. Asher, N., Vieu, L.: Toward a geometry of common sense: A semantics and a complete axiomatization of mereotopology. In: Mellish, C. (ed.) Proceedings of the 14th International Joint Conference on Artificial Intelligence (IJCAI 1995), pp. 846–852 (1995)
5. Birkhoff, G.: Lattice Theory, 3rd edn., vol. XXV. American Mathematical Society Colloquium Publications (1968)
6. Cohn, A.G.: Qualitative spatial representation and reasoning techniques. In: Brewka, G., Habel, C., Nebel, B. (eds.) KI 1997. LNCS, vol. 1303, pp. 1–30. Springer, Heidelberg (1997)
7. Cohn, A.G., Bennett, B., Gooday, J., Gotts, N.M.: Representing and reasoning with qualitative spatial relations about regions. In: Stock, O. (ed.) Spatial and Temporal Reasoning, pp. 97–134. Kluwer, IRST (1997)
8. Clarke, B.L.: A calculus of individuals based on 'connection'. Notre Dame J. of Formal Logic 22, 204–218 (1981)
9. de Morgan, A.: On the syllogism: IV, and on the logic of relations. Transactions of the Cambridge Philosophical Society 10, 331–358 (1860)
10. De Laguna, T.: Point, line and surface as sets of solids. Journal of Philosophy 19, 449–461 (1922)
11. Dimov, G., Vakarelov, D.: Contact algebras and region-based theory of space: A proximity approach - II. Fundamenta Informaticae 74, 251–282 (2006)
12. Düntsch, I.: Relation algebras and their application in temporal and spatial reasoning. Artificial Intelligence Review 23, 315–357 (2005)
13. Düntsch, I., Winter, M.: A representation theorem for boolean contact algebras. Theoretical Computer Science 347, 498–512 (2003)
14. Düntsch, I., Winter, M.: Weak contact structures. In: MacCaull, W., Winter, M., Düntsch, I. (eds.) RelMiCS 2005. LNCS, vol. 3929, pp. 73–82. Springer, Heidelberg (2006)
15. Düntsch, I., Winter, M.: The lattice of contact relations on a boolean algebra. In: Berghammer, R., Möller, B., Struth, G. (eds.) RelMiCS/AKA 2008. LNCS, vol. 4988, pp. 99–109. Springer, Heidelberg (2008)
16. Düntsch, I., Schmidt, G., Winter, M.: A necessary relation algebra for mereotopology. Studia Logica 69, 381–409 (2001)
17. Düntsch, I., Wang, H., McCloskey, S.: Relations algebras in qualitative spatial reasoning. Fundamenta Informaticae 39, 229–248 (1999)
18. Düntsch, I., Wang, H., McCloskey, S.: A relation-algebraic approach to the region connection calculus. Theoretical Computer Science 255, 63–83 (2001)
19. Egenhofer, M.: Reasoning about binary topological relations. In: Günther, O., Schek, H.-J. (eds.) SSD 1991. LNCS, vol. 525, pp. 141–160. Springer, Heidelberg (1991)
20. Egenhofer, M.: Deriving the composition of binary topological relations. Journal of Visual Languages and Computing 5, 133–149 (1994)

21. Egenhofer, M., Franzosa, R.: Point-set topological spatial relations. International Journal of Geographic Information Systems 5, 161–174 (1991)
22. Egenhofer, M., Sharma, J.: Topological consistency. In: Fifth International Symposium on Spatial Data Handling, Charleston, pp. 335–343 (1992)
23. Egenhofer, M., Sharma, J.: Assessing the consistency of complete and incomplete topological information. Geographical Systems 1, 47–68 (1993)
24. Ghosh, M.: Region Connection Calculus: Composition Tables and Constraint Satisfaction Problems. MSc Thesis, Brock University (2013), https://dr.library.brocku.ca/handle/10464/5109
25. Henkin, L., Monk, J.D., Tarski, A.: Cylindric algebras, Part II. Studies in Logic and the Foundations of Mathematics, vol. 115. North-Holland (1985)
26. Kahl, W., Schmidt, G.: Exploring (Finite) Relation Algebras Using Tools Written in Haskell. Technical Report 2000-02, University of the Federal Armed Forces Munich (2000)
27. Maddux, R.D.: Some varieties containing relation algebras. Transactions of the AMS 272, 501–526 (1982)
28. Maddux, R.D.: Relation algebras. Studies in Logic and the Foundations of Mathematics, vol. 150. Elsevier (2006)
29. Mormann, T.: Holes in the region connection calculus, Oisterwijk (2001) (Preprint presented at RelMiCS 6)
30. Pratt, I., Schoop, D.: A complete axiom system for polygonal mereotopology of the real plane. Journal of Philosophical Logic 27(6), 621–658 (1998)
31. Pratt, I., Schoop, D.: Expressivity in polygonal, plane mereotopology. Journal of Symbolic Logic 65(2), 822–838 (2000)
32. Randell, D.A., Cohn, A.G.: Modelling topological and metrical properties in physical processes. Principles of Knowledge Representation and Reasoning, 357–368 (1989)
33. Randell, D.A., Cui, Z., Cohn, A.G.: A spatial logic based on regions and connection. Principles of Knowledge Representation and Reasoning, 165–176 (1992)
34. Siddavaatam, P., Winter, M.: Splitting atoms in relation algebras. In: de Swart, H. (ed.) RAMICS 2011. LNCS, vol. 6663, pp. 331–346. Springer, Heidelberg (2011)
35. Stell, J.G.: Boolean connection algebras: A new approach to the Region-Connection Calculus. Artificial Intelligence 122, 111–136 (2000)
36. Vakarelov, D., Dimov, G., Düntsch, I., Bennett, B.: A proximity approach to some region-based theories of space. Journal of Applied Non-classical Logics 12, 527–559 (2002)
37. Varzi, A.C.: Parts, wholes, and part-whole relations: The Prospect of Mereotopology. Data & Knowledge Engineering 20, 259–286 (1996)
38. Whitehead, A.N.: Process and Reality. MacMillan (1929)

Fuzzifying Modal Algebra

Jules Desharnais[1] and Bernhard Möller[2]

[1] Département d'informatique et de génie logiciel
Université Laval, Québec, QC, Canada
jules.desharnais@ift.ulaval.ca
[2] Institut für Informatik, Universität Augsburg, 86135 Augsburg, Germany
bernhard.moeller@informatik.uni-augsburg.de

Abstract. Fuzzy relations are mappings from pairs of elements into the interval $[0, 1]$. As a replacement for the complement operation one can use the mapping that sends x to $1 - x$. Together with the concepts of t-norm and t-conorm a weak form of Boolean algebra can be defined. However, to our knowledge so far no notion of domain or codomain has been investigated for fuzzy relations. These might, however, be useful, since fuzzy relations can, e.g., be used to model flow problems and many other things. We give a new axiomatisation of two variants of domain and codomain in the more general setting of idempotent left semirings that avoids complementation and hence is applicable to fuzzy relations. Some applications are sketched as well.

Keywords: fuzzy relations, semirings, domain operator, modal operator.

1 Introduction

The basic idea of the present paper is to bring together the concepts of fuzzy semirings, say in the form of fuzzy relations or matrices, and modal semirings that offer domain and codomain operators and, based on these, algebraic definitions of box and diamond. The latter have been thoroughly studied for more than ten years now (see [DMS06] for an early survey) and applied to many different areas, such as program semantics, knowledge and belief logics [Möl13] or preference queries in databases [MRE12], and many more.

Domain and codomain in a certain sense "measure" enabledness in transition systems. This observation motivated an investigation whether modal semirings might also be interesting for handling fuzzy systems. So the idea is not to invent a new kind of algebraic "meta-system" for all kinds of fuzzy logics, but rather to apply and hence re-use an existing and well established algebraic system to the particular case of fuzzy systems.

Let us briefly recapitulate the theory of fuzzy relations. These are mappings from pairs of elements into the interval $[0, 1]$. The values can be interpreted as transition probabilities or as capacities and in various other ways. Hence the idea was to take up the above idea of measuring and to enrich fuzzy relations semirings with domain/codomain operators and applying the corresponding modal

P. Höfner et al. (Eds.): RAMiCS 2014, LNCS 8428, pp. 395–411, 2014.

operators in the description and derivation of systems or algorithms in that realm. To our knowledge this has not been done so far.

The classical relational operators are adapted as follows:

$$(R \sqcup S)(x, y) = \max(R(x, y), S(x, y)) \ ,$$
$$(R \sqcap S)(x, y) = \min(R(x, y), S(x, y)) \ ,$$
$$(R \,;\, S)(x, y) = \sup_z \min(R(x, z), S(z, y)) \ .$$

Under these operations, fuzzy relations form an idempotent semiring (see below for the precise definition). One can also define a weak notion of complementation by setting $\overline{R}(x, y) = 1 - R(x, y)$. This already shows the main problem one encounters in transferring the concept of domain to fuzzy semirings: the original axiomatisation of domain used a Boolean subring of the overall semiring as the target set of the domain operator, and this generally is not present in the fuzzy case.

However, using the above weak negation and the concepts of t-norm and t-conorm (see again below for the details) a substitute for Boolean algebra can be defined.

We give a new axiomatisation of two variants of domain and codomain in the more general setting of idempotent left semirings that avoids complementation and hence is applicable to fuzzy relations. Such an axiomatisation has been given in [DS11] for idempotent semirings. We study the more general case of idempotent left semirings in which left distributivity of multiplication over addition and right annihilation of zero are not required. At the same time we weaken the domain axioms by requiring only isotony rather than distributivity over addition. Surprisingly still a wealth of properties known from the semiring case persist in the more general setting. However, it is no longer true that complemented subidentities are domain elements. This is not really disturbing, though, because the fuzzy world has its own view of complementation anyway.

In the main part of the paper we develop the theory, involving the new concept of *restrictors*. It tuns out that the axiomatisation we come up with can be parameterised in certain ways to characterise a whole family of domain operators in a uniform way. We then investigate how the domain operators extend to matrices, since this is the application we are after. It turns out that the axioms apart from the the so-called locality axiom extend well from the base left semiring S to the matrix semiring over it, while locality extends only if S is actually a semiring. Finally, some applications are sketched.

2   Preliminaries

We will frequently use the reasoning principle of *indirect equality* for partial orders $(M, \leq)$. For $a, b \in M$ we have $a = b \Leftrightarrow (\forall c \in M : b \leq c \Leftrightarrow a \leq c)$. The implication $(\Rightarrow)$ is trivial. For $(\Leftarrow)$, choosing $c = a$ and $c = b$ yields $b \leq a$ and $a \leq b$, respectively, so that antisymmetry of $\leq$ shows the claim.

Definition 2.1. For elements $a, b \in M$, the *interval* $[a, b]$ is

$$[a, b] =_{df} \{c \mid a \le c \wedge c \le b\} .$$

This entails $[a, b] = \emptyset$ if $a \not\le b$.

Now we define our central algebraic structure.

Definition 2.2. A *left (or lazy) semiring*, briefly an *L-semiring*, is a quintuple $(S, +, 0, \cdot, 1)$ with the following properties:

1. $(S, +, 0)$ is a commutative monoid.
2. $(S, \cdot, 1)$ is a monoid.
3. The $\cdot$ operation is right-distributive over $+$ and *left-strict*, i.e., $(a + b) \cdot c = a \cdot c + b \cdot c$ and $0 \cdot a = 0$. As customary, $\cdot$ binds tighter than $+$.

A right semiring is defined symmetrically. A *semiring* [Van34] is a structure which is both a left and right semiring. In particular, its multiplication is both left and right distributive over its addition and its 0 is a left and right annihilator.

Definition 2.3. An *idempotent* left semiring [Möl07], briefly *IL-semiring*, is an L-semiring $(S, +, 0, \cdot, 1)$ with the following additional requirements.

- Addition is idempotent. Hence it induces an upper semilattice with the *natural order* $\le$ given by $a \le b \Leftrightarrow_{df} a + b = b$, which means that b offers at least all the choices of a, but possibly more.
- Multiplication is right-isotone w.r.t. the natural order. This can be axiomatised as super-disjunctivity $a \cdot b + a \cdot c \le a \cdot (b + c)$.

An *I-semiring* is an idempotent semiring. Finally, an IL-semiring is *bounded* if it has a greatest element $\top$.

3   Predomain and Restrictors

As is well known, predicates on states can be modelled by tests, which are defined involving the Boolean operation of negation. As mentioned in the introduction, we want to avoid that and hence give the following new axiomatisation of a (pre)domain operation whose range will replace the set of tests.

Definition 3.1. A *prepredomain IL-semiring* is a structure $(S, \ulcorner \,)$, where S is an IL-semiring and the *prepredomain operator* $\ulcorner : S \to S$ satisfies, for all $a \in S$,

$$\ulcorner a \le 1 , \tag{sub-id}$$

$$\ulcorner 0 \le 0 , \tag{strict}$$

$$a \le \ulcorner a \cdot a . \tag{d1}$$

By $\ulcorner S$ we denote the image of S under the prepredomain operation. The operator $\ulcorner$ is called a *predomain operator* if additionally, for all $a, b \in S$ and $p \in \ulcorner S$,

$$\ulcorner a \le \ulcorner (a + b) , \tag{isot}$$

$$\ulcorner (p \cdot a) \le p . \tag{d2}$$

Finally, a predomain operator $^\ulcorner$ is called a *domain operator* if additionally it satisfies the *locality axiom*, i.e., for all $a, b \in S$,

$$^\ulcorner(a \cdot {}^\ulcorner b) \le {}^\ulcorner(a \cdot b) . \tag{d3}$$

In the latter cases, $(S, {}^\ulcorner)$ is called a *predomain IL-semiring* and a *domain IL-semiring*, resp. An element of $^\ulcorner S$ is called a *((pre)pre)domain element*. We will consistently write $a, b, c \ldots$ for arbitrary semiring elements and $p, q, r, \ldots$ for elements of $^\ulcorner S$.

Since by definition $^\ulcorner a \le 1$, by isotony of $\cdot$ the reverse inequation to (d1) holds as well, so that (d1) is equivalent to $a = {}^\ulcorner a \cdot a$. To simplify matters we will refer to that equation as (d1), too. Using Mace4 it can be shown that the above axioms are independent. They can be understood as follows. The equational form of Ax. (d1) means that restriction to all starting states is no actual restriction, whereas (d2) means that after restriction the remaining starting states satisfy the restricting test. Ax. (isot) states, by the definition of $\le$, that $^\ulcorner$ is isotone, i.e., monotonically increasing. Ax. (d3), which, as will be shown in Lemma 5.4.1, again strengthens to an equality, states that the domain of $a \cdot b$ is not determined by the inner structure or the final states of b; information about $^\ulcorner b$ in interaction with a suffices.

The auxiliary notion of prepredomain already admits a few useful results.

Lemma 3.2. *Assume a prepredomain IL-semiring $(S, {}^\ulcorner)$. Then for all $a \in S$ and $p \in {}^\ulcorner S$ we have the following properties.*

1. *If $a \le 1$ then $a \le {}^\ulcorner a$.*
2. $^\ulcorner 1 = 1$ *and hence $1 \in {}^\ulcorner S$.*
3. (d1) $\Leftrightarrow$ $({}^\ulcorner a \le p \Rightarrow a \le p \cdot a)$.

Proof.

1. By (d1), the assumption, isotony of $\cdot$ and neutrality of 1, $a \le {}^\ulcorner a \cdot a \le {}^\ulcorner a \cdot 1 = {}^\ulcorner a$.
2. We have $1 \le {}^\ulcorner 1$ by Part 1 and $^\ulcorner 1 \le 1$ by (sub-id).
3. ($\Rightarrow$) Assume (d1) and suppose $^\ulcorner a \le p$. Then by isotony of $\cdot$, $a = {}^\ulcorner a \cdot a \le p \cdot a$.
   ($\Leftarrow$) Set $p = {}^\ulcorner a$ in the right hand side. $\square$

To reason more conveniently about predomain we introduce the following auxiliary notion.

Definition 3.3. A *restrictor* in an IL-semiring is an element $x \in [0, 1]$ such that for all $a, b \in S$ we have

$$a \le x \cdot b \Rightarrow a \le x \cdot a .$$

Note that by $x \le 1$ this is equivalent to

$$a \le x \cdot b \Rightarrow a = x \cdot a . \tag{1}$$

The set of all restrictors of S is denoted by $\mathsf{rest}(S)$. In particular, $0, 1 \in \mathsf{rest}(S)$.

A central result is the following.

Lemma 3.4. *In a predomain IL-semiring* $\ulcorner S \subseteq \mathsf{rest}(S)$.

Proof. Assume $a \leq \ulcorner b \cdot c$. By (isot) and (d2) we infer $\ulcorner a \leq \ulcorner (\ulcorner b \cdot c) \leq \ulcorner b$. Now by (d1) and isotony of multiplication we obtain $a = \ulcorner a \cdot a \leq \ulcorner b \cdot a$. Hence $\ulcorner b \in \mathsf{rest}(S)$. $\square$

To allow comparison with previous approaches we recapitulate the following notion.

Definition 3.5. An element r of an IL-semiring is a *test* if it has a relative complement $s \in S$ with $r + s = 1$ and $r \cdot s = 0 = s \cdot r$. The set of all tests of S is denoted by $\mathsf{test}(S)$. In particular, $0, 1 \in \mathsf{test}(S)$.

In an I-semiring, $\mathsf{test}(S)$ is a Boolean algebra with 0 and 1 as the least and greatest elements and $+$ as join and $\cdot$ as meet; moreover, the relative complements are unique if they exist.

Lemma 3.6. $\mathsf{test}(S) \subseteq \mathsf{rest}(S)$.

Proof. Consider an $r \in \mathsf{test}(S)$ with relative complement s. Assume that $a \leq r \cdot b$. Then by isotony of multiplication, the definition of relative complement and left annihilation of 0,

$$s \cdot a \leq s \cdot r \cdot b = 0 \cdot b = 0 \; . \tag{*}$$

Now, by neutrality of 1, the definition of relative complement, right distributivity, $(*)$ and neutrality of 0,

$$a = 1 \cdot a = (r + s) \cdot a = r \cdot a + s \cdot a = r \cdot a + 0 = r \cdot a \; . \qquad \square$$

4 Properties of Restrictors

Next we show some fundamental properties of restrictors which will be useful in proving the essential laws of predomain. In this section we will, for economy, use p, q for restrictors, since domain elements are not mentioned here.

Lemma 4.1. *Assume an IL-semiring S. Then for all $a, b, c \in S$ and all $p, q \in \mathsf{rest}(S)$ the following properties hold.*

1. $p \cdot p = p$.
2. $p \cdot q \in \mathsf{rest}(S)$.
3. $p \cdot q = q \cdot p$.
4. $p \cdot q$ *is the infimum of p and q.*
5. *If the infimum $a \sqcap b$ exists then* $p \cdot (a \sqcap b) = p \cdot a \sqcap b = p \cdot a \sqcap p \cdot b$.
6. $p \cdot q \cdot a = p \cdot a \sqcap q \cdot a$.
7. $p \cdot q = 0 \Rightarrow p \cdot a \sqcap q \cdot a = 0$.
8. *If $b \leq a$ then $p \cdot b = b \sqcap p \cdot a$.*

Assume now that S is bounded.

9. $p \cdot b = b \sqcap p \cdot \top$. *In particular,* $p = 1 \sqcap p \cdot \top$.
10. $p \leq q \Leftrightarrow p \cdot \top \leq q \cdot \top$.

Proof.

1. Set $x = a = p$ and $b = 1$ in (1).
2. Assume $a \leq p \cdot q \cdot b$. Since p is a restrictor we obtain $a = p \cdot a$. By $p \leq 1$ we also obtain $a \leq p \cdot q \cdot b \leq q \cdot b$, and hence $a = q \cdot a$ since q is a restrictor, too. Altogether, $a = p \cdot a = p \cdot q \cdot a$.
3. By the previous part, Part 1 and $p, q \leq 1$ we have

$$p \cdot q = p \cdot q \cdot p \cdot q \leq q \cdot p \,.$$

The reverse inequation is shown symmetrically.
4. By $p, q \leq 1$ and isotony of multiplication we have $p \cdot q \leq p, q$. Let c be an arbitrary lower bound of p and q. Then by Part 1, p being a restrictor and isotony of multiplication,

$$c \leq p \wedge c \leq q \Leftrightarrow c \leq p \cdot p \wedge c \leq q \Rightarrow c \leq p \cdot c \wedge c \leq q \Rightarrow c \leq p \cdot q \,,$$

which shows that $p \cdot q$ is the greatest lower bound of p and q.
5. We show the first equation. By isotony and $p \leq 1$ we have $p \cdot (a \sqcap b) \leq p \cdot a$ and $p \cdot (a \sqcap b) \leq b$, i.e., $p \cdot (a \sqcap b)$ is a lower bound of $p \cdot a$ and b. Let c be an arbitrary lower bound of $p \cdot a$ and b. Since p is a restrictor, this implies $c = p \cdot c$. Moreover, $p \cdot a \leq a$ implies that c is also a lower bound of a and b and hence $c \leq a \sqcap b$. Now by isotony of multiplication we have $c = p \cdot c \leq p \cdot (a \sqcap b)$. This means that $p \cdot (a \sqcap b)$ is the greatest lower bound of $p \cdot a$ and b. The second equation follows using idempotence of p (Part 1) and applying the first equation twice:

$$
\begin{aligned}
p \cdot (a \sqcap b) &= p \cdot p \cdot (a \sqcap b) = p \cdot (p \cdot a \sqcap b) \\
&= p \cdot (b \sqcap p \cdot a) = p \cdot b \sqcap p \cdot a = p \cdot a \sqcap p \cdot b \,.
\end{aligned}
$$

6. Employ that $a \sqcap a = a$ and use Part 5 with $b = a$:

$$p \cdot q \cdot a = p \cdot q \cdot (a \sqcap a) = p \cdot (q \cdot a \sqcap a) = p \cdot (a \sqcap q \cdot a) = p \cdot a \sqcap q \cdot a \,.$$

7. Immediate from Part 6.
8. Since $b \leq a$ the meet $a \sqcap b$ exists and equals b. Now Part 5 shows the claim.
9. For the first claim substitute $\top$ for a in Part 8. For the second claim substitute 1 for b in the first claim.
10. ($\Rightarrow$) Immediate from isotony of $\cdot$.
    ($\Leftarrow$) Assume $p \cdot \top \leq q \cdot \top$. Then by Part 9 and isotony we have

$$p = 1 \sqcap p \cdot \top \leq 1 \sqcap q \cdot \top = q \,. \qquad \square$$

The restrictor laws will help to obtain smoother and shorter proofs of the predomain properties in the next section. We will make some further observations about restrictors in the parameterised predomain axiomatisation in the appendix.

5   (Pre)domain Calculus

For a further explanation of (d1) and (d2) we show an equivalent characterisation of their conjunction. For this we use the formula

$$\ulcorner a \leq p \Leftrightarrow a \leq p \cdot a. \tag{llp}$$

One half of this bi-implication was already mentioned in Lemma 3.2.3.
   Now we can deal with the second half.

Lemma 5.1.

1. $\forall a \in S, p \in \ulcorner S : (\text{sub-id}) \wedge (\text{d2}) \Rightarrow (a \leq p \cdot a \Rightarrow \ulcorner a \leq p)$.
2. $(\forall a \in S, p \in \ulcorner S : a \leq p \cdot a \Rightarrow \ulcorner a \leq p) \Rightarrow (\forall a \in S, p \in \ulcorner S : (\text{d2}))$.

Proof.

1. Assume (sub-id) and (d2) and suppose $a \leq p \cdot a$. Since $p \leq 1$ this implies $a = p \cdot a$ and by (d2) we get $\ulcorner a = \ulcorner (p \cdot a) \leq p$.
2. Consider an arbitrary $p \in \ulcorner S$. By Lemmas 3.4 and 4.1.1, p is multiplicatively idempotent. Hence, substituting in the left hand side of the antecedent $p \cdot a$ for a makes that true, so that the right hand side of the antecedent, which is (d2) in that case, is true as well. □

Corollary 5.2. *All predomain elements satisfy* (llp), *which states that $\ulcorner a$ is the least left preserver of a in $\ulcorner S$. Hence, if $\ulcorner S$ is fixed then predomain is uniquely characterised by the axioms if it exists.*

Proof. The first part is immediate from Lemmas 3.2.3 and 5.1. The second part holds, because least elements are unique in partial orders. □

   Now we can show a number of important laws for predomain.

Theorem 5.3. *Assume a predomain IL-semiring $(S, \ulcorner)$ and let a, b range over S and p, q over $\ulcorner S$.*

1. $\ulcorner p = p$. (Stability)
2. *The predomain operator is fully strict, i.e., $\ulcorner a = 0 \Leftrightarrow a = 0$.*
3. *Predomain preserves arbitrary existing suprema. More precisely, if a subset $A \subseteq S$ has a supremum b in S then the image set of A under $\ulcorner$ has a supremum in $\ulcorner S$, namely $\ulcorner b$. Note that neither completeness of S nor that of $\ulcorner S$ is required.*
4. $\ulcorner S$ *forms an upper semilattice with supremum operator $\sqcup$ given by $p \sqcup q = \ulcorner (p + q)$. Hence for $r \in \ulcorner S$ we have $p \leq r \wedge q \leq r \Leftrightarrow p \sqcup q \leq r$.*
5. $\ulcorner (a + b) = \ulcorner a \sqcup \ulcorner b$.
6. *We have the absorption laws $p \cdot (p \sqcup q) = p$ and $p \sqcup (p \cdot q) = p$. Hence $(\ulcorner S, \cdot, \sqcup)$ is a lattice.*
7. $\ulcorner (a \cdot b) \leq \ulcorner (a \cdot \ulcorner b)$.
8. $\ulcorner (a \cdot b) \leq \ulcorner a$.

9. *Predomain satisfies the partial import/export law* $\ulcorner(p \cdot a) \leq p \cdot \ulcorner a$.

10. $p \cdot q = \ulcorner(p \cdot q)$.

Assuming that S is bounded, the following additional properties hold.

11. *We have the Galois connection* $\ulcorner a \leq p \Leftrightarrow a \leq p \cdot \top$.

12. $\ulcorner(a \cdot \top) = \ulcorner a$. *Hence also* $\ulcorner(p \cdot \top) = p$, *in particular* $\ulcorner\top = 1$.

Proof.

1. By Lemma 3.2.1 it remains to show ($\leq$). By neutrality of 1 and (d2) we obtain $\ulcorner p = \ulcorner(p \cdot 1) \leq p$.

2. The direction ($\Leftarrow$) is Ax. (strict). ($\Rightarrow$) is immediate from (d1) and left strictness of 0.

3. Let $b = \bigsqcup \{a \mid a \in A\}$ exist for some set $A \subseteq S$. We show that $\ulcorner b$ is a supremum of $\ulcorner A =_{df} \{\ulcorner a \mid a \in A\}$ in $\ulcorner S$. First, by (isot), $\ulcorner b$ is an upper bound of $\ulcorner A$, since b is an upper bound of A.

   Now let p be an arbitrary upper bound of $\ulcorner A$ in $\ulcorner S$. Then for all $a \in A$ we have $\ulcorner a \leq p$, equivalently $a \leq p \cdot a$ by (llp), and therefore $a \leq p \cdot b$ by definition of b and isotony of $\cdot$. Hence $p \cdot b$ is an upper bound of A and therefore $b \leq p \cdot b$. By (llp) this is equivalent to $\ulcorner b \leq p$, so that $\ulcorner b$ is indeed the least upper bound of $\ulcorner A$ in $\ulcorner S$.

4. Consider $p, q \in \ulcorner S$. By Part 1 we know that $\ulcorner p = p$ and $\ulcorner q = q$. Part 3 tells us that $\ulcorner(p + q)$ is the supremum of $\ulcorner p$ and $\ulcorner q$ and hence of p and q.

5. By Part 3, $\ulcorner(a + b)$ is the supremum of $\ulcorner a$ and $\ulcorner b$ which, by Part 4 is $\ulcorner a \sqcup \ulcorner b$.

6. For the first claim, assume $p = \ulcorner a$ and $q = \ulcorner b$. By Part 5, (isot) with Lemma 3.4 and Lemma 4.1.4,

$$p \cdot (p \sqcup q) = \ulcorner a \cdot \ulcorner(a + b) = \ulcorner a = p \,.$$

   The second claim follows by $q \leq 1$ and the definition of supremum.

7. By (llp) and (d1) thrice we obtain

$$\ulcorner(a \cdot b) \leq \ulcorner(a \cdot \ulcorner b) \Leftrightarrow a \cdot b \leq \ulcorner(a \cdot \ulcorner b) \cdot a \cdot b$$

$$\Leftrightarrow a \cdot b \leq \ulcorner(a \cdot \ulcorner b) \cdot a \cdot \ulcorner b \cdot b \Leftrightarrow \text{TRUE} \,.$$

8. By Part 7, $\ulcorner b \leq 1$, isotony of $\ulcorner$ and neutrality of 1 we have $\ulcorner(a \cdot b) \leq \ulcorner(a \cdot \ulcorner b) \leq \ulcorner(a \cdot 1) = \ulcorner a$.

9. By (d2) we know $\ulcorner(p \cdot a) \leq p$. By $p \leq 1$, isotony of $\cdot$ and $\ulcorner$ and neutrality of 1 we obtain $\ulcorner(p \cdot a) \leq \ulcorner(1 \cdot a) = \ulcorner a$. Now the claim follows by isotony of $\cdot$, Lemma 3.4 and idempotence of $\cdot$ on restrictors and hence domain elements.

10. ($\leq$) follows from Lemma 3.2.1, since $p, q \leq 1$ implies $p \cdot q \leq 1$. For ($\geq$) we obtain by Parts 9 and 1 that $\ulcorner(p \cdot q) \leq p \cdot \ulcorner q = p \cdot q$.

11. We calculate, employing (llp), greatestness of $\top$ and isotony of $\cdot$, isotony of $\ulcorner$, and finally (d2),

$$\ulcorner a \leq p \Leftrightarrow a \leq p \cdot a \Rightarrow a \leq p \cdot \top \Rightarrow \ulcorner a \leq \ulcorner(p \cdot \top) \Rightarrow \ulcorner a \leq p \,.$$

12. By Part 8 we know $^\ulcorner(a \cdot \top) \leq {}^\ulcorner a$. The reverse inequation follows from $a = a \cdot 1 \leq a \cdot \top$ and isotony of domain. The remaining claims result by first specialising a to p and using Part 1, and second by further specialising p to 1. $\qquad\qquad\qquad\qquad\qquad\qquad\qquad\qquad\qquad\qquad\qquad\quad$ □

We now show additional properties of a domain operation.

Lemma 5.4. *Assume a domain IL-semiring* $(S, {}^\ulcorner)$ *and let* a, b *range over* S *and* p, q *over* $^\ulcorner S$.

1. (d3) *strengthens to an equality.*
2. *Domain satisfies the full import/export law* $^\ulcorner(p \cdot a) = p \cdot {}^\ulcorner a$.
3. *In an I-semiring, the lattice* $({}^\ulcorner S, \cdot, \sqcup)$ *is distributive.*

Proof.

1. This is immediate from Lemma 5.3.7.
2. By Part 1 and Lemma 5.3.10 we obtain $^\ulcorner(p \cdot a) = {}^\ulcorner(p \cdot {}^\ulcorner a) = p \cdot {}^\ulcorner a$.
3. We show one distributivity law; it is well known that the second one follows from it. By Lemma 5.3.5, Part 2, distributivity of $\cdot$, Lemma 5.3.5 and Part 2 again,

$$^\ulcorner a \cdot ({}^\ulcorner b \sqcup {}^\ulcorner c) = {}^\ulcorner a \cdot {}^\ulcorner(b + c) = {}^\ulcorner({}^\ulcorner a \cdot (b + c)) = {}^\ulcorner({}^\ulcorner a \cdot b + {}^\ulcorner a \cdot c)$$
$$= {}^\ulcorner({}^\ulcorner a \cdot b) \sqcup {}^\ulcorner({}^\ulcorner a \cdot c)) = {}^\ulcorner a \cdot {}^\ulcorner b \sqcup {}^\ulcorner a \cdot {}^\ulcorner c \,. \qquad\qquad □$$

6 Fuzzy Domain Operators

We now present the application of our theory to the setting of fuzzy systems. First we generalise the notion of t-norms (e.g. [EGn03, Haj98]) and pseudo-complementation to general IL-semirings, in particular to semirings that do not just consist of the interval $[0, 1]$ (as, say, a subset of the real numbers) and where that interval is not necessarily linearly ordered.

Definition 6.1. Consider an IL-semiring S with the interval $[0, 1]$ as specified in Def. 2.1. A *t-norm* is a binary operator $\otimes : [0, 1] \times [0, 1] \to [0, 1]$ that is isotone in both arguments, associative and commutative and has 1 as unit.

The definition implies $p \otimes q \leq p, q$, since, e.g., $p \otimes q \leq p \otimes 1 = p$. In an IL-semiring, by the axioms the operator $\cdot$ restricted to $[0, 1]$ is a t-norm.

Definition 6.2. A *weak complement* operator in an IL-semiring is a function $\neg : [0, 1] \to [0, 1]$ that is an order-antiisomorphism, i.e., is bijective and satisfies $p \leq q \Leftrightarrow \neg q \leq \neg p$, such that additionally $\neg\neg p = p$. This implies $\neg 0 = 1$ and $\neg 1 = 0$.

Based on $\neg$ we can define the *weak relative complement* $p - q =_{df} p \otimes \neg q$ and *weak implication* $p \to q =_{df} \neg p + q$. We have $1 - p = \neg p$ and $1 \to p = p$.

Moreover, if the IL-semiring has a t-norm $\otimes$ the associated *t-conorm* $\oslash$ is defined as the analogue of the De Morgan dual of the t-norm:

$$p \oslash q =_{df} \neg(\neg p \otimes \neg q) .$$

Lemma 6.3. *Assume an IL-semiring with weak negation.*

1. $p \leq p \oslash q$.
2. *If $p \otimes q$ is the infimum of p and q then $p \sqcup q = p \oslash u$.*

Proof.

1. By definition of $\oslash$, antitony of complement and $\neg q \leq 1$,

$$p \leq p \oslash q \Leftrightarrow p \leq \neg(\neg p \otimes \neg q) \Leftrightarrow \neg p \otimes \neg q \leq \neg p \Leftrightarrow \text{TRUE} .$$

2. By Part 1 $p \oslash q$ is an upper bound of p and q. Let $r \in [0,1]$ be an arbitrary upper bound of p and q. Then by antitony of $\neg$ we have $\neg r \leq \neg p, \neg q$ and hence, by the assumption that $\otimes$ is the infimum operator, $\neg r \leq \neg p \otimes \neg q$. Again by antitony of $\neg$ this entails $\neg(\neg p \otimes \neg q) \leq \neg\neg r = r$, i.e., $p \oslash q \leq r$ by definition of $\oslash$. Hence $p \oslash q$ is the supremum of p and q. $\qquad\square$

Next, we deal with a special t-norm and its associated t-conorm.

Lemma 6.4. *Consider the sub-interval $I =_{df} [0,1]$ of the real numbers with $x \otimes y =_{df} \min(x,y)$ and $x \oslash y =_{df} \max(x,y)$. Then $(I, \oslash, 0, \otimes, 1)$ is an I-semiring and the identity function is a domain operator on I.*

The proof is straightforward. Since this domain operator is quite boring, in Sect. 8 we will turn to matrices over I, where the behaviour becomes non-trivial.

7   Modal Operators

Following [DMS06], in a predomain IL-semiring we can define a *forward diamond operator* as

$$|a\rangle p =_{df} \ulcorner(a \cdot p) .$$

By right-distributivity, diamond is homomorphic w.r.t. $+$:

$$|a + b\rangle p = |a\rangle p \sqcup |b\rangle p .$$

Hence diamond is isotone in the first argument:

$$a \leq b \Rightarrow |a\rangle p \leq |b\rangle p .$$

Diamond is also isotone in its second argument:

$$p \leq q \Rightarrow |a\rangle p \leq |a\rangle q .$$

For predomain elements p, q we obtain by Thm. 5.3.10 that $|p\rangle q = p \cdot q$. Hence, $|1\rangle$ is the identity function on predomain elements. Moreover, $|0\rangle p = 0$. If the underlying semiring is even a domain semiring, by the property (d3) we obtain multiplicativity of diamond:

$$|a \cdot b\rangle p = |a\rangle |b\rangle p \ .$$

If the semiring has a weak complement the diamond can be dualised to a forward box operator by setting

$$|a]q =_{df} \neg |a\rangle \neg q \ .$$

This De Morgan duality gives the *swapping rule*

$$|a\rangle p \leq |b]q \ \Leftrightarrow \ |b\rangle \neg q \leq |a]\neg p \ .$$

We now study the case where $\cdot$ plays the role of a t-norm $\oslash$ on $[0, 1]$. By right-distributivity, Thm. 5.3.5, Lemma 6.3 and duality then for predomain elements p, q we have

$$|a + b]p = (|a]p) \cdot (|b]p) \ ,$$

i.e., box is anti-homomorphic w.r.t. $+$ and hence antitone in its first argument:

$$a \leq b \ \Rightarrow \ |a]p \geq |b]p \ .$$

Box is also isotone in its second argument:

$$p \leq q \ \Rightarrow \ |a]p \leq |a]q \ .$$

For predomain elements p, q we get by Thm. 5.3.10 and the definition of $\rightarrow$ that

$$|p]q = p \rightarrow q \ .$$

Hence, $|1]$, too, is the identity function on tests. Moreover, $|0]p = 1$. If the underlying semiring is even a domain semiring, by locality (d3) we obtain multiplicativity of box as well:

$$|a \cdot b]p = |a]|b]p \ .$$

One may wonder about the relation of these operators to those in other systems of fuzzy modal logic (e.g [MvA13]). These approaches usually deal only with algebras where the whole carrier set coincides with the interval $[0, 1]$. This would, for instance, rule out the matrix semirings to be discussed in Section 8. On the other hand, it would be interesting to see whether the use of residuated lattices there could be carried over fruitfully to the interval $[0, 1]$ of general semirings. However, this is beyond the scope of the present paper.

8   Predomain and Domain in Matrix Algebras

We can use the elements of an IL-semiring as entries in matrices. With pointwise addition and the usual matrix product the set of $n \times n$ matrices for some $n \in \mathbb{N}$ becomes again an IL-semiring with the zero matrix as 0 and the diagonal unit matrix as 1. The restrictors in the matrix IL-semiring are precisely the diagonal matrices with restrictors in the diagonal.

Let us work out what the characteristic property (llp) of a predomain operator means in the matrix world, assuming a predomain operator on the underlying IL-semiring. We perform our calculations for 2×2 matrices to avoid tedious index notation; they generalise immediately to general matrices.

$$\begin{pmatrix} a & b \\ c & d \end{pmatrix} \leq \begin{pmatrix} p & 0 \\ 0 & q \end{pmatrix} \cdot \begin{pmatrix} a & b \\ c & d \end{pmatrix}$$

$\Leftrightarrow$ $\{$ definition of matrix multiplication $\}$

$$\begin{pmatrix} a & b \\ c & d \end{pmatrix} \leq \begin{pmatrix} p \cdot a & p \cdot b \\ q \cdot c & q \cdot d \end{pmatrix}$$

$\Leftrightarrow$ $\{$ pointwise order $\}$

$a \leq p \cdot a \wedge b \leq p \cdot b \wedge c \leq q \cdot c \wedge d \leq q \cdot d$

$\Leftrightarrow$ $\{$ by (llp) $\}$

$\ulcorner a \leq p \wedge \ulcorner b \leq p \wedge \ulcorner c \leq q \wedge \ulcorner d \leq q$

$\Leftrightarrow$ $\{$ by Th. 5.3.4 $\}$

$\ulcorner a \sqcup \ulcorner b \leq p \wedge \ulcorner c \sqcup \ulcorner d \leq q$.

$\Leftrightarrow$ $\{$ pointwise order $\}$

$$\begin{pmatrix} \ulcorner a \sqcup \ulcorner b & 0 \\ 0 & \ulcorner c \sqcup \ulcorner d \end{pmatrix} \leq \begin{pmatrix} p & 0 \\ 0 & q \end{pmatrix}$$

Since (llp) characterises predomain uniquely for fixed $\ulcorner S$, we conclude, by the principle of indirect equality, that predomain in the matrix IL-semiring must be

$$\ulcorner \begin{pmatrix} a & b \\ c & d \end{pmatrix} =_{df} \begin{pmatrix} \ulcorner a \sqcup \ulcorner b & 0 \\ 0 & \ulcorner c \sqcup \ulcorner d \end{pmatrix} .$$

Next we investigate the behaviour of domain in the matrix case.

Lemma 8.1. *Let S be an I-semiring. If S has a domain operator, then so does the set of $n \times n$ matrices over S.*

Proof. We need to show that the above representation of predomain on matrices satisfies (d3) provided the predomain operator on S does. Again we treat only the case of 2×2-matrices.

$$\ulcorner \left(\begin{pmatrix} a & b \\ c & d \end{pmatrix} \cdot \ulcorner \begin{pmatrix} e & f \\ g & h \end{pmatrix} \right)$$

$=$ $\{$ above representation of predomain $\}$

$$\ulcorner \left(\begin{pmatrix} a & b \\ c & d \end{pmatrix} \cdot \begin{pmatrix} \ulcorner e \sqcup \ulcorner f & 0 \\ 0 & \ulcorner g \sqcup \ulcorner h \end{pmatrix} \right)$$

$=$ { definition of matrix product and right annihilation }

$$\ulcorner\begin{pmatrix} a \cdot (\ulcorner e \sqcup \ulcorner f) & b \cdot (\ulcorner g \sqcup \ulcorner h) \\ c \cdot (\ulcorner e \sqcup \ulcorner f) & d \cdot (\ulcorner g \sqcup \ulcorner h) \end{pmatrix}$$

$=$ { by Lemma 5.3.5 }

$$\ulcorner\begin{pmatrix} a \cdot \ulcorner(e+f) & b \cdot \ulcorner(g+h) \\ c \cdot \ulcorner(e+f) & d \cdot \ulcorner(g+h) \end{pmatrix}$$

$=$ { above representation of predomain }

$$\begin{pmatrix} \ulcorner(a \cdot \ulcorner(e+f)) \sqcup \ulcorner(b \cdot \ulcorner(g+h)) & 0 \\ 0 & \ulcorner(c \cdot \ulcorner(e+f)) \sqcup \ulcorner(d \cdot \ulcorner(g+h)) \end{pmatrix}$$

$=$ { by (d3) }

$$\begin{pmatrix} \ulcorner(a \cdot (e+f)) \sqcup \ulcorner(b \cdot (g+h)) & 0 \\ 0 & \ulcorner(c \cdot (e+f)) \sqcup \ulcorner(d \cdot (g+h)) \end{pmatrix}$$

$=$ { left distributivity }

$$\begin{pmatrix} \ulcorner(a \cdot e + a \cdot f) \sqcup \ulcorner(b \cdot g + b \cdot h) & 0 \\ 0 & \ulcorner(c \cdot e + c \cdot f) \sqcup \ulcorner(d \cdot g + d \cdot h) \end{pmatrix}$$

$=$ { by Lemma 5.3.5 }

$$\begin{pmatrix} \ulcorner(a \cdot e + a \cdot f + b \cdot g + b \cdot h) & 0 \\ 0 & \ulcorner(c \cdot e + c \cdot f + d \cdot g + d \cdot h) \end{pmatrix}$$

$=$ { associativity and commutativity of $+$ }

$$\begin{pmatrix} \ulcorner(a \cdot e + b \cdot g + a \cdot f + b \cdot h) & 0 \\ 0 & \ulcorner(c \cdot e + d \cdot g + c \cdot f + d \cdot h) \end{pmatrix}$$

$=$ { by Lemma 5.3.5 }

$$\begin{pmatrix} \ulcorner(a \cdot e + b \cdot g) \sqcup \ulcorner(a \cdot f + b \cdot h) & 0 \\ 0 & \ulcorner(c \cdot e + d \cdot g) \sqcup \ulcorner(c \cdot f + d \cdot h) \end{pmatrix}$$

$=$ { above representation of predomain }

$$\ulcorner\begin{pmatrix} a \cdot e + b \cdot g & a \cdot f + b \cdot h \\ c \cdot e + d \cdot g & c \cdot f + d \cdot h \end{pmatrix}$$

$=$ { definition of matrix product }

$$\ulcorner\left(\begin{pmatrix} a & b \\ c & d \end{pmatrix} \cdot \begin{pmatrix} e & f \\ g & h \end{pmatrix} \right).$$

$\square$

Finally, we calculate the diamond operator in the matrix IL-semiring.

$$\left| \begin{pmatrix} a & b \\ c & d \end{pmatrix} \right\rangle \begin{pmatrix} p & 0 \\ 0 & q \end{pmatrix}$$

$=$ { definition of diamond }

$$\ulcorner\left(\begin{pmatrix} a & b \\ c & d \end{pmatrix} \cdot \begin{pmatrix} p & 0 \\ 0 & q \end{pmatrix} \right)$$

$=$ { definition of matrix multiplication }

$$\ulcorner\begin{pmatrix} a \cdot p & b \cdot q \\ c \cdot p & d \cdot q \end{pmatrix}$$

$=$ { definition of matrix predomain }

$$\begin{pmatrix} \ulcorner(a \cdot p) \sqcup \ulcorner(b \cdot q) & 0 \\ 0 & \ulcorner(c \cdot p) \sqcup \ulcorner(d \cdot q) \end{pmatrix}$$

$= \quad \{\!\!\{ \text{ definition of predomain }\}\!\!\}$

$$\begin{pmatrix} |a\rangle p \sqcup |b\rangle q & 0 \\ 0 & |c\rangle p \sqcup |d\rangle q \end{pmatrix}.$$

9   Application to Fuzzy Matrices

Assume now that in $[0,1]$ we use the t-norm $p \oslash q = p \cdot q$ and that there is a weak complement operator $\neg$. Then by Lemma 6.3.2 the above formula for the diamond transforms into

$$\left| \begin{pmatrix} a & b \\ c & d \end{pmatrix} \right\rangle \begin{pmatrix} p & 0 \\ 0 & q \end{pmatrix} = \begin{pmatrix} |a\rangle p \oslash |b\rangle q & 0 \\ 0 & |c\rangle p \oslash |d\rangle q \end{pmatrix}$$

and a straightforward calculation shows

$$\left| \begin{pmatrix} a & b \\ c & d \end{pmatrix} \right] \begin{pmatrix} p & 0 \\ 0 & q \end{pmatrix} = \begin{pmatrix} |a]p \oslash |b]q & 0 \\ 0 & |c]p \oslash |d]q \end{pmatrix}.$$

A potential application of this is the following. Using the approach of [Kaw06] one can model a flow network as a matrix with the pipe capacities between the nodes as entries, scaled down to the interval $[0,1]$. Note that the entries may be arbitrary elements of $[0,1]$ and not just 0 or 1. By Lemma 6.4 the algebra with $\oslash = \min$ and $\oslash = \max$ is an I-semiring with domain and hence, by Lemma 8.1 the set of fuzzy $n \times n$ matrices is, too. For such a matrix C the expressions $\ulcorner C$ and $\neg \ulcorner C'$, where C' is the componentwise negation of C, give for each node the maximum and minimum capacity emanating from that node.

To describe network shapes and restriction we can use crisp matrices, i.e., matrices with 0/1 entries only. Using crisp diagonal matrices P, we can express pre-/post-restriction by matrix multiplication on the appropriate side. So if a matrix C gives the pipe capacities in a network, $P \cdot C$ and $C \cdot P$ give the capacities in the network in which all starting/ending points outside P are removed. Hence, if we take again $\oslash = \min$ and $\oslash = \max$, the expression $|C\rangle P$ gives for each node the maximum outgoing capacity in the output restricted network $C \cdot P$. To explain the significance of $|C]P$, we take a slightly different view of the fuzzy matrix model for flow analysis. Scaling down the capacities to $[0,1]$ could be done relative to a top capacity (not necessarily occurring in the network). Then $p \in [0,1]$ would indicate how close the flow level is to the top flow. Then $|C]P$ would indicate the level of "non-leaking" outside of P. If for instance $|C]P = 0$, then the maximal flow outside of P is 1, i.e., leaking is maximal.

Since on crisp matrices weak negation coincides with standard Boolean negation, we can, additionally, use these ideas to replay the algebraic derivation of the Floyd/Warshall and Dijkstra algorithms in [HM12].

Elaborating on these examples will be the subject of further papers.

10   Conclusion

Despite the weakness in assumptions, the generalised theory of predomain and domain has turned out to be surprisingly rich in results. Concerning applications, we certainly have just skimmed the surface and hope that others will join our further investigations.

Acknowledgement. We are grateful for valuable comments by Han-Hing Dang and the anonymous referees.

References

[DMS06] Desharnais, J., Möller, B., Struth, G.: Kleene Algebra with Domain. ACM Transactions on Computational Logic 7, 798–833 (2006)

[DS11]   Desharnais, J., Struth, G.: Internal axioms for domain semirings. Sci. Comput. Program. 76, 181–203 (2011)

[EGn03] Esteva, F., Godo, L., García-Cerdaña, À.: On the hierarchy of t-norm based residuated fuzzy logics. In: Fitting, M., Orłowska, E. (eds.) Beyond Two, pp. 251–272. Physica (2003)

[Haj98] Hajek, P.: The Metamathematics of Fuzzy Logic. Kluwer (1998)

[HM12]   Höfner, P., Möller, B.: Dijkstra, Floyd and Warshall meet Kleene. Formal Asp. Comput. 24, 459–476 (2012)

[Kaw06] Kawahara, Y.: On the Cardinality of Relations. In: Schmidt, R.A. (ed.) RelMiCS/AKA 2006. LNCS, vol. 4136, pp. 251–265. Springer, Heidelberg (2006)

[Möl07] Möller, B.: Kleene getting lazy. Sci. Comput. Program. 65, 195–214 (2007)

[MRE12] Möller, B., Roocks, P., Endres, M.: An Algebraic Calculus of Database Preferences. In: Gibbons, J., Nogueira, P. (eds.) MPC 2012. LNCS, vol. 7342, pp. 241–262. Springer, Heidelberg (2012)

[Möl13] Möller, B.: Modal Knowledge and Game Semirings. Computer Journal 56, 53–69 (2013)

[MvA13] Morton, W., van Alten, C.: Modal MTL-algebras. Fuzzy Sets and Systems 222, 58–77 (2013)

[Van34] Vandiver, H.: Note on a simple type of algebra in which the cancellation law of addition does not hold. Bulletin of the American Mathematical Society 40, 914–920 (1934)

Appendix: A Parametrised Axiomatisation of Predomain

Experiments have shown that the ((pre)pre)domain axioms of Sect. 3 can be formulated in a more general way, leading to a whole family of ((pre)pre)domain operators. The key is to factor out the set over which p is quantified in Ax. (d2) and make that into a parameter. This leads to the following definition.

Definition 10.1. By a *parameterised prepredomain IL-semiring* we mean a structure $(S, \ulcorner)$ with an IL-semiring S and the *prepredomain operator* $\ulcorner : S \to S$ satisfying, for all $a \in S$,

$$a \le \ulcorner a \cdot a \,. \tag{pd1}$$

A *parameterised predomain IL-semiring* is a structure $(S, \ulcorner, T)$ with a subset $T \subseteq S$ such that $(S, \ulcorner)$ is a parameterised prepredomain IL-semiring and for all $a, p \in S$,

$$p \in T \Rightarrow \ulcorner(p \cdot a) \le p \,, \tag{pd2}$$

$$\ulcorner a \le \ulcorner(a + b) \,. \tag{p-isot}$$

We will impose varying conditions on the set T using the following formulas.

$$T \subseteq [0, 1] \,, \tag{T-sub-id}$$

$$\ulcorner S \subseteq T \,, \tag{dom-in-T}$$

$$T \text{ is closed under } +, \tag{T-plus-closed}$$

$$T \text{ is closed under } \cdot. \tag{T-dot-closed}$$

Using Prover9/Mace4 it is now an easy albeit somewhat tedious task to investigate which of the properties in Sects. 3 and 5 follow from which subsets of the parameterised axioms and the restrictions on T. We list the results below in Table 1. All of the proofs and counterexamples are generated quite fast. The table is to be understood as follows: "Proved with set A of axioms" means that for all proper subsets of A Mace4 finds counterexamples to the formulas listed.

The strictness property $\ulcorner 0 = 0$ does not follow from any subset of the above formulas; it would need to be an extra axiom.

Using Prover 9 one can also show that (T-sub-id), (dom-in-T), (pd1), (pd2) and (p-isot) determine predomain uniquely: use two copies of these axiom sets with two names for the predomain operator, say d_1 and d_2, and use the goal $d_1(a) = d_2(a)$.

There remains the question whether there are any interesting sets T that meet a relevant subset of the restricting conditions. We can offer four candidates:

- $T = [0, 1]$. This trivially satisfies (T-sub-id), and also (T-plus-closed). Moreover, we have $0 \in T$. So if we stipulate $\ulcorner a \le 1$ as an additional axiom we obtain the full set of properties in the table above, plus the full strictness property $\ulcorner a = 0 \Leftrightarrow a = 0$.
- $T = \ulcorner S$. This choice trivially satisfies (dom-in-T), but to obtain (T-sub-id) we need again the additional axiom $\ulcorner a \le 1$. Since nothing else is known about $\ulcorner S$, we cannot assume (T-plus-closed), and so we only get the properties of the table above the last row, which still is quite a rich set.
- $T = \mathrm{rest}(S)$. This choice trivially satisfies (T-sub-id). But as the table shows, (dom-in-T) is needed in the proof of $\ulcorner S \subseteq T$, so that things get circular here. For that reason this choice does not lead as many results as the two before.
- $T = \mathrm{test}(S)$. This satisfies (T-sub-id), but not necessarily (dom-in-T); the question is the subject of ongoing investigation.

Table 1. Proof results

Properties Proved	Interpretation
with (pd1)	
$⌜a = 0 \Rightarrow a = 0$	strictness
$a \leq 1 \Rightarrow a \leq ⌜1$	sub-Identity I
$a \leq 1 \Rightarrow a \leq ⌜a$	sub-Identity II
$1 \leq ⌜1$	sub-Identity III
with (pd2)	
$1 \in T \Rightarrow ⌜a \leq 1$	1 dominates predomain
with (pd1), (pd2)	
$a \in T \Rightarrow ⌜a \leq a$	predomain is contracting
with (T-sub-id), (pd1), (pd2)	
$⌜a \leq ⌜b \Rightarrow a \leq ⌜b \cdot a$	first half of (llp)
$a \in T \Rightarrow (⌜a \leq b \Rightarrow a \leq b \cdot a)$	analogue of first half of (llp)
$a \in T \Rightarrow a = ⌜a \cdot a$	equational form of (pd1)
$a \in T \Rightarrow a = ⌜a$	stability
$a \in T \Rightarrow (⌜a \leq b \Leftarrow a \leq b \cdot a)$	analogue of second half of (llp)
with (T-sub-id), (pd1), (pd2), (p-isot)	
$b \in T \wedge a \leq b \cdot c \Rightarrow a \leq b \cdot a$	$T \subseteq \mathsf{rest}(S)$
with (T-sub-id), (dom-in-T), (pd1), (pd2)	
$⌜a \leq ⌜b \Leftarrow a \leq ⌜b \cdot a$	second half of (llp)
with (T-sub-id), (dom-in-T), (T-plus-closed), (pd1), (pd2), (p-isot)	
$⌜(a + b) \leq ⌜a + ⌜b$	additivity
$⌜a \cdot (⌜b + ⌜c) = ⌜a \cdot ⌜b + ⌜a \cdot ⌜c$	left distributivity
$⌜a + ⌜b \cdot ⌜c = (⌜a + ⌜b) \cdot (⌜a + ⌜c)$	distributivity II
with (T-sub-id), (dom-in-T), (pd1), (pd2), (p-isot)	
$a \leq ⌜b \cdot c \Rightarrow a \leq ⌜b \cdot a$	$⌜S \subseteq \mathsf{rest}(S)$
$a \leq ⌜b \cdot c \Rightarrow ⌜a \leq ⌜b$	analogue of second half of (llp)
$⌜a + ⌜a \cdot ⌜b = ⌜a$	absorption I
$⌜a \cdot (⌜a + ⌜b) = ⌜a$	absorption II
$(⌜a + ⌜b) \cdot ⌜a = ⌜a$	absorption III
$⌜a \cdot ⌜b = ⌜b \cdot ⌜a$	predomain elements commute
$⌜a \cdot ⌜a = ⌜a$	predomain elements are idempotent
$⌜a \cdot 0 = 0$	0 is a right annihilator on $⌜S$
$⌜1 = 1$	
$⌜⌜a = ⌜a$	stability
$⌜(a \cdot ⌜b) = ⌜a \cdot ⌜b$	domain of product
$r \in T \wedge (r \leq ⌜a \wedge r \leq ⌜b) \Rightarrow r \leq ⌜a \cdot ⌜b$	infimum I
$(⌜c \leq ⌜a \wedge ⌜c \leq ⌜b) \Rightarrow ⌜c \leq ⌜a \cdot ⌜b$	infimum II
$r \in T \wedge (⌜a \leq r \wedge ⌜b \leq r) \Rightarrow ⌜(a + b) \leq r$	supremum I
$(⌜a \leq ⌜c \wedge ⌜b \leq ⌜c) \Rightarrow ⌜(a + b) \leq ⌜c$	supremum II
(T-dot-closed)	T closed under $\cdot$
$0 \in T \Rightarrow ⌜0 = 0$	strictness (not valid without the premise)

Tableau Development
for a Bi-intuitionistic Tense Logic*

John G. Stell[1], Renate A. Schmidt[2], and David Rydeheard[2]

[1] School of Computing, University of Leeds, Leeds, UK
[2] School of Computer Science, University of Manchester, Manchester, UK

Abstract. The paper introduces a bi-intuitionistic logic with two modal operators and their tense versions. The semantics is defined by Kripke models in which the set of worlds carries a pre-order relation as well as an accessibility relation, and the two relations are linked by a stability condition. A special case of these models arises from graphs in which the worlds are interpreted as nodes and edges of graphs, and formulae represent subgraphs. The pre-order is the incidence structure of the graphs. These examples provide an account of time including both time points and intervals, with the accessibility relation providing the order on the time structure. The logic we present is decidable and has the effective finite model property. We present a tableau calculus for the logic which is sound, complete and terminating. The MetTel system has been used to generate a prover from this tableau calculus.

1 Introduction

We start by reviewing the motivation for developing a theory of 'relations on graphs' which generalizes that of 'relations on sets'. One novel feature of relations on graphs is a pair of adjoint converse operations instead of the involution found with relations on sets. One half of this pair (the 'left converse') is used later in defining a relational semantics for a novel bi-intuitionistic modal logic.

Relations on sets underlie the most fundamental of the operations used in mathematical morphology [Ser82, BHR07]. Using $\mathbb{Z}^2$ to model a grid of pixels, binary (i.e., black and white) images are modelled by subsets of $\mathbb{Z}^2$. One aim of processing images is to accentuate significant features and to lessen the visual impact of the less important aspects. Several basic transformations on images are parameterized by small patterns of pixels called structuring elements. These structuring elements generate relations which transform subsets of $\mathbb{Z}^2$ via the correspondence between relations $R \subseteq \mathbb{Z}^2 \times \mathbb{Z}^2$ and union-preserving operations on the powerset $\mathscr{P}\mathbb{Z}^2$. Several fundamental properties of image processing operations can be derived using only properties of these relations.

There have been various proposals for developing a version of mathematical morphology for graphs, one of the earliest being [HV93]. However, most work in this area does not use a relational approach, probably because a theory of relations on graphs may be constructed in several different ways. In one way [Ste12],

* This research was supported by UK EPSRC research grant EP/H043748/1.

P. Höfner et al. (Eds.): RAMiCS 2014, LNCS 8428, pp. 412–428, 2014.

the set of relations on a graph, or more generally a hypergraph, forms a generalization of a relation algebra where, in particular, the usual involutive converse operation becomes a pair of operations (the left converse and the right converse) forming an adjoint pair. In the present paper, we use these relations to give a semantics for a bi-intuitionistic modal logic in which propositions are interpreted over subgraphs as opposed to subsets of worlds as in standard Kripke semantics.

Accessibility relations with additional structure are already well-known in intuitionistic modal logic [ZWC01]. However, the semantics for the logic we present is distinguished both from this work, and from other related work we discuss in Section 5, by the use of the left-converse operation. This leads to a logic with novel features which include $\Diamond\varphi$ being equivalent to $\lnot\Box\lnot\varphi$, where $\lnot$ and $\lnot$ are respectively the co-intuitionistic and the intuitionistic negation.

Connections between mathematical morphology and modal logic, have been developed by Aiello and Ottens [AO07], who implemented a hybrid modal logic for spatial reasoning, and by Aiello and van Benthem [Av02] who pointed out connections with linear logic. Bloch [Blo02], also motivated by applications to spatial reasoning, exploited connections between relational semantics for modal logic and mathematical morphology. These approaches used morphology operations on sets, and one motivation for our own work is to extend these techniques to relations on graphs. This has potential for applications to spatial reasoning about discrete spaces based on graphs.

In this paper we restrict our attention to the logic itself, rather than its applications, and the semantic setting we use is more general than that arising from relations on graphs or hypergraphs.

The main contribution of the paper is a bi-intuitionistic tense logic, called BISKT, for which a Kripke frame consists of a pre-order H interacting with an accessibility relation R via a stability condition. The semantics interprets formulae as H-sets, the downwardly closed sets of the pre-order. A particular case arises when the worlds represent the edges and nodes of a graph and formulae are interpreted as subgraphs. We show that BISKT is decidable and has the effective finite model property, by showing that BISKT can be mapped to the guarded fragment which is known to be decidable and has the effective finite model property [ANvB98, Grä99].

The semantic setting for BISKT is a relational setting in which it is not difficult to develop deduction calculi. Semantic tableau deduction calculi in particular are easy to develop. In this paper we follow the methodology of tableau calculus synthesis and refinement as introduced in [ST11, TS13] to develop a tableau calculus for the logic. We give soundness, completeness and termination results as consequences of results of tableau synthesis and that BISKT has the effective finite model property.

Implementing a prover is normally a time-consuming undertaking but MetTeL is software for automatically generating a tableau prover from a set of tableau rules given by the user [Met, TSK12]. For us using MetTeL turned out to be useful because we could experiment with implementations of different initial versions of the calculus. In combination with the tableau synthesis method it was easy to run tests on a growing collection of problems with different provers for several preliminary versions of formalisations of bi-intuitionistic tense logics before settling on the definition given in this paper. MetTeL has also allowed to us experiment with different refinements of the rules and different forms of blocking. Blocking is a technique for forcing termination of tableau calculi for decidable logics.

The paper is structured as follows. Section 2 presents the basic notions of bi-Heyting algebras and relations on downwardly closed sets as well as graphs. Section 3 defines the logic BISKT as a bi-intuitionistic stable tense logic in which subgraphs are represented as downwardly closed sets. In Section 4 we present a terminating labelled tableau calculus for BISKT. With a MetTeL generated prover we have tested several formulae for validity and invalidity in BISKT; a selection of the validities shown are given in the section. Connections to other work is discussed in Section 5.

The MetTeL specification of the tableau calculus for BISKT and the generated prover can be downloaded from the accompanying website: `http://staff.cs.manchester.ac.uk/~schmidt/publications/biskt13/`. There also our current set of problems and performance graphs can be found.

2   Relations on Pre-orders and on Graphs

2.1   The Bi-Heyting Algebra of H-Sets

Let U be a set with a subset $X \subseteq U$, and let $R \subseteq U \times U$ be a binary relation. We recall the definitions of the key operations used in mathematical morphology.

Definition 1. *The **dilation**, $\oplus$, and the **erosion**, $\ominus$, are given by:*

$$X \oplus R = \{u \in U : \exists x \, ((x, u) \in R \wedge x \in X)\},$$
$$R \ominus X = \{u \in U : \forall x \, ((u, x) \in R \rightarrow x \in X)\}.$$

For a fixed R these operations form an adjunction from the lattice $\mathscr{P}U$ to itself in the following sense, with $\_ \oplus R$ being left adjoint to $R \ominus \_$.

Definition 2. *Let $(X, \leq_X), (Y, \leq_Y)$ be partially ordered sets. An **adjunction** between X and Y consists of a pair of order-preserving functions $f : X \to Y$ and $g : Y \to X$ such that $x \leq_X g(y)$ iff $f(x) \leq_Y y$ for all $x \in X$ and all $y \in Y$. The function f is said to be **left adjoint to** g, and g is **right adjoint to** f.*

Erosion and dilation interact with composition of relations as follows.

Lemma 1. *If R and S are any binary relations on U and $X \subseteq U$, then*

$$(S\,;R) \ominus X = S \ominus (R \ominus X) \qquad and \qquad X \oplus (R\,;S) = (X \oplus R) \oplus S.$$

The operations $\oplus$ and $\ominus$ can be applied to subgraphs of a graph when relations on graphs are defined. We see how this works in Section 2.3, but first work in a more general setting of H-sets which we now define. Let U be a set and H a pre-order on U (i.e., a reflexive and transitive relation).

Definition 3. *A subset* $X \subseteq U$ *is an **H-set** if* $X \oplus H \subseteq X$.

Since H is reflexive, the condition is equivalent to $X \oplus H = X$, and if we were to write H as $\geqslant$ these would be downsets. It follows from the adjunction between dilation and erosion that X satisfies $X = X \oplus H$ iff it satisfies $X = H \ominus X$.

The set of all H-sets forms a lattice which is a bi-Heyting algebra. The H-sets are closed under unions and intersections but not under complements. When A and B are H-sets, we can construct the following H-sets where $-$ denotes the complement of a subset of U.

$$
\begin{aligned}
A \to B &= H \ominus (-A \cup B) & &\text{relative pseudocomplement} \\
A \succ B &= (A \cap -B) \oplus H & &\text{dual relative pseudocomplement} \\
\neg A &= H \ominus (-A) & &\text{pseudocomplement} \\
\lnot A &= (-A) \oplus H & &\text{dual pseudocomplement}
\end{aligned}
$$

2.2   Relations on H-Sets

Relations on a set U can be identified with the union-preserving functions on the lattice of subsets. When U carries a pre-order H, the union-preserving functions on the lattice of H-sets correspond to relations on U which are stable:

Definition 4. *A binary relation R on U is **stable** if* $H\,;R\,;H \subseteq R$.

Stable relations are closed under composition, with H as the identity element for this operation, but they are not closed under converse. They do however support an adjoint pair of operations, the left and the right converse, denoted by $\smile$ and $\frown$ respectively. Properties of these include $\smile\frown R \subseteq R \subseteq \frown\smile R$ for any stable relation R.

Definition 5. *The **left converse** of a stable relation R is* $\smile R = H\,;\check{R}\,;H$ *where $\check{R}$ is the (ordinary) converse of R.*

The stability of $\smile R$ is immediate since H is a pre-order, and the left converse can be characterized as the smallest stable relation which contains $\check{R}$. The right converse is characterized as the largest stable relation contained in $\check{R}$, but it plays no role in this paper, so we omit an explicit construction (see [Ste12] for details).

The connection between erosion, dilation, complementation and converse in the lemma below is well-known [BHR07]. We need it to prove Theorem 3 below which generalizes the lemma to the case of a stable relation acting on an H-set.

Lemma 2. *For any relation R on U and any $X \subseteq U$, $X \oplus \check{R} = -(R \ominus (-X))$.*

The following was proved in [Ste12] for the special case of hypergraphs, but we give a direct proof of the general case as it underlies one of the novel features of the logic we consider.

Theorem 3. *For any stable relation R and any H-set A,*

$$
A \oplus (\smile R) = \lnot (R \ominus (\neg A)) \quad \text{and} \quad (\smile R) \ominus A = \neg ((\lnot A) \oplus R).
$$

Proof.

$$
\begin{aligned}
A \oplus (\smile R) &= A \oplus (H \,;\, \check{R} \,;\, H) \\
&= ((A \oplus H) \oplus \check{R}) \oplus H \\
&= (A \oplus \check{R}) \oplus H \\
&= (-(R \ominus (-A))) \oplus H \\
&= \lrcorner (R \ominus (-A))) \\
&= \lrcorner ((R \,;\, H) \ominus (-A))) \\
&= \lrcorner (R \ominus (H \ominus (-A))) \\
&= \lrcorner (R \ominus (\neg A))
\end{aligned}
\qquad
\begin{aligned}
(\smile R) \ominus A &= (H \,;\, \check{R} \,;\, H) \ominus A \\
&= H \ominus (\check{R} \ominus (H \ominus A)) \\
&= H \ominus (\check{R} \ominus A) \\
&= H \ominus -((-A) \oplus R) \\
&= \neg((-A) \oplus R) \\
&= \neg((-A) \oplus (H \,;\, R)) \\
&= \neg(((-A) \oplus H) \oplus R) \\
&= \neg((\lrcorner A) \oplus R)
\end{aligned}
$$

2.3   Relations on Graphs

A special case of the above constructions is when U is the set of all edges and nodes of a graph (that is, an undirected multigraph with multiple loops permitted). For an edge e and a node n we put $(e, n) \in H$ iff e is incident with n, and otherwise $(u, v) \in H$ holds only when $u = v$.

In this setting, the H-sets are exactly the subgraphs, that is sets of nodes and edges which include the incident nodes of every edge in the set. The importance of the bi-Heyting algebra of subgraphs of a directed graph has been highlighted by Lawvere as explained in [RZ96].

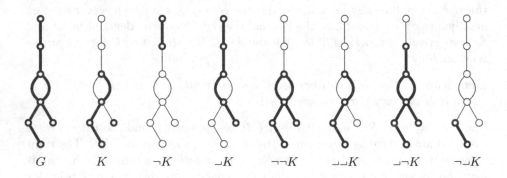

G K $\neg K$ $\lrcorner K$ $\neg\neg K$ $\lrcorner\lrcorner K$ $\lrcorner\neg K$ $\neg\lrcorner K$

Fig. 1. Graph G with subgraph K and the pseudocomplement operation and its dual

We give an example for an undirected graph of the operations $\neg$ and $\lrcorner$ of Section 2.1 since these motivate the semantics for the two negations in our logic. Figure 1 shows various subgraphs of a graph where the subgraphs are distinguished by depicting the edges and nodes in bold. For a subgraph K, the operations $\neg$ and $\lrcorner$ yield respectively the largest subgraph disjoint from K and the smallest subgraph containing all the edges and nodes not present in K. In Figure 1 it can be seen that neither $\neg\neg K$ nor $\lrcorner\lrcorner K$ is equal to K. The subgraph $\neg\neg K$ consists of K completed by the addition of any edges all of whose incident nodes are in K. The subgraph $\lrcorner\lrcorner K$ is K with the removal of any nodes that have incident edges none of which is present in K. The subgraph $\lrcorner\neg K$ can be interpreted as the expansion of K to include things up to one edge away from K, and $\neg\lrcorner K$ is a kind of contraction, removing any nodes on the boundary of K and any edges incident on them.

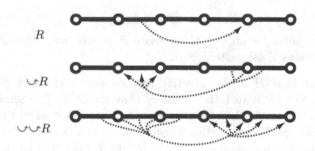

The graph includes edges v,w,x,y,z and nodes a,b,c. The relation is the set $\{(z,a),(x,z),(b,z),(v,z),(w,z),(x,a),$ $(b,a),(v,a),(w,a),(y,b),(y,v),(y,c)\}$.

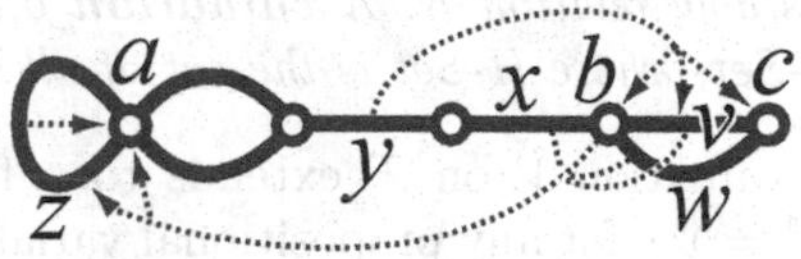

Fig. 2. Relation on a graph drawn as arrows with multiple heads and tails

The stable relations on a graph can be visualized as in Figure 2. The arrows used may have multiple heads and multiple tails; the meaning is that every node or edge at a tail is related to all the edges and nodes at the various heads. The stability condition implies that if a node, n, is related to something, u say, then every edge incident with n is also related to u. Stablility also implies that if u, which may be an edge or a node, is related to an edge e, then u is also related to every node incident with e.

Fig. 3. A relation R, its left converse, and the left converse of the left converse of R

The left converse operation is illustrated in Figure 3 showing that iterating this operation can lead to successively larger relations.

3   Bi-intuitionistic Stable Tense Logic

We now propose a modal logic BISKT for which a Kripke frame is a pre-ordered set (U, H) together with a stable relation. The semantics in this section interprets formulae as H-sets. A particular case is when the worlds are the edges and nodes of a graph and formulae are interpreted as subgraphs.

Definition 6. *The language of* BISKT *consists of a set Vars of propositional variables: $p, q, \ldots$, a constant: $\bot$, unary connectives: $\neg$ and $\lrcorner$, binary connectives: $\wedge, \vee, \rightarrow, \succ$, and unary modal operators: $\Box, \blacklozenge, \Diamond,$ and $\blacksquare$. The set Form of formulae is defined in the usual way.*

Definition 7. *An H-frame $\mathcal{F} = (U, H, R)$ is a pre-order (U, H) together with a stable relation R. A **valuation** on an H-frame $\mathcal{F}$ is a function $\mathcal{V} : Vars \rightarrow H\text{-Set}$, where H-Set is the set of all H-sets.*

A valuation $\mathcal{V}$ on $\mathcal{F}$ extends to a function $[\![\ ]\!] : Form \rightarrow H\text{-Set}$ by putting $[\![v]\!] = \mathcal{V}v$ for any propositional variable v, and making the following definitions.

$$[\![\bot]\!] = \varnothing \qquad\qquad\qquad [\![\top]\!] = U$$

$$[\![\alpha \vee \beta]\!] = [\![\alpha]\!] \cup [\![\beta]\!] \qquad\qquad [\![\alpha \wedge \beta]\!] = [\![\alpha]\!] \cap [\![\beta]\!]$$

$$[\![\neg\alpha]\!] = \neg[\![\alpha]\!] \qquad\qquad\qquad [\![\lrcorner\alpha]\!] = \lrcorner[\![\alpha]\!]$$

$$[\![\alpha \rightarrow \beta]\!] = [\![\alpha]\!] \rightarrow [\![\beta]\!] \qquad\qquad [\![\alpha \succ \beta]\!] = [\![\alpha]\!] \succ [\![\beta]\!]$$

$$[\![\Box\alpha]\!] = R \ominus [\![\alpha]\!] \qquad\qquad [\![\Diamond\alpha]\!] = [\![\alpha]\!] \oplus (\smallsmile R)$$

$$[\![\blacklozenge\alpha]\!] = [\![\alpha]\!] \oplus R \qquad\qquad [\![\blacksquare\alpha]\!] = (\smallsmile R) \ominus [\![\alpha]\!]$$

From Theorem 3 we have $[\![\Diamond\alpha]\!] = [\![\lrcorner\Box\neg\alpha]\!]$ and $[\![\blacksquare\alpha]\!] = [\![\neg\blacklozenge\lrcorner\alpha]\!]$.

Definition 8. *Given a frame $\mathcal{F}$, a valuation $\mathcal{V}$ on $\mathcal{F}$ and a world $w \in U$ we define $\Vdash$ by*

$$\mathcal{F}, \mathcal{V}, w \Vdash \alpha \text{ iff } w \in [\![\alpha]\!].$$

When $\mathcal{F}, \mathcal{V}, w \Vdash \alpha$ holds for all $w \in U$ we write $\mathcal{F}, \mathcal{V} \Vdash \alpha$, and when $\mathcal{F}, \mathcal{V} \Vdash \alpha$ holds for all valuations $\mathcal{V}$ we write $\mathcal{F} \Vdash \alpha$.

In the special case that H is the identity relation on U, stability places no restriction on R and $\smallsmile R$ is just the ordinary converse of R. The semantics is then equivalent to the usual relational semantics for tense logic when time is not assumed to have any specific properties. Figure 4 illustrates how the semantics of $\Diamond$ and $\blacksquare$ can differ from this usual case. In the figure, R is denoted by the broken lines and H is determined by the graph. The H-sets are shown in bold. We can give a temporal interpretation to the example by taking the nodes to be time points, the edges to be open intervals, and R to relate each open interval to all instants that either end the interval or end some later interval. The times when $\Diamond p$ holds are then the open intervals for which p holds at some later instant, together with both endpoints of those intervals. The times when $\blacksquare q$ holds are all the closed intervals where q holds at all times and has always held.

Theorem 4. *The logic* BISKT *is decidable and has the effective finite model property.*

The proof involves an embedding of BISKT into a traditional modal logic with forward and backward looking modal operators defined by H and R as accessibility relations. The frame conditions are reflexivity and transitivity of H, and

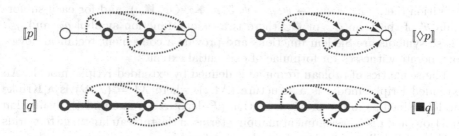

Fig. 4. Example of semantics for ◊ and ■

the stability of R with respect to H. Monotonicity also needs to be suitably ensured. This logic can be shown to be decidable and have the effective finite model property by mapping it to the guarded fragment. This can be done using the axiomatic translation principle introduced in [SH07]. As the guarded fragment is decidable and has the effective finite model property [ANvB98, Grä99] these properties are inherited by the modal logic and also BISKT. It further follows that the computational complexity of reasoning in BISKT is no worse than EXPTIME. (Due to space constraints the detailed proof has been omitted.)

4   Tableau Calculus for BISKT

Since the accessibility relations in the Kripke models of BISKT involve converse relations it is natural to use a semantic tableau method, which does not place any limitations on how the proof search can be performed. In particular, we use a labelled tableau approach because this ensures proof confluence. This means there is no need for backtracking over the proof search, and there is more flexibility in defining search heuristics in an implementation. These are aspects which make it harder to develop tableau calculi where the rules operate on formulae of the logic and do not include any syntactic entities referring to semantics. An additional advantage of semantic tableau calculi is that they return concrete models for satisfiable formulae.

Since BISKT is based on intuitionistic logic the calculus presented in this section is a labelled *signed* tableau calclulus. Tableau formulae have these forms.

$$\bot \quad s : S\,\varphi \quad H(s,t) \quad R(s,t) \quad s \approx t \quad s \not\approx t$$

S denotes a sign (either T or F for true or false), s and t represent worlds in the models constructed by the tableau calculus, and $\approx$ is the standard equality symbol. Technically, s and t denote terms in the term algebra freely generated from a finite set of constants and a finite set of unary function symbols f_θ, $g_{\theta'}$, $g'_{\theta'}$, which are uniquely associated with subformulae of the input set. Specifically, the f_θ are associated with subformulae involving quantification in their semantic definition (i.e., $\neg\varphi$, $\rightharpoondown\varphi$, $\varphi \rightarrow \psi$, $\varphi \succ \psi$, $\Box\varphi$, $\blacklozenge\varphi$, $\Diamond\varphi$, $\blacksquare\varphi$), and for each subformula θ' of the form $\Diamond\varphi$ or $\blacksquare\varphi$ there is a unique function symbol $g_{\theta'}$ and $g'_{\theta'}$. These symbols are Skolem functions and provide a convenient technical device to generate witnesses for formulae of existential extent.

The semantics of tableau formulae is defined by extended Kripke models. An extended Kripke model is a structure $(\mathcal{M}, \iota)$, where $\mathcal{M} = (\mathcal{F}, \mathcal{V})$ is a Kripke model defined as in the previous section ($\mathcal{F}$ denotes a frame and $\mathcal{V}$ a valuation function) and ι is an assignment mapping terms in the tableau language to worlds in U. Satisfiability of tableau formulae is defined by:

$$\mathcal{M}, \iota \not\Vdash \bot$$

$$\mathcal{M}, \iota \Vdash s : T\,\varphi \quad \text{iff} \quad \mathcal{M}, \iota(s) \Vdash \varphi \qquad \mathcal{M}, \iota \Vdash s : F\,\varphi \quad \text{iff} \quad \mathcal{M}, \iota(s) \not\Vdash \varphi$$

$$\mathcal{M}, \iota \Vdash H(s,t) \quad \text{iff} \quad (\iota(s), \iota(t)) \in H \quad \mathcal{M}, \iota \Vdash R(s,t) \quad \text{iff} \quad (\iota(s), \iota(t)) \in R$$

$$\mathcal{M}, \iota \Vdash s \approx t \quad \text{iff} \quad \iota(s) = \iota(t) \qquad \mathcal{M}, \iota \Vdash s \not\approx t \quad \text{iff} \quad \iota(s) \neq \iota(t)$$

Let Tab_{BISKT} be the calculus consisting of the rules in Figure 5. The rules are to be applied top-down. Starting with set of tableau formulae the rules are used to decompose formulae in a goal-directed way. Since some of the rules are branching the inference process constructs a tree derivation. As soon as a contradiction is derived in a branch (that is, when $\bot$ has been derived) that branch is regarded **closed** and no more rules are applied to it. If a branch is not closed then it is **open**. When in an open branch no more rules are applicable then the derivation can stop because a model for the input set can be read off from the branch.

The way to use the tableau calculus is as follows. Suppose we are interested in the validity of a formula, say φ, in BISKT. Then the input to the tableau derivation is the set $\{a : F\varphi\}$ where a is a constant representing the initial world, and the aim is to find a counter-model for φ. If a counter-model is found then φ is not valid, on the other hand, if a closed tableau is constructed then φ is valid.

The first group of rules in the calculus are the closure rule and the decomposition rules of the operators of bi-intuitionistic logic. The closure rule derives $\bot$ when for a formula φ both $s : T\,\varphi$ and $s : F\,\varphi$ occur on the branch. The branch is then closed. The other rules can be thought of as 'decomposing' labelled formulae and building an ever growing tree derivation. These inference steps basically follow the semantics of the main logical operator of the formula being decomposed. For example, the rule for positive occurrences of implication extends the current branch with $t : F\,\varphi$ and $t : T\,\psi$ thereby creating two branches, if formulae of the form $s : T\,\varphi \rightarrow \psi$ and $H(s,t)$ belong to the current branch. The rule for negative occurrences of implication extends the current branch with the three formulae $H(s, f_{\varphi\rightarrow\psi}(s))$, $f_{\varphi\rightarrow\psi}(s) : T\,\varphi$ and $f_{\varphi\rightarrow\psi}(s) : F\,\psi$, if the formula $s : F\,\varphi \rightarrow \psi$ occurs on the current branch. The effect is that an H-successor world is created for s and ψ is assigned false in this successor, while φ is assigned true.

Rules for operators of bi-intuitionistic logic and the closure rule:

$$\frac{s:T\,\varphi,\ s:F\,\varphi}{\bot}\ \text{pv. 0} \qquad\qquad \frac{s:T\,\bot}{\bot}\ \text{pv. 0}$$

$$\frac{s:T\,\varphi\wedge\psi}{s:T\,\varphi,\ s:T\,\psi}\ \text{pv. 1} \qquad\qquad \frac{s:F\,\varphi\wedge\psi}{s:F\,\varphi\ |\ s:F\,\psi}\ \text{pv. 7}$$

$$\frac{s:F\,\varphi\vee\psi}{s:F\,\varphi,\ s:F\,\psi}\ \text{pv. 1} \qquad\qquad \frac{s:T\,\varphi\vee\psi}{s:T\,\varphi\ |\ s:T\,\psi}\ \text{pv. 7}$$

$$\frac{s:T\,\neg\varphi,\ H(s,t)}{t:F\,\varphi}\ \text{pv. 2} \qquad\qquad \frac{s:F\,\neg\varphi}{H(s,f_{\neg\varphi}(s)),\ f_{\neg\varphi}(s):T\,\varphi}\ \text{pv. 10}$$

$$\frac{s:F\,\dashv\varphi,\ H(t,s)}{t:T\,\varphi}\ \text{pv. 2} \qquad\qquad \frac{s:T\,\dashv\varphi}{H(f_{\dashv\varphi}(s),s),\ f_{\dashv\varphi}(s):F\,\varphi}\ \text{pv. 10}$$

$$\frac{s:T\,\varphi\rightarrow\psi,\ H(s,t)}{t:F\,\varphi\ |\ t:T\,\psi}\ \text{pv. 2} \qquad\qquad \frac{s:F\,\varphi\rightarrow\psi}{H(s,f_{\varphi\rightarrow\psi}(s)),\ f_{\varphi\rightarrow\psi}(s):T\,\varphi,\ f_{\varphi\rightarrow\psi}(s):F\,\psi}\ \text{pv. 10}$$

$$\frac{s:F\,\varphi\succ\psi,\ H(t,s)}{t:F\,\varphi\ |\ t:T\,\psi}\ \text{pv. 2} \qquad\qquad \frac{s:T\,\varphi\succ\psi}{H(f_{\varphi\succ\psi}(s),s),\ f_{\varphi\succ\psi}(s):T\,\varphi,\ f_{\varphi\succ\psi}(s):F\,\psi}\ \text{pv. 10}$$

Rules for the tense operators:

$$\frac{s:T\,\Box\varphi,\ R(s,t)}{t:T\,\varphi}\ \text{pv. 2} \qquad\qquad \frac{s:F\,\Box\varphi}{R(s,f_{\Box\varphi}(s)),\ f_{\Box\varphi}(s):F\,\varphi}\ \text{pv. 10}$$

$$\frac{s:F\,\blacklozenge\varphi,\ R(t,s)}{t:F\,\varphi}\ \text{pv. 2} \qquad\qquad \frac{s:T\,\blacklozenge\varphi}{R(f_{\blacklozenge\varphi}(s),s),\ f_{\blacklozenge\varphi}(s):T\,\varphi}\ \text{pv. 10}$$

$$\frac{s:F\,\Diamond\varphi,\ H(t,s),\ R(t,u),\ H(v,u)}{v:F\,\varphi}\ \text{pv. 4}$$

$$\frac{s:T\,\Diamond\varphi}{H(g_{\Diamond\varphi}(s),s),\ R(g_{\Diamond\varphi}(s),g'_{\Diamond\varphi}(s)),\ H(f_{\Diamond\varphi}(s),g'_{\Diamond\varphi}(s)),\ f_{\Diamond\varphi}(s):T\,\varphi}\ \text{pv. 10}$$

$$\frac{s:T\,\blacksquare\varphi,\ H(s,t),\ R(u,t),\ H(u,v)}{v:T\,\varphi}\ \text{pv. 4}$$

$$\frac{s:F\,\blacksquare\varphi}{H(s,g_{\blacksquare\varphi}(s)),\ R(g'_{\blacksquare\varphi}(s),g_{\blacksquare\varphi}(s)),\ H(g'_{\blacksquare\varphi}(s),f_{\blacksquare\varphi}(s)),\ f_{\blacksquare\varphi}(s):F\,\varphi}\ \text{pv. 10}$$

Rules for frame and model conditions:

$$(\text{refl})\ \frac{}{H(s,s)}\ \text{pv. 3} \qquad (\text{tr})\ \frac{H(s,t),\ H(t,u)}{H(s,u)}\ \text{pv. 2}$$

$$(\text{mon})\ \frac{s:T\,\varphi,\ H(s,t)}{t:T\,\varphi}\ \text{pv. 2} \qquad (\text{stab})\ \frac{H(s,t),\ R(t,u),\ H(u,v)}{R(s,v)}\ \text{pv. 4}$$

Fig. 5. Tableau calculus Tab_{BISKT}. (pv = priority value. Rules of highest priority have pv 0, rules of lowest priority have pv 10.)

The rules for the $\Box$ and $\blacklozenge$ operators are signed versions of the standard rules for tense modalities in semantic tableaux for traditional modal logics. The rules for the $\Diamond$ and $\blacksquare$ operators are more complicated versions, as they refer to the composite relation $H;\breve{R};H$.

The third group of rules ensures the models constructed have the required properties. For example, the (refl)-rule ensures all terms (representing worlds) are reflexive in a fully expanded branch, and (tr)-rule ensures the H-relation is transitively closed. The rule (mon) accommodates the property that the truth sets form downsets. It is justified since we can show monotonicity not only for atomic formulae but any formulae of the logic. The rule (stab) ensures the relation R will be stable with respect to H in any generated model.

The rules have been systematically derived from the definition of the semantics of BISKT as given in Section 3. We first expressed the semantics in first-order logic and then converted the formulae to inference rules following the tableau synthesis method described in [ST11]. We do not describe the conversion here because it is completely analogous to the conversion for intuitionistic logic considered as a case study in [ST11, Section 9]. The subset of the rules in the calculus Tab_{BISKT} restricted to the operators and frame conditions relevant to intuitionistic logic in fact coincides with the tableau calculus derived there for intuitionistic logic (there are just insignificant variations in notation). We just note for the rule refinement step in the synthesis process atomic rule refinement as introduced in [TS13] is sufficient. Atomic rule refinement is a specialization of a general rule refinement technique described in [ST11] with the distinct advantage that it is automatic because separate proofs do no need to be given.

The values accompanying each rule in Figure 5 are the priority values we used in the implementation using the MetTeL tableau prover generator [Met, TSK12]. Lower values mean higher prority.

Theorem 5. *The tableau calculus* Tab_{BISKT} *is sound and (constructively) complete with respect to the semantics of* BISKT.

This follows by the results of the tableau synthesis framework and atomic rule refinement [ST11, TS13], because we can show the semantics of BISKT defined in Section 3 is well-defined in the sense of [ST11].

To obtain a terminating tableau calculus, adding the unrestricted blocking mechanism provides an easy way to obtain a terminating tableau calculus for any logic with the finite model property [ST08, ST11, ST13]. The main ingredient of the unrestricted blocking mechanism is the following rule.

Unrestricted blocking rule: (ub) $\dfrac{}{s \approx t \mid s \not\approx t}$ pv. 9

Since this involves equality $\approx$, provision needs to be made for equality reasoning. This can be achieved, for example, via the inclusion of these paramodulation rules.

$$\frac{s \not\approx s}{\bot} \qquad \frac{s \approx t}{t \approx s} \qquad \frac{s \approx t,\ G[s]_\lambda}{G[\lambda/t]}$$

Here, G denotes any tableau formula. The notation $G[s]_\lambda$ means that s occurs as a subterm at position λ in G, and $G[\lambda/t]$ denotes the formula obtained by replacing s at position λ with t. In MetTeL equality reasoning is provided in the form of ordered rewriting which is more efficient [TSK12].

Unrestricted blocking systematically merges terms (sets them to be equal) in order to find small models if they exists. The intuition of the (ub)-rule is that merging two terms s and t, either leads to a model, or it does not, in which case s and t cannot be equal. In order that small models are found it is crucial that blocking is performed eagerly before the application of any term creating rules. The term creating rules are the rules expanding formulae with implicit existential quantification. As can be seen in our implementation using MetTeL the (ub)-rule has been given higher priority (a lower priority value 9) than all the rule creating new Skolem terms (priority value 10).

We denote the extension of the calculus Tab_{BISKT} by the unrestricted blocking mechanism, including some form of reasoning with equality $\approx$, by $Tab_{\mathrm{BISKT}}(ub)$.

Using unrestricted blocking, because the logic has the finite model property, as we have shown in Theorem 4, the tableau calculus of Figure 5 extended with unrestricted blocking provides the basis for a decision procedure.

Theorem 6. *The tableau calculus $Tab_{\mathrm{BISKT}}(ub)$ is sound, (constructively) complete and terminating for* BISKT.

Implementing a prover requires lots of specialist knowledge and there are various non-trivial obstacles to overcome, but using the MetTeL tableau prover generator requires just feeding in the rules of the calculus into the tool which then fully automatically generates an implemention in Java. The tableau calculus in Figure 5 is in exactly the form as supported by MetTeL and unrestricted blocking is available in MetTeL. We have therefore implemented the calculi using MetTeL. MetTeL turned out to be useful to experiment with several initial versions of the calculus. Moreover, in combination with the tableau synthesis method it was easy to experiment with different tableau provers for several preliminary versions of formalizations of bi-intuitionistic tense logics before settling on BISKT. MetTeL has also allowed us to experiment with different rule refinements, limiting the monotonicity rule to atomic formulae or not, and alternative forms of blocking. A natural variation of the (ub) rule is given by these two rules.

Predecessor blocking rules:
$$\frac{H(s,t)}{s \approx t \mid s \not\approx t}\ \text{pv. 9} \qquad \frac{R(s,t)}{s \approx t \mid s \not\approx t}\ \text{pv. 9}$$

Whereas the (ub)-rule blocks any two distinct terms, these rules restrict blocking to terms related via the H and R relations. The rules implement a form of predecessor blocking, which is known to give decision procedures for basic multi-modal logic $K_{(m)}$. We conjecture it also provides a decision procedure for BISKT. Because Tab_{BISKT} is sound and complete and both rules are sound, basing blocking on these rules instead of (ub) preserves soundness and completeness.

The following are examples of formulae provable using the calculus as have been verified by a MetTeL generated prover.

Lemma 7. *The following hold in* BISKT.

1. (a) $[\![\bot \to \varphi]\!] = [\![\top]\!]$

   (b) $[\![\varphi \to \neg\bot]\!] = [\![\top]\!]$

   (c) $[\![\neg(\varphi \to \psi) \to (\neg\varphi \to \neg\psi)]\!] =$ $[\![\top]\!]$

   (d) $[\![\lrcorner(\varphi \to \psi) \to (\lrcorner\varphi \to \lrcorner\psi)]\!] =$ $[\![\top]\!]$

   (e) $[\![\varphi \wedge \neg\varphi]\!] = [\![\bot]\!]$

   (f) $[\![\varphi \vee \lrcorner\varphi]\!] = [\![\top]\!]$

   (g) $[\![\neg\neg\neg\varphi]\!] = [\![\neg\varphi]\!]$

   (h) $[\![\lrcorner\lrcorner\lrcorner\varphi]\!] = [\![\lrcorner\varphi]\!]$

   (i) $[\![\neg(\varphi \vee \psi)]\!] = [\![\neg\varphi \wedge \neg\psi]\!]$

   (j) $[\![\neg(\varphi \wedge \psi)]\!] = [\![\neg\neg(\neg\varphi \vee \neg\psi)]\!]$

   (k) $[\![\lrcorner(\varphi \wedge \psi)]\!] = [\![\lrcorner\varphi \vee \lrcorner\psi]\!]$

   (l) $[\![\lrcorner(\varphi \vee \psi)]\!] = [\![\lrcorner\lrcorner(\lrcorner\varphi \wedge \lrcorner\psi)]\!]$

   (m) $[\![\varphi]\!] \subseteq [\![\neg\neg\varphi]\!]$

   (n) $[\![\lrcorner\lrcorner\varphi]\!] \subseteq [\![\varphi]\!]$

   (o) $[\![\neg\varphi \vee \neg\psi]\!] \subseteq [\![\neg(\varphi \wedge \psi)]\!]$

   (p) $[\![\lrcorner(\varphi \vee \psi)]\!] \subseteq [\![\lrcorner\varphi \wedge \lrcorner\psi]\!]$

   (q) $[\![\varphi \wedge \psi]\!] \subseteq [\![\neg(\neg\varphi \vee \neg\psi)]\!]$

   (r) $[\![\varphi \vee \psi]\!] \subseteq [\![\neg(\neg\varphi \wedge \neg\psi)]\!]$

   (s) $[\![\neg\neg(\varphi \wedge \psi)]\!] = [\![\neg\neg\varphi \wedge \neg\neg\psi]\!]$

   (t) $[\![\neg\neg(\varphi \to \psi)]\!] = [\![\neg\neg\varphi \to \neg\neg\psi]\!]$

2. (a) $[\![\varphi]\!] \subseteq [\![\lrcorner\neg\varphi]\!]$

   (b) $[\![\neg\varphi]\!] \subseteq [\![\lrcorner\varphi]\!]$

   (c) $[\![\neg\lrcorner\varphi]\!] \subseteq [\![\varphi]\!]$

   (d) $[\![\lrcorner\lrcorner\varphi]\!] \subseteq [\![\lrcorner\neg\varphi]\!]$

   (e) $[\![\neg\lrcorner\varphi]\!] \subseteq [\![\neg\neg\varphi]\!]$

   (f) $[\![\neg\lrcorner\varphi]\!] \subseteq [\![\lrcorner\lrcorner\varphi]\!]$

   (g) $[\![\lrcorner\varphi]\!] = [\![\lrcorner\varphi \vee \bot]\!]$

3. (a) $[\![\Box\neg\bot]\!] = [\![\top]\!]$, $[\![\Box(\varphi \to \psi) \to (\Box\varphi \to \Box\psi)]\!] = [\![\top]\!]$

   (b) $[\![\blacklozenge\bot]\!] = [\![\bot]\!]$, $[\![\neg\blacklozenge(\varphi \to \psi) \to (\neg\blacklozenge\varphi \to \neg\blacklozenge\psi)]\!] = [\![\top]\!]$

   (c) $[\![\Diamond\bot]\!] = [\![\bot]\!]$, $[\![\neg\Diamond(\varphi \to \psi) \to (\neg\Diamond\varphi \to \neg\Diamond\psi)]\!] = [\![\top]\!]$

   (d) $[\![\blacksquare\neg\bot]\!] = [\![\top]\!]$, $[\![\blacksquare(\varphi \to \psi) \to (\blacksquare\varphi \to \blacksquare\psi)]\!] = [\![\top]\!]$

   These properties mean all box and diamond operators are in fact 'modal'.

4. (a) $[\![\Diamond\varphi]\!] = [\![\lrcorner\Box\neg\varphi]\!]$

   (b) $[\![\Diamond\varphi]\!] \subseteq [\![\lrcorner\Box\neg\lrcorner\neg\varphi]\!]$

   (c) $[\![\neg\Box\lrcorner\varphi]\!] \subseteq [\![\Diamond\varphi]\!]$

   (d) $[\![\neg\Diamond\lrcorner\varphi]\!] \subseteq [\![\Box\varphi]\!]$

   (e) $[\![\blacksquare\varphi]\!] = [\![\neg\blacklozenge\lrcorner\varphi]\!]$

5. (a) $[\![\lrcorner\varphi]\!] = [\![\neg\bot \succ \varphi]\!]$

   (b) $\neg(\bot \succ \varphi)$

   (c) $\neg(\varphi \succ \neg\bot)$

   (d) $[\![\Diamond\varphi]\!] \subseteq [\![\neg\bot \succ \Box\neg(\neg\bot \succ \neg\varphi)]\!]$

Since we can show the following hold in BISKT, the inclusions in Lemma 7 also hold as implications.

Lemma 8. *1.* $[\![\varphi]\!] \subseteq [\![\psi]\!]$ *implies* $\Vdash \varphi \to \psi$.

2. $[\![\varphi]\!] = [\![\top]\!]$ *implies* $\Vdash \varphi$.

3. $[\![\varphi]\!] = [\![\bot]\!]$ *implies* φ *is contradictory.*

5   Connections to Other Work

Propositional bi-intuitionistic logic was studied by Rauszer [Rau74] who referred to it as H-B logic, standing for Heyting-Brouwer logic. The co-intuitionistic fragment of H-B logic is one of the propositional paraconsistent logics investigated by Wansing [Wan08], but neither of these papers was concerned with bi-intuitionistic modal logic. The stable relations we used are already well known in intuitionistic modal logic [ZWC01, p219] and they provide a special case of the category-theoretic notion of a distributor [Bén00]. However, as far as we are aware, the left converse operation that we use has not featured in either of these contexts.

Reyes and Zolfaghari [RZ96] present modal operators with semantics in bi-Heyting algebras. Graphs are an important example, as in our work, but the modalities are quite different, arising from iterating alternations of ¬ and ¬.

Goré et al [GPT10] studied a bi-intuitionistic modal logic, BiKt, with the same langauge as BISKT but with a semantics producing no relationship between the box and diamond operators. The four modal operators form two residuated pairs $(\Box, \blacklozenge)$ and $(\blacksquare, \Diamond)$ but without any necessary relationship between $\Box$ and $\Diamond$ or between $\blacksquare$ and $\blacklozenge$. In our case the same pairs are residuated (or adjoint) but we have a different semantics for the $(\blacksquare, \Diamond)$ pair, and consequently we do get relationships between $\Box$ and $\Diamond$ and between $\blacksquare$ and $\blacklozenge$.

We next describe the semantics for BiKt [GPT10, p26], using our notation and terminology to clarify the connection with BISKT.

Definition 9. *A BiKt Frame, $\langle U, H, R, S \rangle$, consists of a set U, a relation H on U which is reflexive and transitive, and two relations R and S on U that satisfy $R \,;\, H \subseteq H \,;\, R$ and $\breve{H} \,;\, S \subseteq S \,;\, \breve{H}$.*

The condition $R \,;\, H \subseteq H \,;\, R$ is strictly more general than $R = H \,;\, R \,;\, H$. For example, if $U = \{a, b\}$ and $H = \{(a, a), (a, b), (b, b)\}$ then taking R to be $\{(b, a), (b, b)\}$ we find that $R \,;\, H \subseteq H \,;\, R$ holds but not $R = H \,;\, R \,;\, H$. However, we shall see shortly that this additional generality is not essential. The semantics interprets formulae by H-sets and the modalities are defined as follows.

$$[\![\Diamond\alpha]\!] = [\![\alpha]\!] \oplus \breve{S} \quad [\![\Box\alpha]\!] = (H\,;R) \ominus [\![\alpha]\!] \quad [\![\blacklozenge\alpha]\!] = [\![\alpha]\!] \oplus R \quad [\![\blacksquare\alpha]\!] = (H\,;\breve{S}) \ominus [\![\alpha]\!]$$

Since $[\![\alpha]\!]$ is an H-set $[\![\alpha]\!] \oplus \breve{S} = [\![\alpha]\!] \oplus (H \,;\, \breve{S})$ and $[\![\alpha]\!] \oplus R = [\![\alpha]\!] \oplus (H \,;\, R)$. Thus the only accessibility relations needed in the semantics are $R' = H \,;\, R$ and $S' = H \,;\, \breve{S}$. The following lemma shows that the constraints on R and S are equivalent to R' and S' being stable with respect to H.

Lemma 9. *Let U be any set, let H be any pre-order on U, and let S be any binary relation on U. Then the following are equivalent.*

1. *$S = H \,;\, S \,;\, H$.*
2. *There is some relation $R \subseteq U \times U$ such that $S = H \,;\, R$ and $R \,;\, H \subseteq H \,;\, R$.*

Thus we can rephrase the semantics in [GPT10, p26] as

1. A frame $\langle U, H, R', S' \rangle$, consists of a set U, a pre-order H on U, and two stable relations R' and S' on U. A valuation assigns an H-set to each propositional variable, and the connectives are interpreted as for BISKT.
2. The semantics of the modal operators is:

$$[\![\Diamond\alpha]\!] = [\![\alpha]\!] \oplus S' \quad [\![\Box\alpha]\!] = R' \ominus [\![\alpha]\!] \quad [\![\blacklozenge\alpha]\!] = [\![\alpha]\!] \oplus R' \quad [\![\blacksquare\alpha]\!] = S' \ominus [\![\alpha]\!]$$

There is no relationship between R' and S' whereas the approach in Section 3 above is the special case in which $S' = \smallsmile R'$. The significance of our semantics for BISKT is that we are able to define all four modalities from a single accessibility relation. In developing, for example, a bi-intuitionistic modal logic of time it seems reasonable that the forward procession of time should not be completely unrelated to the backwards view looking into the past. It is the left converse operation on stable relations that allows us to state what appears to be the appropriate connection between the two directions.

6   Conclusion

Motivated by the theory of relations on graphs and applications to spatial reasoning, we have presented a bi-intuitionistic logic BISKT with tense operators. The need to interact well with a graph structure, and more generally with a preorder, meant that our accessibility relations needed to be stable. The stability condition itself is not novel, but stable relations are not closed under the usual converse operation and our work is the first to show that the weaker left converse, which does respect stability, can be used to define semantics for modalities.

In contrast to other intuitionistic and bi-intuitionistic tense logics where all the modal operators are independent of each other, in BISKT the white diamond can be defined in terms of the white box, although not conversely. Dually the black box can be defined from the black diamond, also by using a pairing of intuitionistic negation and dual intuitionistic negation.

We showed BISKT is decidable and has the effective finite model property. The proof is via a reduction to the guarded fragment, which also gives an upper complexity bound of EXPTIME. Future work includes giving a tight complexity result for BISKT.

We have presented a tableau calculus for BISKT, which was shown to be sound, complete and terminating. This was obtained and refined using the tableau synthesis methodology and refinement techniques of [ST11, TS13]. We used a prover generated with the MetTeL tool [Met, TSK12] to analyse the logic and investigate the properties that hold in it, including relationship between different logical operators.

As the reduction to the guarded fragment is via an embedding into a multi-modal logic, this provides another route to obtaining a tableau calculus and tableau prover for BISKT. Preliminary experiments on about 120 problems have however shown that the performance of this alternative prover is not better than the prover for the calculus presented in this paper. The reason is that the rules are more fine grained and less tailor-made, which allows fewer rule refinements. However deeper analysis and more experiments are needed.

Other future work includes extending BISKT with modalities based on the right converse operator.

References

[Met]   MetTeL website, http://www.mettel-prover.org
[Av02]   Aiello, M., van Benthem, J.: A Modal Walk Through Space. Journal of
           Applied Non-Classical Logics 12, 319–363 (2002)
[AO07]   Aiello, M., Ottens, B.: The Mathematical Morpho-Logical View on Reason-
           ing about Space. In: Veloso, M.M. (ed.) IJCAI 2007, pp. 205–211. AAAI
           Press (2007)
[ANvB98]   Andréka, H., Németi, I., van Benthem, J.: Modal Languages and Bounded
           Fragments of Predicate Logic. Journal of Philosophical Logic 27, 217–274
           (1998)
[Bén00]   Bénabou, J.: Distributors at Work (2000),
           http://www.mathematik.tu-darmstadt.de/ streicher/FIBR/DiWo.pdf
[Blo02]   Bloch, I.: Modal Logics Based on Mathematical Morphology for Qualita-
           tive Spatial Reasoning. Journal of Applied Non-Classical Logics 12, 399–423
           (2002)
[BHR07]   Bloch, I., Heijmans, H.J.A.M., Ronse, C.: Mathematical Morphology. In:
           Aiello, M., Pratt-Hartmann, I., van Benthem, J. (eds.) Handbook of Spatial
           Logics, pp. 857–944. Springer (2007)
[GPT10]   Goré, R., Postniece, L., Tiu, A.: Cut-elimination and Proof Search for Bi-
           Intuitionistic Tense Logic. arXiv e-Print 1006.4793v2 (2010)
[Grä99]   Grädel, E.: On the restraining power of guards. Journal of Symbolic Logic 64,
           1719–1742 (1999)
[HV93]   Heijmans, H., Vincent, L.: Graph Morphology in Image Analysis. In:
           Dougherty, E.R. (ed.) Mathematical Morphology in Image Processing, pp.
           171–203. Marcel Dekker (1993)
[Rau74]   Rauszer, C.: A formalization of the propositional calculus of H-B logic. Stu-
           dia Logica 33, 23–34 (1974)
[RZ96]   Reyes, G.E., Zolfaghari, H.: Bi-Heyting Algebras, Toposes and Modalities.
           Journal of Philosophical Logic 25, 25–43 (1996)
[SH07]   Schmidt, R.A., Hustadt, U.: The Axiomatic Translation Principle for Modal
           Logic. ACM Transactions on Computational Logic 8, 1–55 (2007)
[ST08]   Schmidt, R.A., Tishkovsky, D.: A General Tableau Method for Deciding
           Description Logics, Modal Logics and Related First-Order Fragments. In:
           Armando, A., Baumgartner, P., Dowek, G. (eds.) IJCAR 2008. LNCS,
           vol. 5195, pp. 194–209. Springer, Heidelberg (2008)
[ST11]   Schmidt, R.A., Tishkovsky, D.: Automated Synthesis of Tableau Calculi.
           Logical Methods in Computer Science 7, 1–32 (2011)
[ST13]   Schmidt, R.A., Tishkovsky, D.: Using Tableau to Decide Description Logics
           with Full Role Negation and Identity. To appear in ACM Transactions on
           Computational Logic (2013)
[Ser82]   Serra, J.: Image Analysis and Mathematical Morphology. Academic Press
           (1982)
[Ste12]   Stell, J.G.: Relations on Hypergraphs. In: Kahl, W., Griffin, T.G. (eds.)
           RAMICS 2012. LNCS, vol. 7560, pp. 326–341. Springer, Heidelberg (2012)
[TSK12]   Tishkovsky, D., Schmidt, R.A., Khodadadi, M.: The Tableau Prover Gen-
           erator MetTeL2. In: del Cerro, L.F., Herzig, A., Mengin, J. (eds.) JELIA
           2012. LNCS, vol. 7519, pp. 492–495. Springer, Heidelberg (2012)

[TS13]   Tishkovsky, D., Schmidt, R.A.: Refinement in the Tableau Synthesis Framework. arXiv e-Print 1305.3131v1 (2013)
[Wan08]   Wansing, H.: Constructive negation, implication, and co-implication. Journal of Applied Non-Classical Logics 18, 341–364 (2008)
[ZWC01]   Zakharyaschev, M., Wolter, F., Chagrov, A.: Advanced Modal Logic. In: Gabbay, D., Guenthner, F. (eds.) Handbook of Philosophical Logic, vol. 3, pp. 83–266. Kluwer (2001)

Nominal Sets over Algebraic Atoms*

Joanna Ochremiak

Institute of Informatics, University of Warsaw, Warsaw, Poland
ochremiak@mimuw.edu.pl

Abstract. Nominal sets, introduced to Computer Science by Gabbay and Pitts, are useful for modeling computation on data structures built of atoms that can only be compared for equality. In certain contexts it is useful to consider atoms equipped with some nontrivial structure that can be tested in computation. Here, we study nominal sets over atoms equipped with both relational and algebraic structure. Our main result is a representation theorem for orbit-finite nominal sets over such atoms, a generalization of a previously known result for atoms equipped with relational structure only.

1 Introduction

Nominal sets [Pit13] are sets whose elements depend on *atoms* – elements of a fixed countably infinite set $\mathbb{A}$. Examples include:

- the set $\mathbb{A}$ itself,
- the set $\mathbb{A}^n$ of n-tuples of atoms,
- the set $\mathbb{A}^{(n)}$ of n-tuples of distinct atoms,
- the set $\mathbb{A}^*$ of finite words over $\mathbb{A}$,
- the set of graphs edge-labeled with atoms, etc.

Any such set is acted upon by permutations of the atoms in a natural way, by renaming all atoms that appear in it. We require the result of applying a permutation of atoms to each element of a nominal set to be determined by a finite set of atoms, called a *support* of this element. Sets $\mathbb{A}$, $\mathbb{A}^n$ and $\mathbb{A}^{(n)}$ are nominal, since each tuple of atoms is supported by the finite set of atoms that appear in it. Another example of a nominal set is $\mathbb{A}^*$, where a word is supported by the set of its letters. The set of all cofinite subsets of atoms is also nominal: one of the supports of a cofinite set is simply its complement.

Nominal sets were introduced in 1922 by Fraenkel as an alternative model of set theory. In this context they were further studied by Mostowski, which is why they are sometimes called Fraenkel-Mostowski sets. Rediscovered for the computer science community in the 90s by Gabbay and Pitts [GP02], nominal sets gained a lot of interest in semantics. In this application area atoms, whose

---

* This work was supported by the Polish National Science Centre (NCN) grant 2012/07/B/ST6/01497.

P. Höfner et al. (Eds.): RAMiCS 2014, LNCS 8428, pp. 429–445, 2014.

only structure is equality, are used to describe variable names in programs or logical formulas. Permutations of atoms correspond to renaming of variables.

In parallel, nominal sets were studied in automata theory [Pis99], under the name of *named sets with symmetries*[1], and used to model computation over infinite alphabets that can only be accessed in a limited way.

An example of such a model that predates nominal sets are Francez-Kaminski register automata [KF94] that, over the alphabet of atoms $\mathbb{A}$, recognize languages such as "the first letter does not appear any more":

$$L = \{a_1 \dots a_n \ : \ a_1 \neq a_i \text{ for all } i > 1\}.$$

To this end, after reading the first letter the automaton stores it in its register. Then it reads the rest of the input word and rejects if any letter equals the one in the register. The automaton has one register and three states: $0, 1, \top$, where 0 is initial and $\top$ is rejecting. Alternatively, in the framework of nominal sets, this may be modelled as an automaton with an infinite state space $\{0, \top\} \cup \mathbb{A}$ and the transition relation defined by the graph:

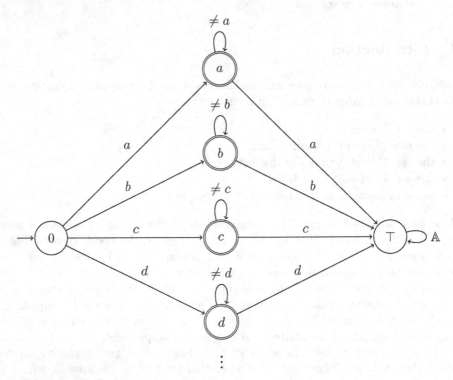

In [BKL11] and [BKL] Bojańczyk, Klin and Lasota showed that automata over infinite alphabets whose letters are built of atoms that can only be tested

---

[1] The equivalence between named sets and nominal sets was proven in [FS06] and [GMM06].

for equality, are essentially automata in the category of nominal sets. As a continuation of this line of research, Turing machines that operate over such alphabets were studied in [BKLT13].

The key notion in the above constructions is *orbit-finiteness* – a more relaxed notion of finiteness provided by nominal sets. A nominal set is considered orbit-finite if it has finitely many elements, up to permutations of atoms. The set 𝔸 of atoms is orbit-finite: in fact, it has only one orbit. This single orbit can be represented by any atom a, because a can be mapped to every other atom by a suitable permutation. Another example of an orbit-finite set is the set of configurations of any register automaton. The automaton described above has infinitely many configurations. However, there are only three of them up to permutations: the initial state 0 with an empty register, state 1 with an atom stored in the register and the rejecting state ⊤ with an empty register.

Atoms turn out to be a good framework to speak of data that can be accessed only in a limited way. Nominal sets, as defined in [Pit13], intuitively correspond to data with no structure except for equality. To model a device with more access to its alphabet one may use atoms with additional structure. An example here are atoms with total order. A typical language recognized by a nominal automaton over such atoms is the language of all monotonic words:

$$L = \{a_1 \ldots a_n \ : \ a_i < a_j \text{ for all } i < j\}.$$

In [BKL] atoms are modelled as countable relational structures. In this setting the definition of a nominal set remains essentially the same. The only change is that we consider only those permutations of atoms that preserve and reflect the relational structure, i.e., we talk about *automorphisms* of atoms. A choice of such automorphisms is called an *atom symmetry*.

Since interesting orbit-finite nominal sets are usually infinite (for example, the transition relation of the automaton above), to manipulate them effectively we need to represent them in a finite way. In [BKL] Bojańczyk et al. provide such a concrete, finite representation of orbit-finite nominal sets for atoms that are *homogeneous* relational structures over finite vocabularies (the corresponding atom symmetries are called *Fraïssé symmetries*). Each element of a nominal set is represented as a finite substructure of atoms modulo some group of local automorphisms. There are two technical assumptions needed for the theorem to hold: existence of least supports and so-called *fungibility* (meaning roughly that one can always find an automorphism that fixes a concrete substructure of atoms without fixing other atoms).

In some contexts, a relational structure of atoms is not enough. In [BL12] Bojańczyk and Lasota use the theory of nominal sets to obtain a machine-independent characterization of the languages recognized by deterministic timed automata. To do so they introduce atoms with a total order and a function symbol +1 and they relate deterministic timed automata to automata over these *timed atoms*. An example of a language recognized by such a nominal automaton is the set of all monotonic words where the distance between any two consecutive letters is smaller than 1:

$$L = \{a_1 \ldots a_n \ : \ a_{i-1} + 1 > a_i > a_{i-1} \text{ for all } i > 1\}.$$

One could easily think of other types of potentially useful functional dependencies on atoms, such as composing two atoms to get another atom. It is therefore natural to ask if the representation theorem can be generalized to cover atoms with algebraic as well as relational structure. This paper gives a positive answer to this question.

The proof of the representation theorem for atoms with both relational and function symbols follows the same pattern as the proof for relational structures given in [BKL]. There are, however, some subtleties, since instead of finite supports one has to consider *finitely generated supports* (which can be infinite) and, as a result, the notion of fungibility becomes less clear.

The structure of this paper is as follows. In Section 2 we define atom symmetries and introduce the category of G-sets. In Section 3, following [BKL11,BKL], we focus on the theory of nominal sets for Fraïssé symmetries, introduce the category of nominal sets, and explain the notion of the least finitely generated support. In Section 4 we define the property of fungibility and finally prove the representation theorem for fungible Fraïssé symmetries that admit least finitely generated supports.

2   Atom Symmetries

A (right) *group action* of a group G on a set X is a binary operator $\cdot : X \times G \to X$ that satisfies following conditions:

for all $x \in X$ $x \cdot e = x$, where e is the neutral element of G,

for all $x \in X$ and $\pi, \sigma \in G$ $x \cdot (\pi\sigma) = (x \cdot \pi) \cdot \sigma$.

The set X equipped with such an action is called a *G-set*.

Example 2.1. For a set X let $\mathrm{Sym}(X)$ denote the symmetric group on X, i.e., the group of all bijections of X. Take any subgroup G of the symmetric group $\mathrm{Sym}(X)$. There is a natural action of the group G on the set X defined by $x \cdot \pi = \pi(x)$.

Definition 2.2. An *atom symmetry* $(\mathbb{A}, G)$ is a set $\mathbb{A}$ of *atoms*, together with a subgroup $G \leq \mathrm{Sym}(\mathbb{A})$ of the symmetric group on $\mathbb{A}$.

Example 2.3. Examples of atom symmetries include:

- the *equality symmetry*, where $\mathbb{A}$ is a countably infinite set, say the natural numbers, and $G = \mathrm{Sym}(\mathbb{A})$ contains all bijections of $\mathbb{A}$,
- the *total order symmetry*, where $\mathbb{A} = \mathbb{Q}$ is the set of rational numbers, and G is the group of all monotone permutations,

– the *timed symmetry*, where $\mathbb{A} = \mathbb{Q}$ is the set of rational numbers, and G is the group of all permutations of rational numbers that preserve the order relation $\leq$ and the successor function $x \mapsto x + 1$.[2]

For any element x of a G-set X the set

$$x \cdot G = \{x \cdot \pi \mid \pi \in G\} \subseteq X$$

is called the *orbit* of x. Orbits form a partition of X. The set X is called *orbit-finite* if the partition has finitely many parts. Each of the orbits can be perceived as a separate G-set. Therefore we can treat any G-set X as a disjoint union of its orbits.

Example 2.4. For any atom symmetry $(\mathbb{A}, G)$ the action of G on $\mathbb{A}$ extends pointwise to an action of G on the set of tuples $\mathbb{A}^n$. In the equality symmetry, the set $\mathbb{A}^2$ has two orbits:

$$\{(a,a) \mid a \in \mathbb{A}\} \qquad \{(a,b) \mid a \neq b \in \mathbb{A}\}.$$

In the timed symmetry, the set $\mathbb{A}^2$ is not orbit-finite. Notice that for any $a \in \mathbb{Q}$ each of the elements $(a, a+1)$, $(a, a+2), \ldots$ is in a different orbit.

Let X be a G-set. A subset $Y \subseteq X$ is *equivariant* if $Y \cdot \pi = Y$ for every $\pi \subset G$, i.e., it is preserved under group action. Considering a pointwise action of a group G on the Cartesian product $X \times Y$ of two G-sets X, Y we can define an *equivariant relation* $R \subseteq X \times Y$. In the special case when the relation is a function $f \colon X \to Y$ we obtain a following definition of an *equivariant function*

$$f(x \cdot \pi) = f(x) \cdot \pi \text{ for any } x \in X,\ \pi \in G.$$

The identity function on any G-set is equivariant, and the composition of two equivariant functions is again equivariant, therefore for any group G, G-sets and equivariant functions form a category, called G-**Set**.

Definition 2.5. For any x in a G-set X, the group

$$G_x = \{\pi \in G \mid x \cdot \pi = x\} \leq G$$

is called the *stabilizer* of x.

Lemma 2.6. *If $G_x \leq G$ is the stabilizer of an element x of a G-set X then $G_{x \cdot \pi} = \pi^{-1} G_x \pi$ for each $\pi \in G$.*

Proof. Obviously $\pi^{-1} G_x \pi \subseteq G_{x \cdot \pi}$. On the other hand, $x \cdot (\pi \sigma \pi^{-1}) = x$ for any $\sigma \in G_{x \cdot \pi}$. Hence $\pi G_{x \cdot \pi} \pi^{-1} \subseteq G_x$, which means that $G_{x \cdot \pi} \subseteq \pi^{-1} G_x \pi$. As a result $G_{x \cdot \pi} = \pi^{-1} G_x \pi$, as required.

---

[2] The timed symmetry was originally defined in [BL12] for $\mathbb{A} = \mathbb{R}$. Considering the rational numbers instead of the reals makes little difference but is essential for our purposes. To fit the Fraïssé theory we need the set of atoms to be countable.

Proposition 2.7. *Let x be an element of a single-orbit G-set X. For any G-set Y equivariant functions from X to Y are in bijective correspondence with those elements $y \in Y$ for which $G_x \leq G_y$.*

Proof. Given an equivariant function $f\colon X \to Y$, let $y = f(x)$. If $\pi \in G_x$ then

$$y \cdot \pi = f(x) \cdot \pi = f(x \cdot \pi) = f(x) = y,$$

hence $G_x \leq G_y$. On the other hand, given $y \in Y$ such that $G_x \leq G_y$, define a function $f\colon X \to Y$ by $f(x \cdot \pi) = y \cdot \pi$. Function f is well-defined. Indeed, if $x \cdot \pi = x \cdot \sigma$ then $\pi\sigma^{-1} \in G_x \subseteq G_y$, hence $y \cdot \pi = y \cdot \sigma$.

It is easy to check that the two above constructions are mutually inverse.

3   Fraïssé Symmetries

In the following, we shall consider atom symmetries that arise as automorphism groups of algebraic structures. Such symmetries behave particularly well if those structures arise as so-called *Fraïssé limits*, which we introduce in this sections.

3.1   Fraïssé Limits

An *algebraic signature* is a set of relation and function names together with (finite) arities. We will consider structures over a fixed finite algebraic signature. For two structures $\mathfrak{A}$ and $\mathfrak{B}$, an *embedding* $f\colon \mathfrak{A} \to \mathfrak{B}$ is an injective function from the carrier of $\mathfrak{A}$ to the carrier of $\mathfrak{B}$ that preserves and reflects all relations and functions in the signature.

Definition 3.1. A class $\mathcal{K}$ of finitely generated structures over some fixed algebraic signature is called a *Fraïssé class* if it:

- is closed under isomorphisms as well as finitely generated substructures and has countably many members up to isomorphism,
- has *joint embedding property*: if $\mathfrak{A}, \mathfrak{B} \in \mathcal{K}$ then there is a structure $\mathfrak{C}$ in $\mathcal{K}$ such that both $\mathfrak{A}$ and $\mathfrak{B}$ are embeddable in $\mathfrak{C}$,
- has *amalgamation*: if $\mathfrak{A}, \mathfrak{B}, \mathfrak{C} \in \mathcal{K}$ and $f_{\mathfrak{B}}\colon \mathfrak{A} \to \mathfrak{B}$, $f_{\mathfrak{C}}\colon \mathfrak{A} \to \mathfrak{C}$ are embeddings then there is a structure $\mathfrak{D}$ in $\mathcal{K}$ together with two embeddings $g_{\mathfrak{B}}\colon \mathfrak{B} \to \mathfrak{D}$ and $g_{\mathfrak{C}}\colon \mathfrak{C} \to \mathfrak{D}$ such that $g_{\mathfrak{B}} \circ f_{\mathfrak{B}} = g_{\mathfrak{C}} \circ f_{\mathfrak{C}}$.

Examples of Fraïssé classes include:

- all finite structures over an empty signature, i.e., finite sets,
- finite total orders,
- all finite structures over a signature with a single binary relation symbol, i.e., directed graphs,
- finite Boolean algebras,
- finite groups,
- finite fields of characteristic p.

Classes that are not Fraïssé include:

- total orders of size at most 7 – due to lack of amalgamation,
- all finite fields – due to lack of joint embedding property.

Some Fraïssé classes admit a stronger version of amalgamation property. We say that a class $\mathcal{K}$ has *strong amalgamation* if it has amalgamation and moreover, $g_{\mathfrak{B}} \circ f_{\mathfrak{B}}(\mathfrak{A}) = g_{\mathfrak{C}} \circ f_{\mathfrak{C}}(\mathfrak{A}) = g_{\mathfrak{B}}(\mathfrak{B}) \cap g_{\mathfrak{C}}(\mathfrak{C})$. It means that we can make amalgamation without identifying any more points than absolutely necessary.

Example 3.2. All the Fraïssé classes listed above, except for the class of finite fields of characteristic p, have the strong amalgamation property.

The *age* of a structure $\mathfrak{U}$ is the class $\mathcal{K}$ of all structures isomorphic to finitely generated substructures of $\mathfrak{U}$. A structure $\mathfrak{U}$ is *homogeneous* if any isomorphism between finitely generated substructures of $\mathfrak{U}$ extends to an automorphism of $\mathfrak{U}$. The following theorem says that for a Fraïssé class $\mathcal{K}$ there exists a so-called *universal* homogeneous structure of age $\mathcal{K}$. We shall refer to it also as the *Fraïssé limit* of the class $\mathcal{K}$ (see e.g. [Hod93]).

Theorem 3.3. *For any Fraïssé class $\mathcal{K}$ there exists a unique, up to isomorphism, countable (finite or infinite) structure $\mathfrak{U}_{\mathcal{K}}$ such that $\mathcal{K}$ is the age of $\mathfrak{U}_{\mathcal{K}}$ and $\mathfrak{U}_{\mathcal{K}}$ is homogeneous.*

Example 3.4. The Fraïssé limit of the class of finite total orders is $\langle \mathbb{Q}, \leq \rangle$. For finite Boolean algebras it is the countable atomless Boolean algebra.

A structure $\mathfrak{U}$ is called *weakly homogeneous* if for any two finitely generated substructures $\mathfrak{A}, \mathfrak{B}$ of $\mathfrak{U}$, such that $\mathfrak{A} \subseteq \mathfrak{B}$, any embedding $f_{\mathfrak{A}} : \mathfrak{A} \to \mathfrak{U}$ extends to an embedding $f_{\mathfrak{B}} : \mathfrak{B} \to \mathfrak{U}$. It turns out that a countable structure $\mathfrak{U}$ is homogeneous if and only if it is weakly homogeneous (see [Hod93]). Hence, one way to obtain a Fraïssé class $\mathcal{K}$ is to take a weakly homogeneous, countable structure $\mathfrak{U}$ and simply consider its age.

Fact 3.5. *Every countable, weakly homogeneous structure $\mathfrak{U}$ is a Fraïssé limit of its age.*

Example 3.6. Consider an algebraic signature with a single binary relation symbol $\leq$, and two unary function symbols $+1$ and -1. It is not difficult to see that the structure $\langle \mathbb{Q}, \leq, +1, -1 \rangle$ is countable and weakly homogeneous. Therefore it is the Fraïssé limit of its age. Observe that its automorphism group contains precisely those permutations of rational numbers which are monotone and preserve the successor funtion $x \mapsto x + 1$.

From a Fraïssé class $\mathcal{K}$ we obtain an atom symmetry $(\mathbb{A}_{\mathcal{K}}, G_{\mathcal{K}})$, where $\mathbb{A}_{\mathcal{K}}$ is the carrier of $\mathfrak{U}_{\mathcal{K}}$ and $G_{\mathcal{K}} = \mathrm{Aut}(\mathfrak{U}_{\mathcal{K}})$ is its group of automorphisms. Such an atom symmetry is called a *Fraïssé symmetry*.

For simplicity we frequently identify the elements of age $\mathcal{K}$ with finitely generated substructures of $\mathfrak{U}_{\mathcal{K}}$.

Example 3.7. All symmetries in Example 2.3 are Fraïssé symmetries. The equality symmetry arises from the class of all finite sets, the total order symmetry – from the class of finite total orders and the timed symmetry – from the class of all finitely generated substructures of $\langle \mathbb{Q}, \leq, +1, -1 \rangle$ (see Example 3.6).

The timed symmetry was originally defined based on a structure without the unary function -1. In the context of [BL12] adding -1 to the signature does not make any difference since the automorphism groups of both structures are the same. As we will show, thanks to this slight modification the timed symmetry satisfies all the conditions of our representation theorem.

3.2   Least Supports

From now on, we focus on G-sets for groups arising from Fraïssé symmetries. Consider such a symmetry $(\mathbb{A}_\mathcal{K}, G_\mathcal{K})$ and a $G_\mathcal{K}$-set X. By $\pi|_C$ we denote the restriction of a permutation π to a subset C of its domain.

Definition 3.8. A set $C \subseteq \mathbb{A}_\mathcal{K}$ *supports* an element $x \in X$ if $x \cdot \pi = x$ for all $\pi \in G_\mathcal{K}$ such that $\pi|_C = \text{id}|_C$. A $G_\mathcal{K}$-set is *nominal* in the symmetry $(\mathbb{A}_\mathcal{K}, G_\mathcal{K})$ if every element in the set is supported by the carrier of a finitely generated substructure $\mathfrak{A}$ of $\mathfrak{U}_\mathcal{K}$. We call $\mathfrak{A}$ a *finitely generated support* of x.

Nominal $G_\mathcal{K}$-sets and equivariant functions between them form a category $G_\mathcal{K}$-**Nom** which is a full subcategory of $G_\mathcal{K}$-**Set**. When the symmetry $(\mathbb{A}_\mathcal{K}, G_\mathcal{K})$ under consideration is the equality symmetry, the category $G_\mathcal{K}$-**Nom** coincides with the category **Nom** defined in [Pit13].

Example 3.9. For any Fraïssé symmetry $(\mathbb{A}_\mathcal{K}, G_\mathcal{K})$ the sets $\mathbb{A}_\mathcal{K}$ and $\mathbb{A}_\mathcal{K}^n$ are nominal. A tuple $(d_1, ..., d_n)$ is supported by the structure generated by its elements.

Lemma 3.10. *The following conditions are equivalent:*
(1) *C supports an element $x \in X$;*
(2) *for any $\pi, \sigma \in G_\mathcal{K}$ if $\pi|_C = \sigma|_C$ then $x \cdot \pi = x \cdot \sigma$.*

Proof. For the implication (1) $\Longrightarrow$ (2), notice that if $\pi|_C = \sigma|_C$, then $\pi\sigma^{-1}$ acts as identity on C, hence $x \cdot \pi\sigma^{-1} = x$ and $x \cdot \pi = x \cdot \sigma$, as required. The opposite implication follows immediately from the definition if we take $\sigma = \text{id}$.

It is easy to see that if an element $x \in X$ has a finitely generated support $\mathfrak{A}$ then it is also supported by the finite set C of its generators. Thus we can equivalently require x to be finitely supported.

Fact 3.11. *A $G_\mathcal{K}$-set is nominal if and only if its every element has a finite support.*

Example 3.12. Consider the structure $\langle \mathbb{Q}, \leq, +1 \rangle$. It is countable and weakly homogeneous, and therefore gives rise to a Fraïssé symmetry. This symmetry is almost the same as the timed symmetry (the carriers and automorphim groups of both $\langle \mathbb{Q}, \leq, +1 \rangle$ and $\langle \mathbb{Q}, \leq, +1, -1 \rangle$ are the same). It has, though, some unwanted properties. Notice that an automorphism π of $\langle \mathbb{Q}, \leq, +1 \rangle$ which preserves an atom $a \in \mathbb{Q}$ necessarily preserves also $a + i$ for any integer i. Therefore, if an element x of a nominal set is supported by a substructure generated e.g. by $\{1, 30\frac{1}{2}, 100\frac{5}{7}\}$ it is also supported by its proper substructure generated by $\{1000, 300\frac{1}{2}, 105\frac{5}{7}\}$. Hence, in this case for any finitely generated support $\mathfrak{A}$ of an element x one can find a finitely generated substructure $\mathfrak{B}$, which is properly contained in $\mathfrak{A}$ and still supports x.

An element of a nominal set has many supports. In particular, supports are closed under adding atoms. If every element of a nominal set X has a unique least finitely generated support, we say that X is *supportable*. As shown in Example 3.12 it is not always the case. It turns out that to check if a single-orbit nominal set is supportable, one just needs to find out if any element of the set has the least finitely generated support.

Lemma 3.13. *If $\mathfrak{A} \subseteq \mathfrak{U}_{\mathcal{K}}$ is the least finitely generated support of an element $x \in X$, then $\mathfrak{A} \cdot \pi$ is the least finitely generated support of $x \cdot \pi$ for any $\pi \in G_{\mathcal{K}}$.*

Proof. First we prove that $\mathfrak{A} \cdot \pi$ supports $x \cdot \pi$. Indeed, if an arbitrary $\rho \in G_{\mathcal{K}}$ is an identity on $\mathfrak{A} \cdot \pi$, then $\pi \rho \pi^{-1}$ is an identity on $\mathfrak{A}$, hence $x \cdot (\pi \rho \pi^{-1}) = x$. As a result $(x \cdot \pi) \cdot \rho = x \cdot \pi$, as required.

Now let $\mathfrak{B} \subseteq \mathfrak{U}_{\mathcal{K}}$ be any finitely generated support of $x \cdot \pi$. We need to show that $\mathfrak{A} \cdot \pi \subseteq \mathfrak{B}$. A reasoning similar to the one above shows that $\mathfrak{B} \cdot \pi^{-1}$ supports x, from which we obtain $\mathfrak{A} \subseteq \mathfrak{B} \cdot \pi^{-1}$. Therefore, since π is a bijection, $\mathfrak{A} \cdot \pi \subseteq \mathfrak{B}$. $\qquad\blacksquare$

Definition 3.14. A Fraïssé symmetry $(\mathbb{A}_{\mathcal{K}}, G_{\mathcal{K}})$ is *supportable* if every nominal $G_{\mathcal{K}}$-set is supportable.

We call a structure $\mathfrak{U}$ *locally finite* if all its finitely generated substructures are finite. Notice that if the universal structure $\mathfrak{U}_K$ is locally finite then being supportable is equivalent to finitely generated supports being closed under finite intersections. The same holds under the weaker assumption that any finitely generated structure has only finitely many finitely generated substructures.

Example 3.15. If we have only relation symbols in the signature it is obvious that any finitely generated structure is finite. One can prove that in the equality symmetry the intersection of two supports is a support itself. Hence the equality symmetry is supportable. The same holds for the total order symmetry. Both facts are proved e.g. in [BKL].

From Example 3.12 we learned that the symmetry arising from the structure $\langle \mathbb{Q}, \leq, +1 \rangle$ is not supportable (even though the finitely generated supports are closed under finite intersections). In the structure $\langle \mathbb{Q}, \leq, +1, -1 \rangle$ all the elements $a + i$ are bound together and, as a result, we obtain a Fraïssé symmetry that is supportable.

Proposition 3.16. *The timed symmetry is supportable.*

Proof. Notice that any finitely generated substructure of $\langle \mathbb{Q}, \leq, +1, -1 \rangle$ has only finitely many substructures. Hence it is enough to show that finitely generated supports are closed under finite intersections.

Take any two finitely generated substructures $\mathfrak{A}$, $\mathfrak{B}$ of $\langle \mathbb{Q}, \leq, +1, -1 \rangle$. Let A and B be the sets of elements of $\mathfrak{A}$ and $\mathfrak{B}$ that are contained in the interval $[0, 1)$. These are (finite) sets of generators. Moreover, the structure $\mathfrak{A} \cap \mathfrak{B}$ is generated by $A \cap B$. Hence, it is enough to show that if an automorphism π acts as identity on $A \cap B$, then π can be decomposed as

$$\pi = \sigma_1 \tau_1 \sigma_2 \tau_2 ... \sigma_n \tau_n,$$

where σ_i acts as identity on A and τ_i acts as identity on B. Indeed, since each σ_i, τ_i acts as identity on $\mathfrak{A}$ and $\mathfrak{B}$ respectively, we have $x \cdot \sigma_i = x$ and $x \cdot \tau_i = x$. As a result $x \cdot \pi = x$.

Let l be the smallest and h the biggest element of the set $A \cup B$. Notice that $h - l < 1$. Take two different open intervals (l_A, h_A), (l_B, h_B) of length 1 such that

$$[l, h] \subseteq (l_A, h_A) \text{ and } [l, h] \subseteq (l_B, h_B).$$

Now, consider sets $A' = A \cup \{l_A, h_A\}$, $B' = B \cup \{l_B, h_B\}$. Take an automorphism π that acts as identity on $A \cap B = A' \cap B'$. Obviously π is a monotone bijection of the set of rational numbers. Therefore, since the total order symmetry is supportable,

$$\pi = \sigma_1' \tau_1' \sigma_2' \tau_2' ... \sigma_n' \tau_n',$$

where σ_i', τ_i' are monotone bijections of $\mathbb{Q}$ and σ_i' act as identity on A', τ_i' act as identity on B'. For each of the permutations σ_i', τ_i' take an automorphism σ_i, τ_i of the universal structure $\langle \mathbb{Q}, \leq, +1, -1 \rangle$, such that

$$\sigma_i'|_{(l_A, h_A)} = \sigma_i|_{(l_A, h_A)}, \quad \tau_i'|_{(l_B, h_B)} = \tau_i|_{(l_B, h_B)}.$$

Then σ_i act as identity on A and τ_i act as identity on B. Moreover $\pi = \sigma_1 \tau_1 \sigma_2 \tau_2 ... \sigma_n \tau_n$, as required.

4   Structure Representation

For any $C \subseteq \mathbb{A}$ and $G \leq \mathrm{Sym}(\mathbb{A})$, the restriction of G to C is defined by

$$G|_C = \{ \pi|_C \mid \pi \in G, \ C \cdot \pi = C \} \leq \mathrm{Sym}(C).$$

Lemma 4.1. *Let $\mathfrak{A} \in \mathcal{K}$ be a finitely generated structure. The set of embeddings $u \colon \mathfrak{A} \to \mathfrak{U}_{\mathcal{K}}$ with the $G_{\mathcal{K}}$-action defined by composition:*

$$u \cdot \pi = u\pi$$

is a single-orbit nominal set.

Proof. First notice that any embedding $u\colon \mathfrak{A} \to \mathfrak{U}_K$ is supported by its image $u(\mathfrak{A})$. Indeed, if an automorphism $\pi \in G_K$ is an identity on $u(\mathfrak{A})$ then obviously $u \cdot \pi = u$. Hence the set of embeddings is a nominal set. Now take any two embeddings u and v. The images $u(\mathfrak{A})$, $v(\mathfrak{A})$ are finitely generated isomorphic substructures of $\mathfrak{U}_K$. By extending any isomorphism between $u(\mathfrak{A})$ and $v(\mathfrak{A})$, we obtain an automorphism $\pi \in G_K$ such that $u \cdot \pi = v$.

As we shall show now, in a supportable symmetry $(\mathbb{A}_K, G_K)$ every single-orbit nominal set is isomorphic to one of the above form, quotiented by some equivariant equivalence relation.

Notice that the quotient of a G-set by an equivariant equivalence relation R has a natural structure of a G-set, with the action defined as follows:

$$[x]_R \cdot \pi = [x \cdot \pi]_R.$$

It is easy to see that if X has one orbit, then so does the quotient X/R. Moreover, any support C of an element $x \in X$ supports the equivalence class $[x]_R$, hence if X is nominal then X/R is also nominal.

Definition 4.2. A *structure representation* is a finitely generated structure $\mathfrak{A} \in \mathcal{K}$ together with a group of automorphisms $S \leq \mathrm{Aut}(\mathfrak{A})$ (the *local symmetry*). Its *semantics* $[\mathfrak{A}, S]$ is the set of embeddings of $u\colon \mathfrak{A} \to \mathfrak{U}_K$, quotiented by the equivalence relation:

$$u \equiv_S v \Leftrightarrow \exists \tau \in S \ \tau u = v.$$

A G_K-action on $[\mathfrak{A}, S]$ is defined by composition:

$$[u]_S \cdot \pi = [u\pi]_S.$$

Proposition 4.3. *(1)* $[\mathfrak{A}, S]$ *is a single-orbit nominal G_K-set.* *(2)* *If a Fraïssé symmetry* $(\mathbb{A}_K, G_K)$ *is supportable then every single-orbit nominal G_K-set X is isomorphic to some* $[\mathfrak{A}, S]$.

Proof. For (1), use Lemma 4.1. The set of embeddings $u\colon \mathfrak{A} \to \mathfrak{U}_K$ is a single-orbit nominal G_K-set, and so is the quotient $[\mathfrak{A}, S]$.

For (2), take a single-orbit nominal set X and let $H \leq G_K$ be the stabilizer of some element $x \in X$. Put $S = H|_{\mathfrak{A}}$ where $\mathfrak{A} \in \mathcal{K}$ is the least finitely generated support of x. Define $f\colon X \to [\mathfrak{A}, S]$ by $f(x \cdot \pi) = [\pi|_{\mathfrak{A}}]_S$. The function f is well defined: if $x \cdot \pi = x \cdot \sigma$ then $\pi\sigma^{-1} \in H$. As $\mathfrak{A} \cdot \pi\sigma^{-1}$ is the least finitely generated support of $x \cdot \pi\sigma^{-1} = x$, we obtain $\mathfrak{A} \cdot \pi\sigma^{-1} = \mathfrak{A}$. Therefore for $\tau = (\pi\sigma^{-1})|_{\mathfrak{A}} \in S$ we have $\tau\sigma|_{\mathfrak{A}} = \pi|_{\mathfrak{A}}$, hence $[\pi|_{\mathfrak{A}}]_S = [\sigma|_{\mathfrak{A}}]_S$. It is easy to check that f is equivariant.

It remains to show that f is bijective. For injectivity, assume $f(x \cdot \pi) = f(x \cdot \sigma)$. This means that there exists $\tau \in S$ such that $\tau\sigma|_{\mathfrak{A}} = \pi|_{\mathfrak{A}}$, then $(\pi\sigma^{-1})|_{\mathfrak{A}} \in S$, hence $(\pi\sigma^{-1})|_{\mathfrak{A}} = \rho|_{\mathfrak{A}}$ for some $\rho \in H$. Therefore $x \cdot \pi\sigma^{-1} = x \cdot \rho = x$, from which we obtain $x \cdot \pi = x \cdot \sigma$. For surjectivity of f, note that by universality of the structure $\mathfrak{U}_K$ any embedding $u\colon \mathfrak{A} \to \mathfrak{U}_K$ can be extended to an automorphism π of $\mathfrak{U}_K$, for which we have $f(x \cdot \pi) = [u]_S$.

Structures over signatures with no function symbols are called *relational structures*. Structure representation was defined by Bojańczyk et al. in the special case of $\mathfrak{U}_K$ being a relational structure. The proposition above generalizes Proposition 11.7 of [BKL].

Example 4.4. Consider the universal structure $\langle \mathbb{Q}, \leq, +1, -1 \rangle$ and its substructure $\mathfrak{A}$ generated by $\{\frac{1}{3}, \frac{1}{2}, \frac{3}{4}\}$. Notice that mapping one of the generators, say $\frac{1}{2}$, to any element of $\mathfrak{A}$, say $\frac{1}{2} \mapsto 3\frac{3}{4}$, uniquely determines an automorphism π of $\mathfrak{A}$. The automorphism can be seen as a shift. It maps $\frac{1}{3}$ to $3\frac{1}{2}$ and $\frac{3}{4}$ to $4\frac{1}{3}$. This observation leads to the conclusion that $\mathrm{Aut}(\mathfrak{A}) = \mathbb{Z}$. Any subgroup S of $\mathrm{Aut}(\mathfrak{A})$ is therefore isomorphic to $\mathbb{Z}$ and generated by a single automorphism π of the form described above. The same holds for any finitely generated substructure $\mathfrak{A}$. In the case of timed symmetry Proposition 4.3 provides a very nice finite representation of single-orbit nominal sets.

4.1 Fungibility

Even if the symmetry is supportable it may happen that some finitely generated structure is not the least finitely generated support of anything. Now we will introduce a condition which ensures that any finitely generated structure is the least finitely generated support of some element of some nominal set.

Definition 4.5. A finitely generated substructure $\mathfrak{A}$ of $\mathfrak{U}_K$ is *fungible* if for every finitely generated substructure $\mathfrak{B} \subsetneq \mathfrak{A}$, there exists $\pi \in G_K$ such that:

- $\pi|_{\mathfrak{B}} = \mathrm{id}|_{\mathfrak{B}}$,
- $\pi(\mathfrak{A}) \neq \mathfrak{A}$.

A Fraïssé symmetry $(\mathbb{A}_K, G_K)$ is fungible if every finitely generated substructure $\mathfrak{A}$ of $\mathfrak{U}_K$ is fungible.

Example 4.6. The equality, total order and timed symmetries are all fungible. The symmetry obtained from the universal structure $\langle \mathbb{Q}, \leq, +1 \rangle$ is not fungible. Take a structure $\mathfrak{A}$ generated by $\{0\}$ and its substructure $\mathfrak{B}$ generated by $\{1\}$. Obviously if an automorphism π acts as identity on $\mathfrak{B}$ then it acts as identity also on $\mathfrak{A}$.

In general, being supportable and being fungible are independent properties of symmetries. Examples are given in [BKL]. The following result generalizes Lemma 10.8. of [BKL].

Lemma 4.7. *(1) If $(\mathbb{A}_K, G_K)$ is supportable then every finitely generated fungible $\mathfrak{A} \subseteq \mathfrak{U}_K$ is the least finitely generated support of $[id|_{\mathfrak{A}}]_S$, for any $S \leq \mathrm{Aut}(\mathfrak{A})$.*

(2) If $(\mathbb{A}_K, G_K)$ is fungible then every finitely generated $\mathfrak{A} \subseteq \mathfrak{U}_K$ is the least finitely generated support of $[id|_{\mathfrak{A}}]_S$, for any $S \leq \mathrm{Aut}(\mathfrak{A})$.

Proof. For (1), recall from Lemma 4.1 that an embedding $u \colon \mathfrak{A} \to \mathfrak{U}_\mathcal{K}$ is supported by its image. Therefore $\mathfrak{A}$ supports $\mathrm{id}\,|_\mathfrak{A}$ and hence also $[\mathrm{id}\,|_\mathfrak{A}]_S$. Now consider any finitely generated structure $\mathfrak{B}$ properly contained in $\mathfrak{A}$. Since $\mathfrak{A}$ is fungible there exists an automorphism π from the Definition 4.5. The automorphism π acts as identity on $\mathfrak{B}$, but $[\mathrm{id}\,|_\mathfrak{A}]_S \cdot \pi = [\pi|_\mathfrak{A}]_S \neq [\mathrm{id}\,|_\mathfrak{A}]_S$ as the image of π is not $\mathfrak{A}$.

For (2), we first show that $\mathfrak{A}$ supports $[\mathrm{id}\,|_\mathfrak{A}]_S$ as in (1) above. Then let $\mathfrak{B}$ be another support of $[\mathrm{id}\,|_\mathfrak{A}]_S$ and assume $\mathfrak{A}$ is not contained in $\mathfrak{B}$, i.e., there exists some $a \in \mathfrak{A} \setminus \mathfrak{B}$. Since the structure $\mathfrak{C}$ generated by $\mathfrak{A} \cup \mathfrak{B}$ is fungible, there exists an automorphism π such that $\pi|_\mathfrak{B} = \mathrm{id}\,|_\mathfrak{B}$ and $\pi(\mathfrak{C}) \neq \mathfrak{C}$, which means that also $\pi(\mathfrak{A}) \neq \mathfrak{A}$. Hence $[\mathrm{id}\,|_\mathfrak{A}]_S \cdot \pi = [\pi|_\mathfrak{A}]_S \neq [\mathrm{id}\,|_\mathfrak{A}]_S$ and we obtain a contradiction as it turns out that $\mathfrak{B}$ does not support $[\mathrm{id}\,|_\mathfrak{A}]_S$.

Let us focus for a moment on relational structures. In this case to obtain a fungible symmetry it is enough to require an existence of π that is not an identity on $\mathfrak{A}$.

Definition 4.8. A finitely generated substructure $\mathfrak{A}$ of $\mathfrak{U}_\mathcal{K}$ is *weakly fungible* if for every finitely generated substructure $\mathfrak{B} \subsetneq \mathfrak{A}$, there exists $\pi \in G_\mathcal{K}$ such that:

- $\pi|_\mathfrak{B} = \mathrm{id}\,|_\mathfrak{B}$,
- $\pi|_\mathfrak{A} \neq \mathrm{id}\,|_\mathfrak{A}$.

A Fraïssé symmetry $(\mathbb{A}_\mathcal{K}, G_\mathcal{K})$ is weakly fungible if every finitely generated substructure $\mathfrak{A}$ of $\mathfrak{U}_\mathcal{K}$ is weakly fungible.

On the other hand, if we restrict ourselves to relational structures, we can also equivalently require an existence of automorphisms π that satisfy a stronger condition.

Definition 4.9. A finitely generated substructure $\mathfrak{A}$ of $\mathfrak{U}_\mathcal{K}$ is *strongly fungible* if for every finitely generated substructure $\mathfrak{B} \subsetneq \mathfrak{A}$, there exists $\pi \in G_\mathcal{K}$ such that:

- $\pi|_\mathfrak{B} = \mathrm{id}\,|_\mathfrak{B}$,
- $\pi(\mathfrak{A}) \cap \mathfrak{A} = \mathfrak{B}$.

A Fraïssé symmetry $(\mathbb{A}_\mathcal{K}, G_\mathcal{K})$ is strongly fungible if every finitely generated substructure $\mathfrak{A}$ of $\mathfrak{U}_\mathcal{K}$ is strongly fungible.

Fact 4.10. *Let $(\mathbb{A}_\mathcal{K}, G_\mathcal{K})$ be a Fraïssé symmetry over a signature containing only relation symbols. The following conditions are equivalent:*
(1) $(\mathbb{A}_\mathcal{K}, G_\mathcal{K})$ is weakly fungible,
(2) $(\mathbb{A}_\mathcal{K}, G_\mathcal{K})$ is fungible,
(3) $(\mathbb{A}_\mathcal{K}, G_\mathcal{K})$ is strongly fungible.

The general picture is more complicated. When we introduce function symbols, the notions of weak fungibility, fungibility and strong fungibility differ from each other. Before showing this let us notice that the condition of strong fungibility is in fact equivalent to the strong amalgamation property.

Proposition 4.11. *A Fraïssé symmetry* $(\mathbb{A}_{\mathcal{K}}, G_{\mathcal{K}})$ *is strongly fungible if and only if the age* $\mathcal{K}$ *of the universal structure* $\mathfrak{U}_{\mathcal{K}}$ *has the strong amalgamation property.*

Proof. The *if* part is easily proved using homogeneity. For the *only if* part take any finitely generated substructures $\mathfrak{A}$, $\mathfrak{B}$, $\mathfrak{C}$ of $\mathfrak{U}_{\mathcal{K}}$ and embeddings $f_{\mathfrak{B}} \colon \mathfrak{A} \to \mathfrak{B}$, $f_{\mathfrak{C}} \colon \mathfrak{A} \to \mathfrak{C}$. Thanks to amalgamation there exists a finitely generated substructure $\mathfrak{D}$ of $\mathfrak{U}_{\mathcal{K}}$ together with two embeddings $g_{\mathfrak{B}} \colon \mathfrak{B} \to \mathfrak{D}$ and $g_{\mathfrak{C}} \colon \mathfrak{C} \to \mathfrak{D}$ such that $g_{\mathfrak{B}} \circ f_{\mathfrak{B}}(\mathfrak{A}) = g_{\mathfrak{C}} \circ f_{\mathfrak{C}}(\mathfrak{A}) = \mathfrak{A}'$. Take $\pi \in G_{\mathcal{K}}$ for which $\pi|_{\mathfrak{A}'} = \mathrm{id}|_{\mathfrak{A}'}$ and $\pi(\mathfrak{D}) \cap \mathfrak{D} = \mathfrak{A}'$. Let $\mathfrak{D}'$ be a substructure generated by $\mathfrak{D} \cup \pi(\mathfrak{D})$. The embeddings $g_{\mathfrak{B}}$ and $g_{\mathfrak{C}}' = \pi \circ g_{\mathfrak{C}}$ into $\mathfrak{D}'$ are as needed:

$$g_{\mathfrak{B}} \circ f_{\mathfrak{B}}(\mathfrak{A}) = g_{\mathfrak{C}}' \circ f_{\mathfrak{C}}(\mathfrak{A}) = g_{\mathfrak{B}}(\mathfrak{B}) \cap g_{\mathfrak{C}}'(\mathfrak{C}).$$

Corollary 4.12. *A Fraïssé symmetry* $(\mathbb{A}_{\mathcal{K}}, G_{\mathcal{K}})$ *over a signature containing only relation symbols is fungible if and only if the age* $\mathcal{K}$ *of the universal structure* $\mathfrak{U}_{\mathcal{K}}$ *has the strong amalgamation property.*

Example 4.13. Consider an algebraic signature with unary function symbols F and G. For any integer i let $\mathbb{A}_i$ be the set of all infinite, binary sequences $\langle a_n \rangle$ defined for $n \geq i$ and equal 0 almost everywhere. Take $\mathbb{A} = \bigcup \mathbb{A}_i$ and define a structure $\mathfrak{U}$ with a carrier $\mathbb{A}$, where

$$F(\langle a_i, a_{i+1}, a_{i+2}, \ldots \rangle) = \langle a_{i+1}, a_{i+2}, \ldots \rangle, \quad G(0w) = 1w, \quad G(1w) = 0w.$$

Since the structure is weakly homogeneous, we obtain a Fraïssé symmetry. The symmetry is weakly fungible, but it is not fungible, as the structure generated by $\{0w, 1w\}$ is not fungible for any $w \in \mathbb{A}$.

Example 4.14. Consider an algebraic signature with a single unary function symbol F. For any integer i let $\mathbb{A}_i$ be the set of all infinite sequences $\langle a_n \rangle$ of natural numbers defined for $n \geq i$ and equal 0 almost everywhere. Take $\mathbb{A} = \bigcup \mathbb{A}_i$ and define a structure $\mathfrak{U}$ with a carrier $\mathbb{A}$, where

$$F(\langle a_i, a_{i+1}, a_{i+2}, \ldots \rangle) = \langle a_{i+1}, a_{i+2}, \ldots \rangle.$$

Notice that the age of $\mathfrak{U}$ is the class $\mathcal{K}$ of all finitely generated structures that satisfy the following axioms

- for any a, b there exist $m, n \in \mathbb{N}$ such that $F^m(a) = F^n(b)$,
- there are no loops, i.e., $F^n(a) \neq a$ for all $n \in \mathbb{N}$.

Since the structure is weakly homogeneous, we obtain a Fraïssé symmetry $(\mathbb{A}_{\mathcal{K}}, G_{\mathcal{K}})$. It is easy to check that the symmetry is fungible.

Now, take any nonempty finitely generated substructure $\mathfrak{A}$ of $\mathfrak{U}$ and the empty substructure $\emptyset \subseteq \mathfrak{A}$. For any automorphism π of $\mathfrak{U}$ and $a \in \mathfrak{A}$ there exist $m, n \in \mathbb{N}$ for which $F^m(a) = F^n(a \cdot \pi)$. Hence there is no π for which $\pi(\mathfrak{A}) \cap \mathfrak{A} = \emptyset$ and the structure $\mathfrak{A}$ is not strongly fungible. Therefore the symmetry $(\mathbb{A}_{\mathcal{K}}, G_{\mathcal{K}})$ is not strongly fungible.

4.2   Representation of Functions

For any finitely generated substructure $\mathfrak{A}$ of $\mathfrak{U}_\mathcal{K}$ and any $S \leq \mathrm{Aut}(\mathfrak{A})$, the $G_\mathcal{K}$-extension of S is

$$ext_{G_\mathcal{K}}(S) = \{\pi \in G_\mathcal{K} \mid \pi|_\mathfrak{A} \in S\} \leq G_\mathcal{K}.$$

Notice that $ext_{G_\mathcal{K}}(S)$ is exactly the stabilizer of $[\mathrm{id}\,|_\mathfrak{A}]_S$ in $G_\mathcal{K}$.

Lemma 4.15. *For each embedding $u\colon \mathfrak{A} \to \mathfrak{U}_\mathcal{K}$ the group $ext_{G_\mathcal{K}}(u^{-1}Su)$, where $u^{-1}Su \leq \mathrm{Aut}(u(\mathfrak{A}))$, is the stabilizer of an element $[u]_S \in [\mathfrak{A}, S]$.*

Proof. For any $\pi \in G_\mathcal{K}$ that extends u we have $[u]_S = [\mathrm{id}\,|_\mathfrak{A}]_S \cdot \pi$. Hence, by Lemma 2.6, the stabilizer of $[u]_S$ is $\pi^{-1}ext_{G_\mathcal{K}}(S)\pi$. It is easy to check that

$$\pi^{-1}ext_{G_\mathcal{K}}(S)\pi = ext_{G_\mathcal{K}}(u^{-1}Su).$$

Lemma 4.16. *For any supportable and fungible Fraïssé symmetry $(\mathbb{A}_\mathcal{K}, G_\mathcal{K})$ let $\mathfrak{A}, \mathfrak{B}$ be finitely generated substructures of $\mathfrak{U}_\mathcal{K}$ and let $S \leq \mathrm{Aut}(\mathfrak{A})$, $T \leq \mathrm{Aut}(\mathfrak{B})$, then $ext_{G_\mathcal{K}}(S) \leq ext_{G_\mathcal{K}}(T)$ if and only if $\mathfrak{B} \subseteq \mathfrak{A}$ and $S|_\mathfrak{B} \leq T$.*

Proof. The *if* part is obvious. For the *only if* part, we first prove that $\mathfrak{B} \subseteq \mathfrak{A}$. Notice that if $\pi|_\mathfrak{A} = \mathrm{id}\,|_\mathfrak{A}$ then $\pi \in ext_{G_\mathcal{K}}(S)$ and hence $\pi \in ext_{G_\mathcal{K}}(T)$, which is the stabilizer of $[\mathrm{id}\,|_\mathfrak{B}]_T$. Therefore $\mathfrak{A}$ supports $[\mathrm{id}\,|_\mathfrak{B}]_T$. By Lemma 4.7 (2) the least support of $[\mathrm{id}\,|_\mathfrak{B}]_T$ is $\mathfrak{B}$. Hence $\mathfrak{B} \subseteq \mathfrak{A}$. Then we have

$$ext_{G_\mathcal{K}}(S) \leq ext_{G_\mathcal{K}}(T)$$

$$\Updownarrow$$

$$\forall \pi \in G_\mathcal{K} \ \ \pi|_\mathfrak{A} \in S \Longrightarrow \pi|_\mathfrak{B} \in T$$

$$\Updownarrow$$

$$\forall \pi \in G_\mathcal{K} \ \ \pi|_\mathfrak{A} \in S \Longrightarrow (\pi|_\mathfrak{A})|_\mathfrak{B} \in T$$

$$\Updownarrow$$

$$\forall \tau \in S \ \ \tau|_\mathfrak{B} \in T.$$

Similar facts about finite substructures of a universal relational structure $\mathfrak{U}_\mathcal{K}$ were proven in [BKL]. The following proposition generalizes Proposition 11.8.

Proposition 4.17. *For any supportable and fungible Fraïssé symmetry $(\mathbb{A}_\mathcal{K}, G_\mathcal{K})$ let $X = [\mathfrak{A}, S]$ and $Y = [\mathfrak{B}, T]$ be single-orbit nominal sets. The set of equivariant functions from X to Y is in one to one correspondence with the set of embeddings $u\colon \mathfrak{B} \to \mathfrak{A}$, for which $uS \subseteq Tu$, quotiented by $\equiv_T$.*

Proof. By Proposition 2.7 and Lemma 4.15 equivariant functions from $[\mathfrak{A}, S]$ to $[\mathfrak{B}, T]$ are in bijective correspondence with those elements $[u]_T \in [\mathfrak{B}, T]$ for which

$$ext_{G_\mathcal{K}}(S) \leq ext_{G_\mathcal{K}}(u^{-1}Tu).$$

Hence, by Lemma 4.16, equivariant functions from $[\mathfrak{A}, S]$ to $[\mathfrak{B}, T]$ correspond to those elements $[u]_T \in [\mathfrak{B}, T]$ for which

$$u(\mathfrak{B}) \subseteq \mathfrak{A} \quad \text{and} \quad S|_{u(\mathfrak{B})} \leq u^{-1}Tu,$$

which means that u is an embedding from $\mathfrak{B}$ to $\mathfrak{A}$ and $uS \subseteq Tu$, as required.

Let $G_\mathcal{K}\text{-}\mathbf{Nom}^1$ denote the category of single-orbit nominal sets and equivariant functions. Propositions 4.3 and 4.17 can be phrased in the language of category theory:

Proposition 4.18. *In a supportable and fungible Fraïssé symmetry, the category* $G_\mathcal{K}\text{-}\mathbf{Nom}^1$ *is equivalent to the category with:*

- *as objects, pairs* $(\mathfrak{A}, S)$ *where* $\mathfrak{A} \in \mathcal{K}$ *and* $S \leq \mathrm{Aut}(\mathfrak{A})$,
- *as morphisms from* $(\mathfrak{A}, S)$ *to* $(\mathfrak{B}, T)$, *those embeddings* $u \colon \mathfrak{B} \to \mathfrak{A}$ *for which* $uS \subseteq Tu$, *quotiented by* $\equiv_T$.

Since a nominal set is a disjoint union of single-orbit sets, this representation extends to orbit-finite sets in an obvious way:

Theorem 4.19. *In a supportable and fungible Fraïssé symmetry, the category* $G_\mathcal{K}\text{-}\mathbf{Nom}$ *is equivalent to the category with:*

- *as objects, finite sets of pairs* $(\mathfrak{A}_i, S_i)$ *where* $\mathfrak{A}_i \in \mathcal{K}$ *and* $S_i \leq \mathrm{Aut}(\mathfrak{A}_i)$,
- *as morphisms from* $\{(\mathfrak{A}_1, S_1), \ldots, (\mathfrak{A}_n, S_n)\}$ *to* $\{(\mathfrak{B}_1, T_m), \ldots, (\mathfrak{B}_m, T_m)\}$, *pairs* $(f, \{[u_i]_{T_{f(i)}}\}_{i=1,\ldots,n})$, *where* $f \colon \{1, \ldots, n\} \to \{1, \ldots, m\}$ *is a function and each* u_i *is an embedding* $u_i \colon \mathfrak{B}_{f(i)} \to \mathfrak{A}_i$ *such that* $u_i S_i \subseteq T_{f(i)} u_i$.

In the special case of relational structures the above theorem was formulated and proved in [BKL].

5   Conclusions and Future Work

Orbit-finite nominal sets can be used to model devices, such as automata or Turing machines, which operate over infinite alphabets. This approach makes sense only if one can treat objects with atoms as data structures and manipulate them using algorithms. To do so the existence of a finite representation of orbit-finite nominal sets is crucial.

In this paper we have generalized the representation theorem due to Bojańczyk et al. to cover atoms with algebraic structure. The result is however not entirely satisfying. Our representation uses automorphism groups of finitely generated substructures of the atoms. If such groups are finitely presentable Theorem 4.19

indeed provides a concrete, finite representation of orbit-finite nominal sets (the timed symmetry being an example). But is it always the case? So far we do not know and we regard it as a field for a further research effort.

Another thing left to be done is a characterization of "well-behaved" atom symmetries in terms of Fraïssé classes that induce them. One might think of algebraic atoms that could be potentially interesting from the point of view of computation theory: strings with the concatenation operator, binary vectors with addition, etc. Yet checking the technical conditions, such as supportability and fungibility, needed for the representation theorem to hold requires each time a lot of effort. This is because these conditions are formulated in terms of Fraïssé limits, and these are not always easy to construct. It would be desirable to have more natural criteria that would be easier to verify.

Acknowledgments. I would like to thank Mikołaj Bojańczyk, Bartek Klin, Sławomir Lasota and Szymon Toruńczyk for inspiring discussions and helpful remarks on earlier drafts of this paper, and the anonymous referees for many insightful comments.

References

[BKL]   Bojańczyk, M., Klin, B., Lasota, S.: Automata Theory in Nominal Sets (to appear)

[BKL11]   Bojańczyk, M., Klin, B., Lasota, Ś.: Automata with Group Actions. In: Proc. LICS 2011, pp. 355–364 (2011)

[BL12]   Bojańczyk, M., Lasota, S.: A Machine-independent Characterization of Timed Languages. In: Czumaj, A., Mehlhorn, K., Pitts, A., Wattenhofer, R. (eds.) ICALP 2012, Part II. LNCS, vol. 7392, pp. 92–103. Springer, Heidelberg (2012)

[BKLT13] Bojańczyk, M., Klin, B., Lasota, S., Toruńczyk, S.: Turing Machines with Atoms. In: Proc. LICS 2013, pp. 183–192 (2013)

[FS06]   Fiore, M., Staton, S.: Comparing Operational Models of Name-passing Process Calculi. Inf. Comput. 204, 524–560 (2006)

[GP02]   Gabbay, M.J., Pitts, A.M.: A new approach to abstract syntax with variable binding. Formal Aspects of Computing 13, 341–363 (2002)

[GMM06]   Gadducci, F., Miculan, M., Montanari, U.: About Permutation Algebras (Pre)Sheaves and Named Sets. Higher Order Symbol. Comput. 19, 283–304 (2006)

[Hod93]   Hodges, W.: Model theory. Cambridge University Press (1993)

[KF94]   Kaminski, M., Francez, N.: Finite-memory Automata. Theor. Comput. Sci. 134, 329–363 (1994)

[Pis99]   Pistore, M.: History Dependent Automata. PhD thesis, Università di Pisa, Dipartimento di Informatica. available at University of Pisa as PhD Thesis TD-5/99 (1999)

[Pit13]   Pitts, A.M.: Nominal Sets: Names and Symmetry in Computer Science. Cambridge Tracts in Theoretical Computer Science, vol. 57. Cambridge University Press (2013)

Fixed-Point Theory in the Varieties $\mathcal{D}_n$

Sabine Frittella and Luigi Santocanale

Laboratoire d'Informatique Fondamentale de Marseille
Aix-Marseille Université, Marseille, France

Abstract. The varieties of lattices $\mathcal{D}_n$, $n \geq 0$, were introduced in [Nat90] and studied later in [Sem05]. These varieties might be considered as generalizations of the variety of distributive lattices which, as a matter of fact, coincides with $\mathcal{D}_0$. It is well known that least and greatest fixed-points of terms are definable on distributive lattices; this is an immediate consequence of the fact that the equation $\phi^2(\bot) = \phi(\bot)$ holds on distributive lattices, for any lattice term $\phi(x)$. In this paper we propose a generalization of this fact by showing that the identity $\phi^{n+2}(x) = \phi^{n+1}(x)$ holds in $\mathcal{D}_n$, for any lattice term $\phi(x)$ and for $x \in \{\top, \bot\}$. Moreover, we prove that the equations $\phi^{n+1}(x) = \phi^n(x)$, $x = \bot, \top$, do not hold in the variety $\mathcal{D}_n$ nor in the variety $\mathcal{D}_n \cap \mathcal{D}_n^{op}$, where $\mathcal{D}_n^{op}$ is the variety containing the lattices L^{op}, for $L \in \mathcal{D}_n$.

1 Introduction

The research that we present in this paper stems from fixed-point theory, as conceived in computer science logic, specifically with logics of computation. It is customary here to add least and greatest fixed-point operators to existing logical or algebraic frameworks to increase their expressive power. The elementary theory of the formal systems so obtained, the μ-calculi, is covered in [AN01]; a main example of a fixed-point logic is the propositional modal μ-calculus [Koz83]. The *lattice μ-calculus* [San02] is among the most elementary μ-calculi. Its μ-terms are built according to the following grammar:

$$\phi := x \mid \top \mid \bot \mid \phi \wedge \phi \mid \phi \vee \phi \mid \mu_x.\phi \mid \nu_x.\phi. \tag{1}$$

Syntactically, the lattice μ-calculus is obtained from the signature of lattice theory (including constants for least and greatest elements) by adding the power of forming terms $\mu_x.\phi, \nu_x.\phi$. Semantically, when all the variables but x are evaluated, a μ-term ϕ gives rise to a monotone function; $\mu_x.\phi$ (resp. $\nu_x.\phi$) is then used to denote its least (resp. greatest) fixed-point.

Applications of the lattice μ-calculus, through its theory of circular proofs [FS13], lie more in the area of functional programming. We tackle here a different problem, explained next, concerning the expressive power of μ-calculi.

The alternation complexity of μ-terms measures the number of nested distinct fixed-point operators; it gives rise to the fixed-point alternation hierarchy of a μ-calculus, see [AN01, §8]. It was shown in [San01] that the alternation hierarchy

P. Höfner et al. (Eds.): RAMiCS 2014, LNCS 8428, pp. 446–462, 2014.
© Springer International Publishing Switzerland 2014

for the lattice μ-calculus is infinite. This means that for each integer n we can find a μ-term of alternation complexity n which is not semantically equivalent to any other μ-term of alternation complexity $m < n$. This result appeared striking if compared with well known folklore of fixed-point theory—see for example [NW96, Lemma 2.2] or [AN01, Proposition 3.1.2]: the alternation hierarchy of the lattice μ-calculus is degenerate if the interpretation of μ-terms is restricted to *distributive* lattices; namely, every μ-term, of arbitrary alternation complexity, can be replaced with a semantically equivalent (on distributive lattices) lattice term, with no fixed-point operators.

The reasons for such a degeneracy are easy to guess for a scientist trained in universal algebra. The variety of distributive lattices has, as its unique sub-directly irreducible member, the two element lattice 2. Its height being 2, the increasing chain

$$\bot \leq \phi(\bot) \leq \phi^2(\bot) \leq \ldots \leq \phi^n(\bot) \leq \ldots \tag{2}$$

must become stationary after one step. Thus, the equation $\phi^2(\bot) = \phi(\bot)$ holds in 2, for every lattice term ϕ, therefore it holds on all the distributive lattices, and thus it forces (the interpretation of) $\phi(\bot)$ to be the least fixed-point of (the interpretation of) $\phi(x)$. A dual fact evidently holds for greatest fixed-points. An algorithm to eliminate all the fixed-point operators from μ-terms can be therefore devised.

For the sake of fixed-point theory it would be desirable to understand the alternation hierarchy uniformly along different classes of models; for example, you would like to exhibit and classify generic reasons for the alternation hierarchy to be degenerate. So, what about the existence of non-trivial equations holding on a class of models and enforcing the increasing chains of approximations (2) to stabilize after a bounded number of steps? Is this a valid generic reason?

In this paper we provide evidence that the answer to the last question is positive. Such an evidence is constructed within lattice theory. We exhibit lattice varieties $\mathcal{D}_n$, $n \geq 0$, such that

$$\mathcal{D}_n \models \phi^{n+2}(\bot) = \phi^{n+1}(\bot), \qquad \text{for each lattice term } \phi. \tag{3}$$

We also prove that the equations $\phi^{n+2}(\top) = \phi^{n+1}(\top)$ hold on $\mathcal{D}_n$. Moreover, we provide examples of terms ψ for which

$$\mathcal{D}_n \not\models \psi^{n+1}(\bot) = \psi^n(\bot), \tag{4}$$

and dually. Thus the phenomenon observed for distributive lattices generalizes in a non-trivial way to each variety $\mathcal{D}_n$: each variety $\mathcal{D}_n$ has a degenerate alternation hierarchy, the alternation hierarchy of $\mathcal{D}_{n+1}$ being, under some respect, less degenerate than the one of $\mathcal{D}_n$.

The varieties $\mathcal{D}_n$, $n \geq 0$, were first considered in [Nat90] and studied further in [Sem05]. Actually, $\mathcal{D}_0$ is exactly the variety of distributive lattices and the identities axiomatizing these varieties (for which we have $\mathcal{D}_n \subseteq \mathcal{D}_{n+1}$) can be thought as increasingly weaker forms of the distributive law.

To see that (3) holds for $n \geq 1$, we cannot use the same argument used for distributive lattices, as already $\mathcal{D}_1$ contains subdirectly irreducible lattices of unbounded height. The reasons for (3) to hold appear deeper, based on some modal view of lattice theory. Roughly speaking, we represent a finite lattice as the set of closed subsets for a closure operator. On the logical side, a closure operator is just a monotone non-normal (that is, it does not distribute over joins) operator satisfying additional equations. The modal logic we have in mind is therefore monotonic modal logic, see for example [Han03, VS10]. Precise connections between lattice theory and monotonic modal logic still need to be fully explored and they are not the main aim of this paper. The preprint [San09] (from which some of the results presented here are borrowed) is just a first step towards establishing these connections and does not cover other attempts [KW99, §6] to establish them. Yet, we freely use ideas and tools from modal logic (semantics, games) and move them to the lattice-theoretic context; these are the tools that allow us to achieve the desired results. From lattice theory, we make heavy use of the notion of OD-graph [Nat90] of a finite lattice, which we consider as the adequate working notion for a dual space of a finite lattice.

The paper is organized as follows. Having introduced some elementary notions and results in Section 2, we present the varieties $\mathcal{D}_n$ in Section 3. Section 4 introduces the main working tool, the semantic relation $\models$ and the associated game. In the following Sections 5 and 6 we prove our main results, relations (3) and the dual ones, and show these relations are the least possible. Finally, in Section 7, we sketch how to obtain relations similar to (4) for the varieties $\mathcal{D}_n \cap \mathcal{D}_n^{op}$.

2 Elementary Notions and Results

For elementary notions about ordered sets and lattices, we invite the reader to consult the standard literature [Bir73, Grä98, DP02]. Similarly, we refer the reader to standard monographs for elementary facts on category theory or universal algebra [Mac98, Awo10, Grä08]. A reference for the lattice-theoretic notions to be introduced in this section is [FJN95, Chapter II].

If P is a poset and $p \in P$, then $\downarrow p$ shall denote the *principal ideal of p*, that is, the set $\{ p' \in P \mid p' \leq p \}$. Similarly, if $X \subseteq P$, then $\downarrow X$ denotes *the downset generated by X*, $\downarrow X = \{ p \in P \mid \exists x \in X \text{ s.t. } p \leq x \}$. A downset of P is a subset $X \subseteq P$ such that $\downarrow X = X$. We use similar notations, $\uparrow p$ and $\uparrow X$, for *principal filter of p* and the *the upset generated by X*.

If L is a lattice, an element $x \in L$ is said to be join-irreducible (resp. meet-irreducible) if $x = \bigvee X$ (resp. $x = \bigwedge X$) implies $x \in X$; $J(L)$ (resp. $M(L)$) denotes the collection of join-irreducible (resp. meet-irreducible) elements of L. If L is finite and $j \in J(L)$, then we denote by j_* the unique lower cover of j in L; namely, j_* is the unique element such that $\{ x \in L \mid j_* \leq x \leq j \} = \{ j_*, j \}$ and $j_* < j$. Dually, for $m \in M(L)$, m^* denotes the unique upper cover of m in L. The *join-dependency relation* $D_L \subseteq J(L) \times J(L)$ is defined as follows:

$$j D_L k \text{ if } j \neq k \text{ and, for some } p \in L, \ j \leq p \vee k \text{ and } j \not\leq p \vee k_*.$$

If $x \in L$ and $X \subseteq L$ is such that $x \leq \bigvee X$, then we say that X is a *join-cover* of x; we write $\mathcal{C}(x)$ for the collection of join-covers of $x \in L$. The set $\mathcal{C}(x)$—and, more in general, the collection of subsets of L—is pre-ordered by the *refinement relation* $\ll$ defined as follows:

$$X \ll Y \quad \text{iff} \quad \forall x \in X \, \exists y \in Y \text{ s.t. } x \leq y.$$

We say that $X \in \mathcal{C}(x)$ is a *minimal join-cover* of x if, for all $Y \in \mathcal{C}(x)$, $Y \ll X$ implies $X \subseteq Y$. The set of minimal join-covers of x is denoted $\mathcal{M}(x)$; we shall also write $x \lhd X$ whenever $X \in \mathcal{M}(x)$. It is easily seen that $X \subseteq J(L)$ for each $x \in L$ and $X \in \mathcal{M}(x)$. A lattice has the *minimal join-cover refinement property* if, for each $x \in L$ and $X \in \mathcal{C}(x)$, there is some $Y \in \mathcal{M}(x)$ such that $Y \ll X$. Every finite lattice has this property. We say that $Y \in \mathcal{C}(x)$ is a *non-trivial* join-cover of x if $x \nleq y$ for each $y \in Y$. We let $\mathcal{M}_-(x)$ denote the set of non-trivial minimal join-covers of x.

It is known [FJN95, Lemma 2.31] that $j D_L k$ iff k belongs to a minimal non-trivial join-cover of j, that is:

$$j D_L k \text{ iff } k \in C, \text{ for some } C \in \mathcal{M}_-(j).$$

Definition 2.1. *The* OD-*graph of a finite lattice L, denoted* $\mathrm{OD}(L)$*, is the structure* $\langle J(L), \leq, \mathcal{M} \rangle$*, where $\leq$ is the restriction of the order of L to the join-irreducible elements, and $\mathcal{M} : J(L) \longrightarrow \mathcal{P}(\mathcal{P}(J(L)))$ is the function sending $j \in J(L)$ to $\mathcal{M}(j)$.*

The notion of OD-graph was introduced in [Nat90]. The letter O stands for *order*, while D stands *dependency*. For example, for the lattice N_5 (see Figure 1), we have $J(N_5) = \{a, b, c\}$, $a \leq c$, $\mathcal{M}(a) = \{\{a\}\}, \mathcal{M}(b) = \{\{b\}\}, \mathcal{M}(c) = \{\{c\}, \{a,b\}\}$. As suggested in [Nat90], the OD-graph of a lattice can be represented as a labelled digraph where a cover of a join-irreducible element is recovered as the set of its successors by a same label; a dotted arrow might be used to code

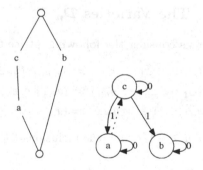

Fig. 1. The lattice N_5 with its OD-graph

the order in $J(L)$. Figure 1 exemplifies this for the lattice N_5 and its OD-graph.

The OD-graph of a finite lattice is a particular instance of a structure that we call in [San09] a presentation.

Definition 2.2. *A presentation is a structure $\langle V, \leq, M \rangle$, where $\langle V, \leq \rangle$ is a poset and $M : V \longrightarrow \mathcal{PP}(V)$. We write $x \lhd C$ if $C \in M(x)$. Given a presentation $\langle V, \leq, M \rangle$, we say that a downset $X \subseteq V$ is closed if, for each $x \in V$ and $C \subseteq V$, $x \lhd C \subseteq X$ implies $x \in X$. The collection of closed subsets of V, ordered by subset inclusion, form a poset that we denote $\mathcal{L}(V, \leq, M)$.*

Let us recall that $\mathcal{L}(V, \leq, M)$ is a lattice, where the meet is given by intersection. For the join operation, see Lemma 4.1 and, more generally, [Bir73, Chapter V].

The following representation Theorem was proved in [Nat90].

Theorem 2.3. *A finite lattice L is isomorphic to $\mathcal{L}(\mathrm{OD}(L))$ via the map ψ defined by $\psi(l) := \{\, j \in J(L) \mid j \leq l \,\}$.*

A presentation $\langle V, \leq, M \rangle$ is *direct* if, for each downset $X \subseteq V$, the set $\{\, x \in V \mid \exists C \in M(x) \text{ s.t. } C \subseteq X \,\}$ belongs to $\mathcal{L}(V, \leq, M)$; it is *atomistic* if the order $\leq$ on V is the identity. The OD-graph of a lattice is a direct presentation. An *atomistic lattice* is a finite lattice for which every join-prime element is an atom. As atoms are not pairwise comparable, the OD-graph of an atomistic lattice is an atomistic presentation. If $\langle V, \leq, M \rangle$ is an atomistic presentation, then we omit the order and write only $\langle V, M \rangle$.

The following Lemma—see [San09, Proposition 6.4]—characterizes OD-graphs of atomistic lattices.

Lemma 2.4. *Suppose that an atomistic presentation $\langle V, M \rangle$ satisfies:*

1. *$\{\, v \,\} \in M(v)$ and is the unique singleton in $M(v)$,*
2. *$M(v)$ is an antichain w.r.t subset inclusion,*
3. *if $v \lhd C$ and $c \lhd D$ for some $c \in C$, then there exists $E \subseteq V$ such that $v \lhd E \subseteq (C \setminus \{\, c \,\}) \cup D$,*

for each $v \in V$. Then $\mathcal{L}(V, M)$ is atomistic and $\langle V, M \rangle$ is isomorphic $\mathrm{OD}(\mathcal{L}(V, M))$. Every OD-graph of an atomistic lattice satisfies the constraints (1), (2), and (3).

3   The Varieties $\mathcal{D}_n$

Let us consider the following lattice terms:

$$t_{-1} := x_{-1}, \qquad t_{n+1} := x_{n+1} \wedge (y_{n+1} \vee t_n),$$
$$s_{-1} := \bot, \qquad s_{n+1} := (x_{n+1} \wedge x_n) \vee (x_{n+1} \wedge (y_{n+1} \vee s_n)),$$
$$w_{-1} := \bot, \qquad w_{n+1} := (x_{n+1} \wedge t_n) \vee (x_{n+1} \wedge y_{n+1}) \vee (x_{n+1} \wedge (y_{n+1} \vee w_n)).$$

For $n \geq -1$, define the inclusions (β_n) and (d_n) as follows:

$$t_n \leq s_n, \qquad (\beta_n) \qquad t_n \leq w_n. \qquad (d_n)$$

Let $\mathcal{B}_n$ (resp. $\mathcal{D}_n$) be the variety determined by (β_n) (resp. (d_n)). The inclusions (β_n) were considered in [Nat90], while the inclusions (d_n) were considered in [Sem05]. We have that $\mathcal{B}_{-1} = \mathcal{D}_{-1}$ is the trivial variety, while $\mathcal{B}_0 = \mathcal{D}_0$ is the variety of distributive lattices. For $n \geq 1$, we can think of the varieties $\mathcal{B}_n$ and $\mathcal{D}_n$ as classes of lattices obeying weaker and weaker forms of the distributive law.

Notice that the inclusions $w_n \leq s_n$ and $s_n \leq t_n$ are valid, for each $n \geq -1$, so that (d_n) implies (β_n) and the variety $\mathcal{D}_n$ is contained in $\mathcal{B}_n$. In order to argue that $\mathcal{B}_n \subseteq \mathcal{D}_n$, let us recall that a variety $\mathcal{V}$ is *locally finite* if every finitely generated algebra in $\mathcal{V}$ is finite, as well as the following Theorem.

Theorem 3.1 (See [Nat90], §5). *The varieties $\mathcal{B}_n$ are locally finite. Thus an identity holds in $\mathcal{B}_n$ if and only if it holds in all the finite lattices in $\mathcal{B}_n$.*

As a subvariety of a locally finite variety is again locally finite, we obtain:

Corollary 3.2. *The varieties $\mathcal{D}_n$ are locally finite.*

In particular, both varieties $\mathcal{B}_n$ and $\mathcal{D}_n$ are determined by the finite lattices in it.

The following Proposition was proved in [Nat90, Section 5].

Proposition 3.3. *A finite lattice L belongs to the variety $\mathcal{B}_n$ if and only if every sequence $j_k D_L j_{k-1} D_L \ldots D_L j_0$ has length $k \leq n$.*

The same statement, with $\mathcal{B}_n$ replaced by $\mathcal{D}_n$, was proved in [Sem05, Proposition 3.4]. As the two varieties have the same finite lattices and are locally finite, we obtain:

Theorem 3.4. *For each $n \geq -1$, the varieties $\mathcal{D}_n$ and $\mathcal{B}_n$ coincide.*

As $\mathcal{D}_n$ is reminiscent of the bound on the length of a sequence of the join-dependency relation, we prefer to keep this naming in the rest of the paper.

To understand the role of the varieties $\mathcal{D}_n$ within lattice theory, let us recall a few definitions and results.

Definition 3.5. *A lattice is:*

- join-semidistributive *if it satisfies the Horn sentence*

$$x \vee y = x \vee z \text{ implies } x \vee y = x \vee (y \wedge z), \qquad (SD_\vee)$$

- meet-semidistributive *if it satisfies the dual Horn sentence*

$$x \wedge y = x \wedge z \text{ implies } x \wedge y = x \wedge (y \vee z), \qquad (SD_\wedge)$$

- semidistributive *, if it is both join $-$ semidistributive and meet$-$semidistributive,*
- lower bounded *if there exists an epimorphism π from a finitely generated free lattice F to L, with a left-adjoint ℓ from L to F:*

$$F \underset{\ell}{\overset{\pi}{\underset{\top}{\rightleftarrows}}} L$$

- bounded[1] *if there exists such an epimorphism π with both a left adjoint ℓ and a right adjoint ρ.*

---

[1] The naming, *bounded*, is a short for *bounded homomorphic image of a free lattice*; it is nowadays widespread among lattice theorists. Unfortunately, it clashes with the notion of a lattice with both a least and a greatest element.

Theorem 3.6. *The following statements hold:*

(*i*) *A finite lattice L is lower bounded iff $L \in \mathcal{D}_n$ for some $n \geq 0$.*
(*ii*) *Every lower bounded lattice is join-semidistributive.*
(*iii*) *A finite lattice is bounded if and only if it is lower bounded and meet-semidistributive.*
(*iv*) *The class of bounded lattices generates the variety of all lattices.*

For (*i*) see [FJN95, Corollary 2.39], for (*ii*) see [FJN95, Theorem 2.20], for (*iii*) see [FJN95, Corollary 2.65]. Finally, (*iv*) was the main result of [Day92], see also [FJN95, Theorem 2.84]

4   Game Semantics

Nation's representation (Theorem 2.3) gives rise to a covering semantics similar to the one developed in [Gol06] for non-commutative linear logic. We give here a game-theoretic account of this semantics, which allows us to have further insights on the problems we wish to tackle.

Lemma 4.1. *If a presentation $\langle V, \leq, M \rangle$ is direct and X is a downset, then the set $\{\, y \in V \mid \exists C \in M(y) \text{ s.t. } C \subseteq X \,\}$ is the least closed downset of V containing X. In particular the join of X_1, X_2 in $\mathcal{L}(V, \leq, M)$ can be computed as*

$$X_1 \vee X_2 = \{\, y \in V \mid \exists C \in M(y) \text{ s.t. } C \subseteq X_1 \cup X_2 \,\}. \tag{5}$$

Let $\mathcal{X}$ be a set of variables, denote by $\mathcal{T}(\mathcal{X})$ the algebra of lattice terms whose free variables are among $\mathcal{X}$, denote by $\mathcal{F}(\mathcal{X})$ the free lattice generated by the set $\mathcal{X}$. Recall that there are natural bijections between the following type of data: valuations $u : \mathcal{X} \longrightarrow \mathcal{L}(V, \leq, M)$, algebra morphisms $u' : \mathcal{T}(\mathcal{X}) \longrightarrow \mathcal{L}(V, \leq, M)$ such that $u'(x) = u(x)$, lattice homomorphisms $\tilde{u} : \mathcal{F}(\mathcal{X}) \longrightarrow \mathcal{L}(V, \leq, M)$ such that $\tilde{u}(x) = u(x)$. In order to simplify the notation, we shall often use the same notation u for the three different kind of data.

We introduce next a semantical relation reminiscent of Kripke semantics in modal logic.

Definition 4.2. *For $u : \mathcal{X} \longrightarrow \mathcal{L}(V, \leq, M)$, $v \in V$ and $t \in \mathcal{T}(\mathcal{X})$, the relation $v \models_u t$ is defined, inductively on t, as follows :*

- $v \models_u x$ *if $v \in u(x)$,*
- $v \models_u \bigwedge_{i \in I} t_i$ *if $v \models t_i$ for each $i \in I$,*
- $v \models_u \bigvee_{i \in I} t_i$ *if there exists $C \in M(v)$ such that, for all $c \in C$, there exists $i \in I$ with $c \models_u t_i$.*

The following statement is quite obvious:

Lemma 4.3. $v \models_u t$ *if and only if $v \in u(t)$.*

Let us denote by P the presentation $\langle V, \leq, M \rangle$. The same relation is characterized by means of a game $\mathcal{G}(P, u, t)$ between two players, Eva and Adam, described as follows. Its set of positions is the disjoint union of three sets,

$$V \times Sub(t), \quad \bigcup_{v \in V} M(v) \times Sub_\vee(t), \quad \text{and} \quad V \times \mathcal{P}(Sub(t)),$$

where $Sub(t)$ denotes the set of subterms of t and $Sub_\vee(t)$ denotes the set of subterms of t that are formal joins. The moves of the game are as follows:

- In position (v, x) Eva wins if $v \in u(x)$ and otherwise Eva loses.
- In position $(v, \bigwedge_{i \in I} t_i)$ Adam chooses $i \in I$ and moves to (v, t_i).
- In position $(v, \bigvee_{i \in I} t_i)$ Eva chooses $C \in M(v)$ and moves to $(C, \bigvee_{i \in I} t_i)$;
  - in $(C, \bigvee_{i \in I} t_i)$ Adam chooses $v' \in C$ and moves to $(v', \{ t_i \mid i \in I \})$;
  - in position $(v', \{ t_i \mid i \in I \})$ Eva chooses $i \in I$ and moves to (v', t_i).

If a player cannot move then he loses.

Proposition 4.4. *Eva has a winning strategy from position (v, t) in the game $\mathcal{G}(P, u, t)$ if and only if $v \in u(t)$, if and only if $v \models_u t$.*

5   Least Fixed-Points

Next, let ϕ be a lattice term possibly containing the variable z. We define terms $\phi^n(\perp)$, $n \geq 0$, inductively by the rules

$$\phi^0(\perp) := \perp, \qquad\qquad \phi^{n+1}(\perp) := \phi[\, \phi^n(\perp)/z \,].$$

Proposition 5.1. *The identity*

$$\phi^{n+2}(\perp) = \phi^{n+1}(\perp)$$

holds on $\mathcal{D}_n$.

Proof. The inclusion $\phi^{n+1}(\perp) \leq \phi^{n+2}(\perp)$ always holds, so that we only need to verify that the inclusion $\phi^{n+2}(\perp) \leq \phi^{n+1}(\perp)$ holds as well in $\mathcal{D}_n$. As $\mathcal{D}_n$ is locally finite, we can focus our attention on the finite lattices in $\mathcal{D}_n$. So, let $L \in \mathcal{D}_n$ be finite; as L and $\mathcal{L}(J(L), \leq, \mathcal{M})$ are isomorphic, we can work with the latter. Consider a valuation $u : \mathcal{X} \longrightarrow \mathcal{L}(J(L), \leq, \mathcal{M})$; we suppose that $j_0 \models_u \phi^{n+2}(\perp)$ and prove that $j_0 \models_u \phi^{n+1}(\perp)$. To this goal, we consider a winning strategy for Eva in the game $\mathcal{G}(\mathrm{OD}(L), u, \phi^{n+2}(\perp))$ from $(j_0, \phi^{n+2}(\perp))$ and transform it into a winning strategy for Eva in $\mathcal{G}(\mathrm{OD}(L), u, \phi^{n+1}(\perp))$ from $(j_0, \phi^{n+1}(\perp))$.

Notice first that if $i \geq 1$, then the structure of the games $\mathcal{G}(\mathrm{OD}(L), u, \phi^{i+1}(\perp))$ and $\mathcal{G}(\mathrm{OD}(L), u, \phi^i(\perp))$ are, at the beginning, the same from the positions of the form $(j, \phi^{i+1}(\perp))$ and $(j, \phi^i(\perp))$. In particular, a winning strategy for Eva from $(j, \phi^{i+1}(\perp))$ in $\mathcal{G}(\mathrm{OD}(L), u, \phi^{i+1}(\perp))$ can be simulated, at the beginning, within the game $\mathcal{G}(\mathrm{OD}(L), u, \phi^i(\perp))$ from position $(j, \phi^i(\perp))$.

Therefore Eva plays as follows: she plays from $(j_0, \phi^{n+1}(\perp))$ as if she was playing in $(j_0, \phi^{n+2}(\perp))$ according to the given winning strategy. If, at some point,

– she hits a position $(j_i, \phi^{n-i+1}(\perp))$, where she is simulating the winning strategy from position $(j_i, \phi^{n-i+2}(\perp))$ in the game $\mathcal{G}(\mathrm{OD}(L), u, \phi^{n+2}(\perp))$, and
– the position $(j_i, \phi^{n-i+1}(\perp))$ is also part of the winning strategy in the game $\mathcal{G}(\mathrm{OD}(L), u, \phi^{n+2}(\perp))$,

then she jumps forward in the simulation: she continues simulating the winning strategy from $(j_i, \phi^{n-i+1}(\perp))$ in the game $\mathcal{G}(\mathrm{OD}(L), u, \phi^{n+2}(\perp))$.

Evidently, this strategy might be losing if Eva cannot jump. That is, Adam might force a play to visit positions $(j_i, \phi^{n-i+1}(\perp))$, $i = 0, \ldots, n+1$, such that all the j_i are distinct. As the $j_i \neq j_{i+1}$, this means that Eva, in the play that has lead from $(j_i, \phi^{n-i+1}(\perp))$ to $(j_{i+1}, \phi^{n-(i+1)+1}(\perp))$, has gone through at least one sequence of choices of the form $C \in M(k)$ and $k' \in C$ with $k \neq k'$, i.e. kD_Lk'; thus we have $j_i D_L^+ j_{i+1}$, where D_L^+ is the transitive closure of the join-dependency relation. Thus, we can depict the play forced by Adam as follows:

$$(j_0, \phi^{n+1}(\perp)) \xrightarrow{D_L^+} (j_1, \phi^n(\perp)) \xrightarrow{D_L^+} \cdots \quad \cdots \xrightarrow{D_L^+} (j_n, \phi(\perp)) \xrightarrow{D_L^+} (j_{n+1}, \perp).$$

The play therefore witnesses the existence of a sequence the join-dependency relation of length at least $n + 1$. By Proposition 3.3, this contradicts the fact that $L \in \mathcal{D}_n$.   □

For each $n \geq 0$, let

$$\sigma_\mu(n) := \min\{ k \geq 0 \mid \mathcal{D}_n \models \phi^{k+1}(\perp) = \phi^k(\perp), \text{ for each lattice term } \phi(z) \},$$

where we write $\mathcal{D}_n \models \phi^{k+1}(\perp) = \phi^k(\perp)$ to mean that the equation $\phi^{k+1}(\perp) = \phi^k(\perp)$ holds on every lattice in $\mathcal{D}_n$. Proposition 5.1 shows that $\sigma_\mu(n) \leq n+1$. Our next goal, achieved with Proposition 5.3, is to show that $\sigma_\mu(n) = n+1$.

Definition 5.2. *For a fixed interger $n \geq 0$, let $A_n = \langle V_n, M \rangle$ be the atomistic presentation where*

$$V_n := \{ v_n, w_n, v_{n-1}, w_{n-1}, \ldots, v_1, w_1, v_0 \},$$
$$M(w_i) := \{\{ w_i \}\}, \qquad i = n, \ldots, 1,$$
$$M(v_i) := \{\{ w_j \mid j = i, \ldots, k+1 \} \cup \{ v_k \} \mid k = i, \ldots, 0 \}, \qquad i = n, \ldots, 0.$$

The presentation A_n satisfies the constraints of Lemma 2.4, so that A_n is isomorphic to $\mathrm{OD}(\mathcal{L}(A_n))$. Figure 2 exhibits generators for the join-dependency relation of $\mathcal{L}(A_n)$ from which it is easily seen that $\mathcal{L}(A_n) \in \mathcal{D}_n$.

Moreover, we consider the valuations $u_n : \{ a, b, c, d, e \} \longrightarrow \mathcal{L}(A_n)$ defined as follows:

$$u_n(a) := \{ v_i \mid i \text{ is even} \}, \qquad u_n(b) := \{ v_i \mid i \text{ is odd} \},$$
$$u_n(c) := \{ w_i \mid i \text{ is even} \}, \qquad u_n(d) := \{ w_i \mid i \text{ is odd} \},$$
$$u_n(e) := \{ v_0 \}.$$

As $u_n(z) = u_m(z) \cap V_n$ for $n \leq m$, we shall abuse of notation and write simply u for any such u_n. Notice that $u(z)$ is indeed a closed subset, for $z \in \{ a, b, c, d, e \}$.

Proposition 5.3. *Let*

$$\psi(z) := (a \wedge (c \vee (b \wedge z))) \vee (b \wedge (d \vee (a \wedge z))) \vee (a \wedge e).$$

For each $n \geq 0$, *the atomistic lattice* $\mathcal{L}(A_n, M)$ *fails the inclusion* $\psi^{n+1}(\bot) \leq \psi^n(\bot)$.

Proof. We shall show that, for each $n \geq 0$,
$v_n \models_u \psi^{n+1}(\bot)$ while $v_n \not\models_u \psi^n(\bot)$.

When $n = 0$ the result is immediate. Let us suppose therefore that $n > 0$, $v_{n-1} \not\models_u \psi^{n-1}(\bot)$, and $v_{n-1} \models_u \psi^n(\bot)$.

Claim. Adam has a winning strategy in the game $\mathcal{G}(A_n, u', \psi(z))$ from position $(v_n, \psi(z))$, where u' is the valuation which extends u by setting $u'(z) := u(\psi^{n-1}(\bot))$.

Proof of Claim. Let us describe Adam's strategy. The starting position is:

$$(v_n, (a \wedge (c \vee (b \wedge z))) \vee (b \wedge (d \vee (a \wedge z))) \vee (a \wedge e)).$$

Recall that w_n satisfies neither a nor b. Thus, if Eva chooses from the starting position a cover of the form $\{w_n, w_{n-1}, ..., w_{k+1}, v_k\}$ with $n > k$, then Adam easily wins by choosing w_n from this cover. If Eva chooses the cover $\{v_n\}$, then we hit position $(\{v_n\}, \psi(z))$ where Adam has the only choice to move to

$$(v_n, \{a \wedge (c \vee (b \wedge z)), \ b \wedge (d \vee (a \wedge z)), \ a \wedge e\}).$$

If Eva moves to $(v_n, a \wedge e)$, then Adam wins because $n > 0$ and $v_n \notin u'(e)$. If n is odd (resp. even) and Eva moves to position $(v_n, a \wedge (c \vee (b \wedge z)))$ (resp. $(v_n, b \wedge (d \vee (a \wedge z)))$), then Adam wins because $v_n \notin u'(a)$ (resp. $v_n \notin u'(b)$).

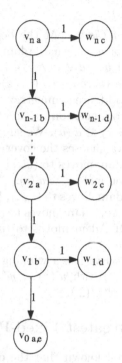

Fig. 2. Dependency generators for the presentation A_n (only the covers of the form $v_i \lhd \{w_i, v_{i-1}\}$ are represented). The valuation u is represented by labels on nodes.

Let us assume that n is even and that Eva moves to position $(v_n, a \wedge (c \vee (b \wedge z)))$ (the case where n is odd is similar and can be treated in a similar way). Here Adam moves to position $(v_n, c \vee (b \wedge z))$ and Eva has to pick a cover of v_n. If she picks the cover $\{v_n\}$ or a cover containing both w_n and w_{n-1}, then Adam wins because, as n is even, $v_n \notin u'(b)$, $v_n \notin u'(c)$, $w_{n-1} \notin u'(c)$ and $w_{n-1} \notin u'(b)$. We suppose therefore that she picks the cover $\{w_n, v_{n-1}\}$ and moves to $(\{w_n, v_{n-1}\}, c \vee (b \wedge z))$. Here Adam goes to position $(v_{n-1}, \{c, b \wedge z\})$. Here if Eva chooses position (v_{n-1}, c), then Adam wins because $v_{n-1} \notin u'(c)$. If otherwise she moves in position $(v_{n-1}, b \wedge z)$, then Adam chooses position (v_{n-1}, z) and wins, since $v_{n-1} \notin u'(z)$. $\square$ *Claim*

As Adam has a winning strategy in the game $\mathcal{G}(A_n, u', \psi(z))$ from $(v_n, \psi(z))$, then we have $v_n \not\models_{u'} \psi(z)$, that is $v_n \not\models_u \psi^n(\bot)$. Our next goal is to prove that $v_n \models_u \psi^{n+1}(\bot)$.

Claim. Eva has a winning strategy in the game $\mathcal{G}(A_n, u', \psi(z))$ from position $(v_n, \psi(z))$, where u' is the valuation which extends u by setting $u'(z) := u(\psi^n(\bot))$.

Proof of Claim. At the beginning of the game Eva chooses $\{v_n\}$ among the covers of v_n, so that Adam can only move from $(\{v_n\}, \psi(z))$ to $(v_n, \{a \wedge (c \vee (b \wedge z)), b \wedge (d \vee (a \wedge z)), a \wedge e\})$. If n is even (resp. odd), then Eva moves to position $(v_n, a \wedge (c \vee (b \wedge z)))$ (resp. $(v_n, b \wedge (d \vee (a \wedge z)))$). As the cases, n even and n odd, are symmetric, we analyse only the situation where n is even.

If, from position $(v_n, a \wedge (c \vee (b \wedge z)))$, Adam moves to (v_n, a), then Eva wins because $v_n \in u'(a)$. We suppose therefore that Adam moves to $(v_n, c \vee (b \wedge z))$; here Eva chooses the cover $\{w_n, v_{n-1}\}$ and moves to $(\{w_n, v_{n-1}\}, c \vee (b \wedge z))$.

If Adam moves to $(w_n, \{c, b \wedge z\})$, then Eva moves to (w_n, c) and wins, since n is even and $w_n \in u'(c)$.

If Adam moves to $(v_{n-1}, \{c, b \wedge z\})$, then Eva moves to position $(v_{n-1}, b \wedge z)$. If, from here, Adam moves to (v_{n-1}, b) then Eva wins since $n-1$ is odd and $v_{n-1} \in u'(b)$. If Adam moves to (v_{n-1}, z), then Eva wins because $v_{n-1} \in u'(\psi^n(\bot))$. $\square$ *Claim*

Thus Eva has a winning strategy in the game $\mathcal{G}(A_n, u', \psi(z))$ from position $(v_n, \psi(z))$ with $u'(z) = \psi^n(\bot)$. Therefore $v_n \models_{u'} \psi(z)$, which is the same as $v_n \models_u \psi^{n+1}(\bot)$. $\square$

6   Greatest Fixed-Points

It is well known that the distributive identity $x \wedge (y \vee z) = (x \wedge y) \vee (x \wedge z)$ implies the dual identity $x \vee (y \wedge z) = (x \vee y) \wedge (x \vee z)$. We can rephrase this by saying that if $L \in \mathcal{D}_0$, then $L^{op} \in \mathcal{D}_0$ as well—where L^{op} is the dual lattice of L, obtained by reversing its order. As an immediate consequence of this fact—and of the results obtained in the previous Section—we observe that the identities of the form $\phi^2(\top) = \phi(\top)$ hold in distributive lattices.

The varieties $\mathcal{D}_n$ are not in general closed under order reversing. For example, the lattice $\mathcal{L}(A_1)$ is not meet-semidistributive, thus its dual $\mathcal{L}(A_1)^{op}$ is not join-semidistributive and therefore it does not belong to any of the varieties $\mathcal{D}_n$, see Theorem 3.6. We shall see later with Proposition 7.1 that meet-semidistributivity is the only requirement needed to obtain closure under order reversing, and actually characterizes it.

Yet, for lattices in $\mathcal{D}_n$ that are not meet-semidistributive, we do not have a shortcuts to derive an identity $\phi^{n+2}(\top) = \phi^{n+1}(\top)$. We prove in this Section that such identities actually hold in $\mathcal{D}_n$.

For a lattice term ϕ, possibly containing the variable z, define the terms $\phi^n(z)$ and $\phi^n(\top)$, $n \geq 0$, as follows:

$$\phi^0(z) := z, \qquad \phi^{n+1}(z) := \phi[\phi^n(z)/z], \qquad \phi^n(\top) := (\phi^n(z))[\top/z].$$

Proposition 6.1. *For every lattice term $\phi(z)$, the identity $\phi^{n+1}(\top) = \phi^{n+2}(\top)$ holds in $\mathcal{D}_n$*

Proof. As $\mathcal{D}_n$ is locally finite, we only have to prove that the identity $\phi^{n+1}(\top) = \phi^{n+2}(\top)$ holds in the finite lattices of $\mathcal{D}_n$. As the inclusion $\phi^{n+2}(\top) \leq \phi^{n+1}(\top)$ always holds, we shall only be concerned with proving the inclusion $\phi^{n+1}(\top) \leq \phi^{n+2}(\top)$.

Since each finite lattice L is isomorphic to $\mathcal{L}(J(L), \leq, \mathcal{M})$, it will be enough to prove that if L is a finite lattice, $u : \mathcal{X} \to \mathcal{L}(J(L), \leq, \mathcal{M})$ is a valuation, and $j_0 \in J(L)$, then $j_0 \models_u \phi^{n+1}(\top)$ implies $j_0 \models_u \phi^{n+2}(\top)$.

Therefore, we suppose that Eva has a winning strategy in the game $\mathcal{G}(\mathrm{OD}(L), u, \phi^{n+1}(\top))$ from position $(j_0, \phi^{n+1}(\top))$ and construct, out of this strategy, a winning strategy for Eva in the game $\mathcal{G}(\mathrm{OD}(L), u, \phi^{n+2}(\top))$ from position $(j_0, \phi^{n+2}(\top))$.

Before presenting the construction, let us have a look at the structure of a complete play in a game of the form $\mathcal{G}(\mathrm{OD}(L), u, \phi^k(\top))$, for some $k \geq 0$. Such a play goes through a maximal sequence of positions of the form

$$(j_0, \phi^k(\top)), (j_1, \phi^{k-1}(\top)), \ldots, (j_{m-1}, \phi^{k-m+1}(\top)), (j_m, \phi^{k-m}(\top)),$$

with $0 \leq m \leq k$. Moreover, for $i = 1, \ldots, m$, either $j_{i-1} = j_i$ or $j_{i-1} D_L^+ j_i$—where D_L^+ is the transitive closure of the join-dependency relation. Thus we can say that such a play *has a repetition* if $j_{i-1} = j_i$ for some i, and that it is *simple* otherwise. Let us observe that, for a simple play, we have $j_0 D_L^+ j_1 D_L^+ \ldots j_{m-1} D_L^+ j_m$; consequenlty we have $m \leq n$ since L is a finite lattice in the variety $\mathcal{D}_n$, see Proposition 3.3.

We can now describe a strategy for Eva in the game $\mathcal{G}(\mathrm{OD}(L), u, \phi^{n+2}(\top))$ from the initial position $(j_0, \phi^{n+2}(\top))$. To this goal, notice that, as $\phi^{n+2}(\top) = \phi^{n+1}[\phi(\top)/z]$, the structure of the terms $\phi^{n+2}(\top)$ and $\phi^{n+1}(\top)$ and therefore of the games $\mathcal{G}(\mathrm{OD}(L), u, \phi^{n+2}(\top))$ and $\mathcal{G}(\mathrm{OD}(L), u, \phi^{n+1}(\top))$ are identical as far as the first part of a play is concerned.

In position $(j_0, \phi^{n+2}(\top))$ of the game $\mathcal{G}(\mathrm{OD}(L), u, \phi^{n+2}(\top))$ Eva plays as if she was playing in the game $\mathcal{G}(\mathrm{OD}(L), u, \phi^{n+1}(\top))$ in position $(j_0, \phi^{n+1}(\top))$, according to the given winning strategy.

Each time the play hits a position of the form $(j_i, \phi^{n+2-i}(\top))$ Eva compares j_i with j_{i-1}. If $j_i \neq j_{i-1}$ and $n + 1 - i > 0$, then she keeps playing from position $(j_i, \phi^{n+2-i}(\top))$ as if she was playing from position $(j_i, \phi^{n+1-i}(\top))$ within the game $\mathcal{G}(\mathrm{OD}(L), u, \phi^{n+1}(\top))$ according to the given winning strategy. If $j_i \neq j_{i-1}$ but $i = n + 1$—let us say that a *critical position* of the form $(j_{n+1}, \phi(\bot))$ is met, where Eva cannot use anymore the given strategy in the game $\mathcal{G}(\mathrm{OD}(L), u, \phi^{n+1}(\top))$ from the corresponding position $(j_{n+1}, \top)$—then she keeps moving randomly until the play reaches a final position (which might be a lost for her).

If instead $j_i = j_{i-1}$, then she recalls that, from position $(j_{i-1}, \phi^{n+2-(i-1)}(\top))$ she has been playing using a winning strategy in position $(j_{i-1}, \phi^{n+1-(i-1)}(\top))$. As $j_i = j_{i-1}$ and $(j_i, \phi^{n+2-i}(\top)) = (j_{i-1}, \phi^{n+1-(i-1)}(\top))$, she realizes that she disposes of a winning strategy from $(j_i, \phi^{n+2-i}(\top))$: she continues using that winning strategy.

Let us argue that Eva's strategy is a *winning* strategy. If a complete play has $j_{i-1} = j_i$ as its first repeat, then Eva has been using a winning strategy from the position $(j_i, \phi^{n+2-i}(\top))$, thus the play is win for Eva.

If on the other hand the play is simple, so that it hits the positions

$$(j_0, \phi^{n+2}(\top)), (j_1, \phi^{n+2-1}(\top)), \ldots, (j_m, \phi^{n+2-m}(\top)),$$

with j_i all distinct, then—by the previous remark on the length of chain of the join-dependency relation—$m \leq n$ and $m < n+1$: a critical position $(j_{n+1}, \phi(\top))$ is not met. This shows that the complete path entirely comes from mimicking a winning play from the game $\mathcal{G}(\mathrm{OD}(L), u, \phi^{n+1}(\top))$, there have been no need of random moves. Thus the play has been played according to a winning strategy and it is a win for Eva.   □

We show next that the identity $\phi^{n+1}(\top) = \phi^n(\top)$ does not hold within the variety $\mathcal{D}_n$.

Proposition 6.2. *Let*

$$\psi(z) := (a \wedge (c \vee (b \wedge z))) \vee (b \wedge (d \vee (a \wedge z))).$$

For each $n \geq 0$ the atomistic lattice $\mathcal{L}(A_n)$ fails the inclusion $\psi^n(\top) \leq \psi^{n+1}(\top)$.

The atomistic presentations A_n and the valuations u_n have been defined in Section 5. As e is not among the variables of $\psi(z)$, we won't be concerned with the value $u_n(e)$. As before, we shall freely write u for an arbitrary u_n.

Proof of Proposition 6.2. We prove, by induction on $n \geq 0$, that $v_n \models_u \psi^n(\top)$ and $v_n \not\models_u \psi^{n+1}(\top)$.

For $n = 0$, we only need to prove that $v_0 \not\models_u \psi(\top)$ and, to this goal, we show that Adam has a winning strategy in the game $\mathcal{G}(A_0, u', \psi(z))$ from the position $(v_0, \psi(z))$, where u' is the valuation extending u by setting $u(z) = V_0 = \{\, v_0 \,\}$.

In the starting position $(v_0, \psi(z))$ Eva can only move to $(\{\, v_0 \,\}, \psi(z))$, and from here Adam can only move to $(v_0, \{\, a \wedge (c \vee (b \wedge z)), b \wedge (d \vee (a \wedge z)) \,\})$. If Eva moves to $(v_0, b \wedge (d \vee (a \wedge z)))$, then Adam chooses (v_0, b) and wins. If Eva moves to $(v_0, a \wedge (c \vee (b \wedge z)))$, then Adam moves to $(v_0, c \vee (b \wedge z))$. From here (as before) the game necessarily hits position $(v_0, \{\, c, b \wedge z \,\})$ where it is Eva's turn to move. As $v_0 \notin u(c)$, Eva looses if she moves to (v_0, c). If she moves to $(v_0, b \wedge z)$, then Adam moves to (v_0, b) and wins.

We suppose now that $n \geq 1$, $v_{n-1} \models_u \psi^{n-1}(\top)$, $v_{n-1} \not\models_u \psi^n(\top)$.

Claim. Eva has a winning strategy in the game $\mathcal{G}(A_n, u', \psi(z))$ from position $(v_n, \psi(z))$, where u' extends u by $u'(z) = u(\psi^{n-1}(\top))$.

Proof of Claim. From the initial position $(v_n, \psi(z))$ Eva moves to $(\{\, v_n \,\}, \psi(z))$ where Adam can only move to $(v_n, \{\, a \wedge (c \vee (b \wedge z)), b \wedge (d \vee (a \wedge z)) \,\})$. From here, if n is even, then Eva moves to $a \wedge (c \vee (b \wedge z))$; if n is odd, then she moves to $b \wedge (d \vee (a \wedge z))$. By symmetry, we only consider the first case. If Adam, moves to (v_n, a), then Eva wins because $v_n \in u'(a)$; if he moves to $(v_n, c \vee (b \wedge z))$, then Eva chooses the cover $\{w_n, v_{n-1}\}$ and moves to $(\{w_n, v_{n-1}\}, c \vee (b \wedge z))$. From here, if Adam moves to $(w_n, \{c, b \wedge z\})$, then Eva moves in position (w_n, c) and wins; if Adam moves to $(v_{n-1}, \{c, b \wedge z\})$, then Eva moves to $(v_{n-1}, b \wedge z)$ and she wins since $v_{n-1} \in u'(b)$ and $v_{n-1} \in u(\psi^{n-1}(\top)) = u'(z)$. $\qquad \Box$ Claim

Thus Eva has a winning strategy in the game $\mathcal{G}(A_n, u', \psi(z))$ from position $(v_n, \psi(z))$ with $u'(z) = u(\psi^{n-1}(\top))$; we have therefore $v_n \models_{u'} \psi(z)$, that is $v_n \models_u \psi^n(\top)$.

Claim. Adam has a winning strategy in the game $\mathcal{G}(A_n, u', \psi(z))$ from position $(v_n, \psi(z))$ where u' extends u by $u'(z) := u(\psi^n(\top))$.

Proof of Claim. If, from the initial position $(v_n, \psi^n(z))$, Eva chooses a position $(\{w_n, w_{n-1}, ..., w_{k+1}, v_k\}, \psi(z))$ with $k < n$, then Adam wins by moving to $(w_n, \{a \wedge (c \vee (b \wedge z)), b \wedge (d \vee (a \wedge z))\})$, since $w_n \notin u'(a)$ and $w_n \notin u'(b)$.

Let us assume that Eva chooses the position $(\{v_n\}, \psi(z))$, from which Adam can only move to $(v_n, \{a \wedge (c \vee (b \wedge z)), b \wedge (d \vee (a \wedge z))\})$.

If n is even (resp. odd) and Eva moves to position $b \wedge (d \vee (a \wedge z))$ (resp. $a \wedge (c \vee (b \wedge z))$), then Adam wins because $v_n \notin u'(b)$ (resp. $v_n \notin u'(a)$). We describe therefore how Adam plays when n is even and Eva moves to position $a \wedge (c \vee (b \wedge z))$. By symmetry, the analysis will cover the case where n is odd and Eva moves to position $b \wedge (d \vee (a \wedge z))$.

From position $(v_n, a \wedge (c \vee (b \wedge z)))$ Adam moves to $(v_n, c \vee (b \wedge z))$, where Eva has to choose a cover of v_n.

If she moves to $(\{v_n\}, c \vee (b \wedge z))$, then Adam can only move to $(v_n, \{c, b \wedge z\})$. If Eva moves to (v_n, c), then Adam wins because $v_n \notin u'(c)$. If Eva moves to $(v_n, b \wedge z)$, then Adam moves to (v_n, b) and wins because $v_n \notin u'(b)$.

If Eva picks a cover C containing both w_n and w_{n-1}, then Adam moves from $(C, c \vee (b \wedge z))$ to position $(w_{n-1}, \{c, b \wedge z\})$. If Eva moves to (w_{n-1}, c), then Adam wins because $n - 1$ is odd and $w_{n-1} \notin u'(c)$. If Eva moves to $(w_{n-1}, b \wedge z)$, then Adam moves to (w_{n-1}, b) and wins because $w_{n-1} \notin u'(b)$.

Finally, let us consider what happens if Eva picks the cover $\{w_n, v_{n-1}\}$: from position $(\{w_n, v_{n-1}\}, c \vee (b \wedge z))$ Adam moves to $(v_{n-1}, \{c, b \wedge z\})$. Therefore, if Eva moves to (v_{n-1}, c), then Adam wins because $v_{n-1} \notin u'(c)$; if she moves to $(v_{n-1}, b \wedge z)$, then Adam wins by moving to (v_{n-1}, z), since $v_{n-1} \notin u(\psi^n(\top)) = u'(z)$. $\qquad \Box$ Claim

We have shown that Adam has a winning strategy in the game $\mathcal{G}(A_n, u', \psi(z))$ from position $(v_n, \psi(z))$ with $u'(z) = u(\psi^n(\top))$: therefore $v_n \not\models_{u'} \psi(z)$, that is $v_n \not\models_u \psi^{n+1}(\top)$. $\qquad \Box$

For each $n \geq 0$, put

$$\sigma_\nu(n) := \min\{k \geq 0 \mid \mathcal{D}_n \models \phi^{k+1}(\top) = \phi^k(\top), \text{ for each lattice term } \phi(z)\}.$$

Thus we have shown that $\sigma_\nu(n) = \sigma_\mu(n) = n + 1$. Another consequence of the results presented up to here is the following Theorem.

Theorem 6.3. *The alternation hierarchy for the lattice μ-calculus is degenerate on each variety $\mathcal{D}_n$.*

Indeed, if we consider the μ-terms defined in (1), it is easily argued that $\mathcal{D}_n \models t = \mathrm{tr}_n(t)$, where $\mathrm{tr}_n(t)$ is a lattice term (with no fixed-point operators) inductively defined by

$$\mathrm{tr}_n(x) := x, \quad \mathrm{tr}_n(\top) := \top \quad \mathrm{tr}_n(\bot) := \bot,$$
$$\mathrm{tr}_n(t \wedge s) := \mathrm{tr}_n(t) \wedge \mathrm{tr}_n(s), \qquad \mathrm{tr}_n(t \vee s) := \mathrm{tr}_n(t) \vee \mathrm{tr}_n(s),$$
$$\mathrm{tr}_n(\mu_z.\phi(z)) := \mathrm{tr}_n(\phi(z))^{n+1}(\bot), \quad \mathrm{tr}_n(\nu_z.\phi(z)) := \mathrm{tr}_n(\phi(z))^{n+1}(\top).$$

7   Further Lower Bounds

We survey in this Section on further results that can be obtained on extremal fixed-points of lattice terms in given varieties.

We have seen that the varieties $\mathcal{D}_n$ are not dual, for $n \geq 1$. What then if we enforce duality? Namely, for a lattice variety $\mathcal{V}$, let $\mathcal{V}^{op} := \{ L^{op} \mid L \in \mathcal{V} \}$. Clearly, if $\mathcal{V}$ is axiomatized by identities $\{ e_1, \dots e_k \}$, then $\mathcal{V}^{op}$ is axiomatized by the identities $\{ e_1^{op}, \dots e_k^{op} \}$, with e^{op} obtained from e by exchanging in terms meets with joins and joins with meets. What can be said if we consider the varieties $\mathcal{D}_n \cap \mathcal{D}_n^{op}$?

Proposition 7.1. *The variety $\mathcal{D}_n \cap \mathcal{D}_n^{op}$ consists of the meet-semidistributive lattices in $\mathcal{D}_n$.*

The interesting question is then whether the lower bounds we found in Sections 5 and 6 are still valid. Namely, for $n \geq 0$, let

$$\tau_\mu(n) := \min\{ k \geq 0 \mid \mathcal{D}_n \cap \mathcal{D}_n^{op} \models \phi^{k+1}(\bot) = \phi^k(\bot), \phi(z) \text{ a lattice term } \}.$$

Notice that, if τ_ν is defined similarly, with $\bot$ replaced by $\top$, duality enforces $\tau_\nu = \tau_\mu$. Clearly we have $\tau_\mu(n) \leq n + 1$; we argue next that we still have $\tau_\mu(n) = n + 1$.

The lower bounds to the functions σ_μ and σ_ν were obtained using the atomistic presentation A_n. When considering meet-semidistributive lattices, we cannot look at atomistic ones, because of the following Lemma.

Lemma 7.2. *A finite meet-semidistributive atomistic lattice in $\mathcal{D}_n$ is a Boolean algebra; in particular it belongs to $\mathcal{D}_0$.*

In order to exhibit lower bounds to the function τ_μ, we lift the atomistic presentations A_n to presentations O_n by adding to them an ordering; we prove then that each O_n arises as the OD-graph of a semidistributive lattice.

Definition 7.3. *For $n \geq 0$, let O_n be the presentation $\langle V_n, \leq, M \rangle$, where V_n and M are as in Definition 5.2, and $\leq$ is the ordering on V_n whose Hasse diagram is given by $v_0 \prec v_1 \prec \dots \prec v_{n-1} \prec v_n$.*

The technical achievement, from a pure lattice-theoretic perspective, is to show that the presentations O_n arise as OD-graphs of meet-semidistributive lattices.

Proposition 7.4. *For each* $n \geq 0$*, the presentation* O_n *is isomorphic to the OD-graphs of a meet-semidistributive lattice* $L_n \in \mathcal{D}_n$*. Thus,* $\mathcal{L}(O_n)$ *is in* $\mathcal{D}_n \cap \mathcal{D}_n^{op}$*.*

Unfortunately, we cannot reuse the results from the preceeding Sections 5 and 6, as neither $u_n(a)$ nor $u_n(b)$ defined in Section 5 are downsets, therefore they do not belong to $\mathcal{L}(O_n)$. In order to achieve our goals, we have to proceed in a completely different—yet similar—way. We let ψ_n and $u_n : \{b_n, \ldots, b_1, a\} \longrightarrow \mathcal{L}(O_n)$ be defined by

$$\psi_n(z) := a \vee \bigvee_{i=1,\ldots,n} (c \wedge (b_i \vee z)),$$

$$u_n(a) := \{v_0\}, \qquad u_n(b_i) := \{w_i\}, \qquad u_n(c) = \{v_n, v_{n-1}, \ldots, v_0\}.$$

Proposition 7.5. *For each* $n \geq 0$*, we have* $v_n \models_{u_n} \psi_n^{n+1}(\bot)$ *and* $v_n \not\models_{u_n} \psi_n^n(\bot)$*. Thus the inclusion* $\psi_n^{n+1}(\bot) \leq \psi_n^n(\bot)$ *does not hold in* $\mathcal{D}_n \cap \mathcal{D}_n^{op}$ *and we have* $\tau_\mu(n) = n + 1$*.*

8 Conclusions and Future Work

A main aim of our research was to identify generic principles for the degeneracy of alternation hierarchies in fixed-point calculi. We argued, in this paper, that validity of some non-trivial equations is among these principles. We aim at finding other principles; connections between uniform interpolation in logics and definability of extremal fixed-points, see for example [DH00], provide a clear direction for our future researches.

Our paper has also made extensive use of the representation theory for finite lattices based on Nation's Theorem [Nat90] and on an approach to lattice theory from the perspective of modal logic. Among the many open paths, we wish to use this representation theory for studying richer mathematical structures such as the residuated lattices and the algebraic models of linear logic. The study of the fixed-point theories of these structures is also of fundamental relevance for the usual μ-calculi. A main difficulty in the proof of the completeness of Kozen's axiomatization of the modal μ-calculus [Wal00] originates from the odd interplay between the classical conjunction and the least fixed-point operator. Understanding the interplay between the linear conjunction and least fixed-points will shed, most probably, an important enlightenment on the classical case.

References

[AN01] Arnold, A., Niwiński, D.: Rudiments of μ-Calculus, Studies in Logic and the Foundations of Mathematics, vol. 146. Elsevier, Amsterdam (2001)

[Awo10] Awodey, S.: Category Theory. In: Oxford Logic Guides, 2nd edn., vol. 52. Oxford University Press, Oxford (2010)

[Bir73]   Birkhoff, G.: Lattice Theory, American Mathematical Society Colloquium Publications, 3rd edn., vol. 25. American Mathematical Society, Providence (1973)

[DH00]   D'Agostino, G., Hollenberg, M.: Logical Questions Concerning the mu-Calculus: Interpolation, Lyndon and Los-Tarski. J. Symb. Log. 65, 310–332 (2000)

[DP02]   Davey, B., Priestley, H.: Introduction to Lattices and Order, 2nd edn. Cambridge University Press, New York (2002)

[Day92]   Day, A.: Doubling constructions in lattice theory. Can. J. Math. 44, 252–269 (1992)

[FS13]   Fortier, J., Santocanale, L.: Cuts for circular proofs: semantics and cut-elimination. In: Rocca, S.R.D. (ed.) CSL. LIPIcs, vol. 23, pp. 248–262. Schloss Dagstuhl - Leibniz-Zentrum für Informatik (2013)

[FJN95]   Freese, R., Ježek, J., Nation, J.: Free Lattices, Mathematical Surveys and Monographs, vol. 42. American Mathematical Society, Providence (1995)

[Gol06]   Goldblatt, R.: A Kripke-Joyal Semantics for Noncommutative Logic in Quantales. In: Governatori, G., Hodkinson, I.M., Venema, Y. (eds.) Advances in Modal Logic, pp. 209–225. College Publications (2006)

[Grä98]   Grätzer, G.: General Lattice Theory, 2nd edn. Birkhäuser Verlag, Basel (1998); New appendices by the author with Davey, B.A., Freese, R., Ganter, B., Greferath, M., Jipsen, P., Priestley, H.A., Rose, H., Schmidt, E.T., Schmidt, S.E., Wehrung, F., Wille, R

[Grä08]   Grätzer, G.: Universal Algebra, 2nd printing of the 2nd edn. Springer, New York (2008)

[Han03]   Hansen, H.: Monotonic Modal Logics. Master's thesis, Institute for Logic, Language and Computation, Universiteit van Amsterdam (2003), Available as: ILLC technical report: PP-2003-24

[Koz83]   Kozen, D.: Results on the Propositional mu-Calculus. Theor. Comput. Sci. 27, 333–354 (1983)

[KW99]   Kracht, M., Wolter, F.: Normal monomodal logics can simulate all others. J. Symb. Log. 64, 99–138 (1999)

[Mac98]   Mac Lane, S.: Categories for the Working Mathematician, Graduate Texts in Mathematics, 2nd edn., vol. 5. Springer, New York (1998)

[Nat90]   Nation, J.B.: An approach to lattice varieties of finite height. Algebra Universalis 27, 521–543 (1990)

[NW96]   Niwinski, D., Walukiewicz, I.: Games for the mu-Calculus. Theor. Comput. Sci. 163, 99–116 (1996)

[San01]   Santocanale, L.: The alternation hierarchy for the theory of μ-lattices. Theory Appl. Categ. 9, 166–197 (2001)

[San02]   Santocanale, L.: Free μ-lattices. J. Pure Appl. Algebra 168, 227–264 (2002)

[San09]   Santocanale, L.: A duality for finite lattices (2009), http://hal.archives-ouvertes.fr/hal-00432113 (unpublished)

[Sem05]   Semenova, M.V.: Lattices that are embeddable in suborder lattices. Algebra and Logic 44 (2005)

[VS10]   Venema, Y., Santocanale, L.: Uniform interpolation for monotone modal logic. In: Beklemishev, L., Goranko, V., Shehtman, V. (eds.) Advances in Modal Logic, vol. 8, pp. 350–370. College Publications (2010)

[Wal00]   Walukiewicz, I.: Completeness of Kozen's Axiomatisation of the Propositional μ-Calculus. Inf. Comput. 157, 142–182 (2000)

Author Index